f	frequency with which a value occurs	$z_{\alpha/2}$	critical value of z
Σ	capital sigma; summation	t	t distribution
Σx	sum of the values	$t_{\alpha/2}$	critical value of t
Σx^2	sum of the squares of the values	df	number of degrees of freedom
$(\Sigma x)^2$	square of the sum of all values	F	F distribution
Σxy	sum of the products of each x value multiplied by the corresponding y value	χ^2	chi-square distribution
		χ_R^2	right-tailed critical value of chi-square
n	number of values in a sample	χ_L^2	left-tailed critical value of chi-square
$n!$	n factorial	p	probability of an event or the population proportion
N	number of scores in a finite population; also used as the size of all samples combined	q	probability or proportion equal to $1 - p$
k	number of samples or populations or categories	$\hat{p}$	sample proportion
		$\hat{q}$	sample proportion equal to $1 - \hat{p}$
$\overline{x}$	mean of the values in a sample	$\overline{p}$	proportion obtained by pooling two samples
$\overline{R}$	mean of the sample ranges	$\overline{q}$	proportion or probability equal to $1 - \overline{p}$
μ	mu; mean of all values in a population	$P(A)$	probability of event A
s	standard deviation of a set of sample values	$P(A|B)$	probability of event A, assuming event B has occurred
σ	lowercase sigma; standard deviation of all values in a population		
s^2	variance of a set of sample values	$_nP_r$	number of permutations of n items selected r at a time
σ^2	variance of all values in a population	$_nC_r$	number of combinations of n items selected r at a time
z	standard score		

ANNOTATED INSTRUCTOR'S EDITION

ELEMENTARY
STATISTICS

SEVENTH EDITION

ELEMENTARY
STATISTICS

SEVENTH EDITION

MARIO F. TRIOLA

 **ADDISON-WESLEY**

An imprint of Addison Wesley Longman, Inc.

Reading, Massachusetts • Menlo Park, California • New York • Harlow, England
Don Mills, Ontario • Sydney • Mexico City • Madrid • Amsterdam

Sponsoring Editor   Jennifer Albanese
Editorial Assistants   Sara Peterson and Carolyn Lee
Development Editor   Jane Parrigin
Senior Marketing Manager   Brenda Bravener
Managing Editor   Karen Guardino
Production Supervisor   Peggy McMahon
Editorial Production Services   Jennifer Bagdigian
Text Designer   Janet Theurer
Cover Designer   Barbara Atkinson
Cover Illustration   Dave Cutler
Art Editor   Jim Roberts
Photo Researcher   Naomi Kornhauser
Prepress Service Manager   Caroline Fell
Senior Manufacturing Manager   Ralph Mattivello
Technical Illustrator   Scientific Illustrators
Essay Illustrator   Terry Presnall
Composition   Typo-Graphics
Printer   R.R. Donnelly & Sons

Many of the designations used by manufacturers and sellers to distinguish their products are claimed as trademarks. Where those designations appear in this book, and Addison-Wesley was aware of a trademark claim, the designations have been printed in initial caps or all caps.

Photo credits: Page 3, United Nations © Arthur Tress/Photo Researchers, Inc.; Page 34, courtesy of Paul Mones; Page 37, new aluminum cans © Comstock; Page 118, courtesy of Anthony M. DiUglio, Jr., P.O. Box 3608, Poughkeepsie, NY 12603; Page 121, lightning © Robert Friedman/Frozen Images/The Image Works; Page 180, courtesy of Drs. Lee Miringoff and Barbara Carvalho, Marist Institute for Public Opinion; Page 183, plane landing © Kees van den Berg/Photo Researchers, Inc.; Page 227, astronaut courtesy of NASA; Page 284, courtesy of Mario F. Triola; Page 287, thermometer © Mehau Kulyk and Victor de Schwanberg/Science Photo Library/Photo Researchers, Inc.; Page 342, courtesy of Barry Cook, A. C. Nielsen Co.; Page 345, athlete cooling off © John Moore/The Image Works; Page 410, courtesy of Jay Dean, Young & Rubicam San Francisco; Page 413, cigarette filters © George Gardner/Stock Boston; Page 475, bear © Voller Ernst/The Image Works; Page 532, courtesy of Boeing Commercial Airplane Group; Page 535, female troops © Norman R. Rowan/Stock Boston; Page 571, Old Faithful © Paolo Koch/Photo Researchers, Inc.; Page 608, courtesy of Lester R. Curtin; Page 611, used aluminum cans © Mark Antman/The Image Works; Page 638, courtesy of Bill James; Page 641, Earth seen from space courtesy of NASA.

Logos used with permission of The Philadelphia 76ers, Motorola Cellular, Motorola Inc., and The Proctor & Gamble Company.

Insert credit: Edward R. Tufte, *The Visual Display of Quantitative Information* (Cheshire, CT: Graphics Press, 1983). Reprinted with permission.

Library of Congress Cataloging-in-Publication Data

The Library of Congress has catalogued the Student Edition of this title as follows:
Triola, Mario F.
    Elementary statistics / Mario F. Triola.—7th ed.
       p.   cm.
    Includes bibliographical references and index.
    ISBN 0–201–85920–3 (hardcover)
    1. Statistics.   I. Title
QA276. 12. T76.   1997
519.5—dc21

                                           97–14499
                                           CIP

Reprinted with corrections, June 1998

4 5 6 7 8 9 10 DOC 9998

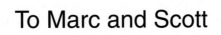

To Marc and Scott

About the Author

Mario F. Triola is a Professor of Mathematics at Dutchess Community College, where he has taught statistics, calculus, linear algebra, technical mathematics, programming, and other courses for 28 years. He is the author of *Mathematics and the Modern World* and *A Survey of Mathematics* and is a coauthor of *Introduction to Technical Mathematics* and *Business Statistics*. His consulting experience includes the mathematical design of casino slot machines, and he has worked with attorneys in determining probabilities in paternity lawsuits and identifying salary inequities based on gender. His interests include golf, running, tennis, and anything that flies. He has a commercial pilot's license with an instrument rating, and has flown airplanes, helicopters, sail planes, hang gliders, and hot air balloons. He has made parachute jumps, has flown in a Goodyear blimp, and has parasailed. He is currently a member of the Project Coalition writing team with NASA and the American Mathematics Association of Two-Year Colleges. He also enjoys travel. With his family, he has visited 49 of the United States, and he hopes to tour Kentucky soon. Other recent trips include a Kenyan photo safari, a rafting trip down the Colorado River through the Grand Canyon, hiking in Yellowstone National Park (and collecting data for the Old Faithful geyser) and travel to interesting sites in Egypt, Greece, Alaska, Scandinavia, and Eastern Europe.

Preface

 Why Study Statistics?

In an interview with the author, David Hall, Division Statistical Manager for the Boeing Commercial Airplane Group, said, "Right now, American industry is crying out for people with an understanding of statistics and the ability to communicate its use." Only a decade ago, students taking a statistics course enjoyed a competitive advantage in the job market, but today's students suffer a competitive disadvantage if they have *not* studied basic concepts of statistics. Fortunately, technological advances in calculators and computers have made the required computations relatively easy, so we can focus more on understanding and interpreting results. It is also fortunate that dedicated educators are committed to the development of an introductory statistics course that is relevant, meaningful, interesting, and enjoyable.

Elementary Statistics, Seventh Edition, reflects the latest technological and teaching advances. Technological advances are reflected in the incorporation of, or reference to, several widely used technologies. For the first time, references to the TI-83® calculator, developed by Texas Instruments especially for use in statistics and business, have been included. The latest version of Minitab® statistical software continues to be featured in the book's exhibits and exercises. Our free, text-specific software, STATDISK®, has been completely reprogrammed for use with the new edition, and it now runs with Windows, Macintosh, and networks. A Web site has been developed to provide additional teaching resources for users of this edition.

Advances in teaching are incorporated through such features as the new **Cooperative Group Activities,** the **From Data to Decision** exercises, and the

Cumulative Review Exercises. We place a strong emphasis on using *real* data throughout the text and exercises, an increased emphasis on methods of data analysis, and an overall focus in examples and exercises on critical thinking and interpretation of results.

As with past editions of this book, the author and the publisher are committed to providing the best possible introductory statistics book for both instructors and students. We are gratified by the book's continued acceptance and by the positive feedback we receive from users. Any additional comments about *Elementary Statistics* are most welcome. Please send them to the Addison-Wesley Statistics Editor or e-mail them to stats@aw.com.

Audience

Elementary Statistics is written for students majoring in any field except mathematics. A strong mathematics background is not necessary, but students should have completed a high school or college elementary algebra course. Although underlying theory is included, this book does not stress the mathematical rigor more suitable for mathematics majors. Because the many examples and exercises cover a wide variety of different and interesting statistical applications, *Elementary Statistics* is appropriate for students pursuing majors in a wide variety of disciplines ranging from the social sciences of psychology and sociology to areas such as education, the allied health fields, business, economics, engineering, the humanities, the physical sciences, journalism, communications, and liberal arts.

Overview of the Seventh Edition

Highlights of the Seventh Edition include:

Content Changes

- A rewritten *Correlation and Regression* chapter (Chapter 9), which facilitates early coverage of these topics for instructors who prefer this approach. Alternative syllabi are also provided in the *Annotated Instructor's Edition* to show how this early coverage can be presented.
- Expanded coverage of design of experiments in Chapter 1.
- Increased emphasis on techniques of data analysis.
- Increased emphasis on conceptual understanding and interpretation throughout every section and exercise set.
- *NEW* sections: Estimating population means with small samples (Section 6-3), finding scores with nonstandard normal distributions

(Section 5-4), the Poisson distribution (Section 4-5), and end-of-semester group projects (Section 14-1).

New Features

- **Cooperative Group Activities** are included for the first time at the end of each chapter to provide activities that increase understanding and interest through active student participation.
- **Cumulative Review Exercises,** also included at the end of each chapter, are designed to reinforce learning and extend conceptual understanding by including topics from earlier chapters.
- **Answers to all Review Exercises and Cumulative Review Exercises** are included in both the *Annotated Instructor's Edition* and the *Student Edition.*
- **TI-83® calculator comments,** where appropriate, are included in the text, and there is a new supplement: *The TI-83® Companion to Elementary Statistics,* by Larry Morgan of Montgomery County Community College.

Exercises

There are 1357 exercises—*60 percent of them new!* Many more of the exercises require interpretation of results. Because exercises are of such core importance to any statistics book, great care has been taken to ensure their usefulness, relevance, and accuracy. Three statisticians have read carefully through the final stages of the book to attest to the precision of the text material, the usefulness of the annotations, and, above all, the accuracy of every exercise answer. Exercises are arranged in order of increasing difficulty by dividing them into two groups: (a) Basic Skills and Concepts and (b) Beyond the Basics. The Beyond the Basics exercises address more difficult concepts or require a somewhat stronger mathematical background. In some cases, these exercises also introduce a new concept.

Hallmark Features

Beyond an interesting and accessible (and sometimes humorous) writing style, great care has been taken to ensure that each chapter of *Elementary Statistics* will help students understand the concepts presented. The following features are designed to help meet that objective:

- **Chapter-opening features:** A list of chapter sections with a brief description of their contents previews the chapter for the student; a chapter-opening problem, using real data, then motivates the chapter material; and a chapter overview provides a statement of the chapter's objectives.

- **End-of-chapter features:** A **Vocabulary List** of important terms (a full glossary is found in Appendix D); a **Chapter Review,** which summarizes the key concepts and topics of the chapter; **Review Exercises** for practice on the chapter concepts and procedures; **Cumulative Review Exercises** to reinforce earlier material; **Computer Projects** for use with Minitab and STATDISK softwares; **From Data to Decision,** a capstone problem requiring critical thinking and a writing component; and **Cooperative Group Activities** to encourage active learning in groups.

- **Margin Essays:** The text includes 116 margin essays, which illustrate uses and abuses of statistics in real, practical applications. Topics include "Predicting the Stock Market," "Statistics: Jobs and Employers," "Statistics in Court," "Census 2000," and "Six Degrees of Separation."

- **Flowcharts:** These appear throughout the text to simplify and clarify more complex concepts and procedures.

- **Real Data Sets:** These are used extensively throughout the entire book. Appendix B lists 16 data sets, 10 of which are new. These data sets are provided in printed form in Appendix B at the back of the book, and in electronic form either on a data disk (Windows or Macintosh formats) available from the publisher, or on the Addison-Wesley Web Site for the new edition. The data sets include such varied topics as eruptions of the Old Faithful geyser, nicotine contents of cigarettes, acid rain, and measurements of wild bears, and they include such topics as how often Old Faithful erupts, the weight of household garbage for one week, and Maryland Lottery results.

- **Interviews:** The text includes nine author-conducted interviews with professionals who use statistics in their day-to-day work.

- **Six appendices:** Appendix A contains tables; Appendix B lists 16 data sets; Appendix C describes the use of the Internet and lists several relevant Web sites; Appendix D is a glossary of important terms; Appendix E is a bibliography of recommended text and reference books; Appendix F contains answers to all the odd-numbered section exercises, as well as *all* answers to Review Exercises and Cumulative Review Exercises.

- **Quick-Reference Endpapers:** A symbol table is included on the front inside cover for quick and easy reference to key symbols. Tables A-2 and A-3 (the normal and *t* distributions) are reproduced on the back inside cover pages.

- **Detachable Formula/Table Card:** This insert, organized by chapter, gives students a quick reference for studying, or for use when taking tests (if allowed by the instructor).

Technology Usage

Elementary Statistics, Seventh Edition, can be used without reference to any specific technology. However, for those who choose to supplement the course with technology, **Computer Projects** are included at the end of each chapter, and particular examples and exercises are designated with a graphing calculator symbol to indicate that they are highlighted in an accompanying TI-83® graphing calculator manual designed specifically for this text.

Two different levels of software are available—STATDISK and Minitab—and their use is discussed with sample displays throughout the book. The data sets in Appendix B (except Data Set 2) are available on disk for use with either software.

STATDISK   STATDISK is an easy-to-use statistical software package developed specifically for use with *Elementary Statistics,* and it is *free* to college-level adopters of the book. The new Version 7.0 of STATDISK is available for IBM PC and compatible computers using Windows and for Macintosh systems.

The *STATDISK Student Laboratory Manual and Workbook* includes instructions on the use of the STATDISK software package, as well as experiments to be conducted by students. The software and manual/workbook have been designed so that instructors can assign computer experiments without using valuable class time.

Minitab   For those who wish to use Minitab, Minitab displays have been included throughout the text, and references are made to Minitab Release 11 for Windows, Windows 95, or Windows NT.

The *Minitab® Student Laboratory Manual and Workbook* is designed specifically for *Elementary Statistics.* It includes instructions on the use of Minitab, as well as experiments to be conducted by students.

Calculators

New to this edition are examples and exercises designated with a graphing calculator icon identifying those examples and exercises that are particularly good for graphing calculator usage. Though many calculators can be used in introductory statistics courses, the TI-83® calculator is the one specifically referred to in this text. The TI-83® is the first calculator to have fairly extensive statistics functions.

The TI-83® Companion to Elementary Statistics, Seventh Edition, by Larry Morgan (Montgomery County Community College) is organized to follow the sequence of the text topics. It includes step-by-step instructions on the use of the TI-83®, along with worked-out examples. Additional TI-83® programs are located on the Data/Program Disk that accompanies the textbook, and these programs are also available on the Web site for the Triola textbook.

Supplements

The student and instructor supplements packages are intended to be the most complete and helpful learning system possible for the introductory statistics course. The following are available from Addison Wesley Longman Publishing Company. Contact your local sales consultant, or call the company directly at 1-800-552-2499 for examination copies.

For the Instructor

- *Annotated Instructor's Edition,* by Mario F. Triola, contains answers to all exercises in the margin, recommended assignments, and teaching suggestions. ISBN: 0-201-85921-1. (Student for-sale edition ISBN: 0-201-85920-3).

- *Instructor's Guide and Solutions Manual,* by Mario F. Triola, contains solutions to all the exercises, quizzes (with answers), transparency masters, and sample course syllabi. ISBN: 0-201-85922-X.

- **Data/Program Disk,** prepared by Mario F. Triola, includes the data sets (except for Data Set 2) from Appendix B in the textbook. These data sets are stored both as text files and as Minitab worksheets. The disk also includes programs for the TI-83® graphing calculator. Windows Data/Program Disk ISBN: 0-201-85931-9; Macintosh Data/Program Disk ISBN: 0-201-85932-7.

- **STATDISK Statistical Software,** developed by Password, Inc., specifically for the Triola text, is a statistical software package (for both Windows and Macintosh) licensed *free* to adopters of *Elementary Statistics,* Seventh Edition. This software is provided to the instructor in a variety of ways—either on disk for loading on to school computers, or for the students to copy onto their own disks. STATDISK Windows Software ISBN: 0-201-85926-2; STATDISK Macintosh Software ISBN: 0-201-85927-0.

- **Testing System:** Great care has been taken to ensure the strongest possible testing system for the new edition of *Elementary Statistics.* Not only is there a printed test bank, there is also a *new* computerized test generator, **TestGen-EQ,** that lets you view and edit testbank questions, transfer them to tests, and print in a variety of formats. The program

also offers many options for organizing and displaying test banks and tests. A built-in random number and test generator makes **TestGen-EQ** ideal for creating multiple versions of tests and provides more possible test items than print testbank questions. Powerful search and sort functions let the instructor easily locate questions and arrange them in the preferred order. Printed Testbank ISBN: 0-201-85930-0; TestGen-EQ Windows ISBN: 0-201-85928-9; TestGen-EQ Macintosh ISBN: 0-201-85929-7.

- **PowerPoint® Lecture Presentation:** Free classroom lecture presentation software geared specifically to the sequence and philosophy of *Elementary Statistics,* prepared by Cheryl Slayden and Bob Maschak of Pellissippi State Technical College. Key graphics from the book are also included. Windows version ISBN: 0-201-31612-9; Macintosh version ISBN: 0-201-33154-3.

- **Triola Elementary Statistics Web Site:** This Web site can be accessed at http://www.triolastats.com, and will be continually changing to provide dynamic resources for the teaching of introductory statistics.

For the Student

- *Student Solutions Manual,* by Milton Loyer, provides detailed, worked-out solutions to all odd-numbered exercises. ISBN: 0-201-85923-8.

- *STATDISK Student Laboratory Manual and Workbook,* written by Mario F. Triola, includes experiments to be conducted by students using STATDISK software, either in the computer lab, or for out-of-class assignments. ISBN: 0-201-85925-4.

- *Minitab® Student Laboratory Manual and Workbook,* written by Mario F. Triola, includes instructions on and examples of Minitab use. It also supplies many computer experiments to be conducted, which allows further exploration of statistical concepts. ISBN: 0-201-85924-6.

- *TI-83® Companion to Elementary Statistics,* Seventh Edition, by Larry Morgan (Montgomery County Community College) is organized to follow the sequence of topics in the text, and is a careful, step-by-step guide to how to use the TI-83® graphing calculator. It provides worked-out examples to help students fully understand and use the graphing calculator. ISBN: 0-201-30727-8.

Also Available from the Publisher . . .

For the last decade, Addison-Wesley Publishing Company has enjoyed an important partnership with Minitab, Inc., through our publication of the popular *Student Editions of Minitab. The Student Editions of Minitab* provide a somewhat reduced version of Minitab professional software (which

allows the user access to 3500 data points and to a full range of Minitab functionalities and graphics capabilities), and an accompanying manual (which includes case studies and hands-on tutorials). Currently available Student Editions are:

Student Edition of Minitab for Windows ISBN: 0-201-59157-X

Student Edition of Minitab, Release 8 for DOS ISBN: 0-201-83590-8

Student Edition of Minitab, Release 8 for the Macintosh ISBN: 0-201-83591-6

Addison-Wesley is also pleased to offer yet another statistical software product, *Data Desk Version 5.0 for Macintosh* and *Data Desk 6.0 for Windows.* Data Desk® is an interactive and highly graphical statistics software program originally developed by Paul Velleman, a recognized leader in statistics education, at Cornell University in the mid-1980s. It is now used by thousands of institutions and individuals around the world in both teaching and research. Combining the philosophy and concepts of Exploratory Data Analysis with a full range of traditional statistical techniques, Data Desk's dynamic displays, drag-and-drop interface, and linked plots allow users to gain exciting insights into the nature and uses of data.

Data Desk Student Version 6.0 for Windows ISBN: 0-201-25831-1

Data Desk Student Version 5.0 for the Macintosh ISBN: 0-201-57124-2

Any of these products can be purchased separately, or bundled with Addison-Wesley texts. Instructors can contact local sales consultants for details or contact the company at 1-800-552-2499 for examination copies of any of these items.

The Addison Wesley Longman publishing team has been truly exceptional, and I thank Julie Berrisford, Jane Parrigin, Jenny Bagdigian, Sara Peterson, Carolyn Lee, Joe Vetere, and the entire Addison Wesley Longman staff.

I would also like to acknowledge the outstanding contributions of Theodore F. Moore, who made many valuable suggestions. Ted's death is a great loss to his family, friends, and the students of Mohawk Valley Community College.

I take great pride and pleasure in thanking my son Marc for developing the Web site, and I thank Marc and Scott for their suggestions and technical assistance. Finally, I thank my wife, Ginny, for her continuing support and encouragement.

M.F.T.
LaGrange, New York
August 1997

Acknowledgments

Writing a textbook is a monumental undertaking, which requires the input of many people beyond that of the author. *Elementary Statistics* has been fortunate in receiving valuable input and feedback from a long list of colleagues, many of whom are listed below, and, of course, from the many students who have used and commented on the text over the years.

For checking the accuracy of answers and solutions, special thanks go to Joe Fred Gonzalez, Milton Loyer, and Jane Parrigin. For agreeing to do interviews, I thank David Burn, Barbara Carvalho, Barry Cook, Lester Curtin, Jay Dean, Anthony DiUglio, David Hall, Bill James, Lee Miringoff, and Paul Mones. For data sets, I thank William Rathje and Masakazu Tani of the University of Arizona; Steve Wasserman, Philip Mackowiak, and Myron Levine of the University of Maryland; Rick Hutchinson and the National Park Service; and the hundreds of researchers who did studies, conducted surveys, and compiled data used in the examples and exercises of this book.

For the beta testing of the new version of STATDISK, I thank:

Jeff Andrews, TSG Associates, Inc.

Denise Brown, Collin County Community College

Dave Kender, Wright State University

Patricia Oakley, Seattle Pacific University

Jane Parrigin

Cheryl Slayden, Pellissippi State University

I extend my sincere thanks for the suggestions made by reviewers and users of this and previous editions of the book to the following:

Mary Abkemeier, Fontbonne College

Jules Albertini, Ulster County Community College

Tim Allen, Delta College

Stu Anderson, College of Du Page

Mary Anne Anthony, Rancho Santiago Community College

William Applebaugh, University of Wisconsin—Eau Claire

James Baker, Jefferson Community College

Anna Bampton, Christopher Newport University

James Beatty, Burlington County College

Philip M. Beckman, Black Hawk College

Marian Bedee, BGSU, Firelands College

Don Benbow, Marshalltown Community College

Michelle Benedict, Augusta College

Kathryn Benjamin, Suffolk County Community College

Ronald Bensema, Joliet Junior College

David Bernklau, Long Island University

Maria Betkowski, Middlesex Community College

Shirley Blatchley, Brookdale Community College

David Blaueuer, University of Findlay

Randy Boan, Aims Community College

John Buchl, John Wood Community College

Michael Butler, Mt. San Antonio College

Jerome J. Cardell, Brevard Community College

Don Chambless, Auburn University

Rodney Chase, Oakland Community College

Bob Chow, Grossmont College

Philip S. Clarke, Los Angeles Valley College

Darrell Clevidence, Carl Sandburg College

Susan Cribelli, Aims Community College

Imad Dakka, Oakland Community College

Arthur Daniel, Macomb Community College

Gregory Davis, University of Wisconsin, Green Bay

Tom E. Davis, III, Daytona Beach Community College

Charles Deeter, Texas Christian University

Joe Dennin, Fairfield University

Richard Dilling, Grace College

Rose Dios, New Jersey Institute of Technology

Paul Duchow, Pasadena City College

Bill Dunn, Las Positas College

Marie Dupuis, Milwaukee Area Technical College

Evelyn Dwyer, Walters State Community College

Jane Early, Manatee Community College

Sharon Emerson-Stonnell, Longwood College

P. Teresa Farnum, Franklin Pierce College

Ruth Feigenbaum, Bergen Community College

Vince Ferlini, Keene State College

Maggie Flint, Northeast State Technical Community College

Bob France, Edmonds Community College

Christine Franklin, University of Georgia

Richard Fritz, Moraine Valley Community College

Maureen Gallagher, Hartwick College

Mahmood Ghamsary, Long Beach City College

Tena Golding, Southeastern Louisiana University

Elizabeth Gray, Southeastern Louisiana University

David Gurney, Southeastern Louisiana University

Francis Hannick, Mankato State University

Sr. Joan Harnett, Molloy College

Leonard Heath, Pikes Peak Community College

Peter Herron, Suffolk County Community College

Mary Hill, College of Du Page

Larry Howe, Rowan College of New Jersey

Lloyd Jaisingh, Morehead State University

Martin Johnson, Gavilan College

Roger Johnson, Carleton College

Herb Jolliff, Oregon Institute of Technology

Francis Jones, Huntington College

Toni Kasper, Borough of Manhattan Community College

Alvin Kaumeyer, Pueblo Community College

William Keane, Boston College

Robert Keever, SUNY, Plattsburgh

Alice J. Kelly, Santa Clara University

Michael Kern, Bismarck State College

Marlene Kovaly, Florida Community College at Jacksonville

Tomas Kozubowski, University of Tennessee

Shantra Krishnamachari, Borough of Manhattan Community College

Richard Kulp, David Lipscomb University

Linda Kurz, SUNY College of Technology

Tommy Leavelle, Mississippi College

R. E. Lentz, Mankato State University

Timothy Lesnick, Grand Valley State University

Dawn Lindquist, College of St. Francis

George Litman, National-Louis University

Benny Lo, Ohlone College

Sergio Loch, Grand View College

Vincent Long, Gaston College

Barbara Loughead, National-Louis University

David Lund, University of Wisconsin—Eau Claire

Rhonda Magel, North Dakota State University

Gene Majors, Fullerton College

Hossein Mansouri, Texas State Technical College

Virgil Marco, Eastern New Mexico University

Joseph Mazonec, Delta College

Caren McClure, Rancho Santiago Community College

Phillip McGill, Illinois Central College

Marjorie McLean, University of Tennessee

Austen Meek, Cañada College

Robert Mignone, College of Charleston

Glen Miller, Borough of Manhattan Community College

Kermit Miller, Florida Community College at Jacksonville

Mitra Moassessi, Santa Monica College

Charlene Moeckel, Polk Community College

Theodore Moore, Mohawk Valley Community College

Gerald Mueller, Columbus State Community College

Sandra Murrell, Shelby State Community College

Faye Muse, Asheville-Buncombe Technical Community College

Gale Nash, Western State College

Felix D. Nieves, Antillean Adventist University

DeWayne Nymann, University of Tennessee

Patricia Odell, Bryant College

James O'Donnell, Bergen Community College

Alan Olinsky, Bryant College

Ron Pacheco, Harding University

Kwadwo Paku, Los Medanos College

Deborah Paschal, Sacramento City College

S. A. Patil, Tennessee Technological University

Robin Pepper, Tri-County Technical College

David C. Perkins, Texas A&M University—Corpus Christi

Anthony Piccolino, Montclair State University

Richard J. Pulskamp, Xavier University

Vance Revennaugh, Northwestern College

C. Richard, Southeastern Michigan College

Sylvester Roebuck, Jr., Olive Harvey College

Kenneth Ross, Broward Community College

Charles M. Roy, Camden County College

Kara Ryan, College of Notre Dame

Richard Schoenecker, University of Wisconsin, Stevens Point

Nancy Schoeps, University of North Carolina, Charlotte

Jean Schrader, Jamestown Community College

A. L. Schroeder, Long Beach City College

Phyllis Schumacher, Bryant College

Sankar Sethuraman, Augusta College

Rosa Seyfried, Harrisburg Area Community College

Calvin Shad, Barstow College

Carole Shapero, Oakton Community College

Lewis Shoemaker, Millersville University

Joan Sholars, Mt. San Antonio College

Galen Shorack, University of Washington

Teresa Siak, Davidson County Community College

Cheryl Slayden, Pellissippi State Technical Community College

Arthur Smith, Rhode Island College

Marty Smith, East Texas Baptist University

Aileen Solomon, Trident Technical College

Sandra Spain, Thomas Nelson Community College

Maria Spinacia, Pasco-Hernandez Community College

Paulette St. Ours, University of New England

W. A. Stanback, Norfolk State University

Carol Stanton, Contra Costa College

Richard Stephens, Western Carolina College

W. E. Stephens, McNeese State University

Terry Stephenson, Spartanburg Methodist College

Consuelo Stewart, Howard Community College

Ellen Stutes, Louisiana State University at Eunice

Sr. Loretta Sullivan, University of Detroit Mercy

Andrew Thomas, Triton College

Evan Thweatt, American River College

Judith A. Tully, Bunker Hill Community College

Gary Van Velsir, Anne Arundel Community College

Randy Villa, Napa Valley College

Hugh Walker, Chattanooga State Technical Community College

Charles Wall, Trident Technical College

Glen Weber, Christopher Newport College

David Weiner, Beaver College

Sue Welsch, Sierra Nevada College

Roger Willig, Montgomery County Community College

Gail Wiltse, St. Johns River Community College

Odell Witherspoon, Western Piedmont Community College

Jean Woody, Tulsa Junior College

Thomas Zachariah, Loyola Marymount University

Elyse Zois, Kean College of New Jersey

Brief Contents

Contents

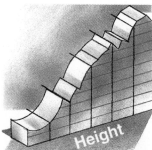

9 Correlation and Regression   474

10 Multinomial Experiments and Contingency Tables   534

11 Analysis of Variance   570

12 Statistical Process Control   610

Suggested Course Syllabus

Almost all instructors will find that the entire textbook cannot be covered in one semester. Because some topics will normally be omitted, we have included samples of course syllabi. The "class hour" referred to in these outlines consists of approximately one hour. In addition to the five syllabi that follow, there are many other configurations that will work very successfully. The particular outline chosen will depend on individual needs and preferences. The following outlines are provided only as a source of some suggestions.

When selecting the sections to be included, consider assigning some sections as optional or required out-of-class independent work. For example, after students understand the basic concepts of confidence intervals and hypothesis testing, consider assigning exercises from Chapter 8 (Inferences from Two Samples) to be done using a computer software package, such as STATDISK. This is an excellent way to include important and useful concepts from Chapter 8 without using valuable and limited class time.

In planning a new course syllabus for the Seventh Edition of *Elementary Statistics*, users of the Sixth Edition should consider these major changes:

- For those professors who prefer *early coverage of correlation and regression*, Sections 9-2 and 9-3 have been rewritten so that they don't require prior coverage of hypothesis testing. Sections 9-2 and 9-3 are now written so that they can follow Chapter 2. See the sample Syllabus Version B.

- Chapter 4 includes a new Section: 4-5 (The Poisson Distribution)

- The former Section 5-3 on Nonstandard Normal Distributions has been broken up into two sections:

    5-3   (Nonstandard Normal Distributions: Finding Probabilities

    5-4   (Nonstandard Normal Distributions: Finding Scores)

- The former Section 6-2 (Estimating a Population Mean) has been broken up into two sections:

    6-2   (Estimating a Population Mean: Large Samples)

    6-3   (Estimating a Population Mean: Small Samples)

- The former Sections 11-2 (One-Way ANOVA With Equal Sample Sizes) and 11-3 (One-Way ANOVA With Unequal Samples Sizes) have been combined into one section: 11-2 (One-Way ANOVA)

- There is a new section in Chapter 14: 14-1 (A Statistics Group Project), which describes an end-of-semester activity with a list of suggested topics.

Syllabus Version A

Class Hour	Text Section	Topic
1	1-1, 1-2, 1-3	Overview; Nature of Data; Uses and Abuses of Statistics
2	1-4	Design of Experiments
3	2-1, 2-2, 2-3	Summarizing Data and Pictures of Data
4	2-4	Measures of Central Tendency
5	2-5	Measures of Variation
6	2-6	Measures of Position
7	2-7	Exploratory Data Analysis
8	Ch. 1, 2	Review
9	Ch. 1, 2	Test 1
10	3-1, 3-2, 3-3	Fundamentals of Probability and Addition Rule
11	3-4	Multiplication Rule
12	4-1, 4-2	Random Variables
13	4-3	Binomial Experiments
14	4-4	Mean and Standard Deviation for the Binomial Distribution
15	Ch. 3, 4	Review
16	Ch. 3, 4	Test 2
17	5-1, 5-2	The Standard Normal Distribution
18	5-3	Nonstandard Normal Distributions: Finding Probabilities
19	5-4	Nonstandard Normal Distributions: Finding Scores
20	5-5	The Central Limit Theorem
21	6-1, 6-2	Estimates and Sample Sizes of Means: Large Samples
22	6-3	Estimates and Sample Sizes of Means: Small Samples
23	6-4	Estimates and Sample Sizes of Proportions
24	Ch. 5, 6	Review
25	Ch. 5, 6	Test 3
26	7-1, 7-2	Fundamentals of Hypothesis Testing
27	7-2, 7-3	Claims about a Mean: Large Samples
28	7-3	Claims about a Mean: Large Samples
29	7-4	Claims about a Mean: Small Samples
30	7-5	Claims about a Proportion
31	7-6	Claims about a Variance or Standard Deviation
32	Ch. 7	Review
33	Ch. 7	Test 4
34	9-1, 9-2	Correlation
35	9-2, 9-3	Correlation and Regression
36	9-3	Regression
37	10-3	Contingency Tables
38	12-1, 12-2	Control Charts for Variation and Mean
39	12-3	Control Charts for Attributes
40	Ch. 9, 10, 12	Review
41	Ch. 9, 10, 12	Test 5
42		Group Project Presentations
43		Review for Comprehensive Final Exam

Syllabus Version B: Early Coverage of Correlation and Regression

Class Hour	Text Section	Topic
1	1-1, 1-2, 1-3	Overview; Nature of Data; Uses and Abuses of Statistics
2	1-4	Design of Experiments
3	2-1, 2-2, 2-3	Summarizing Data and Pictures of Data
4	2-4	Measures of Central Tendency
5	2-5	Measures of Variation
6	2-6	Measures of Position
7	2-7	Exploratory Data Analysis
8	**9-1, 9-2**	**Correlation**
9	**9-3**	**Regression**
10	Ch. 1, 2, 9	Review
11	Ch. 1, 2, 9	Test 1
12	3-1, 3-2, 3-3	Fundamentals of Probability and Addition Rule
13	3-4	Multiplication Rule
14	4-1, 4-2	Random Variables
15	4-3	Binomial Experiments
16	4-4	Mean and Standard Deviation for the Binomial Distribution
17	5-1, 5-2	The Standard Normal Distribution
18	5-3	Nonstandard Normal Distributions: Finding Probabilities
19	5-4	Nonstandard Normal Distributions: Finding Scores
20	5-5	The Central Limit Theorem
21	Ch. 3, 4, 5	Review
22	Ch. 3, 4, 5	Test 2
23	6-1, 6-2	Estimates and Samples Sizes of Means: Large Samples
24	6-3	Estimates and Sample Sizes of Means: Small Samples
25	6-4	Estimates and Sample Sizes of Proportions
26	7-1, 7-2	Fundamentals of Hypothesis Testing
27	7-2, 7-3	Claims about a Mean: Large Samples
28	7-3	Claims about a Mean: Large Samples
29	7-4	Claims about a Mean: Small Samples
30	7-5	Claims about a Proportion
31	7-6	Claims about a Variance or Standard Deviation
32	Ch. 6, 7	Review
33	Ch. 6, 7	Test 3
34	10-1, 10-2	Multinomial Experiments
35	10-3	Contingency Tables
36	12-1, 12-2	Control Charts for Variation and Mean
37	12-3	Control Charts for Attributes
38	Ch. 10, 12	Review
39	Ch. 10, 12	Test 4
40		Review for Comprehensive Final Exam
41		Review for Comprehensive Final Exam

Syllabus Version C: Minimum Coverage of Probability

Class Hour	Text Section	Topic
1	1-1, 1-2, 1-3	Overview, Nature of Data, Uses and Abuses of Statistics
2	1-4	Design of Experiments
3	2-1, 2-2, 2-3	Summarizing Data and Pictures of Data
4	2-4	Measures of Central Tendency
5	2-5	Measures of Variation
6	2-6	Measures of Position
7	2-7	Exploratory Data Analysis
8	Ch. 1, 2	Review
9	Ch. 1, 2	Test 1
10	**3-1, 3-2**	**Fundamentals of Probability**
11	4-1, 4-2	Random Variables
12	4-3	Binomial Experiments
13	4-4	Mean and Standard Deviation for the Binomial Distribution
14	Ch. 3, 4	Review
15	Ch. 3, 4	Test 2
16	5-1, 5-2	The Standard Normal Distribution
17	5-3	Nonstandard Normal Distributions: Finding Probabilities
18	5-4	Nonstandard Normal Distributions: Finding Scores
19	5-5	The Central Limit Theorem
20	6-1, 6-2	Estimates and Sample Sizes of Means: Large Samples
21	6-3	Estimates and Sample Sizes of Means: Small Samples
22	6-4	Estimates and Samples Sizes of Proportions
23	Ch. 5, 6	Review
24	Ch. 5, 6	Test 3
25	7-1, 7-2	Fundamentals of Hypothesis Testing
26	7-3	Claims about a Mean: Large Samples
27	7-3	Claims about a Mean: Large Samples
28	7-4	Claims about a Mean: Small Samples
29	7-5	Claims about a Proportion
30	7-6	Claims about a Variance or Standard Deviation
31	Ch. 7	Review
32	Ch. 7	Test 4
33	9-1, 9-2	Correlation
34	9-2, 9-3	Correlation and Regression
35	9-3	Regression
36	10-1, 10-2	Multinomial Experiments
37	10-3	Contingency Tables
38	12-1, 12-2	Control Charts for Variation and Mean
39	12-3	Control Charts for Attributes
40	Ch. 9, 10, 12	Review
41	Ch. 9, 10, 12	Test 5
42		Review for Comprehensive Final Exam
43		Review for Comprehensive Final Exam

Syllabus Version D: Integration of Some Nonparametric Statistics

Class Hour	Text Section	Topic
1	1-1, 1-2, 1-3	Overview; Nature of Data; Uses and Abuses of Statistics
2	1-4, 13-7	Design of Experiments; **Runs Test for Randomness**
3	2-1, 2-2, 2-3	Summarizing Data and Pictures of Data
4	2-4	Measures of Central Tendency
5	2-5	Measures of Variation
6	2-6	Measures of Position
7	2-7	Exploratory Data Analysis
8	Ch. 1, 2	Review
9	Ch. 1, 2	Test 1
10	3-1, 3-2, 3-3	Fundamentals of Probability and Addition Rule
11	3-4	Multiplication Rule
12	4-1, 4-2	Random Variables
13	4-3	Binomial Experiments
14	4-4	Mean and Standard Deviation for the Binomial Distribution
15	Ch. 3, 4	Review
16	Ch. 3, 4	Test 2
17	5-1, 5-2	The Standard Normal Distribution
18	5-3	Nonstandard Normal Distributions: Finding Probabilities
19	5-4	Nonstandard Normal Distributions: Finding Scores
20	5-5	The Central Limit Theorem
21	6-1, 6-2	Estimates and Sample Sizes of Means: Large Samples
22	6-3	Estimates and Sample Sizes of Means: Small Samples
23	6-4	Estimates and Sample Sizes of Proportions
24	Ch. 5, 6	Review
25	Ch. 5, 6	Test 3
26	7-1, 7-2	Fundamentals of Hypothesis Testing
27	7-2, 7-3	Claims about a Mean: Large Samples
28	7-4	Claims about a Mean: Small Samples
29	7-5	Claims about a Proportion
30	7-6	Claims about a Variance or Standard Deviation
31	Ch. 7	Review
32	Ch. 7	Test 4
33	9-1, 9-2	Correlation and Regression
34	9-3	Regression
35	**13-6**	**Rank Correlation**
36	**13-5**	**Kruskal-Wallis Test**
37	10-3	Contingency Tables
38	12-1, 12-2	Control Charts for Variation and Mean
39	12-3	Control Charts for Attributes
40	Ch. 9, 10, 12, 13	Review
41	Ch. 9, 10, 12, 13	Test 5
42		Review for Comprehensive Final Exam
43		Review for Comprehensive Final Exam

Syllabus Version E: Two Semester Course

A two-semester sequence allows for more time to better develop topics such as computer usage, simulation techniques in probability, and bootstrap resampling techniques in estimating parameters and testing hypotheses.

First Semester:

Chapter 1	(Sections 1-1 through 1-5):	5 classes
Chapter 2	(Sections 2-1 through 2-7):	7 classes
Chapter 3	(Sections 3-1 through 3-6):	5 classes
Chapter 4	(Sections 4-1 through 4-5):	5 classes
Chapter 5	(Sections 5-1 through 5-6):	7 classes
Chapter 6	(Sections 6-1 through 6-5):	7 classes
Chapter 7	(Sections 7-1 through 7-6):	8 classes

Second Semester:

Chapter 8	(Sections 8-1 through 8-6):	8 classes
Chapter 9	(Sections 9-1 through 9-5):	9 classes
Chapter 10	(Sections 10-1 through 10-3):	4 classes
Chapter 11	(Sections 11-1 through 11-3):	6 classes
Chapter 12	(Sections 12-1 through 12-3):	4 classes
Chapter 13	(Sections 13-1 through 13-7):	9 classes
Chapter 14	(Section 14-1):	4 classes

ELEMENTARY

STATISTICS

SEVENTH EDITION

1

Introduction to Statistics

1-1 Overview

The term statistics is defined along with the terms population, sample, parameter, and statistic.

1-2 The Nature of Data

Quantitative data and qualitative data are defined along with discrete data and continuous data. The four levels of measurement (nominal, ordinal, interval, ratio) are also defined.

1-3 Uses and Abuses of Statistics

Examples of beneficial uses of statistics are presented, along with some of the common ways in which statistics are used to deceive. Deceptive uses include small samples, precise numbers, distorted percentages, loaded questions, misleading graphs, and bad samples.

1-4 Design of Experiments

Observational studies and experiments are described, along with good statistical methodology. The importance of good sampling is emphasized. Different sampling methods are defined and described, including random sampling, stratified sampling, systematic sampling, cluster sampling, and convenience sampling.

1-5 Statistics with Calculators and Computers

The importance of calculators and computers is discussed. Calculator usage is discussed along with the statistical software packages of STATDISK and Minitab.

Chapter Problem:

What can we conclude from this survey?

The ABC "Nightline" television show conducted a poll in which viewers were asked to spend 50 cents for a telephone call to a "900" telephone number. They were asked to state whether the United Nations should continue to be located in the United States. Of the 186,000 people who responded, 67% stated that the United Nations should be moved out of the United States. Based on the sample results obtained in this poll, what can we conclude about the way the general population feels about keeping the United Nations in the United States?

1-1 Overview

We begin our study of *statistics* by noting that the word has two basic meanings. In the first sense, the term is used in reference to actual and specific numbers derived from data, as illustrated in these examples:

- In a Bruskin-Goldring Research poll of 1012 people who were asked how a fruitcake should be used, 13% responded that it should be used as a doorstop.
- Among those who are given a drug test for a new job, 3.8% test positive (according to the American Management Association).
- The highest season baseball batting average to date is 0.442, achieved by James O'Neil in 1887.

A second meaning refers to statistics as a method of analysis.

DEFINITION

Statistics is a collection of methods for planning experiments, obtaining data, and then organizing, summarizing, presenting, analyzing, interpreting, and drawing conclusions based on the data.

Statistics involves much more than simply drawing graphs and calculating averages. In this book we will learn how to develop generalized and meaningful conclusions that go beyond the original data. In statistics, we commonly use the terms *population* and *sample*. These terms are at the very core of statistics and we define them now.

DEFINITIONS

A **population** is the complete collection of all elements (scores, people, measurements, and so on) to be studied.

A **census** is the collection of data from *every* element in a population.

A **sample** is a subcollection of elements drawn from a population.

For example, a typical Nielsen television survey uses a *sample* of 4000 households, and the results are used to form conclusions about the *population* of all 97,855,392 households in the United States.

Closely related to the concepts of population and sample are the concepts of *parameter* and *statistic*. The following definitions are easy to remember

---

The State of Statistics

The word *statistics* is derived from the Latin word *status* (meaning "state"). Early uses of statistics involved compilations of data and graphs describing various aspects of a state or country. In 1662, John Graunt published statistical information about births and deaths. Graunt's work was followed by studies of mortality and disease rates, population sizes, incomes, and unemployment rates. Households, governments, and businesses rely heavily on statistical data for guidance. For example, unemployment rates, inflation rates, consumer indexes, and birth and death rates are carefully compiled on a regular basis, and the resulting data are used by business leaders to make decisions affecting future hiring, production levels, and expansion into new markets.

if we recognize the alliteration in "population parameter" and "sample statistic."

Annotations for instructors are included throughout this book, but they are *not* included in the student editions of the book. This entire chapter can be assigned as reading, but the four levels of measurement in Section 1-2 will probably require some explanation.

DEFINITIONS

A **parameter** is a numerical measurement describing some characteristic of a *population*.

A **statistic** is a numerical measurement describing some characteristic of a *sample*.

Let's consider an example. In a Bruskin-Goldring Research survey of 1015 randomly selected people, 269 (or 26.5%) currently own a computer. Because the figure of 26.5% is based on a sample (not the entire population), it is a *statistic* (not a parameter). If a survey of the 50 U.S. governors showed that 42 (or 84%) of them own computers, then the figure of 84% is a *parameter* because it is based on the entire population of governors.

An important aspect of statistics is its obvious applicability to real and relevant situations, and a wide variety of these applications will be found throughout this book.

1-2 The Nature of Data

Some data sets consist of numbers (such as heights), and others are nonnumerical (such as gender). The terms *quantitative data* and *qualitative data* are often used to distinguish between these two types.

The four levels of measurement will require some explanation. Students have the most difficulty distinguishing between the interval and ratio levels of measurement. Suggest a "ratio" test: If one number is twice the other, is the quantity being measured also twice the other quantity? If yes, the data are at the ratio level.

DEFINITIONS

Quantitative data consist of numbers representing counts or measurements.

Qualitative (or **categorical** or **attribute**) **data** can be separated into different categories that are distinguished by some nonnumeric characteristic.

Data Set 4 in Appendix B includes amounts of tar in different cigarettes. Those amounts are quantitative data, but the brand names are qualitative data.

We can further describe quantitative data by distinguishing between the discrete and continuous types.

The concept of "countable" is discussed briefly here, but it could be very difficult for many students to understand well. Consider commenting that "countable" basically means that the item can be counted somehow. See the text example wherein eggs from a hen are counted, but you cannot count the milk from a cow.

DEFINITIONS

Discrete data result from either a finite number of possible values or a countable number of possible values. (That is, the number of possible values is 0, or 1, or 2, and so on.)

Continuous (numerical) data result from infinitely many possible values that can be associated with points on a continuous scale in such a way that there are no gaps or interruptions.

When data represent counts, they are discrete; when they represent measurements, they are continuous. The numbers of eggs that hens lay are *discrete* data because they represent counts; the amounts of milk that cows produce are *continuous* data because they are measurements that can assume any value over a continuous span.

Another common way to classify data is to use four levels of measurement: nominal, ordinal, interval, ratio.

DEFINITION

The **nominal level of measurement** is characterized by data that consist of names, labels, or categories only. The data cannot be arranged in an ordering scheme (such as low to high).

If we associate the term *nominal* with "name only," the meaning becomes easy to remember. An example of nominal data is the political party to which each member of the U.S. Senate belongs.

> **EXAMPLE**   The following are other examples of sample data at the nominal level of measurement.
>
> 1. Survey responses of yes, no, and undecided
> 2. The genders of students in your statistics class

Because the categories lack any ordering or numerical significance, the preceding data cannot be used for calculations. We cannot, for example, "average" 20 women and 15 men. *Caution:* Numbers are sometimes assigned to categories (especially when data are computerized), but these numbers lack any real computational significance and the average calculated with them is usually meaningless. We might find that the Gallup Organization has computerized survey data in which Democrats are assigned 0, Republicans are

assigned 1, and Independents are assigned 2. Even though we now have number labels, the data set continues to be at the nominal level and we shouldn't perform calculations with it.

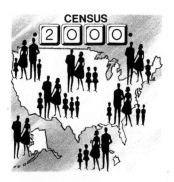

DEFINITION

The **ordinal level of measurement** involves data that may be arranged in some order, but differences between data values either cannot be determined or are meaningless.

EXAMPLE

The following are examples of data at the ordinal level of measurement.

1. An editor rates some manuscripts as "excellent," some as "good," and some as "bad." (We can't find a specific quantitative difference between "good" and "bad.")

2. An Olympic screening committee ranks Gail 3rd, Diana 7th, and Kim 10th. (We can find a difference between ranks of 3 and 7, but the difference of 4 doesn't mean anything.)

This ordinal level provides information about relative comparisons, but the degrees of differences are not usable for calculations. Data at the ordinal level should not be used for calculations.

DEFINITION

The **interval level of measurement** is like the ordinal level, with the additional property that we can determine meaningful amounts of differences between data. However, there is no inherent (natural) zero starting point (where *none* of the quantity is present).

Body temperatures of 98.2°F and 98.6°F are examples of data at this interval level of measurement. Such values are ordered, and we can determine their difference (often called the *distance* between the two values). However, there is no natural starting point. The value of 0°F might seem like a starting point, but it is arbitrary and does not represent "no heat." It is wrong to say that 50°F is twice as hot as 25°F. (Temperature readings on the Kelvin scale are at the ratio level of measurement; that scale has an absolute zero.)

Census 2000

The national census for the year 2000 will be faster, less expensive, and more accurate than the 1990 census. Unlike the 1990 census, Census 2000 will use sampling methods to obtain more accurate results. In 1990, census workers returned up to six times to each of 35 million households that did not mail the census forms, but such missing households will be sampled in 2000. It is expected that sampling will produce more accurate results than attempts to reach each individual household. Census 2000 will cost about $4 billion, which is $1 billion less than the cost of repeating the same methods used in 1990. Census 2000 will be more efficient, although the 1990 census was not quite as inefficient as columnist Dave Barry suggested when he wrote: "Census Bureau mails a hundred million forms, 87 million of which arrive at a single household in Albany."

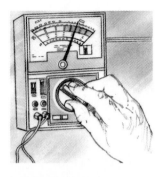

Measuring Disobedience

How are data collected about something that doesn't seem to be measurable, such as people's level of disobedience? Psychologist Stanley Milgram devised the following experiment: A researcher instructed a volunteer subject to operate a control board that gave increasingly painful "electrical shocks" to a third person. Actually, no real shocks were given, and the third person was an actor. The volunteer began with 15 volts and was instructed to increase the shocks by increments of 15 volts. The disobedience level was the point at which the subject refused to increase the voltage. Surprisingly, two-thirds of the subjects obeyed orders even though the actor screamed and faked a heart attack.

EXAMPLE   The following are examples of data at the interval level of measurement.

1. The years 1000, 2000, 1776, and 1944. (Time did not begin in the year 0, so the year 0 is arbitrary instead of being a natural zero starting point.)
2. Average annual temperatures (in degrees Celsius) of the 50 state capitals.

DEFINITION

The **ratio level of measurement** is the interval level modified to include the inherent zero starting point (where zero indicates that *none* of the quantity is present). For values at this level, differences and ratios are both meaningful.

EXAMPLE   The following are examples of data at the ratio level of measurement.

1. Weights of plastic discarded by households (0 lb does represent no plastic discarded, and 10 lb does weigh twice as much as 5 lb)
2. Lengths (in minutes) of movies
3. Distances (in miles) traveled by cars in a test of fuel consumption

Values in each of these data collections can be arranged in order, differences can be computed, and there is an inherent zero starting point. *This level is called the ratio level because the starting point makes ratios meaningful.* Because a 200-lb weight is *twice* as heavy as a 100-lb weight, but 50°F is *not* twice as hot as 25°F, weights are at the ratio level while Fahrenheit temperatures are at the interval level. For a concise comparison and review, study Table 1-1 to see the differences among the four levels of measurement.

In applying statistics to real problems, the level of measurement of the data is an important factor in determining which procedure to use. Our understanding of the four levels of measurement should be supplemented with common sense—an indispensable tool in statistics. For example, it makes no sense to calculate an average of a list of social security numbers, because those numbers don't really *measure* or *count* anything; instead, they simply serve the function of identifying people. Social security numbers are different *names* for people and, as such, they should not be used for calculations. In general, we should not calculate averages for data at the nominal or ordinal levels of measurement.

TABLE 1-1	Levels of Measurement of Data	
Level	Summary	Example
Nominal	Categories only. Data cannot be arranged in an ordering scheme.	Student cars: 10 Corvettes 20 Ferraris 40 Porsches } Categories or names only.
Ordinal	Categories are ordered, but differences cannot be determined or they are meaningless.	Student cars: 10 compact 20 mid-size 40 full size } An order is determined by "compact, mid-size, full-size."
Interval	Differences between values can be found, but there may be no inherent starting point. Ratios are meaningless.	Campus temperatures: 45°F 80°F 90°F } 90°F is not twice as hot as 45°F.
Ratio	Like interval, but with an inherent starting point. Ratios are meaningful.	Weights of college football players: 150 lb 195 lb 300 lb } 300 lb is *twice* 150 lb.

⟨1-2⟩ Exercises A: Basic Skills and Concepts

In Exercises 1–8, identify each number as discrete *or* continuous.

1. Each Camel cigarette has 16.13 mg of tar.   1. Continuous
2. An American Airlines altimeter reports a height of 21,359 ft.   2. Continuous
3. A Bruskin-Goldring Research poll of 1015 people shows that 40 of them now subscribe to an on-line computer service.   3. Discrete
4. Radar indicated that Nolan Ryan pitched the last ball at 82.3 mi/h.
5. Among all SAT scores recorded last year, 27 were perfect.   5. Discrete
6. Among 1000 consumers surveyed, 930 recognized the Campbell's Soup brand name.   6. Discrete
7. The total time a New York City cab driver spends yielding to pedestrians each year is 2.367 sec.   7. Continuous
8. Upon completion of a training program, Shaquille O'Neal weighed 12.44 lb less than when he started the program.   8. Continuous

Recommended exercises: All of 1–18. In Exercise 11, note that social security numbers are really different names for people. In Exercise 15, note that Zip codes are only names for locations; they do not follow a consistent ordering scheme.

4. Continuous

In Exercises 9–18, determine which of the four levels of measurement (nominal, ordinal, interval, ratio) is most appropriate.

9. Ratings of superior, above average, average, below average, or poor for blind dates   9. Ordinal
10. Nicotine contents (in milligrams) of Camel cigarettes   10. Ratio
11. Social security numbers   11. Nominal
12. Temperatures (in degrees Celsius) of a sample of angry taxpayers who are being audited   12. Interval
13. Years in which Democrats won presidential elections   13. Interval
14. Final course grades (A, B, C, D, F) for statistics students   14. Ordinal
15. Zip codes   15. Nominal
16. Annual incomes of nurses   16. Ratio
17. Cars described as subcompact, compact, intermediate, or full-size
18. Colors of a sample of M&M candies   18. Nominal

17. Ordinal

1-2 Exercises B: Beyond the Basics

19. U.S. presidents were assassinated in 1865, 1881, 1901, and 1963. What is the level of measurement for those years? Explain your answer.
20. In the "Born Loser" cartoon strip by Art Sansom, Brutus expresses joy over an increase in temperature from 1° to 2°. When asked what is so good about 2°, he answers that "It's twice as warm as this morning." Why is Brutus wrong once again?

19. Interval. Differences between the years can be determined and are meaningful, but there is no inherent starting point since time did not begin in the year zero.
20. With no natural starting point, the temperatures are at the interval level of measurement; ratios such as "twice" are meaningless.

1-3 Uses and Abuses of Statistics

Uses of Statistics

The applications of statistics have grown so that practically every field of study now benefits in some way from the use of statistical methods. Manufacturers now provide better products at lower costs through the use of statistical quality control techniques. Diseases are controlled through analyses designed to anticipate epidemics. Endangered species of fish and other wildlife are protected through regulations and laws that react to statistical estimates of changing population sizes. By pointing to lower fatality rates, legislators can better justify laws such as those governing air pollution, auto inspections, seat belt and air bag use, and drunk driving. We will cite only these few examples, because a complete compilation of the uses of statistics would easily fill the remainder of this book (a prospect not totally unpleasant to some readers).

Some students choose a statistics course because it's required, but increasing numbers do so voluntarily because they recognize its value and application to whatever field they plan to pursue. Because employers *love* to see a statistics course on the transcript of a job applicant, you will have an advantage

The content of this section is easy enough for students to read on their own, but really stress the importance of good sampling techniques. Discuss the problems with self-selected surveys, such as TV and radio surveys that ask audience members to call in, especially if the caller must pay a fee. Make the important point that if the sampling is not done correctly, even very large samples may be totally worthless.

when seeking employment. Apart from job-motivated or discipline-related reasons, the study of statistics can help you become more critical in your analysis of information, making you less susceptible to misleading or deceptive claims, such as those commonly associated with polls, graphs, and averages. As an educated and responsible member of society, you should sharpen your ability to recognize distorted statistical data; in addition, you should learn to interpret undistorted data intelligently.

Abuses of Statistics

Abuses of statistics have occurred for some time. For example, about a century ago, statesman Benjamin Disraeli famously said, "There are three kinds of lies: lies, damned lies, and statistics." It has also been said that "figures don't lie; liars figure," and that "if you torture the data long enough, they'll admit to anything." Historian Andrew Lang said that some people use statistics "as a drunken man uses lampposts—for support rather than illumination." These statements refer to abuses of statistics in which data are presented in ways that may be misleading. Some abusers of statistics are simply ignorant or careless, whereas others have personal objectives and are willing to suppress unfavorable data while emphasizing supportive data. We will now present a few examples of the many ways in which data can be distorted.

Small Samples   In Chapter 6 we will see that small samples are not necessarily bad, but small sample results are sometimes used as a form of statistical "lying." The toothpaste preferences of only 10 dentists should not be used as a basis for a generalized claim such as "Covariant toothpaste is recommended by 7 out of 10 dentists." Even if a sample is large, it must be unbiased and representative of the population from which it comes. Sometimes a sample might seem relatively large (as in a survey of "2000 randomly selected adult Americans"), but if conclusions are made about subgroups, such as the Catholic male Republicans in the survey, such conclusions might be based on very small samples.

Precise Numbers   Sometimes the numbers themselves can be deceptive. A very precise figure, such as an annual salary of $37,735.29, might be used to sound precise and instill a high degree of confidence in its accuracy. The figure of $37,700 doesn't convey that same sense of precision and accuracy. A statistic that is very precise with many decimal places, however, is not necessarily *accurate*.

Guesstimates   Another source of statistical deception involves estimates that are really guesses and can therefore be in error by substantial amounts. We should consider the source of the estimate and how it was developed. When the Pope visited Miami, officials estimated the crowd size to be 250,000, but the *Miami Herald* used aerial photos and grids to come up with a more accurate estimate of 150,000.

Old Drivers Are Safer than Young?

The American Association of Retired People (AARP) claims that older drivers are involved in fewer crashes than younger drivers. In a recent year, drivers aged 16–19 caused about 1.5 million crashes, compared to only 540,000 crashes caused by drivers aged 70 and over, so the AARP claim seems valid. But because older drivers don't drive as much as younger drivers, that is not the whole picture. Instead of considering only the raw number of accidents, the accident *rates* should also be examined. Here are accident rates per 100 million miles driven: 8.6 for those aged 16–19, 4.6 for those aged 75–79, 8.9 for those aged 80–84, and 20.3 for those aged 85 and over. While the youngest drivers do have more accidents, the oldest drivers have the highest accident rates.

The *Literary Digest* Poll

In the 1936 presidential race, *Literary Digest* magazine ran a poll and predicted an Alf Landon victory, but Franklin D. Roosevelt won by a landslide. Maurice Bryson notes, "Ten million sample ballots were mailed to prospective voters, but only 2.3 million were returned. As everyone ought to know, such samples are practically always biased." He also states, "Voluntary response to mailed questionnaires is perhaps the most common method of social science data collection encountered by statisticians, and perhaps also the worst." (See Bryson's "The *Literary Digest* Poll: Making of a Statistical Myth," *The American Statistician*, Vol. 30, No. 4.)

Distorted Percentages   Misleading or unclear percentages are sometimes used. Continental Airlines ran full-page ads boasting better service. In referring to lost baggage, these ads claimed that this was "an area where we've already improved 100% in the last six months." In an editorial criticizing this statistic, The *New York Times* correctly interpreted the 100% improvement figure to mean that no baggage is now being lost—an accomplishment not yet achieved by Continental Airlines.

Partial Pictures   "Ninety percent of all our cars sold in this country in the last 10 years are still on the road." Millions of consumers heard that commercial message and got the impression that those cars must be well built to last through those long years of driving. What the auto manufacturer failed to mention was that 90% of the cars it sold in this country were sold within the last three years. The claim was technically correct, but it was very misleading in not presenting the complete results.

Deliberate Distortions   In the book *Tainted Truth*, Cynthia Crossen cites an example of the magazine *Corporate Travel* that published results showing that among car rental companies, Avis was the winner in a survey of people who rent cars. When Hertz requested detailed information about the survey, the actual survey responses disappeared and the magazine's survey coordinator resigned. Hertz sued Avis (for false advertising based on the survey) and the magazine; a settlement was reached.

Loaded Questions   Survey questions can be worded to elicit a desired response. A famous case involves presidential candidate Ross Perot, who asked this question in a mail survey: "Should the president have the line item veto to eliminate waste?" The results included 97% "yes" responses. However, 57% said "yes" when subjects were randomly selected and asked this question: "Should the President have the line item veto, or not?" Sometimes questions are unintentionally loaded by such factors as the order of the items being considered. For example, one German poll asked these two questions:

- Would you say that traffic contributes more or less to air pollution than industry?
- Would you say that industry contributes more or less to air pollution than traffic?

When traffic was presented first, 45% blamed traffic and 32% blamed industry; when industry was presented first, those percentages changed dramatically to 24% and 57%, respectively.

Misleading Graphs   Many visual devices—such as bar graphs and pie charts—can be used to exaggerate or deemphasize the true nature of data. (Such devices will be discussed in Chapter 2.) The two graphs in Figure 1-1 depict

(a)

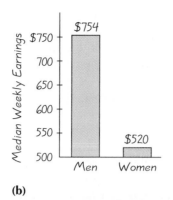

(b)

Figure 1-1   Earnings of Full-Time Professional Workers

the *same data* from the Bureau of Labor Statistics, but part (b) is designed to exaggerate the difference between the earnings of men and women. By not starting the vertical axis at zero, the graph in part (b) tends to produce a misleading subjective impression. Figure 1-1 carries an important lesson: We should analyze the *numerical* information given in the graph, so that we won't be misled by its general shape.

Pictographs   Drawings of objects, called pictographs, may also be misleading. Some objects commonly used to depict data include moneybags, stacks of coins, army tanks (for military expenditures), cows (for dairy production), barrels (for oil production), and houses (for home construction). When drawing such objects, artists can create false impressions that distort differences. If you double each side of a square, the area doesn't merely double; it increases by a factor of four. If you double each side of a cube, the volume doesn't merely double; it increases by a factor of eight. If taxes double over a decade, an artist may depict tax amounts with one moneybag for the first year and a second moneybag that is twice as deep, twice as tall, and twice as wide. Instead of appearing to double, taxes will appear to increase by a factor of eight, so the truth will be distorted by the drawing.

Pollster Pressure   When survey subjects are asked questions, they often provide responses that are favorable to their self-image. In one telephone survey, 94% of the respondents said that they wash their hands after using a bathroom, but observations in such places as Penn Station in New York City and Golden Gate Park in San Francisco showed that the actual rate is only 68%. (Readers are discouraged from replicating these results by lurking about in public restrooms.)

Bad Samples   Another source of deceptive statistics is inappropriate methods of collecting data. It is common for a researcher to analyze data and form conclusions that are wrong because the method of collecting the data was poor.

One typical example is the "Nightline" poll in which 186,000 television viewers each paid 50 cents to call a "900" telephone number with their opinion about keeping the United Nations in the United States. The results showed that 67% of those who called were in favor of moving the United Nations out of the United States. At the beginning of this chapter, we asked what we can conclude about the way the general population feels about keeping the United Nations in the United States. Because the viewers themselves decided whether to be included in the survey, we have an example of a self-selected survey, defined as follows.

DEFINITION

A **self-selected survey** is one in which the respondents themselves decide whether to be included.

In such surveys it often happens that only those people with a strong opinion participate, with the result that the sample of people who respond is not representative of the whole population. Given that 67% of 186,000 respondents favored moving the United Nations out of the United States, *we cannot conclude anything about the general population because of the way the sample was obtained.* In fact, Ted Koppel reported that a "scientific" poll of 500 people showed that 72% of us wanted the United Nations to *stay* in the United States. In this poll of 500 people, respondents were randomly selected by the pollster so the results were much more likely to reflect the true opinion of the general population.

A self-selected survey is only one way in which the method of collecting data can be seriously flawed. Because the method of sampling or collecting data is so important, we will devote the following section to it.

The preceding examples comprise a small sampling of the ways in which statistics can be used deceptively. Entire books have been devoted to this subject, including Darrell Huff's classic *How to Lie with Statistics*, Robert Reichard's *The Figure Finaglers*, and Cynthia Crossen's *Tainted Truth*. Understanding these practices will be extremely helpful in evaluating the statistical data found in everyday situations.

1-3 Exercises A: Basic Skills and Concepts

Recommended exercises: All of 1–12. Students may have some difficulty identifying *two* different problems in Exercise 10.

1. People with unlisted numbers and people without telephones are excluded.

1. You've been hired to research recognition of the Nike brand name, and you must conduct a telephone survey of 1500 consumers in the United States. What is wrong with using telephone directories as the population from which the sample is drawn?

2. Seventy-two percent of Americans squeeze the toothpaste tube from the top. This and other not-so-serious findings are presented in *The First Really Important Survey of American Habits*. Those results are based on 7000 respondents from the 25,000 questionnaires that were mailed. What is wrong with this survey?

3. A report sponsored by the Florida Citrus Commission concluded that cholesterol levels can be lowered by eating citrus products. Why might the conclusion be suspect?

4. An employee has an annual salary of $40,000 but is told that she will be given a 10% cut in pay because of declining company profits. She is also told that next year, she will be given a 10% raise. This doesn't seem too bad because the 10% cut seems to be offset by the 10% raise.

   a. What is the annual income after the 10% cut?   4. a. $36,000

   b. Use the annual income from part a to find the annual income after the 10% raise. Did the 10% cut followed by the 10% raise get the employee back to an annual salary of $40,000?   b. $39,600; no

5. *Glamour* magazine published this survey result: "Seventy-nine percent of those who responded to our August survey say that they believe America has become too lawsuit-happy." The survey question was published in the magazine and readers could respond by mail, fax, or e-mail (Tellus@Galamour.com). How valid is the 79% result?

6. ADT Security Systems advertised that "when you go on vacation, burglars go to work." Their ad stated that "according to FBI statistics, over 26% of home burglaries take place between Memorial Day and Labor Day." What is misleading about this statement?

7. In a study on college campus crimes committed by students high on alcohol or drugs, a mail survey of 1875 students was conducted. A *USA Today* article noted, "Eight percent of the students responding anonymously say they've committed a campus crime. And 62% of that group say they did so under the influence of alcohol or drugs." Assuming that the number of students responding anonymously is 1875, how many actually committed a campus crime while under the influence of alcohol or drugs?

8. A study conducted by the Insurance Institute for Highway Safety found that the Chevrolet Corvette had the highest fatality rate—"5.2 deaths for every 10,000." The car with the lowest fatality rate was the Volvo, with only 0.6 death per 10,000. Does this mean that the Corvette is not as safe as the Volvo?

9. The *Newport Chronicle* claims that pregnant mothers can increase their chances of having healthy babies by eating lobsters. That claim is based on a study showing that babies born to lobster-eating mothers have fewer health problems than babies born to mothers who don't eat lobsters. What is wrong with this claim?

2. The survey is based on voluntary self-selected responses and therefore has serious potential for bias. The respondents are not necessarily representative of all Americans.

3. A study sponsored by the citrus industry is much more likely to reach conclusions favorable to that industry.

5. Because the respondents are self-selected, the survey results are not likely to be valid at all.

6. More than 26% of the year takes place between Memorial Day and Labor Day, so the burglary rate isn't really any higher than at other times during the year.

7. 62% of 8% of 1875 is only 93.

8. Not necessarily. Volvo does have a reputation for safe design, but Corvettes tend to attract disproportionately more reckless drivers.

9. Mothers who eat lobsters tend to be wealthier and can afford better health care.

10. First, it requires a calculation, which will result in some errors. Second, by asking for heights instead of measuring them, we tend to get desired values instead of actual values.

11. A maker of shoe polish has an obvious interest in the importance of the product and there are many ways in which this could affect the survey results.

12. The bars are not drawn in their proper proportions.

13. J. Douglas Carroll wrote in a letter to the editor of the *New York Times* that the mean of 69.5 for all males is measured from birth, whereas males don't become conductors until they have already survived for about 30 years. When this is taken into account, the mean of 73.4 years is not significant.

14. Because the groups consist of 20 mice each, all percentages of success should be multiples of 5. The given percentages cannot be correct.

15. The wording of the question is biased and tends to encourage negative responses. The sample size of 20 is too small. Survey respondents are self-selected instead of being selected by the newspaper. If 20 readers respond, the percentages should be multiples of 5; 87% and 13% are not possible results.

10. A survey includes this item: "Enter your height in inches." It is expected that actual heights of respondents can be obtained and analyzed. Identify the two major problems with this item.

11. "According to a nationwide survey of 250 hiring professionals, scuffed shoes was the most common reason for a male job seeker's failure to make a good first impression." Newspapers carried this statement based on a poll commissioned by Kiwi Brands, producers of shoe polish. Comment on why the results of the survey might be questionable.

12. In an advertising supplement inserted in *Time*, the increases in expenditures for pollution abatement were shown in a graph similar to the one shown. What is wrong with the figure?

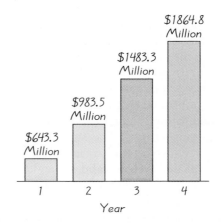

1-3 Exercises B: Beyond the Basics

13. A *New York Times* article noted that the mean life span for 35 male symphony conductors was 73.4 years, in contrast to the mean of 69.5 years for males in the general population. The longer life span was attributed to such factors as fulfillment and motivation. There is a fundamental flaw in concluding that male symphony conductors live longer. What is it?

14. A researcher at the Sloan-Kettering Cancer Research Center was once criticized for falsifying data. Among his data were figures obtained from 6 groups of mice, with 20 individual mice in each group. These values were given for the percentage of successes in each group: 53%, 58%, 63%, 46%, 48%, 67%. What's wrong?

15. Try to identify each of the four major flaws in the following. A daily newspaper ran a survey by asking readers to call in their response to this question: "Do you support the development of atomic weapons that could kill millions of innocent people?" It was reported that 20 readers responded and 87% said "no" while 13% said "yes."

16. A *New York Times* editorial criticized a chart caption that described a dental rinse as one that "reduces plaque on teeth by over 300%."

    a. If you remove 100% of some quantity, how much is left?

    b. What does it mean to reduce plaque by over 300%?

16. a. Nothing is left.
b. According to the *New York Times,* "It would have to remove all the plaque, remove it again and then remove it for a third time plus some more still."

1-4 Design of Experiments

Studies using statistical methods range from those that are well conceived and well executed, producing beneficial results, to those that are poorly designed and poorly executed, leading to conclusions that are misleading and without any real validity or value. Here are some important steps for designing a study that is capable of yielding valid results:

1. Identify the exact question to be answered and clearly identify the relevant population.

2. Develop a plan for collecting data. This plan should describe in detail the conduct of an *observational study* or an *experiment* (both defined below), and it should be carefully designed so that the collected data are representative of the population in question.

3. Collect the data. Be extremely careful to minimize errors that result in a biased collection of data.

4. Analyze the data and draw conclusions. Also, identify possible sources of errors.

Studies that incorporate statistical methods typically arise from two common sources: observational studies and experiments.

Users of previous editions will see that this section has been expanded, as some of them have recommended. Some of the new concepts are included as topics in the Advanced Placement course. In addition to stressing the importance of a good design, again, stress the importance of good sampling techniques.

DEFINITIONS

In an **observational study,** we observe and measure specific characteristics, but we don't attempt to manipulate or modify the subjects being studied.

In an **experiment,** we apply some *treatment* and then proceed to observe its effects on the subjects.

For example, an observational study might involve a survey of citizens to determine what percentage of the population favors the registration of handguns. An experiment might involve a drug treatment given to a group of patients in order to determine its effectiveness as a cure. With the handgun survey we collect data without modifying the people being polled, but the drug treatment consists of modification of the subjects.

Well-planned experiments often involve one group that is given a particular treatment (called a *treatment group*) and a second *control group* that is not

Political Polls Grow

In "Consulting the Oracle," an article for *U.S. News and World Report,* author Stephen Budiansky writes that President Kennedy commissioned 16 polls in his three years as President, Nixon did 233 polls in his six years, and Clinton did between 100 and 150 in his first 2½ years. Clinton's polls cost between $30,000 and $45,000 each, which translates to a cost of about $30 per person. Budiansky reports that polling is complicated by answering machines and people who refuse to cooperate, but good polls include repeated attempts to get responses from those who are not at home and those who refuse to answer. Ignoring those who don't answer or respond could result in a sample that is not representative of the population.

given the treatment. For example, the 1954 polio experiment involved a treatment group of children who were injected with the Salk vaccine and a control group of children who were injected with a placebo that contained no medicine or drug. In experiments of this type, a *placebo effect* occurs when an untreated subject incorrectly believes that he or she is receiving a treatment and reports an improvement in symptoms. The placebo effect can be countered by using *blinding,* a technique in which the subject doesn't know whether he or she is receiving a treatment or a placebo. The polio experiment was *double-blind* in that the children being injected didn't know whether they were given the Salk vaccine or a placebo, and the doctors who gave the injections and evaluated the results didn't know either.

When designing an experiment to test the effectiveness of one or more treatments, you should be careful to assign the *experimental units* (or subjects) to the different groups in such a way that those groups are very similar. (Such similar groups of experimental units are called *blocks.*) One effective approach is to use a *completely randomized experimental design,* which requires that the experimental units be divided into different groups through a process of *random* selection. For example, such a design might involve random assignment of people to a group treated with aspirin and a control group that is not treated. Another approach is to use a *rigorously controlled design,* with experimental units carefully chosen so that the different groups (or blocks) are carefully arranged to be similar. With a rigorously controlled design, you might try to design treatment and control groups to include people similar in age, weight, blood pressure, and so on. It is also important to consider *replication,* which requires that sample sizes be large enough so that effects of chance sample variation are reduced. The polio experiment was a completely randomized experimental design because the subjects in both the treatment group and the control group were selected randomly. It incorporated replication by including very large (200,000) numbers of subjects in each group.

When conducting experiments, results are sometimes ruined because of confounding.

DEFINITION

Confounding occurs when the effects from two or more variables cannot be distinguished from each other.

For example, if you're conducting an experiment to test the effectiveness of a new fire retardant on a brush fire but it begins to rain, confounding occurs because it's impossible to distinguish between the effect from the retardant and the effect from the rain.

One of the worst mistakes is to collect data in a way that is inappropriate. We cannot overstress this very important point:

Data carelessly collected may be so completely useless that no amount of statistical torturing can salvage them.

We noted in Section 1-3 that a self-selected survey is one in which people decide themselves whether to respond. Self-selected surveys are very common, but their results are generally useless for making valid inferences about larger populations.

We now define and describe five of the more common methods of sampling.

DEFINITION

In a **random sample,** members of the population are selected in such a way that each has an *equal chance* of being selected. (A **simple random sample** of n subjects is selected in such a way that every possible sample of size n has the same chance of being chosen.)

Throughout the book, the above calculator image is used to identify places where Larry Morgan's *TI–83 Companion to Elementary Statistics* includes a particularly relevant discussion.

Random samples have been selected by a variety of methods, including using computers to generate random numbers and using tables of random numbers. With random sampling we expect all groups of the population to be (approximately) proportionately represented in the sample. Careless or haphazard sampling can easily result in a biased sample with characteristics very unlike the population from which the sample came. Random sampling, in contrast, is carefully designed to avoid any bias. For example, using telephone directories automatically eliminates anyone with an unlisted number, and ignoring that segment of the population could easily yield misleading results. In Los Angeles, for example, 42.5% of the telephone numbers are unlisted (based on data from Survey Sampling, Inc.). Pollsters commonly circumvent this problem by using computers to generate phone numbers so that all numbers are possible. They must also be careful to include those who are initially unavailable or initially refuse to comment. The Harris polling company has found that the refusal rate for telephone interviews is generally at least 20%. If you ignore those people who initially refuse, you run a real risk of having a biased sample.

DEFINITION

With **stratified sampling,** we subdivide the population into at least two different subpopulations (or strata) that share the same characteristics (such as gender), then we draw a sample from each stratum.

In surveying views on the Equal Rights Amendment to the Constitution, we might use gender as a basis for creating two strata. After obtaining a list of men and a list of women, we use some suitable method (such as random sampling) to select a certain number of people from each list. When the various strata have sample sizes that reflect the general population, we say that we have *proportionate* sampling. If it should happen that some strata are not represented in the proper proportion, then the results can be adjusted or weighted accordingly.

Given a fixed sample size, if you randomly select subjects from different strata, you are likely to get more consistent (and less variable) results than by simply selecting a random sample from the general population. For that reason, stratified sampling is often used to reduce the variation in the results.

DEFINITION

In **systematic sampling,** we select some starting point and then select every kth (such as every 50th) element in the population.

For example, if Motorola wanted to conduct a survey of its 107,000 employees, it could begin with a complete roster, then select every 100th employee to obtain a sample of size 1070. This method is simple and is often used.

DEFINITION

In **cluster sampling,** we first divide the population area into sections (or clusters), then randomly select a few of those sections, and then choose *all* the members from those selected sections.

Stratified sampling and cluster sampling may cause some confusion because they both involve sampling from different groups. Remove the confusion by noting that with cluster sampling, we use *all* members from the selected clusters, whereas stratified sampling uses only a *sample* of members from each strata.

One important way that cluster sampling differs from stratified sampling is that cluster sampling uses *all* members from selected clusters, whereas stratified sampling uses a *sample* of members from each strata. An example of cluster sampling can be found in a pre-election poll, whereby we randomly select 30 election precincts and then survey all the people from each of those chosen precincts. This would be much faster and much less expensive than selecting one person from each of the many precincts in the population area. The results can be adjusted or weighted to correct for any disproportionate representations of groups. Cluster sampling is used extensively by government and private research organizations.

DEFINITION

With **convenience sampling,** we simply use results that are readily available.

In some cases results from convenience sampling may be quite good, but in other cases they may be seriously biased. In investigating the proportion of left-handed people, it would be convenient for a student to survey his or her classmates, because they are readily available. Even though such a sample is not random, the results should be quite good. In contrast, it might be very convenient (and perhaps also profitable) for ABC News to conduct a poll by asking audience members to call a "900" telephone number to register their opinions, but this would be a self-selected survey and the results would likely be biased.

Figure 1-2 on page 22 illustrates the five common methods of sampling just described. The descriptions are intended to be brief and general. Thoroughly understanding these different methods so that you can successfully use them requires much more extensive study than is practical in a single introductory course. To keep this section in perspective, we should note that this text will make frequent reference to "randomly selected" data, and you should understand that such data are carefully selected so that all members of the population have the same chance of being chosen. Although we will not make frequent reference to the other methods of sampling, you should understand that they exist and that the method of sampling requires careful planning and execution. The methods we present throughout this text depend on samples that have been carefully obtained. In addition, the sample size must always be large enough for the required purposes. (Issues of sample size are discussed later in the text, especially in Chapter 6.) Many people incorrectly believe that large samples are good samples, but even large samples may be totally worthless if the data have been collected carelessly. Finally, if you are obtaining measurements of a characteristic (such as height) from people, realize that you will get more accurate results if you do the measuring yourself instead of asking the subjects for the value. Asking tends to yield a disproportionate number of rounded results, as well as many results that reflect *desired* values instead of *actual* values.

No matter how well you plan and execute the sample collection process, there is likely to be some error in the results. For example, randomly select 1000 adults, ask them if they graduated from high school, and record the sample percentage of "yes" responses. If you randomly select another sample of 1000 adults, it is likely that you will obtain a *different* sample percentage.

Meta-Analysis

The term *meta-analysis* refers to a technique of doing a study that essentially combines results of other studies. It has the advantage that separate smaller samples can be combined into one big sample, making the collective results more meaningful. It also has the advantage of using work that has already been done. Meta-analysis has the disadvantage of being only as good as the studies that are used. If the previous studies are flawed, the "garbage in, garbage out" phenomenon can occur. The use of meta-analysis is currently popular in medical research and psychological research. An example: "Reversal of Left Ventricular Hypertrophy in Essential Hypertension: A Meta-analysis of Randomized Double-blind Studies," by Schmieder, Martus, and Klingbeil, *Journal of the American Medical Association,* Vol. 275, No. 19.

Figure 1-2 Common
Sampling Methods

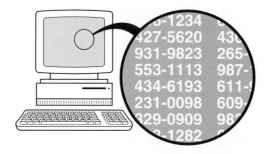

<u>*Random Sampling:*</u>
*Each member of the population has an
equal chance of being selected.
Computers are often used to
generate random telephone numbers.*

<u>*Stratified Sampling:*</u>
*Classify the population into at least
two strata, then draw a sample from
each.*

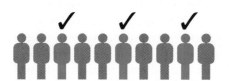

<u>*Systematic Sampling:*</u>
Select every kth member.

<u>*Cluster Sampling:*</u>
*Divide the population area into
sections, randomly select a few of
those sections, and then choose all
members in them.*

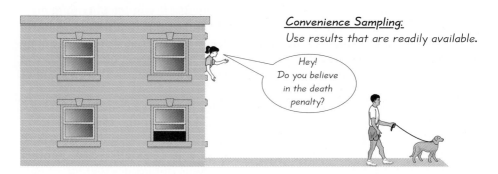

<u>*Convenience Sampling:*</u>
Use results that are readily available.

DEFINITIONS

A **sampling error** is the difference between a sample result and the true population result; such an error results from chance sample fluctuations.

A **nonsampling error** occurs when the sample data are incorrectly collected, recorded, or analyzed. Such an error results from an error other than a chance sample fluctuation, such as selecting a nonrandom and biased sample, using a defective measuring instrument, using a survey question that is biased, obtaining a large number of response refusals, or copying the sample data incorrectly.

If we carefully collect a sample so that it is representative of the population, we can use methods in this book to analyze the sampling error, but we must exercise extreme care so that nonsampling error is minimized.

 Exercises A: Basic Skills and Concepts

Recommended exercises: All of 1–16.

In Exercises 1–4, determine whether the given description corresponds to an observational study or an experiment.

1. Different brands of cigarettes are measured for tar, nicotine, and carbon monoxide (as in Data Set 4 in Appendix B).   1. Observational study
2. People who smoke are asked to halve the number of cigarettes consumed each day so that any effect on pulse rate can be measured.   2. Experiment
3. In a physical education class, the effect of exercise on blood pressure is studied by requiring that half of the students walk a mile each day while the other students run a mile each day.   3. Experiment
4. The relationship between weights of bears and their lengths is studied by measuring bears that have been anesthetized.   4. Observational study

In Exercises 5–16, identify which of these types of sampling is used: random, stratified, systematic, cluster, or convenience.

5. When she wrote *Women and Love: A Cultural Revolution*, author Shere Hite based conclusions on 4500 responses from 100,000 questionnaires distributed to women.   5. Convenience
6. A psychologist at New York University surveys all students from each of 20 randomly selected classes.   6. Cluster
7. A sociologist at the College of Charleston selects 12 men and 12 women from each of 4 English classes.   7. Stratified

Hawthorne and Experimenter Effects

The well-known *placebo effect* occurs when an *untreated* subject incorrectly believes that he or she is receiving a real treatment and reports an improvement in symptoms. The *Hawthorne effect* occurs when *treated* subjects somehow respond differently, simply because they are part of an experiment. (This phenomenon was called the "Hawthorne effect" because it was first observed in a study of factory workers at Western Electric's Hawthorne plant.) An *experimenter effect* (sometimes called a Rosenthall effect) occurs when the researcher or experimenter unintentionally influences subjects through such factors as facial expression, tone of voice, or attitude.

8. Sony selects every 200th compact disk from the assembly line and conducts a thorough test of quality.   8. Systematic

9. A tobacco lobbyist writes the name of each U.S. Senator on a separate card, shuffles the cards, and then draws 10 names.   9. Random

10. The marketing manager for America Online tests a new sales strategy by randomly selecting 250 consumers with less than $50,000 in gross income and 250 consumers with gross income of at least $50,000.   10. Stratified

11. Planned Parenthood polls 500 men and 500 women about their views concerning the use of contraceptives.   11. Stratified

12. A market researcher for American Airlines interviews all passengers on each of 10 randomly selected flights.   12. Cluster

13. A medical researcher from Johns Hopkins University interviews all leukemia patients in each of 20 randomly selected hospitals.   13. Cluster

14. A reporter for *Business Week* magazine interviews every 50th chief executive officer identified in that magazine's listing of the 1000 companies with the highest stock market values.   14. Systematic

15. A reporter for *Business Week* magazine obtains a numbered listing of the 1000 companies with the highest stock market values, then uses a computer to generate 20 random numbers between 1 and 1000, and then interviews the chief executive officers of companies corresponding to these numbers.   15. Random

16. In conducting research for the Boston evening news, a reporter for NBC interviews 15 people as they leave IRS audits.   16. Convenience

1-4   Exercises B: Beyond the Basics

17. Two categories of survey questions are *open* and *closed*. An open question allows a free response, while a closed question allows only a fixed response. Here are examples based on Gallup surveys.

Open question: What do you think can be done to reduce crime?

Closed question: Which of the following approaches would be most effective in reducing crime?

- Hire more police officers.
- Get parents to discipline children more.
- Correct social and economic conditions in slums.
- Improve rehabilitation efforts in jails.
- Give convicted criminals tougher sentences.
- Reform courts.

   a. What are the advantages and disadvantages of open questions?
   b. What are the advantages and disadvantages of closed questions?

(continued)

17. a. An advantage of open questions is that they provide the subject and the interviewer with a much wider variety of responses; a disadvantage is that open questions can be very difficult to analyze.

b. An advantage of closed questions is that they reduce the chance of misinterpreting the topic; a disadvantage is that closed questions prevent the inclusion of valid responses the pollster might not have considered.

   c. Which type is easier to analyze with formal statistical procedures, and why is that type easier?

18. Describe in detail a method that could be used to obtain a simple random sample of the heights of 5 students from your statistics class.

c. Closed questions are easier to analyze with formal statistical procedures.

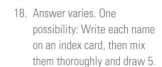

18. Answer varies. One possibility: Write each name on an index card, then mix them thoroughly and draw 5.

1–5 Statistics with Calculators and Computers

One important byproduct of the U.S. space program is the invention of the microprocessor chip—a development that has had a profound influence on the use of statistics. The installation of microprocessor chips in calculators and computers has removed the tremendous burden of monotonous calculations, with the consequence that for many more people the use of statistical methods is now feasible. In this section we briefly discuss the role of calculators and computers in statistics.

Calculators

Statistics students quickly find that a calculator is one of their best friends. Besides having the basic operations ($+$, $-$, $\times$, $\div$, $\sqrt{\ }$, and so on), many calculators now include special statistics features, such as the mean, standard deviation, and correlation/regression results. (These topics will be discussed in later chapters.) In addition to allowing us to evaluate complicated expressions and some special statistics operations, some calculators also allow us to enter and store special programs that can be used throughout the course. The Texas Instruments TI-83 is an excellent example of a calculator that is very suitable for an introductory statistics course. It is programmable, can display graphs, and has an abundance of special statistical functions already included. The *TI-83 Companion to Accompany Elementary Statistics*, written by Larry A. Morgan, is designed to be a supplement to this book. It is strongly recommended for those who use the TI-82 or TI-83 calculator. A separate floppy disk is available with programs written for the TI-82 and TI-83, and these programs can be downloaded from a computer to the calculator. Some statistics professors require that all students use a TI-83 calculator, some require any calculator that can handle two-variable statistics, whereas many others allow the use of almost any calculator. For students who have not yet obtained a calculator, the author recommends one that is capable of dealing with two-variable statistics. Whatever calculator you choose, the manual that comes with it is an extremely valuable guide. If you get stuck, consult your calculator manual and try to follow any examples presented. If that doesn't help, bring your calculator manual to your professor and he or she will be able to help you.

It's important to make an early announcement (preferably in the first class) about calculator requirements. Be sure to notify the class about any calculators that are prohibited during tests. (Many instructors prohibit any calculator with a "QWERTY" keyboard, such as the TI-92.) Comment that calculators capable of handling two-variable statistics will simplify some messy calculations that will be required later in the course.

Computers

Computers now play an important role in almost every aspect of statistical analysis. The widespread availability of computers and software packages has not only made the use of statistics possible for people with many different mathematical backgrounds, but it has also created greater opportunity for the misuse of statistics. It is important to recognize that statistical software and computers have a very serious limitation: They mindlessly follow instructions, even if those instructions are inappropriate or absurd. Computers don't do the necessary human reasoning, and they cannot exercise judgment. An understanding of the principles of statistics is an important prerequisite for correctly interpreting computer results. Even if you don't actually use computers in this course, you should try to develop some skill in interpreting computer displays of statistical analyses, such as those found throughout this text.

We will make frequent reference to two particular software packages: STATDISK and Minitab. STATDISK has two important advantages: (1) It is easy-to-use software designed specifically as a supplement to this textbook, and (2) it's *free* to colleges that adopt this textbook. Minitab is a higher-level statistical software package, but it is also relatively easy to use. Separate workbook/manuals for STATDISK and Minitab are available as supplements to this book.

With STATDISK and Minitab, programs are selected from a main menu bar at the top of the screen, as shown:

STATDISK: `File Edit Analysis Data Help`
Minitab:

`File Edit Manip Calc Stat Graph Editor Window Help`

By using STATDISK or Minitab, we can become more familiar with the general operation of computers. The following examples illustrate some of the basics of STATDISK and Minitab.

To enter a *new* data set:

STATDISK: Select `Data` from the main menu bar, then select the option of `Sample Editor`.

Minitab: Select `File` from the main menu bar, then select the option of `New Worksheet`.

To *save* and name a data set:

STATDISK: Select `File` from the main menu bar, then select the option of `Save As`.

Minitab: Select `File` from the main menu bar, then select the option of `Save Worksheet As...`

STATDISK is free to colleges that adopt this textbook. Students using this book may make their own personal copy of STATDISK. This edition of the book is accompanied by a new version of STATDISK (version 5.0). Comment to the class that even if no statistical software package is used, there is value in interpreting the computer displays that appear throughout this book.

To retrieve (or *open*) a previously stored data set:

STATDISK: Select `File` from the main menu bar, then select the option of `Open`.

Minitab: Select `File` from the main menu bar, then select the option of `Open Worksheet`.

To *print* results:

STATDISK: Select `File` from the main menu bar, then select the option of `Print`.

Minitab: Select `File` from the main menu bar, then select the option of `Print Window`.

To *quit* the program:

STATDISK: Select `File` from the main menu bar, then select the option of `Quit`.

Minitab: Select `File` from the main menu bar, then select the option of `Exit`.

STATDISK and Minitab are each capable of performing almost all of the important operations discussed in this book.

We have presented only a few features of STATDISK and Minitab, but the use of these programs is discussed further in the *STATDISK Student Laboratory Manual and Workbook (7th edition)* and the *Minitab Student Laboratory Manual and Workbook (7th edition)*. Features of these programs and sample displays are also discussed in appropriate places throughout this book.

Some statistics professors prefer to use other software packages, such as SPSS, SAS, BMDP, Execustat, Systat, Mystat, or Statgraphics. Whichever software package is chosen, students will generally benefit by improving the computer skills that have become so important in today's world.

1–5 Exercises A: Basic Skills and Concepts

Recommended exercises: 1–6.
Calculator warmup exercises: In Exercises 1–8, the given expressions are similar to those found in different places throughout this book. Use your calculator to obtain the indicated values.

1. $\dfrac{3.44 + 2.67 + 2.09 + 1.87 + 3.11}{5}$ 1. 2.636

Leave Computer On

Some people turn off their computer whenever they finish an immediate task, whereas others leave it on until they know they won't be using it again until the next day. The computer's circuit board and chips do suffer from repeated on/off electrical power cycles. However, the monitor can be damaged when the same image is left on a screen for very long periods of time. The *mean time between failures* (MTBF) for hard disk drives was once around 5000 hours, but it's now up to about 30,000 hours. Considering the bad effects of on/off cycles on circuit boards and chips and the large MTBF for hard drives, it does make sense to leave computers on until the end of the day, provided the monitor's screen can be protected by making it blank or by using a screen-saver program. Many people use this strategy, which evolved in part from a statistical analysis of past events.

2. $\sqrt{\dfrac{(2-5)^2 + (4-5)^2 + (9-5)^2}{3-1}}$ 2. 3.6055513

3. $\sqrt{\dfrac{3(101) - 15^2}{6}}$ 3. 3.6055513

4. $\dfrac{(12-8.5)^2}{8.5} + \dfrac{(22-25.3)^2}{25.3}$ 4. 1.8716113

5. $\dfrac{1.96^2 \cdot 0.25}{0.03^2}$ 5. 1067.1111

6. $\dfrac{102.7 - 100.0}{\dfrac{14.2}{\sqrt{50}}}$ 6. 1.3444988

7. $\dfrac{15!}{9!6!}$ (*Hint:* $6! = 6 \times 5 \times 4 \times 3 \times 2 \times 1$)   7. 5005

8. $\dfrac{8(56.80) - (14.60)(26)}{\sqrt{8(32.9632) - (14.60)^2} \ \sqrt{8(104) - 26^2}}$ 8. 0.84235978

1-5 Exercises B: Beyond the Basics

9. Load STATDISK or Minitab and retrieve the data set indicated below. Those data sets are already stored. Write the first three values that are listed.   9. STATDISK: 0.838 0.875 0.870 Minitab: 3.22 1 1

STATDISK: BLUE.SDD (weights of blue M&M candies)

Minitab: ALFALFA.MTW (alfalfa yields for different varieties in different fields)

10. Load STATDISK or Minitab and save the following tar amounts (in milligrams per cigarette) for 15 different cigarettes. Save the data with the file name CIGTAR.

16   16   9   8   16   13   15   9   2   15   15   9   14   6   18

Retrieve the data set to verify that these amounts have been saved, then obtain a printed display of the data set.

· ·

Vocabulary List

statistics
population
census
sample
parameter
statistic
quantitative data
qualitative data
discrete data
continuous data
nominal level of measurement
ordinal level of measurement
interval level of measurement

ratio level of measurement
self-selected survey
observational study
experiment
confounding
random sample
simple random sample
stratified sampling
systematic sampling
cluster sampling
convenience sampling
sampling error
nonsampling error

· ·

Review

This chapter began with a general description of the nature of statistics, then discussed different aspects of the nature of data. Uses and abuses of statistics were illustrated with examples. We then discussed the design of experiments, emphasizing the importance of good sampling methods. We concluded the chapter with a brief discussion of the role of calculators and computers. On completing this chapter, you should be able to do the following:

- Distinguish between a population and a sample
- Distinguish between a parameter and a statistic
- Identify the level of measurement (nominal, ordinal, interval, ratio) of a set of data
- Recognize the importance of good sampling methods, as well as the serious deficiency of poor sampling methods
- Recognize that self-selected surveys cannot be used to form valid conclusions about populations

· ·

Review Exercises

1. The Consumer Product Testing Laboratory selects a dozen batteries (labeled 9 volts) from each company that makes them. Each battery is tested for its actual voltage level.

   a. Are the values obtained discrete or continuous?   1. a. Continuous
   b. Identify the level of measurement (nominal, ordinal, interval, ratio) for the voltages.   b. Ratio   *(continued)*

The student edition of this book now includes answers to *all* Review Exercises and Cumulative Review Exercises. For the regular text sections, the student edition continues to include answers for the odd-numbered exercises only.

e. The products that use the batteries may be damaged.

c. Which type of sampling (random, stratified, systematic, cluster, convenience) is being used?   c. Stratified

d. Is this an observational study or an experiment?   d. Observational study

e. What is an important effect of consumers using batteries that are labeled 9 volts when, in reality, the voltage level is very different?

2. Researchers at the Consumer Product Testing Laboratory test samples of electronic surge protectors to find the voltage levels at which computers can be damaged. For each of the following, determine which of the four levels of measurement (nominal, ordinal, interval, ratio) is most appropriate.

a. The measured voltage levels that cause damage   2. a. Ratio

b. Rankings (first, second, third, and so on) in order of quality for a sample of surge protectors   b. Ordinal

c. Ratings of surge protectors as "recommended, acceptable, not acceptable"   c. Ordinal

d. The room temperatures of the rooms in which the surge protectors are tested   d. Interval

e. Nominal

e. The countries in which the surge protectors were manufactured

3. Because it is a mail survey, the respondents are self-selected and will likely include those with strong opinions about the issue. Self-selected respondents do not necessarily represent the views of everyone who invests.

3. *Business Week* magazine conducts a survey by mailing a questionnaire to 5000 people known to invest in securities. On the basis of the results, the magazine editors conclude that most investors in the United States are pessimistic about the economy. What is wrong with that conclusion?

4. Identify each number as discrete or continuous.

a. The Nielsen Media Research Organization surveyed 2027 adults who were watching *Monday Night Football* on ABC.   4. a. Discrete

b. Professor Fisher recorded times used by statistics students to complete a final exam, and the first result was 87.25 minutes.   b. Continuous

c. Kathy Patel weighed her statistics textbook and obtained a value of 1.87 lb.   c. Continuous

5. Identify the type of sampling (random, stratified, systematic, cluster, convenience) used in each of the following.

a. A sample of products is obtained by selecting every 100th item on the assembly line.   5. a. Systematic

d. Stratified

6. Respondents often tend to round off to a nice even number like 50.

7. The sample could be biased by excluding those who work, those who don't eat at school, those who commute, and so on.

8. The figure is very precise, but it is probably not very accurate. The use of such a precise number may incorrectly suggest that it is also accurate.

b. Random numbers generated by a computer are used to select serial numbers of cars to be chosen for sample testing.   b. Random

c. An auto parts supplier obtains a sample of all items from each of 12 different randomly selected retail stores.   c. Cluster

d. A car maker conducts a marketing study involving test drives performed by a sample of 10 men and 10 women in each of 4 different age brackets.

e. A car maker conducts a marketing study by interviewing potential customers who request test drives at a local dealership.   e. Convenience

6. Census takers have found that in obtaining people's ages, they get more people of age 50 than of age 49 or 51. Can you explain how this might occur?

7. You plan to conduct a survey on your campus. What is wrong with selecting every 50th student leaving the cafeteria?

8. The *Southport Chronicle* reports that a pro-choice rally was attended by 8725 people. Comment.

Cumulative Review Exercises

This book's cumulative review exercises are designed to incorporate some material from preceding chapters, a feature that will be implemented in the following chapters. The exercises in this section use concepts learned before beginning the study of this book.

1. The following survey question raised concerns when responses suggested that about 22% of Americans thought that the Holocaust might not have occurred.

   "Does it seem possible or does it seem impossible to you that the Nazi extermination of the Jews never happened?"

   A subsequent poll showed that respondents were probably confused by the double negative in the wording of the question. Here is the wording used in a subsequent Roper poll:

   "Does it seem possible to you that the Nazi extermination of the Jews never happened, or do you feel certain that it happened?"

   Is this second version substantially less confusing? Can you write the question in a way that is clearer than both of these versions?

2. Refer to the accompanying figure on page 32. It is similar to one that Edwin Tufte, author of *The Visual Display of Quantitative Data,* refers to in observing that "This may well be the worst graphic ever to find its way into print." He notes that the graph reports "almost by happenstance, only five pieces of data (since the division within each year adds to 100 percent)." First examine this graph and identify the information it attempts to relate. Then design a new graph that relates the same information.

Computer Project

Refer to Data Set 2 in Appendix B and consider the 106 body temperatures (in degrees Fahrenheit) found in the last column (Day 2, 12 A.M.). (This data set is not included on the data disk that is available as a supplement to this book.) University of Maryland researchers collected body temperature data and found that the average was not 98.6°F, the value most of us assume to be the correct average. Using STATDISK or Minitab, enter the 106 body temperatures and save them under the name BODYTEMP.

   The objective of this computer project is to enter the data and store it on a computer disk. This will allow you to have the data available for use in Chapter 2, and it will also help you develop your ability to enter and store computer data—a skill that is critically important today.

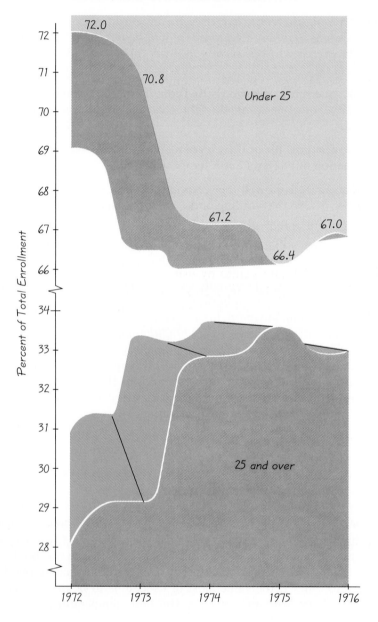

AGE STRUCTURE OF
COLLEGE ENROLLMENT

•••••••• FROM DATA TO DECISION ••••••••

Misrepresented Data

Collect an example from a current newspaper or magazine in which data have been presented in a deceptive manner. Identify the source (including the publication date) from which the example was taken. Explain the way in which the presentation is deceptive, and suggest how the data might be presented more fairly.

••••••••• COOPERATIVE GROUP ACTIVITIES •••••••••

1. *Out-of-class activity:* Divide into groups of 5, then collect 50 data values by using random sampling as described in Section 1-4. Then repeat the collection of 50 data values for each of the other four methods of sampling: stratified, systematic, cluster, convenience. In each case, calculate the "mean." (The mean is defined in Chapter 2 as the average obtained when the scores are added, and the total is then divided by the number of scores.) First describe in detail the procedure used for each method of sampling, then list the scores, then compare the five means. Does it appear that the different methods of sampling produce the same results? The sample data should be selected from a population such as the ages of books in your college library, or the ages of cars in the student parking lot (listed on registration stickers in some states).

2. *In-class-activity:* Divide into groups of three or four and use the data given below to construct a graph that exaggerates the increases in the high points of the Dow Jones Industrial Average. Also construct a graph that deemphasizes those increases, and construct a third graph that represents the data fairly.

Decade	1950s	1960s	1970s	1980s
Dow High	683	995	1052	2796

3. *In-class-activity:* Divide into groups of three or four. Assume that you must conduct a survey of full-time students at your college. Design and describe in detail a procedure for obtaining a *random* sample of 100 students.

interview

Paul Mones

Paul Mones is an attorney, author, and expert consultant. He wrote *When a Child Kills: Abused Children Who Kill Their Parents*. He also wrote *Stalking Justice*, the true story of a detective who first used DNA fingerprinting to catch a serial killer. He has been interviewed on many major shows in the United States, Europe, and Australia, including "60 Minutes," "20/20," and "Larry King Live." His comments have appeared in publications such as the *New York Times* and *Time* magazine, and he was a legal correspondent for "NBC News." He trains doctors, lawyers, and law enforcement officers, and often testifies before legislative groups.

Do you use statistics in your work as an attorney?

I use statistics extensively in my legal work. With DNA fingerprinting, for example, we look at several factors and determine the probability of getting a specific sequence of genotypes in the same people. We once looked at three loci (positions genes occupy on chromosomes), then it went to five, but now it's up to seven. We look at a reference sample and a comparison sample to see how often a particular sequence occurs. If a suspect matches with seven loci, we know we have a pretty good match, then we look at how frequent that sequence is in the population. In *Stalking Justice*, the odds were 1 in 750,000,000 that another person would have the same DNA profile as the defendant in the case. We used hypothesis testing and determined the level of significance for the specific DNA profile. DNA is also very important in paternity and rape cases. With the old traditional blood-enzyme tests, you could get down to about 10% of the population. This means that there is a 1 in 10 chance that the guilty person is someone other than the defendant. With DNA fingerprinting, we might get a 1 in 300,000,000 chance, so we get into a realm of statistical inevitability. This is used not only to convict people, but also to exclude suspects. There's a famous case in North Carolina where two eyewitnesses testified that the defendant was a rapist. He served 11 years, but was freed when DNA was used to show that he was not guilty. In this case, statistics and DNA testing were much more accurate than eyewitness identifications.

I've used statistics in homicide cases, child abuse cases, battered women cases, and paternity cases. With paternity cases today, the DNA results are so accurate that the entire court trial system is being completely short-circuited. People just aren't going to trial when the DNA findings are so clear. Reasonable doubt becomes almost no doubt at all. The greater question is: "Are statistics so powerful that they rob the jury of its decision-making responsibilities?" There are exceptions, but in most cases, the presence of strong statistical evidence is very effective as a plea-bargaining tool.

As an attorney who makes extensive use of statistics, do you believe that *all* attorneys should have an understanding of statistical principles?

They need much more. If you want to effectively marshall your evidence, you need to have some statistical background. The problem is that today, lawyers typically use other experts. In accidental death cases, for example, they often use statisticians for the actuarial data used to find how long someone is likely to live. It's the rare lawyer who even understands what the statistician is saying, so it's a good idea for anyone contemplating a law career to study statistics.

Do you use statistics in your work with child abuse and violence?

I am very interested in the connection between child abuse and violence, and one of the best ways to convince

legislators, or jurors, or audiences, is to use statistics. Among teenagers who kill their parents, we know that one of the biggest risk factors occurs when children see their father beat their mother. The state of Texas did a study and found that among boys who committed homicide, 66 percent had killed someone who had been harming their mother. The more we know about a population, the more we know what statistical inquiries to make.

Do you recommend statistics for today's college students?

Statistics is not just for people in hard sciences. It provides important hands-on tools for people who want to become such things as lawyers, doctors, nurses, or police officers. I've found that I can better appreciate news events, financial news, and profit/loss statements for stocks. The study of statistics is more important than much of the basic math being offered. I use statistics much more than I use geometry or trigonometry.

Which other skills are important for today's college students?

With everything becoming computerized, people are focusing less on communications skills, so the art of the spoken word is becoming somewhat neglected. There might be millions of people who can use a computer, but there are very few who can effectively address an audience. People also need the ability to communicate their ideas effectively in writing.

2

Describing, Exploring, and Comparing Data

2-1 Overview

This chapter presents important tables, graphs, and measurements that can be used to describe a data set, explore a data set, or compare two or more data sets. Later chapters will use many of the important concepts introduced in this chapter.

2-2 Summarizing Data with Frequency Tables

The construction of frequency tables, relative frequency tables, and cumulative frequency tables is described. These tables are useful for condensing a large set of data to a smaller and more manageable summary.

2-3 Pictures of Data

Methods for constructing histograms, relative frequency histograms, dotplots, stem-and-leaf plots, pie charts, Pareto charts, and scatter plots are presented. These graphs are extremely helpful in visually displaying characteristics of data that cannot be seen otherwise.

2-4 Measures of Central Tendency

Measures of central tendency are attempts to find values that are representative of data sets. The following measures of central tendency are defined: mean, median, mode, midrange, and weighted mean. The concept of skewness is also considered.

2-5 Measures of Variation

Measures of variation are numbers that reflect the amount of scattering among the values in a data set. The following measures of variation are defined: range, standard deviation, mean deviation, and variance. Such measures are extremely important in statistical analyses.

2-6 Measures of Position

The standard score (or z score) is defined and used to illustrate how unusual values can be identified. Also defined are percentiles, quartiles, and deciles that are used to compare values within the same data set.

2-7 Exploratory Data Analysis (EDA)

Techniques for exploring data with 5-number-summaries and boxplots are described. Boxplots are especially useful for comparing different data sets.

Chapter Problem:

Can 12-oz aluminum cans be made thinner to save money?

Data Set 15 in Appendix B includes these two samples:

1. 12-oz aluminum cans that are 0.0109-in. thick (reproduced as Table 2-1)

2. 12-oz aluminum cans that are 0.0111-in. thick

We will explore the values in Table 2-1, which lists the axial loads (in pounds) of the sample of aluminum cans that are 0.0109-in. thick. This data set was supplied by a student who used the previous edition of this book. She is an employee of the company that manufactures the cans considered here, and she uses methods she learned in her introductory statistics course. The author is grateful for her contributions.

An axial load of a can is the maximum weight supported by its sides, and it is measured by using a plate to apply increasing pressure to the top of the can until it collapses. It is important to have an axial load high enough so that the can isn't crushed when the top lid is pressed into place. In this particular manufacturing process, the top lids are pressed into place with pressures that vary between 158 pounds and 165 pounds.

Thinner cans have the obvious advantage of using less material so that costs are lower, but thinner cans are probably not as strong as the thicker ones. The company manufacturing these cans is currently using a 0.0111-in. thickness, but is testing to determine if the thinner cans could be used. Using the methods of this chapter, we will explore the data set (reproduced in Table 2-1) for these thinner cans (0.0109-in. thick). Ultimately, we will determine whether these thinner cans could be used.

Table 2-1   Axial Loads of 0.0109-in. Cans

270	273	258	204	254	228	282
278	201	264	265	223	274	230
250	275	281	271	263	277	275
278	260	262	273	274	286	236
290	286	278	283	262	277	295
274	272	265	275	263	251	289
242	284	241	276	200	278	283
269	282	267	282	272	277	261
257	278	295	270	268	286	262
272	268	283	256	206	277	252
265	263	281	268	280	289	283
263	273	209	259	287	269	277
234	282	276	272	257	267	204
270	285	273	269	284	276	286
273	289	263	270	279	206	270
270	268	218	251	252	284	278
277	208	271	208	280	269	270
294	292	289	290	215	284	283
279	275	223	220	281	268	272
268	279	217	259	291	291	281
230	276	225	282	276	289	288
268	242	283	277	285	293	248
278	285	292	282	287	277	266
268	273	270	256	297	280	256
262	268	262	293	290	274	292

2-1   Overview

We sometimes collect data in order to address a specific issue. For example, a safety study of the elevators in the Empire State Building would require data concerning the average weight of the people who ride the elevators. In other cases, we collect or obtain data without having a specific objective, but because we wish to explore the data to see what might be revealed. A geologist may wonder about the time intervals between eruptions of the Old Faithful geyser—are they equally distributed over the range of times or do some time intervals occur more often than others? In both circumstances, we need a variety of tools that will help us *understand* the data set. This chapter presents those tools.

When analyzing a data set, we should first determine whether we have a *sample* or a complete *population*. That determination will affect both the methods we use and the conclusions we form. We use methods of **descriptive statistics** to summarize or describe the important characteristics of a known set of population data, and we use methods of **inferential statistics** when we use sample data to make inferences (or generalizations) about a population. When your professor calculates the final exam average for your statistics class, that result is an example of a descriptive statistic if we consider the population to be the entire class. However, if we state that the result is an estimate of the final exam average for all statistics classes, we are making an inference that goes beyond the known data.

Descriptive statistics and inferential statistics are the two general divisions of the subject of statistics. This chapter deals with the basic concepts of descriptive statistics.

Important Characteristics of Data

We use the tools of descriptive statistics to better understand a data set by learning about its characteristics. The following three characteristics of data are extremely important and can provide considerable insight:

1. The nature or shape of the distribution of the data, such as bell-shaped, uniform, or skewed
2. A representative value, such as an average
3. A measure of scattering or variation

We can learn something about the nature or shape of the distribution by organizing the data and constructing graphs, as in Sections 2-2 and 2-3. In Section 2-4 we will learn how to obtain representative, or average, scores. We will measure the extent of scattering, or variation, among data as we use the tools found in Section 2-5. In Section 2-6 we will determine measures of position so

These three items are typically most important, but not always. For example, a study might be conducted to determine the number of blue whales; here population size would be the only important characteristic. Consider emphasizing this point with a brief class activity: Ask the class to estimate the number of taxicabs a company operates in a town if, while standing on a corner, you observe five of the company cabs with the numbers 4, 9, 12, 20, and 30. A good estimate is 36; students should recognize that there are more than 30 cabs, but not substantially more. The point, though, is that with these five numbers, we don't really care about a representative value or the amount of variation.

that we can better analyze or compare various scores. And in Section 2-7 we will learn about methods for exploring data sets.

2-2 Summarizing Data with Frequency Tables

When investigating large data sets, it is generally helpful to organize and summarize the data by constructing a frequency table.

DEFINITION

A **frequency table** lists categories (or classes) of scores, along with counts (or frequencies) of the number of scores that fall into each category.

Table 2-2 is a frequency table with 10 classes (or categories). The frequency for a particular class is the number of original scores that fall into that class. For example, the first class in Table 2-2 has a frequency of 9, indicating that there are 9 values between 200 and 209 inclusive.

We will first present some standard terms used in discussing frequency tables, and then we will describe a procedure for constructing them. (Many statistics software packages can automatically construct frequency tables.)

DEFINITIONS

Lower class limits are the smallest numbers that can actually belong to the different classes. (Table 2-2 has lower class limits of 200, 210, . . . , 290.)

Upper class limits are the largest numbers that can actually belong to the different classes. (Table 2-2 has upper class limits of 209, 219, . . . , 299.)

Class boundaries are the numbers used to separate classes, but without the gaps created by class limits. They are obtained as follows: Find the size of the gap between the upper class limit of one class and the lower class limit of the next class. Add half of that amount to each upper class limit to find the upper class boundaries; subtract half of that amount from each lower class limit to find the lower class boundaries. (Table 2-2 has class boundaries of 199.5, 209.5, 219.5, . . . , 299.5.)

(continued)

Suggestion: Cover Sections 2-2 and 2-3 in one class. Try to avoid spending too much time on these sections at the expense of some other important topics in later chapters. Careful budgeting of time is very important throughout this chapter. Point out that among the three important characteristics of data listed in Section 2-1, Sections 2-2 and 2-3 include concepts that are helpful in understanding the *distribution* of data.

TABLE 2-2

Frequency Table of Axial Loads of Aluminum Cans

Axial Load	Frequency
200–209	9
210–219	3
220–229	5
230–239	4
240–249	4
250–259	14
260–269	32
270–279	52
280–289	38
290–299	14

> **Class marks** are the midpoints of the classes. (Table 2-2 has class marks of 204.5, 214.5, . . . , 294.5.) Each class mark can be found by adding the lower class limit to the corresponding upper class limit and dividing the sum by 2.
>
> **Class width** is the difference between two consecutive lower class limits or two consecutive lower class boundaries. (Table 2-2 uses a class width of 10.)

The definitions of class marks and class boundaries are tricky. Be careful to avoid the easy mistake of making the class width the difference between the lower class limit and the corresponding upper class limit. See Table 2-2 and note that the class width is 10, not 9. (Students usually have the most difficulty with class boundaries. See the discussion in the following section.) Examine the class limits shown in Table 2-2, and note that there is a gap between 209 and 210, another gap between 219 and 220, and so on. Class boundaries basically split differences and fill in the gaps so that the construction of certain graphs will become easier. Carefully examine the definition of class boundaries, and spend some time until it is understood.

The process of actually constructing a frequency table involves these steps:

Step 1: *Decide on the number of classes your frequency table will contain.* As a guideline, the number of classes should be between 5 and 20. The actual number of classes may be affected by the convenience of using round numbers or other subjective factors. With test grades, for example, it may be convenient to use these 10 classes: 50–54, 55–59, 60–64, . . . , 95–99.

Step 2: *Determine the class width by dividing the range by the number of classes.* (The range is the difference between the highest and the lowest scores.) Round the result *up* to a convenient number. This rounding up (not off) not only is convenient, but also guarantees that all of the data will be included in the frequency table. (If the number of classes divides into the range evenly with no remainder, you will need to add another class for all of the data to be included.)

$$\text{class width} = \text{round } up \text{ of } \frac{\text{range}}{\text{number of classes}}$$

Step 3: *Select as the lower limit of the first class either the lowest score or a convenient value slightly less than the lowest score.* This value serves as the starting point.

Step 4: *Add the class width to the starting point to get the second lower class limit.* Add the class width to the second lower class limit to get the third, and so on.

Step 5:   *List the lower class limits in a vertical column, and enter the upper class limits, which can be easily identified at this stage.*

Step 6:   *Represent each score by a tally in the appropriate class, then use those tally marks to find the total frequency for each class.*

Determination of the number of classes is not yet part of federal law, so it is okay to use a different number of classes which will result in a different frequency table that is also correct. Again, the priority should be on obtaining a table with convenient and understandable values.

EXAMPLE   Construct a frequency table for the 175 axial loads of aluminum cans given in Table 2-1.

SOLUTION   We will list the steps that lead to the development of the frequency table shown in Table 2-2 (see page 39).

Step 1:   We begin by selecting 10 as the number of desired classes. (Many statisticians recommend that we use about 10 classes, but use a smaller number of classes with smaller data sets and a larger number of classes with larger data sets.)

Step 2:   With a minimum of 200 and a maximum of 297, the range is $297 - 200 = 97$, so

$$\text{class width} = \text{round up of } \frac{97}{10}$$
$$= \text{round up of } 9.7$$
$$= 10 \quad \text{(rounded up for the convenience of having whole numbers)}$$

Step 3:   The lowest value is 200. Because it is a convenient value, it becomes the starting point and we use it for the lower limit of the first class.

Step 4:   Add the class width of 10 to the lower limit of 200 to get the next lower limit of 210. Continuing, we get the other lower class limits of 220, 230, and so on.

Step 5:   These lower class limits suggest these upper class limits.

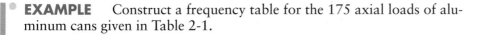

200   209
210   219
etc.

Step 6:   The frequency counts are shown in the right column of Table 2-2.

Table 2-2 provides useful information by making the list of axial loads more intelligible, but we lose the accuracy of the original data. For example, the first class of 200–209 shows 9 scores, but there is no way to determine from the table exactly what those scores are. We cannot reconstruct the original 175 axial loads from the frequency table; we have traded the exactness of the original data for a better understanding of the data.

In constructing frequency tables, the following guidelines should be followed:

1. *Be sure that the classes are mutually exclusive.* That is, each of the original values must belong to exactly one class.
2. *Include all classes,* even if the frequency is zero.
3. *Try to use the same width for all classes,* although sometimes open-ended intervals such as "65 years or older" are impossible to avoid.
4. *Select convenient numbers for class limits.* Round up to use fewer decimal places or use numbers relevant to the situation.
5. *Use between 5 and 20 classes.*
6. *The sum of the class frequencies must equal the number of original data values.*

Relative Frequency Table

An important variation of the basic frequency table uses **relative frequencies,** which are easily found by dividing each class frequency by the total of all frequencies. The **relative frequency table** includes the same class limits as a frequency table, but relative frequencies are used instead of actual frequencies.

$$\text{relative frequency} = \frac{\text{class frequency}}{\text{sum of all frequencies}}$$

Table 2-3 shows the relative frequencies for the 175 axial loads summarized in Table 2-2. The first class has a relative frequency of 9/175 = 0.051. (Relative frequencies can also be given as percentages; that is, 0.051 can be expressed as 5.1%.) The second class has a relative frequency of 3/175 = 0.017, and so on. If constructed correctly, the sum of the relative frequencies should total 1 (or 100%), with small discrepancies allowed for rounding errors.

Relative frequency tables make it easier for us to understand the distribution of the data and to compare different sets of data. For example, we can better understand that 5.1% of the cans have axial loads between 200 lb and 209 lb than to understand that 9 of the 175 cans have axial loads between those values. Also, see Exercise 21 for a situation in which comparison is made easier through the use of relative frequency tables.

TABLE 2-3

Relative Frequency Table of Axial Loads of Aluminum Cans

Axial Load	Relative Frequency
200–209	0.051
210–219	0.017
220–229	0.029
230–239	0.023
240–249	0.023
250–259	0.080
260–269	0.183
270–279	0.297
280–289	0.217
290–299	0.080

Cumulative Frequency Table

Another variation of the standard frequency table is used when cumulative totals are desired. The **cumulative frequency** for a class is the sum of the frequencies for that class and all previous classes. Table 2-4 is an example of a **cumulative frequency table,** with cumulative frequencies used instead of the individual class frequencies. Table 2-4 represents the same 175 aluminum cans summarized in Table 2-2. A comparison of the frequency column of Table 2-2 and the cumulative frequency column of Table 2-4 reveals that the cumulative frequency values are obtained by starting with the frequency for the first class and adding successive frequencies for each class. For example, there are 9 values less than 210, and there are $9 + 3 = 12$ values less than 220, and so on. If constructed correctly, the last cumulative frequency should be equal to the total number of values in the data set.

Frequency tables can be used to identify the general nature of the distribution of data, and they can also be used to construct graphs that visually display the distribution of data. Such graphs are discussed in the next section.

TABLE 2-4	
Cumulative Frequency Table of Axial Loads	
Axial Load	Cumulative Frequency
Less than 210	9
Less than 220	12
Less than 230	17
Less than 240	21
Less than 250	25
Less than 260	39
Less than 270	71
Less than 280	123
Less than 290	161
Less than 300	175

2–2   **Exercises A: Basic Skills and Concepts**

In Exercises 1–4, identify the class width, class marks, and class boundaries for the given frequency table.

1.

Absences	Frequency
0–5	39
6–11	41
12–17	38
18–23	40
24–29	42

2.

Absences	Frequency
0–9	22
10–19	40
20–29	71
30–39	44
40–49	23

3.

Weight (kg)	Frequency
0.0–1.9	20
2.0–3.9	32
4.0–5.9	49
6.0–7.9	31
8.0–9.9	18

4.

Weight (kg)	Frequency
0.0–4.9	60
5.0–9.9	58
10.0–14.9	61
15.0–19.9	62
20.0–24.9	59

Recommended assignment: 1, 2, 5, 6, 9, 10, 13, 15, 20, 21
  Due to space constraints, only answers to even-numbered exercises are listed in this section. See Appendix F for the answers to odd-numbered exercises.

2. Class width: 10. Class marks: 4.5, 14.5, 24.5, 34.5, 44.5. Class boundaries: -0.5, 9.5, 19.5, 29.5, 39.5, 49.5.

4. Class width: 5.0. Class marks: 2.45, 7.45, 12.45, 17.45, 22.45. Class boundaries: -0.05, 4.95, 9.95, 14.95, 19.95, 24.95.

6.

Absences	Relative Frequency
0–9	0.110
10–19	0.200
20–29	0.355
30–39	0.220
40–49	0.115

8.

Weight (kg)	Relative Frequency
0.0–4.9	0.200
5.0–9.9	0.193
10.0–14.9	0.203
15.0–19.9	0.207
20.0–24.9	0.197

10.

Absences	Cumulative Frequency
Less than 10	22
Less than 20	62
Less than 30	133
Less than 40	177
Less than 50	200

12.

Weight (kg)	Cumulative Frequency
Less than 5.0	60
Less than 10.0	118
Less than 15.0	179
Less than 20.0	241
Less than 25.0	300

14. In Exercise 3, the weights have frequencies that start relatively low, increase to a maximum in the middle class, then decrease to the last class. In Exercise 4, the weights are (approximately) evenly spread over the five classes.

16. 0.838–0.854

In Exercises 5–8, construct the relative frequency table that corresponds to the frequency table in the exercise indicated.

5. Exercise 1   **6.** Exercise 2   **7.** Exercise 3   **8.** Exercise 4

In Exercises 9–12, construct the cumulative frequency table that corresponds to the frequency table in the exercise indicated.

9. Exercise 1   **10.** Exercise 2   **11.** Exercise 3   **12.** Exercise 4

13. Compare the *distribution* of the data in Exercise 1 to the distribution of the data in Exercise 2. What is the basic difference?

14. Compare the *distribution* of the data in Exercise 3 to the *distribution* of the data in Exercise 4. What is the basic difference?

In Exercises 15–16, use the given information to find upper and lower limits of the first class. (The data are in Appendix B, but there is no need to refer to the appendix for these exercises.)

15. A data set consists of weights of metal collected from households for one week, and those weights range from 0.26 lb to 4.95 lb. You wish to construct a frequency table with 10 classes.

16. A sample of M&M candies has weights that vary between 0.838 g and 1.033 g. You wish to construct a frequency table with 12 classes.

In Exercises 17–20, construct a frequency table, using the data given.

17. Refer to Data Set 3 in Appendix B and construct a frequency table of the weights of bears. Use 11 classes beginning with a lower class limit of 0.

18. Refer to Data Set 2 in Appendix B and construct a frequency table of the body temperatures for midnight on the second day. Use 8 classes beginning with a lower class limit of 96.5.

19. Refer to Data Set 16 in Appendix B and construct a frequency table of the time intervals between eruptions of the Old Faithful geyser in Yellowstone National Park. Use 7 classes beginning with a lower class limit of 56 min and use a class width of 8 min.

20. Refer to Data Set 11 in Appendix B, and use the weights of all 100 M&Ms to construct a frequency table with 12 classes.

2-2 Exercises B: Beyond the Basics

21. Given below are a frequency table of alcohol consumption prior to arrest for male inmates currently serving sentences for DWI and a corresponding table for women (based on data from the U.S. Department of Justice). First construct the corresponding relative frequency tables, and then use those results to compare the two samples. Note that it is difficult to com-

pare the original frequencies, but it is much easier to compare the relative frequencies.

Ethanol Consumed by Men (oz)	Frequency
0.0–0.9	249
1.0–1.9	929
2.0–2.9	1545
3.0–3.9	2238
4.0–4.9	1139
5.0–9.9	3560
10.0–14.9	1849
15.0 or more	1546

Ethanol Consumed by Women (oz)	Frequency
0.0–0.9	7
1.0–1.9	52
2.0–2.9	125
3.0–3.9	191
4.0–4.9	30
5.0–9.9	201
10.0–14.9	43
15.0 or more	72

18. Temp. (°F)	Frequency
96.5–96.8	1
96.9–97.2	8
97.3–97.6	14
97.7–98.0	22
98.1–98.4	19
98.5–98.8	32
98.9–99.2	6
99.3–99.6	4

20. Weight (g)	Frequency
0.838–0.854	1
0.855–0.871	10
0.872–0.888	13
0.889–0.905	19
0.906–0.922	21
0.923–0.939	15
0.940–0.956	8
0.957–0.973	5
0.974–0.990	5
0.991–1.007	1
1.008–1.024	1
1.025–1.041	1

22. Listed below are two sets of scores that are supposed to be heights (in inches) of randomly selected adult males. One of the sets consists of heights actually obtained from randomly selected adult males, but the other set consists of numbers that were fabricated. Construct a frequency table for each set of heights. By examining the two frequency tables, identify the set of data that you believe to be false, and state your reason.

a. 70 73 70 72 71 73 71 67 68 72 67 72 71 73
　　 72 70 72 68 71 71 71 73 69 73 71 66 77 67

b. 70 73 70 72 71 66 74 76 68 75 67 68 71 77
　　 66 69 72 67 77 75 66 76 76 77 73 74 69 67

23. Data from the U.S. Bureau of the Census are summarized in the accompanying frequency table. Refer to the five guidelines for constructing frequency tables. Which of the guidelines are not followed?

Age	U.S. Population (millions)
Under 15	55
15–24	37
25–44	82
45 and older	79

24. In constructing a frequency table, Sturges' guideline suggests that the ideal number of classes can be approximated by $1 + (\log n) / (\log 2)$, where n is the number of scores. Use this guideline to find the ideal number of classes (rounded off, not rounded up) corresponding to a data collection with the number of scores equal to

a. 50　　**b.** 100　　**c.** 150
d. 500　　**e.** 1000　　**f.** 50,000

22. Height	(a) Frequency	(b) Frequency
66	1	3
67	3	3
68	2	2
69	1	2
70	3	2
71	7	2
72	5	2
73	5	2
74	0	2
75	0	2
76	0	3
77	1	3

24. a. 7 b. 8 c. 8 d. 10
　　e. 11 f. 17

2-3 Pictures of Data

If you require or allow the use of calculators or computers for generating graphs, announce that labeling of axes need not follow the exact formats described in this section (such as class boundaries used in histograms). Suggestion: Discuss histograms and stem-and-leaf plots, and leave the others for reading.

In Section 2-2 we used frequency tables to transform disorganized collections of raw data into organized and understandable summaries. The major objective of this section is to present methods for representing data in pictorial form so that we can easily see the nature of the distribution.

Histograms and the Shape of Data

A common and important graphic device for presenting data is the histogram, an example of which is shown in Figure 2-1. A **histogram** consists of a horizontal scale for values of the data being represented, a vertical scale for frequencies, and bars representing the frequency for each class of values. We generally construct a histogram to represent a set of values after we have first completed a frequency table representing those values. Each bar is marked with its lower class boundary at the left and its upper class boundary at the right. Improved readability, however, is often achieved by using class marks instead of class boundaries. The histogram in Figure 2-1 corresponds directly to the frequency table (Table 2-2 in the previous section).

Before constructing a histogram from a completed frequency table, we must give some consideration to the scales used on the vertical and horizontal axes. The maximum frequency (or the next highest convenient number) should suggest a value for the top of the vertical scale; 0 should be at the bottom. In Figure 2-1 we designed the vertical scale to run from 0 to 60. The horizontal scale should be designed to accommodate all the classes of the frequency table. Ideally, we should try to follow the rule of thumb that the vertical height of the histogram should be about three-fourths of the total width. Both axes should be clearly labeled.

A **relative frequency histogram** will have the same shape and horizontal scale as a histogram, but the vertical scale will be marked with *relative frequencies* instead of actual frequencies, as in Figure 2-2. Figure 2-1 can be modified to be a relative frequency histogram by simply labeling the vertical scale as "relative frequency" and changing the values on that scale to range from 0 to 0.300, as in Figure 2-2. (The highest relative frequency for this data set is 0.297, so it makes sense to use 0.300 as the highest value on the vertical scale; the fact that the highest relative frequency is 0.297 and the maximum score is 297 is coincidental and is one of those little quirks that make life so interesting.) Just as the histogram in Figure 2-1 represents the frequency table in Table 2-2, the relative frequency histogram in Figure 2-2 represents the relative frequency table in Table 2-3.

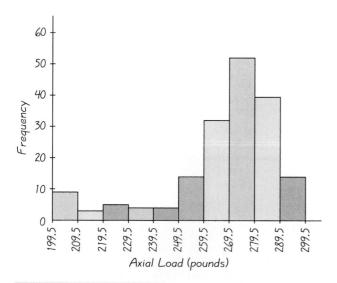

FIGURE 2-1 Histogram of Axial Loads of Aluminum Cans

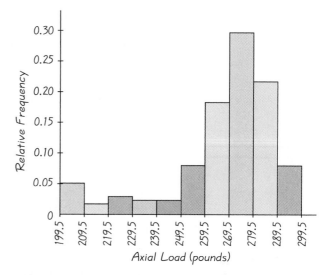

FIGURE 2-2 Relative Frequency Histogram of Axial Loads of Aluminum Cans

Using Calculators and Computers to Generate Histograms

Shown on page 48 is the STATDISK display of a histogram for the axial loads of aluminum cans we are considering in this chapter. The STATDISK display is obtained by using Data from the main menu bar, then entering the data by using the Sample Editor option. Then use the Copy and Paste features to use the data in the Histogram program, which is also found under Data. (The copy and paste features are common to many Windows programs.) The display of the histogram can be obtained from the Windows version of Minitab by first entering the data under column C1 in the data grid. Then use the options of Graph and Histogram, as described in the *Minitab Student Laboratory Manual and Workbook,* which is a supplement to this book. A histogram can also be generated on some graphing calculators, such as the TI-82 and TI-83.

Frequency tables and graphs such as histograms enable us to see how our data are distributed, and the distribution of data is an extremely important characteristic. Figures 2-3 and 2-4 are histograms of real data (see Data Sets 12 and 13 in Appendix B) with fundamentally different distributions.

Strongly consider assigning at least one histogram that should be generated with STATDISK or Minitab.

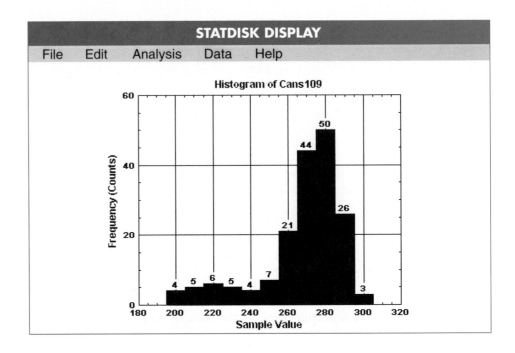

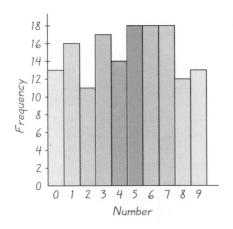

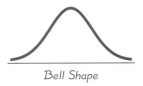

Bell Shape

Figure 2-3 is basically flat or uniform, whereas Figure 2-4 is roughly bell-shaped in the sense that it resembles the shape shown in the margin. Because Figure 2-3 depicts digits selected in Maryland's Pick Three Lottery, we expect that all digits should be equally likely and that the histogram should be basically flat, as in Figure 2-3. Any dramatic departure from a flat or uniform shape would suggest that the lottery isn't working as it should. The bell shape

Figure 2-3 Histogram of Lottery Results

Figure 2-4 Histogram of Weights of Quarters

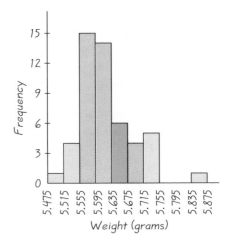

of the weights of quarters in Figure 2-4 is typical of an incredibly wide variety of different circumstances, especially those in manufacturing. Many procedures in statistics require that a distribution have a bell-shaped distribution similar to the one shown in Figure 2-4, and one way to check that requirement is to construct a histogram.

Dotplots

The Minitab display shown below is a dotplot of the same aluminum can data listed in Table 2-1. (Using Minitab, enter the data and then select the options of Graph, Character Graphs, and Dotplot.) From this illustration, it is very easy to see that a **dotplot** consists of a graph in which each data value is plotted as a point (or dot) along a scale of values. The leftmost dot, for example, represents the axial load of 200 lb. When values occur more than once, they are plotted as dots that are stacked vertically above the scale value. For example, the data in Table 2-1 include the load of 204 lb occurring twice, and those values are represented by the two dots stacked above the location corresponding to 204. (This dotplot uses 10 intervals to represent a range of 20 lb, so each dash on the horizontal scale represents two values. The dash above 200 represents the values of 199 and 200.)

The dotplot is similar to the histogram in the sense that we can see the *distribution* of the data.

Dotplots are new to this edition of the book. They are included among the topics recommended for AP statistics.

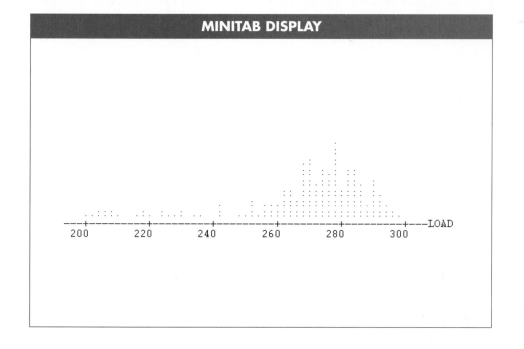

Change from the 6th edition of this book: Stem-and-leaf plots were moved from Section 2-7 to this section.

Stem-and-Leaf Plots

We have noted that the construction of a frequency table and the corresponding histogram give us valuable information about the nature of the distribution of data, but we suffer the disadvantage of losing some information in the process. Generally we cannot reconstruct the original data set from the frequency table or histogram. We will now introduce stem-and-leaf plots that enable us to see the distribution of data *without* losing information in the process.

In a **stem-and-leaf plot** we sort data according to a pattern that reveals the underlying distribution. The pattern involves separating a number (such as 257) into two parts, usually the first one or two digits (25) and the other digits (7). The stem consists of the leftmost digits (in this case, 25), and the leaves consist of the rightmost digits (in this case, 7). The method is illustrated in the following example.

EXAMPLE   Use the axial loads of aluminum cans listed in Table 2-1 to construct a stem-and-leaf plot.

SOLUTION   If we use the two leftmost digits for the stem, the stems consist of 20, 21, . . . , 29. We then draw a vertical line and list the leaves as shown below. The first value in Table 2-1 is 270, and we include that value by entering a 0 in the stem row for 27. We continue to enter all 175 values, and then we arrange the leaves (the digits positioned to the right) so that the numbers are arranged in increasing order. The first row represents the numbers 200, 201, 204, 204, 206, and so on.

Stem	Leaves
20	014466889
21	578
22	03358
23	0046
24	1228
25	01122466677899
26	01222233333455567788888888889999
27	00000000112222233333344445555566667777777788888888999
28	000111122222233333344445556666677899999
29	00011222334557

Consider covering standard stem-and-leaf plots, but assign expanded and condensed types as reading for students to do on their own.

By turning the page on its side, we can see a distribution of these data. Here's the great advantage of the stem-and-leaf plot: We can see the distribution of the data and yet retain all the information in the original list; if necessary, we could reconstruct the original list of values.

You might notice that the rows of digits in a stem-and-leaf plot are similar in nature to the bars in a histogram. One of the guidelines for constructing histograms is that the number of classes should be between 5 and 20, and that same guideline applies to stem-and-leaf plots for the same reasons. Stem-and-leaf plots can be *expanded* to include more rows and can also be *condensed* to include fewer rows. The stem-and-leaf plot of the preceding example can be expanded by subdividing rows into those with the digits 0 through 4 and those with digits 5 through 9. This expanded stem-and-leaf plot is shown here.

When it becomes necessary to *reduce* the number of rows, we can *condense* a stem-and-leaf plot by combining adjacent rows, as in the following illustration. Note that we separate digits in the leaves associated with the numbers in each stem by an asterisk. Every row in the condensed plot must include exactly one asterisk so that the shape of the plot is not distorted.

Stem	Leaves
20	0144
20	66889
21	
21	578
22	033
22	58
23	004
23	6
24	122
24	8
25	011224
25	66677899
26	0122222333334
26	5556778888888889999
27	00000000112222233333334444
27	5555666667777777788888888999
28	0001111222222333334444
28	555666677899999
29	00011222334
29	557

78-79	07*4	← This row represents 780, 787, 794.
80-81	*55	← This row represents 815, 815.
82-83	9*	← This row represents 829.
84-85	*	← This row has no data.
86-87	79*0	← This row represents 867, 869, 870.

Another advantage of stem-and-leaf plots is that their construction provides a fast and easy procedure for *ranking* data (arranging data in order). Data must be ranked for a variety of statistical procedures, such as finding the median (discussed in Section 2-4) and finding percentiles or quartiles (discussed in Section 2-6).

Using Computers for Stem-and-Leaf Plots

STATDISK doesn't do stem-and-leaf plots, but Minitab does. With Minitab, enter the data in column C1, then use the options of `Graph`, `Character Graphs`, and `Stem-and-Leaf`. The Minitab display includes an additional column of cumulative totals. For details, see the *Minitab Student Laboratory Manual and Workbook, seventh edition,* that is a supplement to this book.

Pareto Charts

Consider this statement: Among 75,200 accidental deaths in the United States in a recent year, 43,500 were attributable to motor vehicles, 12,200 to falls, 6400 to poison, 4600 to drowning, 4200 to fire, 2900 to ingestion of food or an object, and 1400 to firearms (based on data from the National Safety

Florence Nightingale

Florence Nightingale (1820–1910) is known to many as the founder of the nursing profession, but she also saved thousands of lives by using statistics. When she encountered an unsanitary and undersupplied hospital, she improved those conditions and then used statistics to convince others of the need for more widespread medical reform. She developed original graphs to illustrate that, during the Crimean War, more soldiers died as a result of unsanitary conditions than were killed in combat. Florence Nightingale pioneered the use of social statistics as well as graphics techniques.

Council). That written statement does a poor job of conveying the relationships among the different categories of qualitative data. A better way to convey relationships among qualitative data is to construct a Pareto chart. (Recall from Section 1-2 that qualitative data represent some nonnumeric characteristic, such as the types of accidental death listed above.) A **Pareto chart** is a bar graph for qualitative data, with the bars arranged in order according to frequencies. As in histograms, vertical scales in Pareto charts can represent frequencies or relative frequencies. The tallest bar is at the left, and the smaller bars are farther to the right, as in Figure 2-5. By arranging the bars in order of frequency, the Pareto chart focuses attention on the more important categories. From Figure 2-5 we can see that motor vehicle accidental deaths pose a problem that is much more serious than the other categories. Although firearm accidental deaths attract considerable media attention, they are a relatively minor problem when compared to the other categories.

Pie Charts

Like Pareto charts, pie charts are used to depict qualitative data in a way that makes them more understandable. Figure 2-6 shows an example of a **pie chart,** which graphically depicts qualitative data as slices of a pie. Construction of such a pie chart involves slicing up the pie into the proper proportions. If the category of motor vehicles represents 57.8% of the total, then the wedge representing motor vehicles should be 57.8% of the total. (The central angle should be $0.578 \times 360° = 208°$.)

The Pareto chart of Figure 2-5 and the pie chart of Figure 2-6 depict the same data in different ways, but a comparison will probably show that the Pareto chart does a better job of showing the relative sizes of the different components.

Scatter Diagrams

Sometimes we have data that are paired in a way that matches each value from one data set with a corresponding value from a second data set. A **scatter diagram** is a plot of the paired (x, y) data with a horizontal x-axis and a vertical y-axis. To manually construct a scatter diagram, construct a horizontal axis for the values of the first variable, construct a vertical axis for the values of the second variable, then plot the points. The pattern of the plotted points is often helpful in determining whether there is some relationship between the two variables. (This issue is discussed at length when the topic of correlation is considered in Section 9-2.) Using the cigarette nicotine and tar data from Data Set 4 in Appendix B, we generated the Minitab scatter diagram to obtain the accompanying result. (To obtain the Minitab graph, first enter or retrieve the two sets of matched data so they are in columns C1 and C2, then use the

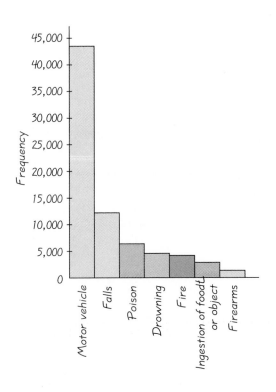

FIGURE 2-5   Pareto Chart of
Accidental Deaths by Type

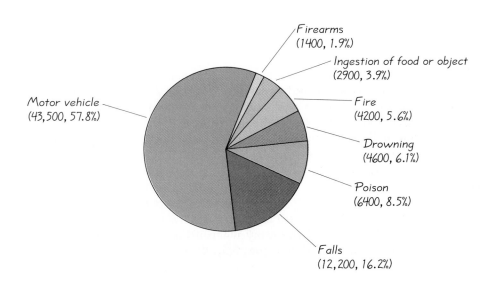

Figure 2-6   Pie Chart of
Accidental Deaths by Type

options of Graph and Plot. STATDISK and the TI-83 calculator are also designed to generate scatter diagrams.) On the basis of that graph, there does appear to be a relationship between the nicotine and tar contents in cigarettes, as shown by the pattern of the points.

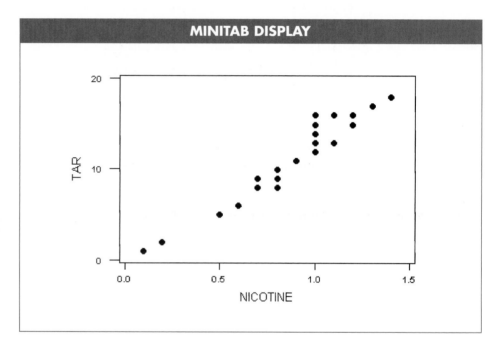

Other Graphs

Numerous pictorial displays other than the ones just described can be used to represent data dramatically and effectively. Exercise 27 involves a frequency polygon, which is a variation of the histogram. Section 2-7 will present boxplots that are very useful for seeing the distribution of data. Pictographs depict data by using pictures of objects, such as soldiers, tanks, airplanes, stacks of coins, or moneybags. Various graphs in Chapter 12 depict patterns of data over time.

See the figure on the insert, which has been described as possibly "the best statistical graphic ever drawn." This figure includes six different variables relevant to the march of Napoleon's army to Moscow in 1812. The thick band at the left depicts the size of the army when it began its invasion of Russia from Poland. The lower band describes Napoleon's retreat, with correspond-

ing temperatures and dates. Although first developed in 1861 by Charles Joseph Minard, this graph is ingenious even by today's standards.

In this section we have focused on the nature or shape of the distribution of data and methods of graphically depicting data. In the following sections we consider ways of measuring other characteristics of data.

Recommended assignment: 2, 3, 5, 7, 10, 13, 14, 18
Due to space constraints, only answers to even-numbered exercises are listed in this section. See Appendix F for the answers to odd-numbered exercises.

2–3 Exercises A: Basic Skills and Concepts

1. Visitors to Yellowstone National Park consider an eruption of the Old Faithful geyser to be a major attraction that should not be missed. The given frequency table summarizes a sample of times (in minutes) between eruptions. Construct a histogram corresponding to the given frequency table. If you're scheduling a bus tour of Yellowstone, what is the minimum time you should allocate to Old Faithful if you want to be reasonably sure that your tourists will see an eruption?

Time	Frequency
40–49	8
50–59	44
60–69	23
70–79	6
80–89	107
90–99	11
100–109	1

2. Samples of student cars and faculty/staff cars were obtained at the author's college, and their ages (in years) are summarized in the accompanying frequency table. Construct a relative frequency histogram for student cars and another relative frequency histogram for faculty cars. Based on the results, what are the noticeable differences between the two samples?

Age	Students	Faculty/Staff
0–2	23	30
3–5	33	47
6–8	63	36
9–11	68	30
12–14	19	8
15–17	10	0
18–20	1	0
21–23	0	1

2. The distribution of faculty/staff cars is weighted slightly more to the left, so their cars are slightly newer.

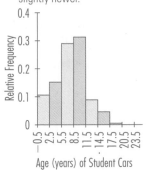

Age (years) of Student Cars

3. The given frequency table describes the speeds of drivers ticketed by the Town of Poughkeepsie police. These drivers were traveling through a 30 mi/h speed zone on Creek Road, which passes the author's college. Construct a histogram corresponding to the given frequency table. What does the distribution suggest about the enforced speed limit compared to the posted speed limit?

Speed	Frequency
42–43	14
44–45	11
46–47	8
48–49	6
50–51	4
52–53	3
54–55	1
56–57	2
58–59	0
60–61	1

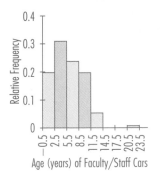

Age (years) of Faculty/Staff Cars

4. Disproportionately more young people are killed by gunfire.

6. 1021 1045 1091 1111
    1132 1177 1183 1204
    1222 1249 1369

8.

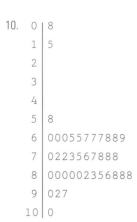

```
  +...:..::..:..:..:....+.....+
1000    1100    1200    1300    1400
```

10.
```
 0 | 8
 1 | 5
 2 |
 3 |
 4 |
 5 | 8
 6 | 00055777889
 7 | 0223567888
 8 | 000002356888
 9 | 027
10 | 0
```

12.
```
 0. |  234
 0. |  66777899999
 1. |  123444444
 1. |  55555566778
 2. |  001112223344
 2. |  7788999
 3. |  00144
 3. |  5
 4. |  4
 4. |  7
 5. |  3
```

4. Insurance companies continually research ages at death and causes of death. Construct a relative frequency histogram that corresponds to the given frequency table. The data are based on a *Time* magazine study of people who died from gunfire in America during one week. What does the histogram suggest about the ages of people who die from gunfire?

Age at Death	Frequency
16–25	22
26–35	10
36–45	6
46–55	2
56–65	4
66–75	5
76–85	1

In Exercises 5 and 6, list the original numbers in the data set represented by the given stem-and-leaf plots.

5.
Stem	Leaves
57	017
58	13349
59	456678
60	23

6.
Stem	Leaves
10	21 45 91
11	11 32 77 83
12	04 22 49
13	69

In Exercises 7–8, construct the dotplot for the data represented by the stem-and-leaf plot in the given exercise.

7. Exercise 5   8. Exercise 6

In Exercises 9–12, construct the stem-and-leaf plots for the given data sets found in Appendix B.

9. The lengths (in inches) of the bears in Data Set 3. (*Hint:* First round the lengths to the nearest inch.)

10. The pulse rates of the female statistics students included in Data Set 8.

11. Weights (in grams) of the 50 quarters listed in Data Set 13. (Use an expanded stem-and-leaf plot with about 8 rows.)

12. Weights (in pounds) of plastic discarded by 62 households: Refer to Data Set 1, and start by rounding the listed weights to the nearest tenth of a pound (or one decimal place). (Use an expanded stem-and-leaf plot with about 11 rows.)

13. A study was conducted to determine how people get jobs. The table on the top of page 57 lists data from 400 randomly selected subjects. The data are based on results from the National Center for Career Strategies. Construct a Pareto chart that corresponds to the given data. If someone would like to get a job, what seems to be the most effective approach?

Job Sources of Survey Respondents	Frequency
Help-wanted ads	56
Executive search firms	44
Networking	280
Mass mailing	20

14. The Pareto chart is more effective in showing the relative importance of job sources.

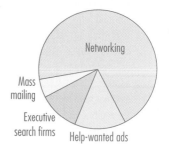

14. Refer to the data given in Exercise 13, and construct a pie chart. Compare the pie chart to the Pareto chart, and determine which graph is more effective in showing the relative importance of job sources.

15. An analysis of train derailment incidents showed that 23 derailments were caused by bad track, 9 were due to faulty equipment, 12 were attributable to human error, and 6 had other causes (based on data from the Federal Railroad Administration). Construct a pie chart representing the given data.

16. Refer to the data given in Exercise 15, and construct a Pareto chart. Compare the Pareto chart to the pie chart, and determine which graph is more effective in showing the relative importance of the causes of train derailments.   16. The Pareto chart appears to be more effective.

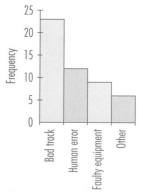

In Exercises 17–18, use the given paired data from Appendix B to construct a scatter diagram.

17. In Data Set 4, use tar for the horizontal scale and use carbon monoxide for the vertical scale. Based on the result, does there appear to be a relationship between cigarette tar and carbon monoxide? If so, describe the relationship.

18. In Data Set 3, use the distances around bear necks for the horizontal scale and use the bear weights for the vertical scale. Based on the result, what is the relationship between a bear's neck size and its weight?

18. Bears with larger neck sizes also tend to weigh more.

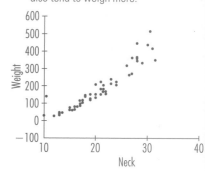

In Exercises 19–22, refer to the data sets in Appendix B.

   a. Construct a histogram.
   b. Describe the general shape of the distribution, such as bell-shaped, uniform, or skewed (lopsided).

19. Data Set 3 in Appendix B: weights of bears (Use 11 classes with a class width of 50 and begin with a lower class boundary of −0.5.)

20. Data Set 11 in Appendix B: weights of 100 M&Ms (Use 12 classes with a class width of 0.017 and begin with a lower class boundary of 0.8375.)

21. Data Set 1 in Appendix B: weights of paper discarded by 62 households in one week (Use 10 classes.)

22. Data Set 12 in Appendix B: the 300 numbers selected in the Maryland Lotto (not the Maryland Pick Three lottery)   22. Uniform

20. Bell-shaped

24. 22,000 of the 50,000 men died; 44%

26. 80,000 (out of 100,000) men died.

28. a. The histogram for π is more uniform.

   b. π is irrational, whereas 22/7 is rational. The decimal form of π is nonrepeating, whereas the decimal form of 22/7 has the digits 142857 repeating.

In Exercises 23–26, refer to the figure on the insert, which describes Napoleon's 1812 campaign to Moscow and back. The thick band at the left depicts the size of the army when it began its invasion of Russia from Poland, and the lower band describes Napoleon's retreat.

23. Find the percentage of men who survived the entire campaign.
24. Find the number of men and the percentage of men who died crossing the Berezina River.
25. How many men died on the return from Moscow during the time when the temperature dropped from 16°F to −6°F?
26. Of the men who made it to Moscow, how many died on the return trip between Moscow and Botr? (Note that 33,000 men did not go to Moscow, but they joined the returning men who did.)

π			22/7	
x	f		x	f
0	8		0	0
1	8		1	17
2	12		2	17
3	11		3	1
4	10		4	17
5	8		5	16
6	9		6	0
7	8		7	16
8	12		8	16
9	14		9	0

2-3 Exercises B: Beyond the Basics

27. A **frequency polygon** is a variation of a histogram that uses line segments connected to points (instead of using bars). Construct a frequency polygon by modifying the histogram in Figure 2-1 as follows: First replace the class boundaries on the horizontal scale with class marks. Second, replace the bars by points located above each class mark at a height equal to the class frequency. Third, connect the points and extend the graph to the left and right so that it begins and ends with a frequency of 0.

28. Frequency tables are given for the first 100 digits in the decimal representation of π and the first 100 digits in the decimal representation of 22/7.
   a. Construct histograms representing the frequency tables, and note any differences.
   b. The numbers π and 22/7 are both real numbers, but how are they fundamentally different?

29. Using a collection of sample data, we construct a frequency table with 10 classes and then construct the corresponding histogram. How is the histogram affected if the number of classes is doubled but the same vertical scale is used?

30. In an insurance study of motor vehicle accidents in New York State, fatal crashes are categorized according to time of day, with the results

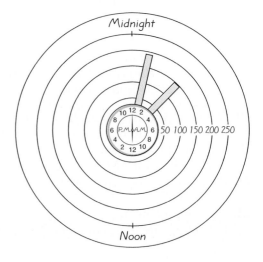

given in the accompanying table (based on data from the New York State Department of Motor Vehicles).

a. Complete the circular bar chart and construct a histogram.

b. Which is more effective in depicting the data? Why?

c. Because the time period 4:00 A.M. to 6:00 A.M. has the lowest number of fatal crashes, is that time period the safest time to drive? Why or why not?

31. In "Ages of Oscar-Winning Best Actors and Actresses" (*Mathematics Teacher* magazine) by Richard Brown and Gretchen Davis, stem-and-leaf plots are used to compare the ages of actors and actresses at the time they won Oscars. Here are the results for 34 recent winners from each category.

Actors:	32	37	36	32	51	53	33	61	35	45	55	39
	76	37	42	40	32	60	38	56	48	48	40
	43	62	43	42	44	41	56	39	46	31	47
Actresses:	50	44	35	80	26	28	41	21	61	38	49	33
	74	30	33	41	31	35	41	42	37	26	34
	34	35	26	61	60	34	24	30	37	31	27

a. Construct a back-to-back stem-and-leaf plot for the above data. The first two scores from each group have been entered below.

Actors' Ages	Stem	Actresses' Ages
	2	
72	3	
	4	4
	5	0
	6	
	7	
	8	

b. Using the results from part a, compare the two different sets of data, and explain any differences.

  2-4 Measures of Central Tendency

The main objective of this section is to present the important measures of central tendency, which are values (or measures) that the data tend to center around.

Time of Day	Number of Fatal Crashes
A.M. 12–2	194
2–4	149
4–6	100
6–8	131
8–10	119
10–12	160
P.M. 12–2	152
2–4	221
4–6	230
6–8	211
8–10	223
10–12	178

30. b. The circular bar chart is more effective because the bar chart hides the pattern created in the time sequence.

c. No, there aren't many cars being driven between 4:00 A.M. and 6:00 A.M. You need to compare the rates of fatal crashes.

This is obviously a very important section. Students must recognize that there are different ways to define an "average." They should thoroughly understand the mean and the median. This is a good time to remind students of the three important characteristics of data: (1) the nature of distribution, (2) center, (3) variation. Sections 2-2 and 2-3 addressed the nature of the distribution; this section addresses measures of central tendency.

> ## DEFINITION
>
> A **measure of central tendency** is a value at the center or middle of a data set.

Whereas Sections 2-2 and 2-3 considered frequency tables and graphs that reveal the nature or shape of the *distribution* of a data set, this section focuses on finding values that are typical or representative of a data set. There are different criteria for determining the center, and so there are different definitions of measures of central tendency, including the mean, median, mode, and midrange. We begin with the mean.

Mean

The (arithmetic) mean is generally the most important of all numerical descriptive measurements, and it is what most people call an average. In Figure 2-7 we illustrate the property that the mean is at the center of the data set in the sense that it is a balance point for the data.

> ## DEFINITION
>
> The **arithmetic mean** of a set of scores is the value obtained by adding the scores and dividing the total by the number of scores. This particular measure of central tendency will be used frequently throughout the remainder of this text, and it will be referred to simply as the **mean**.

This definition can be expressed as Formula 2-1, where the Greek letter Σ (uppercase Greek sigma) indicates *summation* of values, so that Σx represents the sum of all scores. Also, the symbol n denotes the **sample size,** which is the number of scores being considered.

Formula 2-1
$$\text{mean} = \frac{\Sigma x}{n}$$

The mean can be denoted by $\bar{x}$ (pronounced "x-bar") if the available scores are a sample from a larger population; if all scores of the population are available, then we can denote the computed mean by μ (lowercase Greek mu). (Sample statistics are usually represented by English letters, such as $\bar{x}$, whereas population parameters are usually represented by Greek letters, such as μ.) Many calculators can find the mean of a data set: Enter the data, and press the key labeled $\bar{x}$. Entry of the data varies with different calculator models, so refer to the user manual for your calculator.

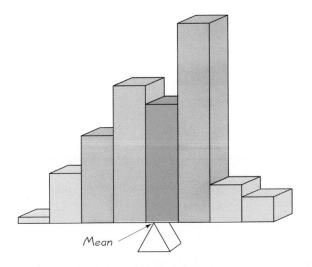

Figure 2-7
Mean as a Balance Point
If a fulcrum is placed at the position of the mean, it will balance the histogram.

Notation	
Σ	denotes *summation* of a set of values.
x	is the *variable* usually used to represent the individual data values.
n	represents the *number of values in a sample*.
N	represents the *number of values in a population*.
$\bar{x} = \dfrac{\Sigma x}{n}$	is the *mean of a set of sample values*.
$\mu = \dfrac{\Sigma x}{N}$	denotes the *mean of all values in a population*.

EXAMPLE   Listed below are the times (in years) that the first ten presidents survived after inauguration). Find the mean for this sample.

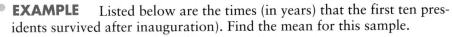

$$10 \quad 29 \quad 26 \quad 28 \quad 15 \quad 23 \quad 17 \quad 25 \quad 0 \quad 20$$

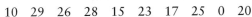

SOLUTION
The mean is computed by using Formula 2-1. First add the scores.

$$\Sigma x = 10 + 29 + 26 + 28 + 15 + 23 + 17 + 25 + 0 + 20 = 193$$

Now divide the total by the number of scores present. Because there are 10 scores, we have $n = 10$ and get

$$\bar{x} = \frac{193}{10} = 19.3$$

The mean value is therefore 19.3 years.

For the 10 values in the preceding example, 19.3 is at the center, according to the definition of the mean. Other definitions of a measure of central tendency involve different perceptions of how the center is determined.

Median

> ### DEFINITION
>
> The **median** of a set of scores is the middle value when the scores are arranged in order of increasing (or decreasing) magnitude. The median is often denoted by $\tilde{x}$ (pronounced "x-tilde").

To find the median, first rank the scores (arrange them in order), then use one of these two procedures:

1. If the number of scores is odd, the median is the number that is located in the exact middle of the list.
2. If the number of scores is even, the median is found by computing the mean of the two middle numbers.

EXAMPLE   Find the median of these survival times (in years after inauguration) for the first five presidents:

$$10 \quad 29 \quad 26 \quad 28 \quad 15$$

SOLUTION   Begin by rearranging the scores so that they are in order:

$$10 \quad 15 \quad 26 \quad 28 \quad 29$$

The number of scores is 5, which is an odd number, so we find the number that is at the exact middle. The number 26 is in the middle, so the median for this data set is 26 years.

EXAMPLE   The following values are the incomes (in dollars) that performers received for one rock concert. The mean is $8900. Find the median.

$$500 \quad 600 \quad 800 \quad 50,000 \quad 1000 \quad 500$$

SOLUTION   Begin by rearranging the values so that they are in order:

$$500 \quad 500 \quad 600 \quad 800 \quad 1000 \quad 50,000$$

The number of scores is 6, which is an even number, so we find the two scores that are at the middle and then find their mean. The two scores at the middle are 600 and 800, so the median is found by adding these two scores and dividing by 2. The median is $700.

For this data set, the mean of $8900 was strongly affected by the extreme value of $50,000, but that extreme score did not affect the median of $700.

Mode

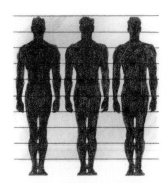

> ### DEFINITION
>
> The **mode** of a data set is the score that occurs most frequently. When two scores occur with the same greatest frequency, each one is a mode and the data set is **bimodal**. When more than two scores occur with the same greatest frequency, each is a mode and the data set is said to be **multimodal**. When no score is repeated, we say that there is no mode. The mode is often denoted by M.

EXAMPLE Find the modes of the following data sets.

a. 5 5 5 3 1 5 1 4 3 5
b. 1 2 2 2 3 4 5 6 6 6 7 9
c. 1 2 3 6 7 8 9 10

SOLUTION

a. The number 5 is the mode because it is the score that occurs most often.
b. The numbers 2 and 6 are both modes because they occur with the same greatest frequency. The data set is bimodal.
c. There is no mode because no score is repeated.

Among the different measures of central tendency we are now considering, the mode is the only one that can be used with data at the nominal level of measurement, as illustrated in the next example.

EXAMPLE A study of reaction times involved 30 left-handed subjects, 50 right-handed subjects, and 20 subjects who are ambidextrous.

An Average Guy

The "average" American male is named Robert. He is 31 years old, 5 ft 9½ in. tall, weighs 172 lb, wears a size 40 suit, wears a size 9½ shoe, and has a 34-in. waist. Each year he eats 12.3 lb of pasta, 26 lb of bananas, 4 lb of potato chips, 18 lb of ice cream, and 79 lb of beef. Each year he also watches television for 2567 hours and gets 585 pieces of mail. After eating his share of potato chips, reading some of his mail, and watching some television, he ends the day with 7.7 hours of sleep. The next day begins with a 21-minute commute to a job at which he will work for 6.1 hours.

Although we can't numerically average these characteristics, we can report that the mode is right-handed, because that is the characteristic with the greatest frequency.

Midrange

The midrange can be easily omitted. It is included to emphasize the point that there are different ways of determining the "center" of a set of data.

DEFINITION

The **midrange** is the value midway between the highest and lowest scores. It is found by adding the highest score to the lowest score and then dividing the sum by 2, as in the following formula.

$$\text{midrange} = \frac{\text{highest score} + \text{lowest score}}{2}$$

EXAMPLE   Find the midrange of these survival times (in years after inauguration) for the first ten presidents.

$$10 \quad 29 \quad 26 \quad 28 \quad 15 \quad 23 \quad 17 \quad 25 \quad 0 \quad 20$$

SOLUTION   The midrange is found as follows:

$$\frac{\text{highest score} + \text{lowest score}}{2} = \frac{29 + 0}{2} = 14.5 \text{ years}$$

The midrange isn't used much, but we include it mainly to emphasize the point that there are several different ways to define the center of a set of data. (See also Exercises 20–22.)

Unfortunately, the term *average* is sometimes used for any measure of central tendency and is sometimes used for the mean. Because of this ambiguity, we should not use the term *average* when referring to a particular measure of central tendency. Instead, we should use the specific term, such as mean, median, mode, or midrange.

EXAMPLE   Refer to Table 2-1, where we list 175 axial loads of aluminum cans. Find the (a) mean, (b) median, (c) mode, and (d) midrange.

SOLUTION

a. Mean: The sum of the 175 values is 46,745, so

$$\bar{x} = \frac{46{,}745}{175} = 267.1 \text{ lb}$$

b. Median: After arranging the scores in increasing order, we find that the 88th score of 273 is in the exact middle, so the median is 273.0 lb. (The scores can easily be arranged in increasing order by constructing a stem-and-leaf plot, as we discussed in Section 2-3, or by using a computer program such as STATDISK or Minitab.) We express the result with an extra decimal place by using the round-off rule that follows this example.

c. Mode: The most frequent axial load is 268 lb, which occurs 9 times. It is therefore the mode.

d. Midrange: We use the following formula to find the midrange:

$$\text{midrange} = \frac{\text{highest score} + \text{lowest score}}{2} = \frac{297 + 200}{2} = 248.5 \text{ lb}$$

We now summarize these results.

mean: 267.1 lb

median: 273.0 lb

mode: 268 lb

midrange: 248.5 lb

Previously, we constructed a frequency table and histogram for the data in Table 2-1 and we saw the distribution of the data. We now have important information about the center of the data.

Round-Off Rule

A simple rule for rounding answers is this:

Carry one more decimal place than is present in the original set of data.

We should round only the final answer and not intermediate values. For example, the mean of 2, 3, 5 is 3.33333333 . . . , and it can be rounded as 3.3. Because the original data were whole numbers, we rounded the answer to the nearest tenth. As another example, the mean of 2.1, 3.4, 5.7 is rounded to 3.73 with two decimal places (one more decimal place than was used for the original values).

Try to encourage this good practice: Don't round off values in the middle of a calculation; instead, carry as many places as the calculator will handle. Only round off at the end. Large errors sometimes occur when intermediate results are rounded too much.

If pressed for time or if you find
that Chapter 2 is taking longer
than expected, consider
omitting weighted mean and
mean from a frequency table, or
simply assign a few problems
for students to do on their own.

Mean from a Frequency Table and the Weighted Mean

When data are summarized in a frequency table, we can approximate the mean by replacing class limits with class marks and assuming that each class mark is repeated a number of times equal to the class frequency. In Table 2-2, for example, the first class of 200–209 contains 9 scores that fall somewhere between those class limits, but we don't know the specific values of those 9 scores. We make calculations possible by assuming that all 9 scores are the class mark of 204.5. With 9 scores of 204.5, we have a total of $9 \times 204.5 = 1840.5$ to contribute toward the grand total of all scores combined. The number of scores is equal to the sum of the frequencies, so Formula 2-2 can be used to find the mean from a frequency table. Formula 2-2 doesn't really involve a fundamentally different concept; it is simply a variation of Formula 2-1.

Formula 2-2 $$\bar{x} = \frac{\Sigma(f \cdot x)}{\Sigma f}$$ mean from frequency table

where x = class mark
                         f = frequency
                         $\Sigma f = n$

The aluminum can axial load data from Frequency Table 2-2 have been entered in Table 2-5, where we apply Formula 2-2. (We can also compute the

TABLE 2-5	Finding Σf and $\Sigma(f \cdot x)$		
Axial Load (lb)	Frequency f	Class Mark x	$f \cdot x$
200–209	9	204.5	1,840.5
210–219	3	214.5	643.5
220–229	5	224.5	1,122.5
230–239	4	234.5	938.0
240–249	4	244.5	978.0
250–259	14	254.5	3,563.0
260–269	32	264.5	8,464.0
270–279	52	274.5	14,274.0
280–289	38	284.5	10,811.0
290–299	14	294.5	4,123.0
Total	$\Sigma f = 175$		$\Sigma(f \cdot x) = 46,757.5$

$$\bar{x} = \frac{\Sigma(f \cdot x)}{\Sigma f} = \frac{46,757.5}{175} = 267.2$$

mean from a frequency table with a TI-83 calculator: Enter the class marks in list L1, enter the frequencies in list L2, then use STAT, CALC, and 1 = Var Stats and enter L1, L2.) When we used the original collection of scores to calculate the mean directly, we obtained a mean of 267.1, so the value of the weighted mean obtained from the frequency table is just a little off.

In some situations the scores vary in their degree of importance, so we may want to compute a **weighted mean,** which is a mean computed with the different scores assigned different weights. In such cases, we can calculate the weighted mean by assigning different weights to different scores, as shown in Formula 2-3.

Formula 2-3   weighted mean: $\bar{x} = \dfrac{\Sigma(w \cdot x)}{\Sigma w}$

For example, suppose we need a mean of 5 test scores (85, 90, 75, 80, 95), but the first 4 tests count for 15% each, while the last score counts for 40%. We can simply assign a weight of 15 to each of the first 4 tests, assign a weight of 40 to the last test, then proceed to calculate the mean by using Formula 2-3 as follows:

$$\bar{x} = \frac{\Sigma(w \cdot x)}{\Sigma w}$$

$$= \frac{(15 \times 85) + (15 \times 90) + (15 \times 75) + (15 \times 80) + (40 \times 95)}{15 + 15 + 15 + 15 + 40}$$

$$= \frac{8750}{100} = 87.5$$

As another example, college grade-point averages can be computed by assigning each letter grade the appropriate number of points (A = 4, B = 3, etc.), then assigning to each number a frequency equal to the number of credit hours. A grade of C in a 3-credit course would be equivalent to a class mark of 2 with a frequency of 3. Again, Formula 2-3 can be used to compute the grade-point average.

The Earned Run Average

A weighted mean is used in baseball's earned run average (ERA), which is a measure of a pitcher's effectiveness. It is computed as follows: Multiply the number of earned runs scored against the pitcher by 9, then divide by the total number of innings pitched. (Earned runs do not include runs scored by players who got on base because of errors.) The ERA represents the total number of runs a pitcher would yield in nine innings. It has the advantage of being weighted to correspond to a full game, so different pitchers can be compared. However, it is criticized because it does not take into account the very different circumstances faced by starting pitchers and relief pitchers.

The Best Measure of Central Tendency

We have found that for the data in Table 2-1, the mean, median, mode, and midrange have values of 267.1, 273.0, 268, and 248.5, respectively. Which measure of central tendency is best? Unfortunately, there is no single best

TABLE 2-6	Comparison of Mean, Median, Mode, and Midrange					
Average	Definition	How Common?	Existence	Takes Every Score into Account?	Affected by Extreme Scores?	Advantages and Disadvantages
Mean	$\bar{x} = \dfrac{\Sigma x}{n}$	most familiar "average"	always exists	yes	yes	used throughout this book; works well with many statistical methods
Median	middle score	commonly used	always exists	no	no	often a good choice if there are some extreme scores
Mode	most frequent score	sometimes used	might not exist; may be more than one mode	no	no	appropriate for data at the nominal level
Midrange	$\dfrac{\text{high} + \text{low}}{2}$	rarely used	always exists	no	yes	very sensitive to extreme values

General comments:
• For a data collection that is approximately symmetric with one mode, the mean, median, mode, and midrange tend to be about the same.
• For a data collection that is obviously asymmetric, it would be good to report both the mean and median.
• The mean is relatively *reliable*. That is, when samples are drawn from the same population, the sample means tend to be more consistent than the other averages (consistent in the sense that the means of samples drawn from the same population don't vary as much as the other averages).

answer to that question because there are no objective criteria for determining the most representative measure for all data sets. The different measures of central tendency have different advantages and disadvantages, some of which are summarized in Table 2-6. An important advantage of the mean is that it takes every score into account, but an important disadvantage is that it is sometimes dramatically affected by a few extreme scores. This disadvantage can be overcome by using a trimmed mean, as described in Exercise 25.

Exercise 27 in Section 2-5 uses the mean, median, and standard deviation to calculate a measure of skewness, which can be used to determine whether a data set has significant skewness.

Skewness

A comparison of the mean, median, and mode can reveal information about the characteristic of skewness, defined below and illustrated in Figure 2-8.

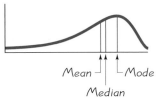 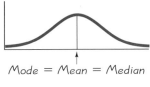 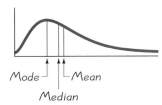

FIGURE 2-8   Skewness

(a) Skewed to the Left (Negatively Skewed): The mean and median are to the *left* of the mode.

(b) Symmetric (Zero Skewness): The mean, median, and mode are the same.

(c) Skewed to the Right (Positively Skewed): The mean and median are to the *right* of the mode.

DEFINITION

A distribution of data is **skewed** if it is not symmetric and extends more to one side than the other. (A distribution of data is **symmetric** if the left half of its histogram is roughly a mirror image of its right half.)

Data skewed to the *left* are said to be **negatively skewed**; the mean and median are to the left of the mode. Although not always predictable, negatively skewed data generally have the mean to the left of the median. (See Figure 2-8(a).) Data skewed to the *right* are said to be **positively skewed**; the mean and median are to the right of the mode. Again, although not always predictable, positively skewed data generally have the mean to the right of the median. (See Figure 2-8(c).)

If we examine the histogram in Figure 2-1 for the axial loads of aluminum cans we are considering in this chapter, we see a graph that appears to be skewed to the left. In practice, many distributions of data are symmetric and without skewness. Distributions skewed to the right are more common than those skewed to the left because it's often easier to get exceptionally large values than values that are exceptionally small. With annual incomes, for example, it's impossible to get values below the lower limit of zero, but there are a few people who earn millions of dollars in a year. Annual incomes therefore tend to be skewed to the right, as in Figure 2-8(c).

Exercises A: Basic Skills and Concepts

In Exercises 1–4, find the (a) mean, (b) median, (c) mode, and (d) midrange for the given sample data.

Recommended assignment: 1, 4, 6, 7, 10, 14, 19 parts a, b, c (part d requires logs), 26

1. $\bar{x}$ = 19.3 oz; median = 19.5
oz; mode = 20 oz;
midrange = 19.0 oz; no

2. $\bar{x}$ = 4.9; median = 6.0;
mode = 6; midrange = 4.5

3. $\bar{x}$ = 6.020; median = 5.780;
mode: none;
midrange = 6.445

4. $\bar{x}$ = 0.187; median = 0.170;
mode = 0.16, 0.17;
midrange = 0.205

5. Jefferson Valley: $\bar{x}$ = 7.15;
median = 7.20; mode = 7.7;
midrange = 7.10
Providence: same results as
Jefferson Valley

6. Students: $\bar{x}$ = 8.2;
median = 8.0; mode = 4, 8;
midrange = 9.0
Faculty/staff: $\bar{x}$ = 7.8;
median = 6.5; mode = 4, 6,
7; midrange = 12.5

7. 4000 B.C.: $\bar{x}$ = 128.7;
median = 128.5;
mode = 131;
midrange = 128.5
150 A.D.: $\bar{x}$ = 133.3;
median = 133.5;
mode = 126;
midrange = 133.5

1. The given values are weights (ounces) of steaks listed on a restaurant menu as "20-ounce Porterhouse" steaks (based on data collected by a student of the author). The weights are supposed to be 21 oz because the steaks supposedly lose an ounce when cooked. Do these steaks appear to weigh enough?

17 20 21 18 20 20 20 18 19 19
20 19 21 20 18 20 20 19 18 19

2. Digits selected in the Maryland Pick Three lottery:

0 7 3 6 2 7 6 6 6 3 8 1 7 8 7
1 6 8 6 9 5 2 1 5 0 3 9 9 0 7

3. Nitrate deposits (in kg per hectare) as part of acid rain for Massachusetts from July through September for recent years (based on data from the U.S. Department of Agriculture):

6.40 5.21 4.66 5.24 6.96 5.53 8.23 6.80 5.78 6.00 5.41

4. Blood-alcohol concentrations of 15 drivers involved in fatal accidents and then convicted with jail sentences (based on data from the U.S. Department of Justice):

0.27 0.17 0.17 0.16 0.13 0.24 0.29 0.24
0.14 0.16 0.12 0.16 0.21 0.17 0.18

In Exercises 5–8, find the mean, median, mode, and midrange for each of the two samples, then compare the two sets of results.

5. Waiting times of customers at the Jefferson Valley Bank (where all customers enter a single waiting line) and the Bank of Providence (where customers wait in individual lines at three different teller windows):

Jefferson Valley: 6.5 6.6 6.7 6.8 7.1 7.3 7.4 7.7 7.7 7.7
Providence: 4.2 5.4 5.8 6.2 6.7 7.7 7.7 8.5 9.3 10.0

6. Samples of the ages (in years) of student cars and faculty/staff cars obtained at the author's college:

Students: 10 4 5 2 9 7 8 8 16 4 13 12
Faculty/staff: 7 10 4 13 23 2 7 6 6 3 9 4

7. Maximum breadth of samples of male Egyptian skulls from 4000 B.C. and 150 A.D. (based on data from *Ancient Races of the Thebaid* by Thomson and Randall-Maciver):

4000 B.C.: 131 119 138 125 129 126 131 132 126 128 128 131
150 A.D.: 136 130 126 126 139 141 137 138 133 131 134 129

8. Weights (in pounds) of paper and plastic discarded by households in one week (data collected for the Garbage Project at the University of Arizona):

Paper: 9.55   6.38   2.80   6.98   6.33   6.16   10.00   12.29
Plastic: 2.19   2.10   1.41   0.63   0.92   1.40   1.74   2.87

In Exercises 9–12, refer to the data set in Appendix B, and find the (a) mean, (b) median, (c) mode, and (d) midrange.

9. Data Set 2 in Appendix B: the body temperatures for 8:00 A.M. on day 1
10. Data Set 4 in Appendix B: the nicotine contents of all cigarettes listed
11. Data Set 3 in Appendix B: the weights of the bears
12. Data Set 11 in Appendix B: the weights of the red M&M plain candies

In Exercises 13–16, find the mean of the data summarized in the given frequency table.

Time	Frequency
40–49	8
50–59	44
60–69	23
70–79	6
80–89	107
90–99	11
100–109	1

13. Visitors to Yellowstone National Park consider an eruption of the Old Faithful geyser to be a major attraction that should not be missed. The given frequency table summarizes a sample of times (in minutes) between eruptions.

14. Samples of student cars and faculty/staff cars were obtained at the author's college, and their ages (in years) are summarized in the accompanying frequency table. Find the mean age of student cars and find the mean age of faculty/staff cars. Based on the results, are there any noticeable differences between the two samples? If so, what are they?

Age	Students	Faculty/Staff
0–2	23	30
3–5	33	47
6–8	63	36
9–11	68	30
12–14	19	8
15–17	10	0
18–20	1	0
21–23	0	1

15. The given frequency table describes the speeds of drivers ticketed by the Town of Poughkeepsie police. These drivers were traveling through a 30 mi/h speed zone on Creek Road, which passes the author's college. How does the mean compare to the posted speed limit of 30 mi/h?   15. 46.7 mi/h

Speed	Frequency
42–43	14
44–45	11
46–47	8
48–49	6
50–51	4
52–53	3
54–55	1
56–57	2
58–59	0
60–61	1

8. Paper: $\bar{x} = 7.561$;
   median = 6.680; mode: none;
   midrange = 7.545
   Plastic: $\bar{x} = 1.658$;
   median = 1.575; mode: none;
   midrange = 1.750
9. $\bar{x} = 98.13$; median = 98.20;
   mode = 97.4, 98.0, 98.2,
   98.8; midrange = 97.80
10. $\bar{x} = 0.94$; median = 1.00;
    mode = 1.0;
    midrange = 0.75
11. $\bar{x} = 182.9$; median = 150.0;
    mode = 140, 150, 166, 202,
    204, 220; midrange = 270.0
12. $\bar{x} = 0.9097$;
    median = 0.9080;
    mode = 0.908;
    midrange = 0.9265
13. 74.4 min
14. Students: $\bar{x} = 7.8$ yr;
    faculty/staff : $\bar{x} = 5.9$ yr
    The student cars average
    about two years older.

16. 35.5 yr

16. Insurance companies continually research ages at death and causes of death. The data are based on a *Time* magazine study of people who died from gunfire in America during one week. What do you conclude from the result?

Age at Death	Frequency
16–25	22
26–35	10
36–45	6
46–55	2
56–65	4
66–75	5
76–85	1

2-4 Exercises B: Beyond the Basics

17. 82.0

17. A student receives quiz grades of 60, 84, and 90. The final exam grade is 88. Find the weighted mean if the quizzes each count for 20% and the final exam counts for 40% of the final grade.

18. 3.000

18. A student's transcript shows an A in a 4-credit course, an A in a 3-credit course, a C in a 3-credit course, and a D in a 2-credit course. Grade points are assigned as follows: A = 4, B = 3, C = 2, D = 1, F = 0. If grade points are weighted according to the number of credit hours, find the weighted mean (grade-point average) rounded to three decimal places.

19. a. $\bar{x}$ = 193,000;
   median = 206,000;
   mode = 236,000;
   midrange = 172,000
   b. Each result is increased by k.
   c. Each result is multiplied by k.
   d. 5.269 ≠ 5.286, so they're not equal.

19. a. Find the mean, median, mode, and midrange for the following annual incomes (in dollars) of self-employed doctors (based on data from the American Medical Association):

$$108,000 \quad 236,000 \quad 179,000 \quad 206,000 \quad 236,000$$

b. If a constant value k is added to each income, how are the results from part (a) affected?

c. If each income given in part (a) is multiplied by a constant k, how are the results from part (a) affected?

d. Data are sometimes transformed by replacing each score x with log x. For the given x values, determine whether the mean of the log x values is equal to log $\bar{x}$.

20. a. 48.0 mi/h
    b. 40.57 mi/h

20. The **harmonic mean** is often used as a measure of central tendency for data sets consisting of rates of change, such as speeds. It is found by dividing the number of scores n by the sum of the *reciprocals* of all scores, expressed as

$$\frac{n}{\sum \frac{1}{x}}$$

(No score can be zero.) For example, the harmonic mean of 2, 4, 10 is

$$\frac{n}{\sum \frac{1}{x}} = \frac{3}{\frac{1}{2} + \frac{1}{4} + \frac{1}{10}} = \frac{3}{0.85} = 3.5$$

(continued)

a. Four students drive from New York to Florida (1200 miles) at a speed of 40 mi/h (yeah, right!) and return at a speed of 60 mi/h. What is their average speed for the round trip? (The harmonic mean is used in averaging speeds.)

b. A dispatcher for the Kramden Bus Company calculates the average round-trip speed (in miles per hour) for the route between Boston and Providence. The results obtained for 14 different runs are listed below. Based on these values, what is the average speed of a bus assigned to this route?

$$
\begin{array}{ccccccc}
42.6 & 41.3 & 38.2 & 42.9 & 43.4 & 43.7 & 40.8 \\
34.2 & 40.1 & 41.2 & 40.5 & 41.7 & 39.8 & 39.6
\end{array}
$$

21. The **geometric mean** is often used in business and economics for finding average rates of change, average rates of growth, or average ratios. Given n scores (all of which are positive), the geometric mean is the nth root of their product. For example, the geometric mean of 2, 4, 10 is found by first multiplying the scores to get 80, then taking the cube root (because there are 3 scores) of the product to get 4.3. The *average growth factor* for money compounded at annual interest rates of 10%, 8%, 9%, 12%, and 7% can be found by computing the geometric mean of 1.10, 1.08, 1.09, 1.12, and 1.07. Find that average growth factor.

22. The **quadratic mean** (or **root mean square**, or **R.M.S.**) is usually used in physical applications. In power distribution systems, for example, voltages and currents are usually referred to in terms of their R.M.S. values. The quadratic mean of a set of scores is obtained by squaring each score, adding the results, dividing by the number of scores n, and then taking the square root of that result. For example, the quadratic mean of 2, 4, 10 is

$$
\sqrt{\frac{\Sigma x^2}{n}} = \sqrt{\frac{4 + 16 + 100}{3}} = \sqrt{\frac{120}{3}} = \sqrt{40} = 6.3
$$

Find the R.M.S. of these power supplies (in volts): 151, 162, 0, 81, −68.

23. Frequency tables often have open-ended classes, such as the accompanying table that summarizes amounts of time spent studying by college freshmen (based on data from *The American Freshman* as reported in *USA Today*). Formula 2-2 cannot be directly applied because we can't determine a class mark for the class of "more than 20." Calculate the mean by assuming that this class is really (a) 21–25, (b) 21–30, and (c) 21–40. What can you conclude?

Hours Studying per Week	Frequency
0	5
1–5	96
6–10	57
11–15	25
16–20	11
More than 20	6

21. 1.092
22. 109.8 volts
23. a. 7.0
    b. 7.1
    c. 7.3 In this case, the open-ended class doesn't have too much effect on the mean. The mean is likely to be around 7.1, give or take about 0.1.

24. When data are summarized in a frequency table, the median can be found by first identifying the *median class* (the class that contains the median). We then assume that the scores in that class are evenly distributed and we can interpolate. This process can be described by

$$\text{(lower limit of median class)} + \text{(class width)}\left(\frac{\left(\dfrac{n+1}{2}\right) - (m+1)}{\text{frequency of median class}}\right)$$

where n is the sum of all class frequencies and m is the sum of the class frequencies that *precede* the median class. Use this procedure and the data in Frequency Table 2-2 to find the median axial load.

25. Because the mean is very sensitive to extreme scores, it is accused of not being a *resistant* measure of central tendency. The **trimmed mean** is more resistant. To find the 10% trimmed mean for a data set, first arrange the data in order, then delete the bottom 10% of the scores and the top 10% of the scores, and calculate the mean of the remaining scores. For the weights of the bears in Data Set 3 from Appendix B, find (a) the mean; (b) the 10% trimmed mean; (c) the 20% trimmed mean. How do the results compare?

26. Using an almanac, a researcher finds the average teacher's salary for each state. He adds those 50 values, then divides by 50 to obtain their mean. Is the result equal to the national average teacher's salary? Why or why not?

  Measures of Variation

Because this section deals with the characteristic of variation, which is so important in statistics, this is one of the most important sections in the entire book. The following key concepts should be mastered: (1) Variation refers to the amount that the scores vary among themselves, and it can be measured with specific numbers; (2) scores that are relatively close together have low measures of variation, whereas scores that are spread farther apart have measures of variation that are larger; (3) the standard deviation is a particularly important measure of variation, and we should be able to find its value from a set of scores; (4) the values of standard deviations must be *interpreted* correctly.

Many banks once required that customers wait in separate lines at each teller's window, but most have now changed to one single main waiting line. Why did they make that change? The mean waiting time didn't change, because the waiting-line configuration doesn't affect the efficiency of the tellers. They changed to the single line because customers prefer waiting times

that are more *consistent* with less variation. In this case, thousands of banks made a change that resulted in lower variation (and happier customers), even though the mean was not affected. Let's now consider the same bank sample data used in Exercise 5 from the preceding section. The listed values are waiting times (in minutes) of customers.

Jefferson Valley Bank (Single waiting line)	6.5	6.6	6.7	6.8	7.1	7.3	7.4	7.7	7.7	7.7
Bank of Providence (Multiple waiting lines)	4.2	5.4	5.8	6.2	6.7	7.7	7.7	8.5	9.3	10.0

Customers at the Jefferson Valley Bank enter a single waiting line that feeds three teller windows. Customers at the Bank of Providence can enter any one of three different lines that have formed at three different teller windows. If we do Exercise 5 from Section 2-4, we find that both banks have the same mean of 7.15, the same median of 7.20, the same mode of 7.7, and the same midrange of 7.10. On the basis of a consideration of only these measures of central tendency, we would think that waiting times at the two banks are pretty much the same. Yet, scanning the original waiting times should reveal a fundamental difference: The Jefferson Valley Bank has waiting times with much less *variation* than the times for the Bank of Providence. If all other characteristics are equal, customers are likely to prefer the Jefferson Valley Bank, where they won't become annoyed by being caught in an individual line that is much slower than the others.

By subjectively comparing the differences among waiting times at the two banks, we can see the characteristic of variation. Let's now proceed to develop some specific ways of actually *measuring* variation. We begin with the range.

Range

The **range** of a set of data is the difference between the highest value and the lowest value. To compute it, simply subtract the lowest score from the highest score. For the Jefferson Valley Bank customers, the range is 7.7 − 6.5 = 1.2 min. The Bank of Providence has waiting times with a range of 5.8 min, and this larger value suggests greater variation.

The range is very easy to compute, but because it depends on only the highest and the lowest scores, it is often inferior to other measures of variation that use the value of every score. (See Exercise 25 for an example in which the range is misleading.)

Good Advice for Journalists

Columnist Max Frankel wrote in *The New York Times* that "Most schools of journalism give statistics short shrift and some let students graduate without any numbers training at all. How can such reporters write sensibly about trade and welfare and crime, or air fares, health care and nutrition? The media's sloppy use of numbers about the incidence of accidents or disease frightens people and leaves them vulnerable to journalistic hype, political demagoguery, and commercial fraud." He cites several cases, including an example of a full-page article about New York City's deficit with a promise by the mayor of New York City to close a budget gap of $2.7 billion; the entire article never once mentioned the *total* size of the budget, so the $2.7 billion figure had no context.

Standard Deviation and Variance

The standard deviation is the measure of variation that is generally the most important and useful. Unlike the range, the standard deviation takes every value into account, but this advantage makes it more difficult to compute. We define the standard deviation in the following box, but to understand this concept fully you will need to read the remainder of this section very carefully.

DEFINITION

The **standard deviation** of a set of sample scores is a measure of variation of scores about the mean. It is calculated by using Formula 2-4.

The College Board gives this formula for standard deviation in its AP statistics materials:

$$s_x = \sqrt{\frac{1}{n-1} \sum (x_i - \bar{x})^2}$$

Formula 2-4 $s = \sqrt{\dfrac{\Sigma(x - \bar{x})^2}{n - 1}}$ sample standard deviation

Almost every scientific calculator and statistical software package is programmed to automatically calculate standard deviations. The use of calculators and computers is discussed in Section 2-6, but you might want to consult your calculator manual now to find the procedure that yields the value of the standard deviation.

Why define a measure of variation in the way described by Formula 2-4? In measuring variation in a set of sample data, it's reasonable to begin with the individual amounts by which scores deviate from the mean. For a particular score x, the amount of **deviation** is $x - \bar{x}$, which is the difference between the score and the mean. However, the sum of all such deviations is always zero, which really doesn't do anything for us. To get a statistic that measures variation (instead of always being zero), we could take absolute values, as in $\Sigma|x - \bar{x}|$. If we find the mean of that sum, we get the **mean deviation** (or **absolute deviation**) described by the following expression:

$$\text{Mean deviation} = \frac{\Sigma |x - \bar{x}|}{n}$$

Instead of using absolute values, we can obtain a better measure of variation by making all deviations $(x - \bar{x})$ nonnegative by squaring them. Finally, we take the square root to compensate for that squaring. As a result, the standard deviation has the same units of measurement as the original scores. For example, if customer waiting times are in minutes, the standard deviation of those times will also be in minutes. On the basis of the format of Formula

2-4, we can describe the procedure for calculating the standard deviation as follows.

Procedure for Finding the Standard Deviation with Formula 2-4

Step 1:   Find the mean of the scores ($\bar{x}$).

Step 2:   Subtract the mean from each individual score ($x - \bar{x}$).

Step 3:   Square each of the differences obtained from Step 2. That is, multiply each value by itself. [This produces numbers of the form $(x - \bar{x})^2$.]

Step 4:   Add all of the squares obtained from Step 3 to get $\Sigma(x - \bar{x})^2$.

Step 5:   Divide the total from Step 4 by the number ($n - 1$); that is, 1 less than the total number of scores present.

Step 6:   Find the square root of the result of Step 5.

● **EXAMPLE**   Find the standard deviation of the Jefferson Valley Bank customer waiting times. Those times (in minutes) are reproduced below:

6.5   6.6   6.7   6.8   7.1   7.3   7.4   7.7   7.7   7.7

SOLUTION   Many students will find it easy to use the standard deviation function built into their calculators, but understanding is enhanced by following the detailed steps of the calculation. (See Table 2-7, where the following steps are executed.)

Step 1:   Obtain the mean of 7.15 by adding the scores and then dividing by the number of scores:

$$\bar{x} = \frac{\Sigma x}{n} = \frac{71.5}{10} = 7.15 \text{ min}$$

Step 2:   Subtract the mean of 7.15 from each score to get these values of $(x - \bar{x})$: $-0.65, -0.55, \ldots, 0.55$.

Step 3:   Square each value obtained in Step 2 to get these values of $(x - \bar{x})^2$: 0.4225, 0.3025, . . . , 0.3025.

More Stocks, Less Risk

In their book *Investments*, authors Zvi Bodie, Alex Kane, and Alan Marcus state that "the average standard deviation for returns of portfolios composed of only one stock was 0.554. The average portfolio risk fell rapidly as the number of stocks included in the portfolio increased." They note that with 32 stocks, the standard deviation is 0.325, indicating much less variation and risk. They make the point that with only a few stocks, a portfolio has a high degree of "firm-specific" risk, meaning that the risk is attributable to the few stocks involved. With more than 30 stocks, there is very little firm-specific risk; instead, almost all of the risk is "market risk," attributable to the stock market as a whole. They note that these principles are "just an application of the well-known law of averages."

TABLE 2-7	Calculating Standard Deviation for Jefferson Valley Bank Customers	
x	$x - \bar{x}$	$(x - \bar{x})^2$
6.5	-0.65	0.4225
6.6	-0.55	0.3025
6.7	-0.45	0.2025
6.8	-0.35	0.1225
7.1	-0.05	0.0025
7.3	0.15	0.0225
7.4	0.25	0.0625
7.7	0.55	0.3025
7.7	0.55	0.3025
7.7	0.55	0.3025
Totals: 71.5		2.0450

$$\bar{x} = \frac{71.5}{10} = 7.15 \text{ min} \qquad s = \sqrt{\frac{2.0450}{10 - 1}} = \sqrt{0.2272} = 0.48 \text{ min}$$

Step 4:   Sum all of the preceding scores to get the value of

$$\Sigma(x - \bar{x})^2 = 2.0450.$$

Step 5:   There are $n = 10$ scores, so divide by 1 less than 10:

$$2.0450 \div 9 = 0.2272$$

Step 6:   Find the square root of 0.2272. The standard deviation is

$$\sqrt{0.2272} = 0.48 \text{ min}$$

Ideally, we would now interpret the meaning of the resulting standard deviation of 0.48 min, but such interpretations will be discussed a little later in this section. For now, practice calculating a standard deviation by using the customer waiting times just given for the Bank of Providence. Using those times, verify that the standard deviation is 1.82 min. Although the interpretations of those standard deviations will be discussed later, we can now compare them and note that the standard deviation of the times for the Jefferson Valley Bank (0.48 min) is much lower than the standard deviation for the Bank of Providence (1.82 min). This

supports our subjective conclusion that the waiting times at the Jefferson Valley Bank have much less variation than those at the Bank of Providence.

In our definition of standard deviation, we referred to the standard deviation of *sample* data. A slightly different formula is used to calculate the standard deviation σ (lowercase Greek sigma) of a population: Instead of dividing by $n - 1$, divide by the population size N, as in the following expression.

$$\sigma = \sqrt{\frac{\Sigma(x - \mu)^2}{N}} \quad \text{population standard deviation}$$

For example, if the 10 scores in Table 2-7 constitute a *population*, the standard deviation is as follows:

$$\sigma = \sqrt{\frac{\Sigma(x - \mu)^2}{N}} = \sqrt{\frac{2.0450}{10}} = 0.45 \text{ min}$$

Because we generally deal with sample data, we will usually use Formula 2-4, in which we divide by $n - 1$. Many calculators do standard deviations, with division by $n - 1$ corresponding to a key labeled σ_{n-1} or s, while the key labeled σ_n or σ corresponds to division by N. For some creative but strange reason, calculators use a variety of different notations; the following notations, however, are very standard in statistics. These notations include reference to the variance of a set of scores, and we will now proceed to describe that measure of variation.

Notation

s denotes the standard deviation of a set of *sample* data
σ denotes the standard deviation of a set of *population* data
s^2 denotes the variance of a set of *sample* data
σ^2 denotes the variance of a set of *population* data
Note: Articles in professional journals and reports often use SD for standard deviation and Var for variance.

If we omit Step 6 (taking the square root) in the procedure for calculating the standard deviation, we get the **variance**, defined in Formula 2-5.

Formula 2-5
$$s^2 = \frac{\Sigma(x - \bar{x})^2}{n - 1} \quad \text{sample variance}$$

Comment that some later procedures use variance (example: the *F*-test in Section 8-4 or all of Chapter 11).

Similarly, we can express the population variance as

$$\sigma^2 = \frac{\Sigma(x - \mu)^2}{N} \quad \text{population variance}$$

By comparing Formulas 2-4 and 2-5, we see that the variance is the square of the standard deviation. Although the variance will be used later in the book, we should concentrate first on the concept of standard deviation as we try to get some sense of this statistic. A major difficulty with the variance is that it is not in the same units as the original data. For example, a data set might have a standard deviation of $3.00 and a variance of 9.00 square dollars. Because a square dollar is an abstract concept that we can't relate to directly, we find variance difficult to understand.

Round-Off Rule

As in Section 2-4, we use this rule for rounding final results:

Carry one more decimal place than was present in the original data.

We should round only the final answer and not intermediate values. (If it's absolutely necessary to round intermediate results, we should carry at least twice as many decimal places as will be used in the final answer.)

Shortcut Formula and Grouped Data

We will now present two additional formulas for standard deviation. These formulas do not involve a different concept; they are only different versions of Formula 2-4. First, Formula 2-4 can be expressed in the following equivalent form.

Formula 2-6 $$s = \sqrt{\frac{n(\Sigma x^2) - (\Sigma x)^2}{n(n - 1)}}$$ Shortcut formula for standard deviation

For calculating standard deviation, some instructors prefer Formula 2-4, others prefer 2-6, others have no preference. Be sure to inform your students of your position on this issue.

Formulas 2-4 and 2-6 are equivalent in the sense that they will always produce the same results. The reader will be mercifully spared from the algebra showing that they are equal. Formula 2-6 is called the *shortcut* formula because it tends to be more convenient to use with messy numbers or with large sets of data. Formula 2-6 is often used in calculators and computer programs because it requires only three memory registers (for n, Σx, and Σx^2), instead of a separate memory register for every individual score. Also, Formula 2-6 eliminates intermediate rounding errors created when the exact value of the mean is not used. Nevertheless, many instructors prefer to use only Formula 2-4 for calculating standard deviations. They argue that Formula 2-4 reinforces the concept that *the standard deviation is a type of aver-*

age deviation, while Formula 2-6 obscures that idea. Other instructors have no objections to Formula 2-6. We have included the shortcut formula so that it is available for those who choose to use it. We have already presented an example illustrating the calculation of a standard deviation using Formula 2-4, and the following example illustrates the use of Formula 2-6.

EXAMPLE   Find the standard deviation of the following Jefferson Valley Bank customer waiting times (in minutes), using Formula 2-6:

$$6.5 \quad 6.6 \quad 6.7 \quad 6.8 \quad 7.1 \quad 7.3 \quad 7.4 \quad 7.7 \quad 7.7 \quad 7.7$$

SOLUTION   Formula 2-6 requires that we find the values of n, Σx, and Σx^2. Because there are 10 scores, we have $n = 10$. The sum of the 10 scores is 71.5, so $\Sigma x = 71.5$. The third required component is calculated as follows:

$$\begin{aligned}
\Sigma x^2 &= 6.5^2 + 6.6^2 + 6.7^2 + \cdots + 7.7^2 \\
&= 42.25 + 43.56 + 44.89 + \cdots + 59.29 \\
&= 513.27
\end{aligned}$$

Formula 2-6 can now be used to find the value of the standard deviation.

$$s = \sqrt{\frac{n(\Sigma x^2) - (\Sigma x)^2}{n(n-1)}} = \sqrt{\frac{10(513.27) - (71.5)^2}{10(10-1)}}$$

$$= \sqrt{\frac{20.45}{90}} = 0.4766783 = 0.48 \text{ min (rounded)}$$

We can develop a formula for standard deviation when the data are summarized in a frequency table. The result is as follows:

$$s = \sqrt{\frac{\Sigma f \cdot (x - \bar{x})^2}{n - 1}}$$

We will express this formula in an equivalent expression that usually simplifies the actual calculations.

Formula 2-7 $$s = \sqrt{\frac{n[\Sigma (f \cdot x^2)] - [\Sigma (f \cdot x)]^2}{n(n-1)}}$$ standard deviation for frequency table

where x = class mark
             f = class frequency
             n = sample size (or Σf = sum of the frequencies)

EXAMPLE   Estimate the standard deviation of the 175 axial loads of aluminum cans by using Formula 2-7 with Frequency Table 2-2.

Where Are the 0.400 Hitters?

The last baseball player to hit above 0.400 was Ted Williams, who hit 0.406 in 1941. There were averages above 0.400 in 1876, 1879, 1887, 1894, 1895, 1896, 1897, 1899, 1901, 1911, 1920, 1922, 1924, 1925, and 1930, but none since 1941. Are there no longer great hitters? Harvard's Stephen Jay Gould notes that the mean batting average has been steady at 0.260 for about 100 years, but the standard deviation has been decreasing from 0.049 in the 1870s to 0.031, where it is now. He argues that today's stars are as good as those from the past, but consistently better pitchers now keep averages below 0.400. Dr. Gould discusses this in Program 4 of the series *Against All Odds: Inside Statistics.*

SOLUTION   Application of Formula 2-7 requires that we find the values of n, $\Sigma(f \cdot x)$, and $\Sigma(f \cdot x^2)$. After finding those values from Table 2-8, we apply Formula 2-7 as follows:

$$s = \sqrt{\frac{n[\Sigma(f \cdot x^2)] - [\Sigma(f \cdot x)]^2}{n(n-1)}} = \sqrt{\frac{175(12,579,173.75) - (46,757.5)^2}{175(175-1)}}$$

$$= \sqrt{\frac{15,091,600}{30,450}} = \sqrt{495.6190476} = 22.3 \text{ lb}$$

The 175 axial loads have a standard deviation estimated to be 22.3 lb. (The exact value calculated from the original set of data is 22.1 lb, so the result obtained here is quite good.

TABLE 2-8	Calculating Standard Deviation from a Frequency Table			
Axial Load	Frequency f	Class Mark x	$f \cdot x$	$f \cdot x^2$
200–209	9	204.5	1,840.5	376,382.25
210–219	3	214.5	643.5	138,030.75
220–229	5	224.5	1,122.5	252,001.25
230–239	4	234.5	938.0	219,961.00
240–249	4	244.5	978.0	239,121.00
250–259	14	254.5	3,563.0	906,783.50
260–269	32	264.5	8,464.0	2,238,728.00
270–279	52	274.5	14,274.0	3,918,213.00
280–289	38	284.5	10,811.0	3,075,729.50
290–299	14	294.5	4,123.0	1,214,223.50
Total	$\Sigma f = 175$		$\Sigma(f \cdot x) = 46,757.5$	$\Sigma(f \cdot x^2) = 12,579,173.75$

We can also use a TI-83 calculator to compute the standard deviation of scores summarized in a frequency table. First enter the class marks in list L1, then enter the frequencies in list L2, then use STAT, CALC, and 1-VarStats and enter L1, L2 to obtain results that include the mean and standard deviation.

Understanding Standard Deviation

We will now attempt to make some intuitive sense of the standard deviation. First, we should clearly understand that the standard deviation measures the variation among scores. Scores close together will yield a small standard deviation, whereas scores spread farther apart will yield a larger standard deviation. Stop reading and take a moment to study Figure 2-9. Observe that as the data spread farther apart, the values of the standard deviation increase.

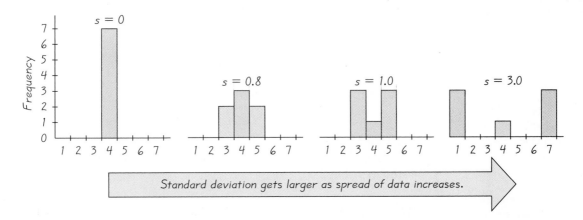

Try to get students to develop an ability for interpreting values of standard deviations, beginning with the range rule of thumb. Randomly select a student and ask him or her to estimate the mean height of all students at your college; they usually do this well.

Because variation is such an important concept and because the standard deviation is such an important tool in measuring variation, we will consider three different ways of developing a sense for values of standard deviations. The first is the **range rule of thumb,** a very rough estimate. (We could improve the accuracy of this rule by taking into account such factors as the size of the sample and the nature of the distribution, but we prefer to sacrifice accuracy for the sake of simplicity. We want a simple rule that will help us interpret values of standard deviations; later methods will produce more accurate results.)

Figure 2-9
Same Means, Different Standard Deviations

Now ask him or her to estimate the standard deviation of heights; comment that it's perfectly natural to have no idea. Then ask him or her to estimate the heights of the shortest and tallest students, and proceed to show how the range rule of thumb can be used to develop an estimate. Women have heights with a standard deviation around 2.5 in., and men have heights with a standard deviation around 2.8 in.

Range Rule of Thumb

For typical data sets, the range of a set of data is approximately 4 standard deviations ($4s$) wide, so the standard deviation can be approximated as follows:

$$\text{standard deviation} \approx \frac{\text{range}}{4} \quad \text{range rule of thumb}$$

This expression provides a rough estimate of the standard deviation when we know the minimum and maximum scores. If we know a value of the standard deviation, we can use it to better understand the data by finding rough estimates of the minimum and maximum scores as follows.

$$\textbf{minimum} \approx (\textbf{mean}) - 2 \times (\textbf{standard deviation})$$
$$\textbf{maximum} \approx (\textbf{mean}) + 2 \times (\textbf{standard deviation})$$

When calculating a standard deviation using Formula 2-4 or 2–6, you can use the range rule of thumb as a check on your result, but realize that although the approximation will get you in the general vicinity of the answer, it can be off by a fairly large amount. For the Jefferson Valley Bank customer waiting times (6.5, 6.6, 6.7, 6.8, 7.1, 7.3, 7.4, 7.7, 7.7, 7.7) we used Formula 2-6 to compute the standard deviation as $s = 0.48$ min. These values have a range of $7.7 - 6.5 = 1.2$, so we could use the range rule of thumb to get a rough estimate of s as follows:

$$s \approx \frac{\text{range}}{4} = \frac{1.2}{4} = 0.3 \text{ min}$$

We found that the standard deviation s is actually 0.48, so the range rule of thumb gives us an estimate of 0.3 that is a bit too low here. However, our estimate does confirm that we are generally in the ballpark, and we would know that a value for s such as 7 is probably not correct.

EXAMPLE   Use the range rule of thumb to find a rough estimate of the standard deviation of the sample of 175 axial loads of aluminum cans listed in Table 2-1.

SOLUTION   In using the range rule of thumb to estimate the standard deviation of sample data, we find the range and divide by 4. By scanning the list of scores, we find that the lowest is 200 and the highest is 297, so the range is $297 - 200 = 97$. The standard deviation s is estimated as follows:

$$s \approx \frac{\text{range}}{4} = \frac{97}{4} = 24.3 \text{ lb}$$

This result is in the ballpark of the correct value of 22.1 lb that is obtained by calculating the exact value of the standard deviation with Formula 2-4 or 2-6.

Because the Table 2-1 axial loads of aluminum cans have a mean of 267.1, a standard deviation of 22.1, and a distribution as shown in Figure 2-1, we conclude that these cans will easily support the pressures of 158 lb–165 lb that are applied when the top lids are pressed into place. Recall from the statement of the Chapter Problem that these cans have a thickness of 0.0109 in., which is thinner than the cans currently used. On the basis of our knowledge of the important characteristics of the data set in Table 2-1, we conclude that money can be saved by using these thinner cans.

The preceding example illustrated how we can use known information about the range to estimate the standard deviation. The following example is

particularly important as an illustration of one way to *interpret* the value of a standard deviation.

EXAMPLE   The Gates Electronics Company makes cordless recharge-able shavers that have lives with a mean of 8.0 years and a standard devi-ation of 3.0 years. Using the range rule of thumb, estimate the longest and shortest lives of these shavers.

SOLUTION   The shortest and longest lives are estimated from the range rule of thumb as follows.

$$\text{minimum} \approx (\text{mean}) - 2 \times (\text{standard deviation})$$
$$= 8.0 - 2(3.0) = 2.0 \text{ years}$$
$$\text{maximum} \approx (\text{mean}) + 2 \times (\text{standard deviation})$$
$$= 8.0 + 2(3.0) = 14.0 \text{ years}$$

We therefore expect that most of the shavers will last between 2.0 years and 14.0 years. Remember, these results are all rough estimates, but by knowing the mean and standard deviation, we are able to approximate the lowest and highest values, thereby developing a much better under-standing of how the data vary.

Empirical (or 68–95–99) Rule for Data

Another rule helpful in interpreting a value for a standard deviation is the **empirical rule,** *which applies only to a data set having a distribution that is approximately bell-shaped,* as in Figure 2-10 on page 86. This figure shows how the mean and standard deviation of the data can be related to the proportion of data falling within certain limits. For exam-ple, data with a bell-shaped distribution will have about 95% of its values within two standard deviations of the mean. The empirical rule is often stated in an abbreviated form, sometimes called the **68–95–99 rule.**

68–95–99 Rule for Data with a Bell-Shaped Distribution

- About 68% of all scores fall within 1 standard deviation of the mean.
- About 95% of all scores fall within 2 standard deviations of the mean.
- About 99.7% of all scores fall within 3 standard deviations of the mean.

Figure 2-10
The Empirical Rule

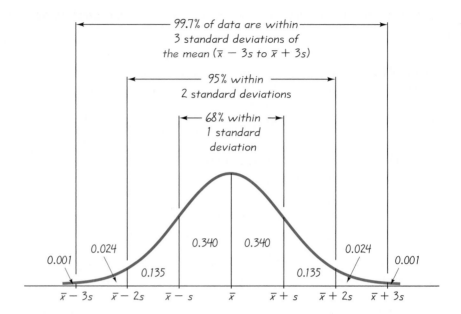

EXAMPLE   Adult IQ scores have a bell-shaped distribution with a mean of 100 and a standard deviation of 15. Use the empirical rule to find the percentage of adults with IQ scores between 55 and 145.

SOLUTION   The key to solving this problem is to recognize that 55 and 145 are each exactly 3 standard deviations away from the mean of 100. (Because the standard deviation is $s = 15$, it follows that $3s = 45$, so 3 standard deviations below the mean is $100 - 45 = 55$, and 3 standard deviations above the mean is $100 + 45 = 145$.) The empirical rule states that 99.7% of all scores are within 3 standard deviations of the mean, so it follows that 99.7% of adults should have IQ scores between 55 and 145. Because values outside of that range are so rare, someone with an IQ above 145 or below 55 is considered exceptional.

A third concept helpful in understanding or interpreting a value of a standard deviation is **Chebyshev's theorem.** The preceding empirical rule applies only to data sets with a bell-shaped distribution. Chebyshev's theorem applies to any data set, but its results are very approximate.

Chebyshev's Theorem

The proportion (or fraction) of *any* set of data lying within K standard deviations of the mean is always *at least* $1 - 1/K^2$, where K is any positive number greater than 1. For $K = 2$ and $K = 3$, we get the following two specific results.

- At least 3/4 (or 75%) of all scores will fall within the interval from 2 standard deviations below the mean to 2 standard deviations above the mean ($\bar{x} - 2s$ to $\bar{x} + 2s$).
- At least 8/9 (or 89%) of all scores will fall within 3 standard deviations of the mean ($\bar{x} - 3s$ to $\bar{x} + 3s$).

Using IQ scores with a mean of 100 and a standard deviation of 15, Chebyshev's theorem tells us that at least 75% of IQ scores will fall between 70 and 130, and at least 89% of IQ scores will fall between 55 and 145.

After studying this section, you should understand that the standard deviation is a measure of variation among scores. Given sample data, you should be able to compute the value of the standard deviation. You should be able to interpret the values of standard deviations that you compute. You should recognize that for typical data sets, it is unusual for a score to differ from the mean by more than 2 or 3 standard deviations.

2-5 Exercises A: Basic Skills and Concepts

In Exercises 1–4, find the range, variance, and standard deviation for the given data. (The same data were used in Section 2-4 where we found measures of central tendency. Here we find measures of variation.)

Recommended assignment: 1, 4, 6, 7, 10, 14, 17–20, 23 parts a, b, c (part d uses logs)

1. The given values are weights (ounces) of steaks listed on a restaurant menu as "20-ounce Porterhouse" steaks (based on data collected by a student of the author).

   17 20 21 18 20 20 20 18 19 19
   20 19 21 20 18 20 20 19 18 19

2. Digits selected in the Maryland Pick Three lottery:

   0 7 3 6 2 7 6 6 6 3 8 1 7 8 7
   1 6 8 6 9 5 2 1 5 0 3 9 9 0 7

1. range = 4.0 oz; $s^2 = 1.2$ oz^2; $s = 1.1$ oz
2. range = 9.0; $s^2 = 8.8$; $s = 3.0$

3. range = 3.570; s^2 = 1.030;
   s = 1.015
4. range = 0.170; s^2 = 0.003;
   s = 0.051
5. Jefferson Valley:
   range = 1.20; s^2 = 0.23;
   s = 0.48
   Providence: range = 5.80;
   s^2 = 3.32; s = 1.82
6. Students: range = 14.0;
   s^2 = 17.1; s = 4.1
   Faculty/staff: range = 21.0;
   s^2 = 32.5; s = 5.7
7. 4000 B.C.: range = 19.0;
   s^2 = 21.5; s = 4.6
   150 A.D.: range = 15.0;
   s^2 = 25.2; s = 5.0
8. Paper: range = 9.490;
   s^2 = 8.592; s = 2.931
   Plastic: range = 2.240;
   s^2 = 0.526; s = 0.725
9. 0.76
10. 0.31
11. 121.8

12. 0.0275

3. Nitrate deposits (in kg per hectare) as part of acid rain for Massachusetts from July through September for recent years (based on data from the U.S. Department of Agriculture):

6.40   5.21   4.66   5.24   6.96   5.53   8.23   6.80   5.78   6.00   5.41

4. Blood-alcohol concentrations of 15 drivers involved in fatal accidents and then convicted with jail sentences (based on data from the U.S. Department of Justice):

0.27   0.17   0.17   0.16   0.13   0.24   0.29   0.24
0.14   0.16   0.12   0.16   0.21   0.17   0.18

In Exercises 5–8, find the range, variance, and standard deviation for each of the two samples, then compare the two sets of results. (The same data were used in Section 2-4.)

5. Waiting times of customers at the Jefferson Valley Bank (where all customers enter a single waiting line) and the Bank of Providence (where customers wait in individual lines at three different teller windows). These data sets were discussed in this section.

Jefferson Valley: 6.5   6.6   6.7   6.8   7.1   7.3   7.4   7.7   7.7   7.7
Providence: 4.2   5.4   5.8   6.2   6.7   7.7   7.7   8.5   9.3   10.0

6. Samples of the ages (in years) of student cars and faculty/staff cars obtained at the author's college.

Students: 10   4   5   2   9   7   8   8   16   4   13   12
Faculty/staff: 7   10   4   13   23   2   7   6   6   3   9   4

7. Maximum breadth of samples of male Egyptian skulls from 4000 B.C. and 150 A.D. (based on data from *Ancient Races of the Thebaid* by Thomson and Randall-Maciver):

4000 B.C.:   131 119 138 125 129 126 131 132 126 128 128 131
150 A.D.:   136 130 126 126 139 141 137 138 133 131 134 129

8. Weights (in pounds) of paper and plastic discarded by households in one week (data collected for the Garbage Project at the University of Arizona):

Paper: 9.55   6.38   2.80   6.98   6.33   6.16   10.00   12.29
Plastic: 2.19   2.10   1.41   0.63   0.92   1.40   1.74   2.87

In Exercises 9–12, refer to the data set in Appendix B and find the standard deviation.

9. Data Set 2 in Appendix B: the body temperatures for 8:00 A.M. on day 1
10. Data Set 4 in Appendix B: the nicotine contents of all cigarettes listed
11. Data Set 3 in Appendix B: the weights of the bears
12. Data Set 11 in Appendix B: the weights of the red M&M plain candies

In Exercises 13–16, find the standard deviation of the data summarized in the given frequency table.

13. Visitors to Yellowstone National Park consider an eruption of the Old Faithful geyser to be a major attraction that should not be missed. The given frequency table summarizes a sample of times (in minutes) between eruptions.

Time	Frequency
40–49	8
50–59	44
60–69	23
70–79	6
80–89	107
90–99	11
100–109	1

13. 14.7

14. Samples of student cars and faculty/staff cars were obtained at the author's college, and their ages (in years) are summarized in the accompanying frequency table. Find the standard deviation of student cars and of faculty/staff cars. Based on the results, are there any noticeable differences between the two samples? If so, what are they?

Age	Students	Faculty/Staff
0–2	23	30
3–5	33	47
6–8	63	36
9–11	68	30
12–14	19	8
15–17	10	0
18–20	1	0
21–23	0	1

14. Students: 3.9; Faculty/staff: 3.7

15. The given frequency table describes the speeds of drivers ticketed by the Town of Poughkeepsie police. These drivers were traveling through a 30 mi/h speed zone on Creek Road, which passes the author's college.

Speed	Frequency
42–43	14
44–45	11
46–47	8
48–49	6
50–51	4
52–53	3
54–55	1
56–57	2
58–59	0
60–61	1

15. 4.3
16. 18.3

16. Insurance companies continually research ages at death and causes of death. The data are based on a *Time* magazine study of people who died from gunfire in America during one week.

Age at Death	Frequency
16–25	22
26–35	10
36–45	6
46–55	2
56–65	4
66–75	5
76–85	1

17. The population of batteries with $\sigma = 1$ month are much more consistent, so they have a smaller chance of failing much earlier than expected.

18. The bulbs with the lower standard deviation are more consistent and it is easier to plan for their systematic replacement.

19. Use $s \approx$ (tallest − shortest) $\div 4$

20. Use $s \approx$ (highest − lowest) $\div 4$

21. a. 68%

    b. 95%

    c. 50, 110

22. At least 75% of heights should fall between 58.6 in. and 68.6 in.; at least 89% of all heights should fall between 56.1 in. and 71.1 in.

23. a. range = 128,000; $s = 53,122$

    b. The results will be the same.

    c. The range and standard deviation will be multiplied by k.

    d. The standard deviation of the log x values is 0.1409, but log $s = 4.7253$.

    e. $\bar{x} = 36.78$; $s = 0.34$

24. a. 28.9

    b. $s = \sqrt{n(n+1)/12}$

    c. $\mu = 0.499999995$; $\sigma = 0.288675135$

17. If you must purchase a replacement battery for your car, would you prefer one that comes from a population with $\sigma = 1$ month or one that comes from a population with $\sigma = 1$ year? (Assume that both populations have the same mean and price.) Explain your choice.

18. As a manager, you must purchase lightbulbs to be used in a hospital. Should you choose Ultralight bulbs that have lives with $\mu = 3000$ h and $\sigma = 200$ h, or should you choose Electrolyte bulbs with $\mu = 3000$ h and $\sigma = 250$ h? Explain.

19. Use the range rule of thumb to estimate the standard deviation of the heights of students in your statistics class.

20. Use the range rule of thumb to estimate the standard deviation of grades on the last statistics final examination.

2-5 Exercises B: Beyond the Basics

21. A typing-competency test yields scores with $\bar{x} = 80.0$ and $s = 10.0$, and a histogram shows that the distribution of the scores is roughly bell-shaped. Use the empirical rule to answer the following:
    a. What percentage of the scores should fall between 70 and 90?
    b. What percentage of the scores should fall within 20 points of the mean?
    c. About 99.7% of the scores should fall between what two values? (The mean of 80.0 should be midway between those two values.)

22. Heights of adult women have a mean of 63.6 in. and a standard deviation of 2.5 in. What does Chebyshev's theorem say about the percentage of women with heights between 58.6 in. and 68.6 in.? Between 56.1 in. and 71.1 in.?

23. a. Find the range and standard deviation s for the following sample of annual incomes (in dollars) of self-employed doctors (based on data from the American Medical Association):

    108,000   236,000   179,000   206,000   236,000

    b. If a constant value k is added to each income, how are the results from part (a) affected?
    c. If each income given in part (a) is multiplied by a constant k, how are the results from part (a) affected?
    d. Data are sometimes transformed by replacing each score x with log x. For the given x values, determine whether the standard deviation of the log x values is equal to log s.
    e. For the body temperature data listed in Data Set 2 of Appendix B (12 A.M. on day 2), $\bar{x} = 98.20°$F and $s = 0.62°$F. Find the values of $\bar{x}$ and s for the data after each temperature has been converted to the Celsius scale. [*Hint:* C = 5(F − 32)/9.]

24. If we consider the values 1, 2, 3, . . . , n to be a population, the standard deviation can be calculated by the formula

$$\sigma = \sqrt{\frac{n^2 - 1}{12}}$$

This formula is equivalent to Formula 2-4 modified for division by n instead of $n - 1$, where the data set consists of the values 1, 2, 3, . . . , n.

a. Find the standard deviation of the population 1, 2, 3, . . . , 100.

b. Find an expression for calculating the sample standard deviation s for the *sample* values 1, 2, 3, . . . , n.

c. Computers and calculators commonly use a random-number generator which produces values between 0.00000000 and 0.99999999. In the long run, all values occur with the same relative frequency. Find the mean and standard deviation for the *population* of those values.

25. Two different sections of a statistics class take the same surprise quiz and the scores are recorded below. Find the range and standard deviation for each section. What do the range values lead you to conclude about the variation in the two sections? Why is the range misleading in this case? What do the standard deviation values lead you to conclude about the variation in the two sections?

Section 1:	1	20	20	20	20	20	20	20	20	20	20
Section 2:	2	3	4	5	6	14	15	16	17	18	19

26. a. The **coefficient of variation,** expressed as a percent, is used to describe the standard deviation relative to the mean. It allows us to compare variability of data sets with different measurement units (such as feet versus minutes), and it is calculated as follows:

$$\frac{s}{x} \cdot 100 \quad \text{or} \quad \frac{\sigma}{\mu} \cdot 100$$

Find the coefficient of variation for the following sample of car ages (in years):

0   1   3   3   5   6   6   6   6   8   12

b. Genichi Taguchi developed a method of improving quality and reducing manufacturing costs through a combination of engineering and statistics. A key tool in the Taguchi method is the **signal-to-noise ratio.** The simplest way to calculate this ratio is to divide the mean by the standard deviation. Find the signal-to-noise ratio for the sample data given in part (a).

27. In Section 2-4 we introduced the general concept of skewness. Skewness can be measured by **Pearson's index of skewness:**

$$I = \frac{3(\overline{x} - \text{median})}{s}$$

(continued)

25. Section 1: range = 19.0;
    $s = 5.7$
    Section 2: range = 17.0;
    $s = 6.7$
    The ranges suggest that Section 2 has less variation, but the standard deviations suggest that Section 1 has less variation.

26. a. 65%   b. 1.5

27. $I = -0.80$; there is not significant skewness.

If $I \geq 1.00$ or $I \leq -1.00$, the data can be considered to be *significantly skewed*. Find Pearson's index of skewness for the axial loads of the aluminum cans listed in Table 2-1, and then determine whether there is significant skewness.

28. a. A sample consists of 6 scores that fall between 1 and 9 inclusive. What is the largest possible standard deviation?
   b. For any data set of n scores with standard deviation s, every score must be within $s\sqrt{n-1}$ of the mean. A statistics teacher reports that the test scores in her class of 17 students had a mean of 75.0 and a standard deviation of 5.0. Kelly, the class's self-proclaimed best student, claims that she received a grade of 97. Could Kelly be telling the truth?

28. a. 4.3817805
   b. No, every score must be within 20 points of the mean, so the highest possible score is 95.

2-6 Measures of Position

Students can generally read and understand *z* scores on their own; discuss percentiles, quartiles, and deciles.

In this section we introduce z scores, which enable us to standardize values so that they can be compared more easily. We also introduce quartiles, percentiles, and deciles, which help us better understand data by showing their positions relative to the whole data set. The quartiles introduced in this section will also be used for boxplots, discussed in the following section.

z Scores

Most of us are reasonably familiar with IQ scores, and we recognize that an IQ of 102 is fairly common, whereas an IQ of 170 is rare. The IQ of 102 is fairly common because it is very close to the mean of 100, but the IQ of 170 is rare because it is so far above 100. This might suggest that we can differentiate between typical scores and rare scores on the basis of the difference between the score and the mean $(x - \bar{x})$. However, the size of such differences is relative to the scale being used. With IQ scores, a 2-point difference is insignificant, but for college grade-point averages, the 2-point difference between 2.00 and 4.00 is very significant, especially to parents. It would be much better if we could use a standard that doesn't require an understanding of the scale being used. With the standard score, we divide the difference $x - \bar{x}$ (or $x - \mu$) by the standard deviation to get such a result.

DEFINITION

The **standard score,** or **z score,** is the number of standard deviations that a given value x is above or below the mean. It is found by using

Sample	Population

$$z = \frac{x - \bar{x}}{s} \quad \text{or} \quad z = \frac{x - \mu}{\sigma}$$

(Round z to two decimal places.)

EXAMPLE   Heights of all adult males have a mean of $\mu = 69.0$ in., a standard deviation of $\sigma = 2.8$ in., and a distribution that is bell-shaped. Basketball player Michael Jordan earned a giant reputation for his skills, but at a height of 78 in., is he exceptionally tall when compared to the general population of adult males? Find the z score for his 78-in. height.

SOLUTION   Because we are dealing with population parameters, the z score is calculated as follows:

$$z = \frac{x - \mu}{\sigma} = \frac{78 - 69.0}{2.8} = 3.21$$

We can interpret this result by stating that Michael Jordan's height of 78 in. is 3.21 standard deviations above the mean.

The role of z scores in statistics is extremely important because they can be used to differentiate between ordinary values and unusual values. Values with standard scores between -2.00 and 2.00 are ordinary, and values with z scores less than -2.00 or greater than 2.00 are unusual. (See Figure 2-11.) Michael Jordan's height converts to a z score of 3.21, so we consider it unusual because it is greater than 2.00. In comparison to the general population, Michael Jordan is exceptionally tall.

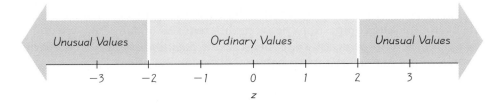

Figure 2-11
Interpreting z Scores
Unusual values are those with z scores less than $z = -2.00$ or greater than $z = 2.00$.

Our criterion for unusual z scores follows from the empirical rule and Chebyshev's theorem. Recall from the empirical rule that for data with a bell-shaped distribution, about 95% of the values are within 2 standard deviations of the mean. (See Figure 2-10 in the preceding section.) Also, Chebyshev's theorem states that for any data set, at least 75% of the values are within 2 standard deviations of the mean.

We noted earlier that z scores are also useful for comparing scores from different populations with different means and different standard deviations. The following example illustrates this use of z scores.

Buying Cars

For buying a new or used car, an excellent reference is the reliability data compiled and reported by *Consumer Reports* magazine. Frequency-of-repair data are based on 10 million pieces of data collected from thousands of readers. Statisticians analyze the data for patterns that lead to lists of both reliable cars and cars that should be avoided. Consumers Union President Rhoda Karpatkin writes, "Because numbers describe so much of our work, it should be no surprise that statisticians are key to that process."

EXAMPLE   A statistics professor gives two different tests to two sections of her course. The statistics are given below. Which score is relatively better: an 82 on the Section 1 test, or a 46 on the Section 2 test?

$$\text{Section 1:} \quad \bar{x} = 75 \text{ and } s = 14$$
$$\text{Section 2:} \quad \bar{x} = 40 \text{ and } s = 8$$

SOLUTION   We can't directly compare the scores of 82 and 46 because they come from different scales. Instead, we convert them both to z scores. For the score of 82 on the Section 1 test, we get a z score of 0.50, because

$$z = \frac{x - \bar{x}}{s} = \frac{82 - 75}{14} = 0.50$$

For the score of 46 on Test B, we get a z score of 0.75, because

$$z = \frac{x - \bar{x}}{s} = \frac{46 - 40}{8} = 0.75$$

That is, a score of 82 on the Section 1 test is 0.50 standard deviation above the mean, whereas a score of 46 on the Section 2 test is 0.75 standard deviation above the mean. This implies that the 46 on the Section 2 test is the better relative score. Although 46 is less than 82, it has a better relative position when considered in the context of the other test results. Later, we will make extensive use of these standard, or z, scores.

The preceding example illustrated that z scores provide useful measurements for making comparisons between different sets of data. Likewise, quartiles, deciles, and percentiles are measures of position useful for comparing scores within one set of data or between different sets of data.

Quartiles, Deciles, and Percentiles

Just as the median divides the data into two equal parts, the three **quartiles,** denoted by Q_1, Q_2, and Q_3, divide the *ranked* scores into four equal parts. (Scores are ranked when they are arranged in order.) Roughly speaking, Q_1 separates the bottom 25% of the ranked scores from the top 75%, Q_2 is the median, and Q_3 separates the top 25% from the bottom 75%. To be more precise, at least 25% of the data will be less than or equal to Q_1, and at least 75% will be greater than or equal to Q_1. At least 75% of the data will be less than or equal to Q_3, while at least 25% will be equal to or greater than Q_3.

Similarly, there are nine **deciles**, denoted by $D_1, D_2, D_3, \ldots, D_9$, which partition the data into 10 groups with about 10% of the data in each group. There are also 99 **percentiles**, which partition the data into 100 groups with about 1% of the scores in each group. (Quartiles, deciles, and percentiles are examples of *fractiles,* which partition data into parts that are approximately equal.) A student taking a competitive college entrance examination might learn that he or she scored in the 92nd percentile. This does not mean that the student received a grade of 92% on the test; it indicates instead that whatever score he or she did achieve was higher than 92% of the scores of those who took a similar test (and also lower than 8% of his or her colleagues). The 92nd percentile is therefore an excellent score relative to the others who took the test.

The process of finding the percentile that corresponds to a particular score x is fairly simple, as indicated in the following expression.

$$\text{percentile of score } x = \frac{\text{number of scores less than } x}{\text{total number of scores}} \cdot 100$$

EXAMPLE   Table 2-9 lists the 175 axial loads of aluminum cans, ranked from lowest to highest. Find the percentile corresponding to 241.

SOLUTION
From Table 2-9 we see that there are 21 values less than 241, so

$$\text{percentile of } 241 = \frac{21}{175} \cdot 100 = 12$$

The axial load of 241 is the 12th percentile.

TABLE 2-9   *Ranked* Axial Loads of Aluminum Cans

200	201	204	204	206	206	208	208	209	215	217	218	220	223	223
225	228	230	230	234	236	241	242	242	248	250	251	251	252	252
254	256	256	256	257	257	258	259	259	260	261	262	262	262	262
262	263	263	263	263	263	264	265	265	265	266	267	267	268	268
268	268	268	268	268	268	268	269	269	269	269	270	270	270	270
270	270	270	270	271	271	272	272	272	272	272	273	273	273	273
273	273	274	274	274	274	275	275	275	275	276	276	276	276	276
277	277	277	277	277	277	277	277	278	278	278	278	278	278	278
279	279	279	280	280	280	281	281	281	281	282	282	282	282	282
282	283	283	283	283	283	283	284	284	284	284	285	285	285	286
286	286	286	287	287	288	289	289	289	289	289	290	290	290	291
291	292	292	292	293	293	294	295	295	297					

Cost of Laughing Index

There really is a Cost of Laughing Index (CLI), which tracks costs of such items as rubber chickens, Groucho Marx glasses, admission to comedy clubs, and 13 other leading humor indicators. This is the same basic approach used in developing the Consumer Price Index (CPI), which is based on a weighted average of goods and services purchased by typical consumers. While standard scores and percentiles allow us to compare different values, they ignore any element of time. Index numbers, such as the CLI and CPI, allow us to compare the value of some variable to its value at some base time period. The value of an index number is the current value, divided by the base value, multiplied by 100.

The preceding example illustrated the procedure for finding the percentile corresponding to a given score. There are several different methods for the reverse procedure of finding the score corresponding to a particular percentile, but the one we will use is summarized in Figure 2-12, which uses the following notation.

Notation	
n	number of scores in the data set
k	percentile being used
L	locator that gives the *position* of a score
P_k	kth percentile

EXAMPLE   Refer to the 175 axial loads of aluminum cans in Table 2-9, and find the score corresponding to the 25th percentile. That is, find the value of P_{25}.

SOLUTION   We refer to Figure 2-12 and observe that the data are already ranked from lowest to highest. We now compute the locator L as follows:

$$L = \left(\frac{k}{100}\right)n = \left(\frac{25}{100}\right) \cdot 175 = 43.75$$

We answer no when asked in Figure 2-12 if 43.75 is a whole number, so we are directed to round L *up* (not off) to 44. (In this particular procedure we round L up to the next higher integer, but in most other situations in this book we generally follow the usual process for rounding.) The 25th percentile, denoted by P_{25}, is the 44th score, counting from the lowest. Beginning with the lowest score of 200, we count through the list to find the 44th score of 262, so $P_{25} = 262$.

Suppose we want to find the percentile corresponding to a score of 262. Verify that there are 41 scores below 262; be sure to count each individual score, including duplicates. Finding the percentile for 262 therefore yields $(41/175) \cdot 100 = 23$ (rounded). There is a small discrepancy: In the preceding

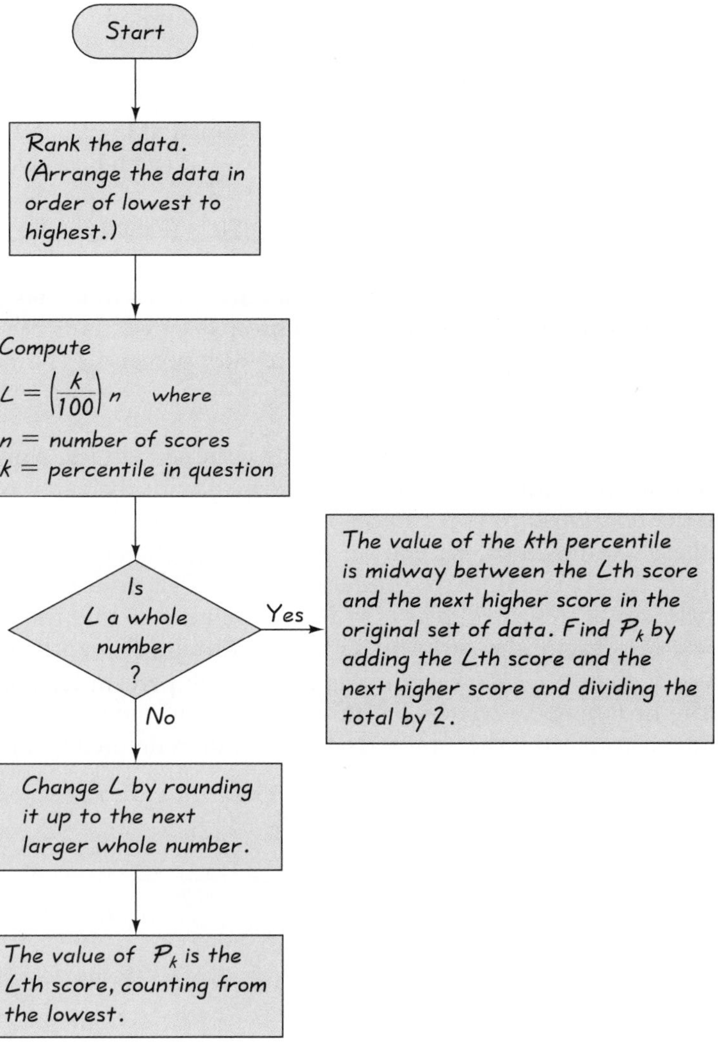

Figure 2-12 Finding the Value of the *k*th Percentile

Start

Rank the data. (Arrange the data in order of lowest to highest.)

Compute

$$L = \left(\frac{k}{100}\right) n \quad \text{where}$$

n = number of scores
k = percentile in question

Is L a whole number ?

Yes → The value of the *k*th percentile is midway between the Lth score and the next higher score in the original set of data. Find P_k by adding the Lth score and the next higher score and dividing the total by 2.

No

Change L by rounding it up to the next larger whole number.

The value of P_k is the Lth score, counting from the lowest.

example we found the 25th percentile to be 262, but when we reverse the process we find that 262 is the 23rd percentile. As the amount of data increases, such discrepancies become smaller. We could eliminate the discrepancy by using a more complicated procedure that includes interpolations instead of rounding.

Because of the sample size in the preceding example, the locator L first became 43.75, which was rounded to 44 because L was not originally a whole number. In the next example we illustrate a case in which L does begin as a whole number. This condition will cause us to branch to the right in Figure 2-12.

EXAMPLE   Refer to the axial load data for aluminum cans, as listed in Table 2-9. Find P_{40}, which denotes the 40th percentile.

SOLUTION   Following the procedure outlined in Figure 2-12 and noting that the data are already ranked from lowest to highest, we compute

$$L = \left(\frac{k}{100}\right)n = \left(\frac{40}{100}\right) \cdot 175 = 70 \quad \text{(exactly)}$$

We note that 70 is a whole number, and Figure 2-12 indicates that P_{40} is midway between the 70th and 71st scores. Because the 70th and 71st scores are both 269, we conclude that the 40th percentile is 269.

Once you have mastered these calculations with percentiles, similar calculations for quartiles and deciles can be performed with the same procedures by noting the relationships given in the margin.

Using these relationships, we can see that finding Q_1 is equivalent to finding P_{25}. In an earlier example we found that $P_{25} = 262$, so it follows that the first quartile can be described by $Q_1 = 262$. If we need to find the third quartile, Q_3, we can restate the problem as that of finding P_{75}, and we can then proceed to use Figure 2-12.

In addition to the measures of central tendency and the measures of variation already introduced, other statistics are sometimes defined using quartiles, deciles, or percentiles, as in the following.

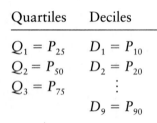

Quartiles	Deciles
$Q_1 = P_{25}$	$D_1 = P_{10}$
$Q_2 = P_{50}$	$D_2 = P_{20}$
$Q_3 = P_{75}$	$\vdots$
	$D_9 = P_{90}$

$$\text{interquartile range} = Q_3 - Q_1$$

$$\text{semi-interquartile range} = \frac{Q_3 - Q_1}{2}$$

$$\text{midquartile} = \frac{Q_1 + Q_3}{2}$$

$$\text{10–90 percentile range} = P_{90} - P_{10}$$

Using Calculators and Computers for Descriptive Statistics

When working with large collections of data, statistical software packages should be used to obtain fast, easy, and reliable results. The following STATDISK and Minitab computer displays are based on the 175 axial loads listed in Table 2-1. These displays are examples of results that are produced almost as quickly as the data can be entered.

Calculators can also be used to obtain descriptive statistics. Most scientific calculators provide at least the mean and standard deviation. With a TI-83 calculator, use STAT and Edit to enter a set of data in a column, such as L1, then use STAT and CALC to get the option of 1-Var Stats. The TI-83 results will include the mean, the sum of the scores, the sum of the squares,

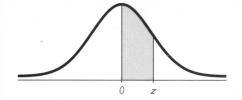

TABLE A-2	Standard Normal (z) Distribution									
z	.00	.01	.02	.03	.04	.05	.06	.07	.08	.09
0.0	.0000	.0040	.0080	.0120	.0160	.0199	.0239	.0279	.0319	.0359
0.1	.0398	.0438	.0478	.0517	.0557	.0596	.0636	.0675	.0714	.0753
0.2	.0793	.0832	.0871	.0910	.0948	.0987	.1026	.1064	.1103	.1141
0.3	.1179	.1217	.1255	.1293	.1331	.1368	.1406	.1443	.1480	.1517
0.4	.1554	.1591	.1628	.1664	.1700	.1736	.1772	.1808	.1844	.1879
0.5	.1915	.1950	.1985	.2019	.2054	.2088	.2123	.2157	.2190	.2224
0.6	.2257	.2291	.2324	.2357	.2389	.2422	.2454	.2486	.2517	.2549
0.7	.2580	.2611	.2642	.2673	.2704	.2734	.2764	.2794	.2823	.2852
0.8	.2881	.2910	.2939	.2967	.2995	.3023	.3051	.3078	.3106	.3133
0.9	.3159	.3186	.3212	.3238	.3264	.3289	.3315	.3340	.3365	.3389
1.0	.3413	.3438	.3461	.3485	.3508	.3531	.3554	.3577	.3599	.3621
1.1	.3643	.3665	.3686	.3708	.3729	.3749	.3770	.3790	.3810	.3830
1.2	.3849	.3869	.3888	.3907	.3925	.3944	.3962	.3980	.3997	.4015
1.3	.4032	.4049	.4066	.4082	.4099	.4115	.4131	.4147	.4162	.4177
1.4	.4192	.4207	.4222	.4236	.4251	.4265	.4279	.4292	.4306	.4319
1.5	.4332	.4345	.4357	.4370	.4382	.4394	.4406	.4418	.4429	.4441
1.6	.4452	.4463	.4474	.4484	.4495 *	.4505	.4515	.4525	.4535	.4545
1.7	.4554	.4564	.4573	.4582	.4591	.4599	.4608	.4616	.4625	.4633
1.8	.4641	.4649	.4656	.4664	.4671	.4678	.4686	.4693	.4699	.4706
1.9	.4713	.4719	.4726	.4732	.4738	.4744	.4750	.4756	.4761	.4767
2.0	.4772	.4778	.4783	.4788	.4793	.4798	.4803	.4808	.4812	.4817
2.1	.4821	.4826	.4830	.4834	.4838	.4842	.4846	.4850	.4854	.4857
2.2	.4861	.4864	.4868	.4871	.4875	.4878	.4881	.4884	.4887	.4890
2.3	.4893	.4896	.4898	.4901	.4904	.4906	.4909	.4911	.4913	.4916
2.4	.4918	.4920	.4922	.4925	.4927	.4929	.4931	.4932	.4934	.4936
2.5	.4938	.4940	.4941	.4943	.4945	.4946	.4948	.4949 *	.4951	.4952
2.6	.4953	.4955	.4956	.4957	.4959	.4960	.4961	.4962	.4963	.4964
2.7	.4965	.4966	.4967	.4968	.4969	.4970	.4971	.4972	.4973	.4974
2.8	.4974	.4975	.4976	.4977	.4977	.4978	.4979	.4979	.4980	.4981
2.9	.4981	.4982	.4982	.4983	.4984	.4984	.4985	.4985	.4986	.4986
3.0	.4987	.4987	.4987	.4988	.4988	.4989	.4989	.4989	.4990	.4990
3.10 and higher	.4999									

NOTE: For values of z above 3.09, use 0.4999 for the area.
*Use these common values that result from interpolation:

z score	Area
1.645	0.4500
2.575	0.4950

From Frederick C. Mosteller and Robert E. K. Rourke, *Sturdy Statistics*, 1973, Addison-Wesley Publishing Co., Reading, MA. Reprinted with permission of Frederick Mosteller.

TRADITIONAL METHOD OF HYPOTHESIS TESTING

1. Identify the specific claim or hypothesis to be tested and put it in symbolic form.
2. Give the symbolic form that must be true when the original claim is false.
3. Of the two symbolic expressions obtained so far, let the null hypothesis H_0 be the one that contains the condition of equality; H_1 is the other statement.
4. Select the significance level α based on the seriousness of a type I error. Make α small if the consequences of rejecting a true H_0 are severe. The values of 0.05 and 0.01 are very common.
5. Identify the statistic that is relevant to this test, and identify its sampling distribution.
6. Determine the test statistic, the critical values, and the critical region. Draw a graph and include the test statistic, critical value(s), and critical region.
7. Reject H_0 if the test statistic is in the critical region. Fail to reject H_0 if the test statistic is not in the critical region.
8. Restate this previous conclusion in simple, nontechnical terms.

FINDING P-VALUES

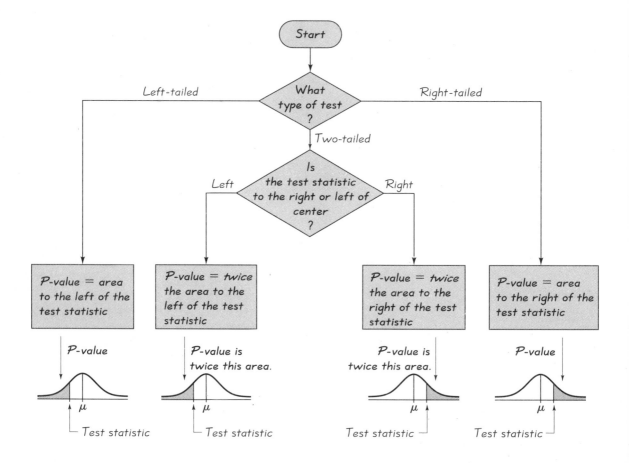

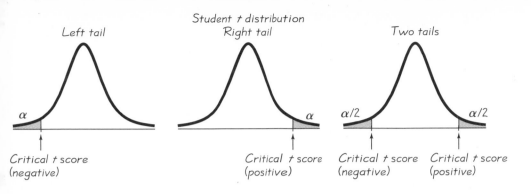

Student t distribution

Left tail

α

Critical t score
(negative)

Right tail

α

Critical t score
(positive)

Two tails

α/2   α/2

Critical t score   Critical t score
(negative)   (positive)

TABLE A-3	t Distribution					

				α		
Degrees of freedom	.005 (one tail) .01 (two tails)	.01 (one tail) .02 (two tails)	.025 (one tail) .05 (two tails)	.05 (one tail) .10 (two tails)	.10 (one tail) .20 (two tails)	.25 (one tail) .50 (two tails)
1	63.657	31.821	12.706	6.314	3.078	1.000
2	9.925	6.965	4.303	2.920	1.886	.816
3	5.841	4.541	3.182	2.353	1.638	.765
4	4.604	3.747	2.776	2.132	1.533	.741
5	4.032	3.365	2.571	2.015	1.476	.727
6	3.707	3.143	2.447	1.943	1.440	.718
7	3.500	2.998	2.365	1.895	1.415	.711
8	3.355	2.896	2.306	1.860	1.397	.706
9	3.250	2.821	2.262	1.833	1.383	.703
10	3.169	2.764	2.228	1.812	1.372	.700
11	3.106	2.718	2.201	1.796	1.363	.697
12	3.054	2.681	2.179	1.782	1.356	.696
13	3.012	2.650	2.160	1.771	1.350	.694
14	2.977	2.625	2.145	1.761	1.345	.692
15	2.947	2.602	2.132	1.753	1.341	.691
16	2.921	2.584	2.120	1.746	1.337	.690
17	2.898	2.567	2.110	1.740	1.333	.689
18	2.878	2.552	2.101	1.734	1.330	.688
19	2.861	2.540	2.093	1.729	1.328	.688
20	2.845	2.528	2.086	1.725	1.325	.687
21	2.831	2.518	2.080	1.721	1.323	.686
22	2.819	2.508	2.074	1.717	1.321	.686
23	2.807	2.500	2.069	1.714	1.320	.685
24	2.797	2.492	2.064	1.711	1.318	.685
25	2.787	2.485	2.060	1.708	1.316	.684
26	2.779	2.479	2.056	1.706	1.315	.684
27	2.771	2.473	2.052	1.703	1.314	.684
28	2.763	2.467	2.048	1.701	1.313	.683
29	2.756	2.462	2.045	1.699	1.311	.683
Large (z)	2.575	2.327	1.960	1.645	1.282	.675

Formulas and Tables

for *Elementary Statistics, Seventh Edition*, by Mario F. Triola
©1998 by Addison Wesley Longman Publishing Company, Inc.

Ch. 13: Nonparametric Tests

$$z = \frac{(x + 0.5) - (n/2)}{\sqrt{n}/2} \quad \text{Sign test for } n > 25$$

$$z = \frac{T - n(n + 1)/4}{\sqrt{\dfrac{n(n + 1)(2n + 1)}{24}}} \quad \begin{array}{l}\text{Wilcoxon signed ranks (two}\\\text{dependent samples for } n > 30)\end{array}$$

$$z = \frac{R - \mu_R}{\sigma_R} = \frac{R - \dfrac{n_1(n_1 + n_2 + 1)}{2}}{\sqrt{\dfrac{n_1 n_2(n_1 + n_2 + 1)}{12}}} \quad \begin{array}{l}\text{Wilcoxon rank-sum}\\\text{(two independent}\\\text{samples)}\end{array}$$

$$H = \frac{12}{n(n + 1)}\left(\frac{R_1^2}{n_1} + \frac{R_2^2}{n_2} + \ldots + \frac{R_k^2}{n_k}\right) - 3(n + 1)$$

Kruskal-Wallis (chi-square df $= k - 1$)

$$r_s = 1 - \frac{6\Sigma d^2}{n(n^2 - 1)} \quad \text{Rank correlation}$$

$$\left(\text{critical value for } n > 30: \frac{\pm z}{\sqrt{n - 1}}\right)$$

$$z = \frac{G - \mu_G}{\sigma_G} = \frac{G - \dfrac{2n_1 n_2}{n_1 + n_2} + 1}{\sqrt{\dfrac{(2n_1 n_2)(2n_1 n_2 - n_1 - n_2)}{(n_1 + n_2)^2(n_1 + n_2 - 1)}}} \quad \begin{array}{l}\text{Runs test}\\\text{for } n > 20\end{array}$$

Ch. 12: Control Charts

R chart: Plot sample ranges

  UCL: $D_4\overline{R}$

  Centerline: $\overline{R}$

  LCL: $D_3\overline{R}$

$\overline{x}$ chart: Plot sample means

  UCL: $\overline{\overline{x}} + A_2\overline{R}$

  Centerline: $\overline{\overline{x}}$

  LCL: $\overline{\overline{x}} - A_2\overline{R}$

p chart: Plot sample proportions

  UCL: $\overline{p} + 3\sqrt{\dfrac{\overline{p}\,\overline{q}}{n}}$

  Centerline: $\overline{p}$

  LCL: $\overline{p} - 3\sqrt{\dfrac{\overline{p}\,\overline{q}}{n}}$

TABLE A-6
Critical Values of the Pearson Correlation Coefficient r

n	$\alpha = .05$	$\alpha = .01$
4	.950	.999
5	.878	.959
6	.811	.917
7	.754	.875
8	.707	.834
9	.666	.798
10	.632	.765
11	.602	.735
12	.576	.708
13	.553	.684
14	.532	.661
15	.514	.641
16	.497	.623
17	.482	.606
18	.468	.590
19	.456	.575
20	.444	.561
25	.396	.505
30	.361	.463
35	.335	.430
40	.312	.402
45	.294	.378
50	.279	.361
60	.254	.330
70	.236	.305
80	.220	.286
90	.207	.269
100	.196	.256

NOTE: To test $H_0: \rho = 0$ against $H_1: \rho \neq 0$, reject H_0 if the absolute value of r is greater than the critical value in the table.

Control Chart Constants			
Subgroup Size n	A_2	D_3	D_4
2	1.880	0.000	3.267
3	1.023	0.000	2.574
4	0.729	0.000	2.282
5	0.577	0.000	2.114
6	0.483	0.000	2.004
7	0.419	0.076	1.924

Formulas and Tables

for *Elementary Statistics, Seventh Edition*, by Mario F. Triola
©1998 by Addison Wesley Longman Publishing Company, Inc.

TABLE A-4 — Chi-Square (χ^2) Distribution

Area to the Right of the Critical Value

Degrees of freedom	0.995	0.99	0.975	0.95	0.90	0.10	0.05	0.025	0.01	0.005
1	—	—	0.001	0.004	0.016	2.706	3.841	5.024	6.635	7.879
2	0.010	0.020	0.051	0.103	0.211	4.605	5.991	7.378	9.210	10.597
3	0.072	0.115	0.216	0.352	0.584	6.251	7.815	9.348	11.345	12.838
4	0.207	0.297	0.484	0.711	1.064	7.779	9.488	11.143	13.277	14.860
5	0.412	0.554	0.831	1.145	1.610	9.236	11.071	12.833	15.086	16.750
6	0.676	0.872	1.237	1.635	2.204	10.645	12.592	14.449	16.812	18.548
7	0.989	1.239	1.690	2.167	2.833	12.017	14.067	16.013	18.475	20.278
8	1.344	1.646	2.180	2.733	3.490	13.362	15.507	17.535	20.090	21.955
9	1.735	2.088	2.700	3.325	4.168	14.684	16.919	19.023	21.666	23.589
10	2.156	2.558	3.247	3.940	4.865	15.987	18.307	20.483	23.209	25.188
11	2.603	3.053	3.816	4.575	5.578	17.275	19.675	21.920	24.725	26.757
12	3.074	3.571	4.404	5.226	6.304	18.549	21.026	23.337	26.217	28.299
13	3.565	4.107	5.009	5.892	7.042	19.812	22.362	24.736	27.688	29.819
14	4.075	4.660	5.629	6.571	7.790	21.064	23.685	26.119	29.141	31.319
15	4.601	5.229	6.262	7.261	8.547	22.307	24.996	27.488	30.578	32.801
16	5.142	5.812	6.908	7.962	9.312	23.542	26.296	28.845	32.000	34.267
17	5.697	6.408	7.564	8.672	10.085	24.769	27.587	30.191	33.409	35.718
18	6.265	7.015	8.231	9.390	10.865	25.989	28.869	31.526	34.805	37.156
19	6.844	7.633	8.907	10.117	11.651	27.204	30.144	32.852	36.191	38.582
20	7.434	8.260	9.591	10.851	12.443	28.412	31.410	34.170	37.566	39.997
21	8.034	8.897	10.283	11.591	13.240	29.615	32.671	35.479	38.932	41.401
22	8.643	9.542	10.982	12.338	14.042	30.813	33.924	36.781	40.289	42.796
23	9.260	10.196	11.689	13.091	14.848	32.007	35.172	38.076	41.638	44.181
24	9.886	10.856	12.401	13.848	15.659	33.196	36.415	39.364	42.980	45.559
25	10.520	11.524	13.120	14.611	16.473	34.382	37.652	40.646	44.314	46.928
26	11.160	12.198	13.844	15.379	17.292	35.563	38.885	41.923	45.642	48.290
27	11.808	12.879	14.573	16.151	18.114	36.741	40.113	43.194	46.963	49.645
28	12.461	13.565	15.308	16.928	18.939	37.916	41.337	44.461	48.278	50.993
29	13.121	14.257	16.047	17.708	19.768	39.087	42.557	45.722	49.588	52.336
30	13.787	14.954	16.791	18.493	20.599	40.256	43.773	46.979	50.892	53.672
40	20.707	22.164	24.433	26.509	29.051	51.805	55.758	59.342	63.691	66.766
50	27.991	29.707	32.357	34.764	37.689	63.167	67.505	71.420	76.154	79.490
60	35.534	37.485	40.482	43.188	46.459	74.397	79.082	83.298	88.379	91.952
70	43.275	45.442	48.758	51.739	55.329	85.527	90.531	95.023	100.425	104.215
80	51.172	53.540	57.153	60.391	64.278	96.578	101.879	106.629	112.329	116.321
90	59.196	61.754	65.647	69.126	73.291	107.565	113.145	118.136	124.116	128.299
100	67.328	70.065	74.222	77.929	82.358	118.498	124.342	129.561	135.807	140.169

From Donald B. Owen, *Handbook of Statistical Tables*, © 1962 Addison-Wesley Publishing Co., Reading, MA. Reprinted with permission of the publisher.

Formulas and Tables

for *Elementary Statistics, Seventh Edition*, by Mario F. Triola
©1998 by Addison Wesley Longman Publishing Company, Inc.

Ch. 7: Test Statistics (one population)

$z = \dfrac{\bar{x} - \mu}{\sigma/\sqrt{n}}$ Mean—one population (σ known or $n > 30$)

$t = \dfrac{\bar{x} - \mu}{s/\sqrt{n}}$ Mean—one population (σ unknown and $n \le 30$)

$z = \dfrac{\hat{p} - p}{\sqrt{\dfrac{pq}{n}}}$ Proportion—one population

$\chi^2 = \dfrac{(n-1)s^2}{\sigma^2}$ Standard deviation or variance—one population

Ch. 8: Test Statistics (two populations)

$t = \dfrac{\bar{d} - \mu_d}{s_d/\sqrt{n}}$ Two means—dependent (df $= n - 1$)

$z = \dfrac{(\bar{x}_1 - \bar{x}_2) - (\mu_1 - \mu_2)}{\sqrt{\dfrac{\sigma_1^2}{n_1} + \dfrac{\sigma_2^2}{n_2}}}$ Two means—independent (σ_1, σ_2 known or $n_1 > 30$ and $n_2 > 30$)

$F = \dfrac{s_1^2}{s_2^2}$ Standard deviation or variance—two populations (where $s_1^2 \ge s_2^2$)

$t = \dfrac{(\bar{x}_1 - \bar{x}_2) - (\mu_1 - \mu_2)}{\sqrt{\dfrac{s_1^2}{n_1} + \dfrac{s_2^2}{n_2}}}$ (df $=$ smaller of $n_1 - 1$, $n_2 - 1$)

Two means—independent (reject $\sigma_1^2 = \sigma_2^2$ and $n_1 \le 30$ or $n_2 \le 30$)

$t = \dfrac{(\bar{x}_1 - \bar{x}_2) - (\mu_1 - \mu_2)}{\sqrt{\dfrac{s_p^2}{n_1} + \dfrac{s_p^2}{n_2}}}$ (df $= n_1 + n_2 - 2$)

where $s_p^2 = \dfrac{(n_1 - 1)s_1^2 + (n_2 - 1)s_2^2}{n_1 + n_2 - 2}$

Two means—independent (fail to reject $\sigma_1^2 = \sigma_2^2$ and $n_1 \le 30$ or $n_2 \le 30$)

$z = \dfrac{(\hat{p}_1 - \hat{p}_2) - (p_1 - p_2)}{\sqrt{\dfrac{\bar{p}\bar{q}}{n_1} + \dfrac{\bar{p}\bar{q}}{n_2}}}$ Two proportions

Ch. 10: Multinomial and Contingency Tables

$\chi^2 = \sum \dfrac{(O - E)^2}{E}$ Multinomial (df $= k - 1$)

$\chi^2 = \sum \dfrac{(O - E)^2}{E}$ Contingency table [df $= (r - 1)(c - 1)$]

where $E = \dfrac{(\text{row total})(\text{column total})}{(\text{grand total})}$

Ch. 9: Linear Correlation/Regression

Correlation $r = \dfrac{n\Sigma xy - (\Sigma x)(\Sigma y)}{\sqrt{n(\Sigma x^2) - (\Sigma x)^2}\sqrt{n(\Sigma y^2) - (\Sigma y)^2}}$

$b_1 = \dfrac{n\Sigma xy - (\Sigma x)(\Sigma y)}{n(\Sigma x^2) - (\Sigma x)^2}$

$b_0 = \bar{y} - b_1\bar{x}$ or $b_0 = \dfrac{(\Sigma y)(\Sigma x^2) - (\Sigma x)(\Sigma xy)}{n(\Sigma x^2) - (\Sigma x)^2}$

$\hat{y} = b_0 + b_1 x$ Estimated eq. of regression line

$r^2 = \dfrac{\text{explained variation}}{\text{total variation}}$

$s_e = \sqrt{\dfrac{\Sigma(y - \hat{y})^2}{n - 2}}$ or $\sqrt{\dfrac{\Sigma y^2 - b_0\Sigma y - b_1\Sigma xy}{n - 2}}$

$\hat{y} - E < y < \hat{y} + E$

where $E = t_{\alpha/2}s_e\sqrt{1 + \dfrac{1}{n} + \dfrac{n(x_0 - \bar{x})^2}{n(\Sigma x^2) - (\Sigma x)^2}}$

Ch. 11: One-Way Analysis of a Variance

$F = \dfrac{ns_{\bar{x}}^2}{s_p^2}$ k samples each of size n (num. df $= k - 1$; den. df $= k(n - 1)$)

$F = \dfrac{\text{MS(treatment)}}{\text{MS(error)}}$ $\leftarrow$ df $= k - 1$ $\leftarrow$ df $= N - k$

$\text{MS(treatment)} = \dfrac{\text{SS(treatment)}}{k - 1}$

$\text{MS(error)} = \dfrac{\text{SS(error)}}{N - k}$ $\text{MS(total)} = \dfrac{\text{SS(total)}}{N - 1}$

$\text{SS(treatment)} = n_1(\bar{x}_1 - \bar{\bar{x}})^2 + \cdots + n_k(\bar{x}_k - \bar{\bar{x}})^2$

$\text{SS(error)} = (n_1 - 1)s_1^2 + \cdots + (n_k - 1)s_k^2$

$\text{SS(total)} = \Sigma(x - \bar{\bar{x}})^2$

$\text{SS(total)} = \text{SS(treatment)} + \text{SS(error)}$

Ch. 11: Two-Way Analysis of Variance

Interaction: $F = \dfrac{\text{MS(interaction)}}{\text{MS(error)}}$

Row Factor: $F = \dfrac{\text{MS(row factor)}}{\text{MS(error)}}$

Column Factor: $F = \dfrac{\text{MS(column factor)}}{\text{MS(error)}}$

Formulas and Tables

for *Elementary Statistics, Seventh Edition*, by Mario F. Triola
©1998 by Addison Wesley Longman Publishing Company, Inc.

Ch. 2: Descriptive Statistics

$$\bar{x} = \frac{\Sigma x}{n} \quad \text{Mean}$$

$$\bar{x} = \frac{\Sigma f \cdot x}{\Sigma f} \quad \text{Mean (frequency table)}$$

$$s = \sqrt{\frac{\Sigma (x - \bar{x})^2}{n - 1}} \quad \text{Standard deviation}$$

$$s = \sqrt{\frac{n(\Sigma x^2) - (\Sigma x)^2}{n(n - 1)}} \quad \begin{array}{l}\text{Standard deviation}\\\text{(shortcut)}\end{array}$$

$$s = \sqrt{\frac{n[\Sigma (f \cdot x^2)] - [\Sigma (f \cdot x)]^2}{n(n - 1)}} \quad \begin{array}{l}\text{Standard deviation}\\\text{(frequency table)}\end{array}$$

$$\text{variance} = s^2$$

Ch. 3: Probability

$P(A \text{ or } B) = P(A) + P(B)$ if A, B are mutually exclusive
$P(A \text{ or } B) = P(A) + P(B) - P(A \text{ and } B)$
    if A, B are not mutually exclusive
$P(A \text{ and } B) = P(A) \cdot P(B)$ if A, B are independent
$P(A \text{ and } B) = P(A) \cdot P(B|A)$ if A, B are dependent
$P(\bar{A}) = 1 - P(A)$ Rule of complements

$$_nP_r = \frac{n!}{(n - r)!} \quad \text{Permutations (no elements alike)}$$

$$\frac{n!}{n_1! \, n_2! \ldots n_k!} \quad \text{Permutations } (n_1 \text{ alike, ...})$$

$$_nC_r = \frac{n!}{(n - r)! \, r!} \quad \text{Combinations}$$

Ch. 4: Probability Distributions

$\mu = \Sigma x \cdot P(x)$ Mean (prob. dist.)
$\sigma = \sqrt{[\Sigma x^2 \cdot P(x)] - \mu^2}$ Standard deviation (prob. dist.)
$P(x) = \dfrac{n!}{(n - x)! \, x!} \cdot p^x \cdot q^{n-x}$ Binomial probability
$\mu = n \cdot p$ Mean (binomial)
$\sigma^2 = n \cdot p \cdot q$ Variance (binomial)
$\sigma = \sqrt{n \cdot p \cdot q}$ Standard deviation (binomial)
$P(x) = \dfrac{\mu^x \cdot e^{-\mu}}{x!}$ Poisson Distribution where $e \approx 2.71828$

Ch. 5: Normal Distribution

$$z = \frac{x - \bar{x}}{s} \text{ or } \frac{x - \mu}{\sigma} \quad \text{Standard score}$$

$\mu_{\bar{x}} = \mu$ Central limit theorem

$\sigma_{\bar{x}} = \dfrac{\sigma}{\sqrt{n}}$ Central limit theorem (Standard error)

Ch. 6: Confidence Intervals (one population)

$$\bar{x} - E < \mu < \bar{x} + E \quad \text{Mean}$$

$$\text{where } E = z_{\alpha/2} \frac{\sigma}{\sqrt{n}} \quad (\sigma \text{ known or } n > 30)$$

$$\text{or } E = t_{\alpha/2} \frac{s}{\sqrt{n}} \quad (\sigma \text{ unknown and } n \leq 30)$$

$$\hat{p} - E < p < \hat{p} + E \quad \text{Proportion}$$

$$\text{where } E = z_{\alpha/2} \sqrt{\frac{\hat{p}\hat{q}}{n}}$$

$$\frac{(n - 1)s^2}{\chi_R^2} < \sigma^2 < \frac{(n - 1)s^2}{\chi_L^2} \quad \text{Variance}$$

Ch. 6: Sample Size Determination

$$n = \left[\frac{z_{\alpha/2}\sigma}{E}\right]^2 \quad \text{Mean}$$

$$n = \frac{[z_{\alpha/2}]^2 \cdot 0.25}{E^2} \quad \text{Proportion}$$

$$n = \frac{[z_{\alpha/2}]^2 \hat{p}\hat{q}}{E^2} \quad \text{Proportion } (\hat{p} \text{ and } \hat{q} \text{ are known})$$

Ch. 8: Confidence Intervals (two populations)

$$\bar{d} - E < \mu_d < \bar{d} + E \quad \text{(Dependent)}$$

$$\text{where } E = t_{\alpha/2} \frac{s_d}{\sqrt{n}} \quad (\text{df} = n - 1)$$

$$(\bar{x}_1 - \bar{x}_2) - E < (\mu_1 - \mu_2) < (\bar{x}_1 - \bar{x}_2) + E \quad \text{(Indep.)}$$

$$\text{where } E = z_{\alpha/2} \sqrt{\frac{\sigma_1^2}{n_1} + \frac{\sigma_2^2}{n_2}} \longleftarrow$$

$$(\sigma_1, \sigma_2 \text{ known or } n_1 > 30 \text{ and } n_2 > 30) \longrightarrow$$

$$E = t_{\alpha/2} \sqrt{\frac{s_1^2}{n_1} + \frac{s_2^2}{n_2}} \quad \begin{array}{l}(\text{df} = \text{smaller of}\\ n_1 - 1, n_2 - 1)\end{array} \longleftarrow$$

$$(\text{reject } \sigma_1^2 = \sigma_2^2 \text{ and } n_1 \leq 30 \text{ or } n_2 \leq 30) \longrightarrow$$

$$E = t_{\alpha/2} \sqrt{\frac{s_p^2}{n_1} + \frac{s_p^2}{n_2}} \quad (\text{df} = n_1 + n_2 - 2) \longleftarrow$$

$$s_p^2 = \frac{(n_1 - 1)s_1^2 + (n_2 - 1)s_2^2}{(n_1 - 1) + (n_2 - 1)}$$

$$(\text{fail to reject } \sigma_1^2 = \sigma_2^2 \text{ and } n_1 \leq 30 \text{ or } n_2 \leq 30) \longrightarrow$$

$$(\hat{p}_1 - \hat{p}_2) - E < (p_1 - p_2) < (\hat{p}_1 - \hat{p}_2) + E$$

$$\text{where } E = z_{\alpha/2} \sqrt{\frac{\hat{p}_1\hat{q}_1}{n_1} + \frac{\hat{p}_2\hat{q}_2}{n_2}}$$

HYPOTHESIS TEST: WORDING
OF FINAL CONCLUSION

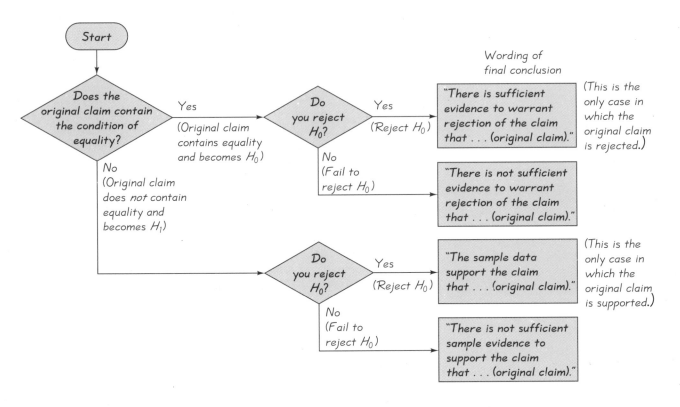

STATDISK DISPLAY FOR CANS 109

File   Edit   Analysis   Data   Help

Cans109

Sample Size, n	175
Mean, $\bar{x}$	267.11
Median	273.00
Midrange	248.50
RMS	268.02
Variance, s^2	488.95
St Dev, s	22.112
Mean Dev	16.019
Range	97.000
Minimum	200.00
1st Quartile	262.00
2nd Quartile	273.00
3rd Quartile	282.00
Maximum	297.00
$\sum x$	46745
$\sum x^2$	12571335

MINITAB DISPLAY FOR CANS 109

Variable	N	Mean	Median	Tr Mean	StDev	SE Mean
CANS109	175	267.11	273.00	269.15	22.11	1.67

Variable	Min	Max	Q1	Q3
CANS109	200.00	297.00	262.00	282.00

---

the standard deviation, the number of scores, the minimum, maximum, median, and quartiles. The TI-83 and Minitab calculate quartiles in a way that is slightly different from the one described in this book, so there may be some discrepancies.

  2–6   Exercises A: Basic Skills and Concepts

In Exercises 1–4, express all z scores with two decimal places.

1. Adult males have heights with a mean of 69.0 in. and a standard deviation of 2.8 in. Find the z scores corresponding to the following.

Recommended assignment (it's not really as long as it appears): 1, 2, 5, 6, 8, 10, 13–24. (Exercises 25–36 use an Appendix B data set that requires ranking of 54 values.)

1. a. −2.14
   b. 5.71
   c. 0.26

a. Basketball player Mugsy Bogues, who is 5 ft. 3 in. tall
   b. Basketball player Shaquille O'Neal, who is 7 ft. 1 in. tall
   c. The author, who is a 69.72-in.-tall golf and tennis "player"
2. Student cars at the author's college have ages with a mean of 7.90 years and a standard deviation of 3.67 years. Find the z scores for cars with the given ages.
   a. A 12-year-old Corvette
   b. A 2-year-old Ferrari
   c. A brand-new Porsche
3. The numbers of hours that college freshmen spend studying each week have a mean of 7.06 h and a standard deviation of 5.32 h (based on data from *The American Freshman*). Find the z score for a freshman who studies 20.00 hours weekly.   3. 2.43
4. The amounts of time that high school students spend working at jobs each week have a mean of 10.7 h and a standard deviation of 11.2 h (based on data from the National Federation of State High School Associations). Find the z score corresponding to a high school student who works 8.0 h each week.   4. −0.24

In Exercises 5–8, express all z scores with two decimal places. Consider a score to be unusual if its z score is less than −2.00 or greater than 2.00.

5. The Beanstalk Club is limited to women and men who are very tall. The minimum height requirement for women is 70 in. Women's heights have a mean of 63.6 in. and a standard deviation of 2.5 in. Find the z score corresponding to a woman with a height of 70 in. and determine whether that height is unusual.   5. 2.56; unusual
6. A woman wrote to *Dear Abby* and claimed that she gave birth 308 days after a visit from her husband, who was in the Navy. Lengths of pregnancies have a mean of 268 days and a standard deviation of 15 days. Find the z score for 308 days. Is such a length unusual? What do you conclude?
7. One of the few working vending machines is located and is found to accept quarters with weights that are not unusual. Find the z score for a quarter weighing 5.50 g. Will it be accepted by this vending machine? (Weights of quarters have a mean of 5.67 g and a standard deviation of 0.070 g.)
8. For men aged between 18 and 24 years, serum cholesterol levels (in mg/100 mL) have a mean of 178.1 and a standard deviation of 40.7 (based on data from the National Health Survey). Find the z score corresponding to a male, aged 18–24 years, who has a serum cholesterol level of 275.2 mg/100 mL. Is this level unusually high?
9. Which of the following two scores has the better relative position?
   a. A score of 60 on a test for which $\bar{x} = 50$ and $s = 5$
   b. A score of 250 on a test for which $\bar{x} = 200$ and $s = 20$
10. Two similar groups of students took equivalent language facility tests. Which of the following results indicates the higher relative level of language facility?

2. a. 1.12
   b. −1.61
   c. −2.15

6. 2.67; unusual
7. −2.43; no
8. 2.39; unusually high
9. 2.00; 2.50; 250 is better
10. −0.50; −0.56; 65 is higher

a. A score of 65 on a test for which $\bar{x} = 70$ and $s = 10$
b. A score of 455 on a test for which $\bar{x} = 500$ and $s = 80$

11. Three prospective employees take equivalent tests of critical thinking. Which of the following scores corresponds to the highest relative position?
    a. A score of 37 on a test for which $\bar{x} = 28$ and $s = 6$
    b. A score of 398 on a test for which $\bar{x} = 312$ and $s = 56$
    c. A score of 4.10 on a test for which $\bar{x} = 2.75$ and $s = 0.92$

12. Three students take equivalent tests of a sense of humor and, after the laughter dies down, their scores are calculated. Which is the highest relative score?
    a. A score of 2.7 on a test for which $\bar{x} = 3.2$ and $s = 1.1$
    b. A score of 27 on a test for which $\bar{x} = 35$ and $s = 12$
    c. A score of 850 on a test for which $\bar{x} = 921$ and $s = 87$

In Exercises 13–16, use the 175 ranked axial loads of aluminum cans listed in Table 2-9. Find the percentile corresponding to the given value.

13. 254   **14.** 265   **15.** 277   **16.** 288

In Exercises 17–24, use the 175 ranked axial loads of aluminum cans listed in Table 2-9. Find the indicated percentile, quartile, or decile.

17. P_{70} **18.** P_{20} **19.** D_6 **20.** D_3
21. Q_3 **22.** Q_1 **23.** D_1 **24.** P_1

In Exercises 25–28, use the weights (in pounds) of bears listed in Data Set 3 of Appendix B. Find the percentile corresponding to the given weight.

25. 144   **26.** 212   **27.** 316   **28.** 90

In Exercises 29–36, use the weights (in pounds) of bears listed in Data Set 3 of Appendix B. Find the indicated percentile, quartile, or decile.

29. P_{85} **30.** P_{35} **31.** Q_1 **32.** Q_3
33. D_9 **34.** D_3 **35.** P_{50} **36.** P_{95}

2–6 Exercises B: Beyond the Basics

37. Use the ranked axial loads of aluminum cans listed in Table 2-9.
    a. Find the interquartile range.
    b. Find the midquartile.
    c. Find the 10–90 percentile range.
    d. Does $P_{50} = Q_2$? If so, does P_{50} *always* equal Q_2?
    e. Does $Q_2 = (Q_1 + Q_3)/2$? If so, does Q_2 *always* equal $(Q_1 + Q_3)/2$?

11. 1.50; 1.54; 1.47; 398 is highest
12. −0.45; −0.67; −0.82; 2.7 is highest
13. 17
14. 30
15. 60
16. 89
17. 279
18. 257
19. 276.5
20. 265
21. 282
22. 262
23. 230
24. 201
25. 44
26. 69
27. 80
28. 26
29. 344
30. 116
31. 86
32. 236
33. 360
34. 105
35. 150
36. 436
37. a. 20
    b. 272
    c. 59
    d. Yes; yes
    e. No

38. $P_{35} = 115.8$
    $Q_1 = 83$
    $D_3 = 96.2$
39. 222.9, 311.3
40. 0; 1; yes

38. When finding percentiles using Figure 2-12, if the locator L is not a whole number, we round it up to the next larger whole number. An alternative to this procedure is to interpolate so that a locator of 23.75 leads to a value that is 0.75 (or 3/4) of the way between the 23rd and 24th scores. Use this method of interpolation to find P_{35}, Q_1, and D_3 for the weights of bears listed in Data Set 3 of Appendix B.

39. For the 175 axial loads of aluminum cans given in Table 2-1, the mean is 267.1 and the standard deviation is 22.1. Find the two cutoff values separating ordinary values from unusual values.

40. Using the scores 2, 5, 8, 9, and 16, first find $\bar{x}$ and s, then replace each score by its corresponding z score. (Don't round the z scores; carry as many decimal places as your calculator can handle.) Now find the mean and standard deviation of the five z scores. Will these new values of the mean and standard deviation result from every set of z scores?

2-7 Exploratory Data Analysis (EDA)

Sometimes we observe or collect data with a specific goal in mind—for example, verifying the effectiveness of a new treatment for insomnia. But sometimes we have no specific goal and simply want to explore the data to see what they reveal. In exploring data, we can use many of the techniques already presented in this chapter. Recall that in Section 2-1 we listed three very important characteristics of data: (1) the nature or shape of the distribution; (2) a representative value; and (3) a measure of variation. The distribution of data should definitely be considered because it may affect the statistical methods we use, as well as the conclusions we draw. In the spirit of **exploratory data analysis** (or **EDA**), we should not simply view a histogram and think that we understand the nature of the distribution—we should *explore*. As an example, we show two STATDISK displays of histograms of the 175 axial loads of the aluminum cans listed in Table 2-1. The first histogram represents the 175 values with one change: The first entry of 270 is incorrectly entered as 2700. The second histogram is correct. Note how one simple error in only one of 175 values has such a dramatic effect on the shape of the histogram. In this case, the incorrect extreme value of 2700 causes a severe distortion of the histogram. In other cases, such extreme values (often called outliers) may be correct but may disguise the true nature of the distribution when illustrated through a histogram. If we didn't further explore the data, we might draw conclusions from the histogram that are seriously wrong.

With EDA, the emphasis is on original explorations with the goals of simplifying the way the data are described and gaining deeper insight into the nature of the data. The table on page 104 compares EDA and traditional statistics in three major areas.

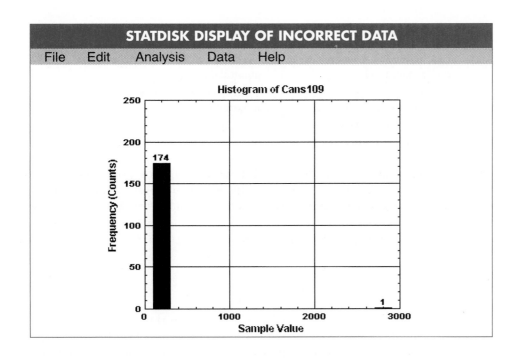

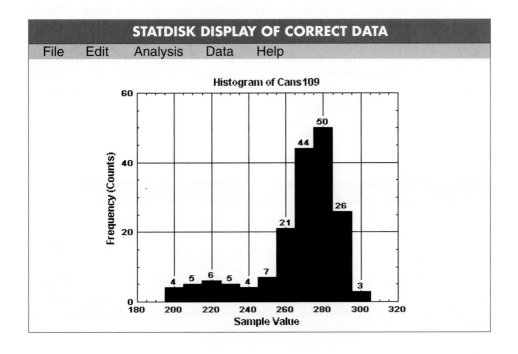

Exploratory Data Analysis	Traditional Statistics
Used to *explore* data at a preliminary level	Used to *confirm* final conclusions about data
Few or no assumptions are made about the data	Typically requires some very important assumptions about the data
Tends to involve relatively simple calculations and graphs	Calculations are often complex and graphs are often unnecessary

In Section 2-3 we already considered stem-and-leaf plots, one of the tools commonly used in EDA. We will now introduce boxplots, which weren't included earlier because they require quartiles, which were discussed in the preceding section.

Boxplots

Changes to this edition: Boxplots are now constructed with *quartiles* instead of hinges as in the previous editions. Also, stem-and-leaf plots have been moved from this section to Section 2-3. The new Windows version of STATDISK now includes an ability to generate boxplots.

Boxplots are graphs that are useful for revealing central tendency, the spread of the data, the distribution of the data, and the presence of outliers (extreme scores). The construction of a boxplot requires that we obtain the minimum score, the first quartile Q_1, the median (or second quartile Q_2), the third quartile Q_3, and the maximum score. Because medians are used to reveal central tendency and quartiles are used to reveal the spread of data, boxplots have the advantage of not being as sensitive to extreme values as other devices based on the mean and standard deviation. Boxplots don't show as much detailed information as histograms or stem-and-leaf plots, so they might not be the best choice when dealing with a single data set. However, boxplots are often more useful when comparing two or more data sets. When using two or more boxplots for comparing different data sets, it is important to use the same scale so that the comparisons can be made.

DEFINITIONS

The minimum score, the first quartile Q_1, the median, the third quartile Q_3, and the maximum score constitute a **5-number summary** of a set of data.

A **boxplot** (or **box-and-whisker diagram**) is a graph of data that consists of a line extending from the lowest score to the highest, and a box with lines drawn at the first quartile Q_1, the median, and the third quartile Q_3.

● **EXAMPLE**   Refer to Data Set 8 of Appendix B and use the pulse rates of smokers.

a.   Find the values constituting the 5-number summary.
b.   Construct a boxplot for the pulse rates of smokers.

SOLUTION

a. The 5-number summary consists of the minimum, Q_1, median, Q_3, and maximum. To find those values, we should first arrange the pulse rates of smokers in order from low to high. Here is the *ranked* list of the 22 pulse rates of smokers from Data Set 8:

$$52 \quad 52 \quad 60 \quad 60 \quad 60 \quad 60 \quad 63 \quad 63 \quad 66 \quad 67 \quad 68$$
$$69 \quad 71 \quad 72 \quad 73 \quad 75 \quad 78 \quad 80 \quad 82 \quad 83 \quad 88 \quad 90$$

From this ranked list, it is easy to identify the minimum of 52 and the maximum of 90. Using the flowchart of Figure 2-12, we find that the first quartile Q_1 (or P_{25}) is 60, which is located by calculating $L = (25/100)22 = 5.5$ and rounding up the result to 6. Q_1 is the 6th score in the ranked list, namely 60. The median is 68.5, which is the value midway between the 11th and 12th scores. We also find that $Q_3 = 78$ by using Figure 2-12 for the 75th percentile. The 5-number summary is therefore 52, 60, 68.5, 78, and 90.

b. In Figure 2-13 we graph the boxplot for the data. We use the minimum (52) and the maximum (90) to determine a scale of values, then we plot the values from the 5-number summary as shown.

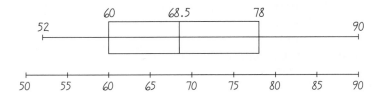

Figure 2-13
Boxplot of Pulse Rates (Beats per Minute) of Smokers

   In Figure 2-14 we show some generic boxplots along with common distribution shapes.

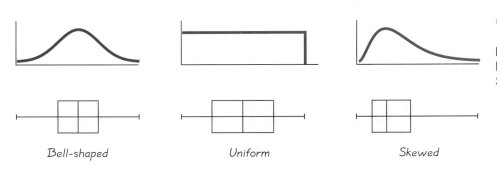

Figure 2-14
Boxplots Corresponding to Bell-Shaped, Uniform, and Skewed Distributions

Outliers

In the course of finding a 5-number summary and graphing a boxplot, it becomes easy to identify **outliers,** which are values that are highly unusual in the sense that they are very far away from most of the data. When exploring a data set, outliers should be considered because they may reveal important information. For example, consider the complete list of pulse rates in Data Set 8. By simply ranking the pulse rates it is easy to see that the low values of 8 and 15 are outliers. Are these outliers exceptional pulse rates or are they errors? Although there were a couple of students whose physical condition could be loosely described as comatose, it is highly unlikely that someone with a pulse rate of 8 or 15 would be capable of entering and leaving a classroom under his or her own power. We therefore conclude that 8 and 15 are errors, and it makes sense to delete them from the data set. Should we delete the maximum pulse rate of 100? No, that value isn't too far away from the others and it most likely came from someone who was excited about being in statistics class. In general, we should delete outliers if they are obvious errors, but they are often interesting anomalies that should be investigated further. In fact, in some data sets, outliers are the most important feature. One study of eggs and cholesterol included a man who had consumed several eggs every day for many years. His egg-consumption rate was an outlier, but it was very important to see that the excess of eggs didn't seem to affect his cholesterol level, which was average. In exploring data, we might study the effects of outliers by constructing graphs and calculating measures with and without the outliers included. (See Exercise 12 for a way to depict outliers on boxplots.)

Using Computers and Calculators for Boxplots

STATDISK, Minitab, and the TI-83 calculator can be used to create boxplots. With STATDISK, choose the main menu item of `Data` and use the `Sample Editor` to enter the data and click on `COPY`, then select `Data/Boxplot` and click on `PASTE`, then `Evaluate`. With Minitab, use the options of `File/ New Worksheet/Graph/Boxplot`. Quartile values calculated by Minitab and the TI-83 calculator may differ slightly from those calculated by applying Figure 2-12, so the boxplots may be slightly different.

We noted that boxplots are useful for comparing sets of data, so we will illustrate Minitab boxplots for the pulse rates of smokers and nonsmokers, as listed in Data Set 8 in Appendix B. After deleting the outliers of 8 and 15 by editing the data grid, the Minitab result will be as shown.

A comparison of the two Minitab boxplots reveals that there aren't any substantial differences. The nonsmokers had more extreme values, but the medi-

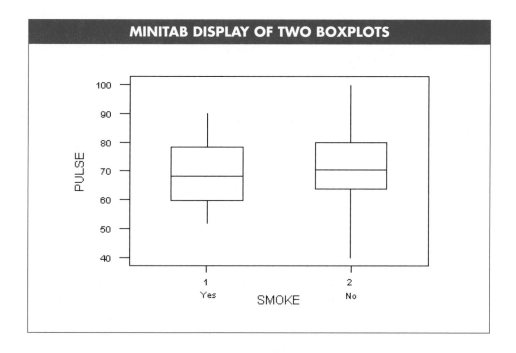

MINITAB DISPLAY OF TWO BOXPLOTS

ans seem to be about the same and the spread of the data is roughly the same. For the population of college students taking statistics, it appears that there are no notable differences in pulse rates between smokers and nonsmokers.

Recommended assignment: 2, 3, 4, 10, 11. Note that if students generate boxplots using calculators or computers, the quartile values may differ from those given as answers in this text.

 2–7   Exercises A: Basic Skills and Concepts

Include values of the 5-number summary in all boxplots.

1. Refer to Data Set 4 in Appendix B and construct a boxplot for the nicotine contents of cigarettes.
2. Refer to Data Set 4 in Appendix B and construct a boxplot for the tar contents of cigarettes.
3. In "Ages of Oscar-Winning Best Actors and Actresses" (*Mathematics Teacher* magazine) by Richard Brown and Gretchen Davis, boxplots are used to compare the ages of actors and actresses at the time they won Oscars. The results for 34 recent winners from each category are listed at the top of the following page. Use boxplots to compare the two data sets.

2.
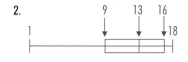

4. Smokers:

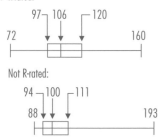

Nonsmokers:

Actors:

	32	37	36	32	51	53	33	61	35	45	55	39
	76	37	42	40	32	60	38	56	48	48	40	
	43	62	43	42	44	41	56	39	46	31	47	

Actresses:

	50	44	35	80	26	28	41	21	61	38	49	33
	74	30	33	41	31	35	41	42	37	26	34	
	34	35	26	61	60	34	24	30	37	31	27	

4. Refer to Data Set 8 in Appendix B for these two data sets: pulse rates of those who smoke and pulse rates of those who don't smoke. Construct a boxplot for each data set. Based on the results, do pulse rates of the two groups appear to be different? If so, how? Is this the result you would expect? (Exclude the 8 and 15, which must be errors.)

5. Refer to Data Set 8 in Appendix B for these two data sets: pulse rates of males and pulse rates of females. Construct a boxplot for each data set. Based on the results, do pulse rates of the two groups appear to be different? If so, how? (Exclude the 8 and 15, which must be errors.)

6. R-rated:

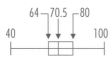

Not R-rated:

6. Refer to Data Set 10 in Appendix B. Use boxplots to compare the lengths of R-rated movies to the lengths of movies with ratings other than R.

7. Refer to Data Set 11 in Appendix B. Use boxplots to compare the weights of red M&M candies to the weights of yellow M&M candies.

8. Refer to Data Set 13 in Appendix B. Construct a boxplot for the weights of quarters. Compare the shape of the resulting boxplot to the generic shapes shown in Figure 2-14. Based on the boxplot, what do you conclude about the nature of the distribution?

8.

9. Refer to Data Set 12 in Appendix B. Construct a boxplot for the 150 digits from the Maryland Pick Three lottery. Compare the shape of the resulting boxplot to the generic shapes shown in Figure 2-14. Based on the boxplot, do the Maryland lottery results appear to be occurring as expected?

10. Paper:

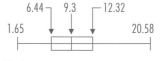

Plastic:

10. Refer to Data Set 1 in Appendix B. Use boxplots to compare the weights of discarded paper to the weights of discarded plastic.

2-7 Exercises B: Beyond the Basics

11. As supervisor of maintenance for a fleet of cars, you must purchase replacement batteries from one of three suppliers. Random samples from those suppliers are tested for longevity and the lives (in months) are summarized in the Minitab boxplots at the top of page 109. Which boxplot corresponds to the brand you will purchase? Why?

12. The boxplots discussed in this section are often called *skeletal* boxplots. In investigating outliers, a useful modification is to construct boxplots as follows:

12.

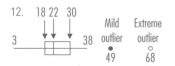

a. Calculate the difference between the quartiles Q_3 and Q_1 and denote it as D, so that $D = Q_3 - Q_1$.

(continued)

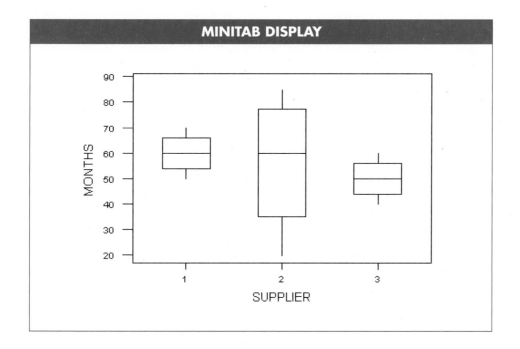

MINITAB DISPLAY

b. Draw the box with the median and quartiles as usual, but when extending the lines that branch out from the box, go only as far as the scores that are within $1.5D$ of the box.

c. **Mild outliers** are scores above Q_3 by an amount of $1.5D$ to $3D$, or below Q_1 by an amount of $1.5D$ to $3D$. Plot mild outliers as solid dots.

d. **Extreme outliers** are scores that exceed Q_3 by more than $3D$ or are below Q_1 by an amount more than $3D$. Plot extreme outliers as small hollow circles.

The accompanying figure is an example of the boxplot described here. Use this procedure to construct the boxplot for the given scores, and identify any mild outliers or extreme outliers.

3   15   17   18   21   21   22   25   27   30   38   49   68

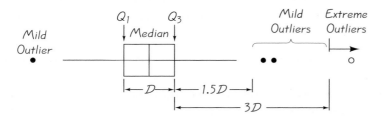

Vocabulary List

descriptive statistics	multimodal
inferential statistics	midrange
frequency table	weighted mean
frequency	skewed
lower class limits	symmetric
upper class limits	negatively skewed
class boundaries	positively skewed
class marks	range
class width	standard deviation
relative frequency	deviation
relative frequency table	mean (or absolute) deviation
cumulative frequency	variance
cumulative frequency table	range rule of thumb
histogram	empirical rule
relative frequency histogram	68–95–99 rule
dotplot	Chebyshev's theorem
stem-and-leaf plot	standard score
Pareto chart	z score
pie chart	quartiles
scatter diagram	deciles
measure of central tendency	percentiles
arithmetic mean	exploratory data analysis (EDA)
mean	5-number summary
sample size	boxplot (or box-and whisker
median	diagram)
mode	outlier
bimodal	

Review

Chapter 2 dealt mainly with the methods and techniques of summarizing, describing, exploring, and comparing data. We noted that three of the most important characteristics of data are (1) the nature or shape of the distribution; (2) a representative value; and (3) a measure of variation. These characteristics can be investigated and described by using the tools of Chapter 2. Specifically, when given a set of data, we should be able to

- Summarize the data by constructing a frequency table or relative frequency table (Section 2-2)
- Visually display the nature of the distribution by constructing a histogram, dotplot, stem-and-leaf plot, pie chart, or Pareto chart (Section 2-3)

- Calculate measures of central tendency by finding the mean, median, mode, and midrange (Section 2-4)
- Calculate measures of variation by finding the standard deviation, variance, and range (Section 2-5)
- Compare individual scores by using z scores, quartiles, deciles, or percentiles (Section 2-6)
- Investigate and explore the spread of data, the center of the data, and the range of values by constructing a boxplot (Section 2-7)

In addition to obtaining the above tables, graphs, and measures, we should *understand* and *interpret* those results. For example, we should clearly understand that the standard deviation is a measure of how much the data vary, and we should be able to use the standard deviation to distinguish between scores that are usual and those that are unusual.

Review Exercises

1. The NCAA was considering ways to speed up the end of college basketball games. The following values are the elapsed times (in seconds) that it took to play the last two minutes of regulation time in the 60 games of the first four rounds of an NCAA basketball tournament (based on data reported in *USA Today*). Using the minimum time as the lower class limit of the first class, construct a frequency table with 9 classes.

756	587	929	871	378	503	564	1128	693	748
448	670	1023	335	540	853	852	495	666	474
443	325	514	404	820	915	793	778	627	483
861	337	292	1070	625	457	676	494	420	862
991	615	609	723	794	447	704	396	235	552
626	688	506	700	240	363	860	670	396	345

2. Construct a relative frequency table (with 9 classes) for the data in Exercise 1.
3. Construct a histogram that corresponds to the frequency table from Exercise 1.
4. For the data in Exercise 1, find (a) Q_1, (b) P_{45}, and (c) the percentile corresponding to the time of 335 s.   4. a. 447.5; b. 575.5; c. 7
5. Use the range rule of thumb to estimate the standard deviation of the data in Exercise 1.
6. Use the frequency table from Exercise 1 to find the mean and standard deviation for the times.
7. Use the data from Exercise 1 to construct a stem-and-leaf plot with 10 rows.

1. Time

Time	Frequency
235–334	4
335–434	9
435–534	11
535–634	9
635–734	9
735–834	6
835–934	8
935–1034	2
1035–1134	2

2. Time

Time	Rel. Freq.
235–334	0.067
335–434	0.150
435–534	0.183
535–634	0.150
635–734	0.150
735–834	0.100
835–934	0.133
935–1034	0.033
1035–1134	0.033

3.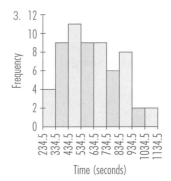

5. $s \approx (1128 - 235)/4 = 223$
6. 621.2; 210.7

The student edition of this book now includes answers to *all* Review Exercises and Cumulative Review Exercises. For the regular text sections, the student edition continues to include answers for the odd-numbered exercises only.

8. Use the data from Exercise 1 to construct a boxplot.

9. Given below are times (in seconds) between an order being placed and the food being received at a McDonald's drive-thru window. Find the (a) mean; (b) median; (c) mode; (d) midrange; (e) range; (f) standard deviation; (g) variance.

$$135 \quad 90 \quad 85 \quad 121 \quad 83 \quad 69 \quad 87 \quad 159 \quad 177 \quad 135 \quad 227$$

10. Given below are the ages of U.S. presidents when they were inaugurated. Find the (a) mean; (b) median; (c) mode; (d) midrange; (e) range; (f) standard deviation; (g) variance; (h) Q_1; (i) P_{30}; (j) D_7.

$$
\begin{array}{cccccccccccc}
57 & 61 & 57 & 57 & 58 & 57 & 61 & 54 & 68 & 51 & 49 & 64 & 50 & 48 \\
65 & 52 & 56 & 46 & 54 & 49 & 51 & 47 & 55 & 55 & 54 & 42 & 51 & 56 \\
55 & 51 & 54 & 51 & 60 & 62 & 43 & 55 & 56 & 61 & 52 & 69 & 64 & 46 \\
\end{array}
$$

11. Scores on a test of depth perception have a mean of 200 and a standard deviation of 40.
   a. Is a score of 260 unusually high? Explain.
   b. What is the z score corresponding to 185?
   c. Assuming that the scores have a bell-shaped distribution, what does the empirical rule say about the percentage of scores between 120 and 280?
   d. What is the mean after 20 points have been added to every score?
   e. What is the standard deviation after 20 points have been added to every score?

12. The accompanying table lists times (in years) required to earn a bachelor's degree for a sample of undergraduate students (based on data from the National Center for Education Statistics). Use the table to find the mean and standard deviation. Based on the results, is it unusual for an undergraduate to require 8 years to earn a bachelor's degree? Explain.

Time (years)	Number
4	147
5	81
6	27
7	15
7.5–11.5	30

13. Using the frequency table given in Exercise 12, construct the corresponding relative frequency histogram.

14. An industrial psychologist gave a subject two different tests designed to measure employee satisfaction. Which score is better: a score of 57 on the first test, which has a mean of 72 and a standard deviation of 20, or a score of 450 on the second test, which has a mean of 500 and a standard deviation of 80? Explain.

15. Refer to the top of page 113 for the Minitab display of two boxplots. The first boxplot represents a sample of skulls from male Egyptians from about 4000 B.C., whereas the second boxplot represents a sample of male Egyptian skulls from about 150 A.D. (based on data from *Ancient Races of the Thebaid* by Thomson and Randall-Maciver). A shift in head sizes would suggest some

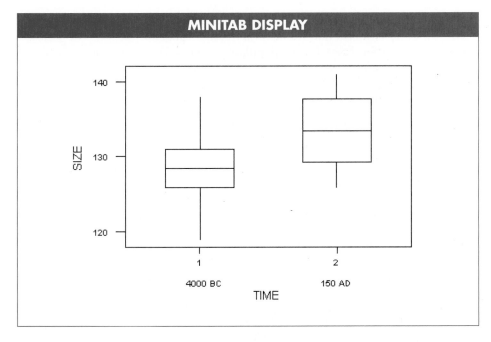

MINITAB DISPLAY

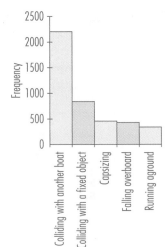

societal changes, such as interbreeding with other cultures. By comparing the two boxplots, is there a shift in the maximum skull breadth? Explain.

16. The United States Coast Guard collected data on serious boating accidents and listed the categories as shown below, with their frequencies given in parentheses. Construct a Pareto chart summarizing the given data.

Colliding with another boat (2203)   Person falling overboard (431)

Colliding with a fixed object (839)   Capsizing (458)

Running aground (341)

• •

Cumulative Review Exercises

1. The amounts of time (in hours) spent on paperwork in one day were obtained from a sample of office managers with the results given below (based on data from Adia Personnel Services):

3.7	2.9	3.4	0.0	1.5	1.8	2.3	2.4	1.0	2.0
4.4	2.0	4.5	0.0	1.7	4.4	3.3	2.4	2.1	2.1

(continued)

New to this edition: Cumulative Review Exercises. The student edition of this book includes answers to all of the Cumulative Review Exercises. For the regular text sections, the student edition continues to include answers for the odd-numbered exercises only.

1. a. $\bar{x} = 2.40$; median $= 2.20$;
      mode: 0.0, 2.0, 2.1, 2.4, 4.4;
      midrange $= 2.25$
   b. $s = 1.29$; $s^2 = 1.67$;
      range $= 4.50$
   c. Continuous
   d. Ratio
2. a. Mode, because the other
      measures of central
      tendencies require
      calculations that cannot (or
      should not) be done with
      data at the nominal level
      of measurement.
   b. Convenience
   c. Cluster
3. No, the 50 values should be
   weighted according to the
   corresponding populations. A
   weighted mean should be
   computed using the
   populations as
   weights.

a. Find the mean, median, mode, and midrange.
b. Find the standard deviation, variance, and range.
c. Are the given scores from a population that is discrete or continuous?
d. What is the level of measurement of these scores? (Nominal, ordinal, interval, ratio)

2. a. A set of data is at the nominal level of measurement and you want to obtain a representative data value. Which of the following is most appropriate: mean, median, mode, or midrange? Why?
   b. A sample is obtained by telephoning the first 250 people listed in the local telephone directory. What type of sampling is being used? (random, stratified, systematic, cluster, convenience)
   c. An exit poll is conducted by surveying everyone who leaves the polling booth at 50 randomly selected election precincts. What type of sampling is being used? (random, stratified, systematic, cluster, convenience)

3. Each year, the United States Energy Department publishes an *Annual Energy Review* that includes the per capita energy consumption (in millions of Btu) for each of the 50 states. If you calculate the mean of these 50 values, is the result the mean per capita energy consumption for the population in all 50 states combined? If it is not, explain how you would calculate the mean per capita energy consumption for all 50 states combined.

Computer Project

It is commonly believed that the mean body temperature of healthy adults is 98.6°F. Refer to Data Set 2 in Appendix B and consider the body temperatures taken at midnight on the second day. Data Set 2 is not stored as a STATDISK or Minitab file, so use STATDISK or Minitab to enter the 106 temperatures and save them as a file named BODYTEMP. Proceed to obtain a histogram, boxplot, measures of central tendency, measures of variation, Q_1, Q_3, the minimum, and the maximum values. Use the results to describe important characteristics of the data set. Based on this sample, what do you conclude about the common belief that the mean body temperature is 98.6°F? Is this the result you would have expected?

•••••••• FROM DATA TO DECISION ••••••••

Garbage In, Insight Out

Refer to Data Set 1 in Appendix B. That data set consists of weights of different categories of garbage discarded by a sample of 62 households. The data were collected as part of the Garbage Project at the University of Arizona. With such data sets, there are often several different issues that can be addressed. In Chapter 9 we will consider the issue of whether there is some relationship between household size and amount of waste discarded so that we might be able to predict the size of the population of a region by analyzing the disposed garbage. For now, we will work with descriptive statistics based on the data.

a. Construct a Pareto chart and a pie chart depicting the relative amounts of the total weights of metal, paper, plastic, glass, food, yard waste, textile waste, and other waste. (Instead of frequencies, use the total weights.) Based on the results, which categories appear to be the largest components of the total amount of waste? Is there any single category that stands out as being the largest component?

b. A pie chart in *USA Today* depicted metal, paper, plastic, glass, food, yard waste, and other waste with the percents of 14%, 38%, 18%, 2%, 4%, 11%, and 13%, respectively. Do these percentages appear to be consistent with Data Set 1 in Appendix B?

c. For each category, find the mean and standard deviation, and construct a histogram of the 62 weights. Enter the results in the table below.

d. The amounts of garbage discarded are listed by weight. Many regions have household waste collected by commercial trucks that compress it, and charges at the destination are based on weight. Under these conditions, is the volume of the garbage relevant to the problem of community waste disposal? What other factors are relevant?

e. Based on the preceding results, if you had to institute conservation or recycling efforts because your region's waste facility was almost at full capacity, what would you do?

	Metal	Paper	Plastic	Glass	Food	Yard	Textile	Other
Mean								
Standard deviation								
Shape of distribution								

········ COOPERATIVE GROUP ACTIVITIES ········

1. *Out-of-class activity:* Are Estimates Influenced by Anchoring Numbers? In the article "Weighing Anchors" in *Omni* magazine, author John Rubin observed that when people estimate a value, their estimate is often "anchored" to (or influenced by) a preceding number, even if that preceding number is totally unrelated to the quantity being estimated. To demonstrate this, he asked people to give a quick estimate of the value of $8 \times 7 \times 6 \times 5 \times 4 \times 3 \times 2 \times 1$. The average answer given was 2250, but when the order of the numbers was reversed, the average became 512. Rubin explained that when we begin calculations with larger numbers (as in $8 \times 7 \times 6$), our estimates tend to be larger. He noted that both 2250 and 512 are far below the correct product, 40,320. The article suggests that irrelevant numbers can play a role in influencing real estate appraisals, estimates of car values, and estimates of the likelihood of nuclear war.

Conduct an experiment to test this theory. Select some subjects and ask them to quickly estimate the value of

$$8 \times 7 \times 6 \times 5 \times 4 \times 3 \times 2 \times 1$$

Then select other subjects and ask them to quickly estimate the value of

$$1 \times 2 \times 3 \times 4 \times 5 \times 6 \times 7 \times 8$$

Record the estimates along with the particular order used. Carefully design the experiment so that conditions are uniform and the two sample groups are selected in a way that minimizes any bias. Don't describe the theory to subjects until after they have provided their estimates. Compare the two sets of sample results by using the methods of this chapter. Provide a typewritten report that includes the data collected, the detailed methods used, the method of analysis, any relevant graphs and/or statistics, and a statement of conclusions. Include a critique of reasons why the results might not be correct and describe ways in which the experiment could be improved.

A variation of the preceding experiment is to survey people about their knowledge of the population of Kenya. First ask half of the subjects whether they think the population is above 5 million or below 5 million, then ask them to estimate the population with an actual number. Ask the other half of the subjects whether they think the population is above 80 million or below 80 million, then ask them to estimate the population. (Kenya's population is 28 million.) Compare the two sets of results and identify the "anchoring" effect of the initial number that the survey subjects are given.

(continued)

•••••••• COOPERATIVE GROUP ACTIVITIES ••••••••

2. *In-class activity:* In each group of three or four students, find the total value of the coins possessed by each individual member. Find the group mean and standard deviation, then exchange those statistics with the other groups. Using the group means as a separate data set, find the mean, standard deviation, and shape of the distribution. How do these results compare to the mean and standard deviation originally found in the group?

3. *In-class activity:* Given below are the ages of motorcyclists at the time they were fatally injured in traffic accidents (based on data from the U.S. Department of Transportation). If your objective is to dramatize the dangers of motorcycles for young people, which would be most effective: Histogram, Pareto chart, pie chart, dotplot, mean, median, . . . ? Construct the graph and find the statistic that best meets that objective. Is it okay to deliberately distort data if the objective is one such as saving lives of motorcyclists?

```
17  38  27  14  18  34  16  42  28  24  40  20  23  31
37  21  30  25  17  28  33  25  23  19  51  18  29
```

interview

Anthony DiUglio

Nuclear Analyst, Probabilistic Risk Assessment, Consolidated Edison Company of New York, Inc.

Anthony DiUglio works in the Probabilistic Risk Assessment (PRA) Group at Consolidated Edison's Indian Point Unit #2 nuclear generating facility in Buchanan, New York. In his work as a Nuclear Analyst, Tony develops probabilities that are used in quantifying various aspects of the plant-specific risk assessment. He is a former student of the author.

What is your job?

In PRA we are concerned with three basic questions about risk: What can happen, how likely is it to happen, and what are the consequences of its happening? We apply these questions about risk to the safe, reliable, and continuous operation of our power plant. When we quantify risk, we obtain numbers that are probabilities. If someone suggests a modification to a plant safety system, we analyze it from a risk perspective. Is the modification better for the system? Does it affect the operation of the plant or put the public health and safety at risk?

How do you use probability and/or statistics?

They're our primary tools. Our PRA requires that we quantify plant-specific repair rates for all safety-related components in our plant. In developing component-repair rates for pumps and valves, we look at industrywide data (generic) and our plant-specific data. We combine this information together, under uncertainty, and end up with component-specific repair probabilities.

How do you use probability and/or statistics in other departments at Indian Point?

Our Performance Department measures various plant parameters, such as heat rate, megawatt generation,

> ## "You can't effectively communicate here unless you use a common language, and that common language happens to be statistics."

cost per kilowatt of generation, etc. These parameters are all done by use of statistics. The statistical tools they use are data trending, normal statistical curves, standard deviations, histograms, etc. Financial Planning makes extensive use of statistics in projecting budget needs and determining its constraints. Our corporate forecasters use probability theory to predict power demands at different times during the year (e.g., winter and summer, one, three, and five years down the road). We have so many people using statistics in their everyday work that statistics has now become a tool for engineers, planners, forecasters, and those of us in Risk Assessment.

In terms of statistics, what would you recommend for prospective employees?

They should have a good understanding of probability, statistics, and their applications. Because PRA is still a relatively new area, we often deal with problems that haven't been addressed before, so many of the problems we address require creative problem solving. Once you have the basic tools, your time is efficiently spent. You can't effectively communicate here unless you use a common language, and that common language happens to be statistics.

Has your work been helpful in convincing the public that your plant is safe?

Safety is always our first concern. In the early 1980s there was a series of public hearings conducted by the Nuclear

Regulatory Commission (NRC) to discuss whether or not the plant should continue operation. Con Edison maintained that the plant was safe, and we were able to help justify the continued operation of our plant through the use of our PRA. At the conclusion of those hearings the NRC agreed with our position, and we continued to operate.

Who was your best math teacher?

Professor Mario Triola.

Is your use of probability and statistics increasing, decreasing, or remaining stable?

It's increasing all the time. We are very much involved with plant performance indicators as parameters for efficient plant operation. With PRA we now have a tool that allows us to focus our attention on the more important plant components and functions. In the case where three components all need maintenance, PRA allows us to identify which component should be returned to service first. In engineering, if we have several components that should be improved, PRA allows us to identify which component should be improved first. We can quantify the effects and thereby target our resources better, thus making the plant safer.

3

Probability

3-1 Overview

Chapter objectives are identified. The importance of probability is discussed, along with its role in basic statistical methods.

3-2 Fundamentals

The relative frequency and classical definitions of probability are both presented and illustrated. Methods are given for finding probabilities of simple events. The law of large numbers is described, and the complement of an event is defined and illustrated. Odds are also considered.

3-3 Addition Rule

The addition rule is described as a method for finding the probability that either one event *or* another event (or both events) occurs when an experiment is conducted. Mutually exclusive events are defined. In applying the addition rule, we avoid or correct for double counting of events that are not mutually exclusive. The rule of complementary events is also introduced.

3-4 Multiplication Rule

The multiplication rule is described as a method for finding the probability that one event occurs in one experiment *and* some other event occurs in another experiment. Independent events are defined. The probability of getting at least one outcome of some event is described. Conditional probability is defined and illustrated.

3-5 Probabilities Through Simulations

Probabilities can often be estimated from simulations that emulate experiments. Methods for creating simulations are described and illustrated.

3-6 Counting

The following important counting techniques are described: the fundamental counting principle, factorial rule, permutations rule (when all items are different), permutations rule (when some items are identical to others), and combinations rule. Such counting devices are used to find the total number of outcomes.

Chapter Problem

Do you have a better chance of being hit by lightning or winning the lottery?

Many of us take actions based on the likelihood of events occurring. Some of us fly in airplanes recognizing that while there is a chance of a crash, the likelihood of this happening is really quite small. Some of us sprint to our cars during a thunderstorm knowing that we could be struck by lightning, but, again, the likelihood of that event is actually quite small. Many of us buy lottery tickets knowing that we could win, although the likelihood of that event is very small, but do we really understand just how small? Is the chance of winning the lottery actually smaller than the chance of being struck by lightning? In this chapter we discuss *probability* as we determine specific ways of measuring the chance of various events. In particular, we will find probabilities for being struck by lightning and for winning a lottery. We will then know which event is more likely.

Some statistics instructors prefer to cover this chapter thoroughly, but some prefer to include only minimum coverage. It is strongly recommended that at least Section 3-2 be included. Normally, Sections 3-1 through 3-4 could be covered, with Sections 3-5 and 3-6 optional. No sections are marked as "optional" because some instructors experienced noticeable student resistance to studying sections labeled that way.

Many students find this chapter to be the most difficult one that they encounter in the book. Section 3-2 covers basic definitions and simple concepts of probability. Emphasize the importance of *understanding* the available data, because some students tend to develop this rule for finding probabilities: Find two numbers, then divide the smaller number by the larger one. For class examples, include at least one example where that rule doesn't work. For example, if a quality control test shows that there are 5 defective printers and 15 that are good, the probability of randomly selecting one that is defective is 5/20, not 5/15.

After covering the definitions given here, you might discuss the one die and two dice examples in detail. Carefully explain how a pair of dice yields the simple events of 1-1, 1-2, . . . , 6-6. Explain the difference between the simple event of 1-6 and the event of getting a total of 7. Students sometimes have difficulty understanding that 3-4 and 4-3 are different simple events; suggest that they consider one die to be green and the other red.

3-1 Overview

In Chapter 2 we described the method of *inferential statistics*, which uses sample evidence to make inferences or conclusions about a whole population. Inferential decisions are based on probabilities—or likelihoods—of events. Suppose, for example, that a check of your college's employee records reveals that of the last 100 people hired, all are men. The probability that 100 consecutive men would be selected with a hiring policy that is unbiased is so low that most reasonable people would conclude that men were favored. This example illustrates the following important principle, which will be a foundation of our reasoning in several of the chapters to come.

> **If, under a given assumption (such as a fair hiring practice), the probability of a particular sample (such as 100 hired men) is exceptionally small, we conclude that the assumption is probably not correct.**

Apart from its use in statistical methodology, probability theory is also becoming an increasingly important analytical tool in a society that must attempt to measure uncertainties. For example, before firing up a nuclear power plant, we should analyze the probability of a meltdown. Before arming a nuclear warhead, we should try to establish the probability of an accidental detonation. And before raising the speed limit on our nation's highways, we should try to estimate the probability of increased fatalities.

The primary objective of this chapter is to develop a sound understanding of probability values that will be used in subsequent chapters. A secondary objective is to develop the basic skills necessary to solve simple probability problems. These skills will be valuable in their own right as decision-making tools that better enable us to understand our world.

3-2 Fundamentals

In considering probability problems, we deal with experiments, events, and the collection of all possible outcomes.

DEFINITIONS

An **experiment** is any process that allows researchers to obtain observations.

An **event** is any collection of results or outcomes of an experiment.

A **simple event** is an outcome or an event that cannot be broken down any further.

The **sample space** for an experiment consists of all possible simple events. That is, the sample space consists of all outcomes that cannot be broken down any further.

For example, the rolling of a single die is an *experiment,* and the result of 3 is an *event.* The outcome of 3 is a *simple event* because it cannot be broken down any further, and the *sample space* consists of these simple events: 1, 2, 3, 4, 5, 6. As another example, the rolling of a pair of dice is an experiment, the result of 7 is an event, but 7 is not itself a simple event because it can be broken down into simpler events, such as 3-4 and 6-1. In the experiment of rolling a pair of dice, the sample space consists of 36 simple events: 1-1, 1-2, . . . , 6-6.

Although there is no universal agreement on how to define the probability of an event, two particular definitions are in common use. We first list some basic notation, then we present the two definitions of probability.

Notation for Probabilities

P denotes a probability. A, B, and C denote specific events. $P(A)$ denotes the probability of event A occurring.

Rule 1: Relative Frequency Approximation of Probability

Conduct (or observe) an experiment a large number of times, and count the number of times that event A actually occurs. Then $P(A)$ is *estimated* as follows:

$$P(A) = \frac{\text{number of times } A \text{ occurred}}{\text{number of times experiment was repeated}}$$

Rule 2: Classical Approach to Probability

Assume that a given experiment has n different simple events, each of which has an *equal chance* of occurring. If event A can occur in s of these n ways, then

$$P(A) = \frac{\text{number of ways } A \text{ can occur}}{\text{number of different simple events}} = \frac{s}{n}$$

It is very important to note that *the classical approach requires equally likely outcomes.* If the outcomes are not equally likely, we must use the relative frequency estimate. Figure 3-1 illustrates this important distinction.

How Probable?

How do we interpret such terms as *probable, improbable,* or *extremely improbable?* The FAA interprets these terms as follows. *Probable:* A probability on the order of 0.00001 or greater for each hour of flight. Such events are expected to occur several times during the operational life of each airplane. *Improbable:* A probability on the order of 0.00001 or less. Such events are not expected to occur during the total operational life of a single airplane of a particular type, but may occur during the total operational life of all airplanes of a particular type. *Extremely improbable:* A probability on the order of 0.000000001 or less. Such events are so unlikely that they need not be considered to ever occur.

Figure 3-1 Comparison of Relative Frequency and Classical Approaches

(a) Relative Frequency Approach (Rule 1): When trying to determine P(the tack lands point up), we must repeat the experiment of tossing the tack many times and then find the ratio of the number of times the tack lands with the point up to the number of tosses. That ratio is our estimate of the probability.

(b) Classical Approach (Rule 2): When trying to determine $P(2)$ with a balanced and fair die, each of the six faces has an equal chance of occurring.

$$P(2) = \frac{\text{number of ways 2 can occur}}{\text{total number of simple events}}$$

$$= \frac{1}{6}$$

Consider doing a class illustration of the law of large numbers: Simulate a family of four children by flipping a coin four times, with male = heads and female = tails. Ask each student to flip a coin four times and report the number of females in his or her simulated family. Some individual students will have all males or all females, but when the class results are combined, the proportion of females should be close to 0.5.

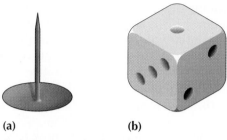

(a)   **(b)**

When finding probabilities with the relative frequency approach (Rule 1), we obtain an *approximation* instead of an exact value. As the total number of observations increases, the corresponding approximations tend to get closer to the actual probability. This property is stated as a theorem commonly referred to as the *law of large numbers*.

Law of Large Numbers
As an experiment is repeated again and again, the relative frequency probability (from Rule 1) of an event tends to approach the actual probability.

The law of large numbers tells us that the relative frequency approximations from Rule 1 tend to get better with more observations. This law reflects a simple notion supported by common sense: A probability estimate based on only a few trials can be off by substantial amounts, but with a very large number of trials, the estimate tends to be much more accurate. For example, if you conduct an opinion poll of only a dozen people your results could easily be in error by large amounts, but if you poll thousands of *randomly selected* people your sample results will be much closer to the true population values.

Figure 3-2 illustrates the law of large numbers by showing computer-simulated results. Note that as the number of births increases, the proportion of girls approaches the 0.5 value.

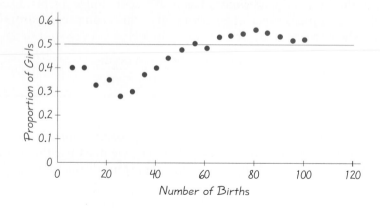

Figure 3-2 Illustration of the Law of Large Numbers

After examining Rules 1 and 2, it might seem that we should always use Rule 2 when an experiment has equally likely outcomes. It frequently happens, though, that such experiments are so complicated that the classical approach (Rule 2) is impractical to use. Instead, we can more easily get estimates of the desired probabilities by using the relative frequency approach (Rule 1). In such cases simulations are often helpful. (A *simulation* of an experiment is a process that behaves in the same ways as the experiment itself, thus producing similar results.) For example, it's much easier to use Rule 1 for estimating the probability of winning at solitaire—that is, to play the game many times (or to run a computer simulation)—than to perform the extremely complex calculations required with Rule 2.

The examples that follow are intended to illustrate the use of Rules 1 and 2. In some of these examples we use the term *random*. Recall these definitions from Section 1-4: In a **random sample of one element** from a population, all elements available for selection have the same chance of being chosen; a sample of *n* items is a **random sample** (or a simple random sample) if it is selected in such a way that every possible sample of *n* items from the population has the same chance of being chosen. The general concept of randomness is extremely important in statistics. When making inferences based on samples, we must have a sampling process that is representative, impartial, and unbiased. If a sample is not carefully selected, it may be totally worthless.

Shakespeare's Vocabulary

According to Bradley Efron and Ronald Thisted, Shakespeare's writings included 31,534 different words. They used probability theory to conclude that Shakespeare probably knew at least another 35,000 words that he didn't use in his writings. The problem of estimating the size of a population is an important problem often encountered in ecology studies, but the result given here is another interesting application. (See "Estimating the Number of Unseen Species: How Many Words Did Shakespeare Know?", in *Biometrika*, Vol. 63, No. 3.)

EXAMPLE   Find the probability that a randomly selected person will be struck by lightning this year.

SOLUTION   The sample space consists of these two simple events: The selected person is struck by lightning this year or is not. Because these simple events are not equally likely, we must use a relative frequency approximation, as described in Rule 1. It isn't practical to conduct experiments, but we can research past events. In a recent year, 371 people were struck by lightning in the United States. In a population of about 260 million, the probability of being struck by lightning in a year is estimated to be

$$\frac{371}{260,000,000} \approx \frac{1}{701,000}.$$

EXAMPLE   On an ACT or SAT test, a typical multiple-choice question has 5 possible answers. If you make a random guess on one such question, what is the probability that your response is wrong?   (*continued*)

SOLUTION   There are 5 possible outcomes or answers, and there are 4 ways to answer incorrectly. Random guessing implies that the outcomes in the sample space are equally likely, so we apply the classical approach (Rule 2) to get

$$P(\text{wrong answer}) = \frac{4}{5} = 0.8$$

In basic probability problems of the type we are now considering, it is very important to examine the available information carefully and to identify the total number of possible outcomes correctly. In some cases, the total number of possible outcomes is directly available, but in other cases we must process the available information to determine that total. The preceding example included the information that the total number of outcomes is 5, but the following example requires us to calculate the total number of possible outcomes.

EXAMPLE   The American Casualty Insurance Company studied causes of accidental deaths in the home and compiled a file consisting of 160 deaths caused by falls, 120 deaths caused by poisons, and 70 deaths caused by fires and burns. If one of these deaths is randomly selected, find the probability that it resulted from poison.

SOLUTION   The total number of accidental deaths is found by computing $160 + 120 + 70 = 350$. With random selection, the 350 deaths are equally likely and Rule 2 is applied as follows.

$$P(\text{poison}) = \frac{\text{number of poison deaths}}{\text{total number of deaths}} = \frac{120}{350} = 0.343$$

There is a 0.343 probability that when one of the deaths is randomly selected, it will be one caused by poison.

EXAMPLE   Find the probability that a couple with 3 children will have exactly 2 boys. Assume that boys and girls are equally likely and that the gender of any child is not influenced by the gender of any other child.

SOLUTION   We first list the sample space that identifies the 8 outcomes. Those outcomes are equally likely, so we use Rule 2. Of those 8 different possible outcomes, 3 correspond to exactly 2 boys, so

$$P(\text{2 boys in 3 births}) = \frac{3}{8} = 0.375$$

There is a 0.375 probability that if a couple has 3 children, exactly 2 will be boys.

Students often need help in constructing a sample space such as the one listing the genders of babies. Suggest a systematic approach to constructing such a table, possibly referring to these patterns that are present: Going from right to left, the columns alternate every line, then every 2 lines, then every 4 lines, and so on.

<u>1st</u> <u>2nd</u> <u>3rd</u>

boy-boy-boy

boy-boy-girl

boy-girl-boy

boy-girl-girl

girl-boy-boy

girl-boy-girl

girl-girl-boy

girl-girl-girl

exactly 2 boys

EXAMPLE   In choosing among several computer suppliers, a purchasing agent wants to know the probability of a personal computer breaking down during the first two years. What is that probability?

SOLUTION   There are only two outcomes: A personal computer either breaks down during the first two years or it does not. Because those two outcomes are not equally likely, the relative frequency approximation must be used. This requires that we somehow observe a large number of personal computers. A *PC World* survey of 4000 personal computer owners showed that 992 of them broke down during the first two years. (The computers broke down, not the owners.) On the basis of that result, we *estimate* that the probability is 992/4000, or 0.248.

EXAMPLE   If a year is selected at random, find the probability that Thanksgiving Day will be on a (a) Wednesday, (b) Thursday.

SOLUTION

a. Thanksgiving Day always falls on the fourth Thursday in November. It is therefore impossible for Thanksgiving to be on a Wednesday. When an event is impossible, we say that its probability is 0.

b. It is certain that Thanksgiving will be on a Thursday. When an event is certain to occur, we say that its probability is 1.

Because any event imaginable is impossible, certain, or somewhere in between, it is reasonable to conclude that the mathematical probability of any event is 0, 1, or a number between 0 and 1 (see Figure 3-3).

- **The probability of an impossible event is 0.**
- **The probability of an event that is certain to occur is 1.**
- $0 \le P(A) \le 1$ **for any event A.**

In Figure 3-3, the scale of 0 through 1 is shown on the left, whereas the more familiar and common expressions of likelihood are shown on the right.

Complementary Events

Sometimes we need to find the probability that an event A does *not* occur.

DEFINITION
The **complement** of event A, denoted by $\overline{A}$, consists of all outcomes in which event A does *not* occur.

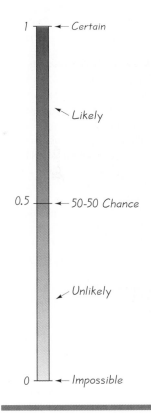

Figure 3-3   Possible Values for Probabilities

EXAMPLE   The Nike Corporation wants to test a new material to be used for making sneakers. A test group consists of 20 men and 30 women. If one person is randomly selected from this test group, find the probability of *not* getting a man.

SOLUTION   First, observe that the total sample space consists of 50 people. Second, because 20 of the 50 subjects are men, it follows that 30 of the 50 subjects are women, so

$$P(\text{not selecting a man}) = P(\overline{\text{man}})$$
$$= P(\text{woman})$$
$$= \frac{30}{50} = 0.6$$

Although it is difficult to develop a universal rule for rounding off probabilities, the following guide will apply to most problems in this text.

Rounding Off Probabilities

When expressing the value of a probability, either give the *exact* fraction or decimal or round off final decimal results to three significant digits. (*Suggestion:* When a probability is not a simple fraction such as 2/3 or 5/9, express it as a decimal.)

All the digits in a number are significant except for the zeros that are included for proper placement of the decimal point.

EXAMPLES

- The probability of 0.0000128506 has six significant digits (128506), and it can be rounded to three significant digits as 0.0000129.
- The probability of 1/3 can be left as a fraction or rounded in decimal form to 0.333, but not 0.3.
- The probability of heads in a coin toss can be expressed as 1/2 or 0.5; because 0.5 is exact, there's no need to express it as 0.500.
- The fraction 7659/32785 is exact, but its value isn't obvious, so express it as the decimal 0.234.

An important concept of this section is the mathematical expression of probability as a number between 0 and 1. This type of expression is fundamental and common in statistical procedures, and we will use it throughout the remainder of this text. A typical computer output, for example, may include a "*P*-value" expression such as "significance less than 0.001." We will discuss the meaning of *P*-values later, but they are essentially probabilities of the type discussed in this section. For now, you should recognize that a probability of 0.001 (equivalent to 1/1000) corresponds to an event so rare that it occurs an average of only once in a thousand trials.

Subjective Probabilities

This section presented the relative frequency approach and the classical approach as two formal methods for finding probabilities of events, but another approach is simply to guess or estimate a probability. The technique of guessing is familiar to those of us who are sometimes not as well prepared for a test as we would like, but it is also used by professionals who set casino odds for sporting events. Subjective probabilities are also used by insurance experts who estimate probabilities for special circumstances, such as the injury of a rock star that causes a tour to be canceled. A guessed or estimated probability based on knowledge of relevant circumstances is commonly referred to as a **subjective probability.** For example, a Las Vegas oddsmaker might estimate that there is a 0.05 probability that the New York Giants will win the Super Bowl next year. That estimate is based on knowledge of relevant factors, such as the abilities of the coach and players.

Odds

Expressions of likelihood are often given as *odds,* such as 50:1 (or "50 to 1"). A serious disadvantage of odds is that they make many calculations extremely difficult. As a result, statisticians, mathematicians, and scientists prefer to use probabilities. The advantage of odds is that they make it easier to deal with money transfers associated with gambling, so they tend to be used in casinos, lotteries, and race tracks. First, we should know that the likelihood of some event can be expressed in terms of the odds against or in favor of that event.

Subjective Probabilities at the Racetrack

Researchers studied the ability of racetrack bettors to develop realistic subjective probabilities. (See "Race-track Betting: Do Bettors Understand the Odds?", by Brown, D'Amato, and Gertner, *Chance* magazine, Vol. 7, No. 3.) After analyzing results for 4400 races, they concluded that although bettors slightly overestimate the winning probabilities of "longshots" and slightly underestimate the winning probabilities of "favorites," their general performance is quite good. The subjective probabilities were calculated from the payoffs, which are based on the amounts bet, and the actual probabilities were calculated from the actual race results.

> ### DEFINITION
>
> The **odds against** event *A* occurring are the ratio $P(\overline{A})/P(A)$, usually expressed in the form of *a:b* (or "*a* to *b*"), where *a* and *b* are integers having no common factors.
>
> The **odds in favor** of event *A* are the reciprocal of the odds against that event. If the odds against *A* are *a:b*, then the odds in favor are *b:a*.

As an example, if $P(A) = 2/5$, then

$$\text{odds against } A = \frac{P(\overline{A})}{P(A)} = \frac{3/5}{2/5} = \frac{3}{2}$$

We express this as 3:2 or "3 to 2." Because the odds against event A are 3:2, the odds in favor are 2:3. See Exercise 31 for conversions from odds into probabilities.

For bets, the odds against an event represent the ratio of net profit to the amount bet.

Odds against event A = (net profit):(amount bet)

Suppose a bet pays 50:1. If the odds aren't specified as being in favor or against, they are probably the odds against the event occurring. If a minor miracle occurs and you win this 50:1 bet, you would make a profit of $50 for each $1 bet. For example, if you bet $2, your net profit would be $100. You would collect a total of $102, which includes the $100 net profit and the original $2 bet.

3-2 Exercises A: Basic Skills and Concepts

1. Which of the following values *cannot* be probabilities?   1. $-0.2, 3/2, \sqrt{2}$
$$0, 0.0001, -0.2, 3/2, 2/3, \sqrt{2}, \sqrt{0.2}$$

2. a.   What is $P(A)$ if event A is that February has 30 days this year?   a. 0
   b.   What is $P(A)$ if event A is that November has 30 days this year?   b. 1
   c.   A sample space consists of 500 separate events that are equally likely. What is the probability of each?   c. 1/500
   d.   On a college entrance exam, each question has 5 possible answers. If you make a random guess on the first question, what is the probability that you are correct?   d. 1/5

3. Find the probability that when a coin is tossed, the result is heads.   3. 1/2

4. We noted in this section that for the experiment of rolling a pair of dice, there are 36 simple events that form the sample space: 1-1, 1-2, . . . , 6-6. Find the probability of rolling a pair of dice and getting a total of 4.

5. Refer to Data Set 11 in Appendix B. Based on those sample results, estimate the probability that when a plain M&M candy is randomly selected, it will be red.   5. 0.210

6. Refer to Data Set 8 in Appendix B. Based on those sample results, estimate the probability that a randomly selected statistics student has at least one credit card.   6. 0.550

Recommended assignment: Exercises 1–4, 6, 8, 10, 13, 14, 20, 24, 30

4. 1/12 (reduced from 3/36)

7. A study of 500 randomly selected American Airlines flights showed that 430 arrived on time (based on data from the Department of Transportation). What is the estimated probability of an American Airlines flight arriving on time? Would you describe the result as being very good?

7. 0.860; yes

8. The Kelly-Lynne Advertising Company is considering a computer campaign that targets teenagers. In a survey of 1066 teens, 181 had a computer on-line service in their household. If a teen is randomly selected, estimate the probability that he or she will have access to an on-line service in his or her household. Would you advise this company to use a computer advertising campaign?

8. 0.170; yes, it could be an inexpensive way to reach 17% of teenagers.

9. In a survey of college students, 1162 stated that they cheated on an exam and 2468 stated that they did not (based on data from the Josephson Institute of Ethics). If one of these college students is randomly selected, find the probability that he or she cheated on an exam.   9. 0.320

10. In a study of Americans over 65 years of age, it is found that 255 have Alzheimer's disease and 2302 do not (based on data from the Alzheimer's Association). If an American over 65 years of age is randomly selected, what is the estimated probability that he or she has Alzheimer's disease? Based on that probability, is Alzheimer's disease a major concern for those over 65 years of age?   10. 0.0997; yes

11. In a study of blood donors, 225 were classified as group O and 275 had a classification other than group O (based on data from the Greater New York Blood Program). What is the approximate probability that a person will have group O blood?   11. 0.450

12. A Nielsen survey of 3857 households shows that 463 have their televisions tuned to CBS on Monday night between 10:00 P.M. and 10:30 P.M. If a household is randomly selected, estimate the probability of getting one tuned to CBS in that time slot.   12. 0.120

13. a. If a person is randomly selected, find the probability that his or her birthday is October 18, which is National Statistics Day in Japan. Ignore leap years.   13. a. 1/365

b. If a person is randomly selected, find the probability that his or her birthday is in November. Ignore leap years.   b. 6/73 (reduced from 30/365)

14. In a study of brand recognition, 831 consumers knew of Campbell's Soup, and 18 did not (based on data from Total Research Corporation). Use these results to estimate the probability that a randomly selected consumer will recognize Campbell's Soup. How do you think this probability value compares to typical values for other brand names?   14. 0.979

15. In a Bruskin-Goldring Research poll, respondents were asked how a fruitcake should be used. One hundred thirty-two respondents indicated that it should be used for a doorstop, and 880 other respondents cited other uses, including birdfeed, landfill, and a gift. If one of these respondents is randomly selected, what is the probability of getting someone who would use a fruitcake for a doorstop?   15. 0.130

16. The U.S. General Accounting Office recently tested the IRS for correctness of answers to taxpayers' questions. For 1733 trials, the IRS was correct

1107 times. Use these results to estimate the probability that a random taxpayer's question will be answered correctly. Based on the result, would you say that the IRS does a good job answering taxpayers' questions correctly?   16. 0.639

17. Among 400 randomly selected drivers in the 20–24 age bracket, 136 were in a car accident during the last year (based on data from the National Safety Council). If a driver in that age bracket is randomly selected, what is the approximate probability that he or she will be in a car accident during the next year? Is the resulting value high enough to be of concern to those in the 20–24 age bracket?   17. 0.340; yes

18. Data provided by the Bureau of Justice Statistics revealed that for a representative sample of convicted burglars, 76,000 were jailed, 25,000 were put on probation, and 2000 received other sentences. Use these results to estimate the probability that a convicted burglar will serve jail time. Does it seem that the result is high enough to deter burglars?18. 0.738; apparently not

19. When the allergy drug Seldane was clinically tested, 70 people experienced drowsiness, and 711 did not (based on data from Merrell Dow Pharmaceuticals, Inc.). Use this sample to estimate the probability of a Seldane user becoming drowsy. Based on the result, is drowsiness a factor that should be considered by Seldane users?   19. 0.0896

20. 0.0000280; no

20. According to the U.S. Department of Transportation, American Airlines boarded 59,377,306 passengers in a recent year. In that same year, 82,796 passengers were voluntarily denied boarding, while 1664 other passengers were involuntarily denied boarding. Find the probability that a randomly selected passenger will be involuntarily denied boarding. Does the resulting value indicate that you should be concerned about being involuntarily denied boarding when you fly to a resort for spring break?

Method of Fraud	Number
Stolen card	243
Counterfeit card	85
Mail/phone order	52
Other	46

21. A study of credit-card fraud was conducted by MasterCard International, and the accompanying table is based on the results. If one case of credit-card fraud is randomly selected from the cases summarized in the table, find the probability that the fraud resulted from a counterfeit card.   21. 0.200

22. A Gallup survey resulted in the sample data in the given table. If one of the respondents is randomly selected, find the probability of getting someone who brushes three times per day, as dentists recommend.

Tooth-Brushings per Day	Number
1	228
2	672
3	240

23. A couple plans to have 2 children.
   a. List the different outcomes according to the gender of each child. Assume that these outcomes are equally likely.   23. a. bb, bg, gb, gg
   b. Find the probability of getting 2 girls.   b. 1/4
   c. Find the probability of getting exactly 1 child of each gender.   c. 1/2

24. A couple plans to have 4 children.
   a. List the 16 different possible outcomes according to the gender of each child. Assume that these outcomes are equally likely.
   b. Find the probability of getting all girls.   24. b. 1/16
   c. Find the probability of getting at least 1 child of each gender.
   d. Find the probability of getting exactly 2 children of each gender.

22. 0.211

24. c. 7/8

24. d. 3/8

25. On a quiz consisting of 3 true/false questions, an unprepared student must guess at each one. The guesses will be random. 25. b. 1/8   c. 1/8   d. 1/2
    a. List the different possible solutions.
    b. What is the probability of answering all 3 questions correctly?
    c. What is the probability of guessing incorrectly for all questions?
    d. What is the probability of passing the quiz by guessing correctly for at least 2 questions?

26. Both parents have the brown/blue pair of eye-color genes, and each parent contributes one gene to a child. Assume that if the child has at least one brown gene, that color will dominate and the eyes will be brown. (Actually, the determination of eye color is somewhat more complex.)
    a. List the different possible outcomes. Assume that these outcomes are equally likely.
    b. What is the probability that a child of these parents will have the blue/blue pair of genes? 26. b. 1/4   c. 3/4
    c. What is the probability that the child will have brown eyes?

27. Find the odds against correctly guessing the answer to a multiple-choice question with five possible answers. 27. 4:1

28. Find the odds against randomly selecting someone who is left-handed, given that 10% of us are left-handed. 28. 9:1

29. a. The probability of a 7 in roulette is 1/38. Find the odds against 7. 29. a. 37:1
    b. If you bet $2 on the number 7 in roulette and you win, the casino gives you $72, which includes the $2 bet. First identify the net profit, then find the odds used for determining the payoff. b. 35:1
    c. How do you explain the discrepancy between the odds in parts (a) and (b)? c. The casino pays off at 35:1 so that it will make a profit, instead of paying off at 37:1, which would not yield a profit.

30. a. In the casino game craps, you can bet that the next roll of the two dice will result in a total of 2. The probability of rolling 2 is 1/36. Find the odds against rolling 2. 30. a. 35:1
    b. If you bet $5 that the next roll of dice will be 2, you will collect $155 (including your $5 bet) if you win. First identify the net profit, then find the odds used for determining the payoff. b. 30:1
    c. How do you explain the discrepancy between the odds in parts (a) and (b)? c. The casino pays off at 30:1 so that it will make a profit, instead of paying off at 35:1, which would not yield a profit.

3–2 Exercises B: Beyond the Basics

31. If the odds against event A are $a:b$, then $P(A) = b/(a + b)$. Find the probability of Horse Cents winning his next race, given that the odds against his winning are 10:3. 31. 3/13

32. The odds against Lazy Lady winning her next race are 9:2. Find the probability that Lazy Lady wins her next race. (See Exercise 31.) 32. 2/11

```
0. | 00
1. | 0578
2. | 00113449
3. | 347
4. | 445
```

IVLIVS CAESAR

Probabilities that Challenge Intuition

In certain cases, our subjective estimates of probability values are dramatically different from the actual probabilities. Here is a classic example: If you take a deep breath, there is better than a 99% chance that you will inhale a molecule that was exhaled in dying Caesar's last breath. In that same morbid and unintuitive spirit, if Socrates' fatal cup of hemlock was mostly water, then the next glass of water you drink will likely contain one of those same molecules. Here's another less morbid example that can be verified: In classes of 25 students, there is better than a 50% chance that at least 2 students will share the same birthday.

33. The stem-and-leaf plot summarizes the time (in hours) that managers spend on paperwork in one day (based on data from Adia Personnel Services). Use this sample to estimate the probability that a randomly selected manager spends more than 2.0 hours per day on paperwork.   33. 0.600

34. After collecting IQ scores from hundreds of subjects, a boxplot is constructed with this 5-number summary: 82, 91, 100, 109, 118. If one of the subjects is randomly selected, find the probability that his or her IQ score is greater than 109.   34. 0.250

35. In part (a) of Exercise 13, leap years were ignored in finding the probability that a randomly selected person will have a birthday on October 18.
    a. Recalculate this probability, assuming that a leap year occurs every 4 years. (Express your answer as an exact fraction.)   35. a. 4/1461
    b. Leap years occur in years evenly divisible by 4, except they are skipped in 3 of every 4 centesimal years (years ending in 00). The years 1700, 1800, and 1900 were not leap years, but 2000 is a leap year. Find the exact probability for this case, and express it as a fraction.   b. 400/146,097

36. a. If 2 flies land on an orange, find the probability that they are on points that are within the same hemisphere.   36. a. 1
    b. Two points along a straight stick are randomly selected. The stick is then broken at those 2 points. Find the probability that the 3 resulting pieces can be arranged to form a triangle. (This is a difficult problem.)   b. 1/4

3-3 Addition Rule

The main objective of this section is to introduce the addition rule as a device for finding $P(A \text{ or } B)$, the probability that either event A occurs or event B occurs (or they both occur) as the single outcome of an experiment. The key word to remember is *or*. (Throughout this text we use the inclusive *or*, which means either one or the other or both. Except for Exercise 27, we will not consider the exclusive *or*, which means either one or the other but not both.) The preceding section considered events categorized as simple because they involved only one outcome, generally denoted by A. Many real situations, however, involve *compound events*, such as the random selection of a consumer who is a woman or who is younger than age 40.

DEFINITION

A **compound event** is any event combining two or more simple events.

Notation for Addition Rule
$P(A \text{ or } B) = P(\text{event } A \text{ occurs or event } B \text{ occurs or they both occur})$

Let's begin with a simple example. Survey subjects are often chosen by using computers to randomly select the last digits of telephone numbers. If we randomly select one of the numbers 0, 1, 2, 3, 4, 5, 6, 7, 8, 9 for the last digit of a telephone number, what is the probability that the selected number will be 0 or 1? Using Rule 2 from Section 3-2, we see that the event of getting 0 or 1 can occur in 2 different ways, and the total number of possible outcomes is 10, so

$$P(0 \text{ or } 1) = \frac{2}{10} = 0.2$$

This example seems to suggest a general rule whereby we simply add the number of outcomes corresponding to each of the events in question. Before declaring any such rule, though, let's consider another example. Let's again make a random selection from {0, 1, 2, . . . , 9}, but let's find the probability of getting an outcome that is odd or above 6. Note that among the 10 possible outcomes, 5 are odd (1, 3, 5, 7, 9) and 3 are above 6 (7, 8, 9). When we count the number of outcomes that are odd or above 6, we must be careful to avoid counting any numbers more than once. There are actually 6 separate outcomes that are odd or above 6: 1, 3, 5, 7, 8, 9. The correct probability is therefore determined as follows.

$$P(\text{odd or above 6}) = \frac{6}{10} = 0.6$$

Here is the key point: When finding the probability that event A occurs or event B occurs, find the total of the number of ways A can occur and the total number of ways B can occur, but *find those totals in such a way that no outcome is counted more than once.* One approach is to combine the number of ways event A can occur with the number of ways event B can occur and, if there is any overlap, compensate by subtracting the number of outcomes that are counted twice, as in the following rule.

Formal Addition Rule
$$P(A \text{ or } B) = P(A) + P(B) - P(A \text{ and } B)$$
where $P(A \text{ and } B)$ denotes the probability that A and B both occur at the same time as an outcome in one experiment.

Although the formal addition rule is presented as a formula, it is generally better to understand the spirit of the rule and apply it intuitively, as follows.

Two comments about notation: In this context, $P(A \text{ and } B)$ means that A and B both occur in the same trial; in Section 3-4 we will use $P(A \text{ and } B)$ for event A on one trial followed by event B on another trial. Also, here is the notation used by The College Board for its AP Statistics exam:

$$P(A \cup B) = P(A) + P(B) - P(A \cap B)$$

Total Area = 1

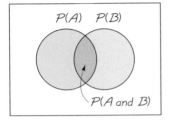

Figure 3-4   Venn Diagram Showing Overlapping Events

Total Area = 1

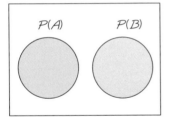

Figure 3-5   Venn Diagram Showing Nonoverlapping Events

Intuitive Addition Rule

To find $P(A \text{ or } B)$, find the sum of the number of ways event A can occur and the number of ways event B can occur, *adding in such a way that every outcome is counted only once*. $P(A \text{ or } B)$ is equal to that sum, divided by the total number of outcomes.

Figure 3-4 shows a Venn diagram that provides a visual illustration of the formal addition rule. In this figure we can see that the probability of A or B equals the probability of A (left circle) plus the probability of B (right circle) minus the probability of A and B (football-shaped middle region). This figure shows that the addition of the areas of the two circles will cause double counting of the football-shaped middle region. This is the basic concept that underlies the addition rule. Because of the relationship between the addition rule and the Venn diagram shown in Figure 3-4, the notation $P(A \cup B)$ is often used in place of $P(A \text{ or } B)$. Similarly, the notation $P(A \cap B)$ is often used in place of $P(A \text{ and } B)$, so the formal addition rule can be expressed as

$$P(A \cup B) = P(A) + P(B) - P(A \cap B)$$

The addition rule is simplified whenever A and B cannot occur simultaneously, so $P(A \text{ and } B)$ becomes zero. Figure 3-5 illustrates that with no overlapping of A and B, we have $P(A \text{ or } B) = P(A) + P(B)$. The following definition formalizes the lack of overlapping shown in Figure 3-5.

DEFINITION

Events A and B are **mutually exclusive** if they cannot occur simultaneously.

The flowchart of Figure 3-6 shows how mutually exclusive events affect the addition rule.

We can sometimes greatly improve our understanding of data as well as our ability to solve problems by reorganizing the data in a more helpful format. Consider this statement: In a test of the allergy drug Seldane, 49 of the 781 Seldane users experienced headaches, 49 of the 665 placebo users experienced headaches, and 24 of the 626 people in the control group experienced headaches. (These results are based on data from Merrell Dow Pharmaceuticals, Inc.) In this verbal statement the data are somewhat difficult to comprehend, but they are much clearer if they are reorganized in tabular form (see Table 3-1). The following examples use the data from Table 3-1.

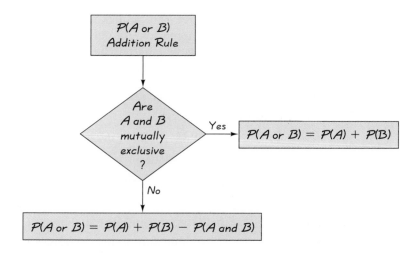

Figure 3-6 Applying the Addition Rule

TABLE 3-1	Test of Seldane			
	Seldane	Placebo	Control Group	**Total**
Headache	49	49	24	**122**
No headache	732	616	602	**1950**
Total	**781**	**665**	**626**	**2072**

Comment that a table arranged in the general format of Table 3-1 is called a "two-way" table or "contingency" table, and such tables are discussed in Section 10-3 of this book. Such tables are very important because they are often used to summarize survey results.

EXAMPLE   If one of the 2072 subjects represented in Table 3-1 is randomly selected, find the probability of getting someone who used a placebo or was in the control group.

SOLUTION   The probability we seek can be denoted by P(placebo or control). Table 3-1 shows that the subjects who used the placebo and those in the control group are mutually exclusive. That is, there is no overlap between those two groups. Consequently, the total number of people who used a placebo or were in the control group is $665 + 626 = 1291$, and the probability we seek is $1291/2072 = 0.623$.

$$P(\text{placebo or control}) = \frac{665}{2072} + \frac{626}{2072} = \frac{1291}{2072} = 0.623$$

EXAMPLE   If one of the 2072 subjects represented in Table 3-1 is randomly selected, find the probability of getting someone who used Seldane or did not experience a headache.   *(continued)*

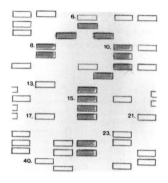

SOLUTION   The probability we seek can be denoted by P(Seldane or no headache). Table 3-1 shows that there is overlap between the group of Seldane users and the group of subjects who had no headache. That is, the events are not mutually exclusive and we must be careful to avoid double counting when we calculate our sums. The intuitive approach is to simply add the Seldane column and the "no headache" row in such a way that the entry of 732 is counted only once.

$$P(\text{Seldane or no headache}) = \frac{49 + 732 + 616 + 602}{2072} = \frac{1999}{2072} = 0.965$$

The same result could be obtained by applying the formal addition rule as follows.

$$P(\text{Seldane or no headache}) = P(\text{Seldane}) + P(\text{no headache})$$
$$- P(\text{Seldane and no headache})$$

$$= \frac{781}{2072} + \frac{1950}{2072} - \frac{732}{2072} = \frac{1999}{2072} = 0.965$$

We can summarize the key points of this section as follows:

1. To find $P(A \text{ or } B)$, begin by associating *or* with addition.
2. Consider whether events A and B are mutually exclusive; in other words, can they happen at the same time? If they are not mutually exclusive (that is, if they can happen at the same time), be sure to avoid (or at least compensate for) double counting when adding the relevant probabilities. If you understand the importance of not double counting when you find $P(A \text{ or } B)$, you don't necessarily have to calculate $P(A) + P(B) - P(A \text{ and } B)$.

Errors made when applying the addition rule often involve double counting. That is, events that are not mutually exclusive are treated as if they were. One indication of such an error is a total probability that exceeds 1; however, errors involving the addition rule do not always cause the total probability to exceed 1.

Complementary Events

In Section 3-2 we defined the complement of event A and denoted it by $\overline{A}$. The definition of complementary events implies that they must be mutually exclusive because it is impossible for an event both to occur and to *not* occur at the same time. Also, we can be absolutely certain that A either does or does not occur. That is, either A or $\overline{A}$ must occur. These observations enable us to apply the addition rule for mutually exclusive events as follows:

$$P(A \text{ or } \overline{A}) = P(A) + P(\overline{A}) = 1$$

We justify $P(A \text{ or } \overline{A}) = P(A) + P(\overline{A})$ by noting that A and $\overline{A}$ are mutually exclusive; we justify the total of 1 by our absolute certainty that A either does or does not occur. This result of the addition rule leads to the following three equivalent forms.

Rule of Complementary Events

$$P(A) + P(\overline{A}) = 1$$
$$P(\overline{A}) = 1 - P(A)$$
$$P(A) = 1 - P(\overline{A})$$

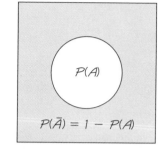

Figure 3-7   Venn Diagram for the Complement of Event A

The first form comes directly from our original result. The second (see Figure 3-7) and third variations involve very simple equation manipulations.

EXAMPLE   If $P(\text{rain}) = 0.4$, find $P(\text{no rain})$.

SOLUTION   Using the rule of complementary events, we get

$$P(\text{no rain}) = 1 - P(\text{rain}) = 1 - 0.4 = 0.6$$

A major advantage of the *rule of complementary events* is that its use can greatly simplify certain problems. We will illustrate this advantage in the following section.

3-3   Exercises A: Basic Skills and Concepts

For each part of Exercises 1 and 2, are the two events mutually exclusive for a single experiment?

Recommended assignment: Exercises 2, 4, 5, 9–12, 17–24.

1. **a.** Selecting a male television viewer   1. a. No
   Selecting someone who rarely uses the TV remote control
   b. Selecting a survey subject who is a registered Democrat   b. No
   Selecting a survey subject opposed to all welfare plans
   c. Spinning a roulette wheel and getting an outcome of 7   c. Yes
   Spinning a roulette wheel and getting an even number
2. **a.** Buying a new Corvette that is free of defects   2. a. Yes
   Buying a car with inoperative headlights
   b. Selecting a math course   b. No
   Selecting a course that is interesting
   c. Selecting a person with blond hair (natural or otherwise)   c. Yes
   Selecting a person who is bald

3. **a.** If $P(A) = 2/5$, find $P(\overline{A})$.   3. a. 3/5
   b. Based on recent data from the U.S. National Center for Health Statistics, the probability of a baby being a boy is 0.513. Find the probability of a baby being a girl.   b. 0.487
4. **a.** Find $P(\overline{A})$, given that $P(A) = 0.228$.   4. a. 0.772
   b. Based on data from the National Conference of Bar Examiners, if you randomly select someone who takes the bar exam, the probability of getting someone who passes is 0.57. Find the probability of getting someone who fails it.   b. 0.43

5. a. 4/13 (reduced from 16/52)
   b. 2/13 (reduced from 8/52)

5. When playing blackjack with a single deck of cards at the Stardust casino in Las Vegas, you are dealt the first card from the top of a shuffled deck. What is the probability that you get (a) a club or an ace? (b) an ace or a 2?
6. If someone is randomly selected, find the probability that his or her birthday is not October 18, which is National Statistics Day in Japan. Ignore leap years.   6. 364/365
7. Refer to Table 3-1 in this section. If one of the 2072 subjects is randomly selected, find the probability of getting someone who used Seldane or a placebo.   7. 0.698
8. Refer to Table 3-1 in this section. If one of the 2072 subjects is randomly selected, find the probability of getting someone who used a placebo or experienced a headache.   8. 0.356
9. Pollsters are concerned about declining levels of cooperation among persons contacted in surveys. A pollster contacts 84 people in the 18–21 age bracket and finds that 73 of them respond and 11 refuse to respond. When 275 people in the 22–29 age bracket are contacted, 255 respond and 20 refuse to respond (based on data from "I Hear You Knocking but You Can't Come In," by Fitzgerald and Fuller, *Sociological Methods and Research*, Vol. 11, No. 1). Assume that 1 of the 359 people is randomly selected. Find the probability of getting someone in the 18–21 age bracket or someone who refused to respond.   9. 0.290

10. 0.944

10. Refer to the same data set from Exercise 9, and find the probability of getting someone who is in the 18–21 age bracket or someone who responded.
11. Problems of sexual harassment received much attention in recent years. In one survey, 420 workers (240 of whom are men) considered a friendly pat on the shoulder to be a form of harassment, whereas 580 workers (380 of whom are men) did not consider that to be a form of harassment (based on data from Bruskin/Goldring Research). If one of the surveyed workers is randomly selected, find the probability of getting someone who does not consider a pat on the shoulder to be a form of harassment.   11. 0.580
12. Refer to the same data set in Exercise 11, and find the probability of randomly selecting a man or someone who does not consider a pat on the shoulder to be a form of harassment.   12. 0.820
13. A study of consumer smoking habits includes 200 married people (54 of whom smoke), 100 divorced people (38 of whom smoke), and 50 adults who never married (11 of whom smoke) (based on data from the U.S.

Department of Health and Human Services). If 1 subject is randomly selected from this sample, find the probability of getting someone who is divorced or smokes.   13. 0.471

14. Refer to the data in Exercise 13, and find the probability of getting someone who was never married or does not smoke.   14. 0.737

In Exercises 15 and 16, use the data in the accompanying table, which summarizes a sample of 200 times (in minutes) between eruptions of the Old Faithful Geyser in Yellowstone National Park.

Time	Frequency
40–49	8
50–59	44
60–69	23
70–79	6
80–89	107
90–99	11
100–109	1

15. Visitors to Yellowstone naturally want to see Old Faithful erupt, so the time interval between eruptions becomes a concern for those with time constraints. If we randomly select one of the times represented in the table, what is the probability that it is at least one hour?   15. 0.740

16. If we randomly select one of the times represented in the table, what is the probability that it is at least 70 min or between 60 and 79 min?   16. 0.740

In Exercises 17–24, refer to the accompanying figure, which describes the blood groups and Rh types of 100 people (based on data from the Greater New York Blood Program). In each case, assume that 1 of the 100 subjects is randomly selected, and find the indicated probability.

17. P(not group O)   17. 0.550
18. P(not type Rh^+)   18. 0.140
19. P(group B or type Rh^-)   19. 0.220
20. P(group O or group A)   20. 0.850
21. P(type Rh^-)   21. 0.140
22. P(group A or type Rh^+)   22. 0.910
23. P(group AB or type Rh^-)   23. 0.180
24. P(group A or B or type Rh^+)
   24. 0.930

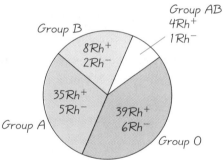

3-3 Exercises B: Beyond the Basics

25. a. If $P(A \text{ or } B) = 1/3$, $P(B) = 1/4$, and $P(A \text{ and } B) = 1/5$, find $P(A)$.   25. a. 17/60
   b. If $P(A) = 0.4$ and $P(B) = 0.5$, what is known about $P(A \text{ or } B)$ if A and B are mutually exclusive events?   b. $P(A \text{ or } B) = 0.9$
   c. If $P(A) = 0.4$ and $P(B) = 0.5$, what is known about $P(A \text{ or } B)$ if A and B are not mutually exclusive events?   c. $P(A \text{ or } B) < 0.9$

26. If events A and B are mutually exclusive and events B and C are mutually exclusive, must events A and C be mutually exclusive? Give an example supporting your answer.   26. No

27. How is the addition rule changed if the *exclusive or* is used instead of the *inclusive or*? Recall that the *exclusive or* means either one or the other, but not both.   27. $P(A \text{ or } B) = P(A) + P(B) - 2P(A \text{ and } B)$

28. $P(A \text{ or } B \text{ or } C) = P(A) +$
    $P(B) + P(C) - P(A \text{ and } B)$
    $- P(A \text{ and } C) - P(B \text{ and } C)$
    $+ P(A \text{ and } B \text{ and } C)$

Key points of this section:
$P(A \text{ and } B)$ suggests multiplication; adjust probabilities for *dependent* events. *Notation:* In this section, $P(A \text{ and } B)$ denotes that event A occurs in one trial and event B occurs in another trial; in Section 3-3 we used $P(A \text{ and } B)$ to denote that events A and B both occur in the same trial.

28. Given that $P(A \text{ or } B) = P(A) + P(B) - P(A \text{ and } B)$, develop a formal rule for $P(A \text{ or } B \text{ or } C)$. (*Hint:* Draw a Venn diagram.)

 Multiplication Rule

The main objective in Section 3-3 was to develop a rule for finding $P(A \text{ or } B)$, the probability that a trial in an experiment has an outcome of A or B or both. The main objective in this section is to develop a rule for finding $P(A \text{ and } B)$, the probability that event A occurs in a first trial and event B occurs in a second trial. In Section 3-3 we associated *or* with addition; in this section we will associate *and* with multiplication. We will see that $P(A \text{ and } B)$ involves multiplication of probabilities and that we must sometimes adjust the probability of event B to reflect the outcome of event A.

Probability theory is used extensively in the analysis and design of standardized tests, such as the SAT, ACT, LSAT (for law), and MCAT (for medicine). For ease of grading, such tests typically use true/false or multiple-choice questions. Let's assume that the first question on a test is a true/false type, while the second question is multiple choice with 5 possible answers (a, b, c, d, e). We will use the following two questions. Try them!

1. True or false: Smoking is one of the leading causes of cancer.

2. The Pearson correlation coefficient is named after

   a. Karl Marx

   b. Carl Friedrich Gauss

   c. Karl Pearson

   d. Carly Simon

   e. Mario Triola

When standardized tests are graded, compensation is usually made for guessing, so let's find the probability that if someone guesses the answers to both questions, the first answer will be correct *and* the second answer will be correct. One way to find that probability is to list the sample space as follows.

T,a	T,b	T,c	T,d	T,e
F,a	F,b	F,c	F,d	F,e

If the answers are random guesses, then the 10 possible outcomes are equally likely. The correct answers are T and c, so

$$P(\text{both correct}) = P(T \text{ and } c) = \frac{1}{10} = 0.1$$

Considering the individual answers of T and c, respectively, we see that with random guesses we have $P(T) = 1/2$ and $P(c) = 1/5$. Recognizing that 1/10 is the product of 1/2 and 1/5, we see that $P(T \text{ and } c) = P(T) \cdot P(c)$. This suggests

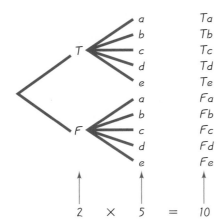

Figure 3-8   Tree Diagram of Test Answers

that, in general, $P(A \text{ and } B) = P(A) \cdot P(B)$, but let's consider another example before making that generalization.

For now, we note that tree diagrams are sometimes helpful in determining the number of possible outcomes in a sample space. A **tree diagram** is a picture of the possible outcomes of an experiment, shown as line segments emanating from one starting point. These diagrams are helpful in counting the number of possible outcomes if the number of possibilities is not too large. The tree diagram shown in Figure 3-8 summarizes the outcomes of the true/false and multiple-choice questions. From Figure 3-8 we see that if both answers are random guesses, all 10 branches are equally likely and the probability of getting the correct pair (T, c) is 1/10. For each response to the first question, there are 5 responses to the second. The total number of outcomes is 5 taken 2 times, or 10. The tree diagram in Figure 3-8 illustrates the reason for the use of multiplication.

Our first example of the true/false and multiple-choice questions suggested that $P(A \text{ and } B) = P(A) \cdot P(B)$, but the next example will introduce another important element. This example involves a deck of cards, so the context will be familiar to most readers, but the principles used can be applied to more meaningful circumstances.

EXAMPLE   When drawing two cards from a shuffled deck, find the probability that the first card is an ace and the second card is a king. That is, find $P(\text{ace and king})$. (Assume that the first card is not replaced before the second card is drawn.)

SOLUTION   We can use Rule 2 from Section 3-2 to find the probability of getting an ace on the first selection. Noting that there are 4 aces among the 52 different cards, we get $P(\text{ace}) = 4/52$. For the second selection we assume that an ace was obtained on the first draw, so we now have

(continued)

Composite Sampling

The U.S. Army once tested for syphilis by giving each inductee an individual blood test that was analyzed separately. One researcher suggested mixing pairs of blood samples. After the mixed pairs were tested, syphilitic inductees could be identified by retesting the few blood samples that were in the pairs that tested positive. The total number of analyses was reduced by pairing blood specimens, so why not put them in groups of three or four or more? Probability theory was used to find the most efficient group size, and a general theory was developed for detecting the defects in any population. This technique is known as *composite sampling.*

4 kings among only 51 cards and we get P(king) = 4/51. The probability of getting an ace on the first draw *and* a king on the second is

$$P(\text{ace and king}) = \frac{4}{52} \cdot \frac{4}{51} = 0.00603$$

We could better justify this result by listing the sample space or providing a tree diagram. The sample space, however, has 2652 different possibilities, and the tree diagram has 2652 branches. Clearly, working out this example would be a bit overwhelming, but the results would show that there are 16 cases among the 2652 possibilities that consist of an ace followed by a king.

This example illustrates the important principle that the probability for event B should take into account the fact that event A has already occurred. This principle is often expressed using the following notation.

Notation for the Multiplication Rule

$P(B|A)$ represents the probability of B occurring after it is assumed that the event A has already occurred. (We can read $B|A$ as "B given A.")

In the preceding example of finding the probability of an ace on the first card and a king on the second card drawn without replacement of the first, we have

$$P(\text{ace on first card}) = \frac{4}{52}$$

$$P(\text{king}|\text{ace}) = \frac{4}{51}$$

where $P(\text{king} | \text{ace})$ denotes the probability of drawing a king on the second card, assuming that the first card drawn was an ace.

DEFINITIONS

Two events A and B are **independent** if the occurrence of one does not affect the probability of the occurrence of the other. (Several events are similarly independent if the occurrence of any does not affect the probabilities of the occurrence of the others.) If A and B are not independent, they are said to be **dependent**.

For example, flipping a coin and then tossing a die are *independent* events because the outcome of the coin has no effect on the probabilities of the out-

Independent Jet Engines

A three-engine jet departed from Miami International Airport en route to South America, but one engine failed immediately after takeoff. While the plane was turning back to the runway, the other two engines also failed, but the pilot was able to make a safe landing. With independent jet engines, the probability of all three failing is only 0.0001^3, or about one chance in a trillion. The FAA found that the same mechanic who replaced the oil in all three engines incorrectly positioned the oil plug sealing rings. A goal in using three separate engines is to increase safety with independent engines, but the use of a single mechanic caused their operation to become dependent. Maintenance procedures now require that the engines be serviced by different mechanics.

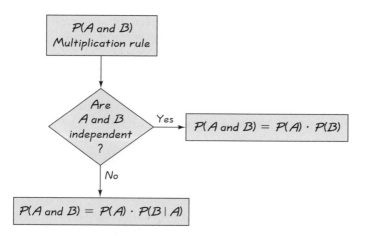

Figure 3-9 Applying the Multiplication Rule

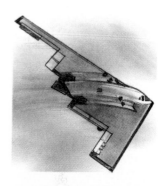

comes of the die. On the other hand, the event of having your car start and the event of driving to class on time are *dependent,* because the outcome of trying to start your car does affect the probability of your getting to class on time (unless you're a dorm resident).

Using the preceding notation and definitions, along with the principles illustrated in the preceding examples, we summarize the key concept of this section in Figure 3-9 and the *multiplication rule.*

Formal Multiplication Rule
$P(A \text{ and } B) = P(A) \cdot P(B)$ if A and B are *independent.* $P(A \text{ and } B) = P(A) \cdot P(B\|A)$ if A and B are *dependent.*

Intuitive Multiplication Rule
When finding the probability that event A occurs in one trial and B occurs in the next trial, multiply the probability of event A by the probability of event B, but be sure that the probability of event B takes into account the previous occurrence of event A.

The multiplication rule is extremely important because it has so many meaningful applications. One area of application involves product testing, as in the following example.

EXAMPLE   A batch of 50 fuel filters is produced by the Detroit Auto Supply Company, and 6 of them are defective. (Optimists would say that

(continued)

Redundancy

Reliability of systems can be greatly improved with redundancy of critical components. Airplanes have two independent electrical systems, and aircraft used for instrument flight typically have two separate radios. The following is from a *Popular Science* article about stealth aircraft: "One plane built largely of carbon fiber was the Lear Fan 2100 which had to carry two radar transponders. That's because if a single transponder failed, the plane was nearly invisible to radar." Such redundancy is an application of the multiplication rule in probability theory. If one component has a 0.001 probability of failure, the probability of two independent components both failing is only 0.000001.

Monkey Typists

A classical claim is that a monkey randomly hitting a keyboard would eventually produce the complete works of Shakespeare, assuming that it continues to type century after century. The multiplication rule for probability has been used to find such estimates. One result of 1,000,000,000,000,000, 000,000,000,000,000,000, 000 years is considered by some to be too short. In the same spirit, Sir Arthur Eddington wrote this poem: "There once was a brainy baboon, who always breathed down a bassoon. For he said, 'It appears that in billions of years, I shall certainly hit on a tune.'"

This is an important note about a common practice used by professional pollsters. Strongly consider discussing it in class.

44 are good.) Two of the filters are selected and tested. Find the probability that the first is good and the second is good if the filters are selected (a) with replacement; (b) without replacement.

SOLUTION

a. If the filters are selected with replacement, the two selections are independent because the second event is not affected by the first outcome. We therefore get

$$P(\text{first good and second good}) = \frac{44}{50} \cdot \frac{44}{50} = 0.774$$

b. If the filters are selected without replacement, the two selections are dependent because the second event is affected by the first outcome. We therefore get

$$P(\text{first good and second good}) = \frac{44}{50} \cdot \frac{43}{49} = 0.772$$

Note that in this case, we adjust the second probability to take into account the selection of a good filter on the first selection. After selecting a good filter the first time, there would be 43 good filters among the 49 that remain. Also note that without replacement, there is a slightly lower chance of getting 2 good filters. If we want to develop a procedure for using samples to test batches of products, we should sample without replacement for two reasons: First, there is a lower chance of getting only good items when some defects are present; second, it doesn't make sense to sample with replacement because it becomes possible to test the same item more than once, and that is a waste.

So far we have discussed two events, but the multiplication rule can be easily extended to several events. In general, the probability of any sequence of independent events is simply the product of their corresponding probabilities. For example, the probability of tossing a coin three times and getting all heads is $0.5 \cdot 0.5 \cdot 0.5 = 0.125$. We can also extend the multiplication rule so that it applies to several dependent events; simply adjust the probabilities as you go along. For example, the probability of getting three aces when three cards are selected without replacement is given by

$$\frac{4}{52} \cdot \frac{3}{51} \cdot \frac{2}{50} = 0.000181$$

In this last example involving three aces, we assumed that the events were dependent because the selections were made without replacement. However, it is a common practice to treat events as independent when small samples are drawn from large populations. (In such cases, it is rare to select the same item twice.) A common guideline is to assume independence whenever the sample size is no more than 5% of the size of the population. When pollsters survey

1200 adults from a population of millions, they typically assume independence, even though they sample without replacement.

The Probability of "At Least One"

The multiplication rule and the rule of complements can be used together to greatly simplify certain types of problems, such as those in which we want to find the probability that among several trials, *at least 1* will result in some specified outcome. In such cases, these two issues of language should be clearly understood:

- "At least 1" is equivalent to "1 or more."
- The complement of getting at least 1 item of a particular type is that you get no items of that type.

Suppose an employee in San Francisco needs to call any 1 of 5 colleagues at home. Assume that the 5 colleagues are random selections from a population and that 39.5% of San Francisco phone numbers are unlisted (based on data from Survey Sampling, Inc.). We need to find the probability that at least 1 of the 5 fellow workers will have a listed number. See the following interpretations.

> **At least 1 listed number = 1 or more listed numbers**
> **The complement of "at least 1 listed number" = no listed numbers (or "all numbers are unlisted")**

The direct solution to this problem is complex, but a simple, indirect approach is given in the solution to the following example.

Students often have some difficulty with this concept. Begin by clearly explaining the meaning of "at least 1." Then discuss the complement of "at least 1." Then emphasize this key point: To find P(at least 1 something), it's usually better to first find the probability of the complement, then subtract from 1.

EXAMPLE   Find the probability that at least 1 of 5 fellow employees in San Francisco has a listed number (and can therefore be called). Assume that telephone numbers are independent and that for San Francisco, 39.5% of the numbers are unlisted.

SOLUTION

Step 1:   Use a symbol to represent the probability desired. In this case, let L = at least 1 listed number among 5 employees.

Step 2:   Identify the complement of the event denoted in Step 1.

> $\overline{L}$ = *not* getting at least 1 listed number among 5 employees
> = getting 5 unlisted numbers among 5 employees

Step 3:   Find the probability of the complement from Step 2.

> $P(\overline{L}) = P(5 \text{ unlisted numbers among 5 employees})$
> $= 0.395 \cdot 0.395 \cdot 0.395 \cdot 0.395 \cdot 0.395$
> $= 0.395^5 = 0.00962$ *(continued)*

Step 4:   Find the probability of the event desired by subtracting the probability of the complement from 1.

$$P(L) = 1 - P(\overline{L}) = 1 - 0.00962 = 0.990$$

There is a 0.990 probability that at least 1 of the employees has a listed number and can therefore be called.

EXAMPLE   Find the probability of getting at least 1 girl if a couple plans to have 3 children. Assume that boys and girls are equally likely and that the gender of any child is not affected by previous children.

SOLUTION

Step 1:   Use a symbol to represent the probability desired. In this case, let A = at least 1 girl among 3 children.

Step 2:   Identify the complement of the event denoted in Step 1.

$\overline{A}$ = *not* getting at least 1 girl among 3 children

= getting 3 boys among 3 children

Step 3:   Find the probability of the complement from Step 2.

$$P(\overline{A}) = P(3 \text{ boys among 3 children}) = 0.5 \cdot 0.5 \cdot 0.5$$
$$= 0.5^3 = 0.125$$

Step 4:   Find the probability of the event desired by subtracting the probability of the complement from 1.

$$P(A) = 1 - P(\overline{A}) = 1 - 0.125 = 0.875$$

There is a 0.875 probability of getting at least 1 girl among 3 children.

Conditional Probability

Students tend to find conditional probability extremely difficult when it is presented in its abstraction. Instead, emphasize careful reading and understanding of the available information and the probability being sought.

The multiplication rule for dependent events can be formally expressed as $P(A \text{ and } B) = P(A) \cdot P(B|A)$. It is easy to solve this equation algebraically for $P(B|A)$; simply divide both sides of the equation by $P(A)$. The result is called the *conditional probability* of event B occurring, given that event A has already occurred.

> **DEFINITION**
>
> The **conditional probability** of B given A is the probability of event B occurring, given that event A has already occurred. It can be found by dividing the probability of events A and B both occurring by the probability of event A, as shown at the top of the following page.

$$P(B|A) = \frac{P(A \text{ and } B)}{P(A)}$$

This formula is a formal expression of conditional probability, but we can often use this intuitive approach:

Intuitive Approach to Conditional Probability

The conditional probability of B given A can be found by assuming that event A has occurred and, operating under that assumption, calculating the probability that event B will occur.

In the multiplication rule for dependent events, if $P(B|A) = P(B)$, then the occurrence of event A has no effect on the probability of event B. This is often used as a test for independence. If $P(B|A) = P(B)$, then A and B are independent events; however, if $P(B|A) \neq P(B)$, then A and B are dependent events. Another test for independence involves checking for the equality of $P(A \text{ and } B)$ and $P(A) \cdot P(B)$. If they are equal, events A and B are independent. If $P(A \text{ and } B) \neq P(A) \cdot P(B)$, then A and B are dependent events. These results are summarized as follows:

Two events A and B are *independent* if
$$P(B|A) = P(B)$$
or $P(A \text{ and } B) = P(A) \cdot P(B)$
Two events A and B are *dependent* if
$$P(B|A) \neq P(B)$$
or $P(A \text{ and } B) \neq P(A) \cdot P(B)$

For example, if $P(B|A) = 0.2$ and $P(B) = 0.2$, then $P(B|A) = P(B)$ and we can conclude that A and B are independent events. Because $P(B|A) = P(B)$, we conclude that the probability of event B is not affected by the occurrence of event A, and this is the definition of independence. However, if $P(B|A) = 0.5$ and $P(B) = 0.6$, then $P(B|A) \neq P(B)$ and we conclude that A and B are dependent events. Here, the different values for $P(B|A)$ and for $P(B)$ show that the probability for event B is affected by the occurrence of event A, so A and B are dependent events.

EXAMPLE   Refer to Table 3-2, and assume that all selections involve the 2000 subjects represented in the table. Find the following:

(continued)

Convicted by Probability

A witness described a Los Angeles robber as a Caucasian woman with blond hair in a ponytail who escaped in a yellow car driven by an African-American male with a mustache and beard. Janet and Malcolm Collins fit this description, and they were convicted based on testimony that there is only about 1 chance in 12 million that any couple would have these characteristics. It was estimated that the probability of a yellow car is 1/10, and the other probabilities were estimated to be 1/4, 1/10, 1/3, 1/10, and 1/1000. The convictions were later overturned when it was noted that no evidence was presented to support the estimated probabilities or the independence of the events. However, because the couple was not randomly selected, a serious error was made in not considering the probability of *other* couples being in the same region with the same characteristics.

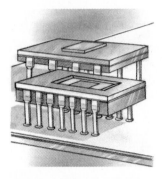

TABLE 3-2	Relationship of Criminal to Victim			
	Homicide	Robbery	Assault	Totals
Stranger	12	379	727	1118
Acquaintance or relative	39	106	642	787
Unknown	18	20	57	95
Totals	69	505	1426	2000

Bayes' Theorem

Thomas Bayes (1702–1761) said that probabilities should be revised when we learn more about an event. Here's one form of Bayes' theorem:

$$P(A|B) = \frac{P(A) \cdot P(B|A)}{P(A) \cdot P(B|A) + P(\overline{A}) \cdot P(B|\overline{A})}$$

Suppose 60% of a company's computer chips are made in one factory (denoted by A) and 40% are made in its other factory (denoted by $\overline{A}$). For a randomly selected chip, the probability it came from factory A is 0.60. Suppose we learn that the chip is defective and the defect rates for the two factories are 35% (for A) and 25% (for $\overline{A}$). We can use the above formula to find that there is a 0.677 probability the defective chip came from factory A.

a.   If one person is randomly selected, what is the likelihood that he or she was victimized by a stranger, given that a robbery victim is selected?

b.   Given that an assault victim is selected, what is the likelihood that the criminal is a stranger?

SOLUTION

a.   We want $P(\text{stranger}|\text{robbery})$. If we assume that the person selected was a robbery victim, we are dealing with the 505 people in the second column of values. Among those 505 people, 379 were victimized by strangers, so

$$P(\text{stranger}|\text{robbery}) = \frac{379}{505} = 0.750$$

The same result can be obtained using this formal approach:

$$P(\text{stranger}|\text{robbery}) = \frac{P(\text{robbery and stranger})}{P(\text{robbery})}$$

$$= \frac{379/2000}{505/2000} = 0.750$$

b.   Here we want $P(\text{stranger}|\text{assault})$. If we assume that the person selected was an assault victim, we are dealing with the 1426 people in the third column. Among those 1426 people, 727 were victimized by strangers, so

$$P(\text{stranger}|\text{assault}) = \frac{727}{1426} = 0.510$$

Again, the same result can be obtained using a formal approach:

$$P(\text{stranger}|\text{assault}) = \frac{P(\text{assault and stranger})}{P(\text{assault})}$$

$$= \frac{727/2000}{1426/2000} = 0.510$$

> By comparing the results from parts (a) and (b) we see that the probability of being victimized by a stranger is very different for robberies than for assaults, so there is a dependency between the type of crime and the relationship of the criminal to the victim.

3-4 Exercises A: Basic Skills and Concepts

In Exercises 1 and 2, for each given pair of events, classify the two events as independent or dependent. Some of the other exercises are based on concepts from earlier sections of this chapter.

Recommended assignment given over two classes: In the first class assign Exercises 2–7, 9, 16–19; in the second class assign Exercises 21, 22, 25–30.

1. **a.** Attending classes in a statistics course   1. a. Dependent
      Passing a statistics course
   b. Getting a flat tire on the way to class   b. Independent
      Sleeping too late for class
   c. Events A and B, where $P(A) = 0.40$, $P(B) = 0.60$, and $P(A \text{ and } B) = 0.20$ c. Dependent

2. **a.** Finding your microwave oven inoperable
      Finding your battery-operated smoke detector inoperable
   b. Finding your kitchen light inoperable   b. Dependent
      Finding your refrigerator inoperable
   c. Events A and B, where $P(A) = 0.90$, $P(B) = 0.80$, and $P(A \text{ and } B) = 0.72$ c. Independent

3. Ten percent of us are left-handed. What is the probability of randomly selecting 2 people who are both left-handed?   3. 0.010

4. Find the probability of answering the first 2 questions on a test correctly if random guesses are made and
   a. The first 2 questions are both true/false types.   4. a. 1/4
   b. The first 2 questions are both multiple choice, each with 5 possible answers.   b. 1/25

5. Find the probability of getting 4 consecutive aces when 4 cards are drawn without replacement from a shuffled deck.   5. 0.00000369

6. Find the probability of getting at least 1 girl when a couple has 5 children. Assume that boys and girls are equally likely and that the gender of any child is independent of the others.   6. 31/32

7. A student experiences difficulties with malfunctioning alarm clocks. Instead of using 1 alarm clock, he decides to use 3. What is the probability that at least 1 alarm clock works correctly if each individual alarm clock has a 98% chance of working correctly?   7. 0.999992

8. We have noted that when rolling a pair of dice, there are 36 different possible outcomes: 1-1, 1-2, . . . , 6-6.
   a. What is the probability of rolling a 7?   8. a. 1/6
   (continued)

8. b. Because the probability is only 0.000000595, the dice are probably loaded.

b. If you have just entered a friendly neighborhood game of craps and the person who brought the dice rolls eight consecutive 7s, what do you conclude? Why?

9. A classic excuse for a missed test is offered by 4 students who claim that their car had a flat tire. On the makeup test, the instructor asks the students to identify the particular tire that went flat. If they really didn't have a flat tire and randomly select one that supposedly went flat, what is the probability that all 4 select the same tire?   9. 1/64

10. Three firms using the same auditor independently and randomly select a month in which to conduct their annual audits. What is the probability that all 3 months are different?   10. 0.764

11. A quality control manager uses test equipment to detect defective computer modems. A sample of 3 different modems is to be randomly selected from a group consisting of 12 that are defective and 18 that have no defects. What is the probability that (a) all 3 selected modems are defective; (b) at least 1 selected modem is defective?   11. a. 0.0542   b. 0.799

12. When driving to class, a student must pass 2 traffic lights that operate independently. For each light, there is a 0.4 probability that it is green. If he must reach both lights when they are green in order to make class on time, what is the probability that he will be on time?   12. 0.160

13. 0.0138; he's probably lying, unless his rate is above 70%.

13. The IRS reports that among all taxpayers audited, 70% end up owing more money. One new auditor randomly selected 12 tax returns, audited them, and boasted that he collected additional taxes from all of them. What is the probability of doing what he boasted about? Based on that result, is it likely that he is being truthful?

14. 0.0000000206

14. One couple attracted media attention when their 3 children, born in different years, were all born on July 4. Ignoring leap years, find the probability that 3 randomly selected people were all born on July 4.

15. If you randomly select a person from the population of people who have died in recent years, there is a 0.0478 probability that the person's death was caused by an accident, according to data from the *Statistical Abstract of the United States*. A Baltimore detective is suspicious about 5 persons whose deaths were categorized as accidental. Find the probability that when 5 dead persons are randomly selected, their deaths were all accidental.   15. 0.000000250

16. 0.263 (with or without replacement)

16. With one method of *acceptance sampling*, a sample of items is randomly selected without replacement and the entire batch is rejected if there is at least one defect. The Niko Electronics Company has just manufactured 5000 CDs, and 3% are defective. If 10 of the CDs are selected and tested, what is the probability that the entire batch will be rejected?

17. A manager can identify employee theft by checking samples of employee shipments. Among 36 employees, 2 are stealing. If the manager checks on 4 different randomly selected employees, find the probability that neither of the thieves will be identified.   17. 0.787

18. In a Riverhead, New York, case, 9 different crime victims listened to voice recordings of 5 different men. All 9 victims identified the same voice as

that of the criminal. If the voice identifications were made by random guesses, find the probability that all 9 victims would select the same person. Does this constitute reasonable doubt?   18. 0.00000256; no

19. As an insurance claims investigator, you are suspicious of 4 brothers who each reported a stolen car in a different region of Houston. If Houston has an annual car-theft rate of 4.5%, find the probability that among 4 randomly selected cars, all are stolen in a given year. (There are 970,000 cars in Houston.) What does the result suggest?   19. 0.00000410; fraud

20. An approved jury list contains 20 women and 20 men. Find the probability of randomly selecting 12 of these people and getting an all-male jury. Under these circumstances, if the defendant is convicted by an all-male jury, is there strong evidence to suggest that the jury was not randomly selected?   20. 0.0000225

21. A blood-testing procedure is made more efficient by combining samples of blood specimens. If samples from 5 people are combined and the mixture tests negative, we know that all 5 individual samples are negative. Find the probability of a positive result for 5 samples combined into 1 mixture, assuming the probability of an individual blood sample testing positive is 0.015.   21. 0.0728

22. An employee claims that a new process for manufacturing VCRs is better because the rate of defects is below 5%, the rate of defects in the past. When 20 VCRs are manufactured with the new process, there are no defects. Assuming that the new method has the same 5% defect rate as in the past, find the probability of getting no defects among the 20 VCRs. Based on the result, is there strong evidence to conclude that the new process is better?   22. 0.358; no

In Exercises 23 and 24, use the data in Table 3-2.

23. **a.** Find the probability that when 1 of the 2000 subjects is randomly selected, the person chosen was victimized by an acquaintance or relative, given that he or she was robbed.   23. a. 0.210

b. Find the probability that when 1 of the 2000 subjects is randomly selected, the person chosen was robbed by an acquaintance or relative.   b. 0.0530

c. Find the probability that when 1 of the 2000 subjects is randomly selected, the person chosen was robbed or was victimized by an acquaintance or relative.   c. 0.593

d. If two different subjects are randomly selected, find the probability that they were both robbed.   d. 0.0637 (not 0.0638)

24. **a.** If one of the crime victims represented in the table is randomly selected, find the probability of getting someone who was victimized by someone unknown to the victim or who was a homicide victim.   24. a. 0.0730

b. If one of the crime victims represented in the table is randomly selected, find the probability of getting someone who was a homicide victim, given that the criminal was a stranger.   b. 0.0107   *(continued)*

d. 0.00223

c. If one of the crime victims represented in the table is randomly selected, find the probability of getting someone who was victimized by a stranger, given that he or she was a homicide victim.   c. 0.174

d. If two different subjects are randomly selected, find the probability that they were both victimized by criminals who were unknown.

In Exercises 25–30, use the following information. The New York State Health Department reports a 10% rate of the HIV virus for the population considered to be "at-risk," and a 0.3% HIV rate for the general population. Lab tests for the HIV virus are currently correct 95% of the time. Based on these results, if we randomly select 5000 people who are at-risk and 20,000 people from the general population, we expect results as summarized in the accompanying tables.

HIV Test Result	Sample from the At-Risk Population		Sample from the General Population	
	Positive	Negative	Positive	Negative
HIV Virus Infected	475	25	57	3
Not HIV Infected	225	4275	997	18,943

25. If a person is randomly selected from the at-risk population, what is the probability that he or she is infected with the HIV virus?   25. 0.100

26. If a person is randomly selected from the general population, what is the probability that he or she tests positive for HIV?   26. 0.0527

27. If a person is randomly selected from the at-risk population, what is the probability that he or she tests positive or is HIV infected?   27. 0.145

28. If a person is randomly selected from the general population, what is the probability that he or she tests positive or is HIV infected?   28. 0.0529

29. a. Consider only the at-risk sample and find the probability that a person has the HIV virus, given that the HIV test is positive.   29. a. 0.679

    b. Consider only the sample from the general population and find the probability that a person has the HIV virus, given that the HIV test is positive.   b. 0.0541

c. Given a positive test result, the likelihood of actually being HIV infected changes dramatically, depending on whether the subject is in the at-risk population.
30. a. 0.321

    c. Compare the results from parts (a) and (b). Why do you suppose a physician asks questions about life style during a consultation after an HIV test?

30. a. Consider only the at-risk sample and find the probability that a person does not have the HIV virus, given that the HIV test is positive.

    b. Consider only the sample from the general population and find the probability that a person does not have the HIV virus, given that the HIV test is positive.   b. 0.946

c. The subject from the general population should be tested again.

    c. Compare the results from parts (a) and (b). If you're the physician in charge, how would you handle a positive HIV test report for a person from each population?

3–4 Exercises B: Beyond the Basics

31. Find the probability that of 25 randomly selected people,
    a. No 2 share the same birthday.   31. a. 0.431
    b. At least 2 share the same birthday.   b. 0.569
32. a. Develop a formula for the probability of not getting either A or B on a single trial. That is, find an expression for $P(\overline{A \text{ or } B})$.
    b. Develop a formula for the probability of not getting A or not getting B on a single trial. That is, find an expression for $P(\overline{A} \text{ or } \overline{B})$.
    c. Compare the results from parts (a) and (b). Are they the same or are they different?   c. Different
33. Two cards are to be randomly selected without replacement from a shuffled deck. Find the probability of getting a 10 on the first card and a club on the second card.   33. 0.0192

32. a. $1 - P(A) - P(B) + P(A \text{ and } B)$
    b. $1 - P(A \text{ and } B)$

3–5 Probabilities Through Simulations

Finding probabilities of events is sometimes very difficult. Sometimes the results, even though correct, just don't seem to be what we might expect. Instead of relying solely on the abstract principles of probability theory, we can often benefit from using a simulation.

For its AP Statistics exam, The College Board states that "the computer facilitates the simulation approach to probability that is emphasized in the AP Statistics course." Carefully discuss simulation methods, such as those described in this section.

DEFINITION

A **simulation** of an experiment is a process that behaves the same way as the experiment, so that similar results are produced.

EXAMPLE   In testing techniques of gender selection, medical researchers need to know probabilities related to the genders of offspring. Assuming that males and females are equally likely, describe an experiment that simulates genders from births.

SOLUTION   One simulation is simply to flip a coin, with heads representing *male* and tails representing *female*. Another approach is to use a computer program, such as STATDISK or Minitab, to generate 0s and 1s, with 0 representing *male* and 1 representing *female*. Those numbers must be generated in such a way that the 0s and 1s are equally likely, as is the case with random-number generators from a uniform distribution. Shown below is a typical computer-generated result. On the basis of this result,

we could use the relative frequency approximation of probability to estimate P(male) as 6/10, because there are 6 males among the 10 births.

0	0	1	0	1	1	1	0	0	0
↓	↓	↓	↓	↓	↓	↓	↓	↓	↓
male	male	female	male	female	female	female	male	male	male

EXAMPLE   One classic exercise in probability is the *birthday problem,* in which we find the probability that in a class of 25 students, at least 2 students have the same birthday. Ignoring leap years, describe a simulation of the experiment that yields the birthdays of 25 students in a class.

SOLUTION   Begin by representing birthdays by integers from 1 through 365 as follows:

$$1 = \text{January 1}$$
$$2 = \text{January 2}$$
$$\vdots$$
$$365 = \text{December 31}$$

Using this representation, we need only generate integers between 1 and 365, instead of actual months and days. Using any source of equally likely integers (such as the uniform random generators in STATDISK, Minitab, or the TI-83 calculator), we can generate a list of 25 random integers between 1 and 365. That list can be ranked so that it becomes easy to determine whether any of the "birthdays" are the same. If we repeat this process many times, we can simulate many different classes, and we can then estimate the probability that in a class of 25 students, at least 2 students share the same birthday.

EXAMPLE   The Delmarva Communications Company manufactures cellular phones and has been experiencing a 6% rate of defects. The quality control manager knows that the phones are produced in batches of 250 and that, on average, there are 15 defects per batch. She wants to know how much the number of defects will typically vary. Describe a simulation of 250 cell phones that are manufactured with a 6% rate of defects.

SOLUTION   Use STATDISK, Minitab, or the TI-83 calculator to generate 250 integers, where each integer is between 1 and 100. Let the integers 1, 2, 3, 4, 5, 6 represent defective cell phones, while 7, 8, 9, . . . ,

100 represent good cell phones. If the 250 integers are ranked, it becomes easy to find the number of "defects," which are the numbers between 1 and 6 inclusive, found at the beginning of the ranked list.

EXAMPLE   Use Minitab to simulate rolling a pair of dice 500 times and, based on the results, estimate $P(7)$.

SOLUTION   The Minitab options of `Calc/Random Data/Integer` can be used to simulate rolling a single die 500 times, with the results stored in column C1. Those same options can then be used for a second die, with the results stored in column C2. The command `LET C3 = C1 + C2` creates a column C3 consisting of the 500 sums of the two dice. Shown in the margin is a summary of the Minitab results (obtained by using `Stat/Tables/Tally`). From these results we see that 7 occurred 84 times among the 500 trials, so we estimate $P(7) = 84/500 = 0.168$. (Using the rules of probability, we get $P(7) = 6/36 = 0.167$.)

C3	Count
2	13
3	29
4	34
5	56
6	73
7	84
8	61
9	55
10	45
11	32
12	18
N=	500

It is extremely important to construct the simulation carefully so that it closely imitates actual circumstances. A serious error would have been created in the preceding example if we had generated the dice totals of 2, 3, 4, . . . , 12 in such a way that they were equally likely. The outcomes would have resembled dice totals in the sense that they would have been integers between 2 and 12 inclusive, but by failing to simulate each individual die and add their values, we would have failed to imitate real dice. That error would have produced very misleading results.

In addition to solving some problems that might otherwise seem unsolvable, simulations can be used for verifying the results of probability calculations. One problem that attracted much attention in recent years is the *Monty Hall problem,* based on the old television game show "Let's Make a Deal," hosted by Monty Hall. Suppose you are a contestant who has selected one of three doors after being told that two of them conceal nothing, but that a new red Corvette is behind one of the three. Next, the host opens one of the doors you didn't select and shows that there is nothing behind it. He then offers you the choice of sticking with your first choice or switching to the other unopened door. Should you stick with your first choice or should you switch? The solution is far from obvious and we're not going to compute it here, but probability theory can be used to show that you should switch because your probability of winning then becomes 2/3. An alternative to the theoretical calculation is to simulate this game by playing it with a friend. The simulation should verify that switching is better than sticking with your first choice because you will win 2/3 of the time. According to *Chance* magazine, business schools at such institutions as Harvard and Stanford use this problem to help students deal with decision making.

The point made in this paragraph is very important and should be emphasized in class.

3-5 Exercises A: Basic Skills and Concepts

Recommended assignment: Exercises 1, 2, 3, 8. (Note that Exercise 8 requires a computer or the TI-83 calculator. If Exercise 8 is assigned, consider combining the student results to develop a probability.)

1. 0.240; the simulated result is too high.
2. 0.300; the simulated result is very good.
3. 0.200; the result is a little low, but it's not dramatically different.
4. 0.420; the result is a little low, but it's not dramatically different.

6. Answer varies, depending on the method of simulation used, but it should be about 0.375; letting even = correct yields 0.280.

1. Simulate an experiment of recording the numbers of girls in families of 3 children by using the Maryland Pick Three lottery results in Data Set 12 in Appendix B. Let each row of 3 digits represent the 3 children in one family, and let an even number represent a male and let an odd number represent a female. Based on the 50 simulated families, what is the estimated probability of getting a family with 3 girls when a family with 3 children is randomly selected? How does the simulated result compare with the actual probability, which is 0.125?

2. Refer to the Maryland Pick Three lottery results in Data Set 12 in Appendix B, but ignore the third digit in each row. Let an even digit represent a male and let an odd digit represent a female. Based on the 50 simulated families, what is the estimated probability of getting a family with 2 girls when a family with 2 children is randomly selected? How does the simulated result compare with the actual probability, which is 0.25?

3. The Telektronic Company manufactures cellular phones in batches of three and has been experiencing a 10% overall rate of defects. Refer to the Maryland Pick Three lottery results in Data Set 12 in Appendix B. Let 0 represent a defective cell phone and let 1, 2, 3, . . . , 9 represent good cell phones. Let each row of three digits represent a simulated batch of cell phones and use the 50 simulated batches to estimate the probability of at least one defect in a batch. How does the estimated probability compare to the theoretical result of 0.271?

4. Repeat Exercise 3 if the overall rate of defects is actually 20%. Let 0 and 1 represent defective cell phones. How does the estimated probability compare to the theoretical result of 0.488?

5. We know that when a single fair die is rolled, the probability of getting a 1 is 1/6, or 0.167. What is the estimated probability obtained when the rolling of a die is simulated with the Maryland Pick Three lottery data (Data Set 12 in Appendix B)? (*Hint:* Let the 150 digits represent simulated rolls of a die, but skip any outcomes that are not 1, 2, 3, 4, 5, or 6.)   5. Approximately 16/94 = 0.170

6. A student guesses answers to each of the three true/false questions on a quiz. Use the Maryland Pick Three lottery data (Data Set 12 in Appendix B) to estimate the probability of getting exactly one correct answer among the three.

3–5 Exercises B: Beyond the Basics

7. Use the Maryland Pick Three lottery data (Data Set 12 in Appendix B) to simulate 50 families with 3 children each. Let even numbers represent male children, and let odd numbers represent female children.
   a. Find the mean number of girls in a family.   7. a. 1.64
   b. Find the standard deviation of the numbers of girls.   b. 0.98

8. The second example in this section describes a method for simulating a class of 25 birthdays. Use Minitab, or STATDISK, or a TI-83 calculator, or any other source of numbers between 1 and 365 (such as a telephone directory) to simulate one class. Rank the results and determine whether there are at least 2 birthdays that are the same. Describe the exact procedure used. If Minitab or STATDISK was used, obtain a printed copy of the display. (See also the Computer Project near the end of this chapter.)

8. Answer varies.

3–6 Counting

This section can be omitted; it is not required for the following chapters. Outline:
1. fundamental counting rule
2. factorial rule
3. permutations rule (all items are different)
4. permutations rule (some items are identical)
5. combinations rule

Let's consider a probability problem pondered seriously by millions of hopeful Americans: What are the chances of winning a lottery? In New York State's lottery, which is typical, you must choose 6 numbers between 1 and 54 inclusive. If you get the same 6-number combination that is randomly drawn, you win millions of dollars. There are some lesser prizes, but they are relatively insignificant. We could use Rule 2 from Section 3-2 to find the probability of winning such a lottery. That rule, which requires equally likely outcomes, states that the probability of an event A can be found by using $P(A) = s/n$, where s is the number of ways A can occur and n is the total number of outcomes. With New York State's lottery, there is only one way to win the grand prize: Choose the same 6-number combination that is drawn in the lottery. Knowing that there is only one way to win, we now need to determine the total number of outcomes; that is, how many 6-number combinations are possible? Writing out a list of the possibilities would take about four years of nonstop work, and that's no fun. We could construct a tree diagram, but it would be about 120 miles high and would violate airspace regulations. We need a more practical way of finding the total number of possibilities. This section introduces efficient methods for finding such numbers. We will return to this lottery problem after we first present some basic principles. We begin with the *fundamental counting rule*.

Fundamental Counting Rule

For a sequence of two events in which the first event can occur m ways and the second event can occur n ways, the events together can occur a total of $m \cdot n$ ways.

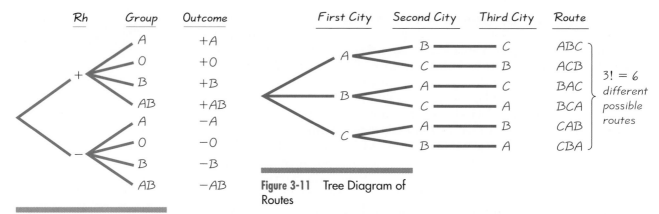

Figure 3-10 Tree Diagram of Blood Types/Rh Factors

Figure 3-11 Tree Diagram of Routes

For example, if a medical researcher must randomly select 1 of the 2 Rh types (positive, negative) and 1 of the 4 blood groups (A, O, B, AB), the total number of possibilities is $2 \cdot 4 = 8$. We can see the reason for multiplication in Figure 3-10 where we use a tree diagram to depict the different possibilities. The fundamental counting rule easily extends to situations involving more than two events, as illustrated in the following example.

EXAMPLE   In designing a computer, if a *byte* is defined to be a sequence of 8 bits and each bit must be a 0 or 1, how many different bytes are possible? (A byte is often used to represent an individual character, such as a letter, digit, or punctuation symbol. For example, one coding system represents the letter *A* as 01000001.)

SOLUTION   Because each bit can occur in two ways (0 or 1) and we have a sequence of 8 bits, the total number of different possibilities is given by

$$2 \cdot 2 \cdot 2 \cdot 2 \cdot 2 \cdot 2 \cdot 2 \cdot 2 = 256$$

There are 256 different possible bytes.

EXAMPLE   When designing surveys, pollsters sometimes try to minimize a *lead-in* effect by rearranging the order in which the questions are presented. (A lead-in effect occurs when some questions influence the responses to the questions that follow.) If Gallup plans to conduct a consumer survey by asking subjects 5 questions, how many different versions of the survey are required if all possible arrangements are included?

SOLUTION   In arranging any individual survey, there are 5 possible choices for the first question, 4 remaining choices for the second question, 3 choices for the third question, 2 choices for the fourth question, and only 1 choice for the fifth question. The total number of possible arrangements is therefore

$$5 \cdot 4 \cdot 3 \cdot 2 \cdot 1 = 120$$

That is, Gallup would need 120 versions of the survey in order to include every possible arrangement.

In the preceding example, we found that 5 survey questions can be arranged $5 \cdot 4 \cdot 3 \cdot 2 \cdot 1 = 120$ different ways. This particular solution can be generalized by using the following notation for the symbol ! and the following *factorial rule*.

Notation
The **factorial symbol !** denotes the product of decreasing positive whole numbers. For example, $4! = 4 \cdot 3 \cdot 2 \cdot 1 = 24$. By special definition, $0! = 1$. (Many calculators have a factorial key.)

Factorial Rule
A collection of n different items can be arranged in order $n!$ different ways. (This **factorial rule** reflects the fact that the first item may be selected n different ways, the second item may be selected $n - 1$ ways, and so on.)

EXAMPLE   Routing problems often involve application of the factorial rule. AT&T wants to route telephone calls through the shortest networks. Federal Express wants to find the shortest routes for its deliveries. Suppose a computer salesperson must visit 3 separate cities denoted by A, B, C. How many routes are possible?

SOLUTION   Using the factorial rule, we see that the 3 different cities (A, B, C) can be arranged in $3! = 6$ different ways. In Figure 3-11 we can see that there are 3 choices for the first city and 2 choices for the second city. This leaves only 1 choice for the third city. The number of possible arrangements for the 3 cities is $3 \cdot 2 \cdot 1 = 6$.

The Random Secretary

One classic problem of probability goes like this: A secretary addresses 50 different letters and envelopes to 50 different people, but the letters are randomly mixed before being put into envelopes. What is the probability that at least one letter gets into the correct envelope? Although the probability might seem like it should be small, it's actually 0.632. Even with a million letters and a million envelopes, the probability is 0.632. The solution is beyond the scope of this text—way beyond.

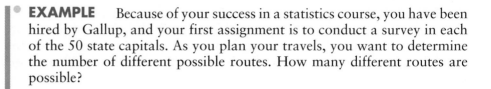

EXAMPLE Because of your success in a statistics course, you have been hired by Gallup, and your first assignment is to conduct a survey in each of the 50 state capitals. As you plan your travels, you want to determine the number of different possible routes. How many different routes are possible?

SOLUTION By applying the factorial rule, we know that 50 items can be arranged in order 50! different ways. That is, the 50 state capitals can be arranged 50! ways, so the number of different routes is 50!, or

30,414,093,201,713,378,043,612,608,166,064,768,844,377,641,568,
960,512,000,000,000,000

As numbers go, this one is very large and is therefore very deserving of the symbol ! used for factorials.

The preceding example is a variation of a classical problem called the *traveling salesman problem*. It is especially interesting because the large number of possibilities means that we can't use a computer to calculate the distance of each route. The time it would take even the fastest computer to calculate the shortest possible route is about

1,000,000,000,000,000,000,000,000,000,000,000,000,000 **centuries,**

which is a bit too long! Considerable effort is currently being applied to finding efficient ways of solving such problems.

Using the factorial counting rule, we determine the number of different possible ways we can arrange a number of items in some type of ordered sequence. The factorial rule tells us how many arrangements are possible when all of n different items are used. Sometimes, however, we want to select only some of the n items. If we must conduct surveys in state capitals, as in the preceding example, but we have time to visit only four of the capitals, the number of different possible routes is $50 \cdot 49 \cdot 48 \cdot 47 = 5,527,200$. Another way to obtain this same result is to evaluate

$$\frac{50!}{46!} = 50 \cdot 49 \cdot 48 \cdot 47 = 5,527,200$$

In this calculation, note that the factors in the numerator divide out with the factors in the denominator, except for the factors of 50, 49, 48, and 47 that remain. We can generalize this result by noting that if we have n different items available and we want to select r of them, the number of different arrangements possible is $n!/(n - r)!$, as in 50!/46!. This generalization is commonly called the *permutations rule*.

Safety in Numbers

Some hotels have abandoned the traditional room key in favor of an electronic key with a number code. A central computer changes the access code to a room as soon as a guest checks out. A typical electronic key has 32 different positions that are either punched or left untouched. This configuration allows for 2^{32}, or 4,294,967,296, different possible codes, so it is impractical to develop a complete set of keys or try to make an illegal entry by trial and error.

> ### Permutations Rule (When items are all different)
>
> The number of **permutations** (or sequences) of r items selected from n available items (not allowing repetition) is
>
> $$_nP_r = \frac{n!}{(n-r)!}$$

Some calculators are designed to automatically calculate values of $_nP_r$. On the TI-83, for example, enter the value for n, then press MATH, PRB, $_nP_r$, and the value for r. If your calculator lacks such a feature, it is still easy to calculate $n!/(n-r)!$ by using the factorial key identified as !.

It is very important to recognize that the permutations rule requires the following conditions:

- We must have a total of n *different* items available. (This rule does not apply if some of the items are identical to others.)
- We must select r of the n items (without repetition).
- We must consider rearrangements of the same items to be different sequences.

When we use the terms permutations, arrangements, or sequences, we imply that *order is taken into account* in the sense that different orderings of the same items are counted separately. The letters *ABC* can be arranged six different ways: *ABC, ACB, BAC, BCA, CAB, CBA*. (Later, we will refer to *combinations*, which do not count such arrangements separately.) In the following example, we are asked to find the total number of different sequences that are possible. That suggests use of the permutations rule.

EXAMPLE   In planning the Monday-night prime-time lineup for the NBC television network, an executive must select 6 shows from 30 that are available. How many different lineups are possible?

SOLUTION   We need to select $r = 6$ shows from $n = 30$ shows available. Order is relevant, because the viewing audience changes later in the evening. Because order counts, we want the number of permutations, which is found as shown:

$$_nP_r = \frac{n!}{(n-r)!} = \frac{30!}{(30-6)!} = 427{,}518{,}000$$

Because there are 427,518,000 possible lineups, there are too many to consider individually.

The permutations rule can be thought of as an extension of the fundamental counting rule. We can also solve the preceding example by using the fundamental counting rule as follows: With 30 shows available and with a stipulation that 6 are to be selected, we know that there are 30 choices for the first time slot, 29 choices for the second time slot, and so on. The total number of possible arrangements is therefore

$$30 \cdot 29 \cdot 28 \cdot 27 \cdot 26 \cdot 25 = 427{,}518{,}000$$

but $30 \cdot 29 \cdot 28 \cdot 27 \cdot 26 \cdot 25$ is actually $30! \div 24!$ [or $30! \div (30 - 6)!$]. In general, whenever we select r items from n available items, the number of different possible arrangements is $n! \div (n - r)!$, and this is expressed in the permutations rule.

We sometimes need to find the number of permutations, but some of the items are identical to others. The following variation of the permutations rule applies to such cases.

Permutations Rule (When some items are identical to others)

If there are n items with n_1 alike, n_2 alike, . . . , n_k alike, the number of permutations of all n items is

$$\frac{n!}{n_1! \, n_2! \, \cdots \, n_k!}$$

EXAMPLE   The classic examples of the permutations rule are those that show that the letters of the word *Mississippi* can be arranged 34,650 different ways and that the letters of the word *statistics* can be arranged 50,400 ways. We will instead consider the letters *DDDDRRRRR*, which are included in a discussion of the runs test for randomness (Section 13-7). Those letters represent a sequence of diet (*D*) and regular (*R*) colas. How many ways can we arrange the letters *DDDDRRRRR*?

SOLUTION   In the sequence *DDDDRRRRR* we have $n = 9$ items, with $n_1 = 4$ alike and $n_2 = 5$ others that are alike. The number of permutations is computed as follows.

$$\frac{n!}{n_1! \, n_2!} = \frac{9!}{4! \, 5!} = \frac{362{,}880}{2880} = 126$$

In Section 13-7 we use the fact that there are 126 different possible sequences of *DDDDRRRRR*, and we can now see how that result is obtained.

The preceding example involved n items, each belonging to one of two categories. When there are only two categories, we can stipulate that x of the items are alike and the other $n - x$ items are alike, so the permutations formula simplifies to

$$\frac{n!}{(n - x)! \, x!}$$

This particular result will be used for binomial experiments, which are introduced in Section 4-3.

When we intend to select r items from n different items but *do not take order into account*, we are really concerned with possible combinations rather than permutations. That is, **when different orderings of the same items are to be counted separately, we have a permutation problem, but when different orderings are not to be counted separately, we have a combination problem** and may apply the following rule.

Combinations Rule

The number of **combinations** of r items selected from n different items is

$$_nC_r = \frac{n!}{(n - r)! \, r!}$$

Some calculators are designed to automatically evaluate $_nC_r$. With the TI-83, for example, enter the value for n, then press MATH, PRB, $_nC_r$, and enter the value for r.

It is very important to recognize that in applying the combinations rule, the following conditions apply:

- We must have a total of n different items available.
- We must select r of the n items (without repetition).
- We must consider rearrangements of the same items to be the same grouping.

Because choosing between the permutations rule and the combinations rule can be confusing, we provide the following example, which is intended to emphasize the difference between them.

EXAMPLE The Board of Trustees at the author's college has 9 members. Each year, they elect a 3-person committee to oversee buildings and grounds. Each year, they also elect a chairperson, vice chairperson, and secretary. *(continued)*

How Many Shuffles?

After conducting extensive research, Harvard mathematician Persi Diaconis found that it takes seven shuffles of a deck of cards to get a complete mixture. The mixture is complete in the sense that all possible arrangements are equally likely. More than seven shuffles will not have a significant effect, and fewer than seven are not enough. Casino dealers rarely shuffle as often as seven times, so the decks are not completely mixed. Some expert card players have been able to take advantage of the incomplete mixtures that result from fewer than seven shuffles.

a. When the board elects the buildings and grounds committee, how many different 3-person committees are possible?
b. When the board elects the 3 officers (chairperson, vice chairperson, and secretary), how many different slates of candidates are possible?

SOLUTION   Note that order is irrelevant when electing the buildings and grounds committee. When electing officers, however, different orders are counted separately.

a. Here we want the number of combinations of $r = 3$ people selected from the $n = 9$ available people. We get

$$_9C_3 = \frac{n!}{(n-r)!\,r!} = \frac{9!}{(9-3)!\,3!} = \frac{362,880}{4320} = 84$$

b. Here we want the number of sequences (or permutations) of $r = 3$ people selected from the $n = 9$ available people. We get

$$_9P_3 = \frac{n!}{(n-r)!} = \frac{9!}{(9-3)!} = \frac{362,880}{720} = 504$$

There are 84 different possible committees of 3 board members, but there are 504 different possible slates of candidates.

The counting techniques presented in this section are sometimes used in probability problems. The following examples illustrate such applications.

EXAMPLE   In the New York State lottery, a player wins first prize by selecting the correct 6-number combination when 6 different numbers from 1 through 54 are drawn. If a player selects one particular 6-number combination, find the probability of winning. (The player need not select the 6 numbers in the same order that they are drawn, so that order is irrelevant.)

SOLUTION   Because 6 different numbers are selected from 54 different possibilities, the total number of combinations is

$$_{54}C_6 = \frac{54!}{(54-6)!\,6!} = \frac{54!}{48!\,6!} = 25,827,165$$

With only one combination selected, the player's probability of winning is only 1/25,827,165.

In the preceding example we found that the probability of winning the New York State lottery is only 1/25,827,165. In Section 3-2 we found that the probability of being struck by lightning this year is estimated to be 1/701,000. A comparison of these two probabilities shows that in a single year, there is a much better chance of being struck by lightning than of winning the lottery in a single try. Of course, you could increase your chances of winning the lottery by buying many tickets—a strategy generally deemed to be unwise.

> **EXAMPLE**   A UPS dispatcher sends a delivery truck to 8 different locations. If the order in which the deliveries are made is randomly determined, find the probability that the resulting route is the shortest possible route.
>
> **SOLUTION**   With 8 locations there are 8!, or 40,320, different possible routes. Among those 40,320 different possibilities, only two routes will be shortest (actually, the same route in two different directions). Therefore, there is a probability of only 2/40,320, or 1/20,160, or 0.0000496 that the selected route will be the shortest one possible.

This section presented five different counting approaches. When deciding which particular one to apply, you should consider a list of questions that address key issues. The following summary may be helpful.

- Is there a sequence of events in which the first can occur m ways, the second can occur n ways, and so on? If so, use the fundamental counting rule and multiply m, n, and so on.
- Are there n *different* items and are *all* of them to be used indifferent arrangements? If so, use the factorial rule and find $n!$.
- Are there n *different* items and are only *some* of them to be used in different arrangements? If so, evaluate $_nP_r$.
- Are there n items with some of them *identical* to each other, and is there a need to find the total number of different arrangements of all of those n items? If so, use the following expression in which n_1 of the items are alike, n_2 are alike, and so on.

$$\frac{n!}{n_1!\, n_2!\, \cdots\, n_k!}$$

- Are there n different items, with *some* of them to be selected, and is there a need to find the total number of combinations (that is, is the order irrelevant)? If so, evaluate $_nC_r$.

The Number Crunch

Every so often telephone companies split regions with one area code into regions with two or more area codes because the increased number of area fax and internet lines has nearly exhausted the possible numbers that can be listed under a single code. A seven-digit telephone number cannot begin with a 0 or 1, but if we allow all other possibilities, we get $8 \cdot 10 \cdot 10 \cdot 10 \cdot 10 \cdot 10 \cdot 10 = 8,000,000$ different possible numbers! Even so, after surviving for 80 years with the single area code of 212, New York City was recently partitioned into the two area codes of 212 and 718. Many other regions have also been assigned split area codes.

3-6 Exercises A: Basic Skills and Concepts

In Exercises 1–16, evaluate the given expressions.

1. $6!$
2. $11!$
3. $100!/97!$
4. $85!/82!$
5. $(10 - 4)!$
6. $(90 - 87)!$
7. $_6C_4$
8. $_6P_4$
9. $_{12}P_9$
10. $_{10}C_9$
11. $_{40}C_6$
12. $_{40}P_6$
13. $_nC_0$
14. $_nP_0$
15. $_nP_n$
16. $_nC_n$

17. The author uses an ADT home security system that has a code consisting of 4 digits (0, 1, . . . , 9) that must be entered in the correct sequence. The digits can be repeated in the code.
  a. How many different possibilities are there?   17. a. 10,000
  b. If it takes a burglar 5 seconds to try a code, how long would it take to try every possibility?   b. 13.9 hours
18. There are 12 members on the board of directors for the Newport General Hospital.
  a. If they must elect a chairperson, first vice chairperson, second vice chairperson, and secretary, how many different slates of candidates are possible?   18. a. 11,880
  b. If they must form an ethics subcommittee of 4 members, how many different subcommittees are possible?   b. 495
19. Each social security number is a sequence of 9 digits. What is the probability of randomly generating 9 digits and getting your social security number?   19. 1/1,000,000,000
20. A typical "combination" lock is opened with the correct sequence of 3 numbers between 0 and 49 inclusive. How many different sequences are possible? (A number can be used more than once.) Are these sequences combinations or are they actually permutations?   20. 125,000 permutations
21. In doing a runs test for randomness (Section 13-7), it is found that genders of survey subjects listed in consecutive order are as follows: MMMMMMMMMMFFFFFFFFF. How many different ways can those letters be arranged?   21. 43,758
22. A Federal Express delivery route must include stops at 5 cities.
  a. How many different routes are possible?   22. a. 120
  b. If the route is randomly selected, what is the probability that the cities will be arranged in alphabetical order?   b. 1/120
23. In attempting to decipher an intercepted message from Libya, a code expert decides to list all different arrangements of the word MGBTQRS. How many different arrangements are there?   23. 5040
24. a. If a couple plans to have 8 children (it happens), how many different gender sequences are possible?   24. a. 256
  b. If a couple has 4 boys and 4 girls, how many different gender sequences are possible?   b. 70

(continued)

c. Based on the results from parts (a) and (b), what is the probability that when a couple has 8 children, the result will consist of 4 boys and 4 girls? c. 70/256 = 0.273

25. The New York State lottery and some other state lotteries once required the selection of 6 numbers between 1 and 40 inclusive.
   a. How many different selections are possible? 25. a. 3,838,380
   b. If you select 6 numbers, what is the probability of winning by getting the same numbers that are drawn in the lottery? b. 1/3,838,380
   c. What are the odds against winning such a lottery? c. 3,838,379:1

26. Refer to the same lottery described in Exercise 25. What is the probability of winning if the rules are changed so that you must select the correct 6 numbers in the same order in which they are drawn? 26. 1/2,763,633,600

27. In Denys Parsons' *Directory of Tunes and Musical Themes,* melodies for more than 14,000 songs are listed according to the following scheme: The first note of every song is represented by an asterisk (*), and successive notes are represented by R (for repeat the previous note), U (for a note that goes up), or D (for a note that goes down). Beethoven's Fifth Symphony begins as *RRD. Classical melodies are represented through the first 16 notes. With this scheme, how many different classical melodies are possible? 27. 14,348,907

28. The Bureau of Fisheries once asked Bell Laboratories for help in finding the shortest route for getting samples from locations in the Gulf of Mexico. How many different routes are possible if samples must be taken from 11 locations? 28. 39,916,800

29. A 4-member FBI investigative team is to be formed from a list of 50 agents not already assigned to a special project.
   a. How many different possible ways can the team be formed?   29. a. 230,300
   b. If the selections are random, what is the probability of getting the 4 agents with the most time in service? b. 1/230,300

30. You become suspicious when a genetics researcher randomly selects groups of 20 newborn babies and seems to consistently get 10 girls and 10 boys. The researcher explains that it is common to get 10 boys and 10 girls in such cases.
   a. If 20 newborn babies are randomly selected, how many different gender sequences are possible? 30. a. 1,048,576
   b. How many different ways can 10 boys and 10 girls be arranged in sequence? b. 184,756
   c. What is the probability of getting 10 boys and 10 girls when 20 babies are born? c. 184,756/1,048,576 = 0.176
   d. Based on the preceding results, do you agree with the researcher's explanation that it is common to get 10 boys and 10 girls when 20 babies are randomly selected?

d. With a probability of 0.176, it is common but it should not happen consistently; it should happen roughly once in every six trials.

31. The Detroit Music Company has purchased the rights to 15 different songs and it plans to release a new CD with 8 of them. Recognizing that the order of the songs is important, how many different CDs are possible?   31. 259,459,200

32. At one Ford Motor Company assembly station, 8 different parts must be attached to a car, but they can be attached in any order. The manager decides to find the most efficient sequence by trying all possibilities. How many different sequences are possible?   32. 40,320

33. **a.** How many different zip codes are possible if each code is a sequence of 5 digits?   33. a. 100,000

    b. If a computer randomly generates 5 digits, what is the probability it will produce your zip code?   b. 1/100,000

34. The programming director for ABC television has decided to schedule 30-minute shows between 8:00 P.M. and 10:00 P.M. on Mondays. If 22 shows are available, how many different lineups are possible for those Monday slots?   34. 175,560

35. In an age-discrimination case against Darmin, Inc., evidence showed that among the last 40 applicants for employment, only the 8 youngest were hired. Find the probability of randomly selecting 8 of 40 people and getting the 8 youngest. Based on the result, does it appear that age discrimination is occurring?   35. 1/76,904,685; yes, if all other factors are equal.

36. The following excerpt is from *The Man Who Cast Two Shadows*, by Carol O'Connell: "The child had only the numbers written on her palm in ink . . . , all but the last four numbers disappeared in a wet smudge of blood . . . . She would put the coins into the public telephones and dial three untried numbers and then the four she knew. If a woman answered, she would say, 'It's Kathy. I'm lost.'" If it costs Kathy 25¢ for each call and she tries every possibility except those beginning with 0 or 1, what is her total cost?   36. $200

37. After testing 12 homes for the presence of radon, an inspector is concerned that her test equipment is defective because the measured radon level at each home was higher than the reading at the preceding home. That is, the 12 readings were arranged in order from low to high. If the homes were randomly selected, what is the probability of getting this particular arrangement? Based on the result, is her concern about the test equipment justified?   37. 1/479,001,600; yes

38. 37,035,180

38. A freight train is to carry 12 coal cars, 5 cars full of lumber, and 4 tankers all carrying kerosene. How many different arrangements are possible?

39. *USA Today* reporter Paul Wiseman described the old rules for telephone area codes by writing about "possible area codes with 1 or 0 in the second digit. (Excluded: codes ending in 00 or 11, for toll-free calls, emergency services, and other special uses.)" Codes beginning with 0 or 1 should also be excluded. How many different area codes were possible under these old rules?   39. 144

40. There are 22 members on the budget committee that is appointed from the 100 members of the U.S. Senate. How many different committees of 22 can be formed from 100 available Senators?   40. 7.33×10^{21}

3-6 Exercises B: Beyond the Basics

41. A common computer programming rule is that names of variables must be between 1 and 8 characters long. The first character can be any of the 26 letters, while successive characters can be any of the 26 letters or any of the 10 digits. For example, allowable variable names are A, BBB, and M3477K. How many different variable names are possible?

41. 2,095,681,645,538 (about 2 trillion)

42. a. Five managers gather for a meeting. If each manager shakes hands with each other manager exactly once, what is the total number of handshakes?   42. a. 10

b. If n managers shake hands with each other exactly once, what is the total number of handshakes?   b. $n(n-1)/2$

c. How many different ways can 5 managers be seated at a round table? (Assume that if everyone moves to the right, the seating arrangement is the same.)   c. 24

d. How many different ways can n managers be seated at a round table?   d. $(n-1)!$

43. Many calculators or computers cannot directly calculate 70! or higher. When n is large, $n!$ can be approximated by $n! = 10^{K}$ where $K = (n + 0.5) \log n + 0.39908993 - 0.43429448n$.

a. Evaluate 50! using the factorial key on a calculator and also by using the approximation given here.

43. a. Calculator: 3.0414093×10^{64}; approximation: 3.0363452×10^{64}

b. The Bureau of Fisheries once asked Bell Laboratories for help in finding the shortest route for getting samples from 300 locations in the Gulf of Mexico. There are 300! different possible routes. If 300! is evaluated, how many digits are used in the result?   b. 615

44. Can computers "think"? According to the *Turing test,* a computer can be considered to think if, when a person communicates with it, the person believes he or she is communicating with another person instead of a computer. In an experiment at Boston's Computer Museum, each of 10 judges communicated with 4 computers and 4 other people and was asked to distinguish between them.

a. Assume that the first judge cannot distinguish between the 4 computers and the 4 people. If this judge makes random guesses, what is the probability of correctly identifying the 4 computers and the 4 people?

44. a. 1/70

b. Assume that all 10 judges cannot distinguish between computers and people, so they make random guesses. Based on the result from part (a), what is the probability that all 10 judges make all correct guesses? (That event would lead us to conclude that computers cannot "think" when, according to the Turing criterion, they can.)   b. 3.54×10^{-19}

Vocabulary List

experiment

event

simple event

sample space

relative frequency approximation of
    probability

classical approach to probability

law of large numbers

random sample of one element

random sample

complement

subjective probability

odds against

odds in favor

compound event

addition rule

mutually exclusive

rule of complementary events

tree diagram

independent events

dependent events

multiplication rule

conditional probability

simulation

fundamental counting rule

factorial symbol

factorial rule

permutations rule

combinations rule

Review

This chapter introduced the basic concepts of probability theory. In Section 3-2 we presented the basic definitions and notation, including the representation of events by letters such as A. We defined probabilities of simple events as

$$P(A) = \frac{\text{number of times } A \text{ occurred}}{\text{number of times experiment was repeated}} \quad \text{(relative frequency)}$$

$$P(A) = \frac{\text{number of ways } A \text{ can occur}}{\text{number of different simple events}} = \frac{s}{n} \quad \text{(for equally likely outcomes)}$$

We noted that the probability of any impossible event is 0, the probability of any certain event is 1, and for any event A, $0 \leq P(A) \leq 1$. Also, $\overline{A}$ denotes the complement of event A. That is, $\overline{A}$ indicates that event A does not occur.

After Section 3-2 we proceeded to consider compound events, which involve more than one event. In general, associate *or* with addition and associate *and* with multiplication. Always keep in mind the following key considerations.

- When conducting one trial, do we want the probability of event *A or B*? If so, use the addition rule, but be careful to avoid counting any outcomes more than once.

- When finding the probability that event *A* occurs on one trial *and* event *B* occurs on a second trial, use the multiplication rule. Multiply the probability of event *A* by the probability of event *B*. *Caution:* When calculating the probability of event *B*, be sure to take into account the fact that event *A* has already occurred.

In some probability problems, the biggest obstacle is finding the total number of possible outcomes. The last section of this chapter was devoted to the following counting techniques, which are briefly summarized at the end of Section 3-6:

- Fundamental counting rule
- Factorial rule
- Permutations rule (when items are all different)
- Permutations rule (when some items are identical to others)
- Combinations rule

Most of the material in the following chapters deals with statistical inferences based on probabilities. As an example of the basic approach used, consider a test of someone's claim that a quarter used in a coin toss is fair. If we flip the quarter 10 times and get 10 consecutive heads, we can make one of two inferences from these sample results:

1. The coin is actually fair, and the string of 10 consecutive heads is a fluke.
2. The coin is not fair.

The statistician's decision as to which inference is correct is based on the probability of getting 10 consecutive heads, which, in this case, is so small (1/1024) that the inference of unfairness is the better choice. Here we can see the important role played by probability in the standard methods of statistical inference.

· ·

Review Exercises

In Exercises 1–8, use the data from Table 3-3 on page 174, which summarize results from a study of 1000 randomly selected deaths of males aged 45–64 (based on data from "Chartbook on Smoking, Tobacco, and Health," USDHEW).

1. If 1 of the 1000 subjects is randomly selected, find the probability of getting a smoker   1. 0.650
2. If 1 of the 1000 subjects is randomly selected, find the probability of getting a smoker or someone who died from heart disease.   2. 0.805
3. If two different subjects are randomly selected, find the probability that they both died from cancer.   3. 0.0359

Reminder: The student version of this book includes *all* answers to Review Exercises and Cumulative Review Exercises. For the regular Section Exercises, answers are given for odd-numbered exercises only.

TABLE 3-3			
	Cause of Death		
	Cancer	Heart Disease	Other
Smoker	135	310	205
Nonsmoker	55	155	140

4. If one subject is randomly selected, find the probability of getting a non-smoker who died from cancer.   4. 0.0550

5. If one subject is randomly selected, find the probability of getting someone who died from cancer or heart disease.   5. 0.655

6. If three different subjects are randomly selected, find the probability that they were all smokers.   6. 0.274

7. If one subject is randomly selected, find the probability of getting a smoker, given that the subject died from cancer.   7. 0.711

8. If one subject is randomly selected, find the probability of getting someone who died from cancer, given that the subject was a smoker. Are smoking and cancer independent events? Why or why not?

9. In setting up a manufacturing process for a new computer memory storage device, the initial configuration has a 16% yield. That is, 16% of the devices are acceptable and 84% are defective. If 12 of the devices are made, what is the probability of getting at least 1 that is good? If it is very important to have at least 1 good unit for testing purposes, is the resulting probability adequate?   9. 0.877

10. On the basis of past experience, a commuting student knows that when he speeds on any given day, there is a 2% chance of being ticketed. What is the probability of not getting a speeding ticket if he speeds every one of the 150 days in one year of college? If this student cannot afford the increased insurance costs resulting from a speeding ticket, what does the resulting probability suggest about a course of action?

11. The board of directors for the Jefferson Valley Bank has 8 members.
    a. If a committee of 3 is formed by random selection, what is the probability that it consists of the 3 wealthiest members?   11. a. 1/56
    b. If the board must elect a chairperson, vice chairperson, and secretary, how many different slates of candidates are possible?   b. 336

12. The New England Life Insurance Company issues one-year policies to 12 men who are all 27 years of age. Based on data from the Department of Health and Human Services, each of these men has a 99.82% chance of living through the year. What is the probability that they all survive the year?   12. 0.979

8. 0.208; no, smokers have a greater probability of having cancer.

10. 0.0483; should not speed.

13. When betting on *even* in roulette, there are 38 equally likely outcomes, but only 2, 4, 6, . . . , 36 are winning outcomes.   13. a. 9/19
    a. Find the probability of winning when betting on even.
    b. Find the odds against winning with a bet on even.   b. 10:9
    c. Casinos pay winning bets according to odds described as 1:1. What is your net profit if you bet $5 on even and you win?   c. $5

14. A question on a history test requires that 5 events be arranged in the proper chronological order. If a random arrangement is selected, what is the probability that it will be correct?   14. 1/120

15. A pollster claims that 12 voters were randomly selected from a population of 200,000 voters (30% of whom are Republicans), and all 12 were Republicans. The pollster claims that this could easily happen by chance. Find the probability of getting 12 Republicans when 12 voters are randomly selected from this population. Based on the result, does it seem that the pollster's claim is correct?   15. 0.000000531; no.

16. In a statistics class of 8 women and 8 men, 2 groups of 8 students are formed by random selection. What is the probability that all of the women are in the first group and all of the men are in the second group? (*Hint:* Find the number of ways of arranging WWWWWWWWMMMMMMMM.)   16. 1/12,870

. . . .

Cumulative Review Exercises

Age	Number
0–4	3,843
5–14	4,226
15–24	19,975
25–44	27,201
45–64	14,733
65–74	8,499
75 and over	16,800

1. Refer to the accompanying frequency table, which describes the age distribution of Americans who died by accident (based on data from the National Safety Council).
    a. Assuming that the class of "75 and over" has a class mark of 80, calculate the mean age of the Americans who died by accident.   1. a. 43.2
    b. Using the same assumption given in part (a), calculate the standard deviation of the ages summarized in the table.   b. 24.1
    c. If one of the 95,277 ages is randomly selected, find the probability that it is below 15 or above 64.   c. 0.350
    d. If one of the 95,277 ages is randomly selected, find the probability that it is below 15 or between 5 and 44.   d. 0.580
    e. If two different ages are randomly selected from those summarized in the table, find the probability that they are both between 0 and 4 years.   e. 0.00163

2. The accompanying boxplot depicts heights (in inches) of a large collection of randomly selected adult women.
    a. If one of these women is randomly selected, find the probability that her height is between 56.1 in. and 62.2 in.   2. a. 1/4
    b. If one of these women is randomly selected, find the probability that her height is below 62.2 in. or above 63.6 in.   b. 3/4
    c. If two women are randomly selected, find the probability that they both have heights between 62.2 in. and 63.6 in.   c. 1/16

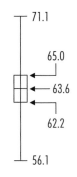

```
10. | 578
11. | 00123447
12. | 099
13. | 3
```

3. An investigation was made of the relationships among the diameters, heights, and volumes of black cherry trees from the Allegheny National Forest in Pennsylvania. The accompanying stem-and-leaf plot represents diameters (in centimeters) for a sample of 15 of the trees used in the study (based on data from *The Minitab Student Handbook*).
   a. Find the mean diameter.   3. a. 11.55 cm
   b. Find the median diameter.   b. 11.30 cm
   c. Find the standard deviation of the diameters.   c. 0.86 cm
   d. Find the variance of the diameters.   d. 0.74 cm²
   e. Which term best identifies the level of measurement of the diameters: nominal, ordinal, interval, ratio?   e. Ratio
   f. If one of these diameters is randomly selected, find the probability that it is less than 11.0 cm.   f. 1/5
   g. If two different diameters are randomly selected, find the probability that they are both above 11.9 cm.   g. 0.0571
   h. If two diameters are randomly selected with replacement, find the probability that they are both above 11.9 cm.   h. 0.0711
   i. If one of these diameters is randomly selected, find the probability that it is less than 11.0 cm or between 10.6 cm and 11.6 cm.   i. 2/3

Computer Project

We can often use computers to find probabilities by simulating an experiment. Recall that a simulation of an experiment is a process that behaves in the same ways as the experiment itself, thus producing similar results. Exercise 31 in Section 3-4 required computation of the probability of getting at least 2 people who share the same birthday when randomly selecting 25 people. Instead of doing theoretical calculations, we will use STATDISK or Minitab to simulate the experiment. Instead of generating actual birthdays, we will generate numbers between 1 and 365, which represent the different possible birthdays. (We will ignore leap years.) For example, a generated number of 5 represents January 5, and 364 represents December 30. We can work with the generated numbers; it isn't necessary to identify the actual day and month. After generating 25 such numbers, we can rank them so that it becomes very easy to see if at least 2 of them are the same "birthday."

Use STATDISK or Minitab to generate 25 "birthdays," then rank them and observe whether at least 2 are the same. Record the result. Repeat this experiment until you are reasonably confident that your estimated probability is approximately correct.

To use STATDISK, first select Data from the main menu bar, then choose Uniform Generator and use Format to set the number of decimal places to 0 (because we want to generate only whole numbers). Proceed to generate a sample size of 25, with a maximum of 365 and a minimum of 1. Use the Sample Editor's Format feature to rank and display the 25 simulated birthdays so

(continued)

you can easily see whether at least 2 are the same.

To use Minitab, select the options of `Calc/Random Data`, `Integer`, and proceed to enter 25 for the number of rows of data, C1 for the column in which to store the results, 1 for the minimum value, 365 for the maximum value, and click on `OK`. Now select `Manip/Sort` and enter C1 for the column, C1 for the column in which to store the results, C1 for the column to sort by, then click on `OK`. The ranked results should be displayed in the first column.

The TI-83 calculator can also be used for this experiment. Press `MATH`, then select `PRB`, then `randInt` and proceed to enter `randInt (1, 365, 25)` to generate 25 simulated birthdays. The birthdays can be stored in list L1 by pressing `STO` and L1. The data can be ranked (or sorted) by pressing `STAT`, then selecting `Sort A`, and entering L1. Now press `STAT` and select `Edit` to view the ranked birthdays.

•••••••• FROM DATA TO DECISION ••••••••

Drug Testing of Job Applicants

According to the American Management Association, most U.S. companies now test at least some employees and job applicants for drug use. The U.S. National Institute on Drug Abuse claims that about 15% of people in the 18–25 age bracket use illegal drugs. Allyn Clark, a 21-year-old college graduate, applied for a job at the Acton Paper Company, took a drug test, and was not offered a job. He suspected that he might have failed the drug test, even though he does not use drugs. In checking with the company's personnel department, he found that the drug test has 99% sensitivity, so only 1% of drug users test negative. Also, the test has 98% specificity, meaning that only 2% of nonusers are incorrectly identified as drug users. Allyn felt relieved by these figures because he believed that they reflected a very reliable test that usually provides good results—but should he be relieved? Can the company feel certain that drug users are not being hired? The accompanying table shows data for Allyn and 1999 other job applicants. Based on those results, find the probability of a "false positive"; that is, find the probability of randomly selecting one of the subjects who tested positive and getting someone who does not use drugs. Also find the probability of a "false negative"; that is, find the probability of randomly selecting someone who tested negative and getting someone who does use drugs. Are the probabilities of these wrong results low enough so that job applicants and the Acton Paper Company need not be concerned?

	Drug Users	Nonusers
Positive test result	297	34
Negative test result	3	1666

········ COOPERATIVE GROUP ACTIVITIES ········

1. *In-class activity:* Divide into groups of three or four and estimate P(2 girls in 3 births) by using a *simulation* with coins. Describe the exact procedure used and the results obtained.

2. *In-class activity:* Divide into groups of three or four and use actual thumbtacks to estimate the probability that when dropped, a thumbtack will land with the point up. How many trials are necessary to get a result that seems to be reasonably accurate?

3. *In-class activity:* Divide into groups of three or four. In each group, agree on the value of a subjective probability for the event that a woman will be elected as President of the United States in the year 2008. Are the various group values approximately the same, or are they very different? Would agreement among the groups mean that the results are accurate?

4. *In-class activity:* Each student should be given a different page torn from the residential section of an old telephone directory. Proceed to randomly select 25 simulated "birthdays" by using the last three digits of telephone numbers, ignoring those above 365. After recording the 25 birthdays, check to determine whether any two of them are the same. The class results can be combined to form an estimate of the probability that when 25 people are randomly selected, at least two of them will have the same birthday.

5. *Out-of-class activity: The Capture-Recapture Method.* Marine biologists often use the capture-recapture method as a way to estimate the size of a population, such as the number of fish in a lake. This method involves capturing a sample from the population, tagging each member in the sample, then returning them to the population. A second sample is later captured and the tagged members are counted along with the total size of this second sample. As an example, suppose a random sample of 50 fish is captured and tagged. Also suppose that a second random sample (obtained later) consists of 100 fish with 20 of them tagged, suggesting that when a fish is captured, the probability of it being tagged is estimated to be 0.20. That is, 20% of the fish in the population have been tagged. Because the original sample of 50 fish were tagged, we estimate that the population size is 250 fish ($50 \div 20/100 = 250$).

It's not easy to actually capture and recapture real fish, but we can simulate an experiment using some uniform collection of items such as BB's, colored beads, M&Ms, or Fruit Loop cereal pieces. "Captured" items in the first sample can be "tagged" with a magic marker (or they can be replaced with similar items of a different color). Illustrate the capture-recapture method by designing and conducting such an experiment. Starting with a large package of M&Ms, for example, collect a sample of 50, then use a magic marker to "tag" each one. Replace the tagged items, mix the whole population, then select a second sample and proceed to estimate the population size. Compare the result to the actual population size obtained by counting all of the items.

6. *In-class activity:* Divide into groups of two. Section 3-5 discussed the "Monte Hall" problem which, according to *Chance* magazine, has been used to study decision making in business schools at Harvard and Stanford. The problem is based on a television game show that was hosted by Monte Hall. Begin by selecting one of the team members to serve as the host. The other team member is the contestant, and there are three doors numbered 1, 2, and 3. The host should *randomly* select one of the doors, and the selection must not be revealed to the contestant. Pretend that the host has put a prize of a new red Corvette behind the door that was randomly selected, but the other two doors have nothing behind them. The contestant should now select one of the three doors. After the contestant reveals which door has been chosen, the host should select an "empty" door and inform the contestant that this particular door has nothing behind it. The host should now offer the contestant a choice of sticking with the original door or switching to the other door that has not been revealed. After the contestant announces his or her decision to stick or switch, the host should announce that the contestant has (or has not) won the Corvette. Record the result along with the contestant's decision to stick or switch. Repeat the game 20 times with the contestant sticking 10 times and switching ten times. Then reverse roles and play the game another 20 times. Find the proportion of times the game was won by sticking and find the proportion of times the game was won by switching. Based on the results, which strategy is better: sticking or switching?

7. *In-class activity:* Divide into groups of two for the purpose of doing an experiment designed to show one approach to dealing with sensitive survey questions, such as those related to drug use, stealing, cheating, or sexual activity. For the purposes of this activity, we will use this innocuous question: "Were you born between January 1 and March 31?" We expect that about 1/4 of all responses should be "yes," but let's assume that the question is very sensitive and subjects are usually reluctant to answer honestly. One team member (the "interviewer") should ask the other (the "survey subject") to flip a coin and write "no" on a piece of paper if the survey subject was *not* born between January 1 and March 31 *and* the coin turns up heads; if the subject was born between those dates or if the coin turns up tails, the response of "yes" should be written. Switch roles so that two responses are obtained from each team. Supposedly, respondents tend to be more honest because the coin flip protects their privacy. Combine all results and analyze them to determine the proportion of people born between January 1 and March 31. The accuracy of the results can be checked against the actual birth dates. The experiment can be repeated with a question that is more sensitive, but such a question is not given here because the author already receives enough mail.

interview

Barbara Carvalho
Director of the Marist College Poll

Lee Miringoff
*Director of the Marist College Institute
for Public Opinion*

Barbara Carvalho and Lee Miringoff
report on their poll results in many
interviews for print and electronic
media, including news programs for
NBC, CBS, ABC, FOX, and public
television. Lee Miringoff appears
regularly on NBC's "Today" show.

What types of polls do you conduct?

We do public polling. We survey public issues, approval
ratings of public officials in New York City, New York
State, and nationwide. We don't do partisan polling for
political parties, political candidates, or lobby groups.
We are independently funded by Marist College, and we
have no outside funding that in any way might suggest
that we are doing research for any particular group on
any one issue. Our program is really an educational
program, but it has wide recognition because the results
are released publicly. Reporters have come to depend on
our results not only for their accuracy and professionalism,
but also because they know that the poll is independent
and is not commissioned by any one media source, as
many polls are.

Who does your interviewing, and what are
their backgrounds?

All of our interviewing is done by paid students who are
trained in interviewing and the specific study they are
working on. Students can take course work in survey
research, public opinion, and data collection. Students in
political science are a natural, but we also get many
communication arts majors, as well as students interested
in statistics, computer analysis, psychology, economics,
sociology, business, and marketing.

Would you recommend statistics for students in fields like history, government, or social science?

Absolutely. They should have at least a research course that deals with the basics of statistical analysis and gives them a foundation by which they can deal with the numbers that they're going to be seeing—no matter what field they go into. The study of statistics is important for understanding one aspect of knowledge, and it's a key to opening up other avenues of pursuit. Statistics cuts across disciplines. Students will inevitably find it in their careers at some point. It might be an evaluation of their work or workplace, or it might involve some marketing aspect or promotional aspect. Surveys are now so pervasive throughout our culture that students are going to see them in their later lives, whether it's in their careers or just as citizens. People are now bombarded with survey information, so it's absolutely vital that as citizens they are in a position to evaluate survey accuracy and worth.

What concepts of statistics do you use?

Statistics comes into play in our sampling, even before we get to data analysis. We use statistics to determine our sample size and to develop an estimate of what would be statistically significant. In the data analaysis most of our studies use basic descriptive statistics. Some of the academic studies get into regression analysis.

How do you select survey respondents?

For a statewide survey we select respondents in proportion to county voter registrations. Different counties have different refusal rates, and if we were to select people at random throughout the state, we would get an uneven model of what the state looks like. We stratify by county and use random digit dialing so that we get listed and unlisted numbers.

What's your typical sample size?

About four people, but they're very carefully selected. Seriously, it can be anywhere from 400 to 1200 or 1500. If we wanted to do an analysis of subgroups within our population group, we would increase our sample size so that we could look at subgroups such as men versus women, or different regional groups, or different income groups.

Is the political process actually influenced by poll results?

Although most polls that people see are public polls, the reality is that the political process is influenced by private polls that the public never sees. No one runs for high office today without using a private poll.

181

4

Probability Distributions

4-1 Overview

The objectives of this chapter are identified. Random variables and probability distributions are described in general, then some special probability distributions are examined.

4-2 Random Variables

Discrete and continuous random variables and probability distributions are described in this section. Methods for finding the mean, variance, and standard deviation for a given probability distribution are also described. The expected value of a probability distribution is defined.

4-3 Binomial Experiments

Binomial experiments are defined. Probabilities in binomial experiments are

calculated using the binomial probability formula, a table of binomial probabilities, and statistical software.

4-4 Mean, Variance, and Standard Deviation for the Binomial Distribution

The mean, variance, and standard deviation for a binomial distribution are calculated. Interpretation of those values is also discussed.

4-5 The Poisson Distribution

The Poisson distribution is described as another special and important example of a discrete probability distribution. A key characteristic of the Poisson distribution is that it applies to occurrences of some event over a specified interval of time, distance, area, or some similar unit.

Chapter Problem:
Are the USAir crashes just a coincidence?

Recently, the media gave much coverage of the fact that USAir jets were involved in four of seven consecutive major air crashes in the United States. At the time of these incidents, USAir was flying 20% of domestic flights. If USAir had 20% of the flights and was as safe as any other airline, we would expect that USAir should have 20% of seven crashes, which is 1.4. Given that USAir had four crashes instead of only one or two, can we conclude that USAir isn't as safe as the other airlines, or is USAir's involvement simply coincidental? That determination depends on the *probability* that the given events occur by chance. We will consider these two questions:

1. Given that USAir flew 20% of all domestic flights, and assuming that USAir is as safe as the other airlines and that plane crashes are independent events that occur at random, what is the probability that USAir will have four of seven consecutive crashes?

2. In deciding whether USAir is unsafe or the victim of coincidence, is the probability described in the preceding question the *relevant* probability? Instead of asking for the probability of USAir having exactly four crashes among seven, is there another question that better addresses the real issue of whether USAir isn't as safe as the other airlines?

The first question can be easily answered using methods presented in this chapter. The second question is more difficult and requires serious thought, but it is extremely important to correctly identify the event that is key to this issue. We will address both questions later in the chapter.

 Overview

In Chapter 2 we saw that we could explore a set of data by using graphs (such as a histogram or boxplot), measures of central tendency (such as the mean), and measures of variation (such as the standard deviation). In Chapter 3 we discussed the basic principles of probability theory. In this chapter we combine those concepts as we develop probability distributions that describe what will *probably* happen instead of what actually did happen. In Chapter 2 we constructed frequency tables and histograms using *observed* real scores, but in this chapter we will construct probability distributions by presenting possible outcomes along with the relative frequencies we *expect*, given an understanding of the relevant circumstances.

Suppose that a casino manager suspects cheating at a dice table. He or she can compare the relative frequency distribution of the actual sample outcomes to a theoretical model that describes the frequency distribution likely to occur with a fair die. A fair die should have a relative frequency histogram similar to the one shown in Figure 4-1(a), but the relative frequency histogram in Figure 4-1(b) depicts a loaded die that favors the outcome of 3. If you get caught sneaking in a die of the type depicted in Figure 4-1(b), you will be in deep trouble.

In Figure 4-1(a) we see relative frequencies based not on actual outcomes, but on our knowledge of the probabilities for the outcomes of a fair die. This figure represents a probability distribution that serves as a model of a theoretically perfect population frequency distribution. In essence, we can describe the frequency table and histogram for a die rolled an infinite number of times. With this knowledge of the population of outcomes, we are able to determine important characteristics, such as the mean and standard deviation. The remainder of this book and the very core of inferential statistics are based on

Figure 4-1   Histograms of Dice Outcomes for (a) a Fair Die and (b) a Loaded Die

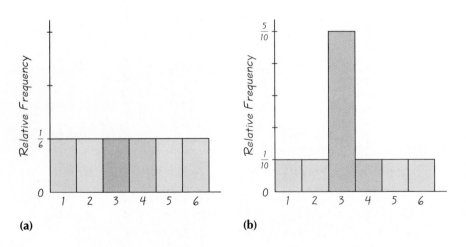

some knowledge of probability distributions. We begin by examining the concept of a random variable, then we consider important distributions that have many real applications.

4-2 Random Variables

In this section we discuss random variable, probability distribution, and procedures for finding the mean and standard deviation for a probability distribution. We will see that a random variable has a number for each outcome of an experiment, and a probability distribution associates a probability value with each numerical outcome of an experiment.

Many everyday situations can be used as experiments that yield outcomes corresponding to some value, and such situations can be described with a random variable.

DEFINITION

A **random variable** is a variable (typically represented by x) that has a single numerical value (determined by chance) for each outcome of an experiment.

Examples of random variables are the following:

x = The number of USAir crashes among seven randomly selected airline crashes

x = The number of women among 10 newly hired employees

x = The number of students absent from statistics class today

x = The height (in inches) of a randomly selected adult male

We use the term *random variable* to describe the value that corresponds to the outcome from a given experiment. The word *random* is used to remind us that we don't usually know what that value is until after the experiment has been conducted.

EXAMPLE   An experiment consists of randomly selecting seven crashes of domestic airline flights and counting those that involve USAir planes. If we let the random variable represent the number of USAir crashes among seven, this experiment has outcomes of 0, 1, 2, 3, 4, 5, 6, 7. (Remember, 0 represents no USAir crashes, 1 represents 1 USAir crash, and so on.) The variable is random in the sense that we don't know the values until after the seven crashes have been selected.

In Section 1-2 we made a distinction between discrete and continuous data. Random variables may also be discrete or continuous, and the following two definitions are consistent with those given in Section 1-2. This chapter deals with discrete random variables, but the following chapters will deal with continuous random variables.

DEFINITIONS

A **discrete random variable** has either a finite number of values or a countable number of values.

A **continuous random variable** has infinitely many values, and those values can be associated with measurements on a continuous scale in such a way that there are no gaps or interruptions.

EXAMPLE

a. The count of the number of patrons viewing a movie is a whole number and is therefore a discrete random variable. The counting device shown in Figure 4-2(a) is capable of indicating only whole numbers. It can therefore be used to obtain values for a discrete random variable.

b. The measure of voltage for a smoke detector battery can be any value between 0 volts and 9 volts and is therefore a continuous random variable. The voltmeter depicted in Figure 4-2(b) is capable of indicating values on a continuous scale, so it can be used to obtain values for a continuous random variable.

(a) Discrete Random Variable: Count of the number of movie patrons.

(b) Continuous Random Variable: The measured voltage of a smoke detector battery.

Figure 4-2 Discrete and Continuous Random Variables

In addition to identifying values of a random variable, we can often identify a probability for each of those values. When we know all values of a random variable along with their corresponding probabilities, we have a probability distribution, defined as follows.

DEFINITION

A **probability distribution** gives the probability for each value of the random variable.

EXAMPLE
Suppose that USAir runs 20% of all flights, and that all flights have the same chance of a crash. If we let the random variable x represent the number of USAir crashes among seven randomly selected crashes, then the probability distribution can be described by Table 4-1.

TABLE 4-1	Probability Distribution for Number of USAir Crashes among Seven	
	x	$P(x)$
	0	0.210
	1	0.367
	2	0.275
	3	0.115
	4	0.029
	5	0.004
	6	0+
	7	0+

(In Section 4-3 we will see how the probabilities in Table 4-1 are found.) In that table, we see that the probability of USAir having no crashes among seven is 0.210, the probability of one crash is 0.367, and so on. The values denoted by 0+ represent positive probabilities that are so small that they become 0.000 when rounded to three decimal places. We prefer to avoid using 0.000 because it incorrectly suggests an impossible event with a probability of 0.

There are various ways to graph a probability distribution, but we present only the **probability histogram.** Figure 4-3 is a probability histogram that resembles the relative frequency histogram from Chapter 2, but the vertical scale delineates *probabilities* instead of actual relative frequencies.

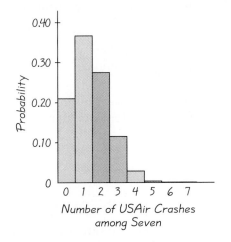

Figure 4-3   Probability Histogram for Number of USAir Crashes among Seven

In Figure 4-3, note that along the horizontal axis, the values of 0, 1, 2, . . . , 7 are located at the centers of the rectangles. This implies that the rectangles are each 1 unit wide, so the areas of the rectangles are 0.210, 0.367, and so on. When the total area of such a probability histogram is 1, the *probabilities* are equal to the corresponding rectangular *areas*. We will see in Chapter 5 and future chapters that this correspondence between area and probability is very useful in statistics.

Every probability distribution must satisfy each of the following two requirements:

Requirements for a Probability Distribution

1. $\Sigma P(x) = 1$ where x assumes all possible values

2. $0 \leq P(x) \leq 1$ for every value of x

The first requirement states that the sum of the individual probabilities must equal 1 and is based on the addition rule for mutually exclusive events. The values of the random variable x represent all possible events in the entire sample space, so we are certain (with probability 1) that one of the events will occur. We use simple addition of the values of $P(x)$ because the different values of x correspond to events that are mutually exclusive. For example, if you randomly select seven airline crashes and let x equal the number from USAir, x cannot be 4 and 5 at the same time. In Table 4-1 we can see that the individual probabilities do result in a sum of 1. Also, the probability rule (see Section 3-2) stating that $0 \leq P(A) \leq 1$ for any event A implies that $P(x)$ must be between 0 and 1 for any value of x. Again, refer to Table 4-1 and note that each individual value of $P(x)$ does fall between 0 and 1. Because Table 4-1 does satisfy both of these requirements, it is an example of a probability distribution. A probability distribution may be described by a table, such as Table 4-1, or a graph, such as Figure 4-3, or a formula, as in the following two examples.

EXAMPLE   Does $P(x) = x/5$ (where x can take on the values of 0, 1, 2, 3) determine a probability distribution?

SOLUTION   If a probability distribution is determined, it must conform to the preceding two requirements. But

$$\Sigma P(x) = P(0) + P(1) + P(2) + P(3)$$

$$= \frac{0}{5} + \frac{1}{5} + \frac{2}{5} + \frac{3}{5}$$

$$= \frac{6}{5} \quad \text{(showing that } \Sigma P(x) \neq 1\text{)}$$

Because the first requirement is not satisfied, we conclude that $P(x)$ given in this example is not a probability distribution.

EXAMPLE   Does $P(x) = x/3$ (where x can be 0, 1, or 2) determine a probability distribution?

SOLUTION   For the given function we find that $P(0) = 0/3$, $P(1) = 1/3$, and $P(2) = 2/3$ so that

1. $\Sigma P(x) = \dfrac{0}{3} + \dfrac{1}{3} + \dfrac{2}{3} = \dfrac{3}{3} = 1$

2. Each of the $P(x)$ values is between 0 and 1 inclusive.

Because the two requirements are both satisfied, the $P(x)$ function given in this example is a probability distribution.

Mean, Variance, and Standard Deviation

In Chapter 2 we saw that there are three extremely important characteristics of data:

1. Representative score, such as an average
2. Measure of scattering or variation, such as a standard deviation
3. Nature or shape of the distribution, such as bell-shaped

The probability histogram can give us insight into the nature or shape of the distribution. Also, we can often find the mean, variance, and standard deviation of data, which provide insight into the other characteristics. The mean, variance, and standard deviation for a probability distribution can be found by applying Formulas 4-1, 4-2, 4-3, and 4-4.

Formula 4-1	$\mu = \Sigma x \cdot P(x)$	Mean for a probability distribution
Formula 4-2	$\sigma^2 = \Sigma[(x - \mu)^2 \cdot P(x)]$	Variance for a probability distribution
Formula 4-3	$\sigma^2 = [\Sigma x^2 \cdot P(x)] - \mu^2$	Variance for a probability distribution
Formula 4-4	$\sigma = \sqrt{[\Sigma x^2 \cdot P(x)] - \mu^2}$	Standard deviation for a probability distribution

σ^2 can be calculated using Formula 4-2 or 4-3. If you have a preference, be sure to inform students of the formula you want them to use.

Caution: The expression $\Sigma x \cdot P(x)$ is the same as $\Sigma[x \cdot P(x)]$. That is, first multiply each value of x by its corresponding probability, and then add the resulting products. Also, evaluate $\Sigma x^2 \cdot P(x)$ by first squaring each value of x, then multiplying each square by the corresponding $P(x)$, then adding. That is, $\Sigma x^2 \cdot P(x) = \Sigma[x^2 \cdot P(x)]$.

   The TI-83 calculator can be used to find the mean and standard deviation for a probability distribution. Enter the values of x in list L1, enter the corresponding probabilities in list L2, then select STAT, CALC, 1-Var Stats, and enter L1, L2 (with the comma). After pressing the ENTER key, the value shown with $\bar{x}$ is actually the mean μ, and the value shown with σx is the value of the standard deviation σ.

> ### Round-off Rule for μ, σ^2, and σ
>
> When using Formulas 4-1 through 4-4, use this rule for rounding results:
>
> **Round results by carrying one more decimal place than the number of decimal places used for the random variable x. If the values of x are integers, round μ, σ^2, and σ to one decimal place.**

It is sometimes necessary to use a different rounding rule because of special circumstances, such as results that require more decimal places to be meaningful.

When we calculate the mean of a probability distribution, we get the average (mean) value that we would expect to get if the trials could be repeated indefinitely. We *don't* get the value we expect to occur most often. In fact, we often get a mean value that cannot occur in any one actual trial (such as 1.5 girls in 3 births). The standard deviation gives us a measure of how much the probability distribution is spread out around the mean. A large standard deviation reflects considerable spread, whereas a smaller standard deviation reflects lower variability with values relatively closer to the mean. The range rule of thumb (see Section 2-5) may also be helpful in interpreting the value of a standard deviation. According to the range rule of thumb, most values should lie within two standard deviations of the mean; it is unusual for a score to differ from the mean by more than two standard deviations.

EXAMPLE   Table 4-1 represents the probability distribution for the number of USAir crashes among seven randomly selected crashes (assuming that USAir has 20% of flights and that crashes are independent and random events). Using the probability distribution described in Table 4-1, assume that we repeat the experiment of randomly selecting seven crashes and each time we find the number of USAir crashes. Find the mean number of USAir crashes (among seven), the variance, and the standard deviation.

SOLUTION   In Table 4-2, the two columns at the left describe the probability distribution given earlier in Table 4-1. We create the three columns at the right for the purposes of the calculations required.

Using Formulas 4-1 and 4-3 and the table results, we get

$$\mu = \Sigma x \cdot P(x) = 1.398 = 1.4 \text{ crashes (rounded)}$$
$$\sigma^2 = [\Sigma x^2 \cdot P(x)] - \mu^2$$
$$= 3.066 - 1.398^2 = 1.111596 = 1.1 \text{ crashes}^2 \text{ (rounded)}$$

TABLE 4-2	Calculating μ, σ^2, and σ for a Probability Distribution				
x	$P(x)$	$x \cdot P(x)$	x^2	$x^2 \cdot P(x)$
0	0.210	0.000	0	0.000
1	0.367	0.367	1	0.367
2	0.275	0.550	4	1.100
3	0.115	0.345	9	1.035
4	0.029	0.116	16	0.464
5	0.004	0.020	25	0.100
6	0+	0.000	36	0.000
7	0+	0.000	49	0.000
Total	1.000	1.398		3.066
	↑	↑		↑
	$\Sigma P(x)$	$\Sigma x \cdot P(x)$		$\Sigma x^2 \cdot P(x)$

If students are performing these calculations with a calculator or computer program, we want the *population* standard deviation and variance (with division by *n*), not the *sample* standard deviation and variance (with division by $n - 1$). Also, here is a hint for using a calculator or computer that allows entry of a frequency table: It may be possible to enter the probabilities as frequencies of 210, 367, 275, 115, 29, 4, 0, 0. The frequency table will have the same μ and σ as the probability distribution.

The standard deviation is the square root of the variance, so

$$\sigma = \sqrt{1.111596} = 1.054323 = 1.1 \text{ crashes (rounded)}$$

We now know that among seven airline crashes, the mean number of USAir crashes is 1.4, the variance is 1.1 "crashes squared," and the standard deviation is 1.1 crashes. Using the range rule of thumb from Section 2-5, we can conclude that most of the time, USAir should have between 0 and 3.6 crashes when seven crashes are randomly selected. (Recall that with the range rule of thumb, we can find rough estimates of minimum and maximum scores by starting with the mean of 1.4 and adding and subtracting 2.2, which is twice the standard deviation.)

Why do Formulas 4-1 through 4-4 work? A probability distribution is actually a model of a theoretically perfect population frequency distribution. The probability distribution is like a relative frequency distribution based on data that behave perfectly, without the usual imperfections of samples. Because the probability distribution allows us to predict the population of outcomes, we are able to determine the values of the mean, variance, and standard deviation. Formula 4-1 accomplishes the same task as the formula for the mean of a frequency table. (Recall that f represents class frequency and N represents population size.) Rewriting the formula for the mean of a frequency table so that it applies to a population and then changing its form, we get

$$\mu = \frac{\Sigma(f \cdot x)}{N} = \Sigma\frac{f \cdot x}{N} = \Sigma x \cdot \frac{f}{N} = \Sigma x \cdot P(x)$$

In the fraction f/N, the value of f is the frequency with which the value x occurs and N is the population size, so f/N is the probability for the value of x.

Similar reasoning enables us to take the variance formula from Chapter 2 and apply it to a random variable for a probability distribution; the result is Formula 4-2. Formula 4-3 is a shortcut version that will always produce the same result as Formula 4-2. Although Formula 4-3 is usually easier to work with, Formula 4-2 is easier to understand directly. On the basis of Formula 4-2, we can express the standard deviation as

$$\sigma = \sqrt{\Sigma(x - \mu)^2 \cdot P(x)}$$

or as the equivalent form given in Formula 4-4.

Expected Value

Many students want to round expected values, such as "1.5 girls." Emphasize that the expected value is not the value expected when an experiment is conducted once; instead, it is a kind of average over a long range. When we say that the expected number of girls in families with three children is 1.5, we don't mean that we expect any one family to have 1.5 girls; we mean that among many families with three children, the mean number of girls is 1.5.

The mean of a discrete random variable is the theoretical mean outcome for infinitely many trials. We can think of that mean as the *expected value* in the sense that it is the average value that we would expect to get if the trials could continue indefinitely. The uses of expected value (also called expectation or mathematical expectation) are extensive and varied, and they play a very important role in an area of application called *decision theory*. (For a discussion of decision theory, see *Business Statistics*, by Triola and Franklin.)

DEFINITION

The **expected value** of a discrete random variable is denoted by E, and it represents the average value of the outcomes. It is found by finding the value of $\Sigma x \cdot P(x)$.

$$E = \Sigma x \cdot P(x)$$

From Formula 4-1 we see that $E = \mu$. That is, the mean of a discrete random variable is the same as its expected value. Repeat the experiment of flipping a coin five times and the *mean* number of heads is 2.5; when flipping a coin five times, the *expected value* of the number of heads is also 2.5.

EXAMPLE   Consider the numbers game started many years ago by organized crime groups and now run legally by many organized governments, as well as some governments that aren't too well organized. Often called a "Pick Three" game, you place a bet that the three-digit number of your choice will be the winning number selected. The typical winning payoff is 499 to 1, meaning that for each winning $1 bet, you would be given $500; your net return is therefore $499. Suppose that you bet $1 on the number 327. What is your expected value of gain or loss?

SOLUTION   For this bet there are two simple outcomes: You win or you lose. Because you have selected the number 327 and there are 1000 possibilities (from 000 to 999), your probability of winning is 1/1000 (or 0.001) and your probability of losing is 999/1000 (or 0.999). Table 4-3 summarizes this situation.

TABLE 4-3	The Numbers Game		
Event	x	$P(x)$	$x \cdot P(x)$
Win	$499	0.001	$0.499
Lose	− $1	0.999	−$0.999
Total			−$0.50 (or −50¢)

From Table 4-3 we can see that when we bet $1 in the numbers game, our expected value is

$$E = \Sigma x \cdot P(x) = -50¢.$$

This means that in the long run, for each $1 bet, we can expect to lose an average of 50¢. This is not a particularly sound investment scheme.

In the preceding example, a player will either lose $1 or win $499; there will never be a loss of 50¢, as the expected value of −50¢ might seem to suggest. The expected value of −50¢ is an average over a long run of bets placed. Even if we're thinking of placing only one bet, the expected value of −50¢ indicates that it's not a good bet. The potential gain is more than offset by the potential loss.

In this section we learned that a random variable has a numerical value associated with each outcome of some chance experiment, and a probability distribution has a probability associated with each value of a random variable. We examined methods for finding the mean, variance, and standard deviation for a probability distribution. We saw that the expected value of a random variable is really the same as the mean. We also learned that lotteries are lousy investments.

4-2 Exercises A: Basic Skills and Concepts

In Exercises 1–4, identify the given random variable as being discrete or continuous.

1. The weight of a randomly selected textbook   1. Continuous
2. The cost of a randomly selected textbook   2. Discrete
3. The number of eggs a hen lays   3. Discrete
4. The amount of milk obtained from a cow   4. Continuous

Cell Phones Cause Brain Cancer?

In *Discover* magazine, mathematician John Allen Paulos cites the case of a man who sued, claiming that his wife's brain cancer was caused by using a cell phone. That lawsuit attracted considerable media attention and caused a drop in stock prices of cell phone manufacturers. Paulos concluded that with an annual brain cancer rate of about 0.007%, and with 10 million cell phone users, we expect about 700 brain tumors each year among those who use cell phones. Paulos's conclusion: "Since only a handful have come to public attention, we should conclude that cellular phones might even effectively ward off brain tumors. Absurd, to be sure, but no more so than the reasoning behind the original hysteria." Paulos refers to this as an example of "the psychological obstacles to rational understanding of statistics."

Recommended assignment: Exercises 2–10 (even), 14, 15, 17, 19, 22.

In Exercises 5–12, determine whether a probability distribution is given. In those cases where a probability distribution is not described, identify the requirement that is not satisfied. In those cases where a probability distribution is described, find its mean, variance, and standard deviation.

5. Probability distribution with
   $\mu = 0.7$, $\sigma^2 = 0.8$, $\sigma = 0.9$

5. When randomly selecting a jail inmate convicted of DWI (driving while intoxicated), the probability distribution for the number x of prior DWI sentences is as described in the accompanying table (based on data from the U.S. Department of Justice).

x	$P(x)$
0	0.512
1	0.301
2	0.132
3	0.055

6. Probability distribution with
   $\mu = 2.0$, $\sigma^2 = 1.0$, $\sigma = 1.0$

6. If your college hires the next 4 employees without regard to gender, and the pool of applicants is large with an equal number of men and women, then the probability distribution for the number x of women hired is described in the accompanying table.

x	$P(x)$
0	0.0625
1	0.2500
2	0.3750
3	0.2500
4	0.0625

7. The Heart Association of Newport plans to open a telephone solicitation office with 8 employees. In planning for the parking area that will serve this office, there is a need to know how many workers will drive their own cars. According to the Hertz Corporation, 69% of all workers commute in their own cars, so the accompanying table describes the probability distribution for the number of workers (among 8 randomly selected workers) who commute in their own cars.   7. Not a probability distribution because $\Sigma P(x) = 0.475 \neq 1$.

x	$P(x)$
0	0.000
1	0.002
2	0.012
3	0.053
4	0.147
5	0.261

8. In assessing credit risks, Jefferson Valley Bank investigates the number of credit cards people have. With x representing the number of credit cards adults have, the accompanying table describes the probability distribution for a population of applicants (based on data from Maritz Marketing Research, Inc.).   8. Probability distribution with $\mu = 2.8$, $\sigma^2 = 6.4$, $\sigma = 2.5$

x	$P(x)$
0	0.26
1	0.16
2	0.12
3	0.09
4	0.07
5	0.09
6	0.07
7	0.14

9. Probability distribution with
   $\mu = 0.8$, $\sigma^2 = 0.5$, $\sigma = 0.7$

9. To settle a paternity suit, two different people are given blood tests. If x is the number having group A blood, then x can be 0, 1, or 2 and the corresponding probabilities are 0.36, 0.48, and 0.16, respectively (based on data from the Greater New York Blood Program).

10. The Baltimore Computer House reports that the probabilities of selling 0, 1, 2, 3, and 4 microcomputers in one day are 0.240, 0.370, 0.205, 0.075, and 0.080, respectively.   10. Not a probability distribution because $\Sigma P(x) = 0.97 \neq 1$.

11. The number of dinners that typical Americans cook in a week are listed with their probabilities (based on data from Millward Brown, reported in *USA Today*): 0 (0.08); 1 (0.05); 2 (0.10); 3 (0.13); 4 (0.15); 5 (0.21); 6 (0.09); 7 (0.19).   11. Probability distribution with $\mu = 4.2$, $\sigma^2 = 4.5$, $\sigma = 2.1$

12. A study of gender bias in media coverage involves the selection of people appearing as the subjects in network TV evening news shows. The subjects are randomly selected in groups of four and the numbers of women are recorded. The probabilities of getting 0, 1, 2, 3, and 4 women are 0.334, 0.421, 0.200, 0.042, and 0.003, respectively (based on data from *USA Today*).   12. Probability distribution with $\mu = 1.0$, $\sigma^2 = 0.7$, $\sigma = 0.9$

13. When you give a casino $5 for a bet on the number 7 in roulette, you have a 1/38 probability of winning $175 and a 37/38 probability of losing $5. What is your expected value? In the long run, how much do you lose for each dollar bet?   13. $-26¢$; $5.26¢$

14. When you give a casino $5 for a bet on the "pass line" in the game of craps, there is a 244/495 probability that you will win $5 and a 251/495 probability that you will lose $5. What is your expected value? In the long run, how much do you lose for each dollar bet?   14. $-7.07¢$; $1.4¢$

15. A 27-year-old woman decides to pay $156 for a one-year life-insurance policy with coverage of $100,000. The probability of her living through the year is 0.9995 (based on data from the U.S. Department of Health and Human Services and AFT Group Life Insurance). What is her expected value for the insurance policy?   15. $-\$106$

16. *Reader's Digest* recently ran a sweepstakes in which prizes were listed along with the chances of winning: $5,000,000 (1 chance in 201,000,000), $150,000 (1 chance in 201,000,000), $100,000 (1 chance in 201,000,000), $25,000 (1 chance in 100,500,000), $10,000 (1 chance in 50,250,000), $5000 (1 chance in 25,125,000), $200 (1 chance in 8,040,000), $125 (1 chance in 1,005,000), and a watch valued at $89 (1 chance in 3774).
   a. Find the expected value of the amount won for one entry.   16. a. $5¢$
   b. Find the expected value if the cost of entering this sweepstakes is the cost of a postage stamp.   b. Mean loss = stamp price $- 5¢$

17. The random variable x represents the number of girls in a family of 3 children. (*Hint:* Assuming that boys and girls are equally likely, we get $P(2) = 3/8$ by examining this sample space: bbb, bbg, bgb, bgg, gbb, gbg, ggb, ggg.) Find the mean, variance, and standard deviation for the random variable x. Also, use the range rule of thumb (from Section 2-5) to approximate the minimum and maximum values of x.

17. $\mu = 1.5$, $\sigma^2 = 0.8$, $\sigma = 0.9$; minimum $= -0.3$ and maximum $= 3.3$, but reality dictates that the minimum and maximum are 0 and 3.

18. The random variable x represents the number of boys in a family of 4 children. (See Exercise 17.) Find the mean, variance, and standard deviation for the random variable x. Also, use the range rule of thumb (from Section 2-5) to approximate the minimum and maximum values of x.

18. $\mu = 2.0$, $\sigma^2 = 1.0$, $\sigma = 1.0$ minimum $= 0$ and maximum $= 4$.

19. $\mu = 0.4$, $\sigma^2 = 0.3$, $\sigma = 0.5$

19. The Menlo Park Electronics Company makes switching devices for traffic signals. One batch of 10 switches includes 2 that are defective. If 2 switches are randomly selected without replacement, let the random variable x represent the number that are defective. Find the mean, variance, and standard deviation for the random variable x.

20. A statistics class includes 3 left-handed students and 24 right-handed students. Two different students are randomly selected for a data collection project and the random variable x represents the number of left-handed students selected. Find the mean, variance, and standard deviation for the random variable x. (*Hint:* Use the multiplication rule of probability to first find $P(0)$ and $P(2)$.)   20. $\mu = 0.2$, $\sigma^2 = 0.2$, $\sigma = 0.4$

23. a. $\mu = 4.5$, $\sigma = 2.872281323$
    b. $\mu = 0$, $\sigma = 1$

24. a.

$$\mu = \Sigma x \cdot P(x) = \left(1 \cdot \frac{1}{n}\right) +$$

$$\left(2 \cdot \frac{1}{n}\right) + \cdots + \left(n \cdot \frac{1}{n}\right)$$

$$= \frac{1}{n}(1 + 2 + \cdots + n)$$

$$= \frac{1}{n} \cdot \frac{n(n + 1)}{2} = \frac{n + 1}{2}$$

b.

$$\sigma^2 = \Sigma x^2 \cdot P(x) - \mu^2$$

$$= \left(1^2 \cdot \frac{1}{n}\right) + \left(2^2 \cdot \frac{1}{n}\right) +$$

$$\cdots + \left(n^2 \cdot \frac{1}{n}\right) - \left(\frac{n + 1}{2}\right)^2$$

$$= \frac{1}{n}(1^2 + 2^2 + \cdots + n^2) -$$

$$\left(\frac{n + 1}{2}\right)^2$$

$$= \frac{1}{n} \cdot \frac{n(n + 1)(2n + 1)}{6} -$$

$$\frac{n^2 + 2n + 1}{4}$$

$$= \frac{2n^2 + 3n + 1}{6} -$$

$$\frac{n^2 + 2n + 1}{4}$$

$$= \frac{4n^2 + 6n + 2 - (3n^2 + 6n + 3)}{12}$$

$$= \frac{n^2 - 1}{12}$$

4-2 Exercises B: Beyond the Basics

21. In each case, determine whether the given function is a probability distribution.
   a. $P(x) = 1/2^x$ where $x = 1, 2, 3, \ldots$ 21. a. Yes
   b. $P(x) = 1/2x$ where $x = 1, 2, 3, \ldots$ b. No, $\Sigma P(x) > 1$
   c. $P(x) = 3/[4(3 - x)!x!]$ where $x = 0, 1, 2, 3$ c. Yes
   d. $P(x) = 0.4(0.6)^{x - 1}$ where $x = 1, 2, 3, \ldots$ d. Yes

22. The mean and standard deviation for a random variable x are 5.0 and 2.0, respectively. Find the mean and standard deviation of the given random variables:
   a. $3 + x$ 22. a. $\mu = 8.0$, $\sigma = 2.0$
   b. $3x$ b. $\mu = 15.0$, $\sigma = 6.0$
   c. $3x + 4$ c. $\mu = 19.0$, $\sigma = 6.0$

23. Digits (0, 1, 2, . . . , 9) are randomly selected for telephone numbers in surveys. The random variable x is the selected digit.
   a. Find the mean and standard deviation of x.
   b. Find the z score for each of the possible values of x, then find the mean and standard deviation of the population of z scores.

24. Assume that the discrete random variable x can assume the values 1, 2, . . . , n, and that those values are equally likely.
   a. Show that $\mu = (n + 1)/2$.
   b. Show that $\sigma^2 = (n^2 - 1)/12$.
   c. An experiment consists of randomly selecting a whole number between 1 and 50, and the random variable x is the value of the number selected. Find the mean and standard deviation for x.
   (*Hint:* $1 + 2 + 3 + \cdots + n = n(n + 1)/2$.
   $1^2 + 2^2 + 3^2 + \cdots + n^2 = n(n + 1)(2n + 1)/6$.)

4-3 Binomial Experiments

In Section 4-2 we learned that a random variable has a numerical value associated with each outcome of some chance experiment, and a probability distribution has a probability associated with each value of a random variable. In most of the examples and exercises in Section 4-2, the probabilities were given for the values of the random variable, but in this section we show how to find probabilities for an important category of probability distributions: *binomial experiments*. Binomial experiments have an element of "twoness": In manufacturing, parts either fail or they do not. In medicine, a patient either survives a year or does not. In advertising, a consumer either recognizes a product or does not.

Change in this edition: The subsection on the Poisson distribution is now included in its own separate section: Section 4-5. Although the binomial distribution is the main focus of this section, Exercises 33–35 deal with the geometric, hypergeometric, and multinomial distributions.

DEFINITION

A **binomial experiment** is one that meets all the following requirements:

1. The experiment must have a *fixed number of trials*.
2. The trials must be independent. (The outcome of any individual trial doesn't affect the probabilities in the other trials.)
3. Each trial must have all outcomes classified into *two categories*.
4. The probabilities must remain *constant* for each trial.

If we conduct a binomial experiment, the distribution of the random variable x is called a *binomial probability distribution* (or *binomial distribution*). The following notation is commonly used.

Notation for Binomial Distributions

S and F (success and failure) denote the two possible categories of all outcomes; p and q will denote the probabilities of S and F, respectively, so

$$P(S) = p$$

$$P(F) = 1 - p = q$$

n denotes the fixed number of trials.

x denotes a specific number of successes in n trials, so x can be any whole number between 0 and n, inclusive.

p denotes the probability of success in *one* of the n trials.

q denotes the probability of failure in *one* of the n trials.

$P(x)$ denotes the probability of getting exactly x successes among the n trials.

Strongly emphasize that x counts *successes* and p is the probability of *success,* so x and p must both refer to the same outcome. A common error is to have x count one category of outcome while p is the probability of the other category of outcome. Students also have some difficulty with the probability p; strongly emphasize that p is the probability of getting a success on *one, single, lone, solitary, individual, isolated* trial.

BOARDING PASSES

The word *success* as used here is arbitrary and does not necessarily describe a desired result. Either of the two possible categories may be called the success S as long as the corresponding probability is identified as p. (The value of q can always be found by subtracting p from 1; if $p = 0.95$, then $q = 1 - 0.95 = 0.05$.) Once a category has been designated as the success S, be sure that p is the probability of a success and x is the number of successes. That is, be sure that the values of p and x refer to the same category designated as a success.

One very common application of statistics involves sampling without replacement, as in testing manufactured items or conducting surveys. Strictly speaking, sampling without replacement involves dependent events, which violates the second requirement in the preceding definition. However, the following rule of thumb is based on the fact that if the sample is very small relative to the population size, the difference in results will be negligible if we treat the trials as independent when they are actually dependent.

> **When sampling without replacement, the events can be considered to be independent if the sample size is no more than 5% of the population size. (That is, $n \leq 0.05N$.)**

EXAMPLE   Given that 10% of us are left-handed, suppose we want to find the probability of getting exactly 3 left-handed students in a class of 15 students. (Some desks are made for left-handed students and the resulting probability could affect the number of such desks that are ordered and placed in classrooms.)

a. Is this a binomial experiment?

b. If this is a binomial experiment, identify the values of n, x, p, and q.

SOLUTION

a. This experiment does satisfy the requirements for a binomial experiment, as shown below.

   1. The number of trials (15) is fixed.

   2. The trials are independent because the left-handedness or right-handedness of any one student doesn't affect the probability of any other student being left-handed.

   3. Each trial has two categories of outcomes: the student either is or is not left-handed.

   4. The probability (0.10) remains constant for the different students.

b. Having concluded that the experiment is binomial, we now proceed to identify the values of n, x, p, and q:

   1. With 15 students in a class, we have $n = 15$.

   2. We want 3 left-handed students (successes), so $x = 3$.

3. The probability of a left-handed student (success) is 0.1, so $p = 0.1$.

4. The probability of failure (not left-handed) is 0.9, so $q = 0.9$.

Again, it is very important to be sure that x and p both refer to the same concept of "success." In this example, we use x to count the desired number of left-handed students, and p is the probability of getting a left-handed student, so x and p do use the same concept of success.

In this section we present three methods for finding probabilities in a binomial experiment. The first method involves calculations using the *binomial probability formula* and is the basis for the other two methods. The second method involves the use of Table A-1, and the third method involves the use of statistical software. We will describe the three methods, illustrate them, and then provide a rationale for each one.

Method 1: Use the Binomial Probability Formula

In a binomial experiment, probabilities can be calculated by using the binomial probability formula.

Formula 4-5
$$P(x) = \frac{n!}{(n - x)!x!} \cdot p^x \cdot q^{n-x} \quad \text{for } x = 0, 1, 2, \ldots, n$$

where
n = number of trials

x = number of successes among n trials

p = probability of success in any one trial

q = probability of failure in any one trial ($q = 1 - p$)

The factorial symbol !, introduced in Section 3-6, denotes the product of decreasing factors. Two examples of factorials are $3! = 3 \cdot 2 \cdot 1 = 6$ and $0! = 1$ (by definition). Many calculators have a factorial key, as well as a key labeled $_nC_r$ that can simplify the computations. For calculators with the $_nC_r$ key, use this version of the binomial probability formula (where n, x, p, and q are the same as in Formula 4-5):

$$P(x) = {_nC_x} \cdot p^x \cdot q^{n-x}$$

The TI-83 calculator has a program for calculating binomial probabilities. The use of the TI-83 calculator will be discussed under Method 3 for finding binomial probabilities.

Note for those planning to take The College Board's AP Statistics exam: The College Board uses $1 - p$ instead of q, and they give the binomial probability formula in this format:

$$P(X = k) = \binom{n}{k} p^k (1 - p)^{n-k}$$

Hint for computations with Formula 4-5: Get a single number for $n!/(n - x)!x!$, a single number for p^x and a single number for q^{n-x}; then simply multiply the three factors together. Also, point out the common calculator error of evaluating a/bc by entering $a \div b \times c$; the correct entry is $a \div b \div c$. Finally, point out that students often get wrong results when they round intermediate results too much.

From Table A–1:

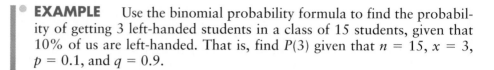

n	x	p .10
15	0	206
	1	343
	2	267
	3	129
	4	043
	5	010
	6	002
	7	0+
	8	0+
	9	0+
	10	0+
	11	0+
	12	0+
	13	0+
	14	0+
	15	0+

Binomial probability
distribution for $n = 15$
and $p = 0.10$

x	P(x)
0	0.206
1	0.343
2	0.267
3	0.129
4	0.043
5	0.010
6	0.002
7	0+
8	0+
9	0+
10	0+
11	0+
12	0+
13	0+
14	0+
15	0+

EXAMPLE   Use the binomial probability formula to find the probability of getting 3 left-handed students in a class of 15 students, given that 10% of us are left-handed. That is, find $P(3)$ given that $n = 15$, $x = 3$, $p = 0.1$, and $q = 0.9$.

SOLUTION   Using the given values of n, x, p, and q in the binomial probability formula (Formula 4-5), we get

$$P(3) = \frac{15!}{(15 - 3)!3!} \cdot 0.1^3 \cdot 0.9^{15-3}$$

$$= \frac{15!}{12!3!} \cdot 0.001 \cdot 0.282429536$$

$$= (455)(0.001)(0.282429536) = 0.129$$

The probability that exactly 3 of the 15 students will be left-handed is 0.129.

Calculation hint: When computing a probability with the binomial probability formula, it's helpful to get a single number for $n!/(n - x)!x!$, a single number for p^x, and a single number for q^{n-x}, then simply multiply the three factors together. Don't round too much when you find those three factors; round only at the end.

Method 2: Use Table A-1 in Appendix A   In some cases, we can easily find binomial probabilities by simply referring to Table A-1 in Appendix A. First locate n and the corresponding value of x that is desired. At this stage, one row of numbers should be isolated. Now align that row with the proper probability of p by using the column across the top. The isolated number represents the desired probability (missing its decimal point at the beginning). A very small probability, such as 0.000000345, is indicated by 0+.

Shown in the margin is part of Table A-1. When $n = 15$ and $p = 0.10$ in a binomial experiment, the probabilities of 0, 1, 2, . . . , 15 successes are 0.206, 0.343, 0.267, . . . , 0+ , respectively.

EXAMPLE   In the preceding example, we used the binomial probability formula to find the probability of 3 successes, given that $n = 15$, $x = 3$, $p = 0.1$, and $q = 0.9$. Use the portion of Table A-1 shown in the margin to find

a. The probability of exactly 3 successes

b. The probability of *at least* 3 successes

SOLUTION

a. The display from Table A-1 shows that when $n = 15$ and $p = 0.1$, $P(3) = 0.129$, which is the same value computed with the binomial probability formula in the preceding example.

b. $P(\text{at least } 3) = P(3 \text{ or } 4 \text{ or } 5 \text{ or } \ldots \text{ or } 15)$

$$= P(3) + P(4) + P(5) + \cdots + P(15)$$
$$= 0.129 + 0.043 + 0.010 + \cdots + 0$$
$$= 0.184$$

In part (b) of the preceding solution, if we wanted to find $P(\text{at least } 3)$ by using the binomial probability formula, we would need to apply that formula 13 times to compute 13 different probabilities, which are then added. (A shortcut would be to calculate $P(0)$, $P(1)$, and $P(2)$ and subtract their sum from 1, but even this shortcut would take much longer than using Table A-1.) Given this choice between the formula and the table, it makes sense to use the table. Note, however, that Table A-1 includes only limited values of n as well as limited values of p, so the table doesn't always work and we must then find the probabilities by using the binomial probability formula or software.

Method 3: Use Computer Software or the TI-83 Calculator   Many computer statistics packages include an option for generating binomial probabilities, and that feature is now available on TI-83 calculators. The following are samples of output obtained from a binomial experiment in which $n = 15$ and $p = 0.1$.

With STATDISK, select Analysis from the main menu, then select the Binomial Probabilities option. Enter the desired values for n and p, and the entire probability distribution will be displayed.

With Minitab, first enter a column C1 of the x values for which you want probabilities (such as 0, 1, 2, 3, 4, 5, 6, 7, 8), then select Calc from the main menu, and proceed to select the submenu items of Probability Distributions and Binomial. Enter the number of trials, the probability of success, and C1 for the input column, then click on OK.

If you're using a TI-83 calculator to find binomial probabilities, press 2nd VARS (to get DISTR, which denotes "distributions"), then select the option identified as binompdf(. Complete the entry of binompdf(n, p, x) with specific values for n, p, and x, then press ENTER, and the result will be the desired probability of $P(x)$.

The following is a good strategy for choosing the best method for finding binomial probabilities:

1. Use computer software or a TI-83 calculator, if available.

2. If a computer or TI-83 calculator cannot be used, use Table A-1 if possible.

3. If a computer or TI-83 calculator cannot be used and the probabilities can't be found using Table A-1, use the binomial probability formula.

Consider pointing out that the selection of "success" is arbitrary. If we stipulate that selection of a *right*-handed student is a "success," then $x = 12$ and $p = 0.9$ and the same solution will be obtained.

In class, make up a few problems that can be solved with the table and be sure that students learn how to use it correctly. Examples: Find the probability of 4 girls in 9 births; find the probability of getting 4 multiple-choice (a, b, c, d, e) questions correct when guesses are made for 15 questions.

Is Statistics Really Worth Anything?

With annual sales exceeding $17 billion, Motorola is one of the largest producers in the world of electronics equipment, including cellular telephones, cordless telephones, pagers, two-way radios, and modules that control car transmissions. In a recent five-year period, Motorola saved approximately $2.5 billion by implementing a quality-improvement plan that makes extensive use of statistical methods. Its pagers and cellular telephones are currently being produced with a target defect rate of only 0.00034%; Motorola is pursuing a goal popularly referred to as the "six-sigma" level of quality, which corresponds to less than 3.4 defects per million parts produced. Motorola has found that the use of statistical methods is necessary for survival in an increasingly competitive market.

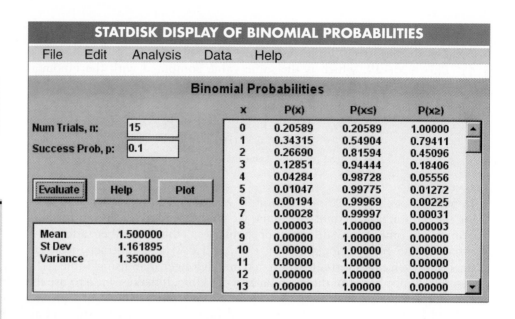

STATDISK DISPLAY OF BINOMIAL PROBABILITIES

File   Edit   Analysis   Data   Help

Binomial Probabilities

Num Trials, n: 15
Success Prob, p: 0.1

Evaluate   Help   Plot

Mean   1.500000
St Dev   1.161895
Variance   1.350000

x	P(x)	P(x≤)	P(x≥)
0	0.20589	0.20589	1.00000
1	0.34315	0.54904	0.79411
2	0.26690	0.81594	0.45096
3	0.12851	0.94444	0.18406
4	0.04284	0.98728	0.05556
5	0.01047	0.99775	0.01272
6	0.00194	0.99969	0.00225
7	0.00028	0.99997	0.00031
8	0.00003	1.00000	0.00003
9	0.00000	1.00000	0.00000
10	0.00000	1.00000	0.00000
11	0.00000	1.00000	0.00000
12	0.00000	1.00000	0.00000
13	0.00000	1.00000	0.00000

MINITAB DISPLAY OF BINOMIAL PROBABILITIES

Binomial with n = 15 and p = 0.100000

x	P(X = x)
0.00	0.2059
1.00	0.3432
2.00	0.2669
3.00	0.1285
4.00	0.0428
5.00	0.0105
6.00	0.0019
7.00	0.0003
8.00	0.0000
9.00	0.0000
10.00	0.0000
11.00	0.0000
12.00	0.0000
13.00	0.0000
14.00	0.0000
15.00	0.0000

In Section 4-2 we presented Table 4-1 as an example of a probability distribution for the USAir crashes discussed at the beginning of the chapter. The following example illustrates how those probabilities were found.

EXAMPLE   At the beginning of this chapter we noted that USAir had 20% of domestic flights and was involved in four of seven consecutive major air crashes in the United States. Assuming that airline crashes are independent and random events, and also assuming that USAir is as safe as the other airlines, find the probability that when seven airliners crash, four of them are from USAir.

SOLUTION   This is a binomial experiment because of the following:

1. We have a fixed number of trials (7).
2. The trials are assumed to be independent.
3. There are two categories: Each crash either does involve a USAir plane or it does not.
4. The probability of a crash involving a plane from USAir (considered a "success" in this example) is 0.20 (because USAir has 20% of domestic flights), and it remains constant for each trial. (We are assuming that the crashes are independent and random).

We begin by identifying the values of n, p, q, and x. We have

$n = 7$ Number of trials (crashes)
$p = 0.20$ Probability of success (when a crash occurs, it is a USAir plane)
$q = 0.80$ Probability of failure (when a crash occurs, it is not a USAir plane)
$x = 4$ Number of USAir crashes among seven

With these particular values of n, p, q, and x, we can obtain the desired probability $P(4)$ from Table A-1 instead of applying the binomial probability formula. (The binomial probability formula could be applied, but using the table is quicker and is less likely to result in an error.) Referring to Table A-1 with $n = 7$, $p = 0.20$, and $x = 4$, we find that $P(4) = 0.029$. That is, there is a 0.029 probability that among seven airline crashes, exactly four of the planes are from USAir.

The low probability (0.029) that we found in the preceding example might seem to suggest that it's unlikely that USAir would be involved in four out of seven crashes just by chance; it seems that USAir's safety isn't as good as the other airlines. But are we really addressing the correct issue? As we asked at the beginning of this chapter, is there another question that better addresses the real issue of whether USAir is unsafe? Here are some other possibilities:

1. What is the probability that USAir has *at least four* out of seven crashes?

How Likely Is an Asteroid Strike?

NASA astronomer David Morrison says that there are about 2000 asteroids with orbits that cross Earth's orbit, yet we have found only 100 of them. It's therefore possible for an undetected asteroid to crash into our planet and cause a global catastrophe that could destroy most life. Morrison says that there is a 1/10,000 probability that within a person's lifetime, there will be an asteroid impact large enough to wipe out all agricultural crops for at least a year, with mass starvation following. Some astronomers recommend a 20-year program aimed at observing asteroids with the goal of early detection of those that are dangerous. Early detection could possibly allow us to alter an asteroid's course so that it isn't a fatal threat.

2. What is the probability that any single airline (with 20% of domestic flights) will have exactly four out of seven crashes?

3. What is the probability that any single airline (with 20% of domestic flights) will have at least four out of seven crashes?

This analysis gets difficult, so consider assigning it as reading to be done very slowly and very carefully. Reduce discomfort levels by informing the class that although it is quite important, the reasoning used here need not be mastered for success in this course.

We eliminate the first question because the real issue concerns the likelihood of *any* airline having four out of seven crashes. To clarify this point, consider your local lottery in which *your* individual chance of winning is extremely small, but the probability of *somebody* winning is fairly high. When someone does win, we don't conclude that this individual had a better chance than anyone else. Similarly, the crash of a USAir plane doesn't necessarily mean that this airline isn't as safe as others; like winning the lottery, the crash might be the result of random events that affect all airlines the same way. We therefore need to find the probability of *any* airline having four out of seven crashes, assuming that each airline has 20% of the flights.

Next, we should eliminate the second question in our list because it refers to *exactly* four crashes out of seven. Our real concern is not the likelihood of getting any specific number of crashes. Instead, the issue rests on the likelihood of getting an outcome *at least as extreme* as the one observed. This is a difficult concept, so let's try to clarify it with another example.

Suppose you were flipping a coin to determine whether it favors heads, and suppose 1000 tosses resulted in 501 heads. Intuition should suggest that this is not evidence that the coin favors heads, because it is quite easy to get 501 heads in 1000 tosses. Yet, the probability of getting exactly 501 heads in 1000 tosses is actually quite small: 0.0252. This low probability reflects the fact that with 1000 tosses, *any specific* number of heads will have a very low probability. However, the result of 501 heads among 1000 tosses is not *unusual* because the probability of getting *at least* 501 heads is high: 0.488. Similarly, we should be asking for the probability of *at least four* crashes, not the probability of exactly four. That is, we need to find the probability that any single airline (with 20% of domestic flights) will have at least four out of seven crashes. If that probability is very low (such as 0.05 or less), it is reasonable to conclude that USAir has a problem, but if it is high (such as greater than 0.05), it is reasonable to conclude that the USAir crashes are a coincidence. First we find the probability of USAir having at least four crashes among seven. We refer to Table A-1 (with $n = 7$, $p = 0.20$, $x = 4, 5, 6, 7$):

$$
\begin{aligned}
P(\text{USAir has at least 4 of 7 crashes}) &= P(4) + P(5) + P(6) + P(7) \\
&= 0.029 + 0.004 + 0^+ + 0^+ \\
&= 0.033
\end{aligned}
$$

Now suppose there are five airlines (denoted by A, B, C, D, E), each with 20% of the flights. The probability of any one of them having at least four of seven crashes is found by using the addition rule for mutually exclusive events. (The events are mutually exclusive because if one airline company has at least four

of seven crashes, no other airline company can also have at least four of seven crashes.)

$$P(A \text{ or } B \text{ or } C \text{ or } D \text{ or } E) = P(A) + P(B) + P(C) + P(D) + P(E)$$
$$= 0.033 + 0.033 + 0.033 + 0.033 + 0.033$$
$$= 0.165$$

To interpret this probability, let's arbitrarily select 0.05 as a cutoff value separating common results from those that are uncommon. (The value of 0.05 is often used. In Chapter 7 we will further discuss the selection of cutoff values.) Because the probability of 0.165 is greater than the cutoff of 0.05, we conclude that it is not uncommon for an airline with 20% of domestic flights to have at least four out of seven crashes. We therefore explain the USAir crashes as a coincidence instead of concluding that USAir has a serious safety problem. This analysis is not simple, but it illustrates an important principle: We should take great care to ask the questions that correctly identify the real issue.

We now have three methods for finding binomial probabilities, but the entries in Table A-1 and statistical software programs both derive their probabilities from the binomial probability formula. We will now explain why that formula is as stated.

In the preceding example we wanted the probability of getting four successes among the seven trials, given a 0.20 probability of success in any one trial. It's correct to reason that for four successes among seven trials, there must be three failures. A common error is to find the probability of four successes and three failures as follows:

$$\underbrace{(0.20 \cdot 0.20 \cdot 0.20 \cdot 0.20)}_{4 \text{ successes}} \cdot \underbrace{(0.80 \cdot 0.80 \cdot 0.80)}_{3 \text{ failures}} = 0.000819$$

This calculation is *wrong* because it contains an implicit assumption that the *first* four outcomes are successes and the *last* three are failures. However, the four successes and three failures can occur in many different sequences, not only the one given above. In fact, there are 35 different sequences of four successes and three failures, each with a probability of 0.000819, so the correct probability is $35 \cdot 0.000819 = 0.029$, as we have found. In general, the number of ways in which it is possible to arrange x successes and $n - x$ failures is shown in Formula 4-6.

Formula 4-6 $\dfrac{n!}{(n - x)!x!}$ Number of outcomes with exactly x successes among n trials

The expression given in Formula 4-6 is from Section 3-6. (Coverage of Section 3-6 is not required for this chapter.) We won't derive Formula 4-6, but its role should be clear: It counts the number of ways we can arrange x successes and

$n - x$ failures. Combining this counting device (Formula 4-6) with the direct application of the multiplication rule for independent events results in the binomial probability formula.

The number of outcomes with exactly x successes among n trials

The probability of x successes among n trials for any 1 particular order

$$P(x) = \frac{n!}{(n - x)!x!} \cdot p^x \cdot q^{n-x}$$

To keep this section in perspective, remember that the binomial probability formula is only one of many probability formulas that can be used for different situations. It is often used in applications such as quality control, voter analysis, medical research, military intelligence, and advertising. Although the main focus of this section is the binomial distribution, other distributions can be found in Exercises 33–35 and Section 4-5.

4-3 Exercises A: Basic Skills and Concepts

Recommended assignment:
Exercises 1–4, 12, 14, 18, 19, 22, 26, 30, 32. These Exercises require the binomial probability formula: 13–16, 21c, 22c, 26, 29, 30, 32.

In Exercises 1–8, determine whether the given experiments are binomial. For those that are not binomial, identify at least one requirement that is not satisfied.

1. Rolling a die 50 times   1. Not binomial; more than two outcomes
2. Tossing an unbiased coin 200 times   2. Binomial
3. Tossing a biased coin 200 times   3. Binomial
4. Surveying 1000 American consumers by asking each one if the brand name of Nike is recognized   4. Binomial
5. Spinning a roulette wheel 500 times   5. Not binomial; more than two outcomes
6. Surveying 1067 people by asking each one if he or she voted in the last election   6. Binomial
7. Sampling (without replacement) a randomly selected group of 12 different tires from a population of 30 tires that includes 5 that are defective
7. Not binomial; trials are not independent
8. Surveying 2000 television viewers to determine whether they can recall a particular product name after watching a commercial   8. Binomial

In Exercises 9–12, assume that in a binomial experiment, a trial is repeated n times. Find the probability of x successes given the probability p of success on a given trial. (Use the given values of n, x, and p and Table A-1.)

9. $n = 3, x = 2, p = 0.9$ 9. 0.243   **10.** $n = 2, x = 0, p = 0.6$ 10. 0.160
11. $n = 8, x = 7, p = 0.99$ 11. 0.075   **12.** $n = 6, x = 1, p = 0.05$ 12. 0.232

In Exercises 13–16, assume that in a binomial experiment, a trial is repeated n times. Find the probability of x successes given the probability p of success

on a single trial. Use the given values of n, x, and p and the binomial probability formula.

13. $n = 3, x = 2, p = 1/4$ 13. 0.141   **14.** $n = 6, x = 2, p = 1/3$ 14. 0.329

15. $n = 10, x = 4, p = 0.35$ **16.** $n = 8, x = 6, p = 0.85$ 16. 0.238   15. 0.238

In Exercises 17–20, refer to the Minitab display in the margin. The probabilities were obtained by entering the values of n = 8 and p = 0.77. When a teenager is randomly selected, there is a 0.77 probability that he or she uses a video game player (based on data from Chilton Research Services). In each case, assume that 8 teenagers are randomly selected and find the indicated probability.

BINOMIAL WITH
N = 8 P = 0.77

K	P(X = K)
0	0.0000
1	0.0002
2	0.0025
3	0.0165
4	0.0689
5	0.1844
6	0.3087
7	0.2953
8	0.1236

17. Find the probability that 6 teenagers use video game players.   17. 0.3087
18. Find the probability that at least 5 teenagers use video game players.
19. Find the probability that fewer than 7 teenagers use video game players.
20. Find the probability that more than 4 teenagers use video game players.

In Exercises 21–32, find the probability requested.

21. Assume that male and female births are equally likely and that the birth of any child does not affect the probability of the gender of any other children. Find the probability of
  a. Exactly 4 girls in 10 births.   21. a. 0.205
  b. At least 4 girls in 10 births.   b. 0.828
  c. Exactly 8 girls in 20 births.   c. 0.120

22. According to Nielsen Media Research, 30% of televisions are tuned to *NFL Monday Night Football* when it is televised. Assuming that this show is being broadcast and that televisions are randomly selected, find the probability that
  a. 5 of 15 televisions are tuned to *NFL Monday Night Football*.
  b. At least 5 of 15 televisions are tuned to *NFL Monday Night Football*.
  c. Exactly 4 of 16 televisions are tuned to *NFL Monday Night Football*.

18. 0.9120
19. 0.5812
20. 0.9120
22. a. 0.206
  b. 0.485
  c. 0.204

23. Mars, Inc., claims that 20% of its plain M&M candies are red. Find the probability that when 15 plain M&M candies are randomly selected, exactly 20% (or 3 candies) are red.   23. 0.250

24. A *Time* magazine article noted that in Los Angeles, for every 100 car accidents in which there is a damage claim, there are 99 claims of injury. If an insurance study begins with the random selection of 12 Los Angeles car accidents with damage claims, find the probability that at least 11 of them have claims of injury.   24. 0.993

25. A statistics quiz consists of 10 multiple-choice questions, each with 5 possible answers. For someone who makes random guesses for all of the answers, find the probability of passing if the minimum passing grade is 60%. Is the probability high enough to make it worth the risk of passing by random guesses instead of studying?   25. 0.007; study

26. Air America has a policy of booking as many as 15 persons on an airplane that can seat only 14. (Past studies have revealed that only 85% of the booked passengers actually arrive for the flight.) Find the probability that if Air America books 15 persons, not enough seats will be available.

27. According to the U.S. Department of Justice, 5% of all U.S. households experienced at least one burglary last year, but Newport police report that a community of 15 homes experienced 4 burglaries last year. By finding the probability of getting 4 or more burglaries in a community of 15 homes, does it seem that this community is just unlucky?

28. A *Computerworld* survey showed that 80% of top executives regularly use personal computers at work. The Telektronic Company plans to move 9 top executives to a new headquarters in Atlanta. If only 7 personal computers are available in Atlanta, find the probability that they will need more. Is that probability high enough so that plans should be initiated to provide more computers?

29. Bill Connors, a quality control manager at the Menlo Park Electronics Company, knows that his company has been making surge protectors with a 10% rate of defective units. He has instituted several measures designed to lower that defect rate. In a test of 20 randomly selected surge protectors, only one is found to be defective. If the 10% defect rate hasn't changed, find the probability that among 20 units, one or none are defective. Based on the result, does it appear that the newly instituted measures are effective?

30. The Telektronic Company purchases large shipments of fluorescent bulbs and uses this *acceptance sampling* plan: Randomly select and test 24 bulbs, then accept the whole batch if there is only one or none that doesn't work. If a particular shipment of thousands of bulbs actually has a 4% rate of defects, what is the probability that this whole shipment will be accepted?   30. 0.751

31. In a study of brand recognition, 95% of consumers recognized Coke (based on data from Total Research Corporation). A pollster reports that among 15 randomly selected consumers, only 10 recognized the Coke brand name. Find the probability that this number will be that low. That is, find the probability of getting 10 or fewer consumers who recognize Coke when 15 are randomly selected. Based on the result, is the pollster's reported result likely to occur by chance?

32. After being rejected for employment, Kim Kelly learns that the Bellevue Advertising Company has hired only two women among the last 20 new employees. She also learns that the pool of applicants is very large, with an approximately equal number of qualified men and women. Help her address the charge of gender discrimination by finding the probability of getting two or fewer women when 20 people are hired, assuming that there is no discrimination based on gender. Does the resulting probability really support such a charge?   32. 0.000201

4-3 Exercises B: Beyond the Basics

33. If a case meets all the conditions of a binomial experiment except that the number of trials is not fixed, then the **geometric distribution** can be used. The probability of getting the first success on the xth trial is given by $P(x) = p(1 - p)^{x - 1}$ where p is the probability of success on any one trial. Assume that the probability of a defective computer component is 0.2. Find the probability that the first defect is found in the seventh component tested.   33. 0.0524

34. If we sample from a small finite population without replacement, the binomial distribution should not be used because the events are not independent. If sampling is done without replacement and the outcomes belong to one of two types, we can use the **hypergeometric distribution.** If a population has A objects of one type, while the remaining B objects are of the other type, and if n objects are sampled without replacement, then the probability of getting x objects of type A and $n - x$ objects of type B is

$$P(x) = \frac{A!}{(A - x)!x!} \cdot \frac{B!}{(B - n + x)!(n - x)!} \div \frac{(A + B)!}{(A + B - n)!n!}$$

In Lotto 54, a bettor selects 6 numbers from 1 to 54 (without repetition), and a winning 6-number combination is later randomly selected. Find the probability of getting
   a. All 6 winning numbers.   34. a. 0.0000000387
   b. Exactly 5 of the winning numbers.   b. 0.0000112
   c. Exactly 3 of the winning numbers.   c. 0.0134
   d. No winning numbers.   d. 0.475

35. The binomial distribution applies only to cases involving 2 types of outcomes, whereas the **multinomial distribution** involves more than 2 categories. Suppose we have 3 types of mutually exclusive outcomes denoted by A, B, and C. Let $P(A) = p_1$, $P(B) = p_2$, and $P(C) = p_3$. In n independent trials, the probability of x_1 outcomes of type A, x_2 outcomes of type B, and x_3 outcomes of type C is given by

$$\frac{n!}{(x_1!)(x_2!)(x_3!)} \cdot p_1{}^{x_1} \cdot p_2{}^{x_2} \cdot p_3{}^{x_3}$$

A genetics experiment involves 6 mutually exclusive genotypes identified as A, B, C, D, E, and F, and they are all equally likely. If 20 offspring are tested, find the probability of getting exactly 5 A's, 4 B's, 3 C's, 2 D's, 3 E's, and 3 F's by expanding the above expression so that it applies to 6 types of outcomes instead of only 3.   35. 0.000535

4-4 Mean, Variance, and Standard Deviation for the Binomial Distribution

In Chapter 2 we explored actual collections of real data and we focused on three important characteristics: (1) the measure of central tendency; (2) the measure of variation; (3) the nature of the distribution. A key point of this chapter is that probability distributions describe what will *probably* happen instead of what actually did happen. In Section 4-2, we learned methods for analyzing probability distributions by finding the mean, the standard deviation, and a probability histogram. In a binomial experiment, the distribution of the random variable x is a binomial probability distribution that describes what will probably happen. With binomial probability distributions, it is again important to investigate the three important characteristics of central tendency, variation, and the nature of the distribution. In Section 4-2 we presented Formulas 4-1, 4-3, and 4-4 for finding the mean, variance, and standard deviation for *any* probability distribution, so those formulas certainly apply to binomial distributions. However, those formulas can be greatly simplified for binomial distributions, as shown below.

For Any Probability Distribution:	**For Binomial Distributions:**
Formula 4-1 $\mu = \Sigma x \cdot P(x)$	Formula 4-7 $\mu = n \cdot p$
Formula 4-3 $\sigma^2 = [\Sigma x^2 \cdot P(x)] - \mu^2$	Formula 4-8 $\sigma^2 = n \cdot p \cdot q$
Formula 4-4 $\sigma = \sqrt{[\Sigma x^2 \cdot P(x)] - \mu^2}$	Formula 4-9 $\sigma = \sqrt{n \cdot p \cdot q}$

Formula 4-7 for the mean makes sense intuitively. If we were to analyze 100 births, we would expect to get about 50 girls, and $n \cdot p$ in this experiment becomes $100 \cdot 1/2$, or 50. In general, if we consider p to be the proportion of successes, then the product $n \cdot p$ will give us the actual number of expected successes in n trials. The variance and standard deviation are not so easily justified, and we prefer to omit the complicated algebraic manipulations that lead to Formulas 4-8 and 4-9. Instead, the following example illustrates that for a binomial experiment, Formulas 4-7, 4-8, and 4-9 will produce the same results as Formulas 4-1, 4-3, and 4-4.

EXAMPLE   At the beginning of this chapter we stated that USAir was involved in four of seven crashes. In Section 4-2 we presented Table 4-1, which represents the probability distribution for the number of USAir crashes among seven randomly selected crashes, assuming that USAir has 20% of flights and that crashes are independent and random events. We then used Formulas 4-1, 4-3, and 4-4 (see Table 4-2) to find these values for the mean, variance, and standard deviation: $\mu = 1.4$ crashes,

$\sigma^2 = 1.1$ "crashes squared," and $\sigma = 1.1$ crashes. In Section 4-3 we found that the USAir crashes could be considered a binomial experiment with $n = 7$, $p = 0.20$, and $q = 0.80$. Given these values, use Formulas 4-7, 4-8, and 4-9 to find the mean, variance, and standard deviation. Verify that the results are the same as those obtained by using Formulas 4-1, 4-3, and 4-4.

SOLUTION   Using the values of $n = 7$, $p = 0.20$, and $q = 0.80$, Formulas 4-7, 4-8, and 4-9 provide these results:

$$\mu = n \cdot p = (7)(0.20) = 1.4$$
$$\sigma^2 = n \cdot p \cdot q = (7)(0.20)(0.80) = 1.1 \quad \text{(rounded)}$$
$$\sigma = \sqrt{n \cdot p \cdot q}$$
$$= \sqrt{(7)(0.20)(0.80)} = \sqrt{1.12} = 1.1 \quad \text{(rounded)}$$

These results show that when we have a binomial experiment, Formulas 4-7, 4-8, and 4-9 will produce the same results as Formulas 4-1, 4-3, and 4-4. (If you compare the results carefully, you may notice small discrepancies in the unrounded values, but these discrepancies are due to rounding errors in Table 4-2.)

EXAMPLE   Some couples prefer to have baby girls because the mothers are carriers of an X-linked recessive disorder that will be inherited by 50% of their sons but none of their daughters. The Ericsson method of gender selection supposedly has a 75% success rate. Suppose 100 couples use the Ericsson method, with the result that among 100 babies, there are 75 girls.

a.   Assuming that the Ericsson method has no effect and assuming that boys and girls are equally likely, find the mean and standard deviation for the number of girls in groups of 100 babies. (The assumption that girls and boys are equally likely is not precisely correct, but it will give us very good results.)

b.   Interpret the values from part (a) to determine whether the result of 75 girls among 100 babies supports a claim that the Ericsson method is effective.

SOLUTION

a.   Let x represent the random variable for the number of girls in 100 births. Assuming that the Ericsson method has no effect and that girls and boys are equally likely, we have $n = 100$, $p = 0.5$, and $q = 0.5$. We can find the mean and standard deviation by using Formulas 4-7 and 4-9 as follows:

(continued)

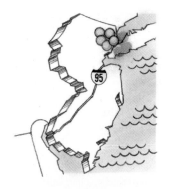

Disease Clusters

Periodically, much media attention is given to a cluster of disease cases in a given community. One New Jersey community had 13 leukemia cases in 5 years, while the normal rate would have been 1 case in 10 years. Research of such clusters can be revealing. A study of a cluster of cancer cases near African asbestos mines led to the discovery that asbestos fibers can be carcinogenic.

When doing such an analysis, we should avoid the mistake of artificially creating a cluster by locating its outer boundary so that it just barely includes cases of disease or death. That would be like gerrymandering, even though it might be unintentional.

$$\mu = n \cdot p = (100)(0.5) = 50$$
$$\sigma = \sqrt{n \cdot p \cdot q} = \sqrt{(100)(0.5)(0.5)} = 5$$

For groups of 100 couples who each have a baby, the mean number of girls is 50 and the standard deviation is 5.

b. We must now interpret the results to determine whether 75 girls among 100 babies is a result that could easily occur by chance, or whether that result is so unlikely that the Ericsson method of gender selection seems to be effective. We will use the range rule of thumb and the empirical rule, both from Section 2-5.

According to the range rule of thumb, rough estimates of the minimum and maximum scores are as follows.

$$\text{minimum} \approx (\text{mean}) - 2 \times (\text{standard deviation}) = 50 - 2(5) = 40$$
$$\text{maximum} \approx (\text{mean}) + 2 \times (\text{standard deviation}) = 50 + 2(5) = 60$$

The range rule of thumb indicates that typical scores are probably between 40 and 60, so 75 girls seems to be a result that is not very likely to occur by chance.

The empirical rule applies only to bell-shaped distributions, and Figure 4-4 shows that this distribution is bell-shaped. According to the empirical rule, 95% of all scores should be within two standard deviations of the mean, and 99.7% of all scores should be within three standard deviations of the mean. The result of 75 girls is five standard deviations away from the mean of 50. (We can calculate the z score as

It is recommended that *interpretation* of standard deviations be stressed and reviewed, as in part (b) of this example. The solution to part (b) incorporates the range rule of thumb, the empirical rule, and the z score.

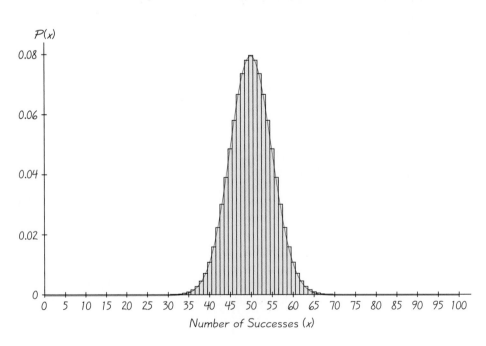

Figure 4-4 Probability Histogram for Binomial Experiment with $n = 100$ and $p = 0.5$

$z = (x - \mu)/\sigma = (75 - 50)/5 = 5$, or we can simply observe that the score of 75 differs from the mean of 50 by 25 units, which is equivalent to five standard deviations.) Again, the result of 75 girls in 100 births is very unusual and is not likely to occur by chance. Instead of concluding that an extremely rare event has occurred, we conclude that the Ericsson method does appear to be effective in increasing the likelihood of a baby being a girl.

It is helpful to develop the technical skill to calculate means and standard deviations, but it is especially important to develop an ability to interpret the significance of values of means and standard deviations, as shown in part (b) of the preceding example.

4-4 Exercises A: Basic Skills and Concepts

In Exercises 1–4, find the mean μ, variance σ^2, and standard deviation σ for the given values of n and p. Assume the binomial conditions are satisfied in each case.

Recommended assignment: Exercises 2, 6, 9–16.

1. $n = 64, p = 0.5$
3. $n = 1068, p = 1/4$
2. $n = 150, p = 0.4$
4. $n = 2001, p = 0.221$

In Exercises 5–8, find the indicated values.

5. Several students are unprepared for a true/false test with 25 questions, and all of their answers are guesses. Find the mean, variance, and standard deviation for the number of correct answers for such students.

6. On a multiple-choice test with 50 questions, each question has possible answers of a, b, c, and d, one of which is correct. For students who guess at all answers, find the mean, variance, and standard deviation for the number of correct answers.

7. The probability of a 7 in roulette is 1/38. In an experiment, the wheel is spun 500 times. If this experiment is repeated many times, find the mean and standard deviation for the number of 7s.   7. $\mu = 13.2, \sigma = 3.6$

8. The probability of winning the New York State lottery is 1/25,827,165. If someone plays twice each week for 50 years (or 5200 times), find the mean and standard deviation for the number of wins. (Express your answers with three significant digits.)   8. $\mu = 0.000201, \sigma = 0.0142$

1. $\mu = 32.0, \sigma^2 = 16.0, \sigma = 4.0$
2. $\mu = 60.0, \sigma^2 = 36.0, \sigma = 6.0$
3. $\mu = 267.0, \sigma^2 = 200.3, \sigma = 14.2$
4. $\mu = 442.2, \sigma^2 = 344.5, \sigma = 18.6$
5. $\mu = 12.5, \sigma^2 = 6.3, \sigma = 2.5$
6. $\mu = 12.5, \sigma^2 = 9.4, \sigma = 3.1$

In Exercises 9–16, consider as unusual any result that differs from the mean by more than two standard deviations. That is, unusual values are either less than $\mu - 2\sigma$ or greater than $\mu + 2\sigma$.

9. When surveyed for brand recognition, 95% of consumers recognize Coke (based on data from Total Research Corporation). A new survey of 1200 randomly selected consumers is to be conducted. For such groups of 1200,
   a. Find the mean and standard deviation for the number who recognize the Coke brand name.   9. a. $\mu = 1140.0$, $\sigma = 7.5$
   b. Is it unusual to get 1170 consumers who recognize the Coke brand name?   b. Yes

10. The New York State Health Department reports a 10% rate of the HIV virus for the "at-risk" population. In one region, an intensive education program is used in an attempt to lower that 10% rate. After running the program, a follow-up study of 200 at-risk individuals is conducted.

10. a. $\mu = 20.0$, $\sigma = 4.2$
    b. 14 is not unusually low and is not strong evidence that the program is effective.

   a. Assuming that the program has no effect, find the mean and standard deviation for the number of HIV cases in groups of 200 at-risk people.
   b. Among the 200 people in the follow-up study, 7% (or 14 people) tested positive for the HIV virus. If the program has no effect, is that rate unusually low? Does this result suggest that the program is effective?

11. Letter frequencies are analyzed by the Central Intelligence Agency in an attempt to decipher intercepted messages. In standard English text, the letter *e* occurs with a relative frequency of 0.130.

11. a. $\mu = 338.0$, $\sigma = 17.1$

   a. Find the mean and standard deviation for the number of times the letter *e* will be found on standard pages of 2600 characters.
   b. In an intercepted message sent to Libya, a standard page of 2600 characters is found to have the letter *e* occurring 307 times. Is this unusual?   b. No

12. b. Yes; the training appears to have been effective.
13. a. $\mu = 1200.0$, $\sigma = 29.0$
    b. 1272 is unusually high, probably due to quality teams or particularly weak programs on other channels.

12. The Newtower Department Store has experienced a 3.2% rate of customer complaints and attempts to lower this rate with an employee retraining program. After the program, 850 customers are observed.
   a. Assuming that the retraining program has no effect, find the mean and standard deviation for the number of complaints in such groups of 850 customers.   12. a. $\mu = 27.2$, $\sigma = 5.1$
   b. In the observed group of 850 customers, 7 complain. Is this result unusual? Does it seem that the retraining was effective?

13. According to Nielsen Media Research, 30% of televisions are tuned to *NFL Monday Night Football* when it is televised. Assume that this show is being broadcast and that 4000 televisions are randomly selected.
   a. For such groups of 4000, find the mean and standard deviation for the number of televisions tuned to *NFL Monday Night Football*.
   b. Is it unusual to find that 1272 of the 4000 TVs are tuned to *NFL Monday Night Football*? What is a likely cause of a rate that is considerably higher than 30%?

14. Mars, Inc., claims that 10% of its M&M plain candies are blue, and a sample of 100 such candies is randomly selected.
   a. Find the mean and standard deviation for the number of blue candies in such groups of 100.   14. a. $\mu = 10.0$, $\sigma = 3.0$
   b. Data Set 11 in Appendix B consists of a random sample of 100 M&Ms in which 5 are blue. Is this result unusual?   b. No

15. A pathologist knows that 14.9% of all deaths are attributable to myocardial infarctions (a type of heart disease).
   a. Find the mean and standard deviation for the number of such deaths that will occur in a typical region with 5000 deaths.   15. a. $\mu = 745.0$, $\sigma = 25.2$
   b. In one region, 5000 death certificates are examined, and it is found that 896 deaths were attributable to myocardial infarction. Is there cause for concern? Why or why not?   b. Yes, unusually high

16. One test of extrasensory perception involves the determination of a shape. Fifty blindfolded subjects are asked to identify the one shape selected from the possibilities of a square, circle, triangle, star, heart, and profile of former president, Millard Fillmore (1800–1874).
   a. Assuming that all 50 subjects make random guesses, find the mean and standard deviation for the number of correct responses in such groups of 50.   16. a. $\mu = 8.3$, $\sigma = 2.6$
   b. If 12 of 50 responses are correct, is this result within the scope of results likely to occur by chance? What would you conclude if 12 of 50 responses are correct?   16. b. Yes. The number of correct guesses is within the realm of expected results; ESP does not appear to be present.

4–4 Exercises B: Beyond the Basics

17. The Providence Computer Supply Company knows that 16% of its computers will require warranty repairs within one month of shipment. In a typical month, 279 computers are shipped.
   a. If x is the random variable representing the number of computers requiring warranty repairs among the 279 sold in one month, find its mean and standard deviation.   17. a. $\mu = 44.6$, $\sigma = 6.1$
   b. For a typical month in which 279 computers are sold, what would be an unusually low figure for the number of computers requiring warranty repair within one month? What would be an unusually high figure? (These values are helpful in determining the number of service technicians that are required.)   b. 32.4, 56.9

18. a. If a company makes a product with an 80% yield (meaning that 80% are good), what is the minimum number of items that must be produced to be at least 99% sure that the company produces at least 5 good items?   18. a. 10
   b. If the company produces batches of items, each with the minimum number determined in part (a), find the mean and standard deviation for the number of good items in such batches.   b. $\mu = 8.0$, $\sigma = 1.3$

5

Normal Probability Distributions

5-1 Overview

Whereas Chapter 4 focused on discrete probability distributions, this chapter focuses on the normal distribution, which is continuous. This entire chapter focuses on the normal distribution because it is the most important single distribution in statistics, and it is used so often throughout the remainder of this book.

5-2 The Standard Normal Distribution

The standard normal distribution is defined as a normal probability distribution having a mean given by $\mu = 0$ and a standard deviation given by $\sigma = 1$. This section presents the basic methods for determining probabilities by using that distribution, as well as for determining standard scores corresponding to given probabilities.

5-3 Nonstandard Normal Distributions: Finding Probabilities

The z score (or standard score) is used to work with normal distributions in which the mean is not 0, the standard deviation is not 1, or both.

5-4 Nonstandard Normal Distributions: Finding Scores

Work with nonstandard normal distributions is continued through methods for finding scores that correspond to given probabilities.

5-5 The Central Limit Theorem

As the sample size increases, the sampling distribution of sample means is shown to approach a normal distribution with mean μ and standard deviation $\sigma/\sqrt{n}$, where n is the sample size and μ and σ represent the mean and standard deviation of the population, respectively. The ideas of this section form a foundation for the important concepts introduced in Chapters 6 and 7.

5-6 Normal Distribution as Approximation to Binomial Distribution

Use of the normal distribution to estimate probabilities in a binomial experiment is described and illustrated.

•••••••• COOPERATIVE GROUP ACTIVITIES ••••••••

1. *In-class activity:* Divide into groups of three or four. Let the random variable x be the value of a coin randomly selected from those in possession of a statistics student. On the basis of the coins belonging to the group members, construct a table (similar to Table 4-1) listing the possible values of x along with their probabilities, then find the mean and standard deviation. What is a practical use of the results?

2. *Out-of-class activity:* Divide into groups of three or four. Conduct a survey that includes the following three questions. (Because the first question may be sensitive to some people, use a procedure that provides for anonymity.)

   • How much do you weigh?

   • Enter three random digits (0, 1, 2, 3, 4, 5, 6, 7, 8, 9). A digit may be chosen more than once.

   • Enter the last three digits of your social security number.

First, compile a list consisting of the *last digit* of each weight. If the weights are accurate and precise, we would expect those *last* digits to have a uniform distribution, so each digit has a probability of 1/10. Assuming that each digit has a probability of 1/10, construct a table describing the probability distribution for the last digits. Next, construct a relative frequency table from the list of recorded *last digits*. Compare the probability distribution to the relative frequency table. (This procedure is often used as a test to determine whether the subjects were actually weighed, or whether they simply reported their weights.) Use a similar procedure to analyze the random digits that were selected. Do those digits seem to be random? Finally, use a similar procedure to analyze the digits in the social security numbers. Do they seem to be random?

5

Normal Probability Distributions

5-1 Overview

Whereas Chapter 4 focused on discrete probability distributions, this chapter focuses on the normal distribution, which is continuous. This entire chapter focuses on the normal distribution because it is the most important single distribution in statistics, and it is used so often throughout the remainder of this book.

5-2 The Standard Normal Distribution

The standard normal distribution is defined as a normal probability distribution having a mean given by $\mu = 0$ and a standard deviation given by $\sigma = 1$. This section presents the basic methods for determining probabilities by using that distribution, as well as for determining standard scores corresponding to given probabilities.

5-3 Nonstandard Normal Distributions: Finding Probabilities

The z score (or standard score) is used to work with normal distributions in which the mean is not 0, the standard deviation is not 1, or both.

5-4 Nonstandard Normal Distributions: Finding Scores

Work with nonstandard normal distributions is continued through methods for finding scores that correspond to given probabilities.

5-5 The Central Limit Theorem

As the sample size increases, the sampling distribution of sample means is shown to approach a normal distribution with mean μ and standard deviation $\sigma/\sqrt{n}$, where n is the sample size and μ and σ represent the mean and standard deviation of the population, respectively. The ideas of this section form a foundation for the important concepts introduced in Chapters 6 and 7.

5-6 Normal Distribution as Approximation to Binomial Distribution

Use of the normal distribution to estimate probabilities in a binomial experiment is described and illustrated.

Distributions, and Binomial. Select the option of Cumulative Probability, enter 350 for the number of trials and 0.95 for the probability of success, enter C1 for the input column, then click on OK. Interpret the results to find the desired probability.

If using a TI-83 calculator, press 2nd VARS (to get DISTR for distributions), then select binomcdf and enter binomcdf(350, 0.95, 340) to get the probability of 340 or fewer successes. The selection of binomcdf calls for the "binomial cumulative density function," which gives the cumulative sum of the probabilities for $x = 0$ up to and including the desired value.

• • • • • • • • FROM DATA TO DECISION • • • • • • • •

Is a Transatlantic Flight Safe with Two Engines?

Statistics is at its best when it is used to benefit humanity in some way. Companies use statistics to become more efficient, increase shareholders' profits, and lower prices. Regulatory agencies use statistics to ensure the safety of workers and clients. This exercise involves a situation in which cost effectiveness and passenger safety are both critical factors. With new aircraft designs and improved engine reliability, airline companies wanted to fly transatlantic routes with twin-engine jets, but the Federal Aviation Administration required at least three engines for transatlantic flights. Lowering this requirement was, of course, of great interest to manufacturers of twin-engine jets (such as the Boeing 767). Also, the two-engine jets use about half the fuel of jets with three or four engines. Obviously, the key issue in approving the lowered requirement is the probability of a twin-engine jet making a safe transatlantic crossing. This probability should be compared to that of three- and four-engine jets. Such a study should involve a thorough understanding of the related probabilities.

A realistic estimate for the probability of an engine failing on a transatlantic flight is 1/14,000. Use this probability and the binomial probability formula to find the probabilities of 0, 1, 2, and 3 engine failures for a three-engine jet and the probabilities of 0, 1, and 2 engine failures for a two-engine jet. Because of the numbers involved, carry all results to as many decimal places as your calculator will allow. Summarize your results by entering the probabilities in the two given tables.

Three Engines			Two Engines	
x	$P(x)$		x	$P(x)$
0	?		0	?
1	?		1	?
2	?		2	?
3	?			

Use the results from the tables and assume that a flight will be completed if at least one engine works. Find the probability of a safe flight with a three-engine jet, and find the probability of a safe flight with a two-engine jet. Write a report for the Federal Aviation Administration that outlines the key issue, and include a recommendation. Support your recommendation with specific results.

f. Find the standard deviation for the number of games for the World Series contests included in the table.   f. 1.1

g. What is the expected number of games for a World Series contest? If a vendor supplies hot dogs to both stadiums involved, and if each stadium averages 30,000 hot dogs sold per game, what is the expected number of hot dogs that will be required?

g. 5.8 games; about 175,000 hot dogs

2. A casino cheat is caught trying to use a pair of loaded dice. At his court trial, physical evidence reveals that some of the black dots were drilled, filled with lead, then repainted to appear normal. In addition to the physical evidence, the dice are rolled in court with these results:

$$
\begin{array}{cccccccccc}
12 & 8 & 9 & 12 & 12 & 9 & 8 & 7 & 12 & 10 \\
12 & 3 & 2 & 12 & 10 & 9 & 12 & 11 & 11 & 12
\end{array}
$$

A probability expert testifies that when fair dice are rolled, the mean should be 7.0 and the standard deviation should be 2.4.

a. Find the mean and standard deviation of the sample values obtained in court.   2. a. $\bar{x} = 9.7$, s = 2.9

b. Based on the outcomes obtained in court, what is the probability of rolling a 12? How does this result compare to the probability of 1/36 (or 0.0278) for fair dice?   b. 0.4 isn't close to 0.0278.

c. If the probability of rolling a 12 with fair dice is 1/36, find the probability of getting at least one 12 when fair dice are rolled 20 times.

c. 0.431

d. If you are the defense attorney, how would you refute the results obtained in court?

d. Claim that the sample of 20 results is too small to obtain meaningful results.

• •

Computer Project

American Air Flight 2705 from New York to San Francisco has seats for 340 passengers. An average of 5% of people with reservations don't show up, so American Air overbooks by accepting 350 reservations for the 340 seats. We can analyze this system by treating it as a binomial experiment with $n = 350$ and $p = 0.95$ (the probability that someone with a reservation does show up).

Find the probability that for a particular flight, there are more passengers than seats. That is, find the probability of at least 341 people showing up with reservations. Because of the value of n, Table A-1 cannot be used and the binomial probability formula would be extremely time consuming and painfully tedious to use. Statistical software or the TI-83 calculator would be the best tool.

With STATDISK, select `Analysis` from the main menu bar, then select the option of `Binomial Probabilities`. Proceed to enter 350 for the Number of Trials and 0.95 for the probability of success on a single trial.

With Minitab, first create a column C1 with the integers 0 through 350 by using the options of `Calc`, then `Set Patterned Data`. Indicate that the result should be stored in column C1, then select a patterned sequence to start at 0 and end at 350 with increment 1. Click on `OK`. Now select `Calc`, `Probability`

c. $\mu = 3.0$, $\sigma = 1.4$

4. a. 0.394
    b. 0.00347
    c. 0.000142
    d. 0.0375

c. **If videotapes are sent to randomly selected college students in many different groups of 10, find the mean and standard deviation for the number (among 10) who own videocassette recorders.**

4. Inability to get along with others is the reason cited in 17% of worker firings (based on data from Robert Half International, Inc.). Concerned about her company's working conditions, the personnel manager at the Flint Fabric Company plans to investigate the 5 employee firings that occurred over the past year. Assuming that the 17% rate applies, find the probability that among those 5 employees, the number fired because of an inability to get along with others is

    a. 0   **b.** 4   **c.** 5   **d.** at least 3

    (Once the actual reasons for the firings have been identified, such probabilities will be helpful in comparing Flint Fabric Company to other companies.)

c. Yes

5. Refer to the data given in Exercise 4. Let the random variable x represent the number of fired employees (among 5) who were let go because of an inability to get along with others.

    a. Find the mean value of x.   5. a. 0.85
    b. Find the standard deviation of the random variable x.   b. 0.84
    c. If we consider as unusual any values that are more than two standard deviations away from the mean, is it unusual to have 4 employees (among 5) fired because of an inability to get along with others?

6. The Washington and Chang Trucking Company operates a large fleet of trucks. Last year, there were 84 breakdowns.

    a. Find the mean number of breakdowns per day.   6. a. $\mu = 0.230$
    b. Find the probability that for a randomly selected day, 2 trucks break down.   b. 0.0210

• •

Cumulative Review Exercises

x	f
4	13
5	22
6	20
7	34

1. The Sports Associates Vending Company supplies refreshments at a baseball stadium and must plan for the possibility of a World Series contest. In the accompanying frequency table (based on past results), x represents the number of baseball games required to complete a World Series contest.

    a. Construct the corresponding relative frequency table.
    b. Does the result from part (a) describe a probability distribution? Why or why not?
    c. Based on the past results, what is the probability that the next World Series contest will last at least 5 games?   c. 0.854
    d. If two different series included in the table are randomly selected, find the probability that they both lasted 7 games.   d. 0.143
    e. Find the mean number of games for the World Series contests included in the table.   e. 5.8

1. b. Yes, each cumulative frequency is a value between 0 and 1, and the sum of the frequencies is 1.

(continued)

- In a *binomial experiment*, probabilities can be found from Table A-1, they can be calculated with Formula 4-5 or a TI-83 calculator, or they can be found with software, such as STATDISK or Minitab.
- In a *binomial experiment*, the mean and standard deviation can be easily found by calculating the values of $\mu = n \cdot p$ and $\sigma = \sqrt{n \cdot p \cdot q}$.
- A *Poisson probability distribution* applies to occurrences of some event over a specified interval, and its probabilities can be computed with Formula 4-10.

· ·

Review Exercises

x	$P(x)$
0	0.0004
1	0.0094
2	0.0870
3	0.3562
4	0.5470

1. **a.** What is a random variable?
   b. What is a probability distribution?
   c. An insurance association's study of home smoke detector use involves homes randomly selected in groups of four. The accompanying table lists values and probabilities for x, the number of homes (in groups of four) that have smoke detectors installed (based on data from the National Fire Protection Association). Does this table describe a probability distribution? Why or why not?
   d. Assuming that the accompanying table does describe a probability distribution, find its mean.
   e. Assuming that the accompanying table does describe a probability distribution, find its standard deviation.

2. Fifteen percent of sport/compact cars are dark green (based on data from DuPont Automotive). Assume that 50 sport/compact cars are randomly selected.
   a. What is the expected number of dark green cars in such a group of 50?
   b. In such groups of 50, what is the mean number of dark green cars?
   c. In such groups of 50, what is the standard deviation for the number of dark green cars?   c. 2.5
   d. Is it unusual to get 15 dark green cars in such a group? Why or why not?
   e. Find the probability that there are exactly 9 dark green cars in such a group of 50.   e. 0.123

3. Thirty percent of college students own videocassette recorders (based on data from the America Passage Media Corporation). The Telektronic Company produced a videotape and sent copies to 10 randomly selected college students as part of a pilot sales program.
   a. Find the probability that exactly one-half of the 10 college students own videocassette recorders.   3. a. 0.103
   b. Find the probability that at least one-half of the 10 college students own videocassette recorders.   b. 0.150

1. a. A random variable is a variable that has a single numerical value (determined by chance) for each outcome of an experiment.
   b. A probability distribution gives the probability for each value of the random variable.
   c. Yes, because each probability value is between 0 and 1 and the sum of the probabilities is 1.
   d. $\mu = 3.4$
   e. $\sigma = 0.7$

2. a. 7.5
   b. 7.5
   d. Yes, using the range rule of thumb, the number is usually between 2.5 and 12.5, so 15 is unusually high.

(continued)

among the 15 trials by using (a) Table A-1 and (b) the Poisson distribution as an approximation to the binomial distribution. Note that the rule of thumb requiring that $n \geq 100$ and $np \leq 10$ suggests that the Poisson distribution might not be a good approximation to the binomial distribution.   13. Table: 0.130; Poisson formula: 0.129

14. The following is a binomial experiment, but the large number of trials involved creates major problems with many calculators. Overcome that obstacle by approximating the binomial distribution by the Poisson distribution.

If you bet on the number 7 for one spin of a roulette wheel, there is a 1/38 probability of winning. Assume that bets are placed on the number 7 in each of 500 different spins.

   14. a. 13.2
       b. 0.110
       c. The approximation is very
          good.

   a. Find the mean number of wins in such experiments.
   b. Find the probability that 7 occurs exactly 13 times.
   c. Compare the result to the probability of 0.111, which is found from the binomial probability formula.

· ·

Vocabulary List

random variable   expected value
discrete random variable   binomial experiment
continuous random variable   binomial probability formula
probability distribution   Poisson distribution
probability histogram

· ·

Review

The central concerns of this chapter were the random variable and the probability distribution. This chapter dealt exclusively with discrete probability distributions. (Chapter 5 will deal with continuous probability distributions.) The following key points were discussed:

• In an experiment yielding numerical results, the *random variable* has numerical values corresponding to different chance outcomes of an experiment.

• A *probability distribution* consists of all values of a random variable, along with their corresponding probabilities. Any probability distribution must satisfy two requirements: $\Sigma P(x) = 1$ and for each value of x, $0 \leq P(x) \leq 1$.

• The important characteristics of a *probability distribution* can be investigated by computing its mean (Formula 4-1) and standard deviation (Formula 4-4) and by constructing a probability histogram.

randomly selected week, the number of accidents requiring medical attention is

    a. 0   **b.** 1   **c.** 2

8. A statistics professor finds that when she schedules an office hour for student help, an average of two students arrive. Find the probability that in
a randomly selected office hour, the number of student arrivals is

    a. 0   **b.** 2   **c.** 5

9. Careful analysis of magnetic computer data tape shows that for each 500
ft of tape, the average number of defects is 2.0. Find the probability of
more than one defect in a randomly selected length of 500 ft of tape.

10. For a recent year, there were 46 aircraft hijackings worldwide (based on
data from the FAA). Using one day as the specified interval required for a
Poisson distribution, we find the mean number of hijackings per day to be
estimated as $\mu = 46/365 = 0.126$. If the United Nations is organizing a
single international hijacking response team, there is a need to know
about the chances of multiple hijackings in one day. Use $\mu = 0.126$ and
find the probability that the number of hijackings (x) in one day is 0 or 1.
Is a single response team sufficient?

11. A classic example of the Poisson distribution involves the number of
deaths caused by horse kicks of men in the Prussian Army between 1875
and 1894. Data for 14 corps were combined for the 20-year period, and
the 280 corps-years included a total of 196 deaths. After finding the mean
number of deaths per corps-year, find the probability that a randomly
selected corps-year has the following numbers of deaths.

    a. 0   **b.** 1   **c.** 2   **d.** 3   **e.** 4

The actual results consisted of these frequencies: 0 deaths (in 144 corps-
years); 1 death (in 91 corps-years); 2 deaths (in 32 corps-years); 3 deaths
(in 11 corps-years); 4 deaths (in 2 corps-years). Compare the actual results
to those expected from the Poisson probabilities. Does the Poisson distribution serve as a good device for predicting the actual results?

12. In a recent year, there were 116 homicide deaths in Richmond, Virginia
(based on "A Classroom Note On the Poisson Distribution: A Model for
Homicidal Deaths In Richmond, VA for 1991," *Mathematics and Computer Education,* by Winston A. Richards). For a randomly selected day,
find the probability that the number of homicide deaths is

    a. 0   **b.** 1   **c.** 2   **d.** 3   **e.** 4

Compare the calculated probabilities to these actual results: 268 days (no
homicides); 79 days (1 homicide); 17 days (2 homicides); 1 day (3 homicides); there were no days with more than 3 homicides.

4–5 Exercises B: Beyond the Basics

13. Assume that a binomial experiment has 15 trials, each with a 0.01 probability of success. Find the probability of getting exactly one success

7. a. 0.819
    b. 0.164
    c. 0.0164

8. a. 0.135
    b. 0.271
    c. 0.0361

9. 0.594

10. 0.993; the single response
team is not sufficient because
there will be about two or
three days each year when
there will be more than two
hijackings.

11. a. 0.497
    b. 0.348
    c. 0.122
    d. 0.0284
    e. 0.00497

The expected frequencies
of 139, 97, 34, 8, and 1.4
compare reasonably well to
the actual frequencies of
144, 91, 32, 11, and 2. The
Poisson distribution does
provide good results.

12. a. 0.728
    b. 0.231
    c. 0.0368
    d. 0.00389
    e. 0.000309

Using the computed proba-
bilities, the expected frequencies are 266, 84, 13,
1.4, and 0.1, and they agree
quite well with the given
actual frequencies.

of not being hit, so we expect that the number of regions with no hits is $576 \cdot 0.395 = 227.5$. (Note that in computing the expected value, we use the number of *regions,* not the number of hits.) This expected value is listed with the others in the third column of Table 4-4. The fourth column describes the results that actually occurred. There were 229 regions that had no hits, 211 regions that were hit once, and so on. We can compare the frequencies *predicted* with the Poisson distribution (third column) to the *actual* frequencies (fourth column) to conclude that there is very good agreement. In this case, the Poisson distribution does a good job of predicting the results that actually occurred. Section 10-2 describes a statistical procedure for determining whether such expected frequencies constitute a good "fit" to the actual frequencies.

The Poisson distribution is sometimes used to approximate the binomial distribution when n is large and p is small. One rule of thumb is to use such an approximation when $n \geq 100$ and $np \leq 10$. When using the Poisson distribution as an approximation to the binomial distribution, the required value of the mean μ can be found from Formula 4-7: $\mu = n \cdot p$. (See Exercises 13 and 14.)

This chapter presented a variety of different *discrete* probability distributions, including binomial (Sections 4-3 and 4-4), Poisson (this section), geometric (Exercise 33 in Section 4-3), hypergeometric (Exercise 34 in Section 4-3), and multinomial (Exercise 35 in Section 4-3). In the following chapter we shift our attention to the extremely important *normal* probability distribution, which is *continuous* instead of discrete.

⁴⁻⁵ Exercises A: Basic Skills and Concepts

Recommended assignment: Exercises 1, 5, 8, 10–13.

In Exercises 1–4, assume that the Poisson distribution has the indicated mean and use Formula 4-10 to find the probability of the value given for the random variable x.

1. $\mu = 2, x = 3$ 1. 0.180
2. $\mu = 4, x = 1$ 2. 0.0733
3. $\mu = 0.845, x = 2$ 3. 0.153
4. $\mu = 0.250, x = 2$ 4. 0.0243

In Exercises 5–12, use the Poisson distribution to find the indicated probabilities.

5. a. 0.105
   b. 0.237
   c. 0.113

6. a. 0.607
   b. 0.303
   c. 0.0758

5. A new hospital is being planned for Newtown, a community that does not yet have its own hospital. If Newtown averages 2.25 births per day, find the probability that in one day, the number of births is
   a. 0   b. 1   c. 4

6. The Scott Auto Park averages 0.5 car sale per day. Find the probability that for a randomly selected day, the number of cars sold is
   a. 0   b. 1   c. 2

7. The Townsend Manufacturing Company experiences a weekly average of 0.2 accident requiring medical attention. Find the probability that in a

The Poisson distribution differs from the binomial distribution in these important ways:

1. The binomial distribution is affected by the sample size n and the probability p, whereas the Poisson distribution is affected only by the mean μ.

2. In a binomial distribution, the possible values of the random variable x are $0, 1, \ldots, n$, but a Poisson distribution has possible x values of $0, 1, 2, \ldots$ with no upper limit.

EXAMPLE   In analyzing hits by V-1 buzz bombs in World War II, South London was subdivided into 576 regions, each with an area of 0.25 km^2. A total of 535 bombs hit the combined area of 576 regions. If a region is randomly selected, find the probability that it was hit exactly twice.

SOLUTION   The Poisson distribution applies because we are dealing with the occurrences of bomb hits over an interval of one region. The mean number of hits per region is $\mu = 535/576 = 0.929$. Because we want the probability of two hits in a region, we let $x = 2$ and use Formula 4-10 as follows.

$$P(x) = \frac{\mu^x \cdot e^{-\mu}}{x!} = \frac{0.929^2 \cdot 2.71828^{-0.929}}{2!} = \frac{0.863 \cdot 0.395}{2} = 0.170$$

The probability of a particular region being hit exactly twice is $P(2) = 0.170$.

In the preceding example, we can also calculate the probabilities for 0, 1, 3, 4, and 5 hits. (We stop at $x = 5$ because no region was hit more than five times and the probabilities for $x > 5$ are 0.000 when rounded to three decimal places.) Those probabilities are listed in Table 4-4. Using the calculated probabilities, we are able to find the expected number of regions with no hits, 1 hit, and so on. For example, the 576 regions each have a probability of 0.395

Table 4-4	V-1 Buzz Bomb Hits for 576 Regions in South London		
Number of Bomb Hits	Probability	Expected Number of Regions	Actual Number of Regions
0	0.395	227.5	229
1	0.367	211.4	211
2	0.170	97.9	93
3	0.053	30.5	35
4	0.012	6.9	7
5	0.002	1.2	1

14. Mars, Inc., claims that 10% of its M&M plain candies are blue, and a sample of 100 such candies is randomly selected.
    a. Find the mean and standard deviation for the number of blue candies in such groups of 100.   14. a. $\mu = 10.0$, $\sigma = 3.0$
    b. Data Set 11 in Appendix B consists of a random sample of 100 M&Ms in which 5 are blue. Is this result unusual?   b. No

15. A pathologist knows that 14.9% of all deaths are attributable to myocardial infarctions (a type of heart disease).
    a. Find the mean and standard deviation for the number of such deaths that will occur in a typical region with 5000 deaths.   15. a. $\mu = 745.0$, $\sigma = 25.2$
    b. In one region, 5000 death certificates are examined, and it is found that 896 deaths were attributable to myocardial infarction. Is there cause for concern? Why or why not?   b. Yes, unusually high

16. One test of extrasensory perception involves the determination of a shape. Fifty blindfolded subjects are asked to identify the one shape selected from the possibilities of a square, circle, triangle, star, heart, and profile of former president, Millard Fillmore (1800–1874).
    a. Assuming that all 50 subjects make random guesses, find the mean and standard deviation for the number of correct responses in such groups of 50.   16. a. $\mu = 8.3$, $\sigma = 2.6$
    b. If 12 of 50 responses are correct, is this result within the scope of results likely to occur by chance? What would you conclude if 12 of 50 responses are correct?   16. b. Yes. The number of correct guesses is within the realm of expected results; ESP does not appear to be present.

4-4 Exercises B: Beyond the Basics

17. The Providence Computer Supply Company knows that 16% of its computers will require warranty repairs within one month of shipment. In a typical month, 279 computers are shipped.
    a. If x is the random variable representing the number of computers requiring warranty repairs among the 279 sold in one month, find its mean and standard deviation.   17. a. $\mu = 44.6$, $\sigma = 6.1$
    b. For a typical month in which 279 computers are sold, what would be an unusually low figure for the number of computers requiring warranty repair within one month? What would be an unusually high figure? (These values are helpful in determining the number of service technicians that are required.)   b. 32.4, 56.9

18. a. If a company makes a product with an 80% yield (meaning that 80% are good), what is the minimum number of items that must be produced to be at least 99% sure that the company produces at least 5 good items?   18. a. 10
    b. If the company produces batches of items, each with the minimum number determined in part (a), find the mean and standard deviation for the number of good items in such batches.   b. $\mu = 8.0$, $\sigma = 1.3$

Speed Kills

In 1987, federal legislation allowed states to raise the maximum speed limit from 55 mi/h to 65 mi/h on certain interstates. A statistical analysis showed that in Iowa, a 20% increase in fatal crashes is attributable to the increased speed limit. (See "Evaluating the Impact of the 65 mph Maximum Speed Limit on Iowa Rural Interstates," by Ledolter and Chan, *The American Statistician*, Vol. 50, No. 1.) This study used the Poisson distribution to find margins of errors for estimated changes in fatalities for rural interstates, urban interstates, rural primary roads, and rural secondary roads. Using the Poisson distribution, the increase in fatalities on rural interstates was found to be statistically significant, but the modest changes on roads with no speed limit changes were not significant.

4-5 The Poisson Distribution

If you've ever waited in a line at Disney World, it's likely that your behavior was analyzed with the Poisson distribution, which is a probability distribution often used as a mathematical model describing arrivals of people in a line. Other applications include the study of vehicle crashes, shoppers arriving at a checkout counter, cars arriving at a gas station, and computer users logging onto the Internet. The Poisson distribution is defined as follows.

DEFINITION

The **Poisson distribution** is a discrete probability distribution that applies to occurrences of some event *over a specified interval*. The random variable x is the number of occurrences of the event in an interval. The interval can be time, distance, area, volume, or some similar unit. The probability of the event occurring x times over an interval is given by Formula 4-10.

Formula 4-10 $$P(x) = \frac{\mu^x \cdot e^{-\mu}}{x!} \quad \text{where } e \approx 2.71828$$

The TI–83 calculator has a program for finding probabilities using Formula 4-10. Press 2nd VARS (to get DISTR, which denotes "distributions"), then select the option identified as poissonpdf(. Complete the entry of poissonpdf(μ, x) with specific values for μ and x, then press ENTER. STAT-DISK also finds probabilities for the Poisson distribution.

Some examples of a random variable in a Poisson distribution include the number of cars arriving at a gas station *during one minute*, the number of aircraft hijackings *in a day*, and the number of defective parts replaced on a new Corvette *during the first year* of warranty.

The Poisson distribution has the following requirements:

- The random variable x is the number of occurrences of an event *over some interval*.
- The occurrences must be *random*.
- The occurrences must be *independent* of each other.
- The occurrences must be *uniformly distributed* over the interval being used.

The Poisson distribution has these parameters:

- The mean is μ.
- The standard deviation is $\sigma = \sqrt{\mu}$.

Chapter Problem:
So, you want to be an astronaut?

If you're 6 ft 8 in. tall and weigh 350 lb, you might make a terrific defensive lineman for the San Francisco 49ers football team, but you wouldn't be selected as an astronaut. Recently, cooperation between the American and Russian space programs highlighted a problem of astronaut size. A Russian Soyuz spacecraft carries astronauts into space and serves as an emergency escape from the Russian Mir space station, but the Soyuz design requires that astronauts have heights between 64.5 in. and 72 in. They must also weigh less than 188 lb. Our American space shuttles are much roomier and require that astronauts have heights between 58.5 in. and 76 in. Because of the greater restrictions imposed by Soyuz spacecraft, NASA faced a shortage of astronauts qualified to participate in joint efforts with Russia. NASA

was therefore considering a recruitment campaign with the goal of finding more qualified astronauts who could meet the more restrictive Russian size limitations.

Height restrictions are important in forming astronaut teams, selecting Rockette dancers, allowing access to amusement park rides, and entering military service. In such cases, it is very helpful to understand the distribution of heights as well as other relevant physical measurements. For example, can we calculate the percentage of American men who have heights between 64.5 in. and 72.0 in.? Can we calculate the percentage of women who fall between those limits? This chapter will introduce methods for finding such percentages.

This chapter is critically important for two major reasons: (1) It discusses the most important distribution in statistics; (2) it introduces concepts that serve as an important foundation in the chapters that follow.

Note to users of the 6th edition and previous editions: Section 5-3 (Nonstandard Normal Distributions) from the 6th edition is now divided into Section 5-3 (for finding probabilities) and Section 5-4 (for finding scores from probabilities). Reviewers and users generally prefer this organization, which had been used in earlier editions.

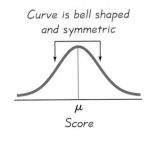

Figure 5-1
The Normal Distribution

5-1 Overview

In Chapter 4 we introduced the concept of the *random variable* as a variable having a single numerical value (determined by chance) for each outcome of an experiment. We noted that a *probability distribution* gives the probability for each value of the random variable. Chapter 4 included only *discrete* random variables, such as those in binomial distributions that have a finite number of possible values. The number of quarters produced by a U.S. mint each day is an example of a discrete distribution. There are also many different *continuous* probability distributions, such as the weights of the quarters produced at the mint. Distributions can be either discrete or continuous, and they can be described by their *shape,* such as a bell shape. This chapter focuses on normal distributions, which are extremely important because they occur so often in real applications. Heights of adult women, weights of adult men, and third-grade reading test scores are some examples of normally distributed populations.

DEFINITION

A continuous random variable has a **normal distribution** if that distribution is symmetric and bell shaped, as in Figure 5-1, and the distribution fits the equation given as Formula 5-1.

Formula 5-1

$$y = \frac{e^{-\frac{1}{2}\left(\frac{x-\mu}{\sigma}\right)^2}}{\sigma\sqrt{2\pi}}$$

The complexity of Formula 5-1 is overwhelming for most students, but we can take comfort in knowing that it isn't really necessary for us to actually use it. What it shows is that any particular normal distribution is determined by two parameters: the mean μ and standard deviation σ. Once specific values are selected for μ and σ, we can graph Formula 5-1 as we would graph any equation relating x and y; the result is a probability distribution with a bell shape. We will see that this normal distribution has many real applications, and we will use it often throughout the remaining chapters.

5-2 The Standard Normal Distribution

We begin this section by using a continuous probability distribution to illustrate the important concept that there is a correspondence between area and probability. It's easier to see that correspondence if we use a distribution such as the one shown in Figure 5-2. That figure shows the probability distribution for temperatures in a manufacturing process. Those temperatures are con-

trolled so that they range between 0°C and 5°C, with all values equally likely. This particular type of distribution occurs so often that we give it a name: *uniform distribution.*

DEFINITION

A **uniform distribution** is a probability distribution in which every value of the random variable is equally likely.

In Section 4-2 we identified two requirements for a probability distribution: (1) $\Sigma P(x) = 1$; and (2) $0 \le P(x) \le 1$ for all values of x. Also in Section 4-2, we stated that the graph of a discrete probability distribution is called a probability histogram. The graph of a continuous probability distribution, such as Figure 5-2, is called a *density curve,* and it must satisfy two properties similar to the requirements for discrete probability distributions, as listed in the Definition.

DEFINITION

A **density curve** is a graph of a continuous probability distribution. It must satisfy the following properties:

1. The total area under the curve must be 1.
2. Every point on the curve must have a vertical height that is 0 or greater.

For users of the TI-83 calculator, it might be helpful to point out the use of the term *density,* as in `normalpdf` (for "normal probability density function," which computes values using Formula 5-1), and `normalcdf` (for "normal cumulative density function," which gives areas under the normal curve).

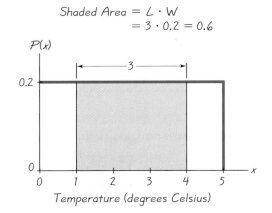

Figure 5-2   Uniform Distribution of Temperatures

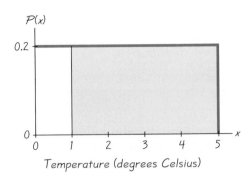

Figure 5-3   Temperatures
Greater than 1°C

The temperatures depicted in Figure 5-2 are controlled so that they range between 0° C and 5° C, and every possible value is equally likely. By setting the height of the rectangle in Figure 5-2 to be 0.2, we force the enclosed area to be $5 \times 0.2 = 1$, as required. This property (area = 1) makes it very easy to solve probability problems. For example, given the uniform distribution of Figure 5-2, the probability of randomly selecting a temperature between 1° C and 4° C is 0.6, which is the area of the shaded region. Here's an important point: **A density curve is the graph of a continuous random variable, so the area under the curve is 1 and there is a correspondence between area and probability.** In Figure 5-2, the probability of a value between 1° C and 4° C is 0.6, and it can be found by finding the area of the corresponding shaded rectangular region with dimensions 0.2 by 3.

EXAMPLE   A uniform distribution of temperatures ranges from 0° C to 5° C. If one temperature is randomly selected, find the probability that it is greater than 1° C.

SOLUTION   In Figure 5-3 the temperatures greater than 1° C are represented by the shaded rectangle that has an area of $4 \cdot 0.2 = 0.8$. Because there is a correspondence here between area and probability, we can conclude that there is a 0.8 probability that a randomly selected temperature will be greater than 1° C.

The density curve of a uniform distribution results in a rectangle, so it's easy to find any area by simply multiplying width and height. The density curve of a normal distribution has the more complicated bell shape shown in Figure 5-1, so it's more difficult to find areas, but the basic principle is the same: There is a correspondence between area and probability.

Just as there are many different uniform distributions, there are also many different normal distributions, with each one depending on two parameters: the population mean μ and the population standard deviation σ. Figure 5-4 shows density curves for heights of adult women and men. Because men have a larger mean height, the density curve for men is farther to the right. Because

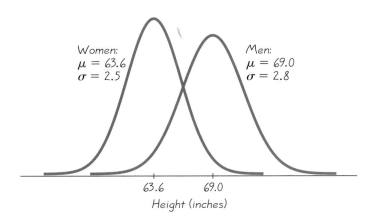

Figure 5-4   Heights of Adult Women and Men

men have a slightly larger standard deviation, the density curve for men is slightly wider. Among the infinite possibilities, there is one normal distribution that is of special interest.

DEFINITION

The **standard normal distribution** is a normal probability distribution that has a mean of 0 and a standard deviation of 1. (See Figure 5-5.)

Suppose that somehow we were forced to perform calculations using Formula 5-1. We would quickly see that the most workable values for μ and σ are $\mu = 0$ and $\sigma = 1$. By letting $\mu = 0$ and $\sigma = 1$, mathematicians have calculated areas under the curve. As shown in Figure 5-5 on the following page, the area under the curve bounded by the mean of 0 and the score of 1 is 0.3413. Remember, the total area under the curve is always 1; this allows us to make the correspondence between area and probability, as we did in the preceding example with the uniform distribution.

Finding Probabilities When Given z Scores

Figure 5-5 (see page 232) shows that the area bounded by the curve, the horizontal axis, and the scores of 0 and 1 is an area of 0.3413. Although the figure shows only one area, Table A-2 (in Appendix A and the *Formulas and Tables* insert card) includes areas (or probabilities) for many different regions.

Table A-2 gives the probability corresponding to the area under the curve bounded on the left by a vertical line above the mean of 0 and bounded on the right by a vertical line above any specific positive score denoted by z (as in Figure 5-6). (The TI-83 calculator can provide areas of the type found in Table A-2; refer to the function identified as `normalcdf`, which represents a normal cumulative density function.) Note that when you use Table A-2, the part of the z score denoting hundredths is found across the top row. The following

Before defining the standard normal distribution, ask the students which values of μ and σ they would select if they had to work with Formula 5-1. Students are usually quite good at recognizing the advantages of choosing $\mu = 0$ and $\sigma = 1$.

For students planning to take the College Board's AP Statistics exam, point out that unlike the table of standard normal probabilities found in most textbooks, the College Board uses a table of *cumulative* probabilities. Be sure to note the difference between the formats of Table A-2 in this book and the table used by the College Board, in which the "table entry for z is the probability lying *below* z." Consider using a reproduction of the College Board's table throughout this chapter.

For students using TI-83 calculators: Point out that instead of using Table A-2, they can find the probabilities by using `normalcdf`.

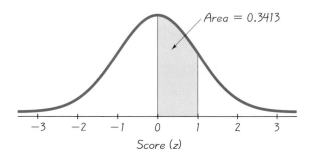

Area = 0.3413

Score (z)

Figure 5-5 Standard Normal Distribution, with Mean $\mu = 0$ and Standard Deviation $\sigma = 1$

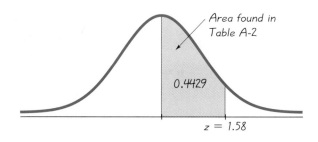

Area found in Table A-2

0.4429

$z = 1.58$

Figure 5-6 The Standard Normal Distribution

The area of the shaded region bounded by the mean of 0 and the positive number z can be found in Table A-2.

example requires that we find the probability associated with a score between 0 and 1.58. Begin with the z score of 1.58 by locating 1.5 in the left column; now find the value in the adjoining row of probabilities that is directly below 0.08, as shown in this excerpt from Table A-2.

z	$\ldots$	.08
.		.
.		.
.		.
1.5	$\ldots$	.4429

Suggestion: Give several examples in class and clarify the use of Table A-2. Comment that use of that table must be mastered sooner or later, and the sooner the better. Also, constantly stress the importance of understanding that the numbers in the body of the table are *areas* or probabilities, whereas the *z* scores are *distances* representing the number of standard deviations that a score is away from the mean.

The area (or probability) value of 0.4429 indicates that there is a probability of 0.4429 of randomly selecting a score between 0 and 1.58. It is essential to remember that Table A-2 is designed only for the standard normal distribution, which has a mean of 0 and a standard deviation of 1. (Nonstandard cases will be considered in the following sections.) It is also essential to avoid confusing z scores and areas. (Remember, a z score measures the number of standard deviations that a value is away from the mean.)

z score: *distance* along horizontal scale on graph; refer to the *leftmost column (and top row)* in Table A-2

Area (or probability): *region* under the curve; refer to the numbers in the *body* of Table A-2

EXAMPLE   The Precision Scientific Instrument Company manufactures thermometers that are supposed to give readings of 0°C at the freezing point of water. Tests on a large sample of these instruments reveal that at the freezing point of water, some thermometers give readings below 0°

(denoted by negative numbers) and some give readings above 0° (denoted by positive numbers). Assume that the mean reading is 0° C and the standard deviation of the readings is 1.00° C. Also assume that the frequency distribution of errors closely resembles the normal distribution. If one thermometer is randomly selected, find the probability that, at the freezing point of water, the reading is between 0° and +1.58°.

SOLUTION   The probability distribution of readings is a standard normal distribution because the readings are normally distributed with $\mu = 0$ and $\sigma = 1$. We need to find the area between 0 and z (the shaded region) in Figure 5-6 with $z = 1.58$. From Table A-2 we find that this area is 0.4429. The probability of randomly selecting a thermometer with an error between 0° and +1.58° is therefore 0.4429. Another way to interpret this result is to conclude that 44.29% of the thermometers will have errors between 0° and +1.58°.

EXAMPLE   Using the thermometers from the preceding example, find the probability of randomly selecting one thermometer that reads (at the freezing point of water) between −2.43° and 0°.

SOLUTION   We are looking for the region shaded in Figure 5-7(a), but Table A-2 is designed to apply only to regions to the right of the mean (0) as in Figure 5-7(b). But by comparing the shaded area in Figure 5-7(a) to the shaded area in Figure 5-7(b), we can see that those two areas are identical because the density curve is symmetric. Referring to Table A-2, we can easily determine that the shaded area of Figure 5-7(b) is 0.4925, so the shaded area of Figure 5-7(a) must also be 0.4925. That is, the probability of randomly selecting a thermometer with an error between −2.43° and 0° is 0.4925. In other words, 49.25% of the thermometers have errors between −2.43° and 0°.

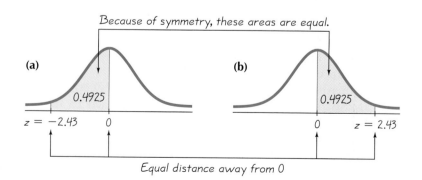

Because of symmetry, these areas are equal.

(a)

0.4925

$z = -2.43$ 0

(b)

0.4925

0    $z = 2.43$

Equal distance away from 0

Figure 5-7   Using Symmetry to Find the Area to the Left of the Mean

The preceding solution illustrates an important principle:

Although a *z* score can be negative, the area under the curve (or the corresponding probability) can never be negative.

Section 2-5 presented the empirical rule, which states that for bell-shaped distributions,

- about 68% of all scores fall within 1 standard deviation of the mean.
- about 95% of all scores fall within 2 standard deviations of the mean.
- about 99.7% of all scores fall within 3 standard deviations of the mean.

If we refer to Figure 5-5 with $z = 1$, Table A-2 shows us that the shaded area is 0.3413. It follows that the proportion of scores between $z = -1$ and $z = 1$ will be $0.3413 + 0.3413 = 0.6826$. That is, about 68% of all scores fall within 1 standard deviation of the mean. A similar calculation with $z = 2$ yields the values of $0.4772 + 0.4772 = 0.9544$ (or about 95%) as the proportion of scores between $z = -2$ and $z = 2$. Similarly, the proportion of scores between $z = -3$ and $z = 3$ is given by $0.4987 + 0.4987 = 0.9974$ (or about 99.7%). These exact values correspond very closely to those given in the empirical rule. In fact, the values of the empirical rule were found directly from the probabilities in Table A-2 and have been slightly rounded for convenience. The empirical rule is sometimes called the *68-95-99 rule*; using exact values from Table A-2, it would be called the *68.26-95.44-99.74 rule*, but then it wouldn't sound as snappy.

Because we are dealing with a density curve for a probability distribution, the total area under the curve must be 1. Now refer to Figure 5-8 and see that a vertical line directly above the mean of 0 divides the area under the curve into two equal parts, each containing an area of 0.5. The following example uses this observation.

In this example, the Table A-2 area must be subtracted from 0.5. Students sometimes try to develop rote methods for finding such areas, but convince them that the best approach is to draw a graph, shade the area of interest, and *understand* how to find the desired area given the limitations of Table A-2.

EXAMPLE   Once again, make a random selection from the same sample of thermometers. Find the probability that the chosen thermometer reads (at the freezing point of water) greater than +1.27°.

SOLUTION   We are again dealing with normally distributed values having a mean of 0° and a standard deviation of 1°. The probability of selecting a thermometer that reads greater than +1.27° corresponds to the shaded area of Figure 5-8. Table A-2 cannot be used to find that area directly, but we can use the table to find that $z = 1.27$ corresponds to the area of 0.3980, as shown in the figure. We now reason that because the area to the right of zero is one-half of the total area, it has an area of 0.5 and the shaded area is $0.5 - 0.3980$, or 0.1020. We conclude that there

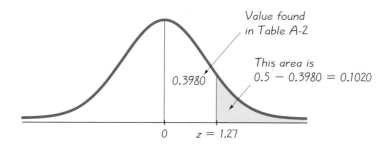

Figure 5-8 Finding the Area to the Right of $z = 1.27$

is a probability of 0.1020 of randomly selecting one of the thermometers with a reading greater than $+1.27°$. Another way to interpret this result is to state that if many thermometers are selected and tested, then 0.1020 (or 10.20%) of them will read greater than $+1.27°$.

We are able to determine the area of the shaded region in Figure 5-8 by an *indirect* application of Table A-2. The following example illustrates yet another indirect use.

EXAMPLE Assuming that one thermometer in our sample is randomly selected, find the probability that it reads (at the freezing point of water) between $1.20°$ and $2.30°$.

SOLUTION The probability of selecting a thermometer that reads between $+1.20°$ and $+2.30°$ corresponds to the shaded area of Figure 5-9. However, Table A-2 is designed to provide only for regions bounded on the left by the vertical line above 0. We can use the table to find that $z = 1.20$ corresponds to an area of 0.3849 and that $z = 2.30$ corresponds

(continued)

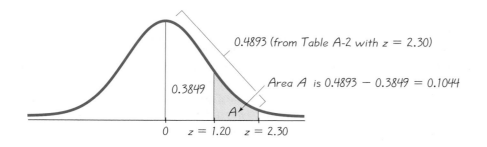

Figure 5-9 Finding the Area Between $z = 1.20$ and $z = 2.30$

to an area of 0.4893, as shown in the figure. If we denote the area of the shaded region by A, we can see from Figure 5-9 that

$$0.3849 + A = 0.4893$$

so

$$A = 0.4893 - 0.3849 = 0.1044$$

If one thermometer is randomly selected, the probability that it reads (at the freezing point of water) between 1.20° and 2.30° is therefore 0.1044.

The preceding example concluded with the statement that the probability of a reading between 1.20° and 2.30° is 0.1044. Such probabilities can also be expressed with the following notation.

Notation	
$P(a < z < b)$	denotes the probability that the z score is between a and b.
$P(z > a)$	denotes the probability that the z score is greater than a.
$P(z < a)$	denotes the probability that the z score is less than a.

Using this notation, we can express the result of the last example as $P(1.20 < z < 2.30) = 0.1044$, which states in symbols that the probability of a z score falling between 1.20 and 2.30 is 0.1044. With a continuous probability distribution such as the normal distribution, the probability of getting any single *exact* value is 0. That is, $P(z = a) = 0$. For example, there is a 0 probability of randomly selecting someone and getting a height of exactly 68.16243357 in. In the normal distribution, any single point on the horizontal scale is represented not by a region under the curve, but by a vertical line above the point. For $P(z = 1.33)$, we have a vertical line above $z = 1.33$, but that vertical line by itself contains no area, so $P(z = 1.33) = 0$. With any continuous random variable, the probability of any one exact value is 0, and it follows that $P(a \leq z \leq b) = P(a < z < b)$. It also follows that the probability of getting a z score of *at most* b is equal to the probability of getting a z score of *less than* b. It is important to interpret correctly key phrases such as *at most, at least, more than, no more than*, and so on. The illustrations

Many of us assume that terms such as "at most 2" or "at least 2" are clearly understood by all students, but that is often a wrong assumption. It could be very helpful to emphasize the importance of understanding and interpreting such terms. Ask students which job offer they would prefer: One that pays at most $50,000, one that pays no more than $50,000, or one that pays at least $50,000.

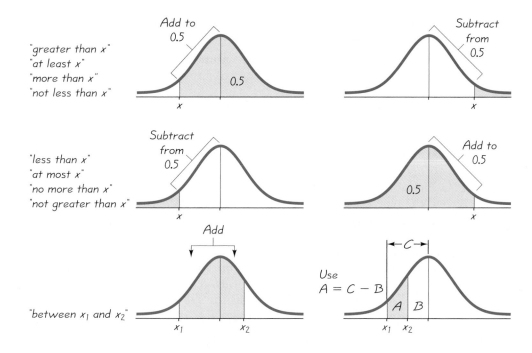

Figure 5-10   Interpreting Areas Correctly

in Figure 5-10 provide an aid to interpreting several of the most common phrases.

Finding z Scores When Given Probabilities

So far, the examples of this section involving the standard normal distribution have all followed the same format: Given some value, we used Table A-2 to find a probability. Many other circumstances exist in which we already know the probability, but we need to find the corresponding z score. In such cases, it is very important to avoid confusion between z scores and areas. Remember, the Table A-2 numbers in the extreme left column and across the top are z scores, which are *distances* along the horizontal scale, whereas the numbers in the body of Table A-2 are *areas* (or probabilities). Also, z scores to the left of the centerline are always negative (as in Figure 5-7a). If we already know a probability and want to determine the corresponding z score, we find it as follows:

1. Identify the probability representing an area bounded by the centerline; this may be the original probability, or it may be found by using the

There are two important points that should be emphasized in this subsection: (1) The numbers in the body of Table A-2 are areas, and the numbers in the extreme left column (and across the top row) are z scores that are actually distances; (2) if a z score is to the *left* of the centerline, it must be negative.

It's always a good idea to check answers to be sure that they are reasonable, but that check becomes particularly important with the problems in this subsection. It is easy to forget that a z score must be made negative when it is to the left of the centerline, but if a check shows that the score is to the right when it should be to the left, it is easy to correct the solution.

original probability. (That is, be sure you're working with a region bounded by the centerline.)

2. Using the probability representing the area bounded by the centerline, locate the closest probability in the *body* of Table A-2 and identify the corresponding z score.

3. If the z score is positioned to the left of the centerline, make it negative.

EXAMPLE   Use the same thermometers with temperature readings that are normally distributed with a mean of $0°\,C$ and a standard deviation of $1°\,C$. Find the temperature corresponding to P_{95}, the 95th percentile. That is, find the temperature separating the bottom 95% from the top 5%. See Figure 5-11.

SOLUTION   Figure 5-11 shows the z score that is the 95th percentile, separating the top 5% from the bottom 95%. We must refer to Table A-2 to find that z score, and we must use a region bounded by the centerline (where $\mu = 0$) on one side, such as the shaded region of 0.45 in Figure 5-11. (Remember, Table A-2 is designed to directly provide only those areas that are bounded on the left by the centerline and on the right by the z score.) We first search for the area of 0.45 *in the body of the table* and then find the corresponding z score. In Table A-2 the area of 0.45 is between the table values of 0.4495 and 0.4505, but there's an asterisk with a special note indicating that 0.4500 corresponds to a z score of 1.645. We can now conclude that the z score in Figure 5-11 is 1.645, so the 95th percentile is the temperature reading of $1.645°\,C$. When tested at freezing, 95% of the readings will be less than or equal to $1.645°\,C$, and 5% of them will be greater than or equal to $1.645°\,C$.

Note that in the preceding solution, Table A-2 led to a z score of 1.645, which is midway between 1.64 and 1.65. When using Table A-2, we can usually avoid interpolation by simply selecting the closest value. However, there

Figure 5-11 Finding the 95th Percentile

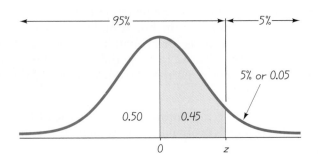

are two special cases involving values that are important because they are used so often in a wide variety of applications (see the accompanying table). Except in these two special cases, we can select the closest value in the table. (If a desired value is midway between two table values, select the larger value.) Also, for z scores above 3.09, we can use 0.4999 as an approximation of the corresponding area.

z score	Area
1.645	0.4500
2.575	0.4950

EXAMPLE   Using the same thermometers, find P_{10}, the 10th percentile. That is, find the temperature reading separating the bottom 10% of all temperatures from the top 90%.

SOLUTION   Refer to Figure 5-12, where the 10th percentile is shown as the z score separating the bottom 10% from the top 90%. Table A-2 is designed for areas bounded by the centerline, so we refer to the shaded area of 0.40 (corresponding to 50% − 10%). *In the body of the table,* we select the closest value of 0.3997 and find that it corresponds to $z = 1.28$. However, because the z score is below the mean of 0, it must be negative. The 10th percentile is therefore −1.28° C. When tested at freezing, 10% of the thermometer readings will be equal to or less than −1.28° C, and 90% of the readings will be equal to or greater than −1.28° C.

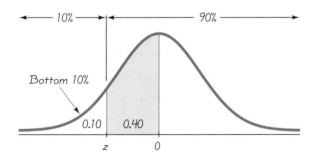

Figure 5-12   Finding the 10th Percentile

The examples in this section were contrived so that the mean of 0 and the standard deviation of 1 coincided exactly with the parameters of the standard normal distribution described in Table A-2. In reality, it is unusual to find such convenient parameters because typical normal distributions involve means different from 0 and standard deviations different from 1. The next section introduces methods for working with such nonstandard normal distributions.

5-2 Exercises A: Basic Skills and Concepts

Recommended assignment: Exercises 1–36. This is actually a reasonable assignment.

In Exercises 1–4, refer to the continuous uniform *distribution depicted in Figure 5-2, assume that one temperature reading is randomly selected, and find the probability of each reading in degrees.*

1. Greater than 2   1. 0.6
3. Between 2 and 4   3. 0.4

2. Less than 3   2. 0.6
4. Between 0.8 and 4.7   4. 0.78

In Exercises 5–24, assume that the readings on the thermometers are normally distributed with a mean of 0° and a standard deviation of 1.00°. A thermometer is randomly selected and tested. In each case, draw a sketch, and find the probability of each reading in degrees.

5. Between 0 and 3.00   5. 0.4987
7. Between 0 and −2.33   7. 0.4901
9. Greater than 2.58   9. 0.0049
11. Less than −2.09   11. 0.0183
13. Between 1.34 and 2.67
15. Between −2.22 and −1.11
17. Less than 0.08   17. 0.5319
19. Greater than −2.29   19. 0.9890
21. Between −1.99 and 2.01
23. Between −1.00 and 4.00

6. Between 0 and 1.96   6. 0.4750
8. Between 0 and −1.28   8. 0.3997
10. Less than −1.47   10. 0.0708
12. Greater than 0.25   12. 0.4013
14. Between −1.72 and −0.31
16. Between 0.89 and 1.78
18. Less than 3.01   18. 0.9987
20. Greater than −1.05   20. 0.8531
22. Between −0.07 and 2.19
24. Between −5.00 and 2.00

13. 0.0863
14. 0.3356
15. 0.1203
16. 0.1492
21. 0.9545
22. 0.5136
23. 0.8412
24. 0.9771

In Exercises 25–28, assume that the readings on the thermometers are normally distributed with a mean of 0° and a standard deviation of 1.00°. Find the indicated probability, where z is the reading in degrees.

25. $P(z > 2.33)$ 25. 0.0099
27. $P(-3.00 < z < 2.00)$

26. $P(2.00 < z < 2.50)$ 26. 0.0166
28. $P(z < -1.44)$ 28. 0.0749

27. 0.9759

In Exercises 29–36, assume that the readings on the thermometers are normally distributed with a mean of 0° and a standard deviation of 1.00°. A thermometer is randomly selected and tested. In each case, draw a sketch, and find the temperature reading corresponding to the given information.

29. Find P_{90}, the 90th percentile. This is the temperature reading separating the bottom 90% from the top 10%.   29. 1.28°
30. Find P_{30}, the 30th percentile.   30. −0.52°
31. Find Q_1, the temperature reading that is the first quartile.   31. −0.67°
32. Find D_1, the temperature reading that is the first decile.   32. −1.28°
33. If 4% of the thermometers are rejected because they have readings that are too high, but all other thermometers are acceptable, find the reading that separates the rejected thermometers from the others.   33. 1.75°

34. If 8% of the thermometers are rejected because they have readings that are too low, but all other thermometers are acceptable, find the reading that separates the rejected thermometers from the others.   34. −1.41°

35. A quality control analyst wants to examine thermometers that give readings in the bottom 2%. What reading separates the bottom 2% from the others?   35. −2.05°

36. If 2.5% of the thermometers are rejected because they have readings that are too high and another 2.5% are rejected because they have readings that are too low, find the two readings that are cutoff values separating the rejected thermometers from the others.   36. −1.96°, 1.96°

5–2 Exercises B: Beyond the Basics

37. Assume that z scores are normally distributed with a mean of 0 and a standard deviation of 1.
    a. If $P(0 < z < a) = 0.3212$, find a.   37. a. 0.92
    b. If $P(-b < z < b) = 0.3182$, find b.   b. 0.41
    c. If $P(z > c) = 0.2358$, find c.   c. 0.72
    d. If $P(z > d) = 0.7517$, find d.   d. −0.68
    e. If $P(z < e) = 0.4090$, find e.   e. −0.23

38. For a standard normal distribution, find the percentage of data that are
    a. within 1 standard deviation of the mean   38. a. 68.26%
    b. within 1.96 standard deviations of the mean   b. 95.00%
    c. between $\mu - 3\sigma$ and $\mu + 3\sigma$ c. 99.74%
    d. between 1 standard deviation below the mean and 2 standard deviations above the mean   d. 81.85%
    e. more than 2 standard deviations away from the mean   e. 4.56%

39. In Formula 5-1, if we let $\mu = 0$ and $\sigma = 1$, and if we approximate e by 2.7 and $\sqrt{2\pi}$ by 2.5, we get

$$y = \frac{2.7^{-x^2/2}}{2.5}$$

Using a scale of 1 in. = 1 unit on the x-axis and 1 in. = 1 unit on the y-axis, graph this equation after finding the y coordinates that correspond to these x coordinates: −4, −3, −2, −1, 0, 1, 2, 3, and 4. Estimate the area (in in.²) bounded by the curve, the x-axis, the vertical line passing through 0 on the x-axis, and the vertical line passing through 1 on the x-axis. Compare this result to the value in Table A-2.

39. The heights at $x = 0$, 1 are 0.4000 and 0.2434. Using a trapezoid, we approximate the area as 0.3217. The value from Table A-2 is 0.3413.

40. Assume that the random variable x has a continuous probability distribution that is *uniform* with $\mu = 0$ and $\sigma = 1$. For this random variable, the minimum is $-\sqrt{3}$ and the maximum is $\sqrt{3}$.
    a. Find $P(x > 1)$.   40. a. 0.2113
    b. Find $P(x > 1)$ if you incorrectly assume that the distribution is normal instead of uniform.   b. 0.1587   *(continued)*

c. Compare the correct result from part (a) to the incorrect result from part (b). Does using the wrong distribution have much of an effect on the result?   c. Yes

5-3 Nonstandard Normal Distributions: Finding Probabilities

Although Section 5-2 introduced important methods for dealing with normal distributions, the examples and exercises included in that section are generally unrealistic because most normally distributed populations have a nonzero mean, a standard deviation different from 1, or both. In this section we include many real and important nonstandard normal distributions. We are able to standardize nonstandard cases by using Formula 5-2.

Formula 5-2
$$z = \frac{x - \mu}{\sigma}$$

See Figure 5-13 where we illustrate the important principle that the area bounded by a score and the population mean is the same as the area bounded by the corresponding z score and the mean of 0. Once we convert a nonstandard score to a z score, we can use Table A-2 the same way it was used in Section 5-2. We therefore recommend the following procedure for finding probabilities for values of a random variable with a normal probability distribution:

1. Draw a normal curve, label the mean and any relevant scores, then *shade* the region representing the desired probability.
2. For each relevant score x that is a boundary for the shaded region, use Formula 5-2 to find the equivalent z score.
3. Refer to Table A-2 to find the area of the shaded region. This area is the desired probability.

The following example uses these three steps. Note that this example focuses on a height of 68.6 in., which is exactly 2 standard deviations above the mean, so the z score is 2.00. The z score measures the number of standard deviations that a score is away from the mean.

EXAMPLE   Heights of women are normally distributed with a mean of 63.6 in. and a standard deviation of 2.5 in. (based on data from the National Health Survey). If a woman is randomly selected, find the probability that her height is between 63.6 in. and 68.6 in.

SOLUTION

Step 1: See Figure 5-14 where we enter the mean of 63.6 and the score of 68.6, and we shade the area representing the probability we want.

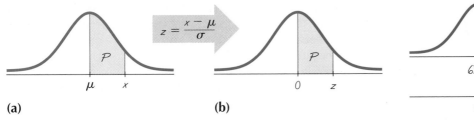

(a)   (b)

Figure 5-13 Converting from a Nonstandard Normal Distribution to the Standard Normal Distribution

Figure 5-14 Probability of Height Between 63.6 in. and 68.6 in.

Step 2: To use Table A-2, we must use Formula 5-2 to convert the non-standard distribution of heights to the standard normal distribution. The height of 68.6 in. is converted to a z score as follows.

$$z = \frac{x - \mu}{\sigma} = \frac{68.6 - 63.6}{2.5} = \frac{5}{2.5} = 2.00$$

This result shows that the height of 68.6 in. differs from the mean of 63.6 in. by 2.00 standard deviations.

Step 3: Referring to Table A-2, we find that $z = 2.00$ corresponds to an area of 0.4772.

There is a probability of 0.4772 of randomly selecting a woman with a height between 63.6 in. and 68.6 in. This can be expressed in symbols as

$$P(63.6 < x < 68.6) = P(0 < z < 2.00) = 0.4772$$

Another way to interpret this result is to conclude that 47.72% of women have heights between 63.6 in. and 68.6 in.

EXAMPLE   At the beginning of this chapter we noted that in order to fit into the Russian Soyuz spacecraft, an astronaut must have a height between 64.5 in. and 72 in.

a. Find the percentage of American women who meet that height requirement.

b. Among 500 randomly selected American women, how many meet the Soyuz height requirement?

SOLUTION   As in the preceding example, we are dealing with normally distributed heights having a mean of 63.6 in. and a standard deviation of 2.5 in.

(continued)

a. The shaded region B in Figure 5-15 represents the proportion of women who meet the Soyuz spacecraft height requirement because their heights are between 64.5 in. and 72 in. We can't find that shaded region directly because Table A-2 isn't designed for such cases, but we can find it indirectly by using the same basic procedures presented in Section 5-2. Here's how we can proceed: Find the shaded area B by finding the difference between region A and the total area of regions A and B combined. That is,

$$B = (A \text{ and } B \text{ combined}) - A.$$

For areas A and B combined:

$$z = \frac{x - \mu}{\sigma} = \frac{72 - 63.6}{2.5} = 3.36$$

Using Table A-2, we find that $z = 3.36$ is off the chart, so we use 0.4999. (See the footnote appearing with Table A-2.)

For region A:

$$z = \frac{x - \mu}{\sigma} = \frac{64.5 - 63.6}{2.5} = 0.36$$

Again using Table A-2, we find that $z = 0.36$ corresponds to an area of 0.1406. Region A has an area of 0.1406.

The shaded area is the difference between 0.4999 and 0.1406:

$$\begin{aligned} \text{Area } B &= (\text{areas of } A \text{ and } B \text{ combined}) - (\text{area } A) \\ &= 0.4999 - 0.1406 = 0.3593 \end{aligned}$$

If American women are recruited as astronauts without regard to height, about 35.93% of them will meet the Soyuz height require-ment.

b. Among 500 randomly selected women, we expect that 35.93% of them meet the Soyuz height requirement. The actual number is there-fore

$$500 \cdot 0.3593 = 179.65 \text{ women}$$

EXAMPLE   The U.S. Army requires that women's heights be between 58 in. and 80 in. Find the percentage of women satisfying that require-ment. Again assume that women have heights that are normally distrib-uted with a mean of 63.6 in. and a standard deviation of 2.5 in.

SOLUTION   Figure 5-16 shows the normal distribution of women's heights, with the shaded region representing heights between 58 in. and 80 in., as required by the U.S. Army. The method for finding the area of the shaded region involves breaking it up into parts A and B as shown. We can use Formula 5-2 and Table A-2 to find the areas of those regions sep-arately; then we can add the results.

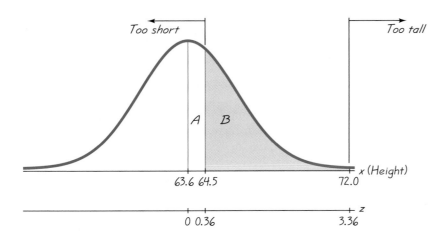

Figure 5-15   Women Fitting into Soyuz Spacecraft

For area *A* only:

$$z = \frac{x - \mu}{\sigma} = \frac{58.0 - 63.6}{2.5} = -2.24$$

We can use Table A-2 to find that $z = -2.24$ corresponds to 0.4875, so area *A* is 0.4875.

For area *B* only:

$$z = \frac{80.0 - 63.6}{2.5} = 6.56$$

Table A-2 does not include *z* scores above 3.09, but it does include a note that for values of *z* above 3.09, we should use 0.4999 for the area. (If necessary, more accurate results can be obtained by using special tables or software.) Area *B* is 0.4999.

The shaded region consists of areas *A* and *B* combined, so

Area of regions *A* and *B* combined = 0.4875 + 0.4999 = 0.9874

Comment on this feature of Table A-2: For values of *z* above 3.09, we use an area of 0.4999. It might be helpful to draw a graph and show that there is more area (or probability) as you go farther to the right, but it becomes negligible once you get past 3.09.

Also, users of a TI-83 calculator can find more exact values. For example, an entry of `normal-cdf(0,6.56)` will show that for $z = 6.56$, the area is 0.4999999994. In this case, our use of 0.4999 instead of 0.4999999994 results in a very small error.

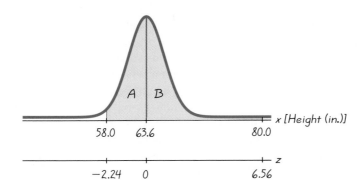

Figure 5-16   Women Who Meet the Height Requirement for the U.S. Army

That is, the proportion of women who meet the height requirement for the U.S. Army is 0.9874. The statement of the problem indicated that a percentage should be found, so we express the result as 98.74%.

If we repeat the preceding example for men, we find that 99.98% of men meet the height requirement. (Men have normally distributed heights with a mean of 69.0 in. and a standard deviation of 2.8 in., and the U.S. Army requires that they have heights between 60 in. and 80 in.) The percentage of eligible men exceeds the percentage of eligible women, but by a very small amount. Both percentages indicate that the U.S. Army rejects very few candidates because they fail to meet its height requirements.

In this section we have extended the concepts of Section 5-2 to include more realistic nonstandard normal probability distributions. However, all of the examples we have considered so far are of the same general type: A probability (or percentage) is found by using the normal distribution (with values listed in Table A-2) when given the values of the population mean, standard deviation, and the relevant score(s). In many practical and real cases, the probability (or percentage) is known and we must find the relevant score(s). Problems of this type are discussed in the next section.

5-3 Exercises A: Basic Skills and Concepts

Recommended assignment: Exercises 1, 3, 6, 7, 10, 12, 18, 20. Also consider assigning Exercise 22, especially if students have access to a computer with STATDISK or Minitab; the data set in Exercise 22 is available on disk for STATDISK and Minitab.

In Exercises 1–6, assume that women's heights are normally distributed with a mean given by $\mu = 63.6$ in. and a standard deviation given by $\sigma = 2.5$ in. (based on data from the National Health Survey). Also assume that a woman is randomly selected. Draw a graph, and find the indicated probability.

1. $P(63.6 \text{ in.} < x < 65.0 \text{ in.})$
2. $P(x < 70.0 \text{ in.})$ 2. 0.9948
3. $P(x > 58.1 \text{ in.})$ 3. 0.9861
4. $P(59.1 \text{ in.} < x < 66.6 \text{ in.})$
5. The heights of the Rockette dancers at New York City's Radio City Music Hall must be between 65.5 in. and 68.0 in. If a woman is randomly selected, find the probability that she meets the height requirement to be a Rockette.   5. 0.1844
6. The Beanstalk Club, a social organization for tall people, has a requirement that women must be at least 70 in. (or 5 ft 10 in.) tall. Suppose you are trying to decide whether to open a branch of the Beanstalk Club in a metropolitan area with 500,000 adult women.
   a. Find the percentage of adult women who are eligible for membership because they meet the minimum height requirement of 70 in.
   b. Among the 500,000 adult women living in this metropolitan area, how many are eligible for Beanstalk Club membership?   b. 2600
   c. Will you open a branch of the Beanstalk Club?
7. Replacement times for TV sets are normally distributed with a mean of 8.2 years and a standard deviation of 1.1 years (based on data from "Get-

1. 0.2123
4. 0.8490
6. a. 0.52%
   c. No; too busy studying statistics.

7. 0.1379

ting Things Fixed," *Consumer Reports*). Find the probability that a randomly selected TV set will have a replacement time less than 7.0 years.

8. Replacement times for CD players are normally distributed with a mean of 7.1 years and a standard deviation of 1.4 years (based on data from "Getting Things Fixed," *Consumer Reports*). Find the probability that a randomly selected CD player will have a replacement time less than 8.0 years.   8. 0.7389

9. Assume that the weights of paper discarded by households each week are normally distributed with a mean of 9.4 lb and a standard deviation of 4.2 lb (based on data from the Garbage Project at the University of Arizona). Find the probability of randomly selecting a household and getting one that discards between 5.0 lb and 8.0 lb of paper in a week.   9. 0.2238

10. Based on the sample results in Data Set 2 of Appendix B, assume that human body temperatures are normally distributed with a mean of 98.20° F and a standard deviation of 0.62° F. If we define a fever to be a body temperature above 100° F, what percentage of normal and healthy persons would be considered to have a fever? Does this percentage suggest that a cutoff of 100° F is appropriate?   10. 0.19%; yes

11. One classic use of the normal distribution is inspired by a letter to *Dear Abby* in which a wife claimed to have given birth 308 days after a brief visit from her husband, who was serving in the Navy. The lengths of pregnancies are normally distributed with a mean of 268 days and a standard deviation of 15 days. Given this information, find the probability of a pregnancy lasting 308 days or longer. What does the result suggest?   11. 0.0038; either a very rare event has occurred or the husband is not the father.

12. The lengths of pregnancies are normally distributed with a mean of 268 days and a standard deviation of 15 days. If we stipulate that a baby is *premature* if born at least three weeks early, what percentage of babies are born prematurely? (Such information is important to hospital administrators, who need to ensure that correct equipment is available to handle premature babies' special needs.)   12. 8.08%

13. According to the Opinion Research Corporation, men spend an average of 11.4 min in the shower. Assume that the times are normally distributed with a standard deviation of 1.8 min. If a man is randomly selected, find the probability that he spends at least 10.0 min in the shower.   13. 0.7823

14. According to the International Mass Retail Association, girls aged 13 to 17 spend an average of $31.20 on shopping trips in a month. Assume that the amounts are normally distributed with a standard deviation of $8.27. If a girl in that age category is randomly selected, what is the probability that she spends between $35.00 and $40.00 in one month?   14. 0.1782

15. IQ scores are normally distributed with a mean of 100 and a standard deviation of 15. Mensa is an organization for people with high IQs, and eligibility requires an IQ above 131.5.
    a. If someone is randomly selected, find the probability that he or she meets the Mensa requirement.   15. a. 0.0179
    b. In a typical region of 75,000 people, how many are eligible for Mensa?   b. 1343

16. 0.39%; yes

19. 96.32%

20. 3.89%

16. An IBM subcontractor was hired to make ceramic substrates which are used to distribute power and signals to and from computer silicon chips. Specifications require resistance between 1.500 ohms and 2.500 ohms, but the population has normally distributed resistances with a mean of 1.978 ohms and a standard deviation of 0.172 ohm. What percentage of the ceramic substrates will not meet the manufacturer specifications? Does this manufacturing process appear to be working well?

17. The serum cholesterol levels in men aged 18 to 24 are normally distributed with a mean of 178.1 and a standard deviation of 40.7. All units are in mg/100 mL, and the data are based on the National Health Survey. If a man aged 18 to 24 is randomly selected, find the probability that his serum cholesterol level is between 200 and 250.   17. 0.2562

18. Measurements of human skulls from different epochs are analyzed to determine whether they change over time. The maximum breadth is measured for skulls from Egyptian males who lived around 3300 B.C. Results show that those breadths are normally distributed with a mean of 132.6 mm and a standard deviation of 5.4 mm (based on data from *Ancient Races of the Thebaid* by Thomson and Randall-Maciver). An archeologist discovers a male Egyptian skull and a field measurement reveals a maximum breadth of 119 mm. Find the probability of getting a value of 119 or less if a skull is randomly selected from the period around 3300 B.C. Is the newly found skull likely to come from that era?   18. 0.0059; no

19. The U.S. Marine Corps requires that men have heights between 64 in. and 78 in. Find the percentage of men meeting those height requirements. (The National Health Survey shows that heights of men are normally distributed with a mean of 69.0 in. and a standard deviation of 2.8 in.)

20. Some vending machines are designed so that their owners can adjust the weights of the quarters that are accepted. If many counterfeit coins are found, adjustments are made to reject more coins, with the effect that most of the counterfeit coins are rejected along with many legal coins. Assume that quarters have weights that are normally distributed with a mean of 5.67 g and a standard deviation of 0.070 g. If a vending machine is adjusted to reject quarters weighing less than 5.50 g or more than 5.80 g, what is the percentage of legal quarters that are rejected?

5-3 Exercises B: Beyond the Basics

In Exercises 21–24, refer to the indicated data set in Appendix B.

   a. Construct a histogram to determine whether the data set has a normal distribution.

   b. Find the sample mean $\bar{x}$ and sample standard deviation s.

c.   Use the sample mean as an estimate of the population mean μ, use the sample standard deviation as an estimate of the population standard deviation σ, and use the methods of this section to find the indicated probability.

21. Use the combined list of 100 weights of M&M plain candies listed in Data Set 11, and estimate the probability of randomly selecting one M&M candy and getting one with a weight greater than 1.000 g.

22. Use the head lengths of bears in Data Set 3, and estimate the probability of randomly selecting one bear with a head length between 12.0 in. and 13.0 in.

23. Use the Iowa precipitation amounts in Data Set 7, and estimate the probability of randomly selecting a year with a precipitation amount less than 40.0 in.

24. Use the total weights of discarded garbage in Data Set 1, and estimate the probability of randomly selecting a household that discards more than 20.0 lb of garbage in a week.

21. a. Normal distribution
    b. $\bar{x} = 0.9147$,
    $s = 0.0369$
    c.   0.0104

22. a.   Normal distribution
    b. $\bar{x} = 12.95$, $s = 2.14$
    c.   0.1780

23. a.   Normal distribution
    b. $\bar{x} = 32.473$, $s = 5.601$
    c.   0.9099

24. a.   Normal distribution
    b. $\bar{x} = 27.443$,
    $s = 12.458$
    c.   0.7257

Nonstandard Normal Distributions: Finding Scores

In this section we consider problems such as this: If women's heights are normally distributed with $\mu = 63.6$ in. and $\sigma = 2.5$ in., find the height separating the top 10% from the others. This problem has a given probability (0.10), and we need to find the appropriate height x. In the examples and exercises in Section 5-3 we used a given score to find a probability. In this section we follow the reverse procedure—we use a given probability to find a score.

In considering problems of finding scores when given probabilities, there are three important cautions to keep in mind.

1. *Don't confuse z scores and areas.* Remember, z scores are *distances* along the horizontal scale, but areas represent *regions* under the normal curve. Table A-2 lists z scores in the left column and across the top row, but areas are found in the body of the table.

2. *Choose the correct (right/left) side of the graph.* A score separating the top 10% from the others will be located on the right side of the graph, but a score separating the bottom 10% will be located on the left side of the graph.

3. *A z score must be negative whenever it is located to the left of the centerline of 0.*

As in Section 5-3, graphs are extremely helpful and they are strongly recommended. They are included as the first step in this recommended procedure:

Reminder to users of the previous edition: Section 5-3 in the 6th edition is now divided into Sections 5-3 and 5-4. There are three important points that should be emphasized once again in this section: (1) The numbers in the body of Table A-2 are areas, and the numbers in the extreme left column (and across the top row) are z scores that are actually distances. (2) If a z score is to the *left* of the centerline, it must be negative. (3) While it's always a good idea to check answers to be sure that they are reasonable, that check becomes particularly important with the problems of this section; if a check shows that the score is above the mean when it should be below it, it is easy to correct the solution.

1. Starting with a rough sketch that bears at least some resemblance to a bell, enter the given probability (or percentage) in the appropriate region of the graph and identify the x value(s) being sought.

2. Use Table A-2 to find the z score corresponding to the region bounded by x and the centerline of 0. Observe the following cautions:

   • Refer to the *body* of Table A-2 to find the closest area, then identify the corresponding z score.

   • Make the z score *negative* if it is located to the left of the centerline.

3. Using Formula 5-2, enter the values for μ, σ, and the z score found in Step 2, then solve for x. On the basis of the format of Formula 5-2, we can solve for x as follows.

$$x = \mu + (z \cdot \sigma) \quad \text{(Another form of Formula 5-2)}$$

4. Refer to the sketch of the curve to verify that the solution makes sense in the context of the graph and in the context of the problem.

EXAMPLE   In Section 5-3 we noted that heights of women are normally distributed with a mean of 63.6 in. and a standard deviation of 2.5 in. (based on data from the National Health Survey). Find the value of P_{90}. That is, find the height separating the bottom 90% from the top 10%.

SOLUTION

Step 1: We begin with the graph shown in Figure 5-17. We have entered the mean of 63.6, shaded the area representing the top 10%, and identified the desired value as x. The area between 63.6 in. and x in. must be 40% of the total area (because the right half of the area must combine to be 50% of the total). Because the total area is 1, the area constituting 40% of the total must be 0.4.

Step 2: We refer to Table A-2, but we look for an area of 0.4000 in the body of the table. (Remember, Table A-2 is designed to list areas only for those regions bounded on the left by the mean and on the right by some score, so we use the area of $0.90 - 0.50 = 0.40$ as shown in Figure 5-17.) The area closest to 0.4000 is 0.3997, and it corresponds to a z score of 1.28.

Step 3: With $z = 1.28$, $\mu = 63.6$, and $\sigma = 2.5$, we solve for x either by using Formula 5-2 directly or by using the following version of Formula 5-2:

$$x = \mu + (z \cdot \sigma) = 63.6 + (1.28 \cdot 2.5) = 66.8$$

Step 4: If we let $x = 66.8$ in Figure 5-17, we see that this solution is reasonable because it should be greater than the mean of 63.6.

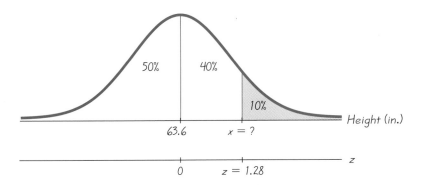

Figure 5-17   Finding P_{90} for Heights of Women

The height of 66.8 in. separates the shortest 90% of women from the tallest 10%.

The preceding solution could be found by using a TI-83 calculator. Press 2nd/DISTR, then use the function invNorm (for inverse normal) with the probability of 0.9, the mean of 63.6, and the standard deviation of 2.5 entered in this format: invNorm(0.9, 63.6, 2.5) to get the result of 66.8.

The next example includes a pitfall not present in the preceding example. Closely follow what happens when the desired score is *below* the mean, as in this example.

EXAMPLE   A study compared the facial behavior of nonparanoid schizophrenic persons with that of a control group of normal persons. The control group was timed for eye contact during a period of 5 minutes, or 300 seconds. The eye-contact times were normally distributed with a mean of 184 s and a standard deviation of 55 s (based on data from "Ethological Study of Facial Behavior in Nonparanoid and Paranoid Schizophrenic Patients," by Pitman, Kolb, Orr, and Singh, *Psychiatry*, 144:1). Because results showed that nonparanoid schizophrenic patients had much lower eye-contact times than did the control group, you have decided to further analyze people in the control group who are in the bottom 5%. For the control group, find P_5, the 5th percentile. That is, find the eye-contact time separating the bottom 5% from the rest.

SOLUTION

Step 1:   The graph is shown in Figure 5-18 on the following page. The mean of 184 is entered, the region representing the bottom 5% is shaded, and the desired score x is identified.

Step 2:   The region bounded by x and the centerline is identified as region A, and it must have an area of 0.45. (The left tail area of

(continued)

Figure 5-18 Finding the 5th Percentile for Eye-Contact Times

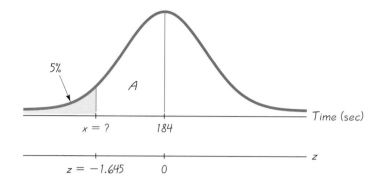

0.05 and region A must combine to form an area of 0.5.) Referring to Table A-2, we search the *body* of the table for an area of 0.45, and we find the corresponding z score of 1.645. *Because the z score is negative whenever it is below (to the left of) the mean, we let $z = -1.645$.*

Step 3:   We let $z = -1.645$, $\mu = 184$, and $\sigma = 55$, then solve for x as follows:

$$x = \mu + (z \cdot \sigma) = 184 + (-1.645 \cdot 55) = 93.5$$

Step 4:   The solution of $x = 93.5$ s seems reasonable in Figure 5-18, because the value of x should be below the mean of 184 s. (If we had forgotten to make the z score negative in Step 2, we would have obtained the unreasonable result of 274.5 s. It should be obvious from Figure 5-18 that x cannot be 274.5 s.)

The result indicates that $P_5 = 93.5$ s. That is, 5% of the times are less than 93.5 s. If you want to further analyze people in the bottom 5% of the control group, then choose people with eye-contact times less than 93.5 s.

5-4 Exercises A: Basic Skills and Concepts

Recommended assignment: Exercises 1, 3, 5, 8, 9, 12, 14.

In Exercises 1–4, assume that women have heights that are normally distributed with a mean of 63.6 in. and a standard deviation of 2.5 in. (based on data from the National Health Survey). Find the height for the given percentile.

2. 64.6 in.
3. 61.0 in.
4. 62.6 in.

1. P_{85} 1. 66.2 in.   **2.** P_{66} **3.** P_{15} **4.** P_{35}

5. Replacement times for TV sets are normally distributed with a mean of 8.2 years and a standard deviation of 1.1 years (based on data from "Getting Things Fixed," *Consumer Reports*). Find the replacement time that separates the top 20% from the bottom 80%. This result will be helpful to an appliance service company that wants to offer service contracts for TV sets.   5. 9.1 years

6. Replacement times for CD players are normally distributed with a mean of 7.1 years and a standard deviation of 1.4 years (based on data from "Getting Things Fixed," *Consumer Reports*). Find the replacement time separating the top 45% from the bottom 55%.   6. 7.3 years

7. Weights of paper discarded by households each week are normally distributed with a mean of 9.4 lb and a standard deviation of 4.2 lb (based on data from the Garbage Project at the University of Arizona). Find the weight that separates the bottom 33% from the top 67%.   7. 7.6 lb

8. Based on the sample results in Data Set 2 of Appendix B, assume that human body temperatures are normally distributed with a mean of 98.20° F and a standard deviation of 0.62° F. What two temperature levels separate the bottom 2% and the top 2%? These values could serve as reasonable limits that could be used to identify people who are likely to be ill.   8. 96.93° F, 99.47° F

9. The lengths of pregnancies are normally distributed with a mean of 268 days and a standard deviation of 15 days. If we stipulate that a baby is *premature* if the length of pregnancy is in the lowest 4%, find the length that separates premature babies from those who are not premature. Premature babies often require special care, and this result could be helpful to hospital administrators in planning for that care.   9. 242 days

10. According to the Opinion Research Corporation, men spend an average of 11.4 min in the shower. Assume that the times are normally distributed with a standard deviation of 1.8 min. Find the values of the quartiles Q_1 and Q_3.   10. 10.2 min, 12.6 min

11. IQ scores are normally distributed with a mean of 100 and a standard deviation of 15. If we define a genius to be someone in the top 1% of IQ scores, find the score separating geniuses from the rest of us. This score could be used by a "think tank" company as one criterion for employment.   11. 135

12. A subcontractor manufactured ceramic substrates for IBM. These devices have resistances that are normally distributed with a mean of 1.978 ohms and a standard deviation of 0.172 ohm. If the required specifications are to be modified so that 3% of the devices are rejected because their resistances are too low and another 3% are rejected because their resistances are too high, find the cutoff values for the acceptable devices.   12. 1.655 ohms, 2.301 ohms

13. The serum cholesterol levels in men aged 18 to 24 are normally distributed with a mean of 178.1 and a standard deviation of 40.7. All units are in mg/100 mL, and the data are based on the National Health Survey.
    a. If a man aged 18 to 24 is randomly selected, find the probability that his serum cholesterol level is less than 200.   13. a. 0.7054
    b. If a serum cholesterol level is deemed to be too high if it is in the top 7%, find the cutoff level for readings that are too high.   13. b. 238.3 mg/100 mL

14. Measurements of human skulls from different epochs are analyzed to determine whether they change over time. The maximum breadth is measured for skulls from Egyptian males who lived around 3300 B.C. Results

show that those breadths are normally distributed with a mean of 132.6 mm and a standard deviation of 5.4 mm (based on data from *Ancient Races of the Thebaid* by Thomson and Randall-Maciver).
a. Find the probability of getting a value greater than 140 mm if a skull is randomly selected from the period of around 3300 B.C.   14. a. 0.0853
b. Find the value that is D_2, the second decile.   b. 128.1 mm

15. a. 98.74%
    b. 57.8 in., 69.4 in.

15. To be eligible for the U.S. Marine Corps, a woman must have a height between 58 in. and 73 in. Recall that heights of women are normally distributed with a mean of 63.6 in. and a standard deviation of 2.5 in.
a. Find the percentage of women who satisfy that requirement.
b. If the requirement is changed to exclude the shortest 1% and the tallest 1%, find the heights that are acceptable.

16. Quarters have weights that are normally distributed with a mean of 5.67 g and a standard deviation of 0.070 g.
a. If a vending machine is adjusted to reject quarters weighing less than 5.53 g or more than 5.81 g, what is the percentage of legal quarters that are rejected?   16. a. 4.56%
b. Find the weights of accepted legal quarters if the machine is readjusted so that the lightest 1.5% are rejected and the heaviest 1.5% are rejected.   b. 5.52 g, 5.82 g

5-4 Exercises B: Beyond the Basics

17. When investigating a data set, construction of a histogram reveals that the distribution is approximately normal and the boxplot is constructed with these quartiles: $Q_1 = 62$, $Q_2 = 70$, $Q_3 = 78$. Estimate the standard deviation.   17. 11.9

18. A: 62.8 and above;
    B: at least 55.2 and below 62.8;
    C: at least 44.8 and below 55.2;
    D: at least 37.2 and below 44.8;
    F: below 37.2

18. A teacher gives a test and gets normally distributed results with a mean of 50 and a standard deviation of 10. If grades are assigned according to the following scheme, find the numerical limits for each letter grade.
A: Top 10%
B: Scores above the bottom 70% and below the top 10%
C: Scores above the bottom 30% and below the top 30%
D: Scores above the bottom 10% and below the top 70%
F: Bottom 10%

19. According to data from the College Entrance Examination Board, the mean math SAT score is 475, and 17.0% of the scores are above 600. Find the standard deviation, and then use that result to find the 99th percentile. (Assume that the scores are normally distributed.)   19. 131.6; 782

20. ≈ 30.9

20. The College Entrance Examination Board writes that "for the SAT Achievement Tests, your score would fall in a range about 30 points above or below your actual ability about two-thirds of the time. This range is called the standard error of measurement (SEM)." Use that statement to estimate the standard deviation for scores of an individual on an SAT Achievement Test. (Assume that the scores are normally distributed.)

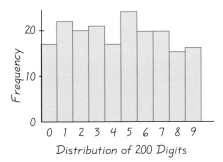

Figure 5-19   Distribution of 200 Digits from Social Security Numbers (Last 4 Digits) of 50 students

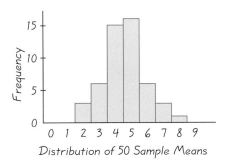

Figure 5-20   Distribution of 50 Sample Means for 50 Students

SSN digits				$\bar{x}$
1	8	6	4	4.75
5	3	3	6	4.25
9	8	8	8	8.25
5	1	2	5	3.25
9	3	3	5	5.00
4	2	6	2	3.50
7	7	1	6	5.25
9	1	5	4	4.75
5	3	3	9	5.00
7	8	4	1	5.00
0	5	6	1	3.00
9	8	2	2	5.25
6	1	5	7	4.75
8	1	3	0	3.00
5	9	6	9	7.25
6	2	3	4	3.75
7	4	0	7	4.50
5	7	5	6	5.75
4	1	5	7	4.25
1	2	0	6	2.25
4	0	2	8	3.50
3	1	2	5	2.75
0	3	4	0	1.75
1	5	1	0	1.75
9	7	4	0	5.00
7	3	1	1	3.00
9	1	1	3	3.50
8	6	5	9	7.00
5	6	4	1	4.00
9	3	9	5	6.50
6	0	7	3	4.00
8	2	9	6	6.25
0	2	8	6	4.00
2	0	9	7	4.50
5	8	9	0	5.50
6	5	4	9	6.00
4	8	7	6	6.25
7	1	2	0	2.50
2	9	5	0	4.00
8	3	2	2	3.75
2	7	1	6	4.00
6	7	7	1	5.25
2	3	3	9	4.25
2	4	7	5	4.50
5	4	3	7	4.75
0	4	3	8	3.75
2	5	8	6	5.25
7	1	3	4	3.75
8	3	7	0	4.50
5	6	6	7	6.00

TABLE 5-1

5-5 The Central Limit Theorem

This section presents the central limit theorem, which is one of the most important and useful concepts in statistics. It forms a foundation for estimating population parameters and hypothesis testing—topics discussed at length in the following chapters. Before considering this theorem, we will first try to develop an intuitive understanding of one of its most important consequences:

As the sample size increases, the sampling distribution of sample means approaches a normal distribution.

We will illustrate the central limit theorem by using the last four digits of social security numbers from each of 50 different students (see Table 5-1). The last four digits of social security numbers are random, unlike the beginning digits, which are used to identify the state and to provide other information. If we combine the four digits from each student into one big collection of 200 numbers, we get a mean of $\bar{x} = 4.5$, a standard deviation of $s = 2.8$, and an approximately uniform distribution with the graph shown in Figure 5-19. Now see what happens when we find the 50 sample means, as shown in Table 5-1. Even though the original collection of data has an approximately *uniform* distribution, the sample means have a distribution that is approximately *normal*. This can be a confusing concept, so you should stop right here and study this paragraph until its major point becomes clear: The original set of 200 individual numbers has a uniform distribution (because the digits 0–9 occur with approximately equal frequencies), but the 50 sample means have a normal distribution. It's a truly fascinating and intriguing phenomenon in statistics that by sampling from any distribution, we can create a distribution that is normal or at least approximately normal.

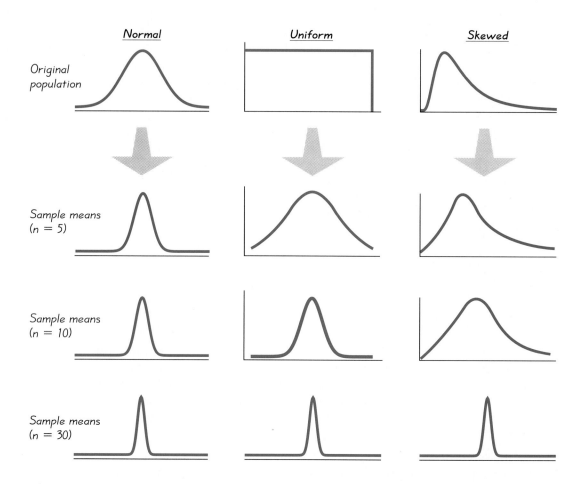

Figure 5-21   Normal, Uniform, and Skewed Distributions

In general, the **sampling distribution of sample means** is the distribution of the sample means obtained when we repeatedly draw samples of the same size from the same population. In other words, if we collect samples of the same size from the same population, compute their means, and then draw a histogram of those means, that histogram will tend to have the bell shape of a normal distribution. This is true regardless of the shape of the distribution of the original population. Figures 5-19 and 5-20 relate to the specific social security digits obtained from 50 students, but Figure 5-21 is a general illustration that also includes normal and skewed distributions. Observations exactly like these led to the formulation of the central limit theorem, which we will now discuss.

Let's assume that the variable x represents scores that may or may not be normally distributed and that the mean of the x values is μ and the standard deviation is σ. Suppose that we collect samples of size n and calculate the sample means. What do we know about the collection of all sample means that we produce by repeating this experiment? The central limit theorem tells us

that as the sample size n increases, the sampling distribution of sample means *approaches* a normal distribution with mean μ and standard deviation $\sigma/\sqrt{n}$. The distribution of sample means approaches a normal distribution in the sense that as n becomes larger, the distribution of sample means gets closer to a normal distribution. This conclusion is not intuitively obvious, and it was arrived at through extensive research and analysis.

Although the formal rigorous proof requires advanced mathematics and is beyond the scope of this text, we can see some justification based on the data in Table 5-1. If we randomly select samples of digits from a uniformly distributed population with mean $\mu = 4.5$, the resulting sample means will also tend to center about 4.5 so that the sample means also have a mean of 4.5; the mean of the 50 sample means in Table 5-1 is in fact 4.5. By visually inspecting the 200 original digits in Table 5-1, we can see that they range from 0 to 9, but the 50 sample means show *less variation* by ranging from 1.75 to 8.25. The original set of 200 digits has a standard deviation of 2.8, but the 50 sample means have a standard deviation of 1.4, which is lower as expected. We now formally state the theorem and give examples of its use.

Central Limit Theorem

Given:

1. The random variable x has a distribution (which may or may not be normal) with mean μ and standard deviation σ.
2. Samples of size n are randomly selected from this population.

Conclusions:

1. The distribution of sample means $\overline{x}$ will, as the sample size increases, approach a normal distribution.
2. The mean of the sample means will be the population mean μ.
3. The standard deviation of the sample means will be $\sigma/\sqrt{n}$.

Practical Rules Commonly Used:

1. For samples of size n larger than 30, the distribution of the sample means can be approximated reasonably well by a normal distribution. The approximation gets better as the sample size n becomes larger.
2. If the original population is itself normally distributed, then the sample means will be normally distributed for *any* sample size n.

Some students instantaneously develop an allergic reaction to this concept because it contains the dreaded word *theorem*. Comment that we do not develop a rigorous proof that must be learned. Instead, we emphasize practical applications of the three important conclusions.

The central limit theorem involves two different distributions: the distribution of the original population and the distribution of the sample means. As in previous chapters, we use the symbols μ and σ to denote the mean and standard deviation of the original population. We now introduce new notation for the mean and standard deviation of the distribution of sample means.

Students can usually see that the mean of the sample means should be μ, but they have trouble seeing that σ should be divided by the square root of *n*. However, they can understand that the standard deviation of sample means should be *smaller* than the standard deviation of the original scores.

Notation for the Central Limit Theorem

If all possible random samples of size n are selected from a population with mean μ and standard deviation σ, the mean of the sample means is denoted by $\mu_{\bar{x}}$, so

$$\mu_{\bar{x}} = \mu$$

Also, the standard deviation of the sample means is denoted by $\sigma_{\bar{x}}$, so

$$\sigma_{\bar{x}} = \frac{\sigma}{\sqrt{n}}$$

$\sigma_{\bar{x}}$ is often called the **standard error of the mean**.

Many important and practical problems can be solved with the central limit theorem. In the following example, part (a) involves an individual score, so we use the methods presented in Section 5-3; those methods apply to the normal distribution of the random variable x. Part (b), however, involves the mean for a group of 36 men, so we must use the central limit theorem in working with the random variable $\bar{x}$.

Instead of doing an example of the central limit theorem alone, it is helpful for students to see how such an example is different from the examples of earlier sections. Consider doing a class example similar to the one given here. Stress that when dealing with the mean for a *group* of scores, it is essential to modify the standard deviation by dividing by the square root of the sample size *n*.

EXAMPLE   In human engineering and product design, it is often important to consider the weights of people so that airplanes or elevators aren't overloaded, chairs don't break, and other such dangerous or embarrassing mishaps do not occur. Given that the population of men has normally distributed weights, with a mean of 173 lb and a standard deviation of 30 lb (based on data from the National Health Survey), find the probability that

a.   if 1 man is randomly selected, his weight is greater than 180 lb.

b.   if 36 different men are randomly selected, their mean weight is greater than 180 lb.

SOLUTION

a.   *Approach: Use the methods presented in Section 5-3* (because we are dealing with an *individual* score from a normally distributed population). We seek the area of the shaded region in Figure 5-22(a).

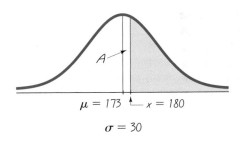

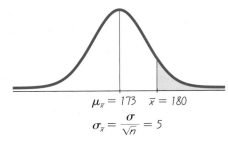

Figure 5-22   Distribution of (a) Individual Men's Weights and (b) Means of Samples of 36 Men's Weights

(a)   (b)

$$z = \frac{x - \mu}{\sigma} = \frac{180 - 173}{30} = 0.23$$

We now refer to Table A-2 to find that region A is 0.0910. The shaded region is therefore $0.5 - 0.0910 = 0.4090$. The probability of the man weighing more than 180 lb is 0.4090.

b. *Approach: Use the central limit theorem* (because we are dealing with the *mean for a group* of 36 scores, not an individual score). Because we are now dealing with a distribution of sample means, we must use the parameters $\mu_{\bar{x}}$ and $\sigma_{\bar{x}}$, which are evaluated as follows:

$$\mu_{\bar{x}} = \mu = 173$$

$$\sigma_{\bar{x}} = \frac{\sigma}{\sqrt{n}} = \frac{30}{\sqrt{36}} = 5$$

We want to determine the shaded area shown in Figure 5-22(b), and the relevant z score is calculated as follows:

$$z = \frac{\bar{x} - \mu_{\bar{x}}}{\sigma_{\bar{x}}} = \frac{180 - 173}{\frac{30}{\sqrt{36}}} = \frac{7}{5} = 1.40$$

Referring to Table A-2, we find that $z = 1.40$ corresponds to an area of 0.4192, so the shaded region is $0.5 - 0.4192 = 0.0808$. The probability that the 36 men have a mean weight greater than 180 lb is 0.0808.

There is a 0.4090 probability that a man will weigh more than 180 lb, but there is only a 0.0808 probability that 36 men will have a mean weight of more than 180 lb. It's much easier for an individual to deviate from the mean than it is for a group of 36. An extreme weight among the 36 weights will lose its impact when it is averaged in with the other 35 weights.

A Professional Speaks About Sampling Error

Daniel Yankelovich, in an essay for *Time*, commented on the sampling error often reported along with poll results. He stated that sampling error refers only to the inaccuracy created by using random sample data to make an inference about a population; the sampling error does not address issues of poorly stated, biased, or emotional questions. He said, "Most important of all, warning labels about sampling error say nothing about whether or not the public is conflict-ridden or has given a subject much thought. This is the most serious source of opinion poll misinterpretation."

The Fuzzy Central Limit Theorem

In *The Cartoon Guide to Statistics* by Gonick and Smith, the authors describe the *Fuzzy Central Limit Theorem* as follows: "Data that are influenced by many small and unrelated random effects are approximately normally distributed. This explains why the normal is everywhere: stock market fluctuations, student weights, yearly temperature averages, SAT scores: All are the result of many different effects." People's heights, for example, are the results of hereditary factors, environmental factors, nutrition, health care, geographic region, and other influences which, when combined, produce normally distributed values.

It would be helpful to review the reasoning here because it is a good lead-in to hypothesis testing. Again, merely obtaining numerical answers might be fine, but the *interpretation* of those results is vital.

EXAMPLE   Assume that the population of human body temperatures has a mean of 98.6° F, as is commonly believed. Also assume that the population standard deviation is 0.62° F (based on data from University of Maryland researchers). If a sample of size $n = 106$ is randomly selected, find the probability of getting a mean of 98.2° F or lower. (The value of 98.2° F was actually obtained; see the midnight temperatures for Day 2 in Data Set 2 of Appendix B.)

SOLUTION   We weren't given the distribution of the population, but because the sample size $n = 106$ exceeds 30, we use the central limit theorem and conclude that the distribution of sample means is a normal distribution with these parameters:

$$\mu_{\bar{x}} = \mu = 98.6$$

$$\sigma_{\bar{x}} = \frac{\sigma}{\sqrt{n}} = \frac{0.62}{\sqrt{106}} = 0.06021972$$

Figure 5-23 shows the shaded area (see the left tail of the graph) corresponding to the probability we seek. Having already found the parameters that apply to the distribution shown in Figure 5-23, we can now find the shaded area by using the same procedures developed in the preceding section. We first find the z score:

$$z = \frac{\bar{x} - \mu_{\bar{x}}}{\sigma_{\bar{x}}} = \frac{98.2 - 98.6}{0.06021972} = -6.64$$

Referring to Table A-2, we find that $z = -6.64$ is off the chart, but for values of z above 3.09, we use an area of 0.4999. We therefore conclude that region A in Figure 5-23 is 0.4999 and the shaded region is $0.5 - 0.4999 = 0.0001$. (More precise tables indicate that the area of the shaded region is closer to 0.000000001.)

The result shows that if the mean of our body temperatures is really 98.6° F, then there is an extremely small probability of getting a sample mean of 98.2° F or lower when 106 subjects are randomly selected. University of Maryland researchers did obtain such a sample mean, and there are two possible explanations: Either the population mean really is 98.6° F and their sample represents a chance event that is extremely rare, or the population mean is actually lower than 98.6° F so their sample is typical. Because the probability is so low, it seems more reasonable to conclude that the population mean is lower than 98.6° F. This is the type of reasoning used in *hypothesis testing*, to be introduced in Chapter 7. For now, we should focus on the use of the central limit theorem for finding the probability of 0.0001, but we should also observe that this theorem will be used later in developing some very important concepts in statistics.

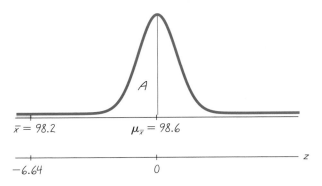

Figure 5-23 Distribution of Sample Mean Body Temperatures ($n = 106$)

In applying the central limit theorem, our use of $\sigma_{\bar{x}} = \sigma/\sqrt{n}$ assumes that the population has infinitely many members. When we sample with replacement (that is, put back each selected item before making the next selection), the population is effectively infinite. Yet many realistic applications involve sampling without replacement, so successive samples depend on previous outcomes. In manufacturing, quality control inspectors typically sample items from a finite production run without replacing them. For such a finite population, we may need to adjust $\sigma_{\bar{x}}$. Here is a common rule of thumb:

> **When sampling without replacement and the sample size n is greater than 5% of the finite population size N (that is, $n > 0.05N$), adjust the standard deviation of sample means $\sigma_{\bar{x}}$ by multiplying it by the *finite population correction factor*:**

$$\sqrt{\frac{N - n}{N - 1}}$$

Except for Exercises 21, 22 and 24, the examples and exercises in this section assume that the finite population correction factor does not apply, because the population is infinite or the sample size doesn't exceed 5% of the population size.

It's very important to note that the central limit theorem applies when we are dealing with the distribution of sample means and either the sample size is greater than 30 or the original population has a normal distribution. Once we have established that the central limit theorem applies, we can determine values for $\mu_{\bar{x}} = \mu$ and $\sigma_{\bar{x}} = \sigma/\sqrt{n}$ and then proceed to use the methods presented in the preceding section.

The central limit theorem is so important because it allows us to use the basic normal distribution methods in a wide variety of different circumstances. In Chapter 6, for example, we will apply the theorem when we use

Because of time limitations, some instructors choose to omit coverage of the finite population correction factor. The exercises that require it are Exercises 21, 22, 24.

sample data to estimate means of populations. In Chapter 7 we will apply it when we use sample data to test claims made about population means. Such applications of estimating population parameters and testing claims are extremely important uses of statistics, and the central limit theorem makes them possible.

5-5 Exercises A: Basic Skills and Concepts

Recommended assignment: Exercises 1, 6, 10, 15, 16, 17, 19, 20. Consider discussing the conclusions reached in Exercises 15, 16, 17, 19, 20, because it is helpful to *interpret* the results in a meaningful and practical way.

In Exercises 1–4, assume that women's heights are normally distributed with a mean given by $\mu = 63.6$ in. and a standard deviation given by $\sigma = 2.5$ in. (based on data from the National Health Survey).

1. **a.** If 1 woman is randomly selected, find the probability that her height is between 63.6 in. and 64.6 in.   1. a. 0.1554
   b. If 36 women are randomly selected, find the probability that they have a mean height between 63.6 in. and 64.6 in.   b. 0.4918
2. **a.** If 1 woman is randomly selected, find the probability that her height is above 63.0 in.   2. a. 0.5948
   b. If 100 women are randomly selected, find the probability that they have a mean height greater than 63.0 in.   b. 0.9918
3. **a.** If 1 woman is randomly selected, find the probability that her height is above 64.0 in.   3. a. 0.4364
   b. If 50 women are randomly selected, find the probability that they have a mean height greater than 64.0 in.   b. 0.1292
4. **a.** If 1 woman is randomly selected, find the probability that her height is between 63.0 in. and 65.0 in.   4. a. 0.3071
   b. If 75 women are randomly selected, find the probability that they have a mean height between 63.0 in. and 65.0 in.   b. 0.9811
5. Replacement times for TV sets are normally distributed with a mean of 8.2 years and a standard deviation of 1.1 years (based on data from "Getting Things Fixed," *Consumer Reports*). Find the probability that 40 randomly selected TV sets will have a mean replacement time less than 8.0 years.   5. 0.1251
6. Replacement times for CD players are normally distributed with a mean of 7.1 years and a standard deviation of 1.4 years (based on data from "Getting Things Fixed," *Consumer Reports*). Find the probability that 45 randomly selected CD players will have a mean replacement time greater than 7.0 years.   6. 0.6844
7. According to the Opinion Research Corporation, men spend an average of 11.4 min in the shower. Assume that the times are normally distributed with a standard deviation of 1.8 min. If 33 men are randomly selected, find the probability that their shower times have a mean between 11.0 min and 12.0 min.   7. 0.8716

8. According to the International Mass Retail Association, girls aged 13 to 17 spend an average of $31.20 on shopping trips in a month. Assume that the amounts have a standard deviation of $8.27. If 85 girls in that age category are randomly selected, what is the probability that their mean monthly shopping expense is between $30.00 and $33.00?   8. 0.8877

9. For women aged 18–24, systolic blood pressures (in mm Hg) are normally distributed with a mean of 114.8 and a standard deviation of 13.1 (based on data from the National Health Survey).
   a. If a woman between the ages of 18 and 24 is randomly selected, find the probability that her systolic blood pressure is above 120.
   b. If 12 women in that age bracket are randomly selected, find the probability that their mean systolic blood pressure is greater than 120.
   c. Given that part (b) involves a sample size that is not larger than 30, why can the central limit theorem be used?

9. a. 0.3446
   b. 0.0838
   c. Because the original population has a normal distribution, samples of *any* size will yield sample means that are normally distributed.

10. The annual precipitation amounts for Iowa appear to be normally distributed with a mean of 32.473 in. and a standard deviation of 5.601 in. (based on data from the U.S. Department of Agriculture).
    a. If one year is randomly selected, find the probability that the annual precipitation is less than 29.000 in.   10. a. 0.2676
    b. If a decade of ten years is randomly selected, find the probability that the annual precipitation amounts have a mean less than 29.000 in.
    c. Given that part (b) involves a sample size that is not larger than 30, why can the central limit theorem be used?

10. b. 0.0250

   c. Because the original population has a normal distribution, samples of *any* size will yield sample means that are normally distributed.

11. The ages of U.S. commercial aircraft have a mean of 13.0 years and a standard deviation of 7.9 years (based on data from Aviation Data Services). If the Federal Aviation Administration randomly selects 35 commercial aircraft for special stress tests, find the probability that the mean age of this sample group is greater than 15.0 years.   11. 0.0668

12. A study of the amounts of time (in hours) college freshmen study each week found that the mean is 7.06 h and the standard deviation is 5.32 h (based on data from *The American Freshman*). If 55 freshmen are randomly selected, find the probability that their mean weekly study time exceeds 7.00 h.   12. 0.5319

13. The typical computer random-number generator yields numbers in a uniform distribution between 0 and 1 with a mean of 0.500 and a standard deviation of 0.289. If 45 random numbers are generated, find the probability that their mean is below 0.565.   13. 0.9345

14. A study was made of seat-belt use among children who were involved in car crashes that caused them to be hospitalized. It was found that children not wearing any restraints had hospital stays with a mean of 7.37 days and a standard deviation of 0.79 day (based on data from "Morbidity Among Pediatric Motor Vehicle Crash Victims: The Effectiveness of Seat Belts," by Osberg and Di Scala, *American Journal of Public Health*, Vol. 82, No. 3). If 40 such children are randomly selected, find the probability that their mean hospital stay is greater than 7.00 days.   14. 0.9985

15. The Town of Newport operates a rubbish waste disposal facility that is overloaded if its 4872 households have total weights with a mean that exceeds 27.88 lb in a week. The total weights are normally distributed with a mean of 27.44 lb and a standard deviation of 12.46 lb (based on data from the Garbage Project at the University of Arizona). What is the proportion of weeks in which the waste disposal facility is overloaded? Is this an acceptable level, or should action be taken to correct a problem of an overloaded system?   15. 0.0069; level is acceptable

16. SAT verbal scores are normally distributed with a mean of 430 and a standard deviation of 120 (based on data from the College Board ATP). Randomly selected SAT verbal scores are obtained from the population of students who took a test preparatory course from the Tillman Training School. Assume that this training course has no effect on test scores.
    a. If 1 of the students is randomly selected, find the probability that he or she obtained a score greater than 440.   16. a. 0.4681
    b. If 100 students are randomly selected, find the probability that their mean score is greater than 440.   b. 0.2033
    c. If 100 Tillman students achieve a sample mean of 440, does it seem reasonable to conclude that the course is effective because the students perform better on the SAT?

c. Because there is a 0.2033 probability that with an ineffective course the mean will be greater than 440, it is not reasonable to conclude that the course is effective. The mean of 440 could have easily occurred by chance.

17. The lengths of pregnancies are normally distributed with a mean of 268 days and a standard deviation of 15 days.
    a. If 1 pregnant woman is randomly selected, find the probability that her length of pregnancy is less than 260 days.   17. a. 0.2981
    b. If 25 randomly selected women are put on a special diet just before they become pregnant, find the probability that their lengths of pregnancy have a mean that is less than 260 days (assuming that the diet has no effect).   b. 0.0038
    c. If the 25 women do have a mean of less than 260 days, should the medical supervisors be concerned?

c. Yes, because it is highly unlikely (with a probability of only 0.0038) that the mean will be that low because of chance.

18. Using a standard measure of satisfaction with salaries, a study finds that college administrators have a mean of 38.9 and a standard deviation of 12.4 (based on data from "Job Satisfaction Among Academic Administrators," by Glick, *Research in Higher Education*, Vol. 33, No. 5). A pollster randomly selects 150 college administrators and measures their levels of satisfaction with their salaries.
    a. Find the probability that the mean is greater than 42.0.   18. a. 0.0011
    b. If a sample of 150 college administrators does yield a mean of 42.0 or greater, is there reason to believe that this sample came from a population with a mean that is higher than 38.9?

b. Yes, because it isn't very likely that such a score would occur by chance.

19. M&M plain candies have a mean weight of 0.9147 g and a standard deviation of 0.0369 g (based on Data Set 11 in Appendix B). The M&M candies used in Data Set 11 came from a package containing 1498 candies, and the package label stated that the net weight is 48.0 oz (3 lb), or

1361 g. (If every package has 1498 candies, the mean weight must exceed 1361/1498 = 0.9085 g for the net contents to weigh at least 1361 g.)

a. If 1 M&M plain candy is randomly selected, find the probability that it weighs more than 0.9085 g.   19. a. 0.5675

b. If 1498 M&M plain candies are randomly selected, find the probability their mean weight is at least 0.9085 g.   b. 0.9999

c. Given these results, does it seem that the Mars Company is providing M&M consumers with the amount claimed on the label?   c. Yes

20. The population of weights of men is normally distributed with a mean of 173 lb and a standard deviation of 30 lb (based on data from the National Health Survey). An elevator in the Dallas Men's Club is limited to 32 occupants, but it will be overloaded if the 32 occupants have a mean weight in excess of 186 lb (yielding a total weight in excess of $32 \cdot 186 = 5952$ lb). If 32 male occupants are the result of a random selection, find the probability that their mean weight will exceed 186 lb, causing the elevator to be overloaded. Based on the value obtained, is there reason for concern?

20. 0.0071; there is reason for concern because even though the probability of an overload is quite small, overloads will occur over the course of thousands of uses.

5–5 Exercises B: Beyond the Basics

21. Repeat Exercise 20, assuming that the population size is $N = 500$ and all sampling is done without replacement. (*Hint:* See the discussion of the finite population correction factor.)   21. 0.0057

22. A *population* consists of these scores: 2, 3, 6, 8, 11, 18.

a. Find μ and σ.   22. a. $\mu = 8.0$, $\sigma = 5.4$

b. List all samples of size $n = 2$ that are obtained without replacement.

c. Find the population of all values of $\bar{x}$ by finding the mean of each sample from part (b).

d. Find the mean $\mu_{\bar{x}}$ and standard deviation $\sigma_{\bar{x}}$ for the population of sample means found in part (c).   d. $\mu_{\bar{x}} = 8.0, \sigma_{\bar{x}} = 3.4$

e. Verify that

$$\mu_{\bar{x}} = \mu \quad \text{and} \quad \sigma_{\bar{x}} = \frac{\sigma}{\sqrt{n}}\sqrt{\frac{N-n}{N-1}}$$

b. 2,3 2,6 2,8 2,11 2,18 3,6 3,8 3,11 3,18 6,8 6,11 6,18 8,11 8,18 11,18

c. 2.5, 4.0, 5.0, 6.5, 10.0, 4.5, 5.5, 7.0, 10.5, 7.0, 8.5, 12.0, 9.5, 13.0, 14.5

23. An education researcher develops an index of academic interest and obtains scores for a randomly selected sample of 350 college students. The results are summarized in the accompanying boxplot. If 15 of the students are randomly selected, find the probability that their mean score is greater than 55.   23. ≈ 0.1949 (using $s \approx 22.4$)

24. The finite population correction factor can be ignored when sampling with replacement or when $n \le 0.05N$. When collecting a sample (without replacement) that is 5% of the population N, what do the values of the finite population correction factor have in common for values of $N \ge 600$?   24. They are all 0.975.

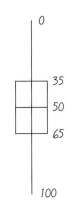

This section is another interesting application of the normal distribution, and it follows up on the earlier work done with the binomial distribution. It is not required for future chapters.

5-6 Normal Distribution as Approximation to Binomial Distribution

In Section 4-3 we introduced the *binomial probability distribution*, which applies to a discrete random variable (rather than to a continuous random variable, as does the normal distribution). We noted that a binomial distribution has these four requirements:

1. The experiment must have a *fixed number of trials*.
2. The trials must be *independent*.
3. Each trial must have all outcomes classified into *two categories*.
4. The probabilities must remain *constant* for each trial.

A typical binomial probability problem involves finding $P(x)$, the probability of x successes among n trials. In Section 4-3 you learned (we hope) how to solve binomial problems by using computer software or Table A-1 or the binomial probability formula. In many cases, however, those methods are not practical. Consider the binomial problem of finding the probability of getting at least 64 men among 100 randomly selected people. Table A-1 doesn't apply because it stops at $n = 15$, and the formula isn't practical because we would have to apply it 37 times (once for each integer from 64 through 100). Such calculations are of the type mathematicians often refer to as "tedious," meaning that they will seriously challenge your sanity. Instead, we will introduce a new method whereby we approximate the binomial distribution by a normal distribution so that the calculations will be greatly simplified. The following box summarizes the key point of this section.

Normal Distribution as Approximation to Binomial Distribution

If $np \geq 5$ and $nq \geq 5$, then the binomial random variable is approximately normally distributed with the mean and standard deviation given as

$$\mu = np$$
$$\sigma = \sqrt{npq}$$

Review Figure 4-4 (which applies to a binomial distribution with $n = 100$, $p = 0.5$, and $q = 0.5$) on page 212, and note that this particular binomial distribution does have a probability histogram with roughly the same shape as that of a normal distribution. The formal justification that allows us to use the normal distribution as an approximation to the binomial distribution results from more advanced mathematics, but Figure 4-4 is a visual argument supporting that approximation.

Reliability and Validity

The reliability of data refers to the consistency with which results occur, whereas the validity of data refers to how well the data measure what they are supposed to measure. The reliability of an IQ test can be judged by comparing scores for the test given on one date to scores for the same test given at another time. To test the validity of an IQ test, we might compare the test scores to another indicator of intelligence, such as academic performance. Many critics charge that IQ tests are reliable, but not valid; they provide consistent results, but don't really measure intelligence.

The normal approximation approach involves the following procedure, which is depicted in the flowchart shown in Figure 5-24:

Normal Approximation Procedure

1. Verify that the binomial probability distribution applies.
2. If available, use software (such as STATDISK or Minitab) or a TI-83 calculator. (Examples of STATDISK and Minitab displays are found in Section 4-3 on page 202.)
3. If software is not available, try to solve the problem by using Table A-1.
4. If Table A-1 cannot be used, consider the binomial probability formula. If the problem can be *easily* solved with the binomial probability formula, use it. Otherwise, continue with Step 5.
5. Establish that the normal distribution is a suitable approximation to the binomial distribution by verifying that $np \geq 5$ and $nq \geq 5$.
6. Find the values of the parameters μ and σ by calculating $\mu = np$ and $\sigma = \sqrt{npq}$.
7. Identify the discrete value x, which represents the number of successes in the binomial experiment. Modify the *discrete* value x by replacing it with the *interval* from $x - 0.5$ to $x + 0.5$. (See the discussion under the subheading *Continuity Corrections* found later in this section.) Draw a normal curve and enter the values of μ, σ, and either $x - 0.5$ or $x + 0.5$ as appropriate.
8. Modify x by replacing it with $x - 0.5$ or $x + 0.5$ as appropriate, then find the z score: $z = (x - \mu)/\sigma$.
9. Using the z score found in Step 8, refer to Table A-1 to find the area between μ and either $x - 0.5$ or $x + 0.5$, as appropriate. Identify the area corresponding to the desired probability.

We will now illustrate this normal approximation procedure with the following example.

She Won the Lottery Twice!

Evelyn Marie Adams won the New Jersey Lottery twice in four months. This happy event was reported as an incredible coincidence with a likelihood of only 1 chance in 17 trillion. But Harvard mathematicians Persi Diaconis and Frederick Mosteller note that there is 1 chance in 17 trillion that a particular person with one ticket in each of two New Jersey lotteries will win both times. However, there is about 1 chance in 30 that someone in the United States will win a lottery twice in a four-month period. Diaconis and Mosteller analyzed coincidences and conclude that "with a large enough sample, any outrageous thing is apt to happen."

> **EXAMPLE** Assume that your college has an equal number of qualified male and female applicants, and assume that 64 of the last 100 newly hired employees are men. Estimate the probability of getting *at least* 64 men if each hiring is done independently and with no gender discrimination. (The probability of getting *exactly* 64 men doesn't really tell us anything, because with 100 trials, the probability of any exact number of men is fairly small. Instead, we need the probability of getting a result *at least* as extreme as the one obtained.) Based on the result, does it seem that the college is discriminating on the basis of gender?

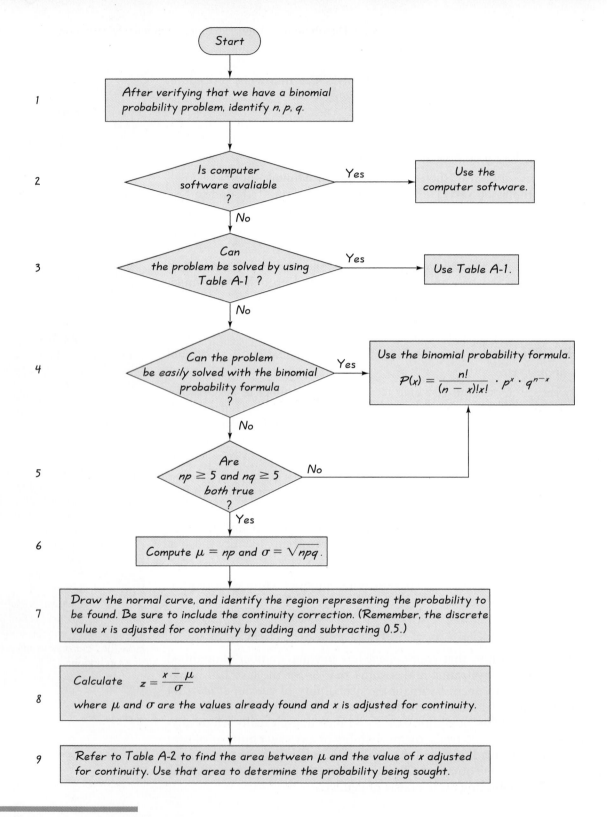

Figure 5-24 Solving Binomial Probability Problems Using a Normal Approximation

SOLUTION

Step 1:   The given problem does involve a binomial distribution with a fixed number of trials ($n = 100$), which are presumably independent, two categories (man, woman) of outcome for each trial, and probabilities that presumably remain constant.

Step 2:   For the purposes of this example, we will assume that computer software is not available.

Step 3:   Table A-1 cannot be used because the value of $n = 100$ exceeds the maximum value of $n = 15$ found in the table.

Step 4:   The formula could be used, but then it would have to be used for each integer from 64 through 100. We prefer to avoid using the binomial probability formula 37 times.

Step 5:   We establish that it is reasonable to approximate the binomial distribution by the normal distribution because $np \geq 5$ and $nq \geq 5$, as verified below.

$$np = 100 \cdot 0.5 = 50 \qquad \text{(Therefore } np \geq 5.\text{)}$$
$$nq = 100 \cdot 0.5 = 50 \qquad \text{(Therefore } nq \geq 5.\text{)}$$

Step 6:   We now proceed to find the values for μ and σ that are needed. We get the following:

$$\mu = np = 100 \cdot 0.5 = 50$$

$$\sigma = \sqrt{npq} = \sqrt{100 \cdot 0.5 \cdot 0.5} = 5$$

Step 7:   In Figure 5-25 we show the discrete value of 64 as being represented by the vertical strip bounded by 63.5 and 64.5. Figure 5-26 shows the area we want: It is the shaded area in the extreme right tail of the graph.

Figure 5-26   Finding the Probability of "At Least" 64 Men Among 100 New Employees

Figure 5-25   Illustration of Continuity Correction

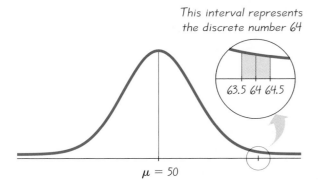

This interval represents the discrete number 64

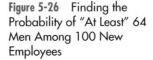

63.5 64 64.5

$\mu = 50$

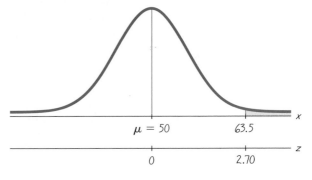

$\mu = 50$

63.5

0

2.70

Students using a TI-83 calculator can solve this problem without using an approximation. Instead, they press 2nd, DISTR, and then proceed to enter `binom-cdf(100,.5,63)`. The result of 0.9966814397 is the probability of 63 or fewer men, so that the probability of at least 64 men is 0.0033185603. The calculator's `binomcdf` (binomial cumulative density function) gives cumulative probabilities for a binomial distribution.

Step 8:   We cannot find the shaded area in Figure 5-26 directly, so we find the area of the region bounded by $\mu = 50$ and the score of 63.5. This area corresponds to the following z score:

$$z = \frac{x - \mu}{\sigma} = \frac{63.5 - 50}{5} = 2.70$$

Step 9:   Using Table A-1, we find that $z = 2.70$ corresponds to an area of 0.4965, so the shaded area we seek is $0.5 - 0.4965 = 0.0035$.

Assuming that men and women have equal chances of being hired, the probability of getting at least 64 men in 100 new employees is approximately 0.0035. (Using the software packages of STATDISK and Minitab and the TI-83 calculator, we get the answer of 0.0033, so the approximation is quite good here.)

Because the probability of getting at least 64 men is so small (0.0035), we conclude that either a very rare event has occurred or the assumption that men and women have the same chance is incorrect. It appears that the college discriminates on the basis of gender. The reasoning here will be considered in more detail in Chapter 7 when we discuss formal methods of testing hypotheses. For now, we should focus on the method of finding the probability by using the normal approximation technique.

Continuity Corrections

Step 7 in the procedure just given is difficult for many people to understand. It is based on the concept of continuity corrections.

DEFINITION

When we use the normal distribution (which is continuous) as an approximation to the binomial distribution (which is discrete), a **continuity correction** is made to a discrete whole number x in the binomial distribution by representing the single value x by the *interval* from $x - 0.5$ to $x + 0.5$ (that is, add and subtract 0.5).

The following practical suggestions should help you use continuity corrections properly.

Procedure for Continuity Corrections

1. When using the normal distribution as an approximation to the binomial distribution, *always* use the continuity correction.

2. In using the continuity correction, first identify the discrete whole number x that is relevant to the binomial probability problem. For example, if you're trying to find the probability of getting at least 64 men in 100 randomly selected people, the discrete whole number of concern is $x = 64$. First focus on the x value itself, and temporarily ignore whether you want at least x, more than x, fewer than x, or whatever.

3. Draw a normal distribution centered about μ, then draw a *vertical strip area* centered over x. Mark the left side of the strip with the number $x - 0.5$, and mark the right side with $x + 0.5$. For $x = 64$, draw a strip from 63.5 to 64.5. *Consider the entire area of the strip to represent the probability of the discrete number x.*

4. Now determine whether the value of x itself should be included in the probability you want. (For example, "at least x" does include x itself, but "more than x" does not include x itself.) Next, determine whether you want the probability of at least x, at most x, more than x, fewer than x, or exactly x. Shade the area to the right or left of the strip, as appropriate; also shade the interior of the strip itself *if and only if* x itself is to be included. This total shaded region corresponds to the probability being sought.

To see how this procedure results in continuity corrections, see the common cases illustrated in Figure 5-27. Those cases correspond to the statements listed below.

Statement	Area
At least 64 (includes 64 and above)	To the *right* of 63.5
More than 64 (doesn't include 64)	To the *right* of 64.5
At most 64 (includes 64 and below)	To the *left* of 64.5
Fewer than 64 (doesn't include 64)	To the *left* of 63.5
Exactly 64	Between 63.5 and 64.5

> Students often have difficulty applying the continuity correction. Carefully describe the four-step procedure given here. It really does help.

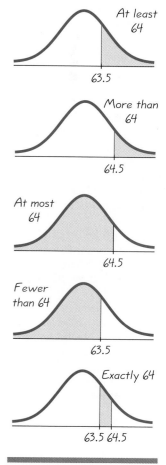

Figure 5-27 Identifying the Correct Area

EXAMPLE   About 4.4% of fatal motor vehicle crashes are caused by defective tires (based on data from the National Safety Council). If a highway safety study begins with the random selection of 750 cases of fatal motor vehicle crashes, estimate the probability that exactly 35 of them were caused by defective tires.   *(continued)*

Picking Lottery Numbers

In a typical state lottery, you select six different numbers. After a random drawing, any entries with the correct combination share in the prize. Since the winning numbers are randomly selected, any choice of six numbers will have the same chance as any other choice, but some combinations are better than others. The combination of 1, 2, 3, 4, 5, 6 is a poor choice because many people tend to select it. In a Florida lottery with a $105 million prize, 52,000 tickets had 1, 2, 3, 4, 5, 6; if that combination had won, the prize would have been only $1000. It's wise to pick combinations not selected by many others. Avoid combinations that form a pattern on the entry card.

SOLUTION   This solution follows the steps outlined in Figure 5-24.

Step 1: The conditions described satisfy the criteria for the binomial distribution with $n = 750$, $p = 0.044$, $q = 0.956$, and $x = 35$. (The value of q is found from $q = 1 - p = 1 - 0.044 = 0.956$.)

Step 2: For the purposes of this example, we assume that computer software is not available.

Step 3: Table A-1 cannot be used because $n = 750$ exceeds the largest table value of $n = 15$.

Step 4: The binomial probability formula applies, but

$$P(35) = \frac{750!}{(750 - 35)!\, 35!} \cdot 0.044^{35} \cdot 0.956^{750 - 35}$$

is difficult to compute because many calculators cannot evaluate 70! or higher.

Step 5: In this problem,

$np = 750 \cdot 0.044 = 33.0$ (Therefore $np \geq 5$.)

$nq = 750 \cdot 0.956 = 717.0$ (Therefore $nq \geq 5$.)

Because np and nq are both at least 5, we conclude that the normal approximation to the binomial distribution is satisfactory.

Step 6: We obtain the values of μ and σ as follows:

$$\mu = np = 750 \cdot 0.044 = 33.0$$
$$\sigma = \sqrt{npq} = \sqrt{(750)(0.044)(0.956)} = 5.6167606$$

Step 7: We draw the normal curve shown in Figure 5-28. The shaded region of the figure represents the probability we want. Use of the continuity correction results in the representation of 35 by the region extending from 34.5 to 35.5.

Step 8: The format of Table A-2 requires that we first find the probability corresponding to the region bounded on the left by the vertical line through the mean of 33.0 and on the right by the vertical line through 35.5. Therefore one of the calculations required in this step is as follows:

$$z = \frac{35.5 - 33.0}{5.6167606} = 0.45$$

We also need the probability corresponding to the region bounded by 33.0 and 34.5, so we calculate

$$z = \frac{34.5 - 33.0}{5.6167606} = 0.27$$

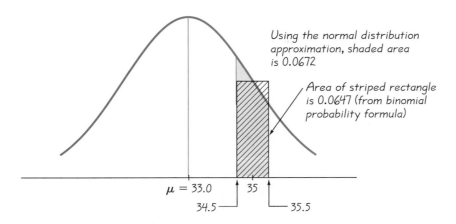

Using the normal distribution approximation, shaded area is 0.0672

Area of striped rectangle is 0.0647 (from binomial probability formula)

$\mu = 33.0$ 35

34.5   35.5

Step 9:   We use Table A-2 to find that $z = 0.45$ corresponds to a probability of 0.1736 and $z = 0.27$ corresponds to a probability of 0.1064. Consequently, the shaded region of Figure 5-28 depicts a probability of $0.1736 - 0.1064 = 0.0672$. The probability of getting exactly 35 cases (out of 750) of fatal crashes caused by defective tires is approximately 0.0672.

If we solve the preceding example using STATDISK, Minitab, or a calculator, we get a result of 0.0647, but the normal approximation method resulted in a value of 0.0672. The discrepancy of 0.0025 occurs because the use of the normal distribution results in an *approximate* value that is the area of the shaded region in Figure 5-28, whereas the correct area is a rectangle centered above 35. (Figure 5-28 illustrates this discrepancy.) The area of the rectangle is 0.0647, but the area of the approximating shaded region is 0.0672.

5-6 Exercises A: Basic Skills and Concepts

In Exercises 1–8, use the continuity correction and describe the region of the normal curve that corresponds to the indicated probability. For example, the probability of "more than 47 successes" corresponds to this area of the normal curve: the area to the right of 47.5.

Recommended assignment: Exercises 1–4, 10, 13, 14, 17–20, 22, 24, 28.

1. Probability of more than 35 defective parts   1. The area to the right of 35.5
2. Probability of at least 175 girls   2. The area to the right of 174.5
3. Probability of fewer than 42 correct answers to multiple-choice questions

3. The area to the left of 41.5

4. The area between 64.5 and 65.5

5. The area to the left of 72.5

6. The area between 34.5 and 45.5

7. The area between 124.5 and 150.5

8. The area between 33.5 and 34.5

10. a. 0.042
    b. Normal approximation is not suitable.

11. a. 0.996
    b. Normal approximation is not suitable.

12. a. 0.515
    b. 0.5199

18. 0.0041; no

20. 0.4602

4. Probability of exactly 65 correct answers to true/false questions

5. Probability of no more than 72 cars with defective brakes

6. Probability that the number of baby girls is between 35 and 45 inclusive

7. Probability that the number of Republican voters is between 125 and 150 inclusive

8. Probability that the number of patients with group A blood is exactly 34

In Exercises 9–12, do the following. (a) *Find the indicated binomial probability by using Table A-1 in Appendix A.* (b) *If $np \geq 5$ and $nq \geq 5$, also estimate the indicated probability by using the normal distribution as an approximation to the binomial distribution; if $np < 5$ or $nq < 5$, then state that the normal approximation is not suitable.*

9. With $n = 14$ and $p = 0.50$, find $P(8)$.   9. a. 0.183   b. 0.1817

10. With $n = 10$ and $p = 0.40$, find $P(7)$.

11. With $n = 15$ and $p = 0.80$, find P(at least 8).

12. With $n = 14$ and $p = 0.60$, find P(fewer than 9).

13. Estimate the probability of getting at least 55 girls in 100 births.   13. 0.1841

14. Estimate the probability of getting exactly 32 boys in 64 births.   14. 0.1034

15. Estimate the probability of passing a true/false test of 50 questions if 60% (or 30 correct responses) represents a passing grade and all responses are random guesses.   15. 0.1020

16. A multiple-choice test consists of 50 questions with possible answers of a, b, c, d, and e. Estimate the probability of getting at most 30% correct if all answers are random guesses.   16. 0.9738

17. There's an 80% chance that a prospective employer will check the educational background of a job applicant (based on data from the Bureau of National Affairs, Inc.). For 100 randomly selected job applicants, estimate the probability that *exactly* 85 have their educational backgrounds checked.   17. 0.0454

18. Susan Stein is the advertising sales director for an NBC TV prime-time police drama, and she wants to persuade advertisers that the viewing audience is actually much larger than the number of people who watch the show at the time of broadcast. A Nielsen survey showed that when viewers use videocassette recorders, 66% of the shows taped are from the major networks. Estimate the probability that among 1000 randomly selected videotaped shows, at least 700 are from major networks, as the sales director claims. Based on the result, does her claim seem plausible?

19. According to a consumer affairs representative from Mars (the candy company, not the planet), 10% of all M&M plain candies are blue. Data Set 11 in Appendix B shows that among 100 M&Ms chosen, 5 are blue. Estimate the probability of randomly selecting 100 M&Ms and getting 5 or fewer that are blue. Assume that the company's 10% blue rate is correct. Based on the result, is it very unusual to get 5 or fewer blue M&Ms when 100 are randomly selected?   19. 0.0668; not *very* unusual

20. Marc Taylor plans to place 200 bets of $1 each on the number 7 at roulette. On any one spin there is a probability of 1/38 that 7 will be the

winning number. For Marc to end up with a profit, the number 7 must occur at least 6 times among the 200 trials. Estimate the probability that Marc finishes with a profit.

21. Based on U.S. Bureau of Justice data, 16% of those arrested are women. In one state, 400 arrest cases are randomly selected. Estimate the probability that the number of women is 38 or lower. If those arrest cases include 38 or fewer women, does it seem plausible that this state arrests women at the 16% rate?   21. 0.0001; no

22. Air America has been experiencing a 7% rate of no-shows on advance reservations. In a test project that requires passengers to confirm reservations, it is found that among 250 randomly selected advance reservations, there are 4 no-shows. Assuming that the confirmation requirement has no effect so the 7% rate applies, estimate the probability of 4 or fewer no-shows among 250 randomly selected reservations. Based on that result, does it appear that the confirmation requirement is effective?   22. 0.0001; yes

23. Currently, about two-thirds of U.S. companies test newly hired employees for drugs, and 3.8% of those prospective employees test positive (based on data from the American Management Association). The Sigma Electronics Company tests 150 prospective employees and finds that 10 of them test positive for drugs. Estimate the probability of 10 or more positive results among 150 subjects. Based on that value, do the 10 positive test results seem to be unusually high?   23. 0.0526; no

24. American Airlines reports that 77.5% of its flights arrive on time (based on data from the U.S. Department of Transportation). A check of 50 randomly selected American Airlines flights shows that 34 (or 68%) of them arrived on time. Estimate the probability that among 50 American Airline flights, 34 or fewer arrive on time. Based on the result, does it seem plausible that the claimed on-time rate of 77.5% could be correct?   24. 0.0749; yes

25. According to the American Medical Association, 18.4% of college graduates smoke. A health study begins with the selection of 300 college graduates, but the number of smokers is 72, which is more than expected. The study is being questioned because the number of smokers does not seem to correspond to the overall rate of 18.4% for the population of college graduates. Estimate the probability of getting at least 72 smokers in a random sample of 300. Based on the result, does it appear that one could easily get 72 smokers by chance, or does it appear that there's something wrong with the sample?   25. 0.0075; something seems wrong with the sample.

26. Some couples have genetic characteristics configured so that one-quarter of all offspring have blue eyes. A study is conducted of 40 couples believed to have those characteristics, with the result that 8 of their 40 offspring have blue eyes. Estimate the probability that among 40 offspring, 8 or fewer have blue eyes. Based on that probability, does it seem that the one-quarter rate is correct?   26. 0.2912; yes

27. Providence Memorial Hospital is conducting a blood drive because its supply of group O blood is low, and it needs 177 donors of group O blood. If 400 volunteers donate blood, estimate the probability that the

number with group O blood is at least 177. Forty-five percent of us have group O blood, according to data provided by the Greater New York Blood Program.   27. 0.6368

28. 0.9505

28. We noted in Section 3-4 that some companies monitor quality by using a method of acceptance sampling whereby an entire batch of items is rejected if, in a random sample, the number of defects is at least some predetermined number. The Dayton Machine Company buys machine bolts in batches of 5000 and rejects a batch if, when 50 of them are sampled, at least 2 defects are found. Estimate the probability of rejecting a batch if the supplier is manufacturing the bolts with a defect rate of 10%.

5-6 Exercises B: Beyond the Basics

29. Replacement times for TV sets are normally distributed with a mean of 8.2 years and a standard deviation of 1.1 years (based on data from "Getting Things Fixed," *Consumer Reports*). Estimate the probability that for 250 randomly selected TV sets, at least 15 of them have replacement times greater than 10.0 years.   29. 0.2946

30. Assume that a baseball player hits .350, so his probability of a hit is 0.350. (Ignore the complications caused by walks). Also assume that his hitting attempts are independent of each other.
  a. Find the probability of at least 1 hit in 4 tries in 1 game.   30. a. 0.821

30. b. 0.9999

  b. Assuming that this batter gets up to bat 4 times each game, estimate the probability of getting a total of at least 56 hits in 56 games.
  c. Assuming that this batter gets up to bat 4 times each game, find the probability of at least 1 hit in each of 56 consecutive games (Joe DiMaggio's 1941 record).   c. 0.0000165
  d. What minimum batting average would be required for the probability in part (c) to be greater than 0.1?   d. 0.552

31. a. 0.4129 − 0.3264 = 0.0865
    b. 0.3192 − 0.2643 = 0.0549
    c. 0.0256 − 0.0228 = 0.0028
    As *n* gets larger, the difference becomes smaller.

31. Find the difference between the answers obtained with and without use of the continuity correction in each of the following. What do you conclude from the results?
  a. Estimate the probability of getting at least 11 girls in 20 births.
  b. Estimate the probability of getting at least 22 girls in 40 births.
  c. Estimate the probability of getting at least 220 girls in 400 births.

32. 262

32. Air America works only with advance reservations and experiences a 7% rate of no-shows. How many reservations could be accepted for an airliner with a capacity of 250 if there is at least a 0.95 probability that all reservation holders who show will be accommodated?

· ·

Vocabulary List

normal distribution

uniform distribution

density curve

standard normal distribution

sampling distribution of sample
    means

central limit theorem

standard error of the mean

finite population correction factor

continuity correction

· ·

Review

Chapter 4 introduced the concept of the probability distribution, but included only discrete types. This chapter introduced continuous probability distributions and focused on the most important type: normal distributions. Normal distributions will be used extensively in the following chapters.

The normal distribution, which appears bell shaped when graphed, can be described algebraically by an equation, but the complexity of that equation usually forces us to use a table of values (Table A-2) instead. Table A-2 gives areas corresponding to specific regions under the standard normal distribution curve, which has a mean of 0 and a standard deviation of 1. Those areas correspond to probability values.

In this chapter we worked with standard procedures for applying Table A-2 to a variety of different situations, including those that involve normal distributions that are nonstandard (with a mean different than 0 or a standard deviation different than 1). Those procedures usually involve use of the standard score: $z = (x - \mu)/\sigma$.

In Section 5-5 we presented the following important points associated with the central limit theorem:

1. The distribution of sample means will, as the sample size n increases, approach a normal distribution.

2. The mean of the sample means is the population mean μ.

3. The standard deviation of the sample means will be $\sigma/\sqrt{n}$.

In Section 5-6 we noted that we can sometimes approximate a binomial probability distribution by a normal distribution. If both $np \geq 5$ and $nq \geq 5$, the binomial random variable x is approximately normally distributed with the mean and standard deviation given as $\mu = np$ and $\sigma = \sqrt{npq}$. Because the binomial probability distribution deals with discrete data and the normal distribution deals with continuous data, we apply the continuity correction, which should be used in normal approximations to binomial distributions.

●●

Review Exercises

In Exercises 1–4, assume that men have normally distributed heights with a mean of 69.0 in. and a standard deviation of 2.8 in. (based on data from the National Health Survey).

1. 64.4 in., 73.6 in.

1. You plan to open a men's clothing store. To minimize start-up costs, you will not stock suits for the tallest 5% and the shortest 5% of men. Find the minimum and maximum heights of the men for whom suits will be stocked.

2. The Beanstalk Club is a social organization for tall people. Men are eligible for membership if they are at least 74 in. tall. What percentage of men are eligible for membership in the Beanstalk Club?   2. 3.67%

3. If a man is randomly selected, what is the probability that he will satisfy the height requirements for the Newport Police Department, which requires men to have heights between 62.0 in. and 76.0 in.?   3. 0.9876

4. If 45 men are randomly selected, what is the probability that their mean height is between 70.0 in. and 71.0 in.?   4. 0.0081

5. *Entertainment Report* magazine runs a sweepstakes as part of a campaign to acquire new subscribers. In the past, 26% of those who received sweepstakes entry materials have ended up entering the contest and subscribing to the magazine (based on data reported in *USA Today*). Estimate the probability that when sweepstakes entry materials are sent to 500 randomly selected households, the resulting number of new subscriptions is between 125 and 150 inclusive.   5. 0.6940

6. The Gleason Supermarket uses a scale to weigh produce, and errors are normally distributed with a mean of 0 oz and a standard deviation of 1 oz. (The errors can be positive or negative.) One item is randomly selected and weighed. Find the probability that the error is
   a. between 0 and 1.25 oz   6. a. 0.3944
   b. greater than 0.50 oz   b. 0.3085
   c. greater than −1.08 oz   c. 0.8599
   d. between −0.50 oz and 1.50 oz   d. 0.6247
   e. between −1.00 oz and −0.25 oz   e. 0.2426

7. Scores on the biology portion of the Medical College Admissions Test are normally distributed with a mean of 8.0 and a standard deviation of 2.6. Among 600 individuals taking this test, how many are expected to score between 6.0 and 7.0?   7. 78.8

8. On the Graduate Record Exam in economics, scores are normally distributed with a mean of 615 and a standard deviation of 107. If a college admissions office requires scores above the 70th percentile, find the cutoff point.   8. 671

9. Among U.S. households, 24% have telephone answering machines (based on data from the U.S. Consumer Electronics Industry). If a telemarketing cam-

paign involves 2500 households, estimate the probability that more than 650
have answering machines.   9. 0.0091
10. The Chemco Company manufactures car tires that last distances that are normally distributed with a mean of 35,600 mi and a standard deviation of 4275
mi.
   a.   If a tire is randomly selected, what is the probability it lasts more than
        30,000 mi?   10. a. 0.9049
   b.   If 40 tires are randomly selected, what is the probability they last dis-
        tances that have a mean greater than 35,000 mi?   b. 0.8133
   c.   If the manufacturer wants to guarantee the tires so that only 3% will be
        replaced because of failure before the guaranteed number of miles, for
        how many miles should the tires be guaranteed?   c. 27,563 mi

Cumulative Review Exercises

1. According to data from the American Medical Association, 10% of us are
left-handed.
   a.   If three people are randomly selected, find the probability that they are all
        left-handed.   1. a. 0.001
   b.   If three people are randomly selected, find the probability that at least one
        of them is left-handed.   b. 0.271
   c.   Why can't we solve the problem in part (b) by using the normal approx-
        imation to the binomial distribution?
   d.   If groups of 50 people are randomly selected, what is the mean number
        of left-handed people in such groups?   d. 5.0
   e.   If groups of 50 people are randomly selected, what is the standard devia-
        tion for the numbers of left-handed people in such groups?
   f.   Would it be unusual to get 8 left-handed people in a randomly selected
        group of 50 people? Why or why not?

c. The requirement that $np \geq 5$ is not satisfied, indicating that the normal approximation would result in errors that are too large.

e. 2.1

f. No, 8 is within two standard deviations of the mean and is within the range of values that could easily occur by chance.

2. The sample scores given below are times (in milliseconds) it took the author's
disk drive to make one revolution. The times were recorded by a diagnostic
software program.

199.7 200.0 200.1 200.1 200.1 200.3 200.3 200.3 200.3 200.3
200.3 200.3 200.4 200.4 200.4 200.4 200.4 200.4 200.4 200.4
200.5 200.5 200.5 200.5 200.5 200.5 200.5 200.5 200.5 200.6
200.6 200.6 200.6 200.6 200.6 200.7 200.8 201.1 201.2 201.2

   a.   Find the mean $\bar{x}$ of the times in this sample.   2. a. 200.46 ms
   b.   Find the median of the times in this sample.   b. 200.45 ms
   c.   Find the mode of the times in this sample.   c. 200.5 ms
   d.   Find the standard deviation s of this sample.   d. 0.29 ms   *(continued)*

e. Convert the time of 200.5 ms to a z score.   e. 0.14
f. Find the actual percentage of these sample scores that are greater than 201.0 ms.   f. 7.5%
g. Assuming a normal distribution, find the percentage of *population* scores greater than 201.0 ms. Use the sample values of $\bar{x}$ and s as estimates of μ and σ.   g. 3.14% (using $\bar{x}$ = 200.46 and s = 0.29)
h. The specifications require times between 197.0 ms and 202.0 ms. Based on these sample results, does the disk drive seem to be rotating at acceptable speeds?   h. Yes

Computer Project

In this chapter we addressed the problem of finding the percentage of women with heights between 64.5 in. and 72 in., as required by Russian Soyuz spacecraft. We found that 35.93% of women have heights between those values. The solution, given in Section 5-3, involves theoretical calculations based on the assumptions that heights of women are normally distributed with a mean of 63.6 in. and a standard deviation of 2.5 in. (based on data from the National Health Survey). This project describes a different method of solution that is based on this simulation technique: We will use a computer or TI-83 calculator to randomly generate 100 heights of women (from a normally distributed population with μ = 63.6 and σ = 2.5), then we will find the percentage of those simulated heights that fall between 64.5 in. and 72 in. The STATDISK, Minitab, and TI-83 procedures are described as follows.

To use STATDISK: Select `Data` from the main menu bar, then choose the option of `Normal Generator`. Proceed to generate 100 values with a mean of 63.6 and a standard deviation of 2.5. (Use the `Format` option to specify 1 decimal place.) Next, rank the data by using the options of `Sample Editor`, then `Format`. With this ranked list, it becomes quite easy to count the number of heights between 64.5 and 72. Divide that number by 100 to find the percentage of these simulated heights that are suitable for Russian Soyuz spacecraft. Compare the result to the theoretical value of 35.93% that was found in Section 5-3.

To use Minitab: Select the options of `Calc`, then `Random Data`, then `Normal`. Proceed to enter 100 for the number of rows, C1 for the column in which to store the data, 63.6 for the value of the mean, and 2.5 for the value of the standard deviation. Now select the option of `Manip`, then `Sort` and proceed to sort column C1, with the sorted column stored in column C1, and with the sorting to be done by column C1. Examine the values in column C1 and count the number of heights between 64.5 and 72, then divide that number by 100 to find the percentage of simulated heights that are suitable for Russian Soyuz spacecraft. Compare the result to the theoretical value of 35.93% that was found in Section 5-3.

To use a TI-83 calculator: Press MATH, then select PRB, then enter `randNorm (63.6, 2.5, 100)` to generate 100 values from a normally distributed population with μ = 63.6 and σ = 2.5. Enter STO→L1 to store the data in list L1. Now press STAT, then enter `SortA(L1)` to arrange the data in order. Examine the

entries in list L1 to find the number of values between 64.5 and 72, then divide that number by 100 to find the percentage of simulated heights that are suitable for Russian Soyuz spacecraft. Compare the result to the theoretical value of 35.93% that was found in Section 5-3.

•••••••• FROM DATA TO DECISION ••••••••

How Can We Outsmart Users of Counterfeit Coins in Vending Machines?

Operators of vending machines are continually devising strategies to combat the use of counterfeit coins and bills. One machine collected dollar bills in a tray and supplied change for use in other vending machines. A rod would pass through the center of the receiving tray and pull the dollar bill into a holding box. A creative but dishonest person found that if he used a dollar bill with a slit through the center, the machine would provide the change but the bill would not be removed. He could empty the machine of all its change, then move on to another machine.

The use of slugs also plagues vending-machine operators. Let's develop a strategy for minimizing losses from slugs that are counterfeit quarters. In Data Set 13 of Appendix B, we have a random sample of weights (in grams) of legal and legitimate quarters. Based on that sample, let's assume that for the population of all such quarters, the distribution is normal, the mean weight is 5.622 g, and the standard deviation is 0.068 g. Let's also assume that a supply of counterfeit quarters has been collected from vending machines in downtown Dallas. Analysis of those counterfeit quarters shows that they have weights that are normally distributed with a mean of 5.450 g and a standard deviation of 0.112 g. A vending machine is designed so that coins are accepted or rejected according to weight. The limits for acceptable coins can be changed, but the machine currently accepts any coin with a weight between 5.500 g and 5.744 g. Find the percentages of legitimate coins and counterfeit coins accepted by the machine.

Here's a dilemma: If we restrict the weight limits on the coins too much, we reduce the proportion of legitimate coins that are accepted and customers are lost because their legitimate coins won't work. If we don't restrict the weight limits on coins enough, too many counterfeit coins are accepted and losses are incurred. What are the percentages of legitimate and counterfeit coins accepted if the lower weight limit is changed to 5.550 g? Experiment with different weight settings and identify limits that seem reasonable in that the machine does not accept too many counterfeit coins but still does accept a reasonable proportion of legitimate coins. It might be helpful to construct a reasonably accurate normal distribution graph for legitimate coins and another one for counterfeit coins; use the same horizontal scale and position one graph above the other so that numbers on the horizontal scale are aligned. With these graphs, you want to find limits that include the most legitimate coins and the fewest counterfeit coins. (There isn't necessarily a unique answer here; to some extent, the choices depend on subjective judgments.) Write a brief report that includes the settings you recommend, along with specific reasons for your choices.

−.204
−.202
−.199
−.196
−.194
−.191
−.188
−.185
−.183
−.180
−.177
−.174
−.171
−.168
−.165
−.161
−.158
−.155
−.151
−.148
−.144
−.140
−.137
−.133
−.129
−.125
−.121
−.116
−.112
−.107
−.102
−.097
−.091
−.085
−.079
−.072
−.065
−.056
−.046
−.032

REACTION TIMER

•••••••• COOPERATIVE GROUP ACTIVITIES ••••••••

1. *Out-of-class activity:* Use the reaction timer by following the instructions given below. Collect reaction times for a sample of at least 40 different subjects taken from a homogeneous group, such as right-handed college students. For each subject, measure the reaction time for each hand. Construct a histogram for the right-hand times and another histogram for the left-hand results. Based on the histogram shapes, do the two sets of times each appear to be normally distributed? Calculate the values of the mean and standard deviation for each of the two data sets and compare the two sets of results. Using the right-hand sample mean $\bar{x}$ as an estimate of the population mean μ, and using the right-hand sample standard deviation s as an estimate of the population standard deviation σ, find the quartiles Q_1, Q_2, and Q_3. Repeat that procedure to find Q_1, Q_2, and Q_3 for the left hand. If the right-hand reaction times are to be used to screen job applicants, what time is the cutoff separating the fastest 5% from the slowest 95%?

Instructions for Using the Reaction Timer

   a. Cut the reaction timer along the dashed line.
   b. Ask the subject to hold his or her thumb and forefinger horizontally; those fingers should be spread apart by a distance that is the same as the width of the reaction timer.
   c. Hold the reaction timer so that the bottom edge is just above the subject's thumb and forefinger.
   d. Ask the subject to catch the reaction timer as quickly as possible, then release it after a few seconds.
   e. Record the reaction time (in seconds) corresponding to the point at which the subject catches it.

2. *In-class activity:* Use the chalkboard to construct two stem-and-leaf plots with stem values of 0, 1, 2, . . . , 9. For the stem-and-leaf plot at the left, each student should enter the last four digits of his or her social security number. For the stem-and-leaf plot at the right, each student should calculate the mean $\bar{x}$ of those same four digits, round the result to the nearest whole number, and enter it on the stem-and-leaf plot. Groups of three or four should now be formed to analyze the results and state how they illustrate the central limit theorem.

COOPERATIVE GROUP ACTIVITIES

3. *In-class activity:* Divide into groups of three or four students. For each group member, find the total value of coins that member has in his or her possession. Next, find the group mean. Share the individual values and the mean value with all of the other groups. What is the distribution of the individual values? What is the distribution of the group means? Do the results illustrate the central limit theorem? How?

4. *In-class activity:* Divide into groups of three or four students. Using a coin to simulate births, each group member should simulate 25 births and record the number of simulated girls. Combine all results in the group and record n = total number of births and x = number of girls. Given batches of n births, compute the mean and standard deviation for the number of girls. Is the simulated result usual or unusual? Why?

interview

David Burn

David Burn is the Director of Statistical Services at Consumers Union. Consumers Union, which publishes *Consumer Reports* magazine, is a nonprofit and independent organization serving consumers. David Burn was formerly a Senior Statistician at Minitab, Inc.

Author's note: The author met with David Burn and the other statisticians at Consumers Union: M. Edna Derderian, Keith Newsom-Stewart, Martin Romm, and Jed Schneider. The author toured the testing facility and saw that Consumers Union meticulously designs, executes, and analyzes objective experiments.

What is the role of statistics in your product evaluation?

The basic role of statistics here at Consumers Union is to ensure that the studies we do are valid, reliable, and understandable. We look for significance in results in the sense that if we conclude that product A is better than product B, that conclusion is based on good, solid statistical theory. But we always remind ourselves that statistical significance does not outweigh practical significance. If there is a statistical significance of 0.01 in the difference between two speakers, but there is a practical significance of 0.1 in the way the human ear hears those speakers, then the statistical significance is irrelevant. That's the way the article would be written. We might say that the tested speakers are of such a high quality that there is really no difference, so you could buy based on price.

Who works on a typical evaluation project?

We work with different departments having people from a variety of backgrounds. Typically, a project leader is from the department where the product belongs. An

electronics engineer might be the project leader for a study of CD players. Other team members typically include a statistician, a market information analyst, a writer, and an editor.

What does the statistician do?

For the statistician, the work really begins when the market survey report comes in and we can see what consumers want to know about. The decision on what to test is made by the project team with input from all members, including the statistician. We design an experiment and method of sampling. For a CD player, we might want to test sound quality and resistance to skipping when the unit is jarred. We work with the project leader on the test protocol, sampling, and other experimental design issues. The statistician begins analyzing the test data by using general exploratory methods with graphs and numerical measurements. We explore the data because we don't necessarily know what we are going to get. At this stage we are using tools like scatter plots, boxplots, and histograms to look for features such as outliers. Then we do analyses that might make some products stand out or appear to be different from the others. Ultimately, we use statistical methods to develop overall scores we call *ratings*. A statistical report is written and checked by one of the other statisticians. The type of analysis is questioned to be sure that it is the best choice, and we check that everything was done correctly and that the results are valid.

How do you ensure objectivity?

We don't accept any type of outside advertising. We don't represent any organizations or companies: We represent the consumer. We buy all of our products on the open market, just the way any consumer would. We have anonymous shoppers throughout the United States in all major areas, and the statistical design for the sampling plan might require shopping in San Francisco, Florida, New York, and Texas—all at the same time for the same product. That way we get a representative sample of the product that the manufacturer is making.

Do you use computers in your work?

We all use state-of-the-art computers, and a variety of statistical and other software packages.

Would you recommend a statistics course to college students?

I would recommend several statistics courses. Certainly, elementary statistics courses are important for an overview of parametric and nonparametric methods. In addition, linear models, sampling, and design of experiment provide the depth and appreciation needed for critical review of many issues involving statistical inquiries. Beyond this, the ability to communicate your ideas is essential. Because we work on several projects simultaneously, personal organization and project management are critical.

6

Estimates and Sample Sizes

6-1 Overview

Chapter objectives are identified. This and the following chapters present topics of *inferential statistics*, characterized by methods of using sample data to form conclusions about population parameters. This chapter focuses on methods of estimating values of population means, proportions, or variances. Methods for determining the sample sizes necessary to estimate those parameters are also presented.

6-2 Estimating a Population Mean: Large Samples

The value of a population mean is approximated with a single value (called a point estimate) and confidence interval, and a method is presented for determining the sample size necessary to estimate a population mean. This section deals with large samples ($n > 30$) only.

6-3 Estimating a Population Mean: Small Samples

This section uses statistics from small samples ($n \leq 30$) to estimate the values of population means. Point estimates and confidence intervals are discussed, and the Student t distribution is introduced.

6-4 Estimating a Population Proportion

The value of a population proportion is estimated with a point estimate and confidence interval. Procedures are described for determining how large a sample must be to estimate a population proportion. This section presents the method commonly used by pollsters to determine how many people must be surveyed in various opinion polls.

6-5 Estimating a Population Variance

The value of a population variance is approximated with a point estimate and confidence interval, and a method for determining the sample size is presented. The chi-square distribution is introduced.

Chapter Problem:

Is the mean body temperature really 98.6°F?

Table 6-1 lists 106 body temperatures (from Data Set 2 in Appendix B) obtained by University of Maryland researchers. Using the methods described in Chapter 2, we can obtain the following important characteristics of the sample data set:

- As revealed by a histogram, the distribution of the data is approximately bell shaped.
- The mean is $\bar{x} = 98.20°F$.
- The standard deviation is $s = 0.62°F$.
- The sample size is $n = 106$.

Most people believe that the mean body temperature is 98.6°F, but the data in Table 6-1 seem to suggest that it is actually 98.20°F. We know that samples tend to vary, so perhaps it is true that the mean body temperature is 98.6°F and the sample mean, $\bar{x} = 98.20°F$, is the result of a chance sample fluctuation. On the other hand, perhaps the sample mean of 98.20°F is correct and the commonly believed value of 98.6°F is wrong.

On the basis of an analysis of the sample data in Table 6-1, we will see whether the mean body temperature is or is not 98.6°F.

Table 6-1   Body Temperatures of 106 Healthy Adults

98.6	98.6	98.0	98.0	99.0	98.4	98.4	98.4	98.4	98.6
98.6	98.8	98.6	97.0	97.0	98.8	97.6	97.7	98.8	98.0
98.0	98.3	98.5	97.3	98.7	97.4	98.9	98.6	99.5	97.5
97.3	97.6	98.2	99.6	98.7	99.4	98.2	98.0	98.6	98.6
97.2	98.4	98.6	98.2	98.0	97.8	98.0	98.4	98.6	98.6
97.8	99.0	96.5	97.6	98.0	96.9	97.6	97.1	97.9	98.4
97.3	98.0	97.5	97.6	98.2	98.5	98.8	98.7	97.8	98.0
97.1	97.4	99.4	98.4	98.6	98.4	98.5	98.6	98.3	98.7
98.8	99.1	98.6	97.9	98.8	98.0	98.7	98.5	98.9	98.4
98.6	97.1	97.9	98.8	98.7	97.6	98.2	99.2	97.8	98.0
98.4	97.8	98.4	97.4	98.0	97.0				

Sample temperatures were obtained by Dr. Philip Mackowiak, Dr. Steven Wasserman, and Dr. Myron Levine, University of Maryland researchers.

Overview

In Chapter 2 we noted that we use *descriptive statistics* to summarize or describe important characteristics of known population data, but with *inferential statistics* we use sample data to make inferences (or generalizations) about a population. The two major applications of inferential statistics involve the use of sample data to (1) estimate the value of a population parameter, and (2) form a conclusion about a population. This chapter presents methods for estimating values of the following population parameters: population means, proportions, and variances. We also present methods for determining the sample sizes necessary to estimate those parameters.

As we proceed with methods for using sample data to form inferences about populations, we should recall an extremely important point first made in Chapter 1:

> **Data collected carelessly can be absolutely worthless, even if the sample is quite large.**

The methods used here and in the following chapters require sound sampling procedures. For example, a news program might encourage television viewers to call a toll-free number to register their preferences among presidential candidates. We could then apply the methods of this chapter to make inferences based on those sample results, but we would have to describe those results with the statistical term *hogwash.* Because the respondents in this case were self-selected, the sample has a strong potential for bias and any inferences made about the population lack any real statistical validity.

 Estimating a Population Mean: Large Samples

Consider the 106 body temperatures given in Table 6-1 at the beginning of this chapter. On the basis of those sample values, we want to estimate the mean of *all* body temperatures. We could use a statistic such as the sample median, midrange, or mode as an estimate of this population mean μ, but the sample mean $\bar{x}$ usually provides the best estimate of a population mean. The choice of $\bar{x}$ is based on careful study and analysis of the distributions of the different statistics that could be used as estimators.

DEFINITIONS

An **estimator** is a sample statistic (such as the sample mean $\bar{x}$) used to approximate a population parameter. An **estimate** is a specific value or range of values used to approximate some population parameter.

For example, on the basis of the data in Table 6-1, we might use the *estimator* $\bar{x}$ to conclude that the *estimate* of the mean body temperature of all healthy adults is 98.20° F.

There are two important reasons why a sample mean is a better estimator of a population mean μ than other estimators such as the median or the mode.

1. For many populations, the distribution of sample means $\bar{x}$ tends to be more consistent (with *less variation*) than the distributions of other sample statistics. (That is, if you use sample means to estimate the population mean μ, those sample means will have a smaller standard deviation than would other sample statistics, such as the median or the mode.)

2. For all populations, we say that the sample mean $\bar{x}$ is an **unbiased estimator** of the population mean μ, meaning that the distribution of sample means tends to center about the value of the population mean μ. (That is, sample means do not systematically tend to overestimate the value of μ, nor do they systematically tend to underestimate μ. Instead, they tend to target the value of μ itself.)

For these reasons, we will use the sample mean $\bar{x}$ as the best estimate of the population mean μ. Because the sample mean $\bar{x}$ is a single value that corresponds to a point on the number scale, we call it a *point estimate*.

DEFINITION

A **point estimate** is a single value (or point) used to approximate a population parameter.

The sample mean $\bar{x}$ is the best point estimate of the population mean μ.

EXAMPLE   Use the sample body temperatures given in Table 6-1 to find the best point estimate of the population mean μ of all body temperatures.

SOLUTION   The sample mean $\bar{x}$ is the best point estimate of the population mean μ, and for the sample data in Table 6-1 we have $\bar{x}$ = 98.20°F. Based on those particular sample values, the best point estimate of the population mean μ of all body temperatures is therefore 98.20° F.

Why Do We Need Confidence Intervals?   In the preceding example we saw that 98.20° F was our *best* point estimate of the population mean μ, but we had

Estimating Wildlife Population Sizes

The National Forest Management Act protects endangered species, including the northern spotted owl, with the result that the forestry industry was not allowed to cut vast regions of trees in the Pacific Northwest. Biologists and statisticians were asked to analyze the problem, and they concluded that survival rates and population sizes were decreasing for the female owls, known to play an important role in species survival. Biologists and statisticians also studied salmon in the Snake and Columbia Rivers in Washington, and penguins in New Zealand. In the article "Sampling Wildlife Populations" (*Chance*, Vol. 9, No. 2), authors Bryan Manly and Lyman McDonald comment that in such studies, "biologists gain through the use of modeling skills that are the hallmark of good statistics. Statisticians gain by being introduced to the reality of problems by biologists who know what the crucial issues are."

no indication of just how good our best estimate was. If we knew only the first four temperatures of 98.6, 98.6, 98.0, and 98.0, the best point estimate of μ would be their mean ($\bar{x} = 98.30°$ F), but we wouldn't expect this point estimate to be very good because it is based on such a small sample. Statisticians have therefore developed another type of estimate that does reveal how good the point estimate is. This estimate, called a confidence interval or interval estimate, consists of a range (or an interval) of values instead of just a single value.

DEFINITION

A **confidence interval** (or **interval estimate**) is a range (or an interval) of values that is *likely* to contain the true value of the population parameter.

A *confidence interval* is associated with a degree of confidence, which is a measure of how certain we are that our interval contains the population parameter. The definition of degree of confidence uses α (lowercase Greek alpha) to describe a probability that corresponds to an area. Refer to Figure 6-1, where the probability α is divided equally between two shaded extreme regions (often called tails) in the standard normal distribution. (We'll describe the role of $z_{\alpha/2}$ later; for now, simply note that α is divided equally between the two tails.)

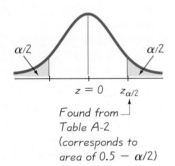

$\alpha/2$ $\alpha/2$

$z = 0$ $z_{\alpha/2}$

Found from Table A-2 (corresponds to area of $0.5 - \alpha/2$)

Figure 6-1 The Standard Normal Distribution: The Critical Value $z_{\alpha/2}$

DEFINITION

The **degree of confidence** is the probability $1 - \alpha$ (often expressed as the equivalent percentage value) that the confidence interval contains the true value of the population parameter. (The degree of confidence is also called the **level of confidence** or the **confidence coefficient**.)

Common choices for the *degree of confidence* are 90% (with $\alpha = 0.10$), 95% (with $\alpha = 0.05$), and 99% (with $\alpha = 0.01$). The choice of 95% is most common because it provides a good balance between precision (as reflected in the width of the confidence interval) and reliability (as expressed by the degree of confidence).

Here's an example of a confidence interval based on the sample data of 106 body temperatures given in Table 6-1:

The 0.95 (or 95%) degree of confidence interval estimate of the population mean μ is 98.08° F $< \mu <$ 98.32° F.

Note that this estimate consists of an interval of values and is associated with a degree of confidence. We interpret this confidence interval as follows: If we were to select many different samples of size $n = 106$ from the population of all healthy people and construct a similar 95% confidence interval estimate for each sample, in the long run 95% of those intervals would actually contain the value of the population mean μ. We should know that μ is a fixed value, not a random variable, and it is therefore *wrong* to say that there is a 95% chance that μ will fall within the interval. Any particular confidence interval either contains μ or does not, and because μ is fixed, there is no probability that μ falls within an interval.

We know from the central limit theorem that sample means $\bar{x}$ tend to be normally distributed, as in Figure 6-1. Sample means have a relatively small chance of falling in one of the extreme tails of Figure 6-1. Denoting the area of each shaded tail by $\alpha/2$, we see that there is a total probability of α that a sample mean will fall in either of the two tails. By the rule of complements (from Chapter 3), it follows that there is a probability of $1 - \alpha$ that a sample mean will fall within the unshaded region of Figure 6-1. The z score separating the right-tail region is commonly denoted by $z_{\alpha/2}$ and is referred to as a *critical value* because it is on the borderline separating sample means that are likely to occur from those that are unlikely to occur.

Notation for Critical Value

$z_{\alpha/2}$ is the positive z value that is at the vertical boundary separating an area of $\alpha/2$ in the right tail of the standard normal distribution. (The value of $-z_{\alpha/2}$ is at the vertical boundary for the area of $\alpha/2$ in the left tail.)

This is a new notation for a concept already introduced in Chapter 5. Some books use the notation $z(\alpha/2)$ instead of $z_{\alpha/2}$, but students sometimes mistake $z(\alpha/2)$ for a product. The notation of $z_{\alpha/2}$ and the concept of a critical value are used in hypothesis testing, which begins in Chapter 7.

DEFINITION

A **critical value** is the number on the borderline separating sample statistics that are likely to occur from those that are unlikely to occur. The number $z_{\alpha/2}$ is a critical value that is a z score with the property that it separates an area of $\alpha/2$ in the right tail of the standard normal distribution. (There is an area of $1 - \alpha$ between the vertical borderlines at $-z_{\alpha/2}$ and $z_{\alpha/2}$.)

Figure 6-2 Finding $z_{\alpha/2}$ for 95% Degree of Confidence

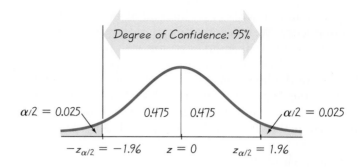

$\alpha/2 = 0.025$ 0.475   0.475 $\alpha/2 = 0.025$

$-z_{\alpha/2} = -1.96$ $z = 0$ $z_{\alpha/2} = 1.96$

EXAMPLE Find the critical value $z_{\alpha/2}$ corresponding to a 95% degree of confidence.

SOLUTION A 95% degree of confidence corresponds to $\alpha = 0.05$. See Figure 6-2, where we show that the area in each of the shaded tails is $\alpha/2 = 0.025$. We find $z_{\alpha/2} = 1.96$ by noting that the region to its left (and bounded by the mean of $z = 0$) must be $0.5 - 0.025$, or 0.475. Now refer to Table A-2 and find that the area of 0.4750 (found in the *body* of the table) corresponds exactly to a z value of 1.96. For a 95% degree of confidence, the critical value is therefore $z_{\alpha/2} = 1.96$.

The preceding example showed that a 95% degree of confidence results in a critical value of $z_{\alpha/2} = 1.96$. This is the most common critical value, and it is listed with two other common values in the table that follows.

Degree of Confidence	α	Critical Value $z_{\alpha/2}$
90%	0.10	1.645
95%	0.05	1.96
99%	0.01	2.575

When we collect a set of sample data, such as the set of 106 body temperatures listed in Table 6-1, we can calculate the sample mean $\bar{x}$, and that sample mean is typically different from the population mean μ. The difference between the sample mean and the population mean can be thought of as an error. In Section 5-5 we saw that $\sigma/\sqrt{n}$ is the standard deviation of sample means. Using $\sigma/\sqrt{n}$ and the $z_{\alpha/2}$ notation, we now define the margin of error E as follows.

> ## DEFINITION
>
> When sample data are used to estimate a population mean μ, the **margin of error,** denoted by E, is the maximum likely (with probability $1 - \alpha$) difference between the observed sample mean $\bar{x}$ and the true value of the population mean μ. The margin of error E is also called the **maximum error of the estimate** and can be found by multiplying the critical value and the standard deviation of sample means, as shown in Formula 6-1.

Formula 6-1
$$E = z_{\alpha/2} \cdot \frac{\sigma}{\sqrt{n}}$$

Given the way that the margin of error E is defined, there is a probability of $1 - \alpha$ that a sample mean will be in error (different from the population mean μ) by no more than E, and there is a probability of α that the sample mean will be in error by more than E. The calculation of the margin of error E as given in Formula 6-1 requires that you know the population standard deviation σ, but in reality it's rare to know σ when the population mean μ is not known. The following method of calculation is common practice.

> ## Calculating E When σ Is Unknown
>
> If $n > 30$, we can replace σ in Formula 6-1 by the sample standard deviation s.
>
> If $n \leq 30$, the population must have a normal distribution and we must know σ to use Formula 6-1. [An alternative method for calculating the margin of error E for small ($n \leq 30$) samples will be discussed more fully in the next section.]

On the basis of the definition of the margin of error E, we can now identify the confidence interval for the population mean μ.

> ## Confidence Interval (or Interval Estimate) for the Population Mean μ (Based on Large Samples: $n > 30$)
>
> $$\bar{x} - E < \mu < \bar{x} + E \quad \text{where} \quad E = z_{\alpha/2} \cdot \frac{\sigma}{\sqrt{n}}$$

Other equivalent forms for the confidence interval are $\mu = \bar{x} \pm E$ and $(\bar{x} - E, \bar{x} + E)$.

The values $\bar{x} - E$ and $\bar{x} + E$ are called **confidence interval limits.**

Point out that for students using a TI-83 calculator, the confidence interval is expressed in a format such as (98.08, 98.32). Also, for students planning to take the AP Statistics exam, the College Board uses the format of $\bar{x} \pm E$. Because of these special interests, this edition uses those other confidence interval formats more than in the previous editions.

Procedure for Constructing a Confidence Interval for μ (Based on a Large Sample: $n > 30$)

1. Find the critical value $z_{\alpha/2}$ that corresponds to the desired degree of confidence. (For example, if the degree of confidence is 95%, the critical value is $z_{\alpha/2} = 1.96$.)

2. Evaluate the margin of error $E = z_{\alpha/2} \cdot \sigma/\sqrt{n}$. If the population standard deviation σ is unknown, use the value of the sample standard deviation s provided that $n > 30$.

3. Using the value of the calculated margin of error E and the value of the sample mean $\bar{x}$, find the values of $\bar{x} - E$ and $\bar{x} + E$. Substitute those values in the general format of the confidence interval:

$$\bar{x} - E < \mu < \bar{x} + E$$

or

$$\mu = \bar{x} \pm E$$

or

$$(\bar{x} - E, \quad \bar{x} + E)$$

4. Round the resulting values by using the following round-off rule.

This is basically the same round-off rule used in Chapter 2 for statistics such as the sample mean.

Round-Off Rule for Confidence Intervals Used to Estimate μ

1. When using the *original set of data* to construct a confidence interval, round the confidence interval limits to one more decimal place than is used for the original set of data.

2. When the original set of data is unknown and only the *summary statistics* $(n, \bar{x}, s)$ are used, round the confidence interval limits to the same number of decimal places used for the sample mean.

The following example clearly illustrates the relatively simple(!) procedure for actually constructing a confidence interval. The original data from Table 6-1 use one decimal place and the summary statistics use two decimal places, so the confidence interval limits will be rounded to two decimal places.

● **EXAMPLE**   For the body temperatures in Table 6-1, we have $n = 106$, $\bar{x} = 98.20$, and $s = 0.62$. For a 0.95 degree of confidence, find both of the following:

a. The margin of error E

b. The confidence interval for μ

SOLUTION

a. The 0.95 degree of confidence implies that $\alpha = 0.05$, so $z_{\alpha/2} = 1.96$, as shown in the preceding example. The margin of error E is calculated by using Formula 6-1 as follows. (Note that σ is unknown, but we can use $s = 0.62$ for the value of σ because $n > 30$.)

$$E = z_{\alpha/2} \cdot \frac{\sigma}{\sqrt{n}} = 1.96 \cdot \frac{0.62}{\sqrt{106}} = 0.12$$

b. With $\bar{x} = 98.20$ and $E = 0.12$, we construct the confidence interval as follows.

$$\bar{x} - E < \mu < \bar{x} + E$$
$$98.20 - 0.12 < \mu < 98.20 + 0.12$$
$$98.08 < \mu < 98.32$$

[This result could also be expressed as $\mu = 98.20 \pm 0.12$ or as (98.08, 98.32).] Based on the sample of 106 body temperatures listed in Table 6-1, the confidence interval for the population mean μ is $98.08°$ F $< \mu < 98.32°$ F, and this interval has a 0.95 degree of confidence. This means that if we were to select many different samples of size 106 and construct the confidence intervals as we did here, 95% of them would actually contain the value of the population mean μ.

Note that the confidence interval limits of $98.08°$ F and $98.32°$ F do not contain $98.6°$ F, the value generally believed to be the mean body temperature. Based on the sample data in Table 6-1, it seems very unlikely that $98.6°$ F is the correct value of μ.

Using Calculators and Computers for Confidence Intervals

STATDISK, Minitab, and many other computer software packages are designed to generate confidence intervals of the type illustrated in the preceding example. With STATDISK, first select Analysis from the main menu bar, then select Confidence Intervals, and proceed to enter the items requested.

Minitab calculates confidence intervals, but it requires an original listing of the raw data. Minitab does not perform calculations using only the summary statistics of n, $\bar{x}$, and s. The *Minitab Student Laboratory Manual and Workbook*, which supplements this textbook, describes a trick for working around this Minitab limitation. To use Minitab, select Stat and Basic Statistics, and then select 1-sample z or 1-sample t. Again, see the Minitab workbook for details on using these options.

The TI-83 calculator can also be used to generate confidence intervals. Either enter the data in list L1 or have the summary statistics available, then press the STAT key. Now select TESTS and choose ZInterval for a confidence interval based on the normal (z) distribution. The TI-83 will require an entry for the population standard deviation, so be prepared to enter the value of s if σ is not known. If you use the data in Table 6-1, the calculator display will include the confidence interval in this format: (98.082, 98.318).

Interpreting a Confidence Interval   We must be careful to interpret confidence intervals correctly. Once we use sample data to find specific limits of $\bar{x} - E$ and $\bar{x} + E$, those limits either enclose the population mean μ or do not, and we cannot determine whether they do or do not without knowing the true value of μ. It is incorrect to state that μ has a 95% chance of falling within the specific limits of 98.08 and 98.32, because μ is a constant, not a random variable. Either μ will fall within these limits or it won't; there's no probability involved. It is correct to say that in the long run these methods will result in confidence intervals that will contain μ in 95% of the cases.

Suppose that in the preceding example, body temperatures really came from a population with a true mean of 98.25° F. Then the confidence interval obtained from the given sample data would contain the population mean, because 98.25 is between 98.08 and 98.32. This is illustrated in Figure 6-3. (See the first confidence interval on the graph.)

Rationale: The basic idea underlying the construction of confidence intervals relates to the central limit theorem, which indicates that with large

The *interpretation* of confidence intervals is discussed here, and this is an important topic that should be discussed. Stress that μ is a fixed value, and it is *not* a variable. It might be helpful to refer to Figure 6-3. Ask the class to imagine 100 confidence intervals instead of the 5 shown in the figure. Ask them to picture 95 of them containing the mean, whereas 5 of them do not contain the mean. This approximates the result we expect when we use 95% for the degree of confidence.

Figure 6-3 Confidence Intervals from Different Samples

The graph shows several confidence intervals, one of which does not contain the population mean μ. For 95% confidence intervals, we expect that among 100 such intervals, 5 will not contain $\mu = 98.25$ while the other 95 will contain it.

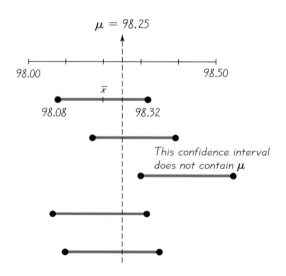

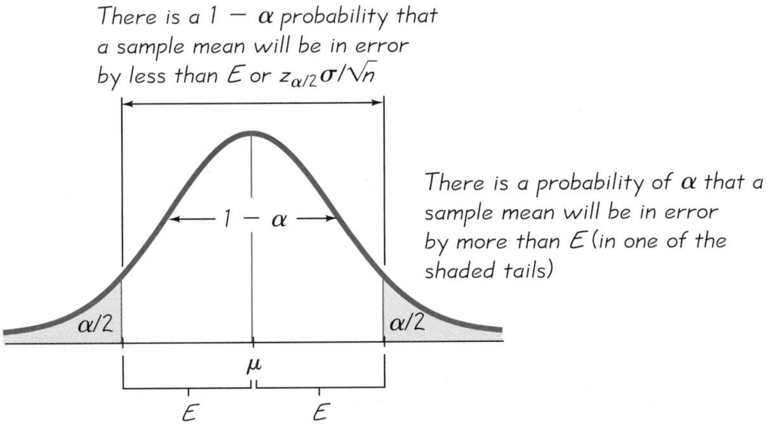

There is a $1 - \alpha$ probability that
a sample mean will be in error
by less than E or $z_{\alpha/2}\sigma/\sqrt{n}$

Figure 6-4   Distribution of
Sample Means

There is a probability of α that a
sample mean will be in error
by more than E (in one of the
shaded tails)

$1 - \alpha$

$\alpha/2$ $\alpha/2$

μ

E E

($n > 30$) samples, the distribution of sample means is approximately normal with mean μ and standard deviation $\sigma/\sqrt{n}$. The confidence interval format is really a variation of the equation

$$z = \frac{\overline{x} - \mu}{\dfrac{\sigma}{\sqrt{n}}}$$

If we solve this equation for μ, we get $\mu = \overline{x} - z\dfrac{\sigma}{\sqrt{n}}$. Using the positive and negative values for z results in the confidence interval limits we are using.

Let's consider the specific case of a 95% degree of confidence, so $\alpha = 0.05$ and $z_{\alpha/2} = 1.96$. For this case there is a probability of 0.05 that a sample mean will be more than 1.96 standard deviations (or $z_{\alpha/2}\sigma/\sqrt{n}$, which we denote by E) away from the population mean μ. Conversely, there is a 0.95 probability that a sample mean will be within 1.96 standard deviations (or $z_{\alpha/2}\sigma/\sqrt{n}$) of μ. (See Figure 6-4.) If the sample mean $\overline{x}$ is within $z_{\alpha/2}\sigma/\sqrt{n}$ of the population mean μ, then μ must be between $\overline{x} - z_{\alpha/2}\sigma/\sqrt{n}$ and $\overline{x} + z_{\alpha/2}\sigma/\sqrt{n}$; this is expressed in the general format of our confidence interval (with $z_{\alpha/2}\sigma/\sqrt{n}$ denoted as E): $\overline{x} - E < \mu < \overline{x} + E$.

Determining Sample Size

So far in this section we have discussed ways to find point estimates and interval estimates of a population mean μ. We based our procedures on known sample data. But suppose that we haven't yet collected the sample. How do we know how many members of the population should be selected? For

For those preparing for the AP Statistics exam, the College Board's "Outline of Topics" apparently omits the determination of sample size, but it is easy to include this very important topic.

example, suppose we want to estimate the mean first-year income of college graduates. How many such incomes must our sample include? Determining the size of a sample is a very important issue, because samples that are needlessly large waste time and money, and samples that are too small may lead to poor results. In many cases we can find the minimum sample size needed to estimate some parameter, such as the population mean μ.

If we begin with the expression for the margin of error E (Formula 6-1) and solve for the sample size n, we get the following.

Discuss the fact that Formula 6-2 does not involve the population size N. The required sample size does not depend on the size of the population, except when sampling is done without replacement from a relatively small finite population (see Exercise 29).

Sample Size for Estimating Mean μ	
Formula 6-2	$$n = \left[\frac{z_{\alpha/2}\sigma}{E}\right]^2$$

The sample size must be a whole number, but the calculations for sample size n often result in a value that is not a whole number. When this happens, we observe the following round-off rule. (It is based on the principle that when rounding is necessary, the required sample size should be rounded *up* so that it is at least adequately large as opposed to slightly too small.)

Round-Off Rule for Sample Size n
When finding the sample size n, if the use of Formula 6-2 does not result in a whole number, always *increase* the value of n to the next *larger* whole number.

Formula 6-2 may be used to determine the sample size necessary to produce results accurate to a desired degree of confidence and margin of error. It should be used when we know the value of σ and want to determine the sample size necessary to establish, with a confidence level of $1 - \alpha$, the value of μ to within $\pm E$. (It's unusual to know σ without knowing μ, but σ might be known from a previous study or it might be estimated from a pilot study or the range rule of thumb.) The existence of such a formula is somewhat remarkable because it implies that the sample size does not depend on the size (N) of the population; the sample size depends on the desired degree of confidence, the desired margin of error, and the value of the standard deviation σ.

EXAMPLE   An economist wants to estimate the mean income for the first year of work for a college graduate who has had the profound wisdom to take a statistics course. How many such incomes must be found if we want to be 95% confident that the sample mean is within $500 of the true population mean? Assume that a previous study has revealed that for such incomes, $\sigma = \$6250$.

SOLUTION   We want to find the sample size n given that $\alpha = 0.05$ (from 95% confidence). We want the sample mean to be within $500 of the population mean, so $E = 500$. Assuming that $\sigma = 6250$, we use Formula 6-2 to get

$$n = \left[\frac{z_{\alpha/2}\sigma}{E} \right]^2 = \left[\frac{1.96 \cdot 6250}{500} \right]^2 = 600.25 = 601 \quad \text{(rounded up)}$$

We should therefore obtain a sample of at least 601 randomly selected first-year incomes of college graduates who have taken a statistics course. With such a sample, we will be 95% confident that the sample mean $\bar{x}$ will be within $500 of the true population mean μ.

If we are willing to settle for less accurate results by using a larger margin of error such as $1000, the sample size drops to 150.0625, which is rounded up to 151. Doubling the margin of error causes the required sample size to decrease to one-fourth its original value. Conversely, halving the margin of error quadruples the sample size. What this implies is that if you want more accurate results, the sample size must be substantially increased. Because large samples generally require more time and money, there is often a need for a trade-off between the sample size and the margin of error E.

What If σ Is Unknown?
Formula 6-2 requires that we substitute some value for the population standard deviation σ, but if it is unknown (as it usually is), we may be able to use a preliminary value obtained from procedures such as these:

1. Using the range rule of thumb (see Section 2-5) to estimate the standard deviation as follows: $\sigma \approx \text{range}/4$.

2. Conducting a pilot study by starting the sample process. Based on the first collection of at least 31 randomly selected sample values, calculate the sample standard deviation s and use it in place of σ. That value can be refined as more sample data are obtained.

Captured Tank Serial Numbers Reveal Population Size

During World War II, Allied intelligence specialists wanted to determine the number of tanks Germany was producing. Traditional spy techniques provided unreliable results, but statisticians obtained accurate estimates by analyzing serial numbers on captured tanks. As one example, records show that Germany actually produced 271 tanks in June 1941. The estimate based on serial numbers was 244, but traditional intelligence methods resulted in the extreme estimate of 1550. (See "An Empirical Approach to Economic Intelligence in World War II," by Ruggles and Brodie, *Journal of the American Statistical Association*, Vol. 42.)

Estimating Sugar in Oranges

In Florida, members of the citrus industry make extensive use of statistical methods. One particular application involves the way in which growers are paid for oranges used to make orange juice. An arriving truckload of oranges is first weighed at the receiving plant, then a sample of about a dozen oranges is randomly selected. The sample is weighed and then squeezed, and the amount of sugar in the juice is measured. Based on the sample results, an estimate is made of the total amount of sugar in the entire truckload. Payment for the load of oranges is based on the estimate of the amount of sugar because sweeter oranges are more valuable than those less sweet, even though the amounts of juice may be the same.

In the two examples that follow, the first uses the range rule of thumb to estimate σ and the second uses a pilot study consisting of sample data found in Appendix B.

EXAMPLE   You plan to estimate the mean selling price of a college textbook. How many textbooks must you sample if you want to be 95% confident that the sample mean is within $2 of the true population mean μ?

SOLUTION   We seek the sample size n given that $\alpha = 0.05$ (from 95% confidence), so $z_{\alpha/2} = 1.96$. We want to be within $2, so $E = 2$. We don't know the standard deviation σ of all textbook selling prices, but we can estimate σ by using the range rule of thumb. If we reason that typical college textbook prices range from $10 to $90, the range becomes $80 so that

$$\sigma \approx \frac{\text{range}}{4} = \frac{(90 - 10)}{4} = 20$$

With $z_{\alpha/2} = 1.96$, $E = 2$, and $\sigma \approx 20$, we use Formula 6-2 as follows.

$$n = \left[\frac{z_{\alpha/2}\sigma}{E}\right]^2 = \left[\frac{1.96 \cdot 20}{2}\right]^2 = 384.16 = 385 \quad \text{(rounded up)}$$

We must randomly select 385 selling prices of college textbooks and then find the value of the sample mean $\bar{x}$. We will be 95% confident that the resulting sample mean is within $2 of the true mean selling price of all college textbooks.

EXAMPLE   If we want to estimate the mean weight of plastic discarded by households in one week, how many households must we randomly select if we want to be 99% confident that the sample mean is within 0.250 lb of the true population mean?

SOLUTION   We seek the sample size n given that $\alpha = 0.01$ (from 99% confidence), so $z_{\alpha/2} = 2.575$. We want to be within 0.250 lb of the true mean, so $E = 0.250$. The value of the population standard deviation σ is unknown, but we can refer to Data Set 1 in Appendix B, which includes the weights of plastic discarded for 62 households. Using that sample as a

pilot study, we can calculate the value of the standard deviation to get $s = 1.065$ lb; because this sample is large ($n > 30$), we can use the value of $s = 1.065$ as an estimate of the population standard deviation σ, as shown below.

$$n = \left[\frac{z_{\alpha/2}\sigma}{E}\right]^2 = \left[\frac{2.575 \cdot 1.065}{0.250}\right]^2 = 120.3 = 121 \quad \text{(rounded up)}$$

On the basis of the population standard deviation estimated from Data Set 1 in Appendix B, we must sample at least 121 randomly selected households to be 99% confident that the sample mean is within 0.250 lb of the true population mean.

This section dealt with the construction of point estimates and confidence interval estimates of population means and presented a method for determining sample sizes needed to estimate population means to the desired degree of accuracy. All of the examples and exercises in this section involve large ($n > 30$) samples. The following section describes procedures to be used when the samples are small ($n \le 30$).

6-2 Exercises A: Basic Skills and Concepts

In Exercises 1–4, find the critical value $z_{\alpha/2}$ that corresponds to the given degree of confidence.

1. 99%   1. 2.575   **2.** 94%   2. 1.88   **3.** 98%   3. 2.33   **4.** 92%   4. 1.75

In Exercises 5–8, use the given degree of confidence and sample data to find (a) the margin of error and (b) the confidence interval for the population mean μ.

5. Heights of women: 95% confidence; $n = 50$, $\bar{x} = 63.4$ in., $s = 2.4$ in.
6. Grade-point averages: 99% confidence; $n = 75$, $\bar{x} = 2.76$, $s = 0.88$
7. Test scores: 90% confidence; $n = 150$, $\bar{x} = 77.6$, $s = 14.2$
8. Police salaries: 92% confidence; $n = 64$, $\bar{x} = \$23{,}228$, $s = \$8779$
9. A sample of 35 skulls is obtained for Egyptian males who lived around 1850 B.C. The maximum breadth of each skull is measured with the result that $\bar{x} = 134.5$ mm and $s = 3.48$ mm (based on data from *Ancient Races*

Recommended assignment: Exercises 1–5, 9–11, 13, 15, 17, 22, 24, 27, 28. (These exercises require determination of sample size: 13, 14, 17, 18, 21, 22, 23, 24.)

5. a. 0.7 in.
  b. 62.7 in. $< \mu <$ 64.1 in.
6. a. 0.26
  b. 2.50 $< \mu <$ 3.02
7. a. 1.9
  b. 75.7 $< \mu <$ 79.5
8. a. $1920
  b. $21,308 $< \mu <$ $25,148

9. 133.3 mm $< \mu <$ 135.7 mm

of the Thebaid by Thomson and Randall-Maciver). Using these sample results, construct a 95% confidence interval for the population mean μ.

10. A sample consists of 75 TV sets purchased several years ago. The replacement times of those TV sets have a mean of 8.2 years and a standard deviation of 1.1 years (based on data from "Getting Things Fixed," *Consumer Reports*). Construct a 90% confidence interval for the mean replacement time of all TV sets from that era. Does the result apply to TV sets currently being sold?   10. 8.0 yr $< \mu <$ 8.4 yr; no

11. The National Center for Education Statistics surveyed 4400 college graduates about the lengths of time required to earn their bachelor's degrees. The mean is 5.15 years, and the standard deviation is 1.68 years. Based on these sample data, construct the 99% confidence interval for the mean time required by all college graduates.   11. 5.08 yr $< \mu <$ 5.22 yr

12. The U.S. Marine Corps is reviewing its orders for uniforms because it has a surplus of uniforms for tall recruits and a shortage for shorter recruits. Its review involves data for 772 men between the ages of 18 and 24. That sample group has a mean height of 69.7 in. with a standard deviation of 2.8 in. (see USDHEW publication 79-1659). Use these sample data to find the 95% confidence interval for the mean height of all men between the ages of 18 and 24.   12. 69.5 in. $< \mu <$ 69.9 in.

13. 543

13. The standard IQ test is designed so that the mean is 100 and the standard deviation is 15 for the population of normal adults. Find the sample size necessary to estimate the mean IQ score of statistics instructors. We want to be 98% confident that our sample mean is within 1.5 IQ points of the true mean. The mean for this population is clearly greater than 100. The standard deviation for this population is probably less than 15 because it is a group with less variation than a group randomly selected from the general population; therefore, if we use $\sigma = 15$, we are being conservative by using a value that will make the sample size at least as large as necessary. Assume then that $\sigma = 15$ and determine the required sample size.

14. The Washington Vending Machine Company must adjust its machines to accept only coins with specified weights. We will obtain a sample of quarters and weigh them to determine the mean. How many quarters must we randomly select and weigh if we want to be 99% confident that the sample mean is within 0.025 g of the true population mean for all quarters? If we use the sample of quarters in Data Set 13 of Appendix B, we can estimate that the population standard deviation is 0.068 g.   14. 50

15. 5.600 g $< \mu <$ 5.644 g; no, the quarters become lighter as they wear down.

15. If we refer to the weights (in grams) of quarters listed in Data Set 13 in Appendix B, we will find 50 weights with a mean of 5.622 g and a standard deviation of 0.068 g. Based on this random sample of quarters in circulation, construct a 98% confidence interval estimate of the population mean of all quarters in circulation. The U.S. Department of the Treasury claims that it mints quarters to yield a mean weight of 5.670 g. Is this claim consistent with the confidence interval? If not, what is a possible explanation for the discrepancy?

16. Data Set 1 in Appendix B includes the *total* weights of garbage discarded by 62 households in one week (based on data collected as part of the Garbage Project at the University of Arizona). For that sample, the mean is 27.44 lb and the standard deviation is 12.46 lb. Construct a 97% confidence interval estimate of the mean weight of garbage discarded by all households. If the city of Providence can handle the garbage as long as the mean is less than 35 lb, is there any cause for concern that there might be too much garbage to handle?   16. 24.01 lb $< \mu <$ 30.87 lb; no

17. A psychologist has developed a new test of spatial perception, and she wants to estimate the mean score achieved by male pilots. How many people must she test if she wants the sample mean to be in error by no more than 2.0 points, with 95% confidence? An earlier study suggests that $\sigma = 21.2$.   17. 432

18. To plan for the proper handling of household garbage, the city of Providence must estimate the mean weight of garbage discarded by households in one week. Find the sample size necessary to estimate that mean if you want to be 96% confident that the sample mean is within 2 lb of the true population mean. For the population standard deviation σ, use the value of 12.46 lb, which is the standard deviation of the sample of 62 households included in the Garbage Project study conducted at the University of Arizona.   18. 164

19. The U.S. Department of Health, Education, and Welfare collected sample data for 1525 women, aged 18 to 24. That sample group has a mean serum cholesterol level (measured in mg/100 mL) of 191.7 with a standard deviation of 41.0 (see USDHEW publication 78-1652). Use these sample data to find the 90% confidence interval for the mean serum cholesterol level of all women in the 18–24 age bracket. If a doctor claims that the mean serum cholesterol for women in that age bracket is 200, does that claim seem to be consistent with the confidence interval?   19. 190.0 $< \mu <$ 193.4; no

20. A random sample of 250 credit-card holders shows that the mean annual credit-card debt for individual accounts is $1592, with a standard deviation of $997 (based on data from *USA Today*). Use these statistics to construct a 94% confidence interval for the mean annual credit-card debt for the population of all accounts.   20. $1473 $< \mu <$ $1711

21. Nielsen Media Research wants to estimate the mean amount of time (in hours) that full-time college students spend watching television each weekday. Find the sample size necessary to estimate that mean with a 0.25 hour (or 15 min) margin of error. Assume that a 96% degree of confidence is desired. Also assume that a pilot study showed that the standard deviation is estimated to be 1.87 hours.   21. 236

22. In deciding whether to attend college, many students are influenced by the increased earnings potential that a college degree is likely to create. Recent data from the Labor Department show that the mean annual income of high-school graduates is $21,652, whereas the mean annual income of college graduates is $40,202. Find the sample size necessary to estimate

304 Chapter 6 Estimates and Sample Sizes

next year's mean annual income of college graduates. Assume that you want 94% confidence that the sample mean will be within $1000 of the true population mean, and assume that the population standard deviation is estimated to be $32,896.   22. 3825

23. You have just been hired by the Boston Marketing Company to conduct a survey to estimate the mean amount of money spent by movie patrons (per movie) in Massachusetts. First use the range rule of thumb to make a rough estimate of the standard deviation of the amounts spent. It is reasonable to assume that typical amounts range from $3 to about $15. Then use that estimated standard deviation to determine the sample size corresponding to 98% confidence and a 25¢ margin of error.   23. 782

24. Answer varies, but using 0 and 8 for minimum and maximum yields 269.

24. Estimate the minimum and maximum ages for typical textbooks currently used in college courses, then use the range rule of thumb to estimate the standard deviation. Next, find the size of the sample required to estimate the mean age (in years) of textbooks currently used in college courses. Assume a 96% degree of confidence that the sample mean will be in error by no more than 0.25 year.

25. $n = 62, \bar{x} = 9.428, s = 4.168; 8.391 \text{ lb} < \mu < 10.466$ lb

25. Refer to Data Set 1 in Appendix B for the 62 weights (in pounds) of *paper* discarded by households (based on data from the Garbage Project at the University of Arizona). Using that sample, construct a 95% confidence interval estimate of the mean weight of paper discarded by all households.

26. $n = 41, \bar{x} = 32.47, s = 5.60;$ $30.22 \text{ in.} < \mu < 34.73 \text{ in.}$

26. Refer to Data Set 7 in Appendix B and construct a 99% confidence interval for the mean total annual precipitation amount in Iowa.

27. $n = 33, \bar{x} = 0.9128,$ $s = 0.0395;$ $0.8979 < \mu < 0.9277;$ no, the sample is small.

27. Refer to Data Set 11 in Appendix B and construct a 97% confidence interval for the mean weight of brown M&M plain candies. Can the methods of this section be used to construct a 97% confidence interval for the mean weight of blue M&M plain candies? Why or why not?

28. $n = 100, \bar{x} = 61.2, s = 116.5;$ $34.1 < \mu < 88.4;$ yes, the populations are different.

28. Refer to Data Set 8 in Appendix B and construct a 98% confidence interval for the mean value of coins in possession of statistics students. Is there reason to believe that this value is different from the mean value of coins in possession of people randomly selected from the general population of adult Americans?

6-2 Exercises B: Beyond the Basics

29. In Formula 6-1 we assume that the population is infinite, that we are sampling with replacement, or that the population is very large. If we have a relatively small population and sample without replacement, we should modify E to include a *finite population correction* factor as follows:

$$E = z_{\alpha/2} \frac{\sigma}{\sqrt{n}} \sqrt{\frac{N - n}{N - 1}}$$

(continued)

where N is the population size. Show that the preceding expression can be solved for n to yield

$$n = \frac{N\sigma^2 (z_{\alpha/2})^2}{(N-1)E^2 + \sigma^2(z_{\alpha/2})^2}$$

Repeat Exercise 13, assuming that the statistics instructors are randomly selected without replacement from a population of $N = 200$ statistics instructors.   29. 147

30. The standard error of the mean is $\sigma/\sqrt{n}$, provided that the population size is infinite. If the population size is finite and is denoted by N, then the correction factor

$$\sqrt{\frac{N-n}{N-1}}$$

should be used whenever $n > 0.05N$. This correction factor multiplies the standard error of the mean, as shown in Exercise 29. Find the 95% confidence interval for the mean of 100 IQ scores if a sample of 31 of those scores produces a mean and standard deviation of 132 and 10, respectively.   30. $129 < \mu < 135$

31. It is found that a sample size of 810 is necessary to estimate the mean weight (in mg) of Bufferin tablets. That sample size is based on a 95% degree of confidence and a population standard deviation that is estimated by the sample standard deviation for Data Set 14 in Appendix B. Find the margin of error E.   31. 0.5 mg

32. A 95% confidence interval for the lives (in minutes) of Kodak AA batteries is $430 < \mu < 470$. (See Program 1 of *Against All Odds: Inside Statistics*.) Assume that this result is based on a sample of size 100.
    a. Construct the 99% confidence interval.   32. a. $424 < \mu < 476$
    b. What is the value of the sample mean?   b. 450
    c. What is the value of the sample standard deviation?   c. 102
    d. If the confidence interval $432 < \mu < 468$ is obtained from the same sample data, what is the degree of confidence?   d. 92%

6–3 Estimating a Population Mean: Small Samples

In Section 6-2 we began our study of inferential statistics by considering point estimates and confidence intervals as methods for estimating the value of a population mean μ. All of the examples and exercises in Section 6-2 involved samples that were large, with sample sizes n greater than 30. Factors such as cost and time often severely limit the size of a sample, so the normal

In the previous edition, the content of this section was included as part of Section 6-2. Users of the previous edition suggested that this separate section would provide for better organization.

distribution may not be a suitable approximation of the distribution of means from small samples. In this section we investigate methods for estimating a population mean μ when the sample size n is small, where *small* is considered to be 30 or fewer.

First, we note that with small samples, the sample mean $\bar{x}$ is generally the best *point estimate* of the population mean μ. Second, confidence intervals can be constructed for small samples by using the normal distribution with the same margin of error from the preceding section, provided that the original population has a normal distribution and the population standard deviation σ is known (a condition not very common in real applications). Later in this section, we will organize and summarize these conditions, but we first consider small sample cases in which the population has a normal distribution, but the value of the population standard deviation σ is not known.

If we have a small ($n \leq 30$) sample and intend to construct a confidence interval but do not know σ, we can sometimes use the Student t distribution developed by William Gosset (1876–1937). Gosset was a Guinness Brewery employee who needed a distribution that could be used with small samples. The Irish brewery where he worked did not allow the publication of research results, so Gosset published under the pseudonym *Student*. As a result of his early experiments and studies of small samples, we can now use the Student t distribution.

We can expect that with smaller samples, the sample means are likely to vary more. The greater variation is accounted for by the t distribution.

Student t Distribution

If the distribution of a population is essentially normal (approximately bell shaped), then the distribution of

$$t = \frac{\bar{x} - \mu}{\frac{s}{\sqrt{n}}}$$

is essentially a **Student t distribution** for all samples of size n. The Student t distribution, often referred to as the t **distribution**, is used to find critical values denoted by $t_{\alpha/2}$.

Table A-3 lists values of the t distribution along with areas denoted by α. Values of $t_{\alpha/2}$ are obtained from Table A-3 by locating the proper value for degrees of freedom in the left column and then proceeding across the corresponding row until reaching the number directly below the applicable value of α for two tails.

> **DEFINITION**
>
> The number of **degrees of freedom** for a data set corresponds to the number of scores that can vary after certain restrictions have been imposed on all scores.

For example, if 10 students have quiz scores with a mean of 80, we can freely assign values to the first 9 scores, but the 10th score is then determined. The sum of the 10 scores must be 800, so the 10th score must equal 800 minus the sum of the first 9 scores. Because those first 9 scores can be freely selected to be any values, we say that there are 9 degrees of freedom available. For the applications of this section, the number of degrees of freedom is simply the sample size minus 1.

$$\text{degrees of freedom} = n - 1$$

Assumption: For the Student t distribution to be applicable, the distribution of the parent population must be essentially normal; it need not be exactly normal, but if it has only one mode and is basically symmetric, we will generally get good results (such as accurate confidence intervals). If there is strong evidence that the population has a very nonnormal distribution, then nonparametric methods (see Chapter 13) or bootstrap resampling methods (see the Computer Project at the end of this chapter) should be used instead.

It might seem a bit strange that with a normally distributed population, we sometimes use the t distribution to find critical values, but when σ is unknown, the use of s from a small sample incorporates another source of error. To maintain the desired degree of confidence, we compensate for the additional variability by widening the confidence interval through a process that replaces the critical value $z_{\alpha/2}$ (found from the standard normal distribution values in Table A-2) with the larger critical value of $t_{\alpha/2}$ (found from the t-distribution values in Table A-3).

Important Properties of the Student t Distribution

1. The Student t distribution is different for different sample sizes. (See Figure 6-5 (on page 308) for the cases $n = 3$ and $n = 12$.)
2. The Student t distribution has the same general symmetric bell shape as the standard normal distribution, but it reflects the greater variability (with wider distributions) that is expected with small samples.
3. The Student t distribution has a mean of $t = 0$ (just as the standard normal distribution has a mean of $z = 0$).

Hispanics Mistreated in Surveys

The Bureau of the Census defines *Hispanic origin* to refer to people of Spanish/Hispanic origin or descent, including Mexicans, Cubans, Puerto Ricans, people from Spanish-speaking countries, and others. The definition includes the comment that "persons of Spanish/Hispanic origin may be of any race." Yet government statistics and summaries sometimes combine race and ethnicity when listing data. In one case, census data are listed under the title of "Race and Hispanic Origin," with categories of whites, blacks, people of Hispanic origin, and others. But how is a white Hispanic classified? The categories are confusing or overlapping. A study by the Office of Management and Budget showed that the combined format of "Race and Hispanic Origin" results in an undercount of Hispanics by 20 to 30 percent. (The author thanks Joseph Diaz-Calderon, a student who used the previous edition of this book, for raising this issue.)

Figure 6-5 Student *t*
Distributions for *n* = 3 and
n = 12
The Student *t* distribution
has the same general shape
and symmetry as the stan-
dard normal distribution,
but it reflects the greater
variability that is expected
with small samples.

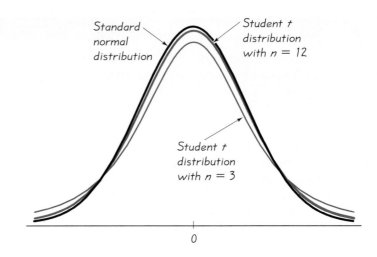

4. The standard deviation of the Student *t* distribution varies with the sample size, but it is greater than 1 (unlike the standard normal distribution, which has $\sigma = 1$).

5. As the sample size *n* gets larger, the Student *t* distribution gets closer to the standard normal distribution. For values of $n > 30$, the differences are so small that we can use the critical *z* values instead of developing a much larger table of critical *t* values. (The values in the bottom row of Table A-3 are equal to the corresponding critical *z* values from the standard normal distribution.)

    Following is a summary of the conditions indicating use of a *t* distribution instead of the standard normal distribution. (These same conditions will also apply in Chapter 7.)

Students tend to oversimplify the
conditions for the *t* distribution,
and they tend to miss these two
aspects: (1) The methods of this
section require that the parent
population has a distribution that
is essentially normal (even when
the *t* distribution is used); (2) if σ
is known, use the normal distribution (not *t*) even if the sample is
small.

Conditions for Using the Student *t* Distribution

1. The sample is small ($n \leq 30$); and
2. σ is unknown; and
3. The parent population has a distribution that is essentially normal. (Because the distribution of the parent population is often unknown, we often estimate it by constructing a histogram of sample data.)

    We can now determine values for the margin of error *E* in estimating μ when a *t* distribution applies. This margin of error can be used for constructing confidence intervals.

Margin of Error for the Estimate of μ
[Based on a Small Sample ($n \leq 30$) and Unknown σ]

Formula 6-3 $E = t_{\alpha/2} \dfrac{s}{\sqrt{n}}$ where $t_{\alpha/2}$ has $n - 1$ degrees of freedom

---

Confidence Interval for the Estimate of μ
[Based on a Small Sample ($n \leq 30$) and Unknown σ]

$$\bar{x} - E < \mu < \bar{x} + E$$

where

$$E = t_{\alpha/2} \dfrac{s}{\sqrt{n}}$$

EXAMPLE   With *destructive testing,* sample items are destroyed in the process of testing them. Crash testing of cars is one very expensive example of destructive testing. If you were responsible for such crash tests, there is no way you would want to tell your supervisor that you must crash and destroy more than 30 cars so that you could use the normal distribution. Let's assume that you have crash tested 12 Dodge Viper sports cars (current list price: $59,300) under a variety of conditions that simulate typical collisions. Analysis of the 12 damaged cars results in repair costs having a distribution that appears to be bell shaped, with a mean of $\bar{x} = \$26{,}227$ and a standard deviation of $s = \$15{,}873$ (based on data from the Highway Loss Data Institute). Find the following.

a. The best point estimate of μ, the mean repair cost for all Dodge Vipers involved in collisions

b. The 95% interval estimate of μ, the mean repair cost for all Dodge Vipers involved in collisions

SOLUTION

a. The best point estimate of the population mean μ is the value of the sample mean $\bar{x}$. In this case, the best point estimate of μ is therefore $26,227.

b. We will proceed to construct a 95% confidence interval by using the t distribution because the following three conditions are met: (1) The sample is small ($n \leq 30$), (2) the population standard deviation σ is

(continued)

unknown, and (3) the population appears to have a normal distribution because the sample data have a bell-shaped distribution.

We begin by finding the value of the margin of error as shown below. Note that the critical value of $t_{\alpha/2} = 2.201$ is found from Table A-3 in the column labeled ".05 two tails" (from 95% confidence) and the row corresponding to 11 degrees of freedom (from $n - 1 = 11$).

$$E = t_{\alpha/2}\frac{s}{\sqrt{n}} = 2.201\frac{15{,}873}{\sqrt{12}} = 10{,}085.29$$

We can now construct the 95% interval estimate of μ by using $E = 10{,}085.29$ and $\bar{x} = 26{,}227$. Because the summary statistics are rounded to the nearest dollar, the confidence interval limits will also be rounded to the nearest dollar.

$$\bar{x} - E < \mu < \bar{x} + E$$
$$26{,}227 - 10{,}085.29 < \mu < 26{,}227 + 10{,}085.29$$
$$\$16{,}142 < \mu < \$36{,}312$$

[This result could also be expressed in the format of $\mu = \$26{,}227 \pm \$10{,}085$ or as ($\$16{,}142$, $\$36{,}312$).]

On the basis of the given sample results, we are 95% confident that the limits of $\$16{,}142$ and $\$36{,}312$ actually do contain the value of the population mean μ. These repair costs appear to be quite high. In fact, the Dodge Viper is currently the most expensive car to repair after a collision. Such information is important to companies that insure Dodge Vipers against collisions.

Using Calculators and Computers for Confidence Intervals

Once you understand the basic theory underlying the use of confidence intervals in estimating parameters and how to construct and interpret them, you can use software that simplifies the manual calculations. With STATDISK, select `Analysis` from the main menu, then select `Confidence Intervals`, and enter the items requested. After you have clicked on `Evaluate`, the confidence interval will be displayed.

Minitab requires an original listing of the raw data and will not work with summary statistics. For a way to get around that limitation and for more details on the use of Minitab, see the *Minitab Student Laboratory Manual and Workbook* that supplements this textbook.

The TI-83 calculator can generate confidence intervals of the type discussed in this section. Either enter the data in list L1 or obtain the summary statistics, then use `TInterval` (after pressing the STAT key and selecting `Tests`).

Using the data from the preceding example, the TI-83 will display a result that includes this confidence interval: (16142, 36312).

Choosing the Appropriate Distribution

It is sometimes difficult to decide whether to use Formula 6-1 (and the standard normal distribution) or Formula 6-3 (and the Student t distribution). The flowchart in Figure 6-6 summarizes the key points to be considered when constructing confidence intervals for estimating μ, the population mean. In Figure 6-6, note that if we have a small ($n \leq 30$) sample drawn from a very nonnormal distribution, we can't use the methods described in this chapter. One alternative is to use nonparametric methods (see Chapter 13), and

If the original population is not normally distributed, the bootstrap method is an interesting alternative, especially for students with great enthusiasm for computers. Consider the Computer Project with an oral and written report as an extra-credit assignment.

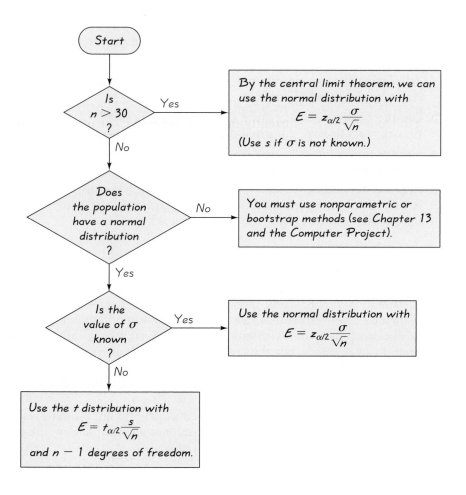

Figure 6-6 Choosing Between Normal (z) and t Distributions

Start

Is $n > 30$? — Yes → By the central limit theorem, we can use the normal distribution with
$$E = z_{\alpha/2}\frac{\sigma}{\sqrt{n}}$$
(Use s if σ is not known.)

No ↓

Does the population have a normal distribution ? — No → You must use nonparametric or bootstrap methods (see Chapter 13 and the Computer Project).

Yes ↓

Is the value of σ known ? — Yes → Use the normal distribution with
$$E = z_{\alpha/2}\frac{\sigma}{\sqrt{n}}$$

No ↓

Use the t distribution with
$$E = t_{\alpha/2}\frac{s}{\sqrt{n}}$$
and $n - 1$ degrees of freedom.

Small Sample

The Children's Defense Fund was organized to promote the welfare of children. The group published *Children Out of School in America,* which reported that in one area, 37.5% of the 16- and 17-year-old children were out of school. This statistic received much press coverage, but it was based on a sample of only 16 children. Another statistic was based on a sample size of only 3 students. (See "Firsthand Report: How Flawed Statistics Can Make an Ugly Picture Look Even Worse," *American School Board Journal,* Vol. 162.)

another alternative is to use the computer-oriented bootstrap method, which makes no assumptions about the original population. This method is described in the Computer Project at the end of this chapter.

EXAMPLE   Assuming that you plan to construct a confidence interval for the population mean μ, use the given data to determine whether the margin of error E should be calculated using the normal distribution, the t distribution, or neither (so that the methods of this chapter cannot be used).

a. $n = 50$, $\bar{x} = 77.6$, $s = 14.2$, and the shape of the distribution is skewed.

b. $n = 25$, $\bar{x} = 77.6$, $s = 14.2$, and the distribution is bell shaped.

c. $n = 25$, $\bar{x} = 77.6$, $\sigma = 14.2$, and the distribution is bell shaped.

d. $n = 25$, $\bar{x} = 77.6$, $\sigma = 14.2$, and the distribution is extremely skewed.

SOLUTION
Refer to Figure 6-6 and use the flowchart to determine the following.

a. Because the sample is large ($n > 30$), use the normal distribution. When calculating the margin of error E, use Formula 6-1, where the sample standard deviation s is used for σ.

b. Use the t distribution because (1) the sample is small, (2) the population appears to have a normal distribution, and (3) σ is unknown.

c. Use the normal distribution because the population appears to have a normal distribution and the value of σ is known.

d. Because the sample is small and the population has a distribution that is very nonnormal, the methods of this chapter do not apply. Neither the normal nor the t distribution can be used.

Section 6-2 presented the three main concepts of point estimate, confidence interval, and determination of sample size. This section extended the concepts of point estimate and confidence interval to small sample cases. This section does not include determination of sample size because with known σ the normal distribution applies, and with unknown σ we need a large sample to justify estimation of σ with s, but the large sample allows us to use the normal distribution again. We therefore base our determinations of sample size on Formula 6-2 only and include no circumstances in which the t distribution is used.

The following section again considers the concepts of point estimate, confidence interval, and determination of sample size, but we will focus on the population proportion instead of the population mean.

 **Exercises A: Basic Skills and Concepts**

In Exercises 1–4, find the critical value $t_{\alpha/2}$ *that corresponds to the given degree of confidence and sample size n.*

1. 99%; *n* = 10   1. 3.250   **2.** 95%; *n* = 16   2. 2.132
3. 98%; *n* = 21   3. 2.528   **4.** 90%; *n* = 8   4. 1.895

Recommended assignment: Exercises 1–5, 9, 13–16, 18, 20. (These exercises include conditions that lead to a normal distribution instead of a *t* distribution: 13, 16, 19, 20 part a.)

In Exercises 5–8, use the given degree of confidence and sample data to find (a) the margin of error and (b) the confidence interval for the population mean μ. In each case, assume that the population has a normal distribution.

5. Heights of women: 95% confidence; *n* = 10, $\bar{x}$ = 63.4 in., *s* = 2.4 in.
6. Grade-point averages: 99% confidence; *n* = 15, $\bar{x}$ = 2.76, *s* = 0.88
7. Test scores: 90% confidence; *n* = 16, $\bar{x}$ = 77.6, *s* = 14.2
8. Police salaries: 98% confidence; *n* = 19, $\bar{x}$ = $23,228, *s* = $8779

5. a. 1.7 in.
   b. 61.7 in. $< \mu <$ 65.1 in.
6. a. 0.68
   b. 2.08 $< \mu <$ 3.44
7. a. 6.2
   b. 71.4 $< \mu <$ 83.8
8. a. $5140
   b. $18,088 $< \mu <$ $28,368

In Exercises 9–20, be sure to determine correctly whether the confidence intervals are calculated with the normal distribution or the t *distribution.*

9. In crash tests of 15 Honda Odyssey minivans, collision repair costs are found to have a distribution that is roughly bell shaped, with a mean of $1786 and a standard deviation of $937 (based on data from the Highway Loss Data Institute). Construct the 99% confidence interval for the mean repair cost in all such vehicle collisions.   9. $1066 $< \mu <$ $2506
10. Data Set 2 in Appendix B includes 106 body temperatures taken at 12:00 A.M. on day 2. Suppose that we have only the first 10 temperatures given below. For these scores, $\bar{x}$ = 98.44 and *s* = 0.30. Construct the 95% confidence interval for the mean of all body temperatures. (A prior study has shown that body temperatures are normally distributed.)

10. 98.23° $< \mu <$ 98.65°

    98.6   98.6   98.0   98.0   99.0   98.4   98.4   98.4   98.4   98.6

11. In a time-use study, 20 randomly selected managers were found to spend a mean of 2.40 h each day on paperwork. The standard deviation of the 20 scores is 1.30 h (based on data from Adia Personnel Services). Also, the sample data appear to have a bell-shaped distribution. Construct the 95% confidence interval for the mean time spent on paperwork by all managers.   11. 1.79 h $< \mu <$ 3.01 h
12. In a study of the amounts of time required for room-service delivery at a newly opened Radisson Hotel, 20 deliveries had a mean time of 24.2 min and a standard deviation of 8.7 min. The sample data appear to have a bell-shaped distribution. Construct the 90% confidence interval for the mean of all deliveries.   12. 20.8 min $< \mu <$ 27.6 min

13. $3.84 < \mu < 4.04$

13. In a study of physical attractiveness and mental disorders, 231 subjects were rated for attractiveness, and the resulting sample mean and standard deviation are 3.94 and 0.75, respectively. (See "Physical Attractiveness and Self-Perception of Mental Disorder," by Burns and Farina, *Journal of Abnormal Psychology,* Vol. 96, No. 2.) Use these sample data to construct the 95% confidence interval for the population mean.

14. A study was conducted to estimate hospital costs for accident victims who wore seat belts. Twenty randomly selected cases have a distribution that appears to be bell shaped with a mean of $9004 and a standard deviation of $5629 (based on data from the U.S. Department of Transportation). Construct the 99% confidence interval for the mean of all such costs.   14. $5403 < \mu < $12,605

15. Construct a 98% confidence interval for the mean income of all full-time workers who have a bachelor's degree. A sample of 25 such workers shows that the distribution of incomes is roughly normal with a mean of $39,271 and a standard deviation of $18,933 (based on data from the U.S. Department of Labor).   15. $29,835 < \mu < $48,707

16. A random sample of 19 women results in a mean height of 63.85 in. Other studies have shown that women's heights are normally distributed with a standard deviation of $\sigma = 2.5$ in. Construct the 98% confidence interval for the mean height of all women.   16. 62.51 in. $< \mu < $ 65.19 in.

17. In a study of the use of hypnosis to relieve pain, sensory ratings were measured for 16 subjects, with the results given below (based on data from "An Analysis of Factors That Contribute to the Efficacy of Hypnotic Analgesia," by Price and Barber, *Journal of Abnormal Psychology,* Vol. 96, No. 1.) Use these sample data to construct the 95% confidence interval for the mean sensory rating for the population from which the sample was drawn.   17. $\bar{x} = 7.96$, $s = 1.60$; $7.11 < \mu < 8.81$

19. $n = 100, \bar{x} = 0.9147$ g, $s =$ 0.0369 g; 0.9075 g $< \mu <$ 0.9219 g; there is reasonable agreement with the result from Exercise 18.

| 8.8 | 6.6 | 8.4 | 6.5 | 8.4 | 7.0 | 9.0 | 10.3 |
| 8.7 | 11.3 | 8.1 | 5.2 | 6.3 | 8.7 | 6.2 | 7.9 |

18. Refer to Data Set 11 in Appendix B and use only the sample of *red* M&M plain candies to construct a 95% confidence interval for the mean weight of all M&Ms.   18. $n = 21$, $\bar{x} = 0.9097$ g, $s = 0.0275$ g; 0.8972 g $< \mu < 0.9222$ g

19. Refer to Data Set 11 in Appendix B and use the entire sample of 100 plain M&M candies to construct a 95% confidence interval for the mean weight of all M&Ms. How does the result compare to the confidence interval for Exercise 18?

20. a. $n = 35, \bar{x} = 111.1$ min, $s =$ 20.5 min; 104.4 min $< \mu <$ 117.9 min

b. $n = 23, \bar{x} = 111.0$ min, $s =$ 25.9 min; 99.8 min $< \mu <$ 122.2 min

c. Part (a) uses a normal distribution, but part (b) uses a *t* distribution. The confidence interval for part (b) is wider, because it is based on a much smaller sample.

20. Refer to Data Set 10 in Appendix B.
  a. Construct a 95% confidence interval for the mean length of movies rated R.
  b. Construct a 95% confidence interval for the mean length of movies rated PG or PG-13.
  c. Compare the methods and results of parts (a) and (b).

6-3 Exercises B: Beyond the Basics

21. Assume that a small ($n \leq 30$) sample is randomly selected from a normally distributed population for which σ is unknown. Construction of a confidence interval should use the t distribution, but how are the confidence interval limits affected if the normal distribution is incorrectly used instead?

22. A confidence interval is constructed for a small sample of temperatures (in degrees Fahrenheit) randomly selected from a normally distributed population for which σ is unknown (such as the data set given in Exercise 10).
    a. How is the margin of error E affected if each temperature is converted to the Celsius scale? [$C = \frac{5}{9}(F - 32)$]   22. a. E is multiplied by 5/9.
    b. If the confidence interval limits are denoted by a and b, find expressions for the confidence interval limits after the original temperatures have been converted to the Celsius scale.
    c. Based on the results from part (b), can confidence interval limits for the Celsius temperatures be found by simply converting the confidence interval limits from the Fahrenheit scale to the Celsius scale?

21. The confidence interval limits are closer than they should be.

22. b. $\frac{5}{9}(a - 32), \frac{5}{9}(b - 32)$
    c. Yes

6-4 Estimating a Population Proportion

In this section we consider the same three concepts discussed in Sections 6-2 and 6-3: (1) point estimate, (2) confidence interval, and (3) determining the required sample size. Whereas Sections 6-2 and 6-3 applied these concepts to estimates of a population mean μ, this section applies them to the population proportion p. For example, Nielsen Media Research might need to estimate the proportion of households that tune in to the *Super Bowl*, and the Hartford Insurance Company might want to estimate the proportion of drunk drivers.

Although this section focuses on the population proportion p, we can also consider a probability or a percentage. Proportions and probabilities are both expressed in decimal or fraction form. When working with percents, convert them to proportions by dropping the percent sign and dividing by 100. For example, the 48.7% rate of people who don't buy any books can be expressed in decimal form as 0.487. The symbol p may therefore represent a proportion, a probability, or the decimal equivalent of a percent. We now introduce the new notation of $\hat{p}$ (called "p hat") for the sample proportion.

For those preparing for the AP Statistics exam, confidence intervals are expressed in the format of $\hat{p} \pm E$ instead of the format of $\hat{p} - E < p < \hat{p} + E$. Also, the College Board uses $1 - \hat{p}$ instead of $\hat{q}$.

Notation for Proportions
$p = $ *population* proportion
$\hat{p} = \dfrac{x}{n}$ *sample* proportion of x successes in a sample of size n

Push Polling

"Push polling" is the practice of political campaigning under the guise of a poll. Its name is derived from its objective of pushing voters away from opposition candidates by asking loaded questions designed to discredit them. Here's an example of one such question that was used: "Please tell me if you would be more likely or less likely to vote for Roy Romer if you knew that Gov. Romer appoints a parole board which has granted early release to an average of four convicted felons per day every day since Romer took office." The National Council on Public Polls characterizes push polls as unethical, but some professional pollsters do not condemn the practice as long as the questions do not include outright lies.

In previous chapters we stipulated that $q = 1 - p$, so it is natural to note here that $\hat{q} = 1 - \hat{p}$. For example, if you survey 1068 Americans and find that 673 of them have answering machines, the sample proportion is $\hat{p} = x/n = 673/1068 = 0.630$, and $\hat{q} = 0.370$ (calculated from $1 - 0.630$). In some cases, the value of $\hat{p}$ may be known because the sample proportion or percentage is given directly. If it is reported that 1068 American television viewers are surveyed and 25% of them are college graduates, then $\hat{p} = 0.25$ and $\hat{q} = 0.75$.

Point Estimate

The sample proportion $\hat{p}$ is the best point estimate of the population proportion p.

Margin of Error of the Estimate of p

Formula 6-4
$$E = z_{\alpha/2} \sqrt{\frac{\hat{p}\hat{q}}{n}}$$

Confidence Interval (or Interval Estimate) for the Population Proportion p

$$\hat{p} - E < p < \hat{p} + E \qquad \text{where } E = z_{\alpha/2} \sqrt{\frac{\hat{p}\hat{q}}{n}}$$

The confidence interval is sometimes expressed in the following formats.
$$p = \hat{p} \pm E$$
or
$$(\hat{p} - E, \hat{p} + E)$$

In Chapter 3 we rounded probabilities expressed in decimal form to three significant digits. We use that same rounding rule here.

Round-Off Rule for Confidence Interval Estimates of p

Round the confidence interval limits to three significant digits.

EXAMPLE   Pollsters are plagued by a variety of confounding factors, such as telephone-answering machines. In a survey of 1068 Americans, 673 stated that they had answering machines (based on data from the International Mass Retail Association, reported in *USA Today*). Using these sample results, find

a. the point estimate of the population proportion of all Americans who have answering machines.

b. the 95% interval estimate of the population proportion of all Americans who have answering machines.

SOLUTION

a. The point estimate of p is

$$\hat{p} = \frac{x}{n} = \frac{673}{1068} = 0.630$$

b. Construction of the confidence interval requires that we first evaluate the margin of error E. The value of E can be found from Formula 6-4. We use $\hat{p} = 0.630$ (found in part a), $\hat{q} = 0.370$ (from $\hat{q} = 1 - \hat{p}$), and $z_{\alpha/2} = 1.96$ (from Table A-2, where 95% converts to $\alpha = 0.05$, which is divided equally between the two tails so that $z = 1.96$ corresponds to an area of 0.4750).

$$E = z_{\alpha/2} \sqrt{\frac{\hat{p}\hat{q}}{n}} = 1.96 \sqrt{\frac{(0.630)(0.370)}{1068}} = 0.0290$$

We can now find the confidence interval by using $\hat{p} = 0.630$ and $E = 0.0290$.

$$\hat{p} - E < p < \hat{p} + E$$

$$0.630 - 0.0290 < p < 0.630 + 0.0290$$

$$0.601 < p < 0.659$$

If we wanted the 95% confidence interval for the true population *percentage,* we could express this result as 60.1% $< p <$ 65.9%. This result is often reported in the following format: "Among Americans, the percentage who have answering machines is estimated to be 63%, with a margin of error of plus or minus 2.9 percentage points." This is a verbal expression of this format for the confidence interval: $p = 63\%$ $\pm$ 2.9%. (The level of confidence should also be reported, but it rarely is in the media. The media typically use a 95% degree of confidence but omit any reference to it.) The confidence interval is also expressed in this format: (0.601, 0.659).

Using Calculators and Computers for Confidence Intervals

STATDISK will construct a confidence interval estimate of a population proportion. Select `Analysis`, then `Confidence Intervals` and proceed to enter the requested items. Minitab can also be used, but it's tricky, so see the *Minitab Student Laboratory Manual and Workbook* for details. The TI-83 calculator can also be used by selecting `1-PropZInt` (after pressing the STAT key and selecting `TESTS`). If using the data from the preceding example, the TI-83 will produce a display that includes the confidence interval in this format: (.6012, .6591).

Rationale: We use $\hat{p}$ as the point estimate of p (just as $\bar{x}$ is used as the point estimate of μ) because it is unbiased and is the most consistent of the estimators that could be used. It is unbiased in the sense that the distribution of sample proportions tends to center about the value of p; that is, sample proportions $\hat{p}$ do not systematically tend to underestimate p, nor do they systematically tend to overestimate p. The sample proportion $\hat{p}$ is the most consistent estimator in the sense that the standard deviation of sample proportions tends to be smaller than the standard deviation of any other unbiased estimators.

Assumptions: Because we are dealing with proportions, we can use the binomial distribution introduced in Section 4-3. We assume in this section that the four conditions given in Section 4-3 for the binomial distribution are essentially satisfied: (1) There is a fixed number of trials, (2) the trials are independent, (3) there are two categories of outcomes, and (4) the probabilities remain constant for each trial. We also assume that the normal distribution can be used as an approximation to the distribution of sample proportions (because the conditions $np \geq 5$ and $nq \geq 5$ are both satisfied). We can therefore use results from Section 5-6 to conclude that μ and σ are given by $\mu = np$ and $\sigma = \sqrt{npq}$. Both of these parameters pertain to n trials, but we convert them to a per trial basis by dividing by n as follows:

$$\text{Mean of sample proportions} = \frac{np}{n} = p$$

$$\text{Standard deviation of sample proportions} = \frac{\sqrt{npq}}{n} = \sqrt{\frac{npq}{n^2}} = \sqrt{\frac{pq}{n}}$$

The first result may seem trivial because we have already stipulated that the true population proportion is p. The second result is nontrivial and is useful in describing the margin of error E, but we replace the product pq by $\hat{p}\hat{q}$ because we don't yet know the value of p (it is the value we are trying to estimate). Formula 6-4 for the margin of error reflects the fact that $\hat{p}$ has a probability of $1 - \alpha$ of being within $z_{\alpha/2}\sqrt{pq/n}$ of p. The confidence interval for p, as given previously, reflects the fact that there is a probability of $1 - \alpha$ that $\hat{p}$ differs from p by less than the margin of error E.

Determining Sample Size

Having discussed point estimates and confidence intervals for p, we will now describe a procedure for determining how large a sample should be when we want to find the approximate value of a population proportion. In the previous section we began with the expression for the margin of error E, then solved for n. Using a similar procedure, we begin with

$$E = z_{\alpha/2} \sqrt{\frac{\hat{p}\hat{q}}{n}}$$

and solve for n to get the sample size, as given in Formula 6-5. Formula 6-5 requires $\hat{p}$ as an estimate of the population proportion p, but if no such estimate is known, we replace $\hat{p}$ by 0.5 and replace $\hat{q}$ by 0.5, with the result given in Formula 6-6.

As in Section 6-2, the sample size does not depend on the population size N, as many people might expect. (Exception: Sampling is done without replacement from a relatively small finite population. See Exercise 29.)

Sample Size for Estimating Proportion p	
When an Estimate $\hat{p}$ Is Known: Formula 6-5	$n = \dfrac{[z_{\alpha/2}]^2 \hat{p}\hat{q}}{E^2}$
When No Estimate $\hat{p}$ Is Known: Formula 6-6	$n = \dfrac{[z_{\alpha/2}]^2 \cdot 0.25}{E^2}$

Round-Off Rule for Determining Sample Size
If the computed sample size is not a whole number, round it up to the next *higher* whole number.

Also consider presenting a geometric demonstration of why the values of $\hat{p} = 0.5$ and $\hat{q} = 0.5$ result in the largest possible product of $\hat{p}\hat{q} = 0.25$. Use a rectangle with perimeter 2, so that $L + W = 1$, as in $\hat{p} + \hat{q} = 1$. Show that given a rectangle with perimeter 2, the area is maximized by using a *square*. It's an interesting link between geometry and statistics.

Use Formula 6-5 when reasonable estimates of $\hat{p}$ can be made by using previous samples, a pilot study, or someone's expert knowledge. When no such guess can be made, we can assign the value of 0.5 to each of $\hat{p}$ and $\hat{q}$, so the resulting sample size will be at least as large as it should be. The underlying reason for the assignment of 0.5 is this: The product $\hat{p} \cdot \hat{q}$ has 0.25 as its largest possible value, which occurs when $\hat{p} = 0.5$ and $\hat{q} = 0.5$. (See the accompanying table, which lists some values of $\hat{p}$ and $\hat{q}$.) Note that Formulas 6-5 and 6-6 do not include the population size N, so the size of the population is irrelevant. (*Exception:* When sampling is without replacement from a relatively small finite population. See Exercise 29.)

$\hat{p}$	$\hat{q}$	$\hat{p} \cdot \hat{q}$
0.1	0.9	0.09
0.2	0.8	0.16
0.3	0.7	0.21
0.4	0.6	0.24
0.5	0.5	0.25
0.6	0.4	0.24
0.7	0.3	0.21
0.8	0.2	0.16
0.9	0.1	0.09

EXAMPLE   Insurance companies are becoming concerned that increased use of cellular telephones is resulting in more car crashes, and they are considering implementing higher rates for drivers who use such phones. We want to estimate, with a margin of error of three percentage points, the percentage of drivers who talk on phones while they are driving. Assuming that we want 95% confidence in our results, how many drivers must we survey?

a. Assume that we have an estimate of $\hat{p}$ based on a prior study that showed that 18% of drivers talk on a phone (based on data from *Prevention* magazine).

b. Assume that we have no prior information suggesting a possible value of $\hat{p}$.

SOLUTION

a. The prior study suggests that $\hat{p} = 0.18$, so $\hat{q} = 0.82$ (found from $\hat{q} = 1 - 0.18$). With a 95% level of confidence, we have $\alpha = 0.05$, so $z_{\alpha/2} = 1.96$. Also, the margin of error is $E = 0.03$ (the decimal equivalent of "three percentage points"). Because we have an estimated value of $\hat{p}$, we use Formula 6-5 as follows.

$$n = \frac{[z_{\alpha/2}]^2 \hat{p}\hat{q}}{E^2} = \frac{[1.96]^2 (0.18)(0.82)}{0.03^2}$$

$$= 630.0224 = 631 \qquad \text{(rounded up)}$$

We must survey at least 631 randomly selected car drivers.

b. As in part (a), we again use $z_{\alpha/2} = 1.96$ and $E = 0.03$, but with no prior knowledge of $\hat{p}$ (or $\hat{q}$), we use Formula 6-6 as follows.

$$n = \frac{[z_{\alpha/2}]^2 \cdot 0.25}{E^2} = \frac{[1.96]^2 \cdot (0.25)}{0.03^2}$$

$$= 1067.1111 = 1068 \qquad \text{(rounded up)}$$

To be 95% confident that our sample percentage is within three percentage points of the true percentage for all car drivers, we should randomly select and survey 1068 drivers. By comparing this result to the sample size of 631 found in part (a), we can see that if we have no knowledge of a prior study, a larger sample is required to achieve the same results as when the value of $\hat{p}$ can be estimated.

Part (b) of the preceding example involved application of Formula 6-6, the same formula frequently used by Nielsen, Gallup, and other professional pollsters. Many people incorrectly believe that we should sample some percentage of the population, but Formula 6-6 shows that the population size is irrelevant. (In reality, the population size is sometimes used, but only in cases in which we sample without replacement from a relatively small population. See

Exercise 29.) Most of the polls featured in newspapers, magazines, and broadcast media involve polls with sample sizes in the range of 1000 to 2000. Even though such polls may involve a very small percentage of the total population, they can provide results that are quite good. When Nielsen surveys 1068 TV households from a population of 93 million, only 0.001% of the households are surveyed; still, we can be 95% confident that the sample percentage will be within three percentage points of the population percentage.

Polls have become very important and prevalent in the United States. They affect the television shows we watch, the leaders we elect, the legislation that governs us, and the products we consume. An understanding of the concepts of this section removes much of the mystery and misunderstanding surrounding polls.

6–4 Exercises A: Basic Skills and Concepts

In Exercises 1–4, assume that a sample is used to estimate a population proportion p. Find the margin of error that corresponds to the given values of n and x and the degree of confidence.

1. $n = 800$, $x = 200$, 95%

2. $n = 1400$, $x = 420$, 99%

3. $n = 4275$, $x = 2576$, 98%

4. $n = 887$, $x = 209$, 90%

Recommended assignment: Exercises 1, 5, 10–16, 19, 21, 24, 25. (These exercises require determination of sample size: 9–12, 15, 16, 21, 25 part b, 26 part b.)

In Exercises 5–8, use the given sample data and degree of confidence to construct the interval estimate of the population proportion p.

5. $n = 800$, $x = 600$, 95% confidence   5. $0.720 < p < 0.780$

6. $n = 2000$, $x = 300$, 99% confidence   6. $0.129 < p < 0.171$

7. $n = 2475$, $x = 992$, 90% confidence   7. $0.385 < p < 0.417$

8. $n = 5200$, $x = 1024$, 98% confidence   8. $0.184 < p < 0.210$

1. 0.0300
2. 0.0315
3. 0.0174
4. 0.0234

In Exercises 9–12, use the given data to find the minimum sample size required to estimate a population proportion or percentage.

9. Margin of error: 0.02; confidence level: 95%; $\hat{p}$ and $\hat{q}$ unknown

10. Margin of error: 0.01; confidence level: 90%; $\hat{p}$ and $\hat{q}$ unknown

11. Margin of error: four percentage points; confidence level: 99%; $\hat{p}$ is estimated to be 0.20 from a prior study   11. 664

12. Margin of error: two percentage points; confidence level: 97%; $\hat{p}$ is estimated to be 0.85 from a prior study   12. 1501

9. 2401
10. 6766

13. The Hartford Insurance Company wants to estimate the percentage of drivers who change tapes or CDs while driving. A random sample of 850 drivers results in 544 who change tapes or CDs while driving (based on data from *Prevention* magazine).

    a. Find the point estimate of the *percentage* of all drivers who change tapes or CDs while driving.   13. a. 64.0%

    b. Find a 90% interval estimate of the *percentage* of all drivers who change tapes or CDs while driving.   b. $61.3\% < p < 66.7\%$

14. When 500 college students are randomly selected and surveyed, it is found that 135 of them own personal computers (based on data from the America Passage Media Corporation).
    a. Find the point estimate of the true proportion of all college students who own personal computers.   14. a. 0.270
    b. Find a 95% confidence interval for the true proportion of all college students who own personal computers.   b. $0.231 < p < 0.309$

15. A reporter for *Byte* magazine wants to conduct a survey to estimate the true proportion of all college students who own personal computers, and she wants 95% confidence that her results have a margin of error of 0.04. How many college students must be surveyed?
    a. Assume that we have an estimate of $\hat{p}$ found from a prior study which revealed a percentage of 27% (based on data from the America Passage Media Corporation).   15. a. 474
    b. Assume that we have no prior information suggesting a possible value of $\hat{p}$.   b. 601

16. As a manufacturer of golf equipment, the Spalding Corporation wants to estimate the proportion of golfers who are left-handed. (The company can use this information in planning for the number of right-handed and left-handed sets of golf clubs to make.) How many golfers must be surveyed if we want 98% confidence that the sample proportion has a margin of error of 0.025?
    a. Assume that there is no available information that could be used as an estimate of $\hat{p}$.   16. a. 2172
    b. Assume that we have an estimate of $\hat{p}$ found from a previous study which suggests that 15% of golfers are left-handed (based on a *USA Today* report).   b. 1108

17. In doing market research for the Ford Motor Company, you find that a random sample of 1220 households includes 1054 in which a vehicle is owned (based on data from the Bureau of the Census). Based on those results, construct a 98% confidence interval for the percentage of all households in which a vehicle is owned.   17. $84.1\% < p < 88.7\%$

18. A *Prime Time Live* television report was based on consumer cheating at a Stew Leonard's grocery store in Connecticut. In speaking for the Connecticut Consumer Protection Agency, Gloria Shaffer stated that out of 1527 packages that were checked, 706 were found to be shortweighted. She commented that this is a very high percentage. Use the sample data to construct a 99% interval estimate of the percentage of all packages that were shortweighted. Based on the result, do you agree with her claim that this is a very high percentage?   18. $42.9\% < p < 49.5\%$

19. $9.53\% < p < 18.5\%$; yes

19. The Greybar Tax Company believes that its clients are selected for audits at a rate substantially higher than the rate for the general population. The IRS reports that it audits 4.3% of those who earn more than $100,000, but a check of 400 randomly selected Greybar returns with earnings above $100,000 shows that 56 of them were audited. Using a 99% level

of confidence, construct a confidence interval for the percentage of Greybar returns with earnings above $100,000 that are audited. Based on the result, does it appear that the high-income Greybar clients are audited at a rate that is substantially higher than the rate for the general population?

20. The Locust Tree Restaurant keeps records of reservations and no-shows. When 150 Saturday reservations are randomly selected, it is found that 70 of them were no-shows (based on data from American Express). Using a 90% degree of confidence, find a confidence interval for the proportion of Saturday no-shows.   20. $0.400 < p < 0.534$

21. How many TV households must Nielsen survey to estimate the percentage that are tuned to *The Late Show with David Letterman*? Assume that you want 97% confidence that your sample percentage has a margin of error of two percentage points. Also assume that nothing is known about the percentage of households tuned in to any television shows after 11 P.M.   21. 2944

22. The West America Communications Company is considering a bid to provide long-distance phone service. You are asked to conduct a poll to estimate the percentage of consumers who are satisfied with their current long-distance phone service. You want to be 90% confident that your sample percentage is within 2.5 percentage points of the true population value, and a Roper poll suggests that this percentage should be about 85%. How large must your sample be?   22. 553

23. The American Resorts hotel chain gives an aptitude test to job applicants and considers a multiple-choice test question to be easy if at least 80% of the responses are correct. A random sample of 6503 responses to one particular question includes 84% correct responses. Construct the 99% confidence interval for the true percentage of correct responses. Is it likely that this question is really easy?   23. $82.8\% < p < 85.2\%$; yes

24. The tobacco industry closely monitors all surveys that involve smoking. One survey showed that among 785 randomly selected subjects who completed four years of college, 18.3% smoke (based on data from the American Medical Association). Construct the 98% confidence interval for the true percentage of smokers among all people who completed four years of college. Based on the result, does the smoking rate for college graduates appear to be substantially different than the 27% rate for the general population?   24. $15.1\% < p < 21.5\%$; yes

25. In a study of store checkout scanners, 1234 items were checked and 20 of them were found to be overcharges (based on data from "UPC Scanner Pricing Systems: Are They Accurate?" by Goodstein, *Journal of Marketing,* Vol. 58).   25. a. $0.00916 < p < 0.0233$
   a. Using the sample data, construct a 95% confidence interval for the proportion of all such scanned items that are overcharges.
   b. Use the sample data as a pilot study and find the sample size necessary to estimate the proportion of scanned items that are overcharges. Assume that you want 99% confidence that the estimate is in error by no more than 0.005.   b. 4229

26. A health study involves 1000 randomly selected deaths, with 331 of them caused by heart disease (based on data from the Centers for Disease Control).
    a. Using the sample data, construct a 99% confidence interval for the proportion of all deaths caused by heart disease.   26. a. $0.293 < p < 0.369$
    b. Use the sample data as a pilot study and find the sample size necessary to estimate the proportion of all deaths caused by heart disease. Assume that you want 98% confidence that the estimate is in error by no more than 0.01.   b. 12,022

27. Columbia Pictures chairman Mark Canton claims that 58% of the movies made are R-rated. Refer to the movie data in Data Set 10 of Appendix B and construct the 95% confidence interval for the percentage of movies with R ratings. Is the resulting confidence interval consistent with Canton's claim?   27. $45.9\% < p < 70.8\%$; yes

28. Refer to Data Set 11 in Appendix B and find the sample proportion of M&Ms that are red. Use that result to construct a 95% confidence interval estimate of the population percentage of M&Ms that are red. Is the result consistent with the 20% rate that is reported by the candy maker Mars?   28. $13.0\% < p < 29.0\%$; yes

6-4 Exercises B: Beyond the Basics

29. 1853

29. This section presented Formulas 6-5 and 6-6, which are used for determining sample size. In both cases we assumed that the population is infinite or very large, or that we are sampling with replacement. When we have a relatively small population with size N and sample without replacement, we modify E to include the *finite population correction factor* shown here, and we can solve for n to obtain the result shown to the right. Use this result to repeat Exercise 21, assuming that we limit our population to a town of 5000 people.

$$E = z_{\alpha/2} \sqrt{\frac{\hat{p}\hat{q}}{n}} \sqrt{\frac{N - n}{N - 1}} \qquad n = \frac{N\hat{p}\hat{q}[z_{\alpha/2}]^2}{\hat{p}\hat{q}\,[z_{\alpha/2}]^2 + (N - 1)E^2}$$

30. A *New York Times* article about poll results states, "In theory, in 19 cases out of 20, the results from such a poll should differ by no more than one percentage point in either direction from what would have been obtained by interviewing all voters in the United States." Find the sample size suggested by this statement.   30. 9604

31. A newspaper article indicates that an estimate of the unemployment rate involves a survey of 47,000 people (a typical sample size for Bureau of Labor Statistics surveys). If the reported unemployment rate must have an error no larger than 0.2 percentage point and the rate is known to be about 8%, find the corresponding confidence level.   31. 89%

32. Heights of women are normally distributed with a mean of 63.6 in. and a standard deviation of 2.5 in. How many women must be surveyed if we want to estimate the percentage who are taller than 5 ft? Assume that we want 98% confidence that the error is no more than 2.5 percentage points. (*Hint:* The answer is substantially smaller than 2172.)   32. 602

33. A *one-sided confidence interval* for p can be written as $p < \hat{p} + E$ or $p > \hat{p} - E$, where the margin of error E is modified by replacing $z_{\alpha/2}$ with z_{α}. If Air America wants to report an on-time performance of at least x percent with 95% confidence, construct the appropriate one-sided confidence interval and then find the percent in question. Assume that a random sample of 750 flights results in 630 that are on time.   33. $p > 0.818$; 81.8%

34. Special tables are available for finding confidence intervals for proportions involving small numbers of cases where the normal distribution approximation cannot be used. For example, given three successes among eight trials, the 95% confidence interval found in *Standard Probability and Statistics Tables and Formulae* (CRC Press) is $0.085 < p < 0.755$. Find the confidence interval that would result if you were to use the normal distribution incorrectly as an approximation to the binomial distribution. Are the results reasonably close?   34. $0.0395 < p < 0.710$; no

6-5   Estimating a Population Variance

In this section we consider the same three concepts discussed in Sections 6-2, 6-3, and 6-4: (1) point estimate, (2) confidence interval, and (3) determining the required sample size. Whereas Sections 6-2, 6-3, and 6-4 applied these concepts to estimates of means and proportions, this section applies them to the population variance σ^2 or standard deviation σ. Many real situations, such as quality control in a manufacturing process, require that we estimate values of population variances or standard deviations. In addition to making products with measurements yielding a desired mean, the manufacturer must make products of *consistent* quality that do not run the gamut from extremely good to extremely poor. As this consistency can often be measured by the variance or standard deviation, these become vital statistics in maintaining the quality of products.

Assumption: In this section we assume that the population has normally distributed values. This assumption was made in earlier sections, but it is more critical here. In using the Student t distribution in Section 6-3, for example, we required that the population of values be approximately normal, but we accepted departures from normality that were not too severe. When we deal with variances using the methods of this section, however, the use of populations with very nonnormal distributions can lead to gross errors. Consequently, the requirement to have a normal distribution is much more strict, and we should check the distribution of the data by constructing a histogram to see if it is symmetrical and bell shaped. We describe this sensitivity to a

Many instructors omit this section because of time limitations. The chi-square distribution is introduced here, but it can be introduced in Chapter 7 instead. (Section 7-6 is written so that the chi-square distribution can be introduced there.) Many other instructors feel strongly that the importance of standard deviations requires inclusion of this section.

normal distribution by saying that inferences about the population variance σ^2 (or population standard deviation σ) made on the basis of the chi-square distribution (defined later) are not *robust*, meaning that the inferences can be very misleading if the population does not have a normal distribution. In contrast, inferences made about the population mean μ based on the Student t distribution are reasonably robust, because departures from normality that are not too extreme will not lead to gross errors.

When we considered estimates of means and proportions in Sections 6-2, 6-3, and 6-4, we used the normal and Student t distributions. When developing estimates of variances or standard deviations, we use another distribution, referred to as the chi-square distribution. We will examine important features of that distribution before proceeding with the development of confidence intervals.

Chi-Square Distribution

In a normally distributed population with variance σ^2, we randomly select independent samples of size n and compute the sample variance s^2 (see Formula 2-5) for each sample. The sample statistic $\chi^2 = (n - 1)s^2/\sigma^2$ has a distribution called the **chi-square distribution.**

Discuss the fact that for a sample with variance s^2 close to the population variance σ^2, the value of χ^2 will be close to the number of degrees of freedom $n-1$ (because the ratio of s^2/σ^2 will be close to 1). Also discuss the fact that s^2 is positive, σ^2 is positive, and $n-1$ is positive, so χ^2 will be positive. This explains why the χ^2 graph begins at 0.

Chi-Square Distribution	
Formula 6-7	$\chi^2 = \dfrac{(n - 1)s^2}{\sigma^2}$
where	n = sample size
	s^2 = sample variance
	σ^2 = population variance

We denote chi-square by χ^2, pronounced "kigh square." (The specific mathematical equations used to define this distribution are not given here because they are beyond the scope of this text.) To find critical values of the chi-square distribution, refer to Table A-4. The chi-square distribution is determined by the number of degrees of freedom, and in this chapter we use $n - 1$ degrees of freedom.

$$\text{degrees of freedom} = n - 1$$

In later chapters we will encounter situations in which the degrees of freedom are not $n - 1$, so we should not universally equate degrees of freedom with $n - 1$.

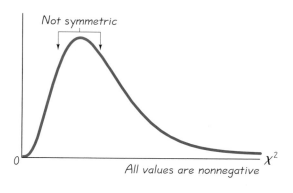

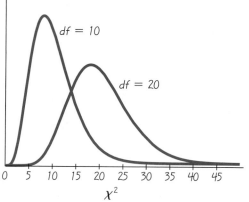

Figure 6-7   Chi-Square
Distribution

Figure 6-8   Chi-Square
Distribution for df = 10 and
df = 20

Properties of the Distribution of the Chi-Square Statistic

1. The chi-square distribution is not symmetric, unlike the normal and Student t distributions (see Figure 6-7). (As the number of degrees of freedom increases, the distribution becomes more symmetric, as Figure 6-8 illustrates.)

2. The values of chi-square can be zero or positive, but they cannot be negative (see Figure 6-7).

3. The chi-square distribution is different for each number of degrees of freedom (see Figure 6-8), which is df = $n - 1$ in this section. As the number of degrees of freedom increases, the chi-square distribution approaches a normal distribution.

Because of the nature of the χ^2 distribution, the methods of this section are very different from those in the preceding sections. One difference can be seen in the procedure for finding critical values, illustrated in the following example. Note the following essential feature of Table A-4:

> **In Table A-4, each critical value of χ^2 corresponds to an area given in the top row of the table, and that area represents the *total region* located *to the right* of the critical value.**

Consider stopping here and illustrating the use of Table A-4 in class.

EXAMPLE   Find the critical values of χ^2 that determine critical regions containing an area of 0.025 in each tail. Assume that the relevant sample size is 10 so that the number of degrees of freedom is 10 − 1, or 9.

(continued)

Figure 6-9 Critical Values of the Chi-Square Distribution

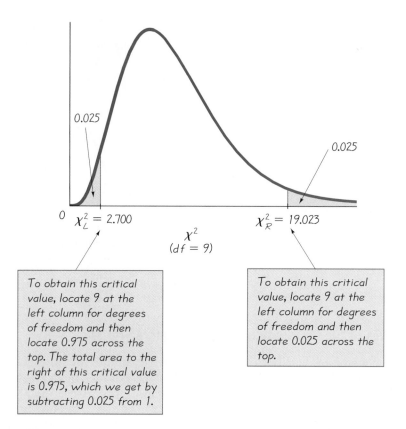

0.025

0.025

0 $\chi_L^2 = 2.700$

$\chi_R^2 = 19.023$

χ^2
$(df = 9)$

To obtain this critical value, locate 9 at the left column for degrees of freedom and then locate 0.975 across the top. The total area to the right of this critical value is 0.975, which we get by subtracting 0.025 from 1.

To obtain this critical value, locate 9 at the left column for degrees of freedom and then locate 0.025 across the top.

SOLUTION   See Figure 6-9 and refer to Table A-4. The critical value to the right ($\chi^2 = 19.023$) is obtained in a straightforward manner by locating 9 in the degrees-of-freedom column at the left and 0.025 across the top. The critical value of $\chi^2 = 2.700$ to the left once again corresponds to 9 in the degrees-of-freedom column, but we must locate 0.975 (found by subtracting 0.025 from 1) across the top because the values in the top row are always *areas to the right* of the critical value. Refer to Figure 6-9 and see that the total area to the right of $\chi^2 = 2.700$ is 0.975.

Figure 6-9 shows that, for a sample of 10 scores taken from a normally distributed population, the chi-square statistic $(n - 1)s^2/\sigma^2$ has a 0.95 probability of falling between the chi-square critical values of 2.700 and 19.023.

When obtaining critical values of chi-square from Table A-4, note that the numbers of degrees of freedom are consecutive integers from 1 to 30, followed by 40, 50, 60, 70, 80, 90, and 100. When a number of degrees of freedom (such as 52) is not found on the table, you can usually use the closest critical value. For example, if the number of degrees of freedom is 52, refer to Table

A-4 and use 50 degrees of freedom. (If the number of degrees of freedom is exactly midway between table values, such as 55, simply find the mean of the two χ^2 values.) For numbers of degrees of freedom greater than 100, use the equation given in Exercise 22, a more detailed table, or a statistical software package.

Estimators of σ^2

Because sample variances s^2 (found by using Formula 2-5) tend to center on the value of the population variance σ^2, we say that s^2 is an *unbiased estimator* of σ^2. That is, sample variances s^2 do not systematically tend to overestimate the value of σ^2, nor do they systematically tend to underestimate σ^2. Instead, they tend to target the value of σ^2 itself. Also, the values of s^2 tend to produce smaller errors by being closer to σ^2 than do other measures of variation. For these reasons, the value of s^2 is generally the best single value (or point estimate) of the various possible statistics we could use to estimate σ^2.

> **The sample variance s^2 is the best point estimate of the population variance σ^2.**

Because s^2 is the best point estimate of σ^2, it would be natural to expect s to be the best point estimate of σ, but this is not the case, because s is a biased estimator of σ. If the sample size is large, however, the bias is so small that we can use s as a reasonably good estimate of σ.

Although s^2 is the best point estimate of σ^2, there is no indication of how good it actually is. To compensate for that deficiency, we develop an interval estimate (or confidence interval) that is more revealing.

Confidence Interval (or Interval Estimate) for the Population Variance σ^2

$$\frac{(n-1)s^2}{\chi_R^2} < \sigma^2 < \frac{(n-1)s^2}{\chi_L^2}$$

This expression is used to find a confidence interval for variance σ^2, but the confidence interval (or interval estimate) for the standard deviation σ is found by taking the square root of each component, as shown below.

$$\sqrt{\frac{(n-1)s^2}{\chi_R^2}} < \sigma < \sqrt{\frac{(n-1)s^2}{\chi_L^2}}$$

The notations χ_R^2 and χ_L^2 in the preceding expressions are described as follows. (Note that some other texts use $\chi_{\alpha/2}^2$ in place of χ_R^2, and they use $\chi_{1-\alpha/2}^2$ in place of χ_L^2.)

Important note about the format of confidence intervals in this section: The format of (0.077, 0.185) is sometimes used instead of $0.077 < \sigma < 0.185$, but the format of $\sigma = s \pm E$ *cannot* be used because s is not at the center of the confidence interval.

Figure 6-10 Chi-Square Distribution with Critical Values χ_L^2 and χ_R^2
The critical values χ_L^2 and χ_R^2 separate the extreme areas corresponding to sample variances that are unlikely (with probability α).

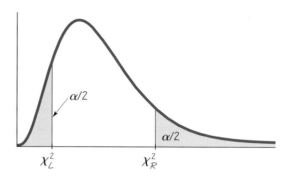

Notation

With a total area of α divided equally between the two tails of a chi-square distribution, χ_L^2 denotes the left-tailed critical value and χ_R^2 denotes the right-tailed critical value. (See Figure 6-10.)

Confidence interval limits for σ^2 and σ should be rounded by using the following round-off rule, which is really the same basic rule given in Section 6-2.

Round-Off Rule for Confidence Interval Estimates of σ or σ^2

1. When using the original set of data to construct a confidence interval, round the confidence interval limits to one more decimal place than is used for the original set of data.

2. When the original set of data is unknown and only the summary statistics (n, s) are used, round the confidence interval limits to the same number of decimal places used for the sample standard deviation or variance.

EXAMPLE The Hudson Valley Bakery makes doughnuts that are packaged in boxes with labels stating that there are 12 doughnuts weighing a total of 42 oz. If the variation among the doughnuts is too large, some boxes will be underweight (cheating consumers) and others will be overweight (lowering profit). A consumer would not be happy with a doughnut so small that it can be seen only with an electron microscope,

nor would a consumer be happy with a doughnut so large that it resembles a tractor tire. The quality control supervisor has found that he can stay out of trouble if the doughnuts have a mean of 3.50 oz and a standard deviation of 0.06 oz or less. Twelve doughnuts are randomly selected from the production line and weighed, with the results given here (in ounces). Construct a 95% confidence interval for σ^2 and a 95% confidence interval for σ, then determine whether the quality control supervisor is in trouble.

3.43   3.37   3.58   3.50   3.68   3.61   3.42   3.52   3.66   3.50   3.36   3.42

SOLUTION   Based on the sample data, the mean of $\overline{x} = 3.504$ seems quite good because it's very close to the desired 3.50 oz. The given scores have a standard deviation of $s = 0.109$, which might seem to be greater than the desired value of 0.06 or less. Let's proceed to find the confidence interval for σ^2.

With a sample of 12 scores we have 11 degrees of freedom. With a 95% degree of confidence, we divide $\alpha = 0.05$ equally between the two tails of the χ^2 distribution and we refer to the values of 0.975 and 0.025 across the top row. The critical values of χ^2 are $\chi_L^2 = 3.816$ and $\chi_R^2 = 21.920$. Using these critical values and the sample standard deviation of $s = 0.109$ and the sample size of 12, we construct the 95% confidence interval by evaluating the following.

$$\frac{(12 - 1)(0.109)^2}{21.920} < \sigma^2 < \frac{(12 - 1)(0.109)^2}{3.816}$$

This becomes $0.006 < \sigma^2 < 0.034$. Taking the square root of each part (before rounding) yields $0.077 < \sigma < 0.185$. Based on the 95% confidence interval for σ, it appears that the standard deviation is greater than the desired value of 0.06 or less, so the quality control supervisor is in trouble and must take corrective action to make the doughnut weights more consistent.

Airlines Sample

Airline companies once used an expensive accounting system to split up income from tickets that involved two or more companies. They now use a sampling method in which a small percentage of these "split" tickets are randomly selected and used as a basis for dividing up all such revenues. The error created by this approach can cause some companies to receive slightly less than their fair share, but these losses are more than offset by the clerical savings accrued by dropping the 100% accounting method. This new system saves companies millions of dollars each year.

The confidence interval $0.077 < \sigma < 0.185$ can also be expressed as (0.077, 0.185), but the format of $\sigma = s \pm E$ *cannot* be used because the confidence interval does not have s at its center.

Using Computer Software for Confidence Intervals about σ or σ^2

STATDISK can be easily used to construct confidence intervals for standard deviations or variances. Select `Analysis` from the main menu, then select `Confidence Intervals`, and proceed to enter the required data.

With Minitab, enter the data in column C1, then select `Editor`, then `Enable Command Language`, and enter the command `%DESCRIBE C1` to obtain output that includes 95% confidence intervals for μ and σ. The degree of confidence can be changed from the default of 95%.

TI-83 calculators do not give confidence intervals for σ or σ^2.

Rationale: Let's now consider why the confidence intervals for σ^2 and σ have the forms just given. If we obtain samples of size n from a population with variance σ^2, the distribution of the $(n-1)s^2/\sigma^2$ values will be as shown in Figure 6-10. For a random sample, there is a probability of $1 - \alpha$ that the statistic $(n-1)s^2/\sigma^2$ will fall between the critical values of χ_L^2 and χ_R^2. In other words (and symbols), there is a $1 - \alpha$ probability that both of the following are true:

$$\frac{(n-1)s^2}{\sigma^2} < \chi_R^2 \qquad \text{and} \qquad \frac{(n-1)s^2}{\sigma^2} > \chi_L^2$$

If we multiply both of the preceding inequalities by σ^2 and divide each inequality by the appropriate critical value of χ^2, we see that the two inequalities can be expressed in the equivalent forms

$$\frac{(n-1)s^2}{\chi_R^2} < \sigma^2 \qquad \text{and} \qquad \frac{(n-1)s^2}{\chi_L^2} > \sigma^2$$

These last two inequalities can be combined into one inequality:

$$\frac{(n-1)s^2}{\chi_R^2} < \sigma^2 < \frac{(n-1)s^2}{\chi_L^2}$$

There is a probability of $1 - \alpha$ that these confidence interval limits contain the population variance σ^2.

Determining Sample Size

Almost every other textbook ignores the topic of sample sizes required to estimate σ or σ^2, even after covering sample sizes required for estimating means and proportions. In many cases, the standard deviation is the most important parameter, so its estimation is critically important. We're going the extra kilometer to make this text the best available.

The procedures for finding the sample size necessary to estimate σ^2 are much more complex than the procedures given earlier for means and proportions. Instead of using very complicated procedures, we will use Table 6-2.

EXAMPLE   With 95% confidence, you wish to estimate σ to within 10%. How large should your sample be? Assume that the population is normally distributed.

SOLUTION   From Table 6-2, we can see that 95% confidence and an error of 10% for σ correspond to a sample of size 191. You should randomly select 191 values from the population.

TABLE 6-2	Sample Size for σ^2		Sample Size for σ	
To be 95% confident that s^2 is within	of the value of σ^2, the sample size n should be at least		To be 95% confident that s is within	of the value of σ, the sample size n should be at least
1%	77,207		1%	19,204
5%	3,148		5%	767
10%	805		10%	191
20%	210		20%	47
30%	97		30%	20
40%	56		40%	11
50%	37		50%	7
To be 99% confident that s^2 is within	of the value of σ^2, the sample size n should be at least		To be 99% confident that s is within	of the value of σ, the sample size n should be at least
1%	133,448		1%	33,218
5%	5,457		5%	1,335
10%	1,401		10%	335
20%	368		20%	84
30%	171		30%	37
40%	100		40%	21
50%	67		50%	13

6-5 Exercises A: Basic Skills and Concepts

In Exercises 1–4, find the critical values χ^2_L and χ^2_R that correspond to the given degree of confidence and sample size.

Recommended assignment: Exercises 1, 5, 9–12, 16–20.

1. 95%; $n = 26$ 1. 13.120, 40.646   **2.** 99%; $n = 17$ 2. 5.142, 34.267
3. 90%; $n = 60$ 3. 43.188, 79.082   **4.** 95%; $n = 50$ 4. 32.357, 71.420

In Exercises 5–8, use the given degree of confidence and sample data to find a confidence interval for the population standard deviation σ. In each case, assume that the population has a normal distribution.

5. Heights of women: 95% confidence; $n = 10$, $\bar{x} = 63.4$ in., $s = 2.4$ in.
6. Grade-point averages: 99% confidence; $n = 15$, $\bar{x} = 2.76$, $s = 0.88$
7. Test scores: 90% confidence; $n = 16$, $\bar{x} = 77.6$, $s = 14.2$
8. Police salaries: 95% confidence; $n = 19$, $\bar{x} = \$23,228$, $s = \$8779$

5. 1.7 in. $< \sigma <$ 4.4 in.
6. 0.59 $< \sigma <$ 1.63
7. 11.0 $< \sigma <$ 20.4
8. \$6634 $< \sigma <$ \$12,982

In Exercises 9–12, assume that each sample is obtained by randomly selecting values from a normally distributed population.

9. Find the minimum sample size needed to be 95% confident that the sample standard deviation s is within 30% of σ.   9. 20

10. Find the minimum sample size needed to be 99% confident that the sample standard deviation s is within 20% of σ.   10. 84

11. Find the minimum sample size needed to be 99% confident that the sample variance is within 30% of the population variance.   11. 171

12. Find the minimum sample size needed to be 95% confident that the sample variance is within 40% of the population variance.   12. 56

In Exercises 13–20, assume that each sample is obtained by randomly selecting values from a population with a normal distribution.

13. 38.2 mL $< \sigma <$ 95.7 mL; no, the fluctation appears to be too high.

13. A container of car antifreeze is supposed to hold 3785 mL of the liquid. Realizing that fluctuations are inevitable, the quality control manager wants to be quite sure that the standard deviation is less than 30 mL. Otherwise, some containers would overflow while others would not have enough of the coolant. She randomly selects a sample, with the results given here. Use these sample results to construct the 99% confidence interval for the true value of σ. Does this confidence interval suggest that the fluctuations are at an acceptable level?

$$
\left.
\begin{array}{l}
3761\ 3861\ 3769\ 3772\ 3675\ 3861 \\
3888\ 3819\ 3788\ 3800\ 3720\ 3748 \\
3753\ 3821\ 3811\ 3740\ 3740\ 3839
\end{array}
\right\}
\quad
\begin{array}{l}
n = 18 \\
\bar{x} = 3787.0 \\
s = 55.4
\end{array}
$$

14. The National Center for Education Statistics surveyed college graduates about the lengths of time required to earn their bachelor's degrees. The mean is 5.15 years, and the standard deviation is 1.68 years. Assume that the sample size is 101. Based on these sample data, construct the 99% confidence interval for the standard deviation of the times required by all college graduates.   14. 1.42 yr $< \sigma <$ 2.05 yr

15. A sample of 35 skulls is obtained for Egyptian males who lived around 1850 B.C. The maximum breadth of each skull is measured with the result that $\bar{x} = 134.5$ mm and $s = 3.5$ mm (based on data from *Ancient Races of the Thebaid* by Thomson and Randall-Maciver). Using these sample results, construct a 95% confidence interval for the population standard deviation σ.   15. 3.0 mm $< \sigma <$ 5.0 mm

16. 1.0 yr $< \sigma <$ 1.3 yr; no, the population parameters have changed.

16. A sample consists of 75 TV sets purchased several years ago. The replacement times of those TV sets have a mean of 8.2 years and a standard deviation of 1.1 years (based on data from "Getting Things Fixed," *Consumer Reports*). Construct a 90% confidence interval for the standard deviation of replacement times for all TV sets from that era. Does the result apply to TV sets being currently sold?

17. The listed values are waiting times (in minutes) of customers at the Jefferson Valley Bank, where customers enter a single waiting line that feeds
three windows. Construct a 95% confidence interval for the population
standard deviation σ.   17. $s = 0.47667832$; 0.33 min $< \sigma <$ 0.87 min

    6.5 6.6 6.7 6.8 7.1 7.3 7.4 7.7 7.7 7.7

18. The listed values are waiting times (in minutes) of customers at the Bank
of Providence, where customers may enter any one of three different lines
that have formed at three different teller windows. Construct a 95% confidence interval for σ and compare the result to the confidence interval for
the data in Exercise 17. Do the confidence intervals suggest a difference in
the variation among waiting times? Which arrangement seems better: the
single line system or the multiple line system?

    4.2 5.4 5.8 6.2 6.7 7.7 7.7 8.5 9.3 10.0

18. $s = 1.8216293$;
1.25 min $< \sigma <$ 3.33 min;
the single line is better
because there is less
variation.

19. Refer to Data Set 7 in Appendix B.
  a. Use the range rule of thumb (see Section 2-5) to estimate σ, the standard deviation of the total annual precipitation amount in Iowa.
  b. Construct a 99% confidence interval for σ.
  c. Does the confidence interval contain your estimated value of σ?

19. a. $s \approx (42.9 - 21.6)/4 =$
5.325
  b. Use $s = 5.6006261$;
4.33 in. $< \sigma <$ 7.78 in.
  c. Yes

20. Refer to Data Set 11 in Appendix B.
  a. Use the range rule of thumb (see Section 2-5) to estimate σ, the standard deviation of the weights of brown M&M plain candies.
  b. Construct a 98% confidence interval for σ.
  c. Does the confidence interval contain your estimated value of σ?

20. a. $s \approx (1.033 - 0.856)/4 =$
0.04425
  b. Use $s = 0.03951697$;
0.0313 g $< \sigma <$ 0.0578 g
  c. Yes

6-5 Exercises B: Beyond the Basics

21. A journal article includes a graph showing that sample data are normally
distributed.
  a. The degree of confidence is inadvertently omitted when this confidence interval is given: $2.8 < \sigma < 6.0$. Find the degree of confidence
for these given sample statistics: $n = 20$, $\bar{x} = 45.2$, and $s = 3.8$.
  b. This 95% confidence interval is given: $19.1 < \sigma < 45.8$. Find the
value of the standard deviation s, which was omitted from the article.

21. a. 98%

  b. 27.0

22. In constructing confidence intervals for σ or σ^2, we use Table A-4 to find
the critical values χ^2_L and χ^2_R, but that table applies only to cases in which
$n \le 101$, so the number of degrees of freedom is 100 or fewer. For larger
numbers of degrees of freedom, we can approximate χ^2_L and χ^2_R by using

22. 2.7 in. $< \sigma <$ 2.9 in.

$$\chi^2 = \frac{1}{2}[\pm z_{\alpha/2} + \sqrt{2k - 1}]^2$$

where k is the number of degrees of freedom. Construct the 95% confidence interval for σ by using the following sample data: The measured
heights of 772 men between the ages of 18 and 24 have a standard deviation of 2.8 in. (based on data from the National Health Survey).

• •

Vocabulary List

estimator	critical value
estimate	margin of error (E)
point estimate	confidence interval limits
confidence interval	Student t distribution
interval estimate	t distribution
degree of confidence	degrees of freedom
level of confidence	chi-square distribution
confidence coefficient	

• •

Review

This chapter and the following chapter introduce the fundamental and important concepts of inferential statistics. This chapter focused on *estimates* of parameters as we considered population means, proportions, and variances to develop procedures for each of the following.

- Identifying a point estimate
- Constructing a confidence interval
- Determining the required sample size

We discussed point estimate (or single-valued estimate) and formed these conclusions:

- The best point of estimate of μ is $\overline{x}$.
- The best point estimate of p is $\hat{p}$.
- The best point estimate of σ^2 is s^2.

As single values, the point estimates don't convey any real sense of how reliable they are, so we introduced confidence intervals (or interval estimates) as more informative estimates. We also considered ways of determining the sample sizes necessary to estimate parameters to within given tolerance factors. This chapter also introduced the Student t and chi-square distributions. We must be careful to use the correct distribution for each set of circumstances.

It is important to know that all of the confidence interval and sample size procedures in this chapter require that we have a population with a distribution that is approximately normal. If the distribution is very nonnormal, we must use other methods, such as the bootstrap method described in the Computer Project at the end of this chapter.

- -

Review Exercises

1. If we refer to the head lengths of bears in Data Set 3 of Appendix B, we will find 54 values with a mean of 12.95 in. and a standard deviation of 2.14 in. and a distribution that is approximately normal.
   a. Based on this sample, construct a 98% confidence interval estimate of the population mean of all bear-head lengths.   1. a. 12.27 in. $< \mu <$ 13.63 in.
   b. Use the sample data to construct a 98% confidence interval estimate of the standard deviation of all bear-head lengths.   b. 1.79 in. $< \sigma <$ 2.86 in.
   c. Using the sample standard deviation as an estimate of the population standard deviation, find the sample size required to estimate the mean head length of all bears. Assume that you want 98% confidence that the margin of error is 0.25 in.   c. 398

2. You have just been hired by General Motors to tour the United States giving randomly selected drivers test rides in a new Corvette (yeah, right). After giving the test drive, you must ask the rider whether he or she would consider buying a Corvette. How many riders must you survey to be 97% confident that the sample percentage is off by no more than two percentage points?   2. 2944

3. A NAPA Auto Parts supplier wants information about how long car owners plan to keep their cars. A random sample of 25 car owners results in $\bar{x} =$ 7.01 years and $s = 3.74$ years, respectively (based on data from a Roper poll). Assuming that the sample is drawn from a normally distributed population, find a 95% confidence interval for the population mean.   3. 5.47 yr $< \mu <$ 8.55 yr

4. Using the same sample data from Exercise 3, find a 95% confidence interval for the population standard deviation.   4. 2.92 yr $< \sigma <$ 5.20 yr

5. Of 1475 transportation workers randomly selected, 32.0% belong to unions (based on data from the U.S. Bureau of Labor Statistics). Construct the 95% confidence interval for the true proportion of all transportation workers who belong to unions. Also, what margin of error (in percentage points) should be reported along with the sample percentage of 32.0%?   5. 0.296 $< p <$ 0.344; 2.4 percentage points

6. In a Gallup poll of 1004 adults, 93% indicated that restaurants and bars should refuse service to patrons who have had too much to drink. If you plan to conduct a new poll to confirm that the percentage continues to be correct, how many randomly selected adults must you survey if you want 98% confidence that the margin of error is four percentage points?   6. 221

7. In designing a new machine to be used on an assembly line at a General Motors plant, an engineer obtains measurements of arm lengths of a random sample of male machine operators. The following values (in centimeters) are obtained. Construct the 95% confidence interval for the mean arm length of all such employees.   7. $n = 16$, $\bar{x} = 72.6625$, $s = 2.5809236$; 71.29 cm $< \mu <$ 74.04 cm

76.8	75.6	69.3	75.7	75.5	71.2	72.5	71.9
70.9	69.4	71.7	72.5	72.2	68.5	75.9	73.0

8. Verbal PSAT scores of a random sample of 40 college-bound high-school juniors have a mean of 40.7, a standard deviation of 10.2, and a distribution that is normal (based on data from Educational Testing Service). Find a 99% confidence interval estimate of the population mean.   8. $36.5 < \mu < 44.9$

9. A medical researcher wishes to estimate the serum cholesterol level (in mg/100 mL) of all women aged 18 to 24. There is strong evidence suggesting that $\sigma = 41.0$ mg/100 mL (based on data from a survey of 1524 women aged 18 to 24, as part of the National Health Survey). If the researcher wants to be 95% confident of obtaining a sample mean that is off by no more than four units, how large must the sample be?   9. 404

10. In a Roper survey of 1,998 randomly selected adults, 24% included loud commercials among the annoying aspects of television. Construct the 99% confidence interval for the percentage of all adults who are annoyed by loud commercials.   10. $21.5\% < p < 26.5\%$

- -

Cumulative Review Exercises

1. The U.S. Marine Corps is reviewing its orders for uniforms because it has a surplus for tall recruits and a shortage for shorter recruits. Its review involves the random sample of heights for male recruits between the ages of 18 and 24, as listed here (in inches).

69.9	69.4	72.6	70.0	70.2	71.8	70.6	72.8
69.0	68.4	68.3	69.6	71.7	69.2	70.8	71.0
70.4	66.8	69.9	69.2	70.5	70.2	70.8	70.0

Find each of the following.

a. mean
b. median
c. mode
d. midrange
e. range
f. variance
g. standard deviation
h. Q_1
i. Q_2
j. Q_3
k. What is the level of measurement of these data? (nominal, ordinal, interval, ratio)
l. Construct a boxplot for the data.
m. Construct a histogram and identify its general shape.
n. Construct a 99% confidence interval for the population mean.
o. Construct a 99% confidence interval for the standard deviation σ.
p. Find the sample size necessary to estimate the mean height so that there is 99% confidence that the sample mean is in error by no more than 0.2 in. Use the sample standard deviation s from part (g) as an estimate of the population standard deviation σ.
q. When men aged 18–24 are randomly selected from the general population, their heights are normally distributed with a mean of 69.7 in. and a standard deviation of 2.8 in. (based on data from the National Health Survey). Based on the preceding results, do the heights of the sample of recruits (as listed above) agree with the population parameters? Explain.

1. a. 70.13 in.   b. 70.10 in.
c. 69.2 in., 69.9 in., 70.0 in., 70.2 in., 70.8 in.
d. 69.80 in.   e. 6.00 in.
f. 1.82 in.^{2} g. 1.35 in.
h. 69.3 in.   i. 70.1 in.
j. 70.8 in.   k. ratio
l.

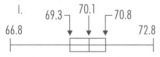

69.3 ┐ 70.1 ┐ ┌─ 70.8
66.8   72.8

m. Answer varies, depending on the choices for the number of classes and the starting point. The histogram is approximately bell shaped.
n. 69.36 in. $< \mu <$ 70.90 in.
o. 0.97 in. $< \sigma <$ 2.13 in.
p. 303
q. The sample mean of $\overline{X} = 70.13$ in. appears to be reasonably close to the population mean of 69.7 in., but the sample standard deviation of $s = 1.35$ in. appears to be considerably smaller than the population standard deviation of $\sigma = 2.8$ in.

2. A genetics expert has determined that for certain couples, there is a 0.25 probability that any child will have an X-linked recessive disorder.

    a. Find the probability that among 200 such children, at least 65 have the X-linked recessive disorder.   2. a. 0.0089

    b. A subsequent study of 200 actual births reveals that 65 of the children have the X-linked recessive disorder. Based on these sample results, construct a 95% confidence interval for the proportion of all such children having the disorder.   b. $0.260 < p < 0.390$

    c. Based on parts (a) and (b), does it appear that the expert's determination of a 0.25 probability is correct? Explain.

c. Because the confidence interval limits do not contain 0.25, it is unlikely that the expert is correct.

Computer Project

The *bootstrap method* can be used to construct confidence intervals for situations in which traditional methods cannot (or should not) be used. For example, the following sample of 10 scores was randomly selected from a very nonnormal distribution, so the methods previously discussed cannot be used.

2.9   564.2   1.4   4.7   67.6   4.8   51.3   3.6   18.0   3.6

The methods of this chapter require that the population have a distribution that is at least approximately normal. The bootstrap method, which makes no assumptions about the original population, typically requires a computer to build a bootstrap population by replicating (duplicating) a sample many times. We can draw from the sample with replacement, thereby creating an approximation of the original population. In this way, we pull the sample up "by its own bootstraps" to simulate the original population. Using the sample data given above, construct a 95% confidence interval estimate of the population mean μ by using the bootstrap method as described in the following Minitab steps.

a. Create 500 new samples, each of size 10, by selecting 10 scores with replacement from the 10 sample scores given above. With Minitab, first enter the sample scores in column C1, then enter probabilities of 0.1, 0.1, . . . , 0.1 (ten times) in column C2. Now select `Calc` from the main menu bar, then select `Random Data` and then `Discrete`. Proceed to generate 500 rows of data, stored in columns C11–C20, with the values in C1 and probabilities in C2, then click OK.

b. Find the means of the 500 bootstrap samples generated in part (a). Select `Calc`, `Row Statistics`, `Mean`, and enter input variables of C11–C20 with results to be stored in C21, then click OK.

c. Rank the 500 means (arrange them in order). Select `Manip` from the main menu bar, choose the option of `Sort`, and proceed to sort column C21, store the sorted column in C21, and sort by column C21. Click OK.

d. Find the percentiles $P_{2.5}$ and $P_{97.5}$ for the ranked means that result from the preceding step. ($P_{2.5}$ is the mean of the 12th and 13th scores in the ranked list of column C21; $P_{97.5}$ is the mean of the 487th and 488th scores in column C21.) Identify the resulting confidence interval by substituting the values for $P_{2.5}$ and $P_{97.5}$ in $P_{2.5} < \mu < P_{97.5}$. Does this confidence interval contain the true value of μ, which is 148?

Now use the bootstrap method to find a 95% confidence interval for the population standard deviation σ. (Use the same steps listed above, but specify *standard deviation* instead of mean in part b.) Compare your result to the interval $318.4 < \sigma < 1079.6$, which was obtained by incorrectly using the methods described in Section 6-5. (Their use is incorrect because the population distribution is very nonnormal.) This incorrect confidence interval for σ does not contain the true value of σ, which is 232.1. Does the bootstrap procedure yield a confidence interval for σ that contains 232.1, verifying that the bootstrap method is effective?

An alternative to using Minitab is to use special software designed specifically for bootstrap resampling methods. The author recommends Resampling Stats, available from Resampling Stats, Inc., 612 N. Jackson St., Arlington, VA 22201.

•••••••• **FROM DATA TO DECISION** ••••••••

He's Angry, But Is He Right?

The following excerpt is taken from a letter written by a corporation president and sent to the Associated Press.

> When you or anyone else attempts to tell me and my associates that 1223 persons account for our opinions and tastes here in America, I get mad as hell! How dare you! When you or anyone else tells me that 1223 people represent America, it is astounding and unfair and should be outlawed.

The writer then goes on to claim that because the sample size of 1223 people represents 120 million people, his letter represents 98,000 people (120 million divided by 1223) who share the same views.

a. Given that the sample size is 1223 and the degree of confidence is 95%, find the margin of error for the proportion. Assume that there is no prior knowledge about the value of that proportion.

b. The writer of the letter is taking the position that a sample size of 1223 taken from a population of 120 million people is too small to be meaningful. Do you agree or disagree? Write a response that either supports or refutes the writer's position that the sample is too small.

c. The writer also makes the claim that because the poll of 1223 people was projected to reflect the opinions of 120 million, any 1 person actually represents 98,000 other people. As the writer is 1 person, he claims to represent 98,000 other people. Is this claim correct? Explain why or why not.

········ COOPERATIVE GROUP ACTIVITIES ········

1. *Out-of-class activity:* Estimate the mean error (in seconds) on wristwatches. First, using a wristwatch that is reasonably accurate, set the time to be exact. Use a radio station or telephone time report which states that "at the tone, the time is . . ." If you cannot set the time to the nearest second, record the error for the watch you are using. Now compare the time on your watch to the time on others. Record the errors with positive signs for watches that are ahead of the actual time and negative signs for those watches that are behind the actual time. Find point estimates and confidence intervals for the mean error and standard deviation of errors. Does the confidence interval for the mean error contain zero? Based on the results, what do you conclude about the accuracy of people's watches? Are the deviations from the correct time random fluctuations, or are there other factors to consider?

2. *In-class activity:* Divide into groups with approximately 10 students in each group. Get the reaction timer from the first Cooperative Group Activity given in Chapter 5 and measure the reaction time of each group member. (Right-handed students should use their right hand, and left-handed students should use their left hand.) Use the methods of this chapter to estimate the mean reaction time for all college students. Construct a 90% confidence interval estimate of that mean. Compare the results to those found in other groups.

3. *In-class activity:* Divide into groups of three or four. Assume that you need to conduct a survey of full-time students at your college with the objective of identifying the percentage of students who are already registered to vote. Identify a margin of error that is reasonable, select an appropriate degree of confidence, then find the minimum sample size. Describe a sampling plan that is likely to yield good results.

4. *In-class activity:* Divide into groups with approximately 10 students in each group. First, each group member should write an estimate of the mean amount of cash being carried by students in the group. Next, each group member should report the actual amount of cash being held. (The amounts should be written anonymously on separate sheets of paper which are mixed, so that nobody's privacy is compromised.) Use the reported values to find $\bar{x}$ and s, then construct a 95% confidence interval estimate of the mean μ. Describe the precise population that is being estimated. Which group member came closest to the value of $\bar{x}$? Compare the results with other groups. Are the results consistent, or are there large variations among the group results?

interview

Barry Cook
*Senior Vice President at
Nielsen Media Research*

Barry Cook is a Senior Vice President
and Chief Research Officer at Nielsen
Media Research. He has taught at
Yale and Hunter College and has
worked for NBC and the USA Cable
Network. He is now in charge of
Nielsen's rating system, doing
research to better understand how the
measurements work, as well as
developing new measurement
systems.

How did you become involved in your current work with media research?

I used to teach statistics myself. Before getting into the
commercial aspect of the research business, I was at Yale
in the psychology department, then at Hunter College in
the sociology department. I taught an introductory
statistics course just about every semester. After teaching
for a number of years, I was at Hunter College in an
applied social research program where I was training
people to go into market research. I looked at where my
students were going and I felt it looked good, so I
followed them. I got into media research at NBC doing
market research on their news programs. I later did
ratings work at NBC. After that, I became involved with
a cable network and then Nielsen.

What major trends do you see in the way Americans watch TV?

In 1985 the average home received 18.8 channels. In
1990 the average home received 33.2 channels. That
obviously has an effect on what people choose to view.

What is your sample size?

For the national survey we use "people meters" in 4000
homes with about 11,000 people. We increased the
sample size because the use of television has changed.
Instead of only three major sources of TV (ABC, NBC,

CBS), there are now dozens of sources of programming, many of which get only a small piece of the audience. In order to measure those smaller pieces with enough precision, a larger sample is needed. In addition we also have meter services in 25 markets; the television sets are metered (but not the people) in 250 to 500 homes per market. Nielsen is still very big in the diary business—not for the national audience, but for measuring audiences in the 200 or so separate markets across the country. Those diaries amount to a combined sample size of 100,000, four times a year.

Have you been experiencing greater resistance to polls and surveys?

There's no question that we have seen a decline in cooperation in both telephone and in-person contacts with people. It's across the entire survey industry. There are concerns about privacy. The data-gathering efforts are being mixed up with sales efforts and that probably is contributing to a decline in the cooperation rate. Also, answering machines make it harder to get through to people.

Do you weight sample results to better reflect population parameters?

We have a policy against that. We try to represent population parameters by doing the sampling correctly in the first place. We sample in a way that gives an equal probability of selection to all housing units in the 50 states. As a result there is a known amount of sampling error and there's also an unknown amount of nonsampling error, but we've done validation research to estimate how close the samples are to measures of the population.

What are some of the specific statistical methods you use?

We use a lot of statistics for our own understanding and our clients' understanding of sampling and trends. We get into hypothesis testing when we try to understand why things change. Confidence intervals are very important in interpreting the estimates of the population.

Could you cite a television programming strategy that is based on survey results?

The most important strategy is called "prime time." The biggest usage of television occurs in the evening hours when most people are home. The most general programming strategy is to put on your best shows when there's the greatest number of people there. With a miniseries what you get on the first episode serves as almost a cap on what you can get after that. You want to get the maximum possible potential audience for the first installment, so Sunday night does that.

7

Hypothesis Testing

7-1 Overview

Chapter objectives are defined. This chapter introduces the basic concepts and procedures used for testing claims made about population parameters. The hypothesis-testing procedures of this chapter and the estimation procedures of Chapter 6 are two of the fundamental and major topics of inferential statistics.

7-2 Fundamentals of Hypothesis Testing

An informal example of a hypothesis test is presented, and its important components are described. The types of errors that can be made are also discussed.

7-3 Testing a Claim about a Mean: Large Samples

The traditional approach to hypothesis testing is presented, along with the P-value approach and a third approach based on confidence intervals. This section is limited to cases involving large $(n > 30)$ samples only.

7-4 Testing a Claim about a Mean: Small Samples

This section presents the procedure for testing a claim made about the mean of a population, given that the data set is small (with 30 or fewer values) and that the value of the population standard deviation is not known. The t distribution will be used in such cases.

7-5 Testing a Claim about a Proportion

The method is described for testing a claim made about a population proportion or percentage. The normal distribution is used for the examples and exercises of this section.

7-6 Testing a Claim about a Standard Deviation or Variance

The method is described for testing a claim made about the standard deviation or variance of a population. The chi-square distribution is used for such tests, and its role is described.

Chapter Problem:

The mean body temperature is 98.6°F, right?

When asked, almost everyone will identify the mean body temperature for healthy adults as 98.6°F. Table 7-1 lists 106 measured temperatures found in Data Set 2 of Appendix B (for 12 A.M. on day 2). Those 106 temperatures, found by University of Maryland researchers, have a mean of $\bar{x} = 98.20°F$ and a standard deviation of $s = 0.62°F$. In Chapter 6 we used the same set of temperatures to estimate μ, the mean body temperature, and we found this 95% confidence interval: $98.08°F < \mu < 98.32°F$. Here's the problem: The confidence interval limits of 98.08°F and 98.32°F do *not* contain 98.6°F, the value generally believed to be the mean body temperature. The confidence interval is an *estimate* of the mean body temperature, but the researchers went further by making a claim that 98.6°F "should be abandoned as a concept having any particular significance for the normal body temperature." Should we reject the common belief that the mean body temperature of healthy adults is 98.6°F? There is a standard procedure for testing such claims, and this chapter will describe that procedure.

Table 7-1   Body Temperatures of 106 Healthy Adults

98.6	98.6	98.0	98.0	99.0	98.4	98.4	98.4	98.4	98.6	98.6
98.8	98.6	97.0	97.0	98.8	97.6	97.7	98.8	98.0	98.0	98.3
98.5	97.3	98.7	97.4	98.9	98.6	99.5	97.5	97.3	97.6	98.2
99.6	98.7	99.4	98.2	98.0	98.6	98.6	97.2	98.4	98.6	98.2
98.0	97.8	98.0	98.4	98.6	98.6	97.8	99.0	96.5	97.6	98.0
96.9	97.6	97.1	97.9	98.4	97.3	98.0	97.5	97.6	98.2	98.5
98.8	98.7	97.8	98.0	97.1	97.4	99.4	98.4	98.6	98.4	98.5
98.6	98.3	98.7	98.8	99.1	98.6	97.9	98.8	98.0	98.7	98.5
98.9	98.4	98.6	97.1	97.9	98.8	98.7	97.6	98.2	99.2	97.8
98.0	98.4	97.8	98.4	97.4	98.0	97.0				

This chapter is designed so that it could be covered before Chapter 6. Earlier editions of this book had hypothesis testing covered before the chapter on estimates and sample sizes, but users generally preferred the current organization.

 Overview

In Chapter 6 we introduced a major topic of inferential statistics: estimating values of population parameters based on sample statistics. This chapter introduces another major topic of inferential statistics: testing claims (or *hypotheses*) made about population parameters.

> **DEFINITION**
>
> In statistics, a **hypothesis** is a claim or statement about a property of a population.

The following statements are examples of hypotheses that will be tested by the procedures we develop in this chapter.

- Medical researchers claim that the mean body temperature of healthy adults is not equal to 98.6° F.
- The percentage of hospitalized drivers is lower for those who crash in cars equipped with air bags than in cars not so equipped.
- When new equipment is used to manufacture aircraft altimeters, the variation in the errors is reduced so that the readings are more consistent.

Before beginning to study this chapter, you should keep in mind this common guideline to reasoning in statistics:

Analyze a sample in an attempt to distinguish between results that can easily occur and results that are highly unlikely.

We can explain the occurrence of highly unlikely results by saying either that a rare event has indeed occurred or that things aren't as they are assumed to be. Let's apply that reasoning in the following example.

Instead of beginning with a sequence of mechanical steps, it is very helpful to begin hypothesis testing with an overview of the basic concept used. Focus on the issue of *significance*: Do the sample results differ from the claim by an amount that is statistically significant? Consider assignment of Section 7-1 for reading to be done before this chapter is discussed in class.

EXAMPLE   ProCare Industries, Ltd., once provided a product called "Gender Choice" which, according to advertising claims, allowed couples to "increase your chances of having a boy up to 85%, a girl up to 80%." Gender Choice was available in blue packages for couples wanting a baby boy and (you guessed it) pink packages for couples wanting a baby girl. Suppose you conduct an experiment that includes 100 couples who want to have baby girls, and they all follow the Gender Choice "easy-to-use in-home system" described in the pink package. Using common sense and no real formal statistical methods, what should you conclude about the effectiveness of Gender Choice if the 100 babies include

a. 52 girls

b. 97 girls

SOLUTION

a. We normally expect around 50 girls in 100 births. The result of 52 girls is close to 50, so we should not conclude that Gender Choice is effective. If the 100 couples used no special methods of gender selection, the result of 52 girls could easily occur by chance.

b. The result of 97 girls in 100 births is extremely unlikely to occur by chance. It could be explained in one of two ways: Either an *extremely* rare event has occurred by chance, or Gender Choice is effective. Because of the extremely low probability of getting 97 girls by chance, the more likely explanation is that the product is effective.

The key point of the preceding example is that we should conclude that the product is effective only if we get *significantly* more girls than we would expect under normal circumstances. Although the outcomes of 52 girls and 97 girls are both "above average," the result of 52 girls is not significant, whereas 97 girls does constitute a significant result.

This brief example illustrates the basic approach used in testing hypotheses. The formal method involves a variety of standard terms and conditions incorporated into an organized procedure. We recommend that you begin the study of this chapter by first reading Sections 7-2 and 7-3 casually to obtain a general idea of their concepts and then rereading Section 7-2 more carefully to become familiar with the terminology.

7-2 Fundamentals of Hypothesis Testing

In this section we begin with an informal example, then we identify the formal components of the standard method of hypothesis testing. In Section 7-3 we will include those components in a formal procedure. Because that procedure involves specific steps, it might be tempting to try to memorize and then mechanically apply them, but if you take the time to *understand* the procedure you will be more likely to remember it, better prepared to apply it to different circumstances, and less likely to make serious errors. You should study the following example until you thoroughly understand it, and you will capture a major concept of statistics.

Large Sample Size Isn't Good Enough

Biased sample data should not be used for inferences, no matter how large the sample is. For example, in *Women and Love: A Cultural Revolution in Progress*, Shere Hite bases her conclusions on 4500 replies that she received after mailing 100,000 questionnaires to various women's groups. A *random* sample of 4500 subjects would usually provide good results, but Hite's sample is biased. It is criticized for overrepresenting women who join groups and women who feel strongly about the issues addressed. Because Hite's sample is biased, her inferences are not valid, even though the sample size of 4500 might seem to be sufficiently large.

EXAMPLE   In the Chapter Problem, we noted that researchers claimed that the mean body temperature of healthy adults is not equal to 98.6° F. The University of Maryland researchers collected sample data with these characteristics: $n = 106$, $\bar{x} = 98.20$, $s = 0.62$, and the shape of the distribution is approximately normal. Here is the key question: *Do the sample data (with $\bar{x} = 98.20$) constitute sufficient evidence to warrant rejection of the common belief that $\mu = 98.6$?*

We conclude that there is sufficient evidence to warrant rejection of the belief that $\mu = 98.6$ because, if the mean is really 98.6, the probability of getting the sample mean of 98.20 is approximately 0.0002, which is too small. (Later, we will show how that probability value of 0.0002 is determined.)

If we were to assume that $\mu = 98.6$ and use the central limit theorem (Section 5-5), we know that sample means tend to be normally distributed with these parameters:

$$\mu_{\bar{x}} = \mu = 98.6 \quad \text{(by assumption)}$$

$$\sigma_{\bar{x}} = \frac{\sigma}{\sqrt{n}} \approx \frac{s}{\sqrt{n}} = \frac{0.62}{\sqrt{106}} = 0.06$$

We construct Figure 7-1 by assuming that $\mu = 98.6$ and by using the parameters just shown. Figure 7-1 also shows that if μ is really 98.6, then 95% of all sample means should fall between 98.48 and 98.72. (The values of 98.48 and 98.72 were found by using the methods of Section 5-5. Specifically, 95% of sample means should fall within 1.96 standard deviations of μ. With $\sigma_{\bar{x}} = 0.06$, 95% of sample means should fall within $1.96 \times 0.06 \approx 0.12$ of 98.6. Falling within 0.12 of 98.6 is equivalent to falling between 98.48 and 98.72.)

Here are the key points:

- The common belief is that $\mu = 98.6$.
- The sample resulted in $\bar{x} = 98.20$.
- Considering the distribution of sample means, the sample size, and the magnitude of the discrepancy between 98.6 and 98.20, we find that a sample mean of 98.20 is unlikely (with less than a 5% chance) to occur if μ is really 98.6.
- There are two reasonable explanations for the sample mean of 98.20: Either a very rare event has occurred, or μ is not really 98.6. Because the probability of getting a sample mean of 98.20 (when $\mu = 98.6$) is so low, we go with the more reasonable explanation: The value of μ is not 98.6 as is commonly believed.

The preceding example illustrates well the basic line of reasoning. Carefully read it several times until you understand it. Try not to dwell on the details of

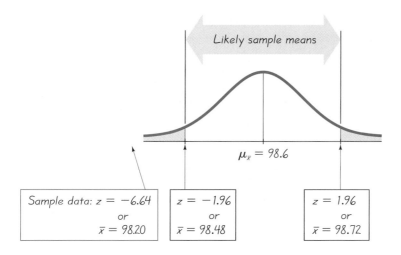

Figure 7-1
Central Limit Theorem
The Expected Distribution of
Sample Means Assuming
that $\mu = 98.6$

the calculations. Instead, focus on the key idea that although there is a common belief that $\mu = 98.6$, the sample mean is $\bar{x} = 98.20$. By using the central limit theorem, we determine that if the mean is really 98.6, then the probability of getting a sample with a mean of 98.20 is very small, which suggests that the belief that $\mu = 98.6$ should be rejected. Section 7-3 will describe the specific procedure used in hypothesis testing, but we will first describe the components of a formal **hypothesis test** or **test of significance.**

Components of a Formal Hypothesis Test

- The **null hypothesis** (denoted by H_0) is a statement about the value of a population parameter (such as the mean), and it must contain the condition of equality and must be written with the symbol $=$, $\leq$, or $\geq$. (When actually conducting the test, we operate under the assumption that the parameter *equals* some specific value.) For the mean, the null hypothesis will be stated in one of these three possible forms:

  $H_0\colon \mu =$ some value $H_0\colon \mu \leq$ some value $H_0\colon \mu \geq$ some value

  For example, the null hypothesis corresponding to the common belief that the mean body temperature is $98.6°\,F$ is expressed as $H_0\colon \mu = 98.6$. We test the null hypothesis directly in the sense that we assume it is true and reach a conclusion to either reject H_0 or fail to reject H_0.

- The **alternative hypothesis** (denoted by H_1) is the statement that must be true if the null hypothesis is false. For the mean, the alternative hypothesis will be stated in only one of three possible forms:

Comment that the special terminology used here is not unique to this book. These terms are commonly used by medical researchers, manufacturers, psychologists, educators, and many other people who use methods of statistics in their professions.

$$H_1: \mu \neq \text{some value} \qquad H_1: \mu > \text{some value} \qquad H_1: \mu < \text{some value}$$

Note that H_1 is the opposite of H_0. For example, if H_0 is given as $\mu = 98.6$, then it follows that the alternative hypothesis is given by $H_1: \mu \neq 98.6$.

Note about Using ≤ or ≥ in H_0: Even though we sometimes express H_0 with the symbol ≤ or ≥ as in $H_0: \mu \leq 98.6$ or $H_0: \mu \geq 98.6$, we conduct the test by assuming that $\mu = 98.6$ is true. We must have a single fixed value for μ so that we can work with a single distribution having a specific mean. (Some textbooks and some software packages use notation in which H_0 *always* contains only the equals symbol. Where this and many other textbooks might use $\mu \leq 98.6$ and $\mu > 98.6$ for H_0 and H_1, respectively, some others might use $\mu = 98.6$ and $\mu > 98.6$ instead.)

Note about Stating Your Own Hypotheses: If you are conducting a research study and you want to use a hypothesis test to *support* your claim, the claim must be stated in such a way that it becomes the alternative hypothesis, so it cannot contain the condition of equality. For example, if you believe that your brand of refrigerator lasts longer than the mean of 14 years for other brands, state the claim that $\mu > 14$, where μ is the mean life of your refrigerators. (In this context of trying to support the goal of the research, the alternative hypothesis is sometimes referred to as the *research hypothesis*. Also in this context, the null hypothesis is assumed true for the purpose of conducting the hypothesis test, but it is hoped that the conclusion will be rejection of the null hypothesis so that the research hypothesis is supported.)

Note about Testing the Validity of Someone Else's Claim: Sometimes we test the validity of someone else's claim, such as the claim of the Coca Cola Bottling Company that "the mean amount of Coke in cans is at least 12 oz," which becomes the null hypothesis of $H_0: \mu \geq 12$. In this context of testing the validity of someone else's claim, their original claim sometimes becomes the null hypothesis (because it contains equality), and it sometimes becomes the alternative hypothesis (because it does not contain equality).

When testing a null hypothesis, we arrive at a conclusion of rejecting it or failing to reject it. Such conclusions are sometimes correct and sometimes wrong (even if we do everything correctly). There are two different types of errors that can be made. Table 7-2 summarizes the different possibilities and shows that we make a correct decision when we either reject a null hypothesis that is false or fail to reject a null hypothesis that is true. However, we make an error when we reject a true null hypothesis or fail to reject a false null hypothesis. The type I and type II errors are described as follows.

- **Type I error:** *The mistake of rejecting the null hypothesis when it is true.* For the preceding informal example, a type I error would be the mistake of rejecting the null hypothesis that the mean body temperature is 98.6 when that mean really is 98.6. The type I error is not a miscalculation or procedural misstep; it is an actual error that can occur when a rare event happens by chance. The *probability* of rejecting the null hypoth-

Some mathematicians acknowledge only the research role of hypothesis testing and insist that *every* claim must be stated as an alternative hypothesis. However, many textbooks and real applications reflect the comments made here about testing the validity of someone else's claim. Some textbooks completely dodge the issue by wording all claims as alternative hypotheses.

Stress that with hypothesis testing, it's not simply a matter of being right or wrong. Different types of errors can have dramatically different consequences, and that is why we distinguish between type I errors and type II errors. Ask the class the difference between these two errors: (1) rejecting a perfectly good parachute and refusing to jump and (2) failing to reject a defective parachute and jumping out of a plane with it.

TABLE 7-2	Type I and Type II Errors		

		True State of Nature	
		The null hypothesis is true.	The null hypothesis is false.
Decision	We decide to reject the null hypothesis.	**Type I error** (rejecting a true null hypothesis)	Correct decision
	We fail to reject the null hypothesis.	Correct decision	**Type II error** (failing to reject a false null hypothesis)

esis when it is true is called the **significance level** and is denoted by the symbol α (alpha). The value of α is typically predetermined, and very common choices are $\alpha = 0.05$ and $\alpha = 0.01$.

- **Type II error**: *The mistake of failing to reject the null hypothesis when it is false.* For the preceding informal example, a type II error would be the mistake of failing to reject the null hypothesis ($\mu = 98.6$) when it is actually false (that is, the mean is not 98.6). The symbol β (beta) is used to represent the probability of a type II error.

The following terms are associated with key components in the hypothesis-testing procedure.

- **Test statistic**: *A sample statistic or a value based on the sample data.* A test statistic is used in making the decision about the rejection of the null hypothesis. For the data in the preceding informal example, we can use the central limit theorem to get a test statistic of $z = -6.64$ as follows:

$$z = \frac{\bar{x} - \mu_{\bar{x}}}{\frac{\sigma}{\sqrt{n}}} = \frac{98.20 - 98.6}{\frac{0.62}{\sqrt{106}}} = -6.64$$

- **Critical region**: *The set of all values of the test statistic that would cause us to reject the null hypothesis.* For the preceding informal example, the critical region is represented by the shaded part of Figure 7-1 and consists of values of the test statistic less than $z = -1.96$ or greater than $z = 1.96$.

- **Critical value:** *The value or values that separate the critical region from the values of the test statistic that would not lead to rejection of the null hypothesis.* The critical values depend on the nature of the null hypothesis, the relevant sampling distribution, and the level of significance α. For the preceding informal example, the critical values of $z = -1.96$ and $z = 1.96$ separate the shaded critical regions. See Figure 7-1.

We will now discuss some important issues in hypothesis testing.

Controlling Type I and Type II Errors   We noted that α is the probability of a type I error (rejecting a true null hypothesis) whereas β is the probability of a type II error (failing to reject a false null hypothesis). One step in our procedure for testing hypotheses involves the selection of the significance level α, which is the probability of a type I error. However, we don't select β [P(type II error)]. It would be great if we could always have $\alpha = 0$ and $\beta = 0$, but in reality that is not possible so we must attempt to manage the α and β error probabilities. Mathematically, it can be shown that α, β, and the sample size n are all related, so when you choose or determine any two of them, the third is automatically determined. We could select both α and β (the required sample size n would then be determined), but the usual practice in research and industry is to determine in advance the values of α and n, so the value of β is determined. Depending on the seriousness of a type I error, try to use the largest α that you can tolerate. For type I errors with more serious consequences, select smaller values of α. Then choose a sample size n as large as is reasonable, based on considerations of time, cost, and other such relevant factors. (Sample size determinations were discussed in Section 6-2.) The following practical considerations may be relevant:

1. For any fixed α, an increase in the sample size n will cause a decrease in β. That is, a larger sample will lessen the chance that you make the error of not rejecting the null hypothesis when it's actually false.

2. For any fixed sample size n, a decrease in α will cause an increase in β. Conversely, an increase in α will cause a decrease in β.

3. To decrease both α and β, increase the sample size.

To make sense of these abstract ideas, let's consider M&Ms (produced by Mars, Inc.) and Bufferin brand aspirin tablets (produced by Bristol-Myers Products). The M&M package contains 1498 candies. The mean weight of the individual candies should be at least 0.9085 g, because the M&M package is labeled as containing 1361 g. The Bufferin package is labeled as holding 30 tablets, each of which contains 325 mg of aspirin. (See the sample data in Data Sets 11 and 14 of Appendix B.) Because M&Ms are candies used for enjoyment whereas Bufferin tablets are drugs used for treatment of health problems, we are dealing with two very different levels of seriousness. If the M&Ms don't have a population mean weight of 0.9085 g, the consequences are not very serious, but if the Bufferin tablets don't have a mean of 325 mg

of aspirin, the consequences could be very serious. If the M&Ms have a mean that is too large, Mars will lose some money but consumers will not complain. In contrast, if the Bufferin tablets have too much aspirin, Bristol-Myers could be faced with consumer lawsuits and Federal Drug Administration actions. Consequently, in testing the claim that $\mu = 0.9085$ g for M&Ms, we might choose $\alpha = 0.05$ and a sample size of $n = 100$; in testing the claim that $\mu = 325$ mg for Bufferin tablets, we might choose $\alpha = 0.01$ and a sample size of $n = 500$. The smaller significance level α and larger sample size n are chosen because of the more serious consequences associated with a commercial drug.

Conclusions in Hypothesis Testing

We have already noted that the original claim sometimes becomes the null hypothesis and at other times becomes the alternative hypothesis. However, our procedure requires that we always test the null hypothesis. In Section 7-3 we will see that our initial conclusion will always be one of the following:

1. Fail to reject the null hypothesis H_0.
2. Reject the null hypothesis H_0.

The conclusion of failing to reject the null hypothesis or rejecting it is fine for those of us with the wisdom to take a statistics course, but we should use simple, nontechnical terms in stating what the conclusion suggests. Students often have difficulty formulating this final nontechnical statement, which describes the practical consequence of the data and computations. It's important to be precise in the language used; the implications of words such as "support" and "fail to reject" are very different. Figure 7-2 on page 354 shows how to formulate the correct wording of the final conclusion. Note that only one case leads to wording indicating that the sample data actually *support* the conclusion. If you want to justify some claim, state it in such a way that it becomes the alternative hypothesis and then hope that the null hypothesis gets rejected. For example, if you want to justify the claim that the mean body temperature is different from 98.6°F, then make the claim that $\mu \neq 98.6$. This claim will be an alternative hypothesis that will be supported if you reject the null hypothesis of H_0: $\mu = 98.6$. If, on the other hand, you claim that $\mu = 98.6$, you will either reject or fail to reject the claim; in either case, you will not *support* the original claim.

Some texts say "accept the null hypothesis" instead of "fail to reject the null hypothesis." Whether we use the term *accept* or *fail to reject*, we should recognize that *we are not proving the null hypothesis;* we are merely saying that the sample evidence is not strong enough to warrant rejection of the null hypothesis. It's like a jury's saying that there is not enough evidence to convict a suspect. The term *accept* is somewhat misleading because it seems to imply incorrectly that the null hypothesis has been proved. The phrase *fail to reject*

Students typically have some difficulty with the correct statement of the final conclusions. Stress that the final conclusion should accurately reflect the work that was done, and differences between terms such as "support" and "fail to reject" are very important. Show how Figure 7-2 can be helpful in forming the wording of the final conclusion.

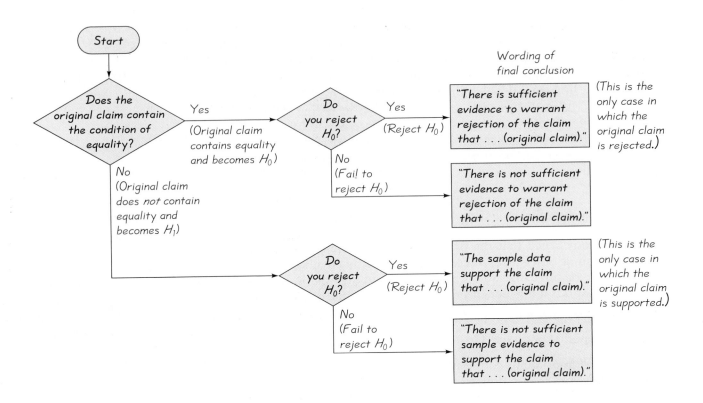

Figure 7-2 Wording of Conclusions in Hypothesis Tests

says more correctly that the available evidence isn't strong enough to warrant rejection of the null hypothesis. In this text we will use the conclusion *fail to reject the null hypothesis,* instead of *accept the null hypothesis.*

Two-Tailed, Left-Tailed, Right-Tailed

The *tails* in a distribution are the extreme regions bounded by critical values. Our informal example of hypothesis testing involved a **two-tailed test** in the sense that the critical region of Figure 7-1 is in the two extreme regions (tails) under the curve. We reject the null hypothesis H_0 if our test statistic is in the critical region because that indicates a significant discrepancy between the null hypothesis and the sample data. Some tests will be **left-tailed,** with the critical region located in the extreme left region under the curve. Other tests may be **right-tailed,** with the critical region in the extreme right region under the curve.

In two-tailed tests, the level of significance α is divided equally between the two tails that constitute the critical region. For example, in a two-tailed test with a significance level of $\alpha = 0.05$, there is an area of 0.025 in each of the two tails. In right- or left-tailed tests , the area of the critical region is α.

Students will like the hint that the symbol $<$ suggests a left-tailed test and the symbol $>$ suggests a right-tailed test. Many will want to use only that criterion, but they should also understand the reasoning used to determine whether a test is left-tailed, right-tailed, or two-tailed. They should not rely blindly on a rote process that makes no logical sense.

By examining the null hypothesis H_0, we should be able to deduce whether a test is right-tailed, left-tailed, or two-tailed. The tail will correspond to the critical region containing the values that would conflict significantly with the null hypothesis. A useful check is summarized in the margin figures, which show how the inequality sign in H_1 points in the direction of the critical region. The symbol $\neq$ is often expressed in programming languages as $< >$, and this reminds us that an alternative hypothesis such as $\mu \neq 98.6$ corresponds to a two-tailed test.

Section 7-3 will discuss formal procedures for testing hypotheses. For now, we will examine the components included in the following examples.

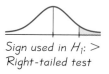

Sign used in H_1: $>$
Right-tailed test

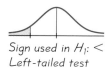

Sign used in H_1: $<$
Left-tailed test

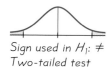

Sign used in H_1: $\neq$
Two-tailed test

EXAMPLE   *Identifying components of a hypothesis test:* After obtaining samples from the gasoline pumps at the Premium Auto Service Center, the Connecticut Consumer Protection Agency claims that consumers are being cheated because of this condition: When the meters indicate 1 gal, the mean amount of gas actually supplied is less than 1 gal.

a. Express in symbolic form the claim that the Premium Auto Service Center is cheating consumers.
b. Identify the null hypothesis H_0.
c. Identify the alternative hypothesis H_1.
d. Identify this test as being two-tailed, left-tailed, or right-tailed.
e. Identify the type I error for this test.
f. Identify the type II error for this test.
g. Assume that the conclusion is to reject the null hypothesis. State the conclusion in nontechnical terms; be sure to address the original claim.
h. Assume that the conclusion is failure to reject the null hypothesis. State the conclusion in nontechnical terms; be sure to address the original claim.

SOLUTION

a. The claim that consumers are being cheated is equivalent to claiming that the mean is less than 1 gal, which is expressed in symbolic form as $\mu < 1$ gal.
b. The original claim of $\mu < 1$ gal does not contain equality, as required by the null hypothesis. The original claim is therefore the alternative hypothesis; the null hypothesis is H_0: $\mu \geq 1$.
c. See part (b). The alternative hypothesis is H_1: $\mu < 1$.
d. This test is left-tailed because the null hypothesis is rejected if the sample mean is significantly less than (or to the left of) 1. (As a double check, note that the alternative hypothesis $\mu < 1$ contains the sign $<$, which points to the left.)   *(continued)*

e. The type I error (rejection of a true null hypothesis) is to reject the null hypothesis $\mu \geq 1$ when the population mean is really equal to or greater than 1. (This is a serious error, because the business will be charged with cheating consumers when no such cheating is actually happening.)

f. The type II error (failure to reject a false null hypothesis) is to fail to reject the null hypothesis $\mu \geq 1$ when the population mean is really less than 1. (That is, we conclude that there isn't sufficient evidence to charge cheating when cheating is actually taking place.)

g. See Figure 7-2 for the case of the original claim not containing equality, but H_0 is rejected. Conclude that there is sufficient evidence to support the claim that the mean is less than 1.

h. See Figure 7-2 for the case of the original claim not containing equality, and we fail to reject H_0. Conclude that there is not sufficient evidence to support the claim that the mean is less than 1.

EXAMPLE   *Finding critical values:* Many passengers on cruise ships wear skin patches that supply dramamine to the body for the purpose of preventing motion sickness. A claim about the mean dosage amount is tested with a significance level of $\alpha = 0.05$. The conditions are such that the standard normal distribution can be used (because the central limit theorem applies). Find the critical value(s) of z if the test is (a) two-tailed, (b) left-tailed, and (c) right-tailed.

SOLUTION

a. In a two-tailed test, the significance level of $\alpha = 0.05$ is divided equally between the two tails, so there is an area of 0.025 in each tail. We can find the critical values in Table A-2 as the values corresponding to areas of 0.4750 (found by subtracting 0.025 from 0.5) to the right and left of the mean. We get critical values of $z = -1.96$ and $z = 1.96$, as shown in Figure 7-3(a).

b. In a left-tailed test, the significance level of $\alpha = 0.05$ is the area of the critical region at the left, so the critical value corresponds to an area of 0.4500 (found from $0.5 - 0.05$) to the left of the mean. Using Table A-2, we get a critical value of $z = -1.645$, as shown in Figure 7-3(b).

c. In a right-tailed test, the significance level of $\alpha = 0.05$ is the area of the critical region to the right, so the critical value corresponds to an area of 0.4500 (found from $0.5 - 0.05$) to the right of the mean. Using Table A-2, we get a critical value of $z = 1.645$, as shown in Figure 7-3(c).

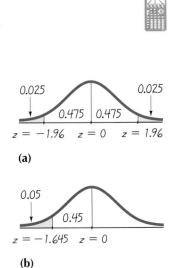

Figure 7-3 Finding Critical Values

7-2 Exercises A: Basic Skills and Concepts

In Exercises 1–8, assume that a hypothesis test of the given claim will be conducted. Use μ for a claim about a mean, p for a claim about a proportion, and σ for a claim about variation.

Recommended assignment: Exercises 1–4 and 9–16.

   a. *Express the claim in symbolic form.*
   b. *Identify the null hypothesis H_0.*
   c. *Identify the alternative hypothesis H_1.*
   d. *Identify the test as being two-tailed, left-tailed, or right-tailed.*
   e. *Identify the type I error for the test.*
   f. *Identify the type II error for the test.*
   g. *Assuming that the conclusion is to reject the null hypothesis, state the conclusion in nontechnical terms; be sure to address the original claim. (See Figure 7-2.)*
   h. *Assuming that the conclusion is failure to reject the null hypothesis, state the conclusion in nontechnical terms; be sure to address the original claim. (See Figure 7-2.)*

1. The attorney for the Food and Drug Administration claims that the Medassist Pharmaceutical Company makes cold caplets that contain amounts of acetaminophen with a mean different from the 650 mg amount indicated on the label.
2. The college Dean of Admissions claims that the times students require to earn a bachelor's degree have a mean less than five years.
3. The Republican presidential candidate claims that she is currently favored by more than 1/2 of all voters.
4. The Home Electronics Supply Company claims that its home circuit breakers trip at levels that have less variation than circuit breakers made by its major competitor, which has variation described by $\sigma = 0.4$ amp.
5. NBC claims that it gets 15% of viewers for the 10:00 P.M. to 11:00 P.M. time slot on Mondays.
6. In trying to attract new business, the Orange County Chamber of Commerce claims that the mean annual income for its region is greater than $50,000.
7. A Ford engineer claims that a new fuel-injection design increases the mean mileage on the Taurus above its current 30 mi/gal level.
8. The buyer for the New England Hospital Supply Company recommends not buying the new digital thermometers because they vary more than the old thermometers with a standard deviation of 0.06° F.

1. a. $\mu \neq 650$ mg
   b. $H_0: \mu = 650$ mg
   c. $H_1: \mu \neq 650$ mg
   d. Two-tailed
2. a. $\mu < 5$ yr
   b. $H_0: \mu \geq 5$ yr
   c. $H_1: \mu < 5$ yr
   d. Left-tailed
3. a. $p > 0.5$
   b. $H_0: p \leq 0.5$
   c. $H_1: p > 0.5$
   d. Right-tailed
4. a. $\sigma < 0.4$ amp
   b. $H_0: \sigma \geq 0.4$ amp
   c. $H_1: \sigma < 0.4$ amp
   d. Left-tailed
5. a. $p = 0.15$
   b. $H_0: p = 0.15$
   c. $H_1: p \neq 0.15$
   d. Two-tailed
6. a. $\mu > \$50,000$
   b. $H_0: \mu \leq \$50,000$
   c. $H_1: \mu > \$50,000$
   d. Right-tailed
7. a. $\mu > 30$ mi/gal
   b. $H_0: \mu \leq 30$ mi/gal
   c. $H_1: \mu > 30$ mi/gal
   d. Right-tailed
8. a. $\sigma > 0.06°F$
   b. $H_0: \sigma \leq 0.06°F$
   c. $H_1: \sigma > 0.06°F$
   d. Right-tailed

In Exercises 9–16, find the critical z values for the given conditions. In each case assume that the normal distribution applies, so Table A-2 can be used. Also, draw a graph showing the critical value and critical region.

9. Right-tailed test; $\alpha = 0.05$
10. Left-tailed test; $\alpha = 0.05$
11. Left-tailed test; $\alpha = 0.01$
12. Two-tailed test; $\alpha = 0.01$
13. Two-tailed test; $\alpha = 0.10$
14. Right-tailed test; $\alpha = 0.025$
15. Left-tailed test; $\alpha = 0.025$
16. Two-tailed test; $\alpha = 0.02$

7-2 Exercises B: Beyond the Basics

17. Someone suggests that in testing hypotheses, you can eliminate a type I error by making $\alpha = 0$. In a two-tailed test, what critical values correspond to $\alpha = 0$? If $\alpha = 0$, will the null hypothesis ever be rejected?

18. Assume that you are using a significance level of $\alpha = 0.05$ to test the claim that $\mu < 2$ and that your sample consists of 50 randomly selected scores. Find the probability of making a type II error, given that the population actually has a normal distribution with $\mu = 1.5$ and $\sigma = 1$. (*Hint:* With $H_0: \mu \geq 2$, begin by finding the values of the sample means that do not lead to rejection of H_0, then find the probability of getting a sample mean with one of those values.)   18. 0.0294

7-3 Testing a Claim about a Mean: Large Samples

In this section we formalize the informal procedure used in the preceding section. We will present three methods of testing hypotheses that appear different but are equivalent in the sense that they always lead to the same conclusions. The first procedure is the traditional method, and it will be the procedure used most often throughout the remainder of this book. The second procedure, based on *P*-values, will be referred to often, and the third procedure, based on confidence intervals, will be used in only a few selected cases.

We begin by identifying the two assumptions that apply to the methods of this section.

Section 7-2 presented the basic components of a hypothesis test. This section describes the formal procedures for testing claims made about a population mean. The examples and exercises of this section all deal with large ($n > 30$) samples, so the test statistic will be based on the normal distribution.

This section presents three methods for testing hypotheses: (1) traditional, (2) *P*-value, (3) confidence intervals. If you are making extensive use of TI-83 calculators or computer software, you might consider emphasizing the *P*-value approach instead of the traditional approach.

Assumptions for Testing a Claim about the Mean of a Single Population

1. The sample is large ($n > 30$), so the central limit theorem applies and we can use the normal distribution.

2. When applying the central limit theorem, we can use the sample standard deviation s as an estimate of the population standard deviation σ whenever σ is unknown and the sample size is large ($n > 30$).

The Traditional Method of Testing Hypotheses

Figure 7-4 on the following page summarizes the steps used in the **traditional** (or **classical**) **method** of testing hypotheses. This procedure uses the components described in Section 7-2 as part of a scheme to identify a sample result that is *significantly* different from the claimed value. The relevant sample statistic (such as $\bar{x}$) is converted to a test statistic, which we compare to a critical value. For the hypothesis tests of this section, the test statistic is a z score used to standardize the sample mean $\bar{x}$, just as the z score was used in Chapter 2 to standardize an individual score x. The test statistic required in Step 6 can be calculated as follows:

Test Statistic for Claims about μ When $n > 30$
$$z = \frac{\bar{x} - \mu_{\bar{x}}}{\frac{\sigma}{\sqrt{n}}}$$

Note that Step 7 in Figure 7-4 uses this decision criterion:

> **Reject the null hypothesis if the test statistic is in the critical region.**

> **Fail to reject the null hypothesis if the test statistic is not in the critical region.**

Section 7-2 presented an informal example of a hypothesis test; the following example formalizes that test.

EXAMPLE   Using the sample data given at the beginning of the chapter ($n = 106$, $\bar{x} = 98.20°$, $s = 0.62$) and a 0.05 significance level, test the claim that the mean body temperature of healthy adults is equal to 98.6° F. Use the traditional method by following the procedure outlined in Figure 7-4.

SOLUTION   Refer to Figure 7-4 and follow these steps.

Step 1:   The claim that the mean is equal to 98.6 is expressed in symbolic form as $\mu = 98.6$.

Step 2:   The alternative (in symbolic form) to the original claim is $\mu \neq 98.6$.

Step 3:   The statement $\mu = 98.6$ contains the condition of equality, and so it becomes the null hypothesis. We have

$$H_0: \mu = 98.6 \text{ (original claim)} \qquad H_1: \mu \neq 98.6$$

This would be a good time to remind students of the importance of good sampling techniques. If a sample is carelessly collected, it might not be a good sample and none of the methods of this chapter will apply. Data carelessly collected may be totally worthless.

The traditional method requires a decision based on a comparison of the test statistic and critical value, but the TI-83 calculator and some software packages (such as Minitab) might not include critical values with their results; they typically do provide P-values, so the P-value approach may be more suitable.

Refer to Figure 7-4 as a general model of the traditional method for testing hypotheses. Describe exactly what you expect in a solution. Recommendation: Require the statements of H_0 and H_1; a graph like Figure 7-5 that shows the critical value(s), critical region, and test statistic; work showing the computation of the test statistic; a statement of either "reject H_0" or "fail to reject H_0"; and a summary statement of the conclusion in nontechnical terms.

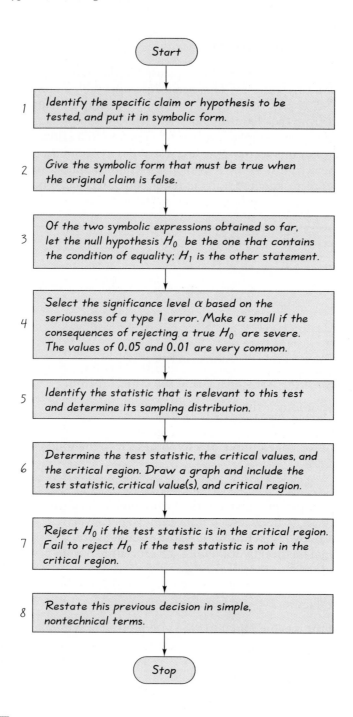

Figure 7-4 Traditional Method
of Hypothesis Testing

Step 4: As specified in the statement of the problem, the significance level is $\alpha = 0.05$.

Step 5: Because the claim is made about the population mean μ, the sample statistic most relevant to this test is $\bar{x} = 98.20$. Because $n > 30$, the central limit theorem indicates that the distribution of sample means can be approximated by a *normal* distribution.

Step 6: In calculating the test statistic, we can use $s = 0.62$ as a reasonable estimate of σ (because $n > 30$), so the test statistic of $z = -6.64$ is found by converting the sample mean of $\bar{x} = 98.20$ to $z = -6.64$ through the following computation.

$$z = \frac{\bar{x} - \mu_{\bar{x}}}{\frac{\sigma}{\sqrt{n}}} = \frac{98.20 - 98.6}{\frac{0.62}{\sqrt{106}}} = -6.64$$

The critical z values are found by first noting that the test is two-tailed because a sample mean significantly less than or greater than 98.6 is strong evidence against the null hypothesis that $\mu = 98.6$. We now divide $\alpha = 0.05$ equally between the two tails to get 0.025 in each tail. We then refer to Table A-2 (because the central limit theorem lets us assume a normal distribution) to find the z value corresponding to $0.5 - 0.025$, or 0.4750. After finding $z = 1.96$, we use the property of symmetry to conclude that the left critical value is $z = -1.96$. The test statistic, critical region, and critical values are shown in Figure 7-5.

(continued)

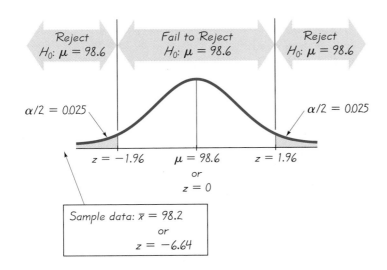

Figure 7-5 Distribution of Means of Body Temperatures Assuming $\mu = 98.6$

Step 7:   The sample mean of $\bar{x} = 98.20$ is converted to a test statistic of $z = -6.64$, which falls within the critical region, so we reject the null hypothesis.

Step 8:   To restate the Step 7 conclusion in nontechnical terms, we might refer to Figure 7-2 in the preceding section. We are rejecting the null hypothesis, which is the original claim. We conclude that there is sufficient evidence to warrant rejection of the claim that the mean body temperature of healthy adults is 98.6° F.

In the preceding example, the sample mean of 98.20 led to a test statistic of $z = -6.64$. That is, the sample mean is 6.64 standard deviations away from the claimed mean of $\mu = 98.6$. In Chapter 2 we saw that a z score of 6.64 is quite large and corresponds to an unusual result if in fact $\mu = 98.6$. This evidence is strong enough to make us believe that the mean is actually different from 98.6. The critical values and critical region clearly identify the range of unusual values that cause us to reject the claimed value of μ.

The preceding example illustrated a two-tailed hypothesis test; the following example illustrates a right-tailed test. The following example also illustrates the difference between *statistical significance* and *practical significance*.

EXAMPLE   The Jack Wilson Health Club claims in advertisements that "you will lose weight after only two days of the Jack Wilson diet and exercise program." The Dade County Bureau of Consumer Affairs conducts a test of that claim by randomly selecting 33 people who have signed up for the program. It was found that the 33 people lost an average of 0.37 lb, with a standard deviation of 0.98 lb. Use a 0.05 level of significance to test the advertised claim.

SOLUTION   We know that measured weights of people naturally fluctuate from day to day, so we need to determine whether the mean weight loss of 0.37 lb represents a significant result. That is, the health club's claim is supported if the mean weight loss is significantly greater than 0 lb. Refer to Figure 7-4 and follow these steps.

Step 1:   We express a mean weight loss significantly greater than 0 as $\mu > 0$.

Step 2:   The opposite (in symbolic form) of the original claim is $\mu \leq 0$.

Step 3:   The statement $\mu \leq 0$ contains the condition of equality, so it becomes the null hypothesis and we have

$$H_0: \mu \leq 0 \qquad H_1: \mu > 0 \qquad \text{(Original claim)}$$

Step 4:   With a 0.05 significance level, we have $\alpha = 0.05$.

Step 5:   The sample mean $\overline{x} = 0.37$ lb should be used to test a claim made about the population mean μ. Because $n > 30$, the central limit theorem indicates that the distribution of sample means can be approximated by a normal distribution.

Step 6:   In calculating the test statistic, we can use $s = 0.98$ as a reasonable estimate of σ (because $n > 30$), so the test statistic of $z = 2.17$ is computed as follows:

$$z = \frac{\overline{x} - \mu_{\overline{x}}}{\dfrac{\sigma}{\sqrt{n}}} = \frac{0.37 - 0}{\dfrac{0.98}{\sqrt{33}}} = 2.17$$

The critical value of $z = 1.645$ is found in Table A-2 as the z score corresponding to an area of 0.4500. (The test is right-tailed because the null hypothesis of $\mu \leq 0$ is rejected if the sample mean $\overline{x}$ is *greater than* 0 by a significant amount.) The test statistic, critical region, and critical value are shown in Figure 7-6.

Step 7:   From Figure 7-6 we see that the sample mean of 0.37 lb does fall within the critical region, so we reject the null hypothesis H_0.

Step 8:   There is sufficient evidence to support the claim that the mean weight loss is greater than 0 lb. That is, the program does appear to be effective. But even though the result is *statistically significant*, it does not appear to have *practical significance* because the mean weight loss of only 0.37 lb is so small.

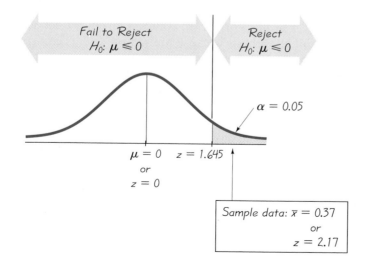

Figure 7-6   Distribution of Means of Weight Losses

In presenting the results of a hypothesis test, it is not always necessary to show all of the steps included in Figure 7-4. However, the results should include the null hypothesis, the alternative hypothesis, the calculation of the test statistic, a graph such as Figure 7-6, the initial conclusion (reject H_0 or fail to reject H_0), and the final conclusion stated in nontechnical terms. The graph should show the test statistic, critical value(s), critical region, and significance level.

The *P*-Value Method of Testing Hypotheses

Current widespread use of computers and TI-83 calculators makes *P*-values an important topic. Also, reconsidering hypothesis testing from another perspective may be helpful in understanding the general approach.

When defining *P*-value, reinforce the importance of getting a value *at least as extreme* as the one found.

Many professional articles and software packages use another approach to hypothesis testing that is based on the calculation of a probability value, or *P*-value. Given a null hypothesis and sample data, the *P*-value reflects the likelihood of getting the sample results obtained assuming that the null hypothesis is actually true. A very small *P*-value (such as 0.05 or lower) suggests that the sample results are very unlikely under the assumption of the null hypothesis, so that such a small *P*-value is evidence against the null hypothesis.

DEFINITION

A **P-value** (or **probability value**) is the probability of getting a value of the sample test statistic that is *at least as extreme* as the one found from the sample data, assuming that the null hypothesis is true.

Whereas the traditional approach results in a "reject/fail to reject" conclusion, *P*-values measure how confident we are in rejecting a null hypothesis. For example, a *P*-value of 0.0002 would lead us to reject the null hypothesis, but it would also suggest that the sample results are extremely unusual if the claimed value of μ is in fact correct. In contrast, given a *P*-value of 0.40, we fail to reject the null hypothesis because the sample results can *easily* occur if the claimed value of μ is correct.

The *P*-value approach uses most of the same basic procedures as the traditional approach, but Steps 6 and 7 are different:

Step 6:   Find the *P*-value.

Step 7:   Report the *P*-value. Some statisticians prefer to simply report the *P*-value and leave the conclusion to the reader. Others prefer to use the following decision criterion:

- *Reject the null hypothesis* if the *P*-value is less than or equal to the significance level α.

- *Fail to reject the null hypothesis* if the *P*-value is greater than the significance level α.

In Step 7, if the conclusion is based on the *P*-value alone, the following guide may be helpful.

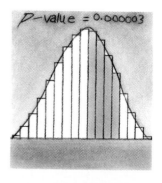

P-Value	Interpretation
Less than 0.01	Highly statistically significant Very strong evidence against the null hypothesis
0.01 to 0.05	Statistically significant Adequate evidence against the null hypothesis
Greater than 0.05	Insufficient evidence against the null hypothesis

Many statisticians consider it good practice to always select a significance level *before* doing a hypothesis test. This is a particularly good procedure when using *P*-values because we may be tempted to adjust the significance level based on the results. For example, with a 0.05 level of significance and a *P*-value of 0.06, we should fail to reject the null hypothesis, but it is sometimes tempting to say that a probability of 0.06 is small enough to warrant rejection of the null hypothesis. Other statisticians believe that prior selection of a significance level reduces the usefulness of *P*-values. They contend that no significance level should be specified and that the conclusion should be left to the reader. We will use the decision criterion that involves a comparison of a significance level and the *P*-value.

Figure 7-7 on the following page outlines key steps and decisions that lead to the *P*-value. The figure shows the following:

Right-tailed test: The *P*-value is the area to the right of the test statistic.

Left-tailed test: The *P*-value is the area to the left of the test statistic.

Two-tailed test: The *P*-value is *twice* the area of the extreme region bounded by the test statistic.

Figure 7-7 shows how to find *P*-values, and Figure 7-8 on page 367 summarizes the *P*-value method for testing hypotheses. A comparison of the traditional method (summarized in Figure 7-4) and the *P*-value method (Figure 7-8) shows that they are essentially the same, but they differ in the decision criterion. The traditional method compares the test statistic to the critical values, whereas the *P*-value method compares the *P*-value to the significance level. However, the traditional and *P*-value methods are equivalent in the sense that they will always result in the same conclusion.

Beware of *P*-Value Misuse

John P. Campbell, editor of the *Journal of Applied Psychology*, wrote the following on the subject of *P*-values. "Books have been written to dissuade people from the notion that smaller *P*-values mean more important results or that statistical significance has anything to do with substantive significance. It is almost impossible to drag authors away from their *P*-values, and the more zeros after the decimal point, the harder people cling to them." Although it might be necessary to provide a statistical analysis of the results of a study, we should place strong emphasis on the significance of the results themselves.

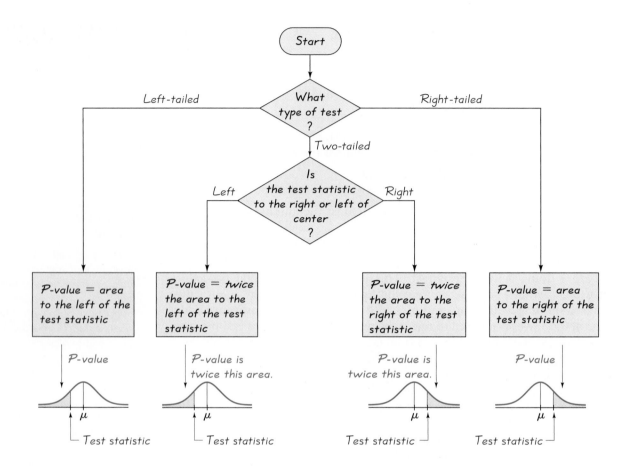

Figure 7-7   Finding *P*-Values

EXAMPLE   Use the *P*-value method to test the claim that the mean body temperature of healthy adults is equal to 98.6° F. As before, use a 0.05 significance level and the sample data summarized in the chapter-opening problem ($n = 106$, $\bar{x} = 98.20$, $s = 0.62$, a bell-shaped distribution).

SOLUTION   Except for Steps 6 and 7, this solution is the same as the one developed earlier in this section using the traditional method. Steps 1, 2, and 3 led to the following hypotheses:

$$H_0: \mu = 98.6 \text{ (original claim)} \qquad H_1: \mu \neq 98.6$$

In Steps 4 and 5 we noted that the significance level is $\alpha = 0.05$ and that the central limit theorem indicates use of the normal distribution. We now proceed to Steps 6 and 7.   *(continued)*

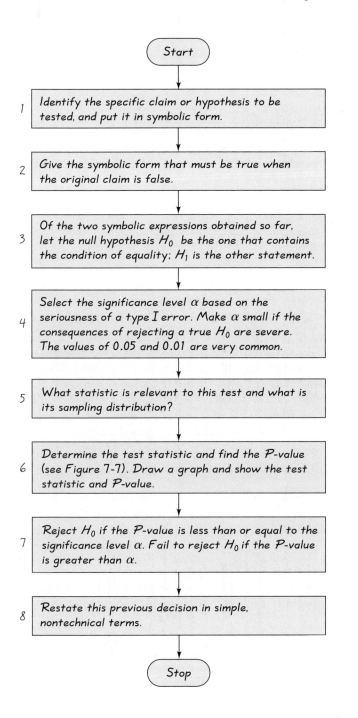

Figure 7-8 *P*-Value Method of Testing Hypotheses

Figure 7-9 *P-Value Method of Testing H_0: $\mu = 98.6$*
Because the test is two-tailed, the *P*-value is *twice* the shaded area.

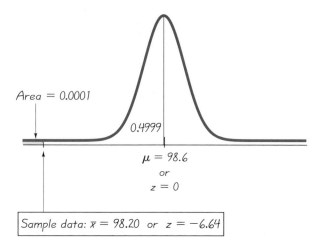

Area = 0.0001

0.4999

$\mu = 98.6$
or
$z = 0$

Sample data: $\bar{x} = 98.20$ or $z = -6.64$

Step 6:   The test statistic of $z = -6.64$ was already found in a preceding example. We can now find the *P*-value by referring to Figure 7-7. Because the test is two-tailed, the *P*-value is twice the area to the left of the test statistic $z = -6.64$. Using Table A-2, we find that the area to the left of $z = -6.64$ is 0.0001, so the *P*-value is $2 \times 0.0001 = 0.0002$ (see Figure 7-9). (Tables more precise than Table A-2 would show that the area to the left of $z = -6.64$ is actually much less than 0.0001.)

Step 7:   Because the *P*-value of 0.0002 is less than the significance level of $\alpha = 0.05$, we reject the null hypothesis.

As with the traditional method, we conclude in Step 8 that there is sufficient evidence to warrant rejection of the claim that the mean body temperature is 98.6° F.

The following example applies the *P*-value approach to the Jack Wilson diet and exercise program discussed in an earlier example in this section.

EXAMPLE   Use the *P*-value approach and a 0.05 significance level to test the claim that the Jack Wilson two-day diet and exercise program will result in a mean weight loss that is greater than 0 lb. Recall that the sample data consist of 33 subjects with a mean loss of 0.37 lb and a standard deviation of 0.98 lb.

SOLUTION   Again, refer to the steps used earlier to solve this problem by the traditional method. Steps 1, 2, and 3 resulted in the following hypotheses:

$$H_0: \mu \leq 0 \text{ lb} \qquad H_1: \mu > 0 \text{ lb (original claim)}$$

Steps 4 and 5 led to a significance level of $\alpha = 0.05$ and the decision that the normal distribution is relevant to this test of a claim about a sample mean. We now proceed to Steps 6 and 7.

Step 6:   The test statistic of $z = 2.17$ was computed in the earlier solution to this same problem. To find the P-value, we refer to Figure 7-7, which indicates that for this right-tailed test, the P-value is the area to the right of the test statistic. Referring to Table A-2, we find that the area to the right of $z = 2.17$ is $0.5 - 0.4850$, or 0.0150. The P-value is therefore 0.0150.

Step 7:   Because the P-value of 0.0150 is less than the significance level of $\alpha = 0.05$, we reject the null hypothesis that $\mu \leq 0$.

As in the previous solution of this same problem, we conclude that there is sufficient evidence to support the claim that the mean weight loss is greater than 0. The P-value of 0.0150 shows us that it is very unlikely that we would get a sample mean of 0.37 if the mean were really 0.

The next procedure for testing hypotheses is based on confidence intervals and therefore requires the concepts discussed in Section 6-2.

Testing Claims with Confidence Intervals

Let's again consider the hypothesis-testing problem described at the beginning of the chapter. We want to test the claim that the mean body temperature of healthy adults is equal to 98.6° F. Sample data consist of $n = 106$ temperatures with mean $\bar{x} = 98.20$ and standard deviation $s = 0.62$. Chapter 6 described methods of constructing confidence intervals. In particular, we used the body temperature sample data to construct the following 95% confidence interval:

$$98.08 < \mu < 98.32$$

We are 95% confident that the limits of 98.08 and 98.32 contain the population mean μ. (This means that if we were to repeat the experiment of collecting a sample of 106 body temperatures, 95% of the samples would result in confidence interval limits that actually do contain the value of the population mean μ.) This confidence interval suggests that it is very unlikely that the population mean is equal to 98.6. That is, on the basis of the confidence interval

This subsection requires Chapter 6 as a prerequisite, so skip it if Chapter 6 has not been covered. This subsection discusses the correspondence between confidence intervals and hypothesis testing, but this approach is not emphasized in the following chapters.

given here, we reject the common belief that the mean body temperature of healthy adults is 98.6° F. We can generalize this procedure as follows: First use the sample data to construct a confidence interval, and then apply the following decision criterion.

> **A confidence interval estimate of a population parameter contains the likely values of that parameter. We should therefore reject a claim that the population parameter has a value that is not included in the confidence interval.**

Using this criterion, we note that the confidence interval given here does not contain the claimed value of 98.6, and we therefore reject the claim that the population mean equals 98.6. (*Note:* We can make a direct correspondence between a confidence interval and a hypothesis test only when the test is two-tailed. A one-tailed hypothesis test with significance level α corresponds to a confidence interval with degree of confidence $1 - 2\alpha$. For example, a right-tailed hypothesis test with a 0.05 significance level corresponds to a 90% confidence interval.)

This use of confidence intervals gives us one method for identifying results that are highly unlikely, so we can determine when there is a *significant* difference between sample results and a claimed value of a parameter.

Using Computers and Calculators to Test Hypotheses

STATDISK, Minitab, and the TI-83 calculator are capable of conducting hypothesis tests for a wide variety of different circumstances. Shown on the next page are the STATDISK and Minitab displays for the test of the claim that $\mu = 98.6$, using the 106 body temperatures listed in Table 7-1. For STATDISK, select the main menu bar item `Analysis`, then select `Hypothesis Testing`, then select `Mean-One Sample`, and proceed to enter the data as requested.

For Minitab, first enter the data in column C1, then use the menu items of `Stat`, `Basic Statistics`, then `1-Sample z`, and enter the required data. The box identified as `alternative` is used to select the form of the alternative hypothesis, and it can include either `not equal`, `less than`, or `greater than`. The default of `not equal` is suitable for the test we are considering, so leave that option as is and click on OK. The Minitab results will be as shown.

If using a TI-83 calculator, press STAT, then select `TESTS` and choose the first option of `Z-Test`. You can use the original data or the summary statistics (`Stats`) by providing the entries indicated in the window display. (Because the sample is large, enter the value of $s = 0.62$ in the line requesting an entry for σ.) The TI-83 results will include the alternative hypothesis of $\mu \neq 98.6$, the test statistic of $z = -6.64$, the P-value ($p = 3.1039\text{E}^{-}11$), the sample mean, and sample size.

STATDISK DISPLAY

File   Edit   Analysis   Data   Help

Claim	$\mu = \mu_{hyp}$
Null Hypothesis	$\mu = \mu_{hyp}$
Sample Size, n	106
Sample Mean, $\bar{x}$	98.20
Sample St Dev, s	0.6229

Large Sample
Test Statistic, z	-6.6115
Critical z	±1.9600
P-Value	0.0000

95% Confidence Interval:
$98.08 < \mu < 98.32$

Reject the Null Hypothesis

Sample provides evidence to reject
the claim

MINITAB DISPLAY

```
TEST OF MU = 98.6000 VS MU N.E. 98.6000
      N      MEAN     STDEV    SE MEAN    T       P VALUE
C1    106    98.2000  0.6229   0.0605    -6.61   0.0000
```

At the beginning of this section, we noted that we are testing a claim made about the mean of a single population and that the sample size is large ($n > 30$). There are many important cases involving claims about a mean in which the sample size is small ($n \leq 30$). Such cases will be considered in the following section. In the following sections and chapters, we will apply methods of hypothesis testing to other circumstances, such as those involving claims about proportions or standard deviations or those involving more than one population. It is easy to become entangled in a complex web of steps without ever understanding the underlying rationale of hypothesis testing. The key to that understanding lies in recognizing the following concept:

When testing a claim, we make an assumption (null hypothesis) that contains equality. We then compare the assumption and the sample results and form one of the following conclusions.

- If the sample results can easily occur when the assumption is true, we attribute the relatively small discrepancy between the assumption and the sample results to chance.
- If the sample results cannot easily occur when the assumption is true, we explain the relatively large discrepancy between the assumption and the sample results by concluding that the assumption is not true.

In interpreting the results of hypothesis tests, we should also keep in mind the distinction between statistical significance and practical significance. With a sufficiently large sample, just about any difference between a sample mean and a claimed population mean becomes significant. As always, we should use common sense to interpret results.

7-3 Exercises A: Basic Skills and Concepts

Recommended assignment: Exercises 1, 5, 6, 9, 16, 20. Also consider giving a second assignment as a follow-up.

In Exercises 1–24, test the given claim using the traditional method of hypothesis testing. Also identify the P-value. Assume that all samples have been randomly selected.

1. Test the claim that the population mean $\mu = 75$, given a sample of $n = 100$ for which $\bar{x} = 78$ and $s = 15$. Test at the $\alpha = 0.05$ significance level.   1. TS: $z = 2.00$. CV: $z = \pm 1.96$.

2. Test the claim that $\mu > 750$, given a sample of $n = 36$ for which $\bar{x} = 800$ and $s = 100$. Use a significance level of $\alpha = 0.01$.   2. TS: $z = 3.00$; CV: $z = 2.33$.

3. TS: $z = -0.50$; CV: $z = -2.05$

3. Test the claim that $\mu < 2.50$, given a sample of $n = 64$ for which $\bar{x} = 2.45$ and $s = 0.80$. Use a significance level of $\alpha = 0.02$.

4. Test the claim that a population mean is different from 32.0, given a sample of $n = 75$ for which $\bar{x} = 31.8$ and $s = 0.85$. Use a significance level of $\alpha = 0.01$.   4. TS: $z = -2.04$. CV: $z = \pm 2.575$.

5. TS: $z = -0.60$; CV: $z = \pm 1.96$.

5. When home runs abound in baseball, there are often charges that the new baseballs are "juiced" to travel farther. Tests of the old balls showed that when dropped 24 ft onto a concrete surface, they bounced an average of 92.84 in. In a test of a sample of 40 new balls, the bounce heights had a mean of 92.67 in. and a standard deviation of 1.79 in. (based on data from Brookhaven National Laboratory and *USA Today*). Use a 0.05 significance level to test the claim that the new balls have bounce heights different from 92.84 in. Are these balls "juiced"?

6. TS: $z = -1.89$; CV: $z = -2.33$.

6. The Hudson Valley Bottling Company distributes root beer in bottles labeled 32 oz. The Bureau of Weights and Measures randomly selects 50 of these bottles, measures their contents, and obtains a sample mean of 31.8 oz and a sample standard deviation of 0.75 oz. Using a 0.01 significance level, test the Bureau's claim that the company is cheating consumers. Should charges be filed?

7. In a study of consumer habits, researchers designed a questionnaire to identify compulsive buyers. For a sample of consumers who identified themselves as compulsive buyers, questionnaire scores have a mean of 0.83 and a standard deviation of 0.24 (based on data from "A Clinical Screener for Compulsive Buying," by Faber and Guinn, *Journal of Consumer Research*, Vol. 19). Assume that the subjects were randomly selected and that the sample size was 32. At the 0.01 level of significance, test the claim that the self-identified compulsive-buyer population has a mean greater than 0.21, the mean for the general population. Does the questionnaire seem to be effective in identifying compulsive buyers?

7. TS: $z = 14.61$; CV: $z = 2.33$.

8. A *New York Times* article noted that the mean life span for 35 male symphony conductors was 73.4 years, in contrast to the mean of 69.5 years for males in the general population. Assuming that the 35 males have life spans with a standard deviation of 8.7 years, use a 0.05 level of significance to test the claim that male symphony conductors have a mean life span that is different from 69.5 years. (See also Exercise 13 in Section 1-3.)   8. TS: $z = 2.65$. CV: $z = \pm1.96$.

9. A study included 123 children who were wearing seat belts when injured in motor vehicle crashes. The amounts of time spent in an intensive care unit have a mean of 0.83 day and a standard deviation of 0.16 day (based on data from "Morbidity Among Pediatric Motor Vehicle Crash Victims: The Effectiveness of Seat Belts," by Osberg and Di Scala, *American Journal of Public Health*, Vol. 82, No. 3). Using a 0.01 significance level, test the claim that the seat belt sample comes from a population with a mean of less than 1.39 days, which is the mean for the population who were not wearing seat belts when injured in motor vehicle crashes. Do the seat belts seem to help?   9. TS: $z = -38.82$; CV: $z = -2.33$.

10. The mean amount of disposable per capita income in Colorado is \$13,901 (based on data from the U.S. Bureau of Economic Analysis). Tom Phelps plans to open a Cadillac car dealership and wants to verify that amount for a particular region of Colorado. He finds results from a recent survey of 200 people, with a mean of \$13,447 and a standard deviation of \$4883. At the 0.05 level of significance, test the claim that the sample was drawn from a population with a mean of \$13,901. Is there any reason Tom should be concerned about the income level in this region?

10. TS: $z = -1.31$; CV: $z = \pm1.96$.

11. The effectiveness of a test preparation course was studied for a random sample of 75 subjects who took the SAT before and after coaching. The differences between the scores resulted in a mean increase of 0.6 and a standard deviation of 3.8. (See "An Analysis of the Impact of Commercial Test Preparation Courses on SAT Scores," by Sesnowitz, Bernhardt, and Kwain, *American Education Research Journal*, Vol. 19, No. 3.) At the 0.05 significance level, test the claim that the population mean increase is greater than 0, indicating that the course is effective in raising scores. Should people take this course?   11. TS: $z = 1.37$. CV: $z = 1.645$

12. In a study of distances traveled by buses before the first major engine failure, a sampling of 191 buses resulted in a mean of 96,700 mi and a standard deviation of 37,500 mi (based on data in *Technometrics*, Vol. 22, No. 4). At the 0.05 level of significance, test the manufacturer's claim that mean distance traveled before a major engine failure is more than 90,000 mi.   12. TS: $z = 2.47$; CV: $z = 1.645$.

13. A poll of 100 randomly selected car owners revealed that the mean length of time that they plan to keep their cars is 7.01 years and the standard deviation is 3.74 years (based on data from a Roper poll). The president of the Newton Car Park is trying to plan a sales campaign targeted at car owners who are ready to buy a different car. Test the claim of the sales manager, who authoritatively states that the mean length of time all car owners plan to keep their cars is less than 7.5 years. Use a 0.05 significance level.   13. TS: $z = -1.31$; CV: $z = -1.645$

14. When 200 convicted embezzlers were randomly selected, the mean length of prison sentence was found to be 22.1 months and the standard deviation was found to be 8.6 months (based on data from the U.S. Department of Justice). Kim Patterson is running for political office on a platform of tougher treatment of convicted criminals. Test her claim that prison terms for convicted embezzlers have a mean of less than 2 years. Use a 0.05 significance level.   14. TS: $z = -3.12$; CV: $z = -1.645$.

15. The nighttime cold medicine Dozenol bears a label indicating the presence of 600 mg of acetaminophen in each fluid ounce of the drug. The Food and Drug Administration randomly selected 65 one-ounce samples and found that the mean acetaminophen content is 589 mg, whereas the standard deviation is 21 mg. Using $\alpha = 0.01$, test the claim of the Medassist Pharmaceutical Company that the population mean is equal to 600 mg. Would you buy this cold medicine?   15. TS: $z = -4.22$; CV: $z = \pm 2.575$.

16. TS: $z = -10.02$; CV: $z = -2.33$.

16. The New England Insurance Company is reviewing the driving habits of women aged 16–24 to determine whether they should continue to pay higher premiums than women in a higher age bracket. In a study of 750 randomly selected women drivers aged 16–24, the mean driving distance for one year is 6047 mi and the standard deviation is 2944 mi (based on data from the Federal Highway Administration). Use a 0.01 significance level to test the claim that the population mean for women in the 16–24 age bracket is less than 7124 mi, which is the known mean for women in the higher age bracket. If women in the 16–24 age bracket drive less, should they be charged lower insurance premiums?

17. TS: $z = 2.44$; CV: $z = 2.575$.

17. The *Late Show with David Letterman* is seen by a relatively large percentage of household members who videotape the show for viewing at a more convenient time. The show's marketing manager claims that the mean income of households with VCRs is greater than $40,000. Test that claim using a 0.005 significance level. A sample of 1700 households with VCRs produces a sample mean of $41,182 and a sample standard deviation of $19,990 (based on data from Nielsen Media Research).

18. The true value of a college degree cannot be quantitatively measured, but there are ways to measure its value in income. Men with only a high-school diploma currently have a mean annual income of $21,652. When 73 men with college degrees are randomly selected, their annual incomes have a mean of $40,202 and a standard deviation of $10,900 (based on data from the U.S. Department of Labor). Use a 0.01 level of significance to test the claim that men with college degrees have a mean annual income that is greater than the mean for men with a high-school diploma. How would you respond to someone who argues that the sample size of 73 is too small in this case?   18. TS: $z = 14.54$; CV: $z = 2.33$.

19. The mean time between failures (in hours) for a Telektronic Company radio used in light aircraft is 420 h. After 35 new radios were modified in an attempt to improve reliability, tests showed that the mean time between failures for this sample is 385 h and the standard deviation is 24 h. Use a 0.05 significance level to test the claim that the modifications improved reliability. (Note that improved reliability should result in a *longer* mean time between failures.)   19. TS: $z = -8.63$; CV: $z = 1.645$

20. As part of a campaign designed to attract farmers, the Iowa Farm Bureau claims that the state's precipitation amount "averages more than two and one-half feet each year." Use a 0.01 level of significance to test the claim that Iowa's mean annual precipitation amount is greater than 2.5 ft. For sample data, use the precipitation amounts listed in Data Set 7 of Appendix B. Is there sufficient evidence to support the advertised claim? Are there grounds for an accusation of false advertising?   20. $n = 41$, $\bar{x} = 32.47$ in., $s = 5.60$ in. TS: $z = 2.82$; CV: $z = 2.33$.

21. Refer to Data Set 1 in Appendix B for the *total* weights of garbage discarded by households in one week (based on data collected as part of the Garbage Project at the University of Arizona). For that data set, the mean is 27.44 lb and the standard deviation is 12.46 lb. At the 0.01 level of significance, test the claim of the city of Providence supervisor that the mean weight of all garbage discarded by households each week is less than 35 lb, the amount that can be handled by the town. Based on the result, is there any cause for concern that there might be too much garbage to handle?   21. TS: $z = -4.78$; CV: $z = -2.33$.

22. If we refer to the weights (in grams) of quarters listed in Data Set 13 in Appendix B, we find 50 weights with a mean of 5.622 g and a standard deviation of 0.068 g. The U.S. Department of the Treasury claims that the procedure it uses to mint quarters yields a mean weight of 5.670 g. Use a 0.01 significance level to test the claim that the mean weight of quarters in circulation is 5.670 g. If the claim is rejected, what is a possible explanation for the discrepancy?   22. TS: $z = -4.99$; CV: $z = \pm 2.575$.

23. Analysis of the last digits in data sometimes reveals whether the data have been accurately measured and reported. If the last digits are uniformly distributed from 0 through 9, the mean should be 4.5. The last digits in the quoted lengths (in miles) of 141 rivers are used to test the claim that they come from a population with a mean of 4.5. When Minitab is used to test

that claim, the display is as shown here (based on data from "Distributions of Final Digits in Data," by Preece, *The Statistician*, Vol. 30). Using a 0.05 significance level, interpret the Minitab results.

```
TEST OF MU  =  4.500  VS  MU N.E.  4.500
  N       MEAN      STDEV    SE MEAN    T      P VALUE
C1 141    2.319     2.899    0.244    -8.93    0.0000
```

24. A package of M&M plain candies is labeled as containing 1361 g, and there are 1498 candies, so the mean weight of the individual candies should be 1361/1498, or 0.9085 g. In a test to determine whether consumers are being cheated, a sample of 33 brown M&Ms is randomly selected. (See Data Set 11 in Appendix B.) When the 33 weights are used with Minitab, the display is as shown here. Interpret those results.

```
TEST OF MU  =  0.90850  VS  MU  <  0.90850
  N       MEAN      STDEV    SE MEAN    T      P VALUE
C1 33    .91282    .03952   .00688    0.63     0.73
```

7-3 Exercises B: Beyond the Basics

25. A journal article reported that a null hypothesis of $\mu = 100$ was rejected because the P-value was less than 0.01. The sample size was given as 62, and the sample mean was given as 103.6. Find the largest possible standard deviation.   25. 11.0

26. In Exercise 17, find the smallest sample mean above $40,000 that will support the claim that the mean is greater than $40,000. (Use the same sample size and sample standard deviation.)   26. $41,248.44

27. For a given hypothesis test, the probability α of a type I error is fixed, whereas the probability β of a type II error depends on the particular value of μ that is used as an alternative to the null hypothesis. For hypothesis tests of the type found in this section, we can find β as follows:

Step 1: Find the value(s) of $\bar{x}$ that correspond to the critical value(s). In

$$z = \frac{\bar{x} - \mu_{\bar{x}}}{\sigma_{\bar{x}}}$$

substitute the critical value(s) for z, enter the values for $\mu_{\bar{x}}$ and $\sigma_{\bar{x}}$, and solve for $\bar{x}$.

Step 2: Given a particular value of μ that is an alternative to the null hypothesis H_0, draw the normal curve with this new value of μ at the center. Also plot the value(s) of $\bar{x}$ found in Step 1.

Step 3: Refer to the graph in Step 2 and find the area of the new critical region bounded by $\bar{x}$. This is the probability of rejecting the null hypothesis, given that the new value of μ is correct.   *(continued)*

Step 4:   The value of β is 1 minus the area from Step 3. This is the probability of failing to reject the null hypothesis, given that the new value of μ is correct.

The preceding steps allow you to find the probability of failing to reject H_0 when it is false. You are determining the area under the curve that excludes the critical region in which you reject H_0; this area corresponds to a failure to reject a false H_0, because we use a particular value of μ that goes against H_0. Refer to the body-temperature example discussed in this section and find the value of β corresponding to the following:

    a. $\mu = 98.7$ **b.** $\mu = 98.4$ b. $\beta = 0.0868$ 27. a. $\beta = 0.6178$

28. The *power* of a test is $1 - \beta$, the probability of rejecting a false null hypothesis. Refer to the example in this section that addressed the claim made by the Jack Wilson Health Club. If the test of that claim has a power of 0.8, find the mean μ (see Exercise 27).   28. $\mu = 0.42393$ lb

7-4 Testing a Claim about a Mean: Small Samples

The hypothesis-testing examples and exercises in Sections 7-2 and 7-3 all use large samples in tests of claims about means, so they all allow the use of the normal distribution. For those large-sample cases, we can use the central limit theorem to conclude that sample means are normally distributed, regardless of the distribution of the original population. However, we cannot use the central limit theorem when the samples are small. Small sample cases were first considered in Section 6-3, where we used the Student t distribution in developing confidence interval estimates of a population mean μ. Figure 7-10 on the following page, based on the same reasoning used in Section 6-3, shows that some claims about means are tested with a normal distribution and some are tested with a Student t distribution. Figure 7-10 summarizes the decisions to be made in choosing between the normal and Student t distributions.

    Figure 7-10 summarizes these observations:

1. According to the central limit theorem, if we obtain large ($n > 30$) samples (from any population with any distribution), the distribution of the sample means can be approximated by the normal distribution.

2. When we obtain samples (of any size) from a population with a normal distribution, the distribution of the sample means will be approximately normal with mean μ and standard deviation $\sigma/\sqrt{n}$. In a hypothesis test, the value of μ corresponds to the null hypothesis, and the value of the population standard deviation σ must be known. If σ is unknown and the samples are large, we can use the sample standard deviation s as a substitute for σ because large random samples tend to be representative of the populations from which they come.

If Chapter 6 has been covered, point out that this section uses the same t distribution introduced there. Also, the criteria for choosing between a normal distribution and a Student t distribution are the same in this chapter as they are in Chapter 6.

Figure 7-10 Choosing Between the Normal and Student t Distributions When Testing a Claim about the Population Mean μ

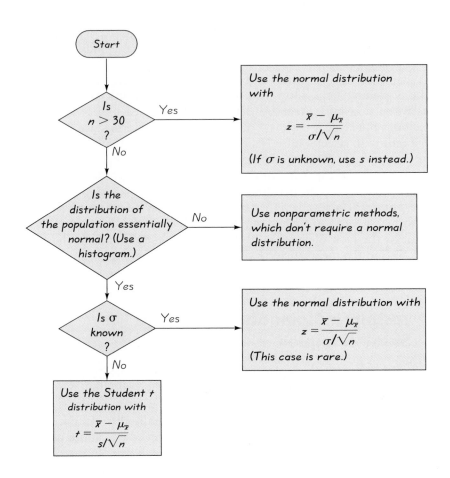

3. The conditions for using the Student t distribution are as follows:

   a. The sample is small ($n \leq 30$); and

   b. σ is unknown; and

   c. The parent population has a distribution that is essentially normal.

4. If our random samples are small, σ is unknown, and the population distribution is grossly nonnormal, we cannot use the methods of this chapter. Instead, we can use nonparametric methods, some of which are discussed in Chapter 13.

When the conditions listed in item 3 are satisfied, we use the Student t distribution with the test statistic and critical values described as follows.

Test Statistic for Claims about μ When $n \le 30$ and σ Is Unknown

If a population is essentially normal, then the distribution of

$$t = \frac{\overline{x} - \mu_{\overline{x}}}{\frac{s}{\sqrt{n}}}$$

is essentially a *Student t distribution* for all samples of size n. (The Student t distribution is often referred to as the t *distribution*.)

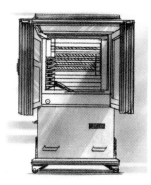

Critical Values

1. Critical values are found in Table A-3.
2. Degrees of freedom = $n - 1$.
3. After finding the number of degrees of freedom, refer to Table A-3 and locate that number in the column at the left. With a particular row of t values now identified, select the critical t value that corresponds to the appropriate column heading. If a critical t value is located at the left tail, be sure to make it negative.

Important Properties of the Student t Distribution

1. The Student t distribution is different for different sample sizes (see Figure 6-5 in Section 6-3).
2. The Student t distribution has the same general bell shape as the standard normal distribution; its wider shape reflects the greater variability that is expected with small samples.
3. The Student t distribution has a mean of $t = 0$ (just as the standard normal distribution has a mean of $z = 0$).
4. The standard deviation of the Student t distribution varies with the sample size and is greater than 1 (unlike the standard normal distribution, which has $\sigma = 1$).
5. As the sample size n gets larger, the Student t distribution gets closer to the standard normal distribution. For values of $n > 30$, the differences are so small that we can use the critical z values instead of developing a much larger table of critical t values. (The values in the bottom row of Table A-3 are equal to the corresponding critical z values from the standard normal distribution.)

The following example involves a small sample drawn from a normally distributed population for which the standard deviation σ is not known. These conditions require that we use the Student t distribution.

EXAMPLE   The seven scores listed here are axial loads (in pounds) for the first sample of seven 12-oz aluminum cans (see Data Set 15 in Appendix B). An axial load of a can is the maximum weight supported by its sides, and it must be greater than 165 pounds, because that is the maximum pressure applied when the top lid is pressed into place. At the 0.01 level of significance, test the claim of the engineering supervisor that this sample comes from a population with a mean that is greater than 165 pounds.

$$270 \quad 273 \quad 258 \quad 204 \quad 254 \quad 228 \quad 282$$

SOLUTION   Using the given sample data, we apply procedures of Chapter 2 to find that $n = 7$, $\bar{x} = 252.7$, and $s = 27.6$. The mean of 252.7 does seem to be well above the required value of 165, but with only seven scores, do we really have enough evidence to support the supervisor's claim? Let's find out by conducting a formal hypothesis test with the same steps outlined in Figure 7-4.

As in the preceding section, Steps 1, 2, and 3 result in the following null and alternative hypotheses:

$$H_0: \mu \leq 165$$
$$H_1: \mu > 165 \quad \text{(supervisor's claim)}$$

Step 4:   The significance level is $\alpha = 0.01$.

Step 5:   In this test of a claim about the population mean, the most relevant statistic is the sample mean. Referring to Figure 7-10, we see that the sample is small (because $n = 7$ and does not exceed 30), it's reasonable to conclude that the population distribution is normal (because we're dealing with physical measurements of a product made under standard conditions), and σ is unknown. Figure 7-10 shows that we should use the Student t distribution (not the normal distribution).

Step 6:   The test statistic is

$$t = \frac{\bar{x} - \mu_{\bar{x}}}{\frac{s}{\sqrt{n}}} = \frac{252.7 - 165}{\frac{27.6}{\sqrt{7}}} = 8.407$$

The critical value of $t = 3.143$ is found by referring to Table A-3. First locate $n - 1 = 6$ degrees of freedom in the column at the left. Because this test is right-tailed with $\alpha = 0.01$, refer to the column with the heading of 0.01 (one tail). The critical value

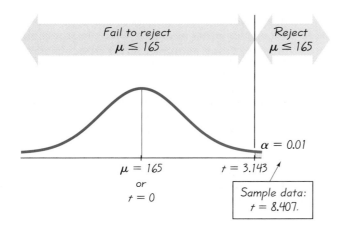

Figure 7-11 *t*-Test of Claim that $\mu > 165$

of $t = 3.143$ corresponds to 6 degrees of freedom and $\alpha = 0.01$ in one tail. The test statistic and critical value are shown in Figure 7-11.

Step 7:   Because the test statistic of $t = 8.407$ falls in the critical region, we reject H_0.

Step 8:   (Refer to Figure 7-2 for help in wording the final conclusion.) There is sufficient evidence to support the supervisor's claim that the sample comes from a population with a mean greater than the required 165 pounds.

The critical value in the preceding example was $t = 3.143$, but if the sample had been large ($n > 30$), the critical value would have been $z = 2.33$. The larger Student t critical value shows that with a small sample, the sample evidence must be *more extreme* before we consider the difference to be significant.

P-Values

The preceding example followed the traditional approach to hypothesis testing, but STATDISK, Minitab, the TI-83 calculator, and much of the literature will display P-values. Because the t distribution table (Table A-3) includes only selected values of the significance level α, we cannot usually find the specific P-value from Table A-3. Instead, we can use that table to identify limits that contain the P-value. In the last example we found the test statistic to be $t = 8.407$, and we know that the test is one-tailed with 6 degrees of freedom. By examining the row of Table A-3 corresponding to 6 degrees of freedom, we see that the test statistic of 8.407 exceeds the largest critical value in that row. Although we cannot pinpoint an exact P-value from Table A-3, we can

conclude that it must be less than 0.005. That is, we conclude that *P*-value < 0.005. (Some calculators and computer programs allow us to find exact *P*-values. For the preceding example, STATDISK displays a *P*-value of 0.00008, Minitab displays a *P*-value of 0.0001, and the TI-83 calculator displays a *P*-value of 0.000077613518.) With a significance level of 0.01 and a *P*-value less than 0.005, we reject the null hypothesis (because the *P*-value is less than the significance level), as we did with the traditional method in the preceding example.

EXAMPLE   Use Table A-3 to find the *P*-value corresponding to these results: The Student *t* distribution is used in a two-tailed test with a sample of *n* = 10 scores, and the test statistic is found to be *t* = 2.567.

SOLUTION   Refer to the row of Table A-3 with 9 degrees of freedom and note that the test statistic of 2.567 falls between the critical values of 2.821 and 2.262. Because the test is two-tailed, we consider the values of α at the top that are identified with "two tails." The critical values of 2.821 and 2.262 correspond to 0.02 (two tails) and 0.05 (two tails), so we express the *P*-value as follows.

$$0.02 < P\text{-value} < 0.05$$

With a true null hypothesis, the chance of getting a sample mean (from 10 sample values) that converts to a test statistic of *t* = 2.567 is somewhere between 0.02 and 0.05.

EXAMPLE   Refer to the 11 amounts of acid rain nitrate deposits listed for Massachusetts (MA) in Data Set 6 from Appendix B (based on data from the U.S. Department of Agriculture). Use a 0.05 significance level to test the claim that the sample comes from a population with a mean equal to 5 kg/hectare.

SOLUTION   In this solution we use the *P*-value approach based on results from a TI-83 calculator. First enter the 11 sample values in list L1, then press STAT and select TESTS. Now choose T-Test and proceed to enter the data required. The TI-83 results will include p = 0.0075851732. (STATDISK or Minitab could also be used to find the *P*-value.) Because this *P*-value is less than the significance level of 0.05, we reject the null hypothesis that the population mean equals 5. There is sufficient evidence to warrant rejection of the claim that the mean equals 5 kg/hectare.

So far, we have discussed tests of hypotheses made about population means only. In the next section we will test hypotheses made about population pro-

portions or percentages, and the last section will consider claims made about standard deviations or variances.

7-4 Exercises A: Basic Skills and Concepts

In Exercises 1 and 2, find the critical t values for the given hypotheses, sample sizes, and significance levels.

1. a. H_0: $\mu = 98.6$
      $n = 7$
      $\alpha = 0.05$

   b. H_0: $\mu \geq 100$
      $n = 12$
      $\alpha = 0.05$

   c. H_1: $\mu > 32$
      $n = 9$
      $\alpha = 0.01$

2. a. H_0: $\mu \leq 1.07$
      $n = 5$
      $\alpha = 0.01$

   b. H_1: $\mu < 75.2$
      $n = 14$
      $\alpha = 0.05$

   c. H_1: $\mu \neq 64$
      $n = 10$
      $\alpha = 0.10$

In Exercises 3 and 4, assume that the sample is randomly selected from a population with a normal distribution. Test the given claim by using the traditional method of testing hypotheses.

3. Use a significance level of $\alpha = 0.05$ to test the claim that $\mu \neq 64.8$. The sample data consist of 12 scores for which $\overline{x} = 59.8$ and $s = 8.7$.

4. Use a significance level of $\alpha = 0.01$ to test the claim that $\mu < 927$. The sample data consist of 10 scores for which $\overline{x} = 874$ and $s = 57.3$.

5. If we use Minitab for the 11 amounts of acid rain *sulfate* deposits listed for Pennsylvania (PA) in Data Set 6 from Appendix B (based on data from the U.S. Department of Agriculture), and we test the claim that the sample comes from a population with a mean greater than 12.00 kg/hectare, the display will be as shown here. Use a 0.01 significance level and interpret the Minitab results from the hypothesis test.   5. TS: $t = 2.23$. *P*-value: 0.025

```
Test of mu = 12.000 vs mu > 12.000
Variable  N    Mean    StDev  SE Mean   T     P-value
C1        11  13.286   1.910   0.576   2.23   0.025
```

6. Refer to the Minitab display in Exercise 5. If the claim is changed from "greater than 12.00" to "not equal to 12.00," how are the values in the bottom row affected?   6. Only the *P*-value changes; it doubles.

For hypothesis tests in Exercises 7–24, use the traditional approach summarized in Figure 7-4. Draw a graph showing the test statistic and critical values. In each case, assume that the population has a distribution that is approximately normal and that the sample is randomly selected. Caution: Some of the exercises require the use of the normal distribution (as described in the preceding section) instead of the Student t distribution (as described in this

Recommended assignment: Exercises 1, 3, 5, 7, 10, 12, 20. These exercises use a normal distribution instead of a *t* distribution: 10, 13, 17, 20.

1. a. ±2.447
   b. −1.796
   c. 2.896
2. a. 3.747
   b. −1.771
   c. ±1.833

3. TS: $t = -1.991$;
   CV: $t = \pm 2.201$
4. TS: $t = -2.925$;
   CV: $t = -2.821$

section); be sure to check the conditions to determine which distribution is appropriate.

7. The Carolina Tobacco Company advertised that its best-selling cigarettes contain at most 40 mg of nicotine, but *Consumer Advocate* magazine ran tests of 10 randomly selected cigarettes and found that $\bar{x}$ = 43.3 mg and s = 3.8 mg. Other evidence suggests that the distribution of nicotine content is a normal distribution. The sample is small because the laboratory work required to extract the nicotine is time consuming and expensive. It's a serious matter to charge that the company advertising is wrong, so the magazine editor chooses a significance level of α = 0.01 in testing her belief that the mean nicotine content is greater than 40 mg. Using a 0.01 significance level, test the editor's belief that the mean is greater than 40 mg.   7. TS: t = 2.746; CV: t = 2.821.

8. TS: t = −1.320; CV: t = −2.718

8. The expense of moving the storage yard for the Consolidated Package Delivery Service (CPDS) is justified only if it can be shown that the daily mean travel distance will be less than 214 mi. In trial runs of 12 delivery trucks, the mean and standard deviation are found to be 198 mi and 42 mi, respectively. At the 0.01 level of significance, test the claim that the mean is less than 214 mi. Should the storage yard be moved?

9. TS: t = −3.214; CV: t = ±2.064

9. Refer to Data Set 2 in Appendix B. Using only the first 25 body temperatures listed for 12 A.M. of Day 2, test the claim that the mean body temperature of all healthy adults is equal to 98.6° F. For the level of significance, use α = 0.05. For the first 25 scores, $\bar{x}$ = 98.24 and s = 0.56.

10. TS: z = −3.00; CV: z = ±1.96

10. Refer to Data Set 2 in Appendix B. Using only the first 35 body temperatures listed for 12 A.M. of Day 2, test the claim that the mean body temperature of all healthy adults is equal to 98.6° F. Use a 0.05 significance level. For the first 35 scores, $\bar{x}$ = 98.27 and s = 0.65.

11. TS: t = 3.344; CV: t = 2.718.

11. For each of 12 organizations, the cost of operation per client was found. The 12 scores have a mean of $2133 and a standard deviation of $345 (based on data from "Organizational Communication and Performance," by Snyder and Morris, *Journal of Applied Psychology,* Vol. 69, No. 3). At the 0.01 significance level, test the claim of a stockholder who complains that the mean for all such organizations exceeds $1800 per client.

12. TS: t = −2.062; CV: t = −2.921

12. The skid properties of a snow tire have been tested, and a mean skid distance of 154 ft has been established for standardized conditions. A new, more expensive tire is developed, but tests on a sample of 17 new tires yield a mean skid distance of 148 ft with a standard deviation of 12 ft. Because of the cost involved, the new tires will be purchased only if it can be shown at the α = 0.005 significance level that they skid less than the current tires. Based on the sample, will the new tires be purchased?

13. TS: z = 1.00; CV: z = 1.96

13. The Bank of New England is concerned about the amount of debt being accrued by customers using its credit cards. The Board of Directors voted to institute an expensive monitoring system if the mean for all of the bank's customers is greater than $2000. The bank randomly selected 50 credit-card holders and determined the amounts they charged. For this sample group, the mean is $2177 and the standard deviation is $1257.

(continued)

Use a 0.025 level of significance to test the claim that the mean amount charged is greater than $2000. Based on the result, will the monitoring system be implemented?

14. A study was conducted to determine whether a standard clerical test would need revision for use on video display terminals (VDTs). The VDT scores of 22 subjects have a mean of 170.2 and a standard deviation of 35.3 (based on data from "Modification of the Minnesota Clerical Test to Predict Performance on Video Display Terminals," by Silver and Bennett, *Journal of Applied Psychology*, Vol. 72, No. 1). At the 0.05 level of significance, test the claim that the mean for all subjects taking the VDT test differs from the mean of 243.5 for the standard printed version of the test. Based on the result, should the VDT test be revised?

14. TS: $t = -9.740$; CV: $t = \pm 2.080$

15. In a study of factors affecting hypnotism, visual analogue scale (VAS) sensory ratings were obtained for 16 subjects. For these sample ratings, the mean is 8.33 while the standard deviation is 1.96 (based on data from "An Analysis of Factors That Contribute to the Efficacy of Hypnotic Analgesia," by Price and Barber, *Journal of Abnormal Psychology*, Vol. 96, No. 1). At the 0.01 level of significance, test the claim that this sample comes from a population with a mean rating of less than 10.00.

15. TS: $t = -3.408$; CV: $t = -2.602$.

16. The following reading test results were obtained for a sample of 15 third-grade students: $\bar{x} = 31.0$, $s = 10.5$. (The data are based on "A Longitudinal Study of the Effects of Retention/Promotion on Academic Achievement," by Peterson et al., *American Educational Research Journal*, Vol. 24, No. 1.) Does this third-grade sample mean differ significantly from a first-grade population mean of 41.9? Assume a 0.01 level of significance. 16. TS: $t = -4.021$; CV: $t = \pm 2.977$

17. Kim Greco is a high-school senior who is concerned about attending college because she knows that many college students require more than 4 years to earn a bachelor's degree. At the 0.10 level of significance, test the claim of her guidance counselor, who states that the mean time is greater than 5 years. Sample data consist of a mean of 5.15 years and a standard deviation of 1.68 years for 80 randomly selected college graduates (based on data from the National Center for Education Statistics).

17. TS: $z = 0.80$; CV: $z = 1.28$

18. Using the weights of the Bufferin tablets in Data Set 14 of Appendix B, test the claim that the mean weight is equal to 650 mg. Use a 0.05 significance level. For the sample data, $\bar{x} = 665.41$ and $s = 7.26$.

18. TS: $t = 11.626$; CV: $t = \pm 2.045$

19. Using the weights of only the *blue* M&Ms listed in Data Set 11 of Appendix B, test the claim that the mean is at least 0.9085 g, the mean value necessary for the 1498 M&Ms to produce a total of 1361 g as the package indicates. Use a 0.05 significance level. For the blue M&Ms, $\bar{x} = 0.9014$ g and $s = 0.0573$ g. Based on the result, can we conclude that the package contents do not agree with the claimed weight printed on the label?

19. TS: $t = -0.277$; CV: $t = -2.132$

20. Using the weights of only the *brown* M&Ms listed in Data Set 11 of Appendix B, test the claim that the mean is greater than 0.9085 g, the mean value necessary for the 1498 M&Ms to produce a total of 1361 g as the package indicates. Use a 0.05 significance level. For the brown

M&Ms, $\bar{x} = 0.9128$ g and $s = 0.0395$ g. Based on the result, can we conclude that the packages contain more than the claimed weight printed on the label?   20. TS: $z = 0.63$; CV: $z = 1.645$

21. Listed here are the total electric energy consumption amounts (in kWh) for the author's home during seven different years.

$$11,943 \quad 11,463 \quad 10,789 \quad 9907 \quad 9012 \quad 9942 \quad 11,153$$

The utility company claims that the mean annual consumption amount is 11,000 kWh and offers a budget payment plan based on that amount. At the 0.05 significance level, test the utility company's claim that the mean is equal to 11,000 kWh.

22. Given here are the birth weights (in kilograms) of male babies born to mothers on a special vitamin supplement (based on data from the New York State Department of Health). At the 0.05 level of significance, test the claim that the mean birth weight for all male babies of mothers given vitamins is equal to 3.39 kg, which is the mean for the population of all males. Based on the result, does the vitamin supplement appear to have an effect on birth weight?

$$3.73 \quad 4.37 \quad 3.73 \quad 4.33 \quad 3.39 \quad 3.68 \quad 4.68 \quad 3.52$$
$$3.02 \quad 4.09 \quad 2.47 \quad 4.13 \quad 4.47 \quad 3.22 \quad 3.43 \quad 2.54$$

23. Rita Gibbons is a stand-up comedian who videotapes her performances and records the total of the times she must wait for audience laughter to subside. Given here are the times (in seconds) for 15 different shows in which she used a new routine. Test the claim that the mean time is greater than 63.2 s, the mean time for her old routine. Based on the results, does her new routine seem to be better than the old one?

$$86 \quad 45 \quad 44 \quad 78 \quad 52 \quad 79 \quad 86 \quad 66 \quad 61 \quad 57 \quad 98 \quad 44 \quad 61 \quad 99 \quad 87$$

24. Refer to the nicotine contents of cigarettes listed in Data Set 4 of Appendix B. Use the sample data to test the claim that the population mean is less than 1.0 mg. Does the conclusion apply to the mean nicotine level of all 100 mm cigarettes (not menthol or light) smoked by consumers? Why or why not?   24. $n = 29, \bar{x} = 0.94, s = 0.31.$ TS: $t = -1.007$; CV: $t = -1.701$

7-4 Exercises B: Beyond the Basics

25. For each given hypothesis test, what can you conclude about the P-value by using only Tables A-2 and A-3?

    a. Exercise 7   **b.** Exercise 8   **c.** Exercise 9   **d.** Exercise 10

26. Because of certain conditions, a hypothesis test requires the Student t distribution, as described in this section. Assume that the standard normal distribution was incorrectly used instead. Does using the standard normal

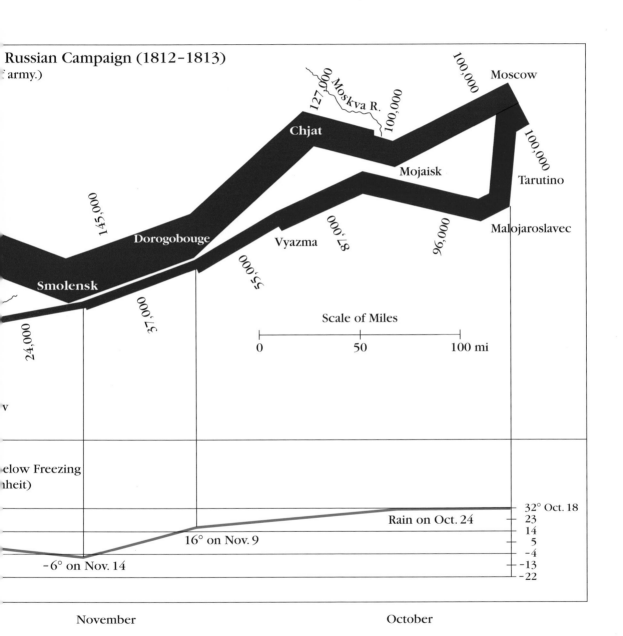

Russian Campaign (1812–1813)
army.)

127,000

Moskva R.

100,000

100,000

Moscow

Chjat

100,000

145,000

Dorogobouge

Mojaisk

100,000

Smolensk

Vyazma

87,000

96,000

Tarutino

55,000

Malojaroslavec

37,000

24,000

Scale of Miles

0   50   100 mi

v

elow Freezing
heit)

32° Oct. 18

23

14

Rain on Oct. 24

16° on Nov. 9

5
-4

-6° on Nov. 14

-13
-22

November

October

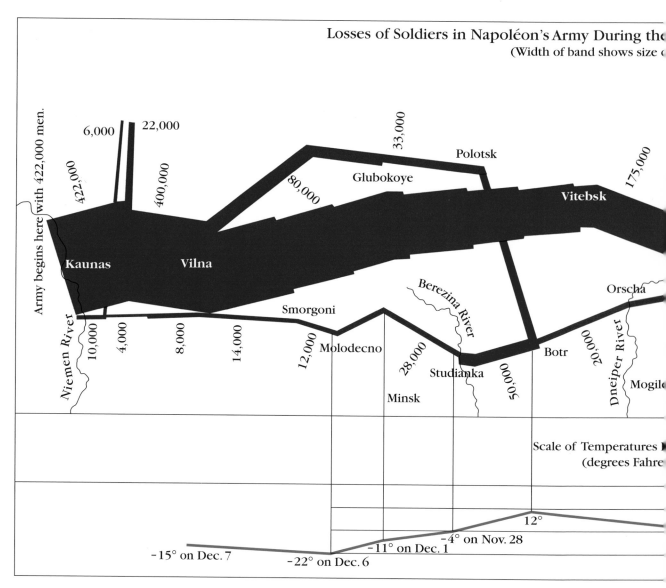

Losses of Soldiers in Napoléon's Army During the
(Width of band shows size of

6,000   22,000

33,000

Polotsk

Glubokoye

Vitebsk

80,000

175,000

Army begins here with 422,000 men.

422,000

400,000

Kaunas   Vilna

Smorgoni

Berezina River

Orscha

Botr

Niemen River

10,000   4,000   8,000   14,000   12,000   Molodecno

Minsk

28,000

Studianka

50,000

20,000

Dneiper River

Mogil

Scale of Temperatures
(degrees Fahre

12°

−4° on Nov. 28

−11° on Dec. 1

−15° on Dec. 7

−22° on Dec. 6

December

Credit: Edward R. Tufte, *The Visual Display of Quantitative Information*
(Cheshire, CT: Graphics Press, 1983). Reprinted with permission.

distribution make you more or less likely to reject the null hypothesis, or does it not make a difference? Explain.   26. More likely to reject the null hypothesis.

27. When finding critical values, we sometimes need significance levels other than those available in Table A-3. Some computer programs approximate critical *t* values by

$$t = \sqrt{\text{df} \cdot (e^{A^2/\text{df}} - 1)}$$

where

$$\text{df} = n - 1$$
$$e = 2.718$$
$$A = z\left(\frac{8\,\text{df} + 3}{8\,\text{df} + 1}\right)$$

and *z* is the critical *z* score. Use this approximation to find the critical *t* score corresponding to $n = 10$ and a significance level of 0.05 in a right-tailed case. Compare the results to the critical *t* value found in Table A-3.

28. Refer to Exercise 7 and assume that you're testing the claim that $\mu > 40$ mg. Find β (the probability of a type II error), given that the actual value of the population mean is $\mu = 45.0518$ mg. (See Exercise 27 from Section 7-3.)   28. $\beta = 0.10$

27. With $z = 1.645$, the table and the approximation both result in $t = 1.833$.

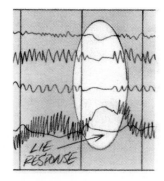

7-5 Testing a Claim about a Proportion

In the preceding sections of this chapter we introduced the basic methods of hypothesis testing, but they were used to address claims made about population *means* only. In this section we see how to apply those same basic methods to claims about population *proportions*. Using the methods of this section, we can test a claim about a proportion, percentage, or probability, as illustrated in these examples:

- Based on a sample survey, fewer than 1/4 of all college graduates smoke.
- The percentage of late-night television viewers who watch *The Late Show with David Letterman* is equal to 18%.
- If a fatal car crash occurs, there is a 0.44 probability that it involves a driver who had been drinking.

Assumptions Used When Testing a Claim about a Population Proportion, Probability, or Percentage

1. The conditions for a *binomial experiment* are satisfied. That is, we have a fixed number of independent trials having constant probabilities, and each trial has two outcome categories, which we classify as "success" and "failure."

2. The conditions $np \geq 5$ and $nq \geq 5$ are both satisfied, so **the binomial distribution of sample proportions can be approximated by a normal distribution with $\mu = np$ and $\sigma = \sqrt{npq}$** (as described in Section 5-6).

Lie Detectors

Why not require all criminal suspects to take lie detector tests and dispense with trials by jury? The Council of Scientific Affairs of the American Medical Association states, "It is established that classification of guilty can be made with 75% to 97% accuracy, but the rate of false positives is often sufficiently high to preclude use of this (polygraph) test as the sole arbiter of guilt or innocence." A "false positive" is an indication of guilt when the subject is actually innocent. Even with accuracy as high as 97%, the percentage of false positive results can be 50%, so half of the innocent subjects incorrectly appear to be guilty.

By now, a pattern for testing hypotheses should be clear. Point out that this section does not introduce dramatically different concepts. Instead, it shows how to adapt the same basic procedure to testing claims about proportions. These are the key questions to be posed in attempting such an adaptation: What distribution applies? What test statistic is appropriate? How are the critical values found?

If these assumptions are not all satisfied, we may be able to use other methods. In this section, however, we consider only those cases in which the assumptions are satisfied, so the sampling distribution of sample proportions can be approximated by the normal distribution. We will use the following notation and test statistic.

Notation

n = number of trials

$\hat{p} = \dfrac{x}{n}$ (*sample* proportion)

p = *population* proportion (used in the null hypothesis)

$q = 1 - p$

Test Statistic for Testing a Claim about a Proportion

$$z = \frac{\hat{p} - p}{\sqrt{\dfrac{pq}{n}}}$$

First, point out the danger of rounding $\hat{p}$ to too few decimal places. Second, note that in Section 5-6 we included a correction for continuity whenever we used the normal distribution to approximate a binomial distribution, but we don't include such a correction in the above test statistic because its effect is negligible with large samples.

The critical value is found from Table A-2 (standard normal distribution) by using the same procedures described in Section 7-2. For example, in a two-tailed test with significance level $\alpha = 0.05$, divide α equally between the two tails, then refer to Table A-2 for the z score corresponding to an area of $0.5 - 0.025 = 0.475$; the result is $z = 1.96$, so the critical values are $z = -1.96$ and $z = 1.96$.

The test statistic is justified by noting that when using the normal distribution to approximate a binomial distribution, we substitute $\mu = np$ and $\sigma = \sqrt{npq}$ to get

$$z = \frac{x - \mu}{\sigma} = \frac{x - np}{\sqrt{npq}}$$

In this expression, x is the number of successes among n trials. Divide the numerator and denominator of this last expression by n, then replace x/n by the symbol $\hat{p}$, and you have the test statistic just given. In other words, the test statistic is simply $z = (x - \mu)/\sigma$ modified for the binomial notation by incorporating the fact that the distribution of sample proportions $\hat{p}$ is a normal distribution with mean p and standard deviation $\sqrt{pq/n}$.

When conducting a test of a claim about a population proportion p, be careful to identify correctly the sample proportion $\hat{p}$. The sample proportion $\hat{p}$ is sometimes given directly, as in the statement that "10% of the observed sports cars are red," which is expressed as $\hat{p} = 0.10$. In other cases, we may need to calculate the sample proportion by using $\hat{p} = x/n$. For example, from the statement that "96 surveyed households have cable TV and 54 do not," we can first find the sample size n to be $96 + 54 = 150$, then we can calculate the value of the sample proportion of households with cable TV as follows:

$$\hat{p} = \frac{x}{n} = \frac{96}{150} = 0.64$$

Caution: The value of 0.64 is exact, but when a calculator or computer display of $\hat{p}$ results in many decimal places, use all of those decimal places when evaluating the z test statistic. Large errors can result from rounding $\hat{p}$ too much.

Let's now consider an example illustrating a practical application of the preceding concepts. This example follows the same basic hypothesis-testing procedures given in Section 7-3. The claim in the following example involves a percentage, but we must use the equivalent decimal form. Although the methods presented in this section can be used to test claims made about proportions, probabilities, or percentages, we must always conduct the actual test using a value of p between 0 and 1, and the sum of p and q must be exactly 1.

● **EXAMPLE**   In a study of air-bag effectiveness, it was found that in 821 crashes of midsize cars equipped with air bags, 46 of the crashes resulted in hospitalization of the drivers (based on data from the Highway Loss Data Institute). Use a 0.01 significance level to test the claim that the air-bag hospitalization rate is lower than the 7.8% rate for crashes of midsize cars equipped with automatic safety belts.

SOLUTION   We will use the traditional method of testing hypotheses as outlined in Figure 7-4. Instead of working with percentages, this solution will use the sample proportion of $\hat{p} = 46/821 = 0.0560292$ and the claimed population proportion of $p = 0.078$.

Step 1:   The original claim is that the air-bag hospitalization rate is lower than 7.8%. We express this in symbolic form as $p < 0.078$.

Step 2:   The opposite of the original claim is $p \geq 0.078$.

Step 3:   Because $p \geq 0.078$ contains equality, we have

$H_0: p \geq 0.078$ (null hypothesis)
$H_1: p < 0.078$ (alternative hypothesis and original claim)

Step 4:   The significance level is $\alpha = 0.01$.

(continued)

Figure 7-12 Hypothesis Test of
Claim that $p < 0.078$

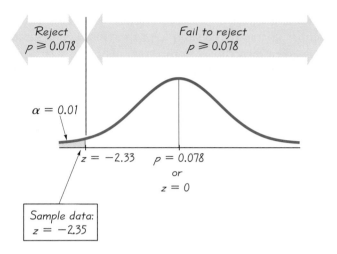

Step 5:   The statistic relevant to this test is $\hat{p} = 46/821 = 0.0560292$. The sampling distribution of sample proportions is approximated by the normal distribution. (The requirements that $np \geq 5$ and $nq \geq 5$ are both satisfied with $n = 821$, $p = 0.078$, and $q = 0.922$.)

Step 6:   The test statistic of $z = -2.35$ is found as follows:

$$z = \frac{\hat{p} - p}{\sqrt{\dfrac{pq}{n}}} = \frac{0.0560292 - 0.078}{\sqrt{\dfrac{(0.078)\,(0.922)}{821}}} = -2.35$$

The critical value of $z = -2.33$ is found from Table A-2. With $\alpha = 0.01$ in the left tail, look for an area of 0.4900 in the body of Table A-2; the closest value is 0.4901 and it corresponds to $z = 2.33$, but it must be negative because it is to the left of the mean. The test statistic and critical value are shown in Figure 7-12.

Step 7:   Because the test statistic does fall within the critical region, reject the null hypothesis.

Step 8:   There is sufficient evidence to support the claim that for midsize car crashes, the air-bag hospitalization rate is lower than the 7.8% rate for automatic safety belts.

EXAMPLE   In a consumer taste test, 100 regular Pepsi drinkers are given blind samples of Coke and Pepsi; 48 of these subjects preferred Coke. At the 0.05 level of significance, test the claim that Coke is preferred by 50% of Pepsi drinkers who participate in such blind taste tests.

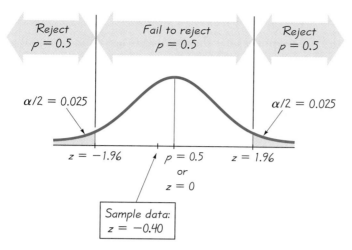

Figure 7-13
Hypothesis Test of $p = 0.5$

SOLUTION   We summarize the key components of the hypothesis test.

$H_0: p = 0.5$ (from the claim that "Coke is preferred by 50%")
$H_1: p \neq 0.5$

Test statistic:

$$z = \frac{\hat{p} - p}{\sqrt{\dfrac{pq}{n}}} = \frac{0.48 - 0.5}{\sqrt{\dfrac{(0.5)(0.5)}{100}}} = -0.40$$

The test statistic, critical values, and critical region are shown in Figure 7-13. Because the test statistic is not in the critical region, we fail to reject the null hypothesis. There is not sufficient evidence to reject the claim that 50% of Pepsi drinkers prefer Coke. (Critics of such taste tests claim that the subjects often cannot observe differences and guess when making their choices.)

The *P*-Value Method

The examples in this section followed the traditional approach to hypothesis testing, but it would be just as easy to use the *P*-value approach. To find the *P*-value corresponding to a z test statistic, use the following procedure, which was described in Section 7-3:

Right-tailed test: *P*-value = area to right of test statistic z

Left-tailed test: *P*-value = area to left of test statistic z

Two-tailed test: *P*-value = *twice* the area of the extreme region bounded by the test statistic z

Discrimination Case Uses Statistics

Statistics often play a key role in discrimination cases. One such case involved Matt Perez and more than 300 other FBI agents who won a class-action suit charging that Hispanics in the FBI were discriminated against in the areas of promotions, assignments, and disciplinary actions. The plaintiff employed statistician Gary Lafree, who showed that the FBI's upper management positions had significantly low proportions of Hispanic employees. Statistics were instrumental in the plaintiff's victory in this case. (See Program 20 from the series *Against All Odds: Inside Statistics* for a discussion of this case.)

As in Section 7-3, we use this decision criterion:

Reject the null hypothesis if the *P*-value is less than or equal to the significance level α.

The last example was two-tailed, so the *P*-value is twice the area to the left of the test statistic $z = -0.40$. Table A-2 indicates that the area between $z = 0$ and $z = -0.40$ is 0.1554, so the area to the left of the test statistic ($z = -0.40$) is $0.5 - 0.1554 = 0.3446$. The *P*-value is $2 \times 0.3446 = 0.6892$. Because the *P*-value of 0.6892 is not less than or equal to the significance level of 0.05, we fail to reject the null hypothesis and again conclude that there is not sufficient evidence to reject the claim that 50% of Pepsi drinkers prefer Coke. Again, the *P*-value method is simply another way of arriving at the same conclusion reached using the traditional approach.

Using Computers and Calculators

STATDISK and the TI-83 calculator can be used to test claims about a population proportion. Minitab can usually be used, but it requires a trick described in the *Minitab Student Laboratory Manual and Workbook*.

Exercises A: Basic Skills and Concepts

Recommended assignment: Exercises 1, 2, 4, 6, 10, 16.

1. TS: $z = 2.19$; CV: $z = \pm 1.96$;
   P-value: 0.0286.

2. TS: $z = -0.93$;
   CV: $z = \pm 1.96$;
   P-value: 0.3524

In Exercises 1–20, use the traditional method to test the given hypothesis. Include the steps listed in Figure 7-4, and draw the appropriate graph. Also identify the P-*value. Assume that all samples have been randomly selected.*

1. In a study of store checkout scanners, 1234 items were checked and 20 of them were found to be overcharges (based on data from "UPC Scanner Pricing Systems: Are They Accurate?" by Goodstein, *Journal of Marketing*, Vol. 58). Use a 0.05 level of significance to test the claim that with scanners, 1% of sales are overcharges. (Before scanners were used, the overcharge rate was estimated to be about 1%.) Based on these results, do scanners appear to help consumers avoid overcharges?

2. Environmental concerns often conflict with modern technology, as is the case with birds that pose a hazard to aircraft during takeoff. An environmental group states that incidents of bird strikes are too rare to justify killing the birds. A pilots' group claims that among aborted takeoffs leading to an aircraft's going off the end of the runway, 10% are due to bird strikes. Use a 0.05 level of significance to test that claim. Sample data consist of 74 aborted takeoffs in which the aircraft overran the runway. Among those 74 cases, 5 were due to bird strikes (based on data from the Air Line Pilots Association and Boeing, as reported in *USA Today*).

3. The quality control manager at the Telektronic Company considers production of telephone-answering machines to be "out of control" when the overall rate of defects exceeds 4%. Testing of a random sample of 150 machines revealed that 9 are defective, so the sample percentage of defects is 6%. The production manager claims that this is only a chance difference, so production is not really out of control and no corrective action is necessary. Use a 0.05 significance level to test the production manager's claim. Does it appear that corrective action is necessary?

3. TS: $z = 1.25$; CV: $z = 1.645$; P-value: 0.1056.

4. In 1990, 5.8% of job applicants who were tested for drugs failed the test. At the 0.01 level, test the claim that the failure rate is now lower if a random sample of 1520 current job applicants results in 58 failures (based on data from the American Management Association). Does the result suggest that fewer job applicants now use drugs?

4. TS: $z = -3.31$; CV: $z = -2.33$; P-value: 0.0001.

5. Ralph Carter is a high-school history teacher who says that if students are not aware of the Holocaust, then the curriculum should be revised to correct that deficiency. A Roper Organization survey of 506 high-school students showed that 268 of them did not know that the term "Holocaust" refers to the Nazi killing of about 6 million Jews during World War II. Using the sample data and a 0.05 significance level, test the claim that most (more than 50%) students don't know what "Holocaust" refers to. Based on these results, should Ralph Carter seek revisions to the curriculum?   5. TS: $z = 1.33$; CV: $z = 1.645$; P-value: 0.0918.

6. Refer to Data Set 8 in Appendix B and consider only those statistics students who are 21 or older. Find the sample percentage of smokers in that age group, then test the claim that statistics students aged 21 or over smoke at a rate that is less than 32%, which is the smoking rate for the general population of persons aged 21 and over (based on data from the U.S. National Institute on Drug Abuse). Is there a reason that statistics students 21 and over would smoke at a rate lower than the rate for the general population in that age group?

6. $\hat{p} = 9/41$; TS: $z = -1.38$; CV: $z = -1.645$; P-value: 0.0838.

7. The Greybar Tax Company claims that its clients are selected for audits at a rate substantially higher than the rate for the general population. The IRS reports that it audits 4.3% of those who earn more than $100,000, but a check of 400 randomly selected Greybar returns with earnings above $100,000 shows that 56 of them were audited. Test the claim using a 0.005 level of significance. If you earn more than $100,000, would you use this tax company?   7. TS: $z = 9.56$; CV: $z = 2.575$; P-value: 0.0001

8. Columbia Pictures chairman Mark Canton claims that 58% of movies made are R-rated. Test that claim using a 0.10 significance level. For sample data, refer to Data Set 10 of Appendix B.

8. $\hat{p} = 35/60$; TS: $z = 0.05$; CV: $z = \pm 1.645$; P-value: 0.9602.

9. A television executive claims that "fewer than half of all adults are annoyed by the violence shown on television." Test this claim by using sample data from a Roper poll in which 48% of 1,998 surveyed adults indicated their annoyance with television violence. Use a 0.05 significance level.   9. TS: $z = -1.79$. CV: $z = -1.645$; P-value: 0.0367

10. TS: $z = -2.06$;
    CV: $z = -1.96$;
    P-value: 0.0197.

12. TS: $z = -4.40$;
    CV: $z = -2.575$;
    P-value: 0.0001

13. TS: $z = 0.99$;
    CV: $z = \pm 1.645$;
    P-value: 0.3222

10. "Bystanders perform CPR correctly less than half the time," according to a *USA Today* article which noted that among 662 cases in which bystanders gave CPR, 46% performed the CPR correctly. Use a 0.025 significance level to test the article's claim.

11. Auto insurance companies are beginning to consider raising rates for those who use telephones while driving. The National Consumers Group claims that the problem really isn't too serious because only 10% of drivers use telephones. The insurance industry conducts a study and finds that among 500 randomly selected drivers, 90 use telephones (based on data from *Prevention* magazine). At the 0.02 level of significance, test the consumer group's claim.   11. TS: $z = 5.96$; CV: $z = \pm 2.33$; *P*-value: 0.0002.

12. Bay Photo is a San Francisco-based photo-development company that wants to use automated telephone messages to solicit customers. A company partner argues that potential customers with unlisted numbers will not be reached, but the marketing manager claims that fewer than one-half of residential telephones in San Francisco have unlisted numbers. A random sample of 400 residential phones in San Francisco resulted in an unlisted rate of 39%. Use a 0.005 level of significance to test the claim that fewer than one-half have unlisted numbers.

13. In clinical studies of the allergy drug Seldane, 70 of the 781 subjects experienced drowsiness (based on data from Merrell Dow Pharmaceuticals, Inc.). A competitor claims that 8% of Seldane users experience drowsiness. Use a 0.10 significance level to test that claim.

14. The Federal Aviation Administration will fund research on spatial disorientation of pilots if there is sufficient sample evidence (at the 0.01 significance level) to conclude that among aircraft accidents involving such disorientation, more than three-fourths result in fatalities. A study of 500 aircraft accidents involving spatial disorientation of the pilot found that 91% of those accidents resulted in fatalities (based on data from the Department of Transportation). Based on these sample results, will the funding be approved?   14. TS: $z = 8.26$; CV: $z = 2.33$; *P*-value: 0.0001

15. Dr. Kelly Roberts is dean of a medical school and must plan courses for incoming students. The college president is encouraging her to increase the emphasis on pediatrics, but Dr. Roberts argues that fewer than 10% of U.S. medical students prefer pediatrics. She refers to sample data indicating that 64 of 1068 randomly selected medical students chose pediatrics (based on data reported by the Association of American Medical Colleges). Is there sufficient sample evidence to support (at the 0.01 significance level) her argument?   15. TS: $z = -4.37$. CV: $z = -2.33$; *P*-value: 0.0001

16. A reservations system for Air America suffered from a 7% rate of no-shows. A new procedure was instituted whereby reservations are confirmed on the day preceding the actual flight, and a study was then made of 5218 randomly selected reservations made under the new system. If 333 no-shows were recorded, test the claim that the no-show rate is lower with the new system. Does the new system appear to be effective in reducing no-shows?   16. TS: $z = -1.75$; CV: $z = -1.645$; *P*-value: 0.0401.

17. In one study of 71 smokers who tried to quit smoking with nicotine patch therapy, 32 were not smoking one year after the treatment (based on data from "High-Dose Nicotine Patch Therapy," by Dale et al., *Journal of the American Medical Association*, Vol. 274, No. 17). Use a 0.10 level of significance to test the claim that among smokers who try to quit with nicotine patch therapy, the majority are smoking a year after the treatment. Do these results suggest that the nicotine patch therapy is not effective?

17. TS: $z = 0.83$; CV: $z = 1.28$; P-value: 0.2033

18. The Kennedy–Nixon presidential race was extremely close. Kennedy won with 34,227,000 votes to Nixon's 34,108,000 votes. The closeness of the results caused some people to speculate that the event was like flipping a coin and Nixon might have won on another day. Although the votes aren't sample data randomly selected from a larger population, assume that they are. At the 0.01 level of significance, test the claim that the true population proportion for Kennedy exceeded 0.5. Does the difference in the votes appear to be a chance difference or a significant difference?

18. TS: $z = 14.40$; CV: $z = 2.33$; P-value: 0.0001

19. A Bruskin-Goldring Research poll of 1012 adults showed that among those who used fruitcakes, 28% ate them and 72% used them for other purposes, such as doorstops, birdfeed, and landfill. This surprised fruitcake producers, who believe that a fruitcake is an appealing food. The president of the Kansas Food Products Company claims that the poll results are a fluke and, in reality, half of all adults eat their fruitcakes. Use a 0.01 level of significance to test that claim. Based on the result, does it appear that fruitcake producers should consider changes to make their product more appealing as a food or better suited for its uses as a doorstop, birdfeed, and so on?   19. TS: $z = -14.00$; CV: $z = \pm 2.575$; P-value: 0.0002

20. A *Prevention* magazine article reported on a poll of 1257 adults. The article noted that among those polled, 27% smoke, and 82% of those who smoke have tried to stop at least once. How many of the surveyed people are smokers who have tried to stop at least once? At the 0.05 significance level, test the claim that less than 25% of adults are smokers who have tried to stop at least once.   20. TS: $z = -2.36$; CV: $z = -1.645$; P-value: 0.0091

7–5   Exercises B: Beyond the Basics

21. A reporter for the *Providence Journal* claims that 10% of the city's residents believe that the mayor is doing a good job. Test her claim if, in a random sample of 15 residents, there are none who believe that the mayor is doing a good job. Use a 0.05 level of significance. Because $np = 1.5$ and is not at least 5, the normal distribution is not a suitable approximation of the distribution of sample proportions, so the test statistic given in this section should not be used.   21. $n = 15$, $p = 0.1$; $P(0) = 0.206$

22. Chemco, a supplier of chemical waste containers, finds that 3% of a sample of 500 units are defective. Being fundamentally dishonest, the Chemco production manager wants to make a claim that the rate of defective units is no more than some specified percentage, and he doesn't want that claim rejected at the 0.05 level of significance if the sample data are used. What is the *lowest* defective rate he can claim under these conditions?

23. Refer to Data Set 11 in Appendix B and find the sample proportion of plain M&Ms that are red. Use that result to test the claim of the Mars candy company that 20% of its plain M&M candies are red. Use a 0.05 significance level.

   a. Use the traditional method.   23. a. $\hat{p} = 21/100$; TS: $z = 0.25$; CV: $z = \pm 1.96$

   b. Use the *P*-value method.   b. *P*-value: 0.8026

   c. Test the claim by constructing and interpreting a 95% confidence interval.   c. $0.130 < p < 0.290$

24. Refer to Exercise 9. If the true value of p is 0.45, find β, the probability of a type II error (see Exercise 27 from Section 7-3). (*Hint:* In Step 3 use the values of $p = 0.45$ and $pq/n = (0.45)(0.55)/1998$.)   24. 0.0023

7-6 Testing a Claim about a Standard Deviation or Variance

The preceding sections of this chapter introduced methods of testing claims made about population means and proportions. This section uses the same basic procedures to test claims made about a population standard deviation σ or variance σ^2. The following requirement must be met for such tests.

Assumption for Testing Claims about σ or σ^2

In testing a hypothesis made about a population standard deviation σ or variance σ^2, we assume that the population has values that are normally distributed.

Other methods of testing hypotheses have also required a normally distributed population, but tests of claims about standard deviations or variances are not as *robust*, meaning that the inferences can be very misleading if the population does not have a normal distribution. In this section, the condition of a normally distributed population is a much stricter requirement. Given the assumption of a normal distribution, the following test statistic has a chi-square distribution with $n - 1$ degrees of freedom and critical values given in Table A-4.

Test Statistic for Testing Hypotheses about σ or σ^2
$$\chi^2 = \frac{(n-1)s^2}{\sigma^2}$$
where n = sample size
s^2 = sample variance
σ^2 = population variance (given in the null hypothesis)

The chi-square distribution was introduced in Section 6-4, where we noted the following important properties.

Properties of the Chi-Square Distribution

1. All values of χ^2 are nonnegative, and the distribution is not symmetric (see Figure 7-14).
2. There is a different distribution for each number of degrees of freedom (see Figure 7-15).
3. The critical values are found in Table A-4 where

$$\text{degrees of freedom} = n - 1$$

One of the difficulties in applying methods of statistics is correctly determining what procedure to apply. Encourage a top-down analysis that begins with this broad question: Is the main objective to test a claim, estimate a parameter, find a probability, or what? Is the parameter of central concern the mean, proportion, standard deviation, or what? What distribution applies? It is helpful to develop a systematic and logical approach.

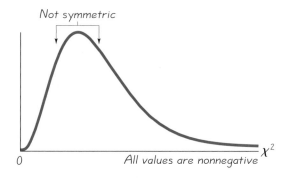

Figure 7-14 Properties of the Chi-Square Distribution

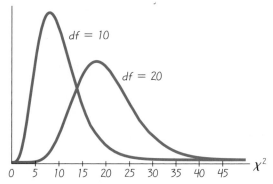

Figure 7-15   Chi-Square Distributions for 10 and 20 Degrees of Freedom
There is a different distribution for each number of degrees of freedom.

Ethics in Experiments

Sample data can often be obtained by simply observing or surveying members selected from the population. Many other situations require that we somehow manipulate circumstances to obtain sample data. In both cases ethical questions may arise. Researchers in Tuskegee, Alabama, withheld the effective penicillin treatment to syphilis victims so that the disease could be studied. This experiment continued for a period of 27 years!

Critical values are found in Table A-4 by first locating the row corresponding to the appropriate number of degrees of freedom (where df $= n - 1$). Next, the significance level α is used to determine the correct column as described in the following list.

Right-tailed test: Locate the area at the top of Table A-4 that is equal to the significance level α.

Left-tailed test: Calculate $1 - \alpha$, then locate the area at the top of Table A-4 that is equal to $1 - \alpha$.

Two-tailed test: Calculate $1 - \alpha/2$ and $\alpha/2$, then locate the area at the top of Table A-4 that equals $1 - \alpha/2$ (for the left critical value) and locate the area at the top equal to $\alpha/2$ (for the right critical value). (See Figure 6-9 and the example on pages 327–328.)

Quality control engineers want to ensure that a product is, on the average, acceptable, but they also want to produce items of *consistent* quality so that there will be few defective products. Consistency is improved by reducing variation. For example, the consistency of aircraft altimeters is governed by Federal Aviation Regulation 91.36, which requires that aircraft altimeters be tested and calibrated to give a reading "within 125 feet (on a 95-percent probability basis)." Even if the mean altitude reading is exactly correct, an excessively large standard deviation will result in individual readings that are dangerously low or high.

EXAMPLE   The Stewart Aviation Products Company has been successfully manufacturing aircraft altimeters with errors normally distributed with a mean of 0 ft (achieved by calibration) and a standard deviation of 43.7 ft. After the installation of new production equipment, 30 altimeters were randomly selected from the new line. This sample group had errors with a standard deviation of $s = 54.7$ ft. Use a 0.05 significance level to test the claim that the new altimeters have a standard deviation different from the old value of 43.7 ft.

SOLUTION   We will use the traditional method of testing hypotheses as outlined in Figure 7-4.

Step 1:   The claim is expressed in symbolic form as $\sigma \neq 43.7$ ft.

Step 2:   If the original claim is false, then $\sigma = 43.7$ ft.

Step 3:   The null hypothesis must contain equality, so we have

$$H_0: \sigma = 43.7 \qquad H_1: \sigma \neq 43.7 \text{ (original claim)}$$

Step 4:   The significance level is $\alpha = 0.05$.

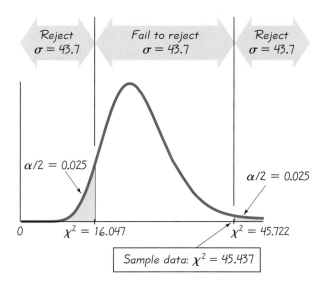

Figure 7-16   Hypothesis Test
of Claim that $\sigma \neq 43.7$

Step 5:   Because the claim is made about σ, we use the chi-square distribution.

Step 6:   The test statistic is

$$\chi^2 = \frac{(n-1)s^2}{\sigma^2} = \frac{(30-1)(54.7)^2}{43.7^2} = 45.437$$

The critical values are 16.047 and 45.722. They are found in Table A-4, in the 29th row (degrees of freedom $= n - 1 = 29$) in the columns corresponding to 0.975 and 0.025. See the test statistic and critical values shown in Figure 7-16.

Step 7:   Because the test statistic is not in the critical region, we fail to reject the null hypothesis.

Step 8:   There is not sufficient evidence to support the claim that the standard deviation is different from 43.7 ft. (For help in wording the final conclusion, refer to Figure 7-2.) However, it would be wise to continue monitoring and testing the new product line.

One aspect of the preceding example that might cause some confusion is the reference to two different distributions. The population of the actual altimeter errors has a *normal* distribution, but in Step 6 of the preceding solution we use the *chi-square* distribution because the distribution of $(n - 1)s^2/\sigma^2$ is a chi-square distribution. The methods of this section require that the original population has a normal distribution, but we use the chi-square distribution in testing claims about standard deviations or variances.

Point out that if the claimed value of σ^2 is actually correct, then samples tend to have variances s^2 that are close to σ^2 so that the ratio of s^2/σ^2 tends to be close to 1 and the test statistic tends to be close to $n - 1$. This follows from simplifying $(n - 1)s^2/\sigma^2$ to $(n - 1)$ when the ratio s^2/σ^2 is approximately 1 because s^2 is close to σ^2.

Recommended assignment: Exercises 1, 3, 4, 5, 10, 12.

1. a. 1.735, 23.589
   b. 45.642
   c. 10.851
2. a. 10.982
   b. 17.275
   c. 6.262, 27.488

The P-Value Method

Instead of the traditional approach to hypothesis testing, we can also use the P-value approach summarized in Figures 7-7 and 7-8. We usually cannot find *exact* P-values because the chi-square distribution table (Table A-4) includes only selected values of α. Instead, we use the table to identify limits that contain the P-value. The test statistic from the last example is $\chi^2 = 45.437$, and we know that the test is two-tailed with 29 degrees of freedom. Refer to the 29th row of Table A-4 and see that the test statistic of 45.437 is between the table critical values of 42.557 and 45.722, indicating that the area to the right of the test statistic is between the corresponding areas of 0.05 and 0.025. Figure 7-7 summarizes the general procedure for finding P-values, and that figure indicates that in a two-tailed test with a test statistic to the right of the center, the P-value is *twice* the area to the right of the test statistic. The P-value is therefore between 0.10 and 0.05, which can be expressed as

$$0.05 < P\text{-value} < 0.10$$

Because the P-value is not less than or equal to the significance level of $\alpha = 0.05$, we again fail to reject the null hypothesis. Again, the traditional method and P-value method are equivalent in the sense that they always lead to the same conclusion.

The P-value approach becomes particularly relevant when using computer programs that provide P-values but do not provide critical values. If the preceding example is run using STATDISK, the P-value of 0.0533 is included in the displayed results, but STATDISK does provide the critical values as well. Minitab and the TI-83 calculator are not designed to test claims involving a standard deviation or variance, but they can be used to find the components needed for such a test.

7-6 Exercises A: Basic Skills and Concepts

In Exercises 1 and 2, use Table A-4 to find the critical values of χ^2 based on the given information.

1. a. $H_0: \sigma = 15$ b. $H_1: \sigma > 0.62$ c. $H_1: \sigma < 14.4$
      $n = 10$ $n = 27$ $n = 21$
      $\alpha = 0.01$ $\alpha = 0.01$ $\alpha = 0.05$

2. a. $H_1: \sigma < 1.22$ b. $H_1: \sigma > 92.5$ c. $H_0: \sigma = 0.237$
      $n = 23$ $n = 12$ $n = 16$
      $\alpha = 0.025$ $\alpha = 0.10$ $\alpha = 0.05$

In Exercises 3–14, use the traditional method to test the given hypotheses. Follow the steps outlined in Figure 7-4, and draw the appropriate graph. In

all cases, assume that the population is normally distributed and that the sample has been randomly selected.

3. The Stewart Aviation Products Company uses a new production method to manufacture aircraft altimeters. A random sample of 81 altimeters resulted in errors with a standard deviation of $s = 52.3$ ft. At the 0.05 level of significance, test the claim that the new production line has errors with a standard deviation different from 43.7 ft, which was the standard deviation for the old production method. If it appears that the standard deviation has changed, does the new production method appear to be better or worse than the old method?   3. TS: $\chi^2 = 114.586$; CV: $\chi^2 = 57.153, 106.629$

4. Tests in the author's past statistics classes have scores with a standard deviation equal to 14.1. One of his current classes now has 27 test scores with a standard deviation of 9.3. Use a 0.01 significance level to test the claim that this current class has less variation than past classes. Does a lower standard deviation suggest that the current class is doing better?   4. TS: $\chi^2 = 11.311$; CV: $\chi^2 = 12.198$

5. With individual lines at its various windows, the Jefferson Valley Bank found that the standard deviation for normally distributed waiting times on Friday afternoons was 6.2 min. The bank experimented with a single main waiting line and found that for a random sample of 25 customers, the waiting times have a standard deviation of 3.8 min. On the basis of previous studies, we can assume that the waiting times are normally distributed. Use a 0.05 significance level to test the claim that a single line causes lower variation among the waiting times. (Customers tend to be happier if the waiting times have lower variation.) Does the use of a single line result in a shorter wait?   5. TS: $\chi^2 = 9.016$; CV: $\chi^2 = 13.848$

6. Do men's heights vary more than the heights of women? The nurse at a women's college has collected data from physical exams. The standard deviation of women's heights is found to be 2.5 in. The college has now become coed and the first 20 male students have heights with a standard deviation of 2.8 in. Use a 0.025 significance level to test the claim that men's heights vary more than the heights of women. If that claim is not supported, does this mean that men's heights do not vary more than the heights of women?   6. TS: $\chi^2 = 23.834$; CV: $\chi^2 = 32.852$

7. The Kansas Farm Products Company uses a machine that fills 50-lb corn seed bags. In the past, the machine has had a standard deviation of 0.75 lb. In an attempt to get more consistent weights, mechanics have replaced some worn machine parts. A random sample of 61 bags taken from the repaired machinery produced a sample mean of 50.13 lb and a sample standard deviation of 0.48 lb. At the 0.05 significance level, test the claim that the weights are "more consistent" with the repaired machinery than in the past. If the weights are more consistent, how is the standard deviation affected?   7. TS: $\chi^2 = 24.576$; CV: $\chi^2 = 43.188$

8. The Collins Investment Company finds that if the standard deviation for the weekly downtimes of their computer is 2 h or less, then the computer is predictable and planning is facilitated. Twelve weekly downtimes for

the computer were randomly selected, and the sample standard deviation was computed to be 2.85 h. The manager of computer operations claims that computer access times are unpredictable because the standard deviation exceeds 2 h. Test her claim, using a 0.025 significance level. Does the variation appear to be too high?   8. TS: $\chi^2 = 22.337$; CV: $\chi^2 = 21.920$

9. The Medassist Pharmaceutical Company uses a machine to pour cold medicine into bottles in such a way that the standard deviation of the weights is 0.15 oz. A new machine was tested on 71 bottles, and the standard deviation for this sample is 0.12 oz. The Dayton Machine Company, which manufactures the new machine, claims that it fills bottles with a lower variation. At the 0.05 significance level, test the claim made by the Dayton Machine Company. If Dayton's machine is being used on a trial basis, should its purchase be considered?   9. TS: $\chi^2 = 44.800$; CV: $\chi^2 = 51.739$

10. TS: $\chi^2 = 10.222$;
    CV: $\chi^2 = 11.689, 38.076$

10. For randomly selected adults, IQ scores are normally distributed with a mean of 100 and a standard deviation of 15. A sample of 24 randomly selected college professors resulted in IQ scores having a standard deviation of 10. A psychologist is quite sure that college professors have IQ scores that have a mean greater than 100. He doesn't understand the concept of standard deviation very well and claims the standard deviation for IQ scores of college professors is equal to 15, the same standard deviation as in the general population. Use a 0.05 level of significance to test that claim. Based on the result, what do you conclude about the standard deviation of IQ scores for college professors?

11. Systolic blood pressure results from contraction of the heart. In comparing systolic blood pressure levels of men and women, Dr. Jane Taylor obtained readings for a random sample of 50 women. The sample mean and standard deviation were found to be 130.7 and 23.4, respectively. If systolic blood pressure levels for men are known to have a mean and standard deviation of 133.4 and 19.7, respectively, test the claim that women have more variation. Use a 0.05 level of significance. (All readings are in millimeters of mercury, and data are based on the National Health Survey.)   11. TS: $\chi^2 = 69.135$; CV: $\chi^2 = 67.505$ (approx.)

12. If we use the body temperatures for 12 A.M. of Day 2 that are listed in Data Set 2 of Appendix B, we reject (at the 0.05 significance level) the claim that $\mu = 98.6°$ F. Those 106 temperatures have a distribution that is approximately normal, their mean is $\bar{x} = 98.20°$ F, and their standard deviation is $s = 0.62°$ F. The test statistic will cause rejection of $\mu = 98.6°$ F as long as the standard deviation is less than 2.11° F. Use the sample statistics and a 0.005 significance level to test the claim that $\sigma < 2.11°$ F.   12. TS: $\chi^2 = 9.066$; CV: $\chi^2 = 67.328$ (approx.)

13. Based on data from the National Health Survey, men aged 25–34 have heights with a standard deviation of 2.9 in. Test the claim that men aged 45–54 have heights with a different standard deviation. The heights of 25 randomly selected men in the 45–54 age bracket are listed at the top of the following page.   13. $s = 2.340$; TS: $\chi^2 = 15.628$; CV: $\chi^2 = 12.401, 39.364$

66.80	71.22	65.80	66.24	69.62	70.49	70.00	71.46	65.72
68.10	72.14	71.58	66.85	69.88	68.69	72.77	67.34	68.40
68.96	68.70	72.69	68.67	67.79	63.97	67.19		

14. Shown below are birth weights (in kilograms) of male babies born to mothers on a special vitamin supplement (based on data from the New York State Department of Health). Test the claim that this sample comes from a population with a standard deviation equal to 0.470 kg, which is the standard deviation for male birth weights in general. Does the vitamin supplement appear to affect the variation among birth weights?

| 3.73 | 4.37 | 3.73 | 4.33 | 3.39 | 3.68 | 4.68 | 3.52 |
| 3.02 | 4.09 | 2.47 | 4.13 | 4.47 | 3.22 | 3.43 | 2.54 |

14. $s = 0.657$; TS: $\chi^2 = 29.339$; CV: $\chi^2 = 6.262, 27.488$

7–6 Exercises B: Beyond the Basics

15. Use Table A-4 to find the range of possible P-values in the given exercises.

   a. Exercise 3   b. Exercise 5   c. Exercise 9

15. a. $0.01 < P\text{-value} < 0.02$
    b. $P\text{-value} < 0.005$
    c. $0.005 < P\text{-value} < 0.01$

16. For large numbers of degrees of freedom, we can approximate critical values of χ^2 as follows:

$$\chi^2 = \frac{1}{2} (z + \sqrt{2k - 1})^2$$

Here k is the number of degrees of freedom and z is the critical value, found in Table A-2. For example, if we want to approximate the two critical values of χ^2 in a two-tailed hypothesis test with $\alpha = 0.05$ and a sample size of 150, we let $k = 149$ with $z = -1.96$, followed by $k = 149$ and $z = 1.96$.

   a. Use this approximation to estimate the critical values of χ^2 in a two-tailed hypothesis test with $n = 101$ and $\alpha = 0.05$. Compare the results to those found in Table A-4.

   b. Use this approximation to estimate the critical values of χ^2 in a two-tailed hypothesis test with $n = 150$ and $\alpha = 0.05$.

16. a. Est. values: 73.772, 129.070; Table A-4 values: 74.222, 129.561
    b. 116.643, 184.199

17. Repeat Exercise 16 using this approximation (with k and z as described in Exercise 16):

$$\chi^2 = k\left(1 - \frac{2}{9k} + z\sqrt{\frac{2}{9k}}\right)^3$$

17. a. Est. values: 74.216, 129.565; Table A-4 values: 74.222, 129.561
    b. $\chi^2 = 117.093, 184.690$

18. Refer to Exercise 5. Assuming that σ is actually 4.0, find β (the probability of a type II error). See Exercise 27 from Section 7-3 and modify the procedure so that it applies to a hypothesis test involving σ instead of μ.

18. Approximately 0.10

● ●

Vocabulary List

hypothesis	test statistic
hypothesis test	critical region
test of significance	critical value
null hypothesis	two-tailed test
alternative hypothesis	left-tailed test
type I error	right-tailed test
type II error	traditional method
significance level	classical method
alpha (α)	P-value
beta (β)	probability value

● ●

Review

Chapters 6 and 7 introduced two important concepts in using sample data to form inferences about population data: estimating values of population parameters (Chapter 6), and testing claims made about population parameters (Chapter 7). The parameters considered in Chapters 6 and 7 are means, proportions, standard deviations, and variances.

Section 7-2 presented the fundamental concepts of a hypothesis test: null hypothesis, alternative hypothesis, type I error, type II error, test statistic, critical region, critical value, and significance level. We also discussed two-tailed tests, left-tailed tests, right-tailed tests, and the statement of conclusions. Section 7-3 used those components in identifying three different methods for testing hypotheses:

1. The traditional method (summarized in Figure 7-4)

2. The P-value method (summarized in Figure 7-8)

3. Confidence intervals (discussed in Chapter 6)

Sections 7-3 through 7-6 discussed specific methods for dealing with the different parameters. Because it is so important to be correct in selecting the distribution and test statistic, we provide Figure 7-17, which summarizes the key decisions to be made.

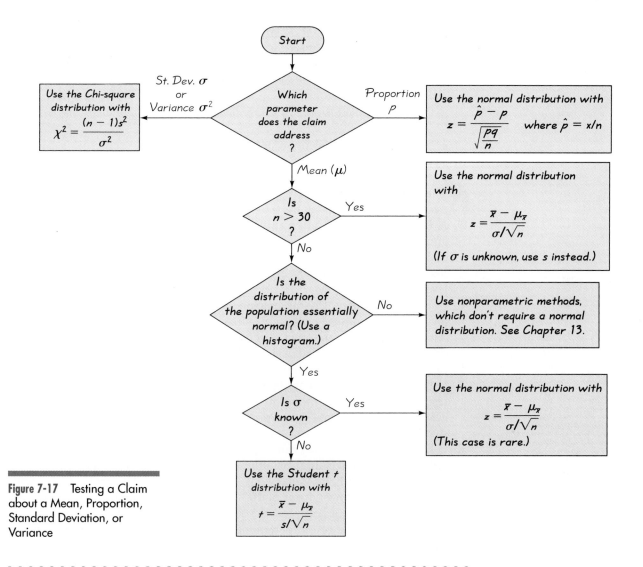

Figure 7-17   Testing a Claim about a Mean, Proportion, Standard Deviation, or Variance

• •

Review Exercises

In Exercises 1 and 2, find the appropriate critical values. In all cases, assume that the population standard deviation σ is unknown.

1. a. $H_1: \mu < 27.3$
  $n = 10$
  $\alpha = 0.05$

b. $H_1: p \neq 0.5$
  $n = 150$
  $\alpha = 0.01$

c. $H_1: \sigma \neq 15$
  $n = 20$
  $\alpha = 0.05$

2. a. $H_0: \mu = 19.9$
  $n = 40$
  $\alpha = 0.10$

b. $H_0: \sigma \leq 0.93$
  $n = 10$
  $\alpha = 0.025$

c. $H_0: p \geq 0.25$
  $n = 540$
  $\alpha = 0.01$

1. a. $t = -1.833$
   b. $z = \pm 2.575$
   c. $\chi = 8.907, 32.852$
2. a. $z = \pm 1.645$
   b. $\chi = 19.023$
   c. $z = -2.33$

In Exercises 3 and 4, respond to each of the following:

   a. Give the null hypothesis in symbolic form.

   b. Is this test left-tailed, right-tailed, or two-tailed?

   c. In simple terms devoid of symbolism and technical language, describe the type I error.

   d. In simple terms devoid of symbolism and technical language, describe the type II error.

   e. Identify the probability of making a type I error.

3. The claim that statistics consultants have a mean income of $90,000 is to be tested at the 0.05 significance level.

4. The claim that lengths of CBS commercials have a standard deviation less than 15 s is to be tested at the 0.01 significance level.   4. a. $\sigma \geq 15$ s, b. Left-tailed

5. A *USA Today*/CNN/Gallup survey was the basis for a report that "57% of gun owners favor stricter gun laws." Assume that the survey group consisted of 504 randomly selected gun owners. Test the claim that the majority (more than 50%) of gun owners favor stricter gun laws. Use a 0.01 significance level.   5. TS: $z = 3.14$; CV: $z = 2.33$

6. A Telektronics dental X-ray machine bears a label stating that the machine gives radiation dosages with a mean of less than 5.00 milliroentgens. Sample data consist of 36 randomly selected observations with a mean of 4.13 milliroentgens and a standard deviation of 1.91 milliroentgens. Using a 0.01 level of significance, test the claim stated on the label.   6. TS: $z = -2.73$; CV: $z = -2.33$

7. Using the same sample data from Exercise 6, use a 0.01 significance level to test the claim that the radiation dosages have a standard deviation less than 2.50 milliroentgens. What is a practical consequence of a standard deviation that is too high in this case?   7. TS: $\chi^2 = 20.429$; CV: between 14.954 and 22.164

8. Is Elvis alive? *USA Today* ran a report about a University of North Carolina poll of 1248 adults from the southern United States. It was reported that 8% of those surveyed believe that Elvis Presley still lives. The article began with the claim that "almost 1 out of 10" Southerners still thinks Elvis is alive. At the 0.01 significance level, test the claim that the true percentage is less than 10%. Based on the result, determine whether the 8% sample result justifies the phrase "almost 1 out of 10."   8. TS: $z = -2.36$; CV: $z = -2.33$

9. The Boston Bottling Company distributes cola in cans labeled 12 oz. The Bureau of Weights and Measures randomly selected 24 cans, measured their contents, and obtained a sample mean of 11.82 oz and a sample standard deviation of 0.38 oz. Use a 0.01 significance level to test the claim that the company is cheating consumers.   9. TS: $t = -2.321$; CV: $t = -2.500$

10. The Medassist Pharmaceutical Company makes a pill intended for children susceptible to seizures. The pill is supposed to contain 20.0 mg of phenobarbital. A random sample of 20 pills yielded the amounts (in mg) listed here. Are these pills acceptable at the $\alpha = 0.01$ significance level?

27.5	26.0	22.9	23.4	23.0
23.9	32.6	20.9	22.9	24.3
24.8	16.1	24.3	17.3	18.9
20.7	33.0	15.6	24.3	23.3

3. a. $\mu = \$90{,}000$

    b. Two-tailed

10. $\bar{x} = 23.29$, $s = 4.53$.

    TS: $t = 3.240$;

    CV: $t = \pm 2.861$

Cumulative Review Exercises

1. For healthy women aged 18–24, systolic blood pressure readings (in mm Hg) are normally distributed with a mean of 114.8 and a standard deviation of 13.1 (based on data from the National Health Survey).
   a. If a healthy woman is randomly selected from the general population of all women aged 18–24, find the probability of getting one with systolic blood pressure greater than 124.23.   1. a. 0.2358
   b. If 16 women are randomly selected, find the probability that the mean of their systolic blood pressure readings is greater than 124.23.   b. 0.0020

2. A medical researcher obtains the systolic blood pressure readings (in mm Hg) in the accompanying list for women aged 18–24 who have a new strain of viral infection. (As in Exercise 1, healthy women in that age group have readings that are normally distributed with a mean of 114.8 and a standard deviation of 13.1.)
   a. Find the sample mean $\bar{x}$ and standard deviation s.   2. a. $\bar{x} = 124.23$, $s = 22.52$
   b. Use a 0.05 significance level to test the claim that the sample comes from a population with a mean equal to 114.8.   b. TS: $t = 1.675$; CV: $t = \pm 2.132$.
   c. Use the sample data to construct a 95% confidence interval for the population mean μ. Do the confidence interval limits contain the value of 114.8, which is the mean for healthy women aged 18–24?   c. $112.23 < \mu < 136.23$; yes
   d. Use a 0.05 significance level to test the claim that the sample comes from a population with a standard deviation equal to 13.1, which is the standard deviation for healthy women aged 18–24.   d. TS: $\chi^2 = 44.339$; CV: $\chi^2 = 6.262, 27.488$.
   e. Based on the preceding results, does it seem that the new strain of viral infection affects systolic blood pressure?

134.9	78.7	108.9	133.0	123.7	96.1	126.9	89.8
132.0	134.7	132.1	121.7	112.3	150.2	158.3	154.4

3. A student majoring in psychology designs an experiment to test for extrasensory perception (ESP). In this experiment, a card is randomly selected from a shuffled deck, and the blindfolded subject must guess the suit (clubs, diamonds, hearts, spades) of the card selected. The experiment is repeated 25 times, with the card replaced and the deck reshuffled each time.
   a. For subjects who make random guesses with no ESP, find the mean number of correct responses.   3. a. 6.3
   b. For subjects who make random guesses with no ESP, find the standard deviation for the numbers of correct responses.   b. 2.2
   c. For subjects who make random guesses with no ESP, find the probability of getting more than 12 correct responses.   c. 0.0024
   d. If a subject gets more than 12 correct responses, test the claim that they made random guesses. Use a 0.05 level of significance.   *(continued)*

4. a.

9.	0005
10.	00000
11.	00055555
12.	05555
13.	000000005555555
14.	00555
15.	055555
16.	00055
17.	0

e. You want to conduct a survey to estimate the percentage of adult Americans who believe that some people have ESP. How many people must you survey if you want 90% confidence that your sample percentage is in error by no more than four percentage points?   4. e. 423

4. An important requirement in Sections 7-4 and 7-6 is that the sample data must come from a population that is normally distributed.

a. Refer to the head lengths of the bears in Data Set 3 of Appendix B and construct a stem-and-leaf plot.

b. Based on the result, does the distribution appear to be normal?   4. b. Yes

c. What other method could be used to determine whether the distribution is normal?   4. c. a histogram

d. Is there anything particularly noteworthy about this data set?   4. d. Yes

Computer Project

Refer to the body temperatures (in °F) summarized in the given stem-and-leaf plot. (The sample is a subset of the body temperatures listed in Table 7-1.) Use STAT-DISK or Minitab (or a TI-83 calculator) and a 0.05 significance level for the following tests.

a. Test the claim that the sample comes from a population with a mean of 98.6° F.

b. Add 5 to each of the original temperatures listed in the stem-and-leaf plot, then test the claim that the sample comes from a population with a mean of 103.6. Compare the results to those found in part (a). In general, what changes when the same constant is added to every score?

c. After multiplying each of the original temperatures by 10, test the claim that the sample comes from a population with a mean of 986. Compare the results to those found in part (a). In general, what changes when every score is multiplied by the same constant?

97.	01
97.	589
98.	00044
98.	5566778
99.	02

d. Fahrenheit temperatures can be converted to the Celsius scale by using $C = (5/9)(F - 32)$. That is, first subtract 32 from each Fahrenheit temperature, then multiply each result by 5/9. Because the conversion from the Fahrenheit scale to the Celsius scale involves adding the same constant (-32) and multiplying by the same constant (5/9), can you predict what will happen when the original temperatures are converted and the claim of $\mu = 37°$ C is tested? (Note that 98.6° F = 37° C.) Verify or disprove your prediction by converting the original temperatures to the Celsius scale and testing the claim that $\mu = 37°$ C.

e. When testing the claim that the mean weight of men is now greater than the known mean from men who lived 100 years ago, does it make any difference if the weights are measured in pounds or kilograms?

•••••••• **FROM DATA TO DECISION** ••••••••

Should the Insurance Discount be Dropped?

Insurance companies are reconsidering the discounts they give for cars equipped with anti-lock braking systems (ABS). Cars worth more than $15,000 and not equipped with ABS have damage amounts with a mean of $8908 when they crash. A random sample of 36 crashed cars worth more than $15,000 and equipped with ABS results in the damage amounts (in dollars) listed here (based on data

from the Highway Loss Data Institute). Use a 0.05 significance level to test the claim that damage amounts are not different when cars have ABS. Based on these results, should insurance companies continue discounts for ABS, or should those discounts be dropped? Based on the results, which course of action would you take as the Chief Executive Officer of a company that insures cars?

```
9982 7509 9359 8934 6512 9761 11226 9332 10047 10200 8417 6948
11491 6902 9866 9227 6060 7260  9026 8651  8974 11184 8079 9371
 9977 7731 8332 8750 6669 9545  7953 7483  7660  8891 7277 7467
```

•••••••• **COOPERATIVE GROUP ACTIVITIES** ••••••••

1. *Out-of-class activity:* A group activity suggested for Chapter 6 involved estimating the mean error (in seconds) on people's wristwatches. Use the same data collected in that activity to test the claim that the mean error of all wristwatches is equal to 0. Recall that some errors are positive (because the watch reads earlier than the actual time) and some are negative (because the watch reads later than the actual time). Describe the details of the test and include a graph showing the test statistic and critical values. Do we collectively run on time, or are we early or late? Also test the claim that the standard deviation of errors is less than one minute. What are the practical implications of a standard deviation that is excessively large?

2. *In-class activity:* In a group of three or four people, conduct an ESP experiment by selecting one of the group members as the subject. Draw a circle on one small piece of paper and

draw a square on another sheet of the same size. Repeat this experiment 20 times: Randomly select the circle or the square and place it in the subject's hand behind his or her back so that it cannot be seen, then ask the subject to identify the shape (without looking at it); record whether the response is correct. Test the claim that the subject has ESP because the proportion of correct responses is greater than 0.5.

3. *In-class activity:* After dividing into groups with sizes between 10 and 20 people, each group member should record the number of heartbeats in one minute. After calculating $\bar{x}$ and s, each group should proceed to test the claim that the mean is greater than 59, which is the author's result. (When people exercise, they tend to have lower pulse rates, and the author runs five miles a few times each week. What a guy.)

interview

Jay Dean

Senior Vice President at the
San Francisco Office of
Young & Rubicam Advertising

Jay Dean is Director of the Consumer Insights Department at the San Francisco office of Young & Rubicam. He has worked with many well-known advertisers, including AT&T, Chevron, Clorox, Coors, Dr. Pepper, Ford Motor Co., General Foods, Gillette, Gulf Oil, H.J. Heinz, Kentucky Fried Chicken, and Warner-Lambert.

How extensive is your use of statistics at Young & Rubicam?

We use statistics every day. We do a lot of consumer research, and we conduct many consumer attitude and usage surveys. For some clients we do product tests, taste tests, basic strategy studies—anything they need. I'm now working on a typical design, an advertising test for Take Heart salad dressing, one of the brands we handle for The Clorox Company. Two commercials were shown to independent samples of consumers. We then asked questions about what was being communicated by each commercial, perceptions of the brand, likes and dislikes, and so on. Significance testing will be used to compare the results and to choose the best commercial.

On larger surveys we use multivariate techniques such as factor analysis and multiple regression. Marketing is becoming tougher and tougher today. There are more brands, increasing price competition, and more sophisticated consumers. A tougher marketing environment requires more marketing research, and that means we're using statistics more often as well.

Could you cite a case in which the use of statistics was instrumental in determining a successful strategy?

We recently conducted a major creative exploratory for Pine-Sol, another Clorox brand. About half a dozen commercials were produced in rough form and shown to independent samples of consumers. Statistics helped us

to identify the best commercial for the new Pine-Sol advertising campaign. Early marketplace results are very encouraging. At Y&R we believe consumer research helps us to produce the most effective advertising. Statistics helps us to make informed decisions, based upon the results of the research.

How do you typically collect data for statistical analysis?

There are many ways to collect data, but typically it's either random telephone sampling for survey work, or if we want to show people something like a commercial, we use a central location interview. For example, respondents are intercepted in a shopping mall by an interviewer, screened for eligibility, asked to go into a testing facility to be shown a commercial, and so on.

Do you find it difficult to obtain representative and unbiased samples?

Yes. The marketing research industry now is very concerned about the growing refusal rate among the general public. That is driven in part by salespeople who use marketing research as a guise for selling. Mail surveys are subject to a huge self-selection bias and response rates are typically quite low. Shopping mall interviews have other problems. Not everyone goes to malls, and there is a certain degree of interviewer bias in approaching prospective respondents.

What is your typical sample size?

For a national survey the rule of thumb is about a thousand people, although you might go with as few as 600 in some cases. For an advertising test a sample on the order of 200 would be a pretty healthy sample size— sometimes as few as a hundred are used.

Do you feel that job applicants in your field are viewed more favorably if they have studied some statistics?

Yes, absolutely. Everyone has to understand and use statistics at some level. It's very important for entry-level people to know statistics well because they do much of our research project work. In the beginning they do a lot of number crunching and a lot of data analysis.

Do you have any advice for today's students?

If I could go back to school, I would certainly study more math, statistics, and computer science. I studied a lot of it, but I would like to know even more. There's an enormous data explosion in business these days. All businesses are becoming much more quantitative than they ever were in the past. Today there is more information than you know what to do with, and you've got to have the analytical tools and knowledge to deal with this information if you want to be successful.

8

Inferences from Two Samples

8-1 Overview

Chapters 6 and 7 presented the two major activities of inferential statistics: estimating a population parameter and testing a hypothesis about a population. Whereas those chapters considered cases involving only a single population, this chapter deals with two populations. Methods for hypothesis testing and the construction of confidence intervals are described for cases involving two populations.

8-2 Inferences about Two Means: Dependent Samples

Two dependent samples (consisting of paired data) are used to test hypotheses about the means of two populations. Methods are presented for constructing confidence intervals as estimates of the difference between two population means.

8-3 Inferences about Two Means: Independent and Large Samples

Independent and large ($n > 30$) samples are used to test hypotheses about the means of two populations. Methods are presented for the construction of confidence intervals used to estimate the difference between two population means.

8-4 Comparing Two Variances

A method is presented for testing hypotheses made about two population variances or standard deviations.

8-5 Inferences about Two Means: Independent and Small Samples

This section deals with cases in which two samples are independent and small ($n \leq 30$). Procedures for testing hypotheses about the difference between two population means are discussed. Also discussed are methods for constructing confidence intervals used to estimate the difference between two population means.

8-6 Inferences about Two Proportions

This section deals with methods of inferential statistics applied to two population proportions. Methods are presented for testing claims about the difference between two population proportions. Confidence intervals are constructed for estimating the difference between two population proportions.

Chapter Problem:

Do cigarette filters really make a difference, or are they just a sales gimmick with little or no effect?

In Table 8-1 we list measured tar, nicotine, and carbon monoxide amounts from randomly selected filtered and nonfiltered king-size cigarettes. All measurements are in milligrams, and the data are from the Federal Trade Commission. In Figure 8-1 we use boxplots to compare the filtered and non-filtered cigarettes. The boxplots suggest that there are distinct differences, but the two samples are relatively small ($n_1 = 21$, $n_2 = 8$) so we might question whether those apparent differences are significant. Later in this section, we will consider the data of Table 8-1 as we test for equality of two population means.

Table 8-1 Tar, Nicotine, and Carbon Monoxide in Cigarettes

Filtered Kings			Nonfiltered Kings		
Tar	Nicotine	CO	Tar	Nicotine	CO
16	1.2	14	23	1.6	14
15	1.3	12	23	1.9	15
16	1.1	14	24	1.6	17
14	1.1	16	26	1.8	17
16	1.0	15	25	1.7	16
1	0.1	2	26	1.7	16
16	1.1	14	21	1.4	14
18	1.0	16	24	1.5	16
10	0.8	11			
14	1.0	13			
12	0.9	13			
11	0.8	12			
14	1.0	13			
13	1.0	12			
13	1.0	13			
13	0.9	14			
16	1.2	14			
16	1.1	14			
8	0.1	9			
16	1.2	17			
11	0.9	12			

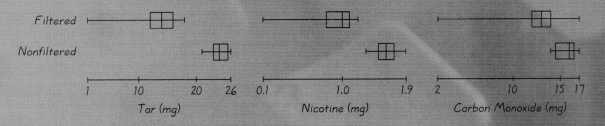

Figure 8-1   Comparisons of Filtered and Nonfiltered Cigarettes

8-1 Overview

Some instructors feel strongly that this chapter should be thoroughly covered, and others skip it. Even if this chapter is skipped, consider assigning some problems from it. Once students understand the basic concepts of hypothesis testing and confidence intervals, they can use software or a TI-83 calculator to do the number crunching. Sections 8-2, 8-3, and 8-6 are especially important, so consider covering at least those sections or assigning problems from them.

Chapter 6 introduced an important activity of inferential statistics: Samples were used to construct confidence intervals, which are used to estimate values of population parameters. Chapter 7 introduced a second important activity of inferential statistics: Samples were used to test hypotheses about population parameters. In both of those chapters all examples and exercises involved the use of *one* sample to form an inference about *one* population. In reality, however, there are many important and meaningful situations in which it becomes necessary to compare *two* sets of sample data. The following are examples typical of those found in this chapter, which presents methods for using data from two samples so that inferences can be made about the populations from which they came.

- Determine whether there is a difference between the mean amounts of nicotine from filtered cigarettes and nonfiltered cigarettes.
- Determine whether women employees exposed to ethyl glycol ethers have a rate of miscarriage that is different from the rate for women not exposed to those chemicals.

8-2 Inferences about Two Means: Dependent Samples

This section requires the *t* distribution introduced in Chapters 6 and 7. It is extremely important to emphasize that different methods of statistics are used for different configurations of data. This section introduces procedures for working with matched or paired data.

In this section we consider methods for testing hypotheses and constructing confidence intervals for two samples that are dependent, which means that they are paired or matched. We begin by formally defining *independent* and *dependent*.

DEFINITION

Two samples are **independent** if the sample selected from one population is not related to the sample selected from the other population. If one sample is related to the other, the samples are **dependent.** Such samples are often referred to as **paired samples** or **matched samples** (because we get two values from each subject, or we get one value from each of two subjects sharing the same characteristic).

Consider the paired sample data shown at the top of the following page. The sample of pretraining weights and the sample of posttraining weights are *dependent samples*, because each pair is matched according to the person involved. "Before/after" data are usually matched and are usually dependent.

Subject	A	B	C	D	E	F
Pretraining weights (kg)	99	62	74	59	70	73
Posttraining weights (kg)	94	62	66	58	70	76

Based on data from the *Journal of Applied Psychology*, Vol. 62, No. 1.

For the following data, however, the two samples are *independent* because the sample of females is not related to the sample of males. The data are not matched as they are in the preceding table.

Weights of females (lb)	115	107	110	128	130	
Weights of males (lb)	128	150	160	140	163	155	175

Assumptions   For the hypothesis tests and confidence intervals described in this section, we make the following assumptions.

1. Two dependent samples must be selected from two populations in a way that is *random*.

2. Each of the two populations must be *normally distributed*. (If they depart radically from normal distributions, we should not use the methods given in this section, in which case we may be able to use nonparametric methods, discussed in Chapter 13.)

In dealing with two dependent samples, we base our calculations on the differences (d) between the pairs of data, as illustrated in the table in the margin. (Comparing the two sample means $\bar{x}$ and $\bar{y}$ would waste important information about the paired data.) The following notation is based on those differences.

It wouldn't hurt to remind students once again about the importance of good sampling methodology. Point out that if the data are not carefully collected, the results may be totally worthless or seriously misleading.

x	10	8	5	20
y	7	2	9	20
d	3	6	−4	0

Notation for Two Dependent Samples

μ_d = mean value of the differences d for the *population* of paired data

$\bar{d}$ = mean value of the differences d for the paired *sample* data (equal to the mean of the $x - y$ values)

s_d = standard deviation of the differences d for the paired *sample* data

n = number of *pairs* of data

For example, using the d values of 3, 6, -4, 0 taken from the table in the margin, we get

$$\bar{d} = \frac{\Sigma d}{n} = \frac{3 + 6 + (-4) + 0}{4} = 1.25$$

$$s_d = \sqrt{\frac{n(\Sigma d^2) - (\Sigma d)^2}{n(n - 1)}} = \sqrt{\frac{4(61) - (5)^2}{4(4 - 1)}} = 4.27$$

$$n = 4$$

Hypothesis Tests

We now use the preceding notation to describe the test statistic to be used in hypothesis tests of claims made about the means of two populations, given that the two samples are dependent. When we randomly select two dependent samples from normally distributed populations in which the population mean of the paired differences is μ_d, the following test statistic possesses a Student t distribution.

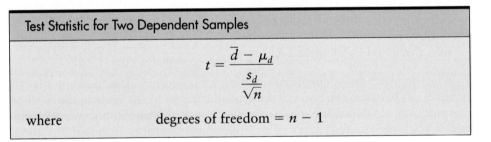

Test Statistic for Two Dependent Samples

$$t = \frac{\bar{d} - \mu_d}{\frac{s_d}{\sqrt{n}}}$$

where   degrees of freedom = $n - 1$

It might be helpful to point out that the test statistic follows the same basic format of many test statistics: The numerator is the difference between the sample statistic and the claimed value, and the denominator is the standard deviation of the sample statistic. It's really a modification of $z = (x - \mu)/\sigma$.

If the number of pairs of data is large ($n > 30$), the number of degrees of freedom will be at least 30, so critical values will be z scores (Table A-2) instead of t scores (Table A-3).

EXAMPLE   Using a reaction timer similar to the one described in the Cooperative Group Activities of Chapter 5, subjects are tested for reaction times with their left and right hands. (Only right-handed subjects were used.) The results (in thousandths of a second) are given in the accompanying table. Use a 0.05 significance level to test the claim that there is a difference between the mean of the right- and left-hand reaction times. If an engineer is designing a fighter-jet cockpit and must locate the ejection-seat activator to be accessible to either the right or the left hand, does it make a difference which hand she chooses?

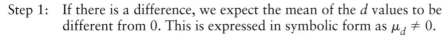

Subject	A	B	C	D	E	F	G	H	I	J	K	L	M	N
Right	191	97	116	165	116	129	171	155	112	102	188	158	121	133
Left	224	171	191	207	196	165	177	165	140	188	155	219	177	174

SOLUTION   Using the traditional method of hypothesis testing, we will test the claim that there is a difference between the right- and left-hand reaction times. Because we are dealing with paired data, begin by finding the differences $d = \text{right} - \text{left}$. Then follow the steps summarized in Figure 7-4.

Step 1: If there is a difference, we expect the mean of the d values to be different from 0. This is expressed in symbolic form as $\mu_d \neq 0$.

Step 2: If the original claim is not true, we have $\mu_d = 0$.

Step 3: The null hypothesis must contain equality, so we have

$$H_0: \mu_d = 0 \qquad H_1: \mu_d \neq 0 \text{ (original claim)}$$

Step 4: The significance level is $\alpha = 0.05$.

Step 5: Because we are testing a claim about the means of paired dependent data, we use the Student t distribution.

Step 6: Before finding the value of the test statistic, we must first find the values of $\bar{d}$ and s_d. When we evaluate the difference d for each subject, we find these differences $d = \text{right} - \text{left}$:

$$-33, \quad -74, \quad -75, \quad -42, \quad -80, \quad -36, \quad -6,$$
$$-10, \quad -28, \quad -86, \quad 33, \quad -61, \quad -56, \quad -41$$

We now use Formulas 2-1 and 2-4 as follows.

$$\bar{d} = \frac{\Sigma d}{n} = \frac{-595}{14} = -42.5$$

$$s_d = \sqrt{\frac{n(\Sigma d^2) - (\Sigma d)^2}{n(n-1)}} = \sqrt{\frac{14(39{,}593) - (-595)^2}{14(14-1)}} = 33.2$$

With these statistics and the assumption that $\mu_d = 0$, we can now find the value of the test statistic.

$$t = \frac{\bar{d} - \mu_d}{\frac{s_d}{\sqrt{n}}} = \frac{-42.5 - 0}{\frac{33.2}{\sqrt{14}}} = -4.790$$

Crest and Dependent Samples

In the late 1950s, Procter & Gamble introduced Crest toothpaste as the first such product with fluoride. To test the effectiveness of Crest in reducing cavities, researchers conducted experiments with several sets of twins. One of the twins in each set was given Crest with fluoride, while the other twin continued to use ordinary toothpaste without fluoride. It was believed that each pair of twins would have similar eating, brushing, and genetic characteristics. Results showed that the twins who used Crest had significantly fewer cavities than those who did not. This use of twins as dependent samples allowed the researchers to control many of the different variables affecting cavities.

(continued)

The critical values of $t = -2.160$ and $t = 2.160$ are found from Table A-3; use the column for 0.05 (two tails), and use the row with degrees of freedom of $n - 1 = 13$. Figure 8-2 shows the test statistic, critical values, and critical region.

Step 7: Because the test statistic does fall in the critical region, we reject the null hypothesis of $\mu_d = 0$.

Step 8: There is sufficient evidence to support the claim of a difference between the right- and left-hand reaction times. Because there does appear to be such a difference, an engineer designing a fighter-jet cockpit should locate the ejection-seat activator so that it is readily accessible to the faster hand, which appears to be the right hand with seemingly lower reaction times. (We could require special training for left-handed pilots if a similar test of left-handed pilots shows that their dominant hand is faster.)

The preceding example is a two-tailed test, but left-tailed tests and right-tailed tests follow the same basic procedure. For example, if we want to test the claim that the right-hand reaction times are lower than the left-hand times, we have a claim that $\mu_d < 0$. That is, the values of "right $-$ left" should be negative. This claim leads to a left-tailed test with a critical value of $t = -1.771$; the test statistic is again $t = -4.790$. With the given test statistic and critical value, we reject the null hypothesis of $\mu_d \geq 0$. That is, this left-tailed test leads to the conclusion that there is sufficient evidence to support the claim that the right hands have lower reaction times.

The preceding two-tailed example used the traditional method, but the P-value approach could be used by modifying Steps 6 and 7. In Step 6, use the test statistic of $t = -4.790$ and refer to the 13th row of Table A-3 to find that 4.790 is beyond the largest table value of 3.012. The P-value in the two-tailed test is therefore less than $2(0.005)$ or 0.01; that is, P-value < 0.01. In Step 7, we again reject the null hypothesis because the P-value is less than the significance level of $\alpha = 0.05$.

We can develop a confidence interval estimate of the population mean difference μ_d by using the sample mean $\overline{d}$, the standard deviation of sample means d (which is $s_d/\sqrt{n}$), and the critical value $t_{\alpha/2}$.

When discussing P-values, remind students that for two-tailed tests, we must *double* the area beyond the test statistic. It might be helpful to conduct a brief but general review of the P-value approach. Students using TI-83 calculators will find the P-value approach easy, because their displays will include P-values.

Because the test statistic and margin of error E both use the t distribution, some students might incorrectly believe that the methods of this section apply to small samples only. Point out that large samples can be handled as well. When it is necessary to find critical t values with more than 29 degrees of freedom, simply use the standard normal distribution (Table A-2).

Confidence Intervals

The confidence interval estimate of the mean difference μ_d is as follows:

$$\overline{d} - E < \mu_d < \overline{d} + E$$

where $E = t_{\alpha/2} \dfrac{s_d}{\sqrt{n}}$ and degrees of freedom $= n - 1$.

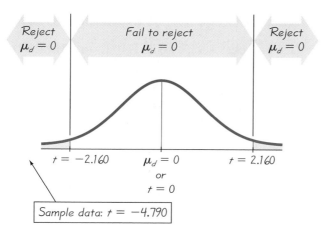

Figure 8-2 Distribution of Differences Between Right- and Left-Hand Reaction Times

EXAMPLE   Use the sample data from the preceding example to construct a 95% confidence interval estimate of μ_d.

SOLUTION   Using the values of $\overline{d} = -42.5$, $s_d = 33.2$, $n = 14$, and $t_{\alpha/2} = 2.160$, we first find the value of the margin of error E.

$$E = t_{\alpha/2} \frac{s_d}{\sqrt{n}} = 2.160 \frac{33.2}{\sqrt{14}} = 19.2$$

The confidence interval can now be found.

$$\overline{d} - E < \mu_d < \overline{d} + E$$
$$-42.5 - 19.2 < \mu_d < -42.5 + 19.2$$
$$-61.7 < \mu_d < -23.3$$

The result is sometimes expressed as $\mu_d = -42.5 \pm 19.2$ or as $(-61.7, -23.3)$. In the long run, 95% of such samples will lead to confidence interval limits that actually do contain the true population mean of the differences. Note that the confidence interval limits do not contain 0, indicating that the true value of μ_d is significantly different from 0. That is, the mean value of the "right − left" differences is different from 0. On the basis of the confidence interval, we conclude that there is sufficient evidence to support the claim that there is a difference between the right- and left-hand reaction times. This conclusion agrees with the conclusion in the preceding example.

It would be helpful to review the interpretation of confidence intervals, as well as the other formats.

Computer and Calculator Usage for Two Dependent Samples

Statistical software packages such as STATDISK and Minitab can be used to test hypotheses and construct confidence intervals for situations involving two dependent samples. Shown here are the STATDISK and Minitab displays for the example of this section. If you are using STATDISK, select Analysis, then Hypothesis Testing, then Mean-Two Dependent Samples. The result will be the display shown here. (STATDISK automatically provides confidence interval limits for two-tailed hypothesis tests.)

STATDISK DISPLAY

File   Edit   Analysis   Data   Help

Claim $\mu_d = 0$
Null Hypothesis $\mu_d = 0$

Sample Size, n   14
Difference Mean, $\bar{x}_d$ -42.50
Difference St Dev, s_d 33.17

Test Statistic, t   -4.7937
Critical t   ±2.1604
P-Value   0.0004
95% Confidence interval:
    $-61.65 < \mu_d < -23.35$

Reject the Null Hypothesis

Sample provides evidence to reject the claim

If you are using Minitab, the basic approach is to create a column of the differences d, then do a t test of the claim that the differences come from a population with a mean of 0. Enter the "right" scores in column C1 and enter the "left" scores in column C2. Now click on Calc, then Calculator, then enter C3 for the column in which to store the result, and enter the expression C1 − C2. (Column C3 now contains the differences d.) Next, select the main menu item of Stat, then Basic Statistics, then 1-Sample t. Enter C3 for the variable, click on the Test mean circle, and enter the value for the desired mean (0 in this case). The box labeled as Alternative has a default of "Not equal," so leave it as is for this example. Click on OK and the result should be as shown. Minitab will also provide a confidence interval if you select Stat, Basic Statistics, 1-Sample t, and the Confidence interval option.

MINITAB DISPLAY						

```
TEST OF MU = 0.00 VS MU N.E. 0.00
        N      MEAN     STDEV     SE MEAN       T    P VALUE
C3     14     -42.50    33.17       8.87     -4.79   0.0004
```

The TI-83 calculator can also be used for the methods of this section. Enter the data for the first variable in list L1, enter the data for the second variable in list L2, then clear the screen and enter L1 − L2 → L3. Now press STAT, then select TESTS and choose the option of T-Test. Using the input option of Data, enter the indicated data and press ENTER when done. If the paired data from the preceding examples are used, the window display will include the test statistic of $t = -4.793724854$, the P-value of 0.0003507665, the value of $\overline{d}$ (indicated as $\overline{x}$), and the value of s_d (indicated as Sx). The TI-83 uses the P-value method of testing hypotheses. A confidence interval can also be found by pressing STAT, then selecting TESTS, then TInterval.

We used Table A-3 to conclude that the P-value is less than 0.01, but from the STATDISK, Minitab, and TI-83 displays we can see that the P-value is actually 0.0004.

8-2 Exercises A: Basic Skills and Concepts

In Exercises 1 and 2, assume that you want to test the claim that the paired sample data come from a population for which the mean difference is $\mu_d = 0$. Assuming a 0.05 level of significance, find (a) $\overline{d}$, (b) s_d, (c) the t test statistic, and (d) the critical values.

Recommended assignment: Exercises 1, 5, 7, 8, 11, 12.

1.

x	8	8	6	9	7
y	3	2	6	4	9

2.

x	20	25	27	27	23	29	30	26
y	20	24	25	29	20	29	32	29

3. Using the sample paired data in Exercise 1, construct a 95% confidence interval for the population mean of all differences $x - y$.

4. Using the sample paired data in Exercise 2, construct a 99% confidence interval for the population mean of all differences $x - y$.

1. a. $\overline{d} = 2.8$
   b. $s_d = 3.6$
   c. $t = 1.757$
   d. $t = \pm 2.776$
2. a. $\overline{d} = -0.125$
   b. $s_d = 2.1$
   c. $t = -0.168$
   d. $t = \pm 2.365$
3. $-1.6 < \mu_d < 7.2$
4. $-2.7 < \mu_d < 2.5$

5. Does it pay to take preparatory courses for standardized tests such as the SAT? Using the sample data in the following table, test the claim that the Allan Preparation Course has an effect on the SAT score. Use a 0.05 level of significance.

Subject	A	B	C	D	E	F	G	H	I	J
SAT score before course (x)	700	840	830	860	840	690	830	1180	930	1070
SAT score after course (y)	720	840	820	900	870	700	800	1200	950	1080

Based on data from the College Board and "An Analysis of the Impact of Commercial Test Preparation Courses on SAT Scores," by Sesnowitz, Bernhardt, and Knain, *American Educational Research Journal*, Vol. 19, No. 3.

6. Use the sample data given in Exercise 5 to construct a 95% confidence interval for the mean difference of the "before" minus "after" scores.

7. Captopril is a drug designed to lower systolic blood pressure. When subjects were tested with this drug, their systolic blood pressure readings (in mm of mercury) were measured before and after the drug was taken, with the results given in the accompanying table. Use the sample data to construct a 99% confidence interval for the mean difference between the before and after readings.

Subject	A	B	C	D	E	F	G	H	I	J	K	L
Before	200	174	198	170	179	182	193	209	185	155	169	210
After	191	170	177	167	159	151	176	183	159	145	146	177

Based on data from "Essential Hypertension: Effect of an Oral Inhibitor of Angiotensin-Converting Enzyme" by MacGregor et al., *British Medical Journal*, Vol. 2.

8. Refer to the sample data given in Exercise 7. Use a 0.05 significance level to test the claim that captopril is effective in lowering systolic blood pressure.

9. A study was conducted to investigate the effectiveness of hypnotism in reducing pain. Results for randomly selected subjects are given in the accompanying table. At the 0.05 significance level, test the claim that the sensory measurements are lower after hypnotism. (The values are before and after hypnosis; the measurements are in centimeters on a pain scale.) Does hypnotism appear to be effective in reducing pain?

Subject	A	B	C	D	E	F	G	H
Before	6.6	6.5	9.0	10.3	11.3	8.1	6.3	11.6
After	6.8	2.4	7.4	8.5	8.1	6.1	3.4	2.0

Based on "An Analysis of Factors That Contribute to the Efficacy of Hypnotic Analgesia," by Price and Barber, *Journal of Abnormal Psychology*, Vol. 96, No. 1.

10. Using the data in Exercise 9, construct a 95% confidence interval for the mean of the "before − after" differences.

11. Mental measurements of young children are often made by giving them blocks and telling them to build a tower as tall as possible. One experiment of block building was repeated a month later, with the times (in seconds) listed in the accompanying table. Use a 0.01 level of significance to test the claim that there is no difference between the two times.

Child	A	B	C	D	E	F	G	H	I	J	K	L	M	N	O
First trial	30	19	19	23	29	178	42	20	12	39	14	81	17	31	52
Second trial	30	6	14	8	14	52	14	22	17	8	11	30	14	17	15

Based on data from "Tower Building," by Johnson and Courtney, *Child Development*, Vol. 3.

12. Refer to the sample data in Exercise 11 and construct a 95% confidence interval for the mean of the differences. Do the confidence interval limits contain 0, indicating that there is not a significant difference between the times of the first and second trials?

13. A study was conducted to investigate some effects of physical training. Sample data are listed below. (See "Effect of Endurance Training on Possible Determinants of VO_2 During Heavy Exercise," by Casaburi et al., *Journal of Applied Physiology*, Vol. 62, No. 1.) At the 0.05 level of significance, test the claim that the mean pretraining weight equals the mean posttraining weight. All weights are given in kilograms. What do you conclude about the effect of training on weight?

Pretraining:	99	57	62	69	74	77	59	92	70	85
Posttraining:	94	57	62	69	66	76	58	88	70	84

14. Use the data from Exercise 13 to construct a 95% confidence interval for the mean of the differences between pretraining and posttraining weights.

15. Refer to Data Set 2 in Appendix B. Use the paired data consisting of body temperatures of women at 8:00 A.M. and at 12:00 A.M. on Day 2. Construct a 95% confidence interval for the mean difference of 8 A.M. temperatures minus 12 A.M. temperatures.

16. Using the sample data described in Exercise 15, and using a 0.05 level of significance, test the claim that for those temperatures, the mean difference is 0. Based on the result, do morning and night body temperatures appear to be about the same?

17. The following Minitab display resulted from an experiment in which 10 subjects were tested for motion sickness before and after taking the drug astemizole. The Minitab data column C3 consists of differences in the number of head movements that the subjects could endure without

10. $0.69 < \mu_d < 5.56$

11. TS: $t = 2.631$.
CV: $t = \pm 2.977$.

12. $4.1 < \mu_d < 40.4$; no

13. TS: $t = 2.301$.
CV: $t = \pm 2.262$.

14. $0.0 < \mu_d < 4.0$

15. $-1.40 < \mu_d < -0.17$

16. TS: $t = -2.840$.
CV: $t = \pm 2.228$.

17. TS: $t = -0.41$.
P-value: 0.69

becoming nauseous. (The differences were obtained by subtracting the "after" values from the "before" values.) Use a 0.05 significance level to test the claim that astemizole has an effect (for better or worse) on vulnerability to motion sickness. Based on the result, would you use astemizole if you were concerned about motion sickness while on a cruise ship?

```
TEST OF MU = 0.0 VS MU N.E. 0.0
        N    MEAN    STDEV    SE MEAN         T    P VALUE
C3 10    -7.5    57.7       18.2    -0.41      0.69
```

18. P-value = 0.345.

18. Refer to the Minitab display in Exercise 17. What P-value would result from a test of the claim that astemizole is effective in preventing motion sickness? What do you conclude?

19. $-48.8 < \mu_d < 33.8$; yes

19. Use the Minitab results from Exercise 17 to construct a 95% confidence interval for the mean difference between the numbers of head movements that the subject could endure without becoming nauseous. Do the confidence interval limits contain 0, indicating that astemizole does not have a significant effect on vulnerability to motion sickness?

8-2 Exercises B: Beyond the Basics

20. a. $0.10 < P$-value < 0.20
b. $0.005 < P$-value < 0.01

20. Refer to the indicated exercise and find the P-value, or find (from Table A-3) limits containing the P-value.

   a. Exercise 5   **b.** Exercise 9

21. Hypothesis test results not affected. Confidence interval limits change.

21. The example in this section used reaction times given in thousandths of a second. Suppose we express all of the times in seconds as 0.191, 0.224, and so on. Is the hypothesis test affected by such a change in units? Is the confidence interval affected by such a change in units? How?

22. 0.025

22. The 95% confidence interval for a collection of paired sample data is $0.0 < \mu_d < 1.2$. Based on this confidence interval, the traditional method of hypothesis testing leads to the conclusion that the claim of $\mu_d > 0$ is supported. What is the smallest possible value of the significance level of the hypothesis test?

8-3 Inferences about Two Means: Independent and Large Samples

Section 8-2 described methods of using inferential statistics for dependent (paired) data, but in this section our two samples are *independent*. In Section 8-2 we defined samples to be independent if the sample selected from one population is not related to the sample selected from the other population.

This section is restricted not only to consideration of independent samples, but also to large ($n > 30$) samples. That is, the sample size n_1 for sample 1 will be greater than 30, and the sample size n_2 for sample 2 will also be greater than 30. In Section 8-5 we will deal with cases involving small independent samples. For the purposes of this section, the following assumptions apply.

Because the methods of this section are relatively easy, some statistics textbooks cover only this section when discussing inferences from two samples.

Assumptions   In testing hypotheses about two population means and constructing confidence intervals for the difference between two population means, we make the following assumptions for the methods used in this section.

1. The two samples are *independent*.
2. The two sample sizes are large. That is, $n_1 > 30$ and $n_2 > 30$.

Boxplots

Before jumping into a formal hypothesis test or construction of a confidence interval, it is often enlightening to construct boxplots that can be used to make a visual comparison of the two samples in question. For example, Figure 8-3 on page 426 includes boxplots for the axial loads of a sample of cans 0.0109 in. thick and another sample of cans 0.0111 in. thick. (An axial load of a can is the maximum weight supported by its sides, and it is measured by using a plate to apply increasing pressure to the top of the can until it collapses. The original data are found in Data Set 15 in Appendix B.)

Inspection of the values in the original two data sets reveals that the axial load of 504 lb (for the cans that are 0.0111 in. thick) is an outlier because it is very far away from all of the other values. If that outlier is deleted, the long right tail (or "whisker") on the bottom boxplot is shortened to the value identified as 317. A comparison of the two boxplots shows that the spread of the data is about the same, but the values for the 0.0111 in. cans appear to be farther to the right, suggesting that those cans have larger axial loads and are stronger. We can now use a hypothesis test to determine whether that apparent difference is significant.

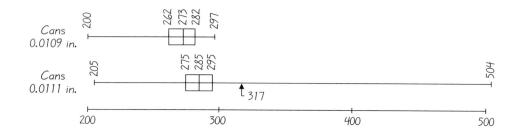

Figure 8-3 Boxplots of Axial Loads (in Pounds) of Cans 0.0109 in. Thick and 0.0111 in. Thick

Hypothesis Tests

One conclusion of the central limit theorem is that sample means tend to be normally distributed. Later in this section we will justify this important property: The *differences* between sample means $(\bar{x}_1 - \bar{x}_2)$ also tend to be normally distributed. Using these properties and the previously listed assumptions, we get the following test statistic to be used for testing hypotheses made about the means of two populations. This test statistic is very similar to several of the test statistics that we have already encountered. Note that it has the same basic format of

$$\frac{\text{(sample statistic)} - \text{(claimed population parameter)}}{\text{(standard deviation of sample statistics)}}$$

Test Statistic for Two Means: Independent and Large Samples
$$z = \frac{(\bar{x}_1 - \bar{x}_2) - (\mu_1 - \mu_2)}{\sqrt{\dfrac{\sigma_1^2}{n_1} + \dfrac{\sigma_2^2}{n_2}}}$$

Some attentive students might wonder why the standard deviation has a sum in it, while the numerator uses a difference. Consider explaining that the variance of the *differences* between two independent random variables equals the variance of the first random variable *plus* the variance of the second random variable. This is discussed later in the section.

As in Chapter 7, if the values of σ_1 and σ_2 are not known, we can use s_1 and s_2 in their places, provided that both samples are large. If σ_1 and σ_2 are known, we use those values in calculating the test statistic, but realistic cases will usually require the use of s_1 and s_2. It is rare that we know the values of population standard deviations without knowing the values of the population means.

EXAMPLE   Use a 0.01 significance level to test the claim that cans 0.0109 in. thick have a lower mean axial load than cans that are 0.0111 in. thick. The original data are listed in Data Set 15 of Appendix B, and the summary statistics are listed in the margin.

Axial Loads (lb)
of 0.0109 in. Cans
_____

$n_1 = 175$

$\bar{x}_1 = 267.1$

$s_1 = 22.1$

SOLUTION

Step 1:   The claim can be expressed symbolically as $\mu_1 < \mu_2$.

Step 2:   If the original claim is false, then $\mu_1 \geq \mu_2$.

Step 3:   The null hypothesis must contain equality, so we have

$$H_0: \mu_1 \geq \mu_2 \qquad H_1: \mu_1 < \mu_2 \text{ (original claim)}$$

We now proceed with the assumption that $\mu_1 = \mu_2$, or $\mu_1 - \mu_2 = 0$.

Axial Loads (lb)
of 0.0111 in. Cans
_____

$n_2 = 175$

$\bar{x}_2 = 281.8$

$s_2 = 27.8$

Step 4:   The significance level is $\alpha = 0.01$.

Step 5:   Because we have two independent and large samples and we are testing a claim about the two population means, we use a normal distribution with the test statistic given earlier in this section.

Step 6:   The values of σ_1 and σ_2 are unknown, but the samples are both large, so we can use the sample standard deviations as estimates of the population standard deviations, and the test statistic is calculated as follows.

$$z = \frac{(\bar{x}_1 - \bar{x}_2) - (\mu_1 - \mu_2)}{\sqrt{\dfrac{\sigma_1^2}{n_1} + \dfrac{\sigma_2^2}{n_2}}} = \frac{(267.1 - 281.8) - 0}{\sqrt{\dfrac{22.1^2}{175} + \dfrac{27.8^2}{175}}} = -5.48$$

Because we are using a normal distribution, the critical value of $z = -2.33$ is found from Table A-2. (With $\alpha = 0.01$ in the left tail, find the z score corresponding to an area of $0.5 - 0.01$, or 0.4900.) The test statistic, critical value, and critical region are shown in Figure 8-4 at the top of the following page.

Step 7:   Because the test statistic does fall within the critical region, reject the null hypothesis $\mu_1 \geq \mu_2$.

Step 8:   There is sufficient evidence to support the claim that the 0.0109 in. cans have a mean axial load that is lower than the mean for the 0.0111 in. cans. We might have expected the thinner cans to be weaker, but we now know that the difference is statistically significant. The *practical* significance of that difference is another issue not addressed by this hypothesis test.

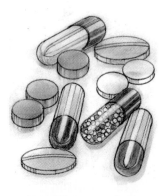

Figure 8-4 Distribution of Differences Between Means of Axial Loads for 0.0109 in. Cans and 0.0111 in. Cans

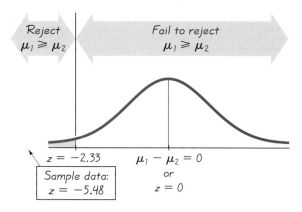

P-Values

Determination of P-values is easy here because the test statistics are z scores from the standard normal distribution. Simply follow the method outlined in Figures 7-7 and 7-8. For the preceding left-tailed hypothesis test, the P-value is the area to the left of the test statistic $z = -5.48$. Referring to Table A-2, we find that the area to the left of $z = -5.48$ is 0.0001. (Because the z score is beyond 3.09, we use an area of 0.4999 as indicated in the footnote to Table A-2.) Because the P-value of 0.0001 is less than the significance level of $\alpha = 0.01$, we again reject the null hypothesis and conclude that the 0.0109 in. cans have a mean that is lower than the mean for the 0.0111 in. cans.

Confidence Intervals

The confidence interval estimate of the difference $\mu_1 - \mu_2$ is as follows:

$$(\bar{x}_1 - \bar{x}_2) - E < (\mu_1 - \mu_2) < (\bar{x}_1 - \bar{x}_2) + E$$

where

$$E = z_{\alpha/2} \sqrt{\frac{\sigma_1^2}{n_1} + \frac{\sigma_2^2}{n_2}}$$

EXAMPLE   Using the sample data given in the preceding example, construct a 99% confidence interval estimate of the difference between the means of the axial loads of the 0.0109 in. cans and the 0.0111 in. cans.

SOLUTION   We first find the value of the margin of error E.

$$E = z_{\alpha/2} \sqrt{\frac{\sigma_1^2}{n_1} + \frac{\sigma_2^2}{n_2}} = 2.575 \sqrt{\frac{22.1^2}{175} + \frac{27.8^2}{175}} = 6.9$$

We now find the desired confidence interval as follows:

$$(\bar{x}_1 - \bar{x}_2) - E < (\mu_1 - \mu_2) < (\bar{x}_1 - \bar{x}_2) + E$$
$$(267.1 - 281.8) - 6.9 < (\mu_1 - \mu_2) < (267.1 - 281.8) + 6.9$$
$$-21.6 < (\mu_1 - \mu_2) < -7.8$$

We are 99% confident that the limits of -21.6 and -7.8 actually do contain the difference between the two population means. This result could be more clearly presented by stating that μ_2 exceeds μ_1 by an amount that is between 7.8 lb and 21.6 lb. Because those limits do not contain 0, it is very unlikely that the two population means are equal.

Why do the test statistic and confidence interval have the particular forms we have presented? Both forms are based on a normal distribution of $\bar{x}_1 - \bar{x}_2$ values with mean $\mu_1 - \mu_2$ and standard deviation $\sqrt{\sigma_1^2/n_1 + \sigma_2^2/n_2}$. This follows from the central limit theorem introduced in Section 5-5. The central limit theorem tells us that sample means $\bar{x}$ are normally distributed with mean μ and standard deviation $\sigma/\sqrt{n}$. Also, when samples have a size of 31 or larger, the normal distribution serves as a reasonable approximation to the distribution of sample means. By similar reasoning, the values of $\bar{x}_1 - \bar{x}_2$ also tend to approach a normal distribution with mean $\mu_1 - \mu_2$. When both samples are large, the following property of variances leads us to conclude that the values of $\bar{x}_1 - \bar{x}_2$ will have a standard deviation of

$$\sqrt{\frac{\sigma_1^2}{n_1} + \frac{\sigma_2^2}{n_2}}$$

The variance of the *differences* between two independent random variables equals the variance of the first random variable *plus* the variance of the second random variable.

That is, the variance of sample values $\bar{x}_1 - \bar{x}_2$ will tend to equal

$$\sigma_{\bar{x}_1}^2 + \sigma_{\bar{x}_2}^2$$

provided that $\bar{x}_1$ and $\bar{x}_2$ are independent. (See Exercise 21.) For large random samples, the standard deviation of sample means is $\sigma/\sqrt{n}$, so the variance of sample means is σ^2/n. If we combine the additive property of variances with

The Placebo Effect

It has been a common belief that when patients are given a placebo (a treatment with no medicinal value), about one-third of them show some improvement. However, a more recent study of 6000 patients showed that for those with mild medical problems, the placebos seemed to result in improvement in about two-thirds of the cases. The placebo effect seems to be strongest when patients are very anxious and they like their physicians. Because it could cloud studies of new treatments, the placebo effect is minimized by using a *double-blind* experiment in which neither the patient nor the physician knows whether the treatment is a placebo or a real medicine.

the central limit theorem's expression for variance of sample means, we get the following:

$$\sigma^2_{\bar{x}_1 - \bar{x}_2} = \sigma^2_{\bar{x}_1} + \sigma^2_{\bar{x}_2} = \frac{\sigma^2_1}{n_1} + \frac{\sigma^2_2}{n_2}$$

Taking the square root, we see that the standard deviation of the differences $\bar{x}_1 - \bar{x}_2$ can be expressed as

$$\sqrt{\frac{\sigma^2_1}{n_1} + \frac{\sigma^2_2}{n_2}}$$

Because z is a standard score that corresponds in general to

$$z = \frac{(\text{sample statistic}) - (\text{population mean})}{(\text{standard deviation of sample statistics})}$$

we get

$$z = \frac{(\bar{x}_1 - \bar{x}_2) - (\mu_1 - \mu_2)}{\sqrt{\frac{\sigma^2_1}{n_1} + \frac{\sigma^2_2}{n_2}}}$$

by noting that the sample values of $\bar{x}_1 - \bar{x}_2$ will have a mean of $\mu_1 - \mu_2$ and the standard deviation previously given.

Using Computers and Calculators for Hypothesis Tests and Confidence Intervals

The hypothesis testing and confidence interval construction of this section can be accomplished by using statistical software packages. To use STATDISK, select Analysis, then Hypothesis Testing, then Mean-Two Indepen-dent Samples. Confidence interval limits are included with hypothesis test results. The result will be as shown on the following page. You can see that the STATDISK results agree quite well with those obtained in this section.

Minitab is designed for raw data and does not deal directly with summary statistics, so the original data must be entered in columns C1 and C2. (See the *Minitab Student Laboratory Manual and Workbook* for a way to circumvent this restriction.) After entering the first data set in column C1 and the second data set in column C2, select Stat/Basic Statistics/2-Sample t. Select the option of Samples in different columns, and enter C1 for the first sample and C2 for the second sample. In the box labeled as Alter-

native, select the option of less than. Enter a confidence level of 99 (for a 0.01 significance level). The Minitab display should be as shown below.

STATDISK DISPLAY

File   Edit   Analysis   Data   Help

Claim	$\mu_1 < \mu_2$
Null Hypothesis	$\mu_1 > $ or $= \mu_2$

Large Samples
Test Statistic, z   -5.4757
Critical z   -2.3263
P-Value   0.0000
99% Confidence Interval:
   $\mu_1 - \mu_2 > -20.95$

Reject the Null Hypothesis

Sample provides evidence to support the claim

Capture-Recapture

Ecologists need to determine population sizes of endangered species. One method is to capture a sample of some species, mark all members of this sample, and then free them. Later, another sample is captured, and the ratio of marked subjects, coupled with the size of the first sample, can be used to estimate the population size. This capture-recapture method was used with other methods to estimate the blue whale population, and the result was alarming: The population was as small as 1000. That led the International Whaling Commission to ban the killing of blue whales to prevent their extinction.

MINITAB DISPLAY

```
Two sample T for C1 vs C2
        N     Mean    StDev   SE Mean
C1    175    267.1    22.1      1.7
C2    175    281.8    27.8      2.1

99% CI for mu C1 − mu C2: (−21.6,   −7.7)
T-Test mu C1 = mu C2 (vs <): T = −5.47  P = 0.0000
DF = 331
```

The TI-83 calculator can also be used for the methods of this section. Begin by entering the data in lists L1 and L2, then press STAT and select TESTS. Choose the option of 2-SampZTest and proceed to enter the data requested. (If summary statistics are known, you can use the Stats option. If both sample sizes are larger than 30, enter the sample standard deviations when

prompted for σ_1 and σ_2.) Using the data in the preceding example will result in a test statistic of $z = -5.475651121$ and a P-value of 2.184799E $^{-}8$, which is really 0.0000000218. To get a confidence interval, press STAT, then select TESTS and choose the option of 2-SampZInt.

8-3 Exercises A: Basic Skills and Concepts

Recommended assignment: Exercises 5–10.

In Exercises 1 and 2, use a 0.05 significance level to test the claim that the two samples come from populations with the same mean. In each case, the two samples are independent and have been randomly selected.

1. TS: $z = 0.79$.
   CV: $z = \pm1.96$.
2. TS: $z = -2.14$.
   CV: $z = \pm1.96$.
3. $-3 < \mu_1 - \mu_2 < 7$; yes.
4. $-1.74 < \mu_1 - \mu_2 < -0.08$; no.
5. TS: $z = 2.88$.
   CV: $z = 2.33$.
6. 23 min $< \mu_1 - \mu_2 <$ 65 min; no.

1.

Control Group	Experimental Group
$n_1 = 50$	$n_2 = 100$
$\bar{x}_1 = 75$	$\bar{x}_2 = 73$
$s_1 = 15$	$s_2 = 14$

2.

Treated Items	Untreated Items
$n_1 = 60$	$n_2 = 75$
$\bar{x}_1 = 8.75$	$\bar{x}_2 = 9.66$
$s_1 = 2.05$	$s_2 = 2.88$

3. Using the sample data given in Exercise 1, construct a 95% confidence interval estimate of the difference between the two population means. Do the confidence interval limits contain 0? What do you conclude about the difference between the control group and experimental group?

4. Using the sample data given in Exercise 2, construct a 95% confidence interval estimate of the difference between the two population means. Do the confidence interval limits contain 0? What do you conclude about the difference between the treated items and those that are untreated?

5. Does stress affect the recall ability of police eyewitnesses? This issue was studied in an experiment that tested eyewitness memory a week after a nonstressful interrogation of a cooperative suspect and a stressful interrogation of an uncooperative and belligerent suspect. The numbers of details recalled a week after the incident are summarized in the margin (based on data from "Eyewitness Memory of Police Trainees for Realistic Role Plays," by Yuille et al., *Journal of Applied Psychology*, Vol. 79, No. 6). Use a 0.01 level of significance to test the claim in the article that "stress decreased the amount recalled."

Nonstress	Stress
$n_1 = 40$	$n_2 = 40$
$\bar{x}_1 = 53.3$	$\bar{x}_2 = 45.3$
$s_1 = 11.6$	$s_2 = 13.2$

6. Nielsen Media Research reported that adult women watch TV an average of 5 hr and 1 min per day, compared to an average of 4 hr and 17 min for adult men. Assume that those results are found from a sample of 100 men and 100 women and that the two groups have the same standard deviation of 57 min. Construct a 99% confidence interval for the difference

$\mu_1 - \mu_2$, where μ_1 is the mean for adult women. Do the confidence interval limits contain 0? Does this suggest that there is or is not a significant difference between the two means?

7. The Medassist Pharmaceutical Company wants to test Dozenol, a new cold medicine intended for night use. Tests for such products often include a "treatment group" of people who use the drug and a "control group" of people who don't use the drug. Fifty people with colds are given Dozenol, and 100 others are not. The systolic blood pressure is measured for each subject, and the sample statistics are as given in the margin. The head of research at Medassist claims that Dozenol does not affect blood pressure—that is, the treatment population mean μ_1 and the control population mean μ_2 are equal. Test that claim using a significance level of 0.01. Based on the result, would you recommend advertising that Dozenol does not affect blood pressure?

8. Refer to the sample data used in Exercise 7 and construct a 99% confidence interval for $\mu_1 - \mu_2$, where μ_1 and μ_2 represent the mean for the treatment group and the control group, respectively.

9. Students at the author's college randomly selected 217 student cars and found that they had ages with a mean of 7.89 years and a standard deviation of 3.67 years. They also randomly selected 152 faculty cars and found that they had ages with a mean of 5.99 years and a standard deviation of 3.65 years. Use a 0.05 significance level to test the claim that student cars are older than faculty cars.

10. Refer to the data from Exercise 9 and construct a 95% confidence interval for the mean $\mu_1 - \mu_2$, where μ_1 is the mean age of student cars.

11. As part of the National Health Survey, data were collected on the weights of men. For 804 men aged 25–34, the mean is 176 lb and the standard deviation is 35.0 lb. For 1657 men aged 65–74, the mean and standard deviation are 164 lb and 27.0 lb, respectively. Test the claim that the older men come from a population with a mean that is less than the mean for men in the 25–34 age bracket. Use a 0.01 level of significance.

12. Use the data from Exercise 11 to construct a 99% confidence interval for the difference between the means of the men in the two age brackets. Do the confidence interval limits contain 0? Does this indicate that there is or is not a significant difference between the two means?

13. In a study of flight attendants, salaries paid by two different airlines were randomly selected. For 40 American Airlines flight attendants, the mean is $23,870 and the standard deviation is $2960. For 35 TWA flight attendants, the mean is $22,025 and the standard deviation is $3065 (based on data from the Association of Flight Attendants). At the 0.10 level of significance, test the claim that American and TWA have the same mean salary for flight attendants. Based on the result, should salary be an important factor for a prospective flight attendant in choosing between American and TWA?

Treatment Group	Control Group
$n_1 = 50$	$n_2 = 100$
$\bar{x}_1 = 203.4$	$\bar{x}_2 = 189.4$
$s_1 = 39.4$	$s_2 = 39.0$

7. TS: $z = 2.06$.
   CV: $z = \pm 2.575$.
8. $-3.5 < \mu_1 - \mu_2 < 31.5$
9. TS: $z = 4.91$.
   CV: $z = 1.645$.
10. $1.14 \text{ yr} < \mu_1 - \mu_2 < 2.66 \text{ yr}$
11. TS: $z = 8.56$.
   CV: $z = 2.33$.
12. $8 \text{ lb} < \mu_1 - \mu_2 < 16 \text{ lb}$; no.
13. TS: $z = 2.64$.
   CV: $z = \pm 1.645$.

14. Use the sample data from Exercise 13 to construct a 90% confidence interval for the difference between the population means ($\mu_1 - \mu_2$), where μ_1 is the mean salary for all American Airlines flight attendants and μ_2 is the mean salary for those at TWA. Do the confidence interval limits contain 0? Does this suggest that there is not a significant difference between the two population means?

15. Data Set 11 in Appendix B contains weights of 100 randomly selected M&M plain candies. Those weights have a mean of 0.9147 g and a standard deviation of 0.0369 g. The previous edition of this book used a different sample of 100 M&M plain candies, with a mean and standard deviation of 0.9160 g and 0.0433 g, respectively. Is the discrepancy between the two sample means significant?

16. A study of seat-belt use involved children who were hospitalized as a result of motor vehicle crashes. For a group of 290 children who were not wearing seat belts, the number of days spent in intensive care units (ICU) has a mean of 1.39 and a standard deviation of 3.06. For a group of 123 children who were wearing seat belts, the number of days in ICU has a mean of 0.83 and a standard deviation of 1.77 (based on data from "Morbidity Among Pediatric Motor Vehicle Crash Victims: The Effectiveness of Seat Belts," by Osberg and Di Scala, *American Journal of Public Health*, Vol. 82, No. 3). At the 0.01 significance level, test the claim that the population of children not wearing seat belts has a higher mean number of days spent in ICU. Based on the result, is there significant evidence in favor of seat-belt use among children?

17. Refer to Data Set 2 of Appendix B and use only the 12:00 A.M. body temperatures for Day 1. We want to test, at the 0.05 significance level, the claim that persons aged 18–24 inclusive have the same mean as those aged 25 and older.
   a. Conduct the hypothesis test using the traditional method.
   b. Conduct the hypothesis test using the *P*-value method.
   c. Conduct the hypothesis test by constructing a 95% confidence interval for the difference $\mu_1 - \mu_2$.
   d. Compare the results from parts (a), (b), and (c).

18. Refer to the pulse rates in Data Set 8 from Appendix B. First identify and delete the two outliers that are unrealistic pulse rates of living and reasonably healthy statistics students.
   a. Use a 0.05 level of significance and the traditional method of testing hypotheses to test the claim that male and female statistics students have the same mean pulse rate.
   b. Use the *P*-value method to test the claim given in part (a).
   c. Construct a 95% confidence interval for the difference $\mu_1 - \mu_2$, where μ_1 and μ_2 are the mean pulse rates for male and female statistics students, respectively.
   d. Compare the results from parts (a), (b), and (c).

8-3 Exercises B: Beyond the Basics

19. The examples in this section used the axial loads of a sample of cans 0.0109 in. thick and a second sample of cans 0.0111 in. thick. (The original data are listed in Data Set 15 of Appendix B.) We noted that the second sample includes an outlier of 504 lb. How are the boxplots of Figure 8-3, the hypothesis test, and the confidence interval affected if we decide that the outlier is actually an error that should be deleted?

20. The examples in this section used the axial loads of cans, where the loads were given in pounds. How are the boxplots, hypothesis test, and confidence interval affected if the loads are all converted to kilograms?

21. a. Find the variance for this *population* of x scores: 5, 10, 15. (See Section 2-5 for the variance σ^2 of a population.)

    b. Find the variance for this *population* of y scores: 1, 2, 3.

    c. List the *population* of all possible differences $x - y$, and find the variance of this population.

    d. Use the results from parts (a), (b), and (c) to verify that $\sigma^2_{x-y} = \sigma^2_x + \sigma^2_y$. (This principle is used to derive the test statistic and confidence interval for this section.)

    e. How is the *range* of the differences $x - y$ related to the range of the x values and the range of the y values?

22. Refer to Exercise 11. If the actual difference between the population means is 6.0, find β, the probability of a type II error. (See Exercise 27 in Section 7-3.) (*Hint:* In Step 1, replace $\bar{x}$ by $(\bar{x}_1 - \bar{x}_2)$, replace $\mu_{\bar{x}}$ by 0, and replace $\sigma_{\bar{x}}$ by

$$\sqrt{\frac{\sigma^2_1}{n_1} + \frac{\sigma^2_2}{n_2}}$$

8-4 Comparing Two Variances

This section is required for coverage of Section 8-5, but it is not required for Section 8-6.

Because the characteristic of variation among data is extremely important, this section presents a method for using two samples to compare the variances of the populations from which the samples are drawn. The method we use requires the following assumptions.

Assumptions   When testing a hypothesis about the *variances* of two populations, we assume that

1. The two populations are *independent* of each other. (Recall from Section 8-2 that two samples are independent if the sample selected from one population is not related to the sample selected from the other population.)

19. The boxplot for the 0.0111 in. cans uses a 5-number summary of 205, 275, 285, 294, 317. Before the deletion of the outlier, the 5-number summary was 205, 275, 285, 295, 504, so the only major change is shortening of the right whisker. In the hypothesis test, the test statistic changes from −5.48 to −5.66, so the conclusions will be the same. After deleting the outlier of 504, the confidence interval limits change from (−21.6, −7.8) to (−19.5, −7.3), so those changes are not substantial.

20. The boxplots will have the same shapes, the hypothesis test will have the same test statistic and conclusions, and the confidence interval limits will be converted to their equivalent values in kilograms.

21. a. 50/3

    b. 2/3

    c. 52/3

    e. The range of the x-y values equals the range of the x values plus the range of the y values.

22. $\beta = 0.0256$

2. The two populations are each *normally distributed*. (This assumption is important, because the test statistic is extremely sensitive to departures from normality.)

The computations of this section will be greatly simplified if we stipulate that s_1^2 represents the *larger* of the two sample variances. It doesn't really make any difference which sample is designated as sample 1, so we use the following notation.

By making s_1^2 the *larger* variance, we avoid the tricky problem of finding critical F values for left-tailed cases. Note, however, that the distribution of s_1^2/s_2^2 is the F distribution illustrated in Figure 8-5 only if we haven't yet imposed the condition that s_1^2 is the larger of the two sample variances. Once we impose that condition, the ratio of s_1^2/s_2^2 must be 1 or greater.

Notation for Hypothesis Tests with Two Variances

$s_1^2 = $ *larger* of the two sample variances

$n_1 = $ size of the sample with the *larger* variance

$\sigma_1^2 = $ variance of the population from which the sample with the *larger* variance was drawn

The symbols s_2^2, n_2, and σ_2^2 are used for the other sample and population.

For two normally distributed populations with equal variances (that is, $\sigma_1^2 = \sigma_2^2$), the sampling distribution of the following test statistic is the **F distribution** shown in Figure 8-5 with critical values listed in Table A-5. If you continue to repeat an experiment of randomly selecting samples from two normally distributed populations with equal variances, the distribution of the ratio s_1^2/s_2^2 of the sample variances is the F distribution. In Figure 8-5, note these properties of the F distribution:

- The F distribution is not symmetric.
- Values of the F distribution cannot be negative.
- The exact shape of the F distribution depends on two different degrees of freedom.

Test Statistic for Hypothesis Tests with Two Variances

$$F = \frac{s_1^2}{s_2^2}$$

If the two populations really do have equal variances, then $F = s_1^2/s_2^2$ tends to be close to 1 because s_1^2 and s_2^2 tend to be close in value. But if the two populations have radically different variances, s_1^2 and s_2^2 tend to be very different numbers. Denoting the larger of the sample variances by s_1^2, we see that the

Figure 8-5   *F* Distribution
There is a different *F* distribution for each different pair of degrees of freedom for numerator and denominator.

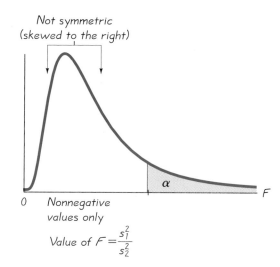

Not symmetric
(skewed to the right)

0   Nonnegative
values only

Value of $F = \dfrac{s_1^2}{s_2^2}$

ratio s_1^2/s_2^2 will be a large number whenever s_1^2 and s_2^2 are far apart in value. Consequently, a value of *F* near 1 will be evidence in favor of the conclusion that $\sigma_1^2 = \sigma_2^2$, but a large value of *F* will be evidence against the conclusion of equality of the population variances.

This test statistic applies to a claim made about two variances, but we can also use it for claims about two population standard deviations. Any claim about two population standard deviations can be restated in terms of the corresponding variances.

Using Table A-5, we obtain critical *F* values that are determined by the following three values:

1. The significance level α. (Table A-5 has six pages of critical values for $\alpha = 0.01$, 0.025, and 0.05.)
2. **Numerator degrees of freedom** $= n_1 - 1$
3. **Denominator degrees of freedom** $= n_2 - 1$

To find a critical value, first refer to the part of Table A-5 corresponding to α (for a one-tailed test) or $\alpha/2$ (for a two-tailed test), then intersect the column representing the degrees of freedom for s_1^2 with the row representing the degrees of freedom for s_2^2. Because we are stipulating that the larger sample variance is s_1^2, all one-tailed tests will be right-tailed and all two-tailed tests will require that we find only the critical value located to the right. Good news: We have no need to find a critical value separating a left-tailed critical region. (Because the *F* distribution is not symmetric and has only nonnegative values, a left-tailed critical value cannot be found by using the negative of the right-tailed critical value; instead, a left-tailed critical value is found by using the reciprocal of the right-tailed value with the numbers of degrees of freedom reversed. See Exercise 15.)

There will certainly be questions about handling cases in which the number of degrees of freedom is not one of those found in the table. We can usually use the nearest values; it usually doesn't make a difference if we use the value above or below the desired missing value. It only makes a difference if the test statistic is between the table values, in which case you could interpolate, but it's best to run the problem on STAT-DISK or Minitab or a TI-83.

Axial Loads (lb) of 0.0109 in. Cans	Axial Loads (lb) of 0.0111 in. Cans
$n = 175$	$n = 175$
$\bar{x} = 267.1$	$\bar{x} = 281.8$
$s = 22.1$	$s = 27.8$

We sometimes have numbers of degrees of freedom that are not included in Table A-5. We could use linear interpolation to approximate the missing values, but in most cases that's not necessary because the F test statistic is either less than the lowest possible critical value or greater than the largest possible critical value. For example, Table A-5 shows that for $\alpha = 0.025$ in the right tail, 20 degrees of freedom for the numerator, and 34 degrees of freedom for the denominator, the critical F value is between 2.0677 and 2.1952. Any F test statistic below 2.0677 will result in failure to reject the null hypothesis, any F test statistic above 2.1952 will result in rejection of the null hypothesis, and interpolation is necessary only if the F test statistic happens to fall between 2.0677 and 2.1952. (See Exercise 13.) The use of a statistical software package such as STATDISK or Minitab eliminates this problem by providing critical values or P-values.

EXAMPLE   In the preceding section we used a sample of 175 aluminum cans that are 0.0109 in. thick and we used a second sample of 175 aluminum cans that are 0.0111 in. thick. Data Set 15 lists the axial loads of the cans in those samples, and the summary statistics are listed in the margin. Use a 0.05 significance level to test the claim that the samples come from populations with the same variance.

SOLUTION   Because we stipulate in this section that the larger variance is denoted by s_1^2, we reverse the subscript notation used in the preceding section, and we let $s_1^2 = 27.8^2$, $n_1 = 175$, $s_2^2 = 22.1^2$, and $n_2 = 175$. We now proceed to use the traditional method of testing hypotheses as outlined in Figure 7-4.

Step 1:   The claim of equal population variances is expressed symbolically as $\sigma_1^2 = \sigma_2^2$.

Step 2:   If the original claim is false, then $\sigma_1^2 \neq \sigma_2^2$.

Step 3:   Because the null hypothesis must contain equality, we have

$$H_0: \sigma_1^2 = \sigma_2^2 \text{ (original claim)} \qquad H_1: \sigma_1^2 \neq \sigma_2^2$$

Step 4:   The significance level is $\alpha = 0.05$.

Step 5:   Because this test involves two population variances, we use the F distribution.

Step 6:   The test statistic is

$$F = \frac{s_1^2}{s_2^2} = \frac{27.8^2}{22.1^2} = 1.5824$$

For the critical values, first note that this is a two-tailed test with 0.025 in each tail. As long as we are stipulating that the larger variance is placed in the numerator of the F test statistic, we need

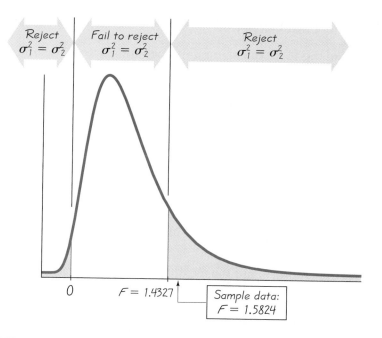

Figure 8-6   Distribution of s_1^2/s_2^2 for Axial Loads of 0.0111 in. Cans and 0.0109 in. Cans

to find only the right-tailed critical value. From Table A-5 we get a critical value of $F = 1.4327$, which corresponds to 0.025 in the right tail, with 174 degrees of freedom for the numerator and 174 degrees of freedom for the denominator. (Actually, the table does not include 174 degrees of freedom, so we chose the closest value of 120 degrees of freedom. If the difference in sample variances is significant based on sample sizes of 121, that difference would certainly be significant with sample sizes of 175.)

Step 7:   Figure 8-6 shows that the test statistic $F = 1.5824$ falls within the critical region, so we reject the null hypothesis.

Step 8:   There is sufficient evidence to warrant rejection of the claim that the two variances are equal.

In the preceding example we used a two-tailed test for the claim of equal variances. A right-tailed test of the claim that the 0.0111 in. cans have a larger variance would yield the same test statistic of $F = 1.5824$, but a different critical value of $F = 1.3519$ (found from Table A-5 with $\alpha = 0.05$). We would have sufficient evidence to support the claim of a larger variance.

We have described the traditional method of testing hypotheses made about two population variances. Exercise 14 deals with the P-value approach, and Exercise 16 deals with the construction of confidence intervals.

Using Computers and Calculators

STATDISK and the TI-83 calculator are programmed to do hypothesis tests of the type described in this section, but Minitab will not do these tests. If using STATDISK, select Analysis from the main menu, then select Hypothesis Testing, then Variance-Two Samples. If using a TI-83 calculator, press the STAT key, then select TESTS, then 2-SampFTEST.

8-4 Exercises A: Basic Skills and Concepts

Recommended assignment:
Exercises 3, 4, 8, 10, 11, 12.

In Exercises 1 and 2, test the given claim. Use a significance level of $\alpha = 0.05$, and assume that all populations are normally distributed. Use the traditional method of testing hypotheses outlined in Figure 7-4, and draw the appropriate graphs.

1. TS: $F = 2.2222$.
   CV: $F = 2.2090$.
2. TS: $F = 6.6859$.
   CV: $F =$ between 2.2664 and 2.2229.

1. **Claim: Populations A and B have different variances ($\sigma_1^2 \neq \sigma_2^2$).**
   Sample A: $n = 25, \bar{x} = 175, s^2 = 900$
   Sample B: $n = 31, \bar{x} = 200, s^2 = 2000$
2. **Claim: Population A has a larger variance than population B.**
   Sample A: $n = 50, \bar{x} = 77.4, s = 18.1$
   Sample B: $n = 15, \bar{x} = 75.7, s = 7.0$

Nonstress	Stress
$n_1 = 40$	$n_2 = 40$
$\bar{x}_1 = 53.3$	$\bar{x}_2 = 45.3$
$s_1 = 11.6$	$s_2 = 13.2$

3. **An experiment was conducted to investigate the effect of stress on the recall ability of police eyewitnesses. The experiment involved a nonstressful interrogation of a cooperative suspect and a stressful interrogation of an uncooperative and belligerent suspect. The numbers of details recalled a week after the incident are summarized in the margin (based on data from "Eyewitness Memory of Police Trainees for Realistic Role Plays," by Yuille et al., *Journal of Applied Psychology,* Vol. 79, No. 6). Use a 0.10 level of significance to test the claim that the samples come from populations with different standard deviations.**

3. TS: $F = 1.2949$.
   CV: $F = 1.6928$ (approx.).
4. TS: $F = 3.8643$.
   CV: $F = 2.9249$.
5. TS: $F = 1.6804$.
   CV: $F < 1.5330$.

4. **Customer waiting times are studied at the Jefferson Valley Bank. When 25 randomly selected customers enter any one of several waiting lines, their times have a mean of 6.896 min and a standard deviation of 3.619 min. When 20 randomly selected customers enter a single main waiting line that feeds the individual teller stations, their waiting times have a mean of 7.460 min and a standard deviation of 1.841 min. Use a 0.01 significance level to test the claim that waiting times for the single line have a lower standard deviation.**

5. **As part of the National Health Survey, data were collected on the weights of men. For 804 men aged 25 to 34, the mean is 176 lb and the standard deviation is 35.0 lb. For 1657 men aged 65 to 74, the mean and standard deviation are 164 lb and 27.0 lb, respectively. At the 0.01 significance level, test the claim that the older men come from a population with a standard deviation less than that for men in the 25 to 34 age bracket.**

6. In an insurance study of pedestrian deaths in New York State, monthly fatalities are totaled for two different time periods. Sample data for the first time period are summarized by these statistics: $n = 12$, $\bar{x} = 46.42$, $s = 11.07$. Sample data for the second time period are summarized by these statistics: $n = 12$, $\bar{x} = 51.00$, $s = 10.39$ (based on data from the New York State Department of Motor Vehicles). At the 0.05 significance level, test the claim that both time periods have the same variance.

7. In a study of flight attendants, salaries paid by two different airlines were randomly selected. For 40 American Airlines flight attendants, the mean is $23,870 and the standard deviation is $2960. For 35 TWA flight attendants, the mean is $22,025 and the standard deviation is $3065 (based on data from the Association of Flight Attendants). At the 0.10 level of significance, test the claim that salaries of American and TWA flight attendants have the same standard deviation.

8. An experiment is devised to study the variability of grading procedures among college professors of organic chemistry. Two different professors are asked to grade the same set of 25 exam solutions, and their grades have variances of 103.4 and 39.7, respectively. At the 0.05 significance level, test the claim that the first professor's grading exhibits greater variance. Given that a student is very weak in organic chemistry, and assuming that both professors give grades with the same mean, is there any advantage in the student's choosing one professor over the other? If so, which one? Why?

9. The effectiveness of a mental training program was tested in a military training program. In an antiaircraft artillery examination, scores for an experimental group and a control group were recorded. Use the given data to test the claim that both groups come from populations with the same variance. Use a 0.05 significance level.

Experimental Group					Control Group			
60.83	117.80	44.71	75.38		122.80	70.02	119.89	138.27
73.46	34.26	82.25	59.77		118.43	54.22	118.58	74.61
69.95	21.37	59.78	92.72		121.70	70.70	99.08	120.76
72.14	57.29	64.05	44.09		104.06	94.23	111.26	121.67
80.03	76.59	74.27	66.87					

Based on "Routinization of Mental Training in Organizations: Effects on Performance and Well-Being," by Larsson, *Journal of Applied Psychology*, Vol. 72, No. 1.

10. Many students have had the unpleasant experience of panicking on a test because the first question was exceptionally difficult. The arrangement of test items was studied for its effect on anxiety. The following scores are measures of "debilitating test anxiety" (which most of us call panic or blanking out). At the 0.05 significance level, test the claim that the two given samples come from populations with the same variance.

(continued)

6. TS: $F = 1.1352$.
   CV: $F =$ between 3.5257 and 3.4296.
7. TS: $F = 1.0722$.
   CV: $F = 1.7444$.
8. TS: $F = 2.6045$.
   CV: $F = 1.9838$.
9. TS: $F = 1.3478$.
   CV: $F = 2.6171$.
10. TS: $F = 2.5908$.
    CV: $F = 2.7006$.

Questions Arranged From Easy to Difficult					Questions Arranged From Difficult to Easy			
24.64	39.29	16.32	32.83	28.02	33.62	34.02	26.63	30.26
33.31	20.60	21.13	26.69	28.90	35.91	26.68	29.49	35.32
26.43	24.23	7.10	32.86	21.06	27.24	32.34	29.34	33.53
28.89	28.71	31.73	30.02	21.96	27.62	42.91	30.20	32.54
25.49	38.81	27.85	30.29	30.72				

Based on data from "Item Arrangement, Cognitive Entry Characteristics, Sex and Test Anxiety as Predictors of Achievement in Examination Performance," by Klimko, *Journal of Experimental Education*, Vol. 52, No. 4.

11. TS: $F = 1.2478$.
    CV: $F = 2.5342$.

11. Sample data were collected in a study of calcium supplements and their effects on blood pressure. A placebo group and a calcium group began the study with blood pressure measurements (based on data from "Blood Pressure and Metabolic Effects of Calcium Supplementation in Normotensive White and Black Men," by Lyle et al., *Journal of the American Medical Association*, Vol. 257, No. 13). At the 0.10 significance level, test the claim that the two sample groups come from populations with the same standard deviation. If the experiment requires groups with equal standard deviations, are these two groups acceptable?

Placebo: 124.6 104.8 96.5 116.3 106.1 128.8 107.2 123.1 118.1 108.5 120.4 122.5 113.6

Calcium: 129.1 123.4 102.7 118.1 114.7 120.9 104.4 116.3 109.6 127.7 108.0 124.3 106.6 121.4 113.2

12. Red: $n = 21$ and $s =$ 0.027474. Yellow: $n = 26$ and $s = 0.033817$. TS: $F = 1.5150$. CV: $F = 2.0825$

13. a. Reject H_0: $\sigma_1^2 = \sigma_2^2$.
    b. Fail to reject H_0: $\sigma_1^2 = \sigma_2^2$.
    c. Reject H_0: $\sigma_1^2 = \sigma_2^2$.

14. P-value < 0.01

12. Refer to Data Set 11 in Appendix B and use a 0.10 level of significance to test the claim that red and yellow M&M plain candies have weights with different standard deviations. If you were responsible for controlling production so that red and yellow M&Ms have weights with the same amount of variation, would you take corrective action?

8-4 Exercises B: Beyond the Basics

13. A hypothesis test of the claim $\sigma_1^2 = \sigma_2^2$ is being conducted with a 0.05 level of significance. What do you conclude in each of the following cases?
  a. Test statistic is $F = 2.0933$; $n_1 = 50$; $n_2 = 35$
  b. Test statistic is $F = 1.8025$; $n_1 = 50$; $n_2 = 35$
  c. Test statistic is $F = 2.3935$; $n_1 = 40$; $n_2 = 20$

14. To test a claim about two population variances by using the *P*-value approach, first find the *F* test statistic, then refer to Table A-5 to determine how it compares to the critical values listed for $\alpha = 0.01$, $\alpha = 0.025$, and

$\alpha = 0.05$. Referring to Exercise 4, what can be concluded about the P-value?

15. For hypothesis tests in this section that were two-tailed, we found only the upper critical value. Let's denote that value by F_R, where the subscript suggests the critical value for the right tail. The lower critical value F_L (for the left tail) can be found as follows: First interchange the degrees of freedom, and then take the reciprocal of the resulting F value found in Table A-5. (F_R is often denoted by $F_{\alpha/2}$, and F_L is often denoted by $F_{1-\alpha/2}$.) Find the critical values F_L and F_R for two-tailed hypothesis tests based on the following values.

a. $n_1 = 10$, $n_2 = 10$, $\alpha = 0.05$ b. $n_1 = 10$, $n_2 = 7$, $\alpha = 0.05$
c. $n_1 = 7$, $n_2 = 10$, $\alpha = 0.05$ d. $n_1 = 25$, $n_2 = 10$, $\alpha = 0.02$
e. $n_1 = 10$, $n_2 = 25$, $\alpha = 0.02$

15. a. $F_L = 0.2484$, $F_R = 4.0260$
    b. $F_L = 0.2315$, $F_R = 5.5234$
    c. $F_L = 0.1810$, $F_R = 4.3197$
    d. $F_L = 0.3071$, $F_R = 4.7290$
    e. $F_L = 0.2115$, $F_R = 3.2560$

16. In addition to testing claims involving σ_1^2 and σ_2^2, we can also construct interval estimates of the ratio σ_1^2/σ_2^2 using the following expression.

16. $0.2692 < \sigma_1^2/\sigma_2^2 < 1.9418$

$$\left(\frac{s_1^2}{s_2^2} \cdot \frac{1}{F_R}\right) < \frac{\sigma_1^2}{\sigma_2^2} < \left(\frac{s_1^2}{s_2^2} \cdot \frac{1}{F_L}\right)$$

Here F_L and F_R are as described in Exercise 15. Construct the 95% confidence interval estimate for the ratio of the experimental group variance to the control group variance for the data in Exercise 9.

17. Sample data consist of temperatures recorded for two different groups of items that were produced by two different production techniques. A quality control specialist plans to analyze the results. She begins by testing for equality of the two population standard deviations.
a. If she adds the same constant to every temperature from both groups, how is the value of the test statistic F affected?
b. If she multiplies every score from both groups by the same constant, how is the value of the test statistic F affected?
c. If she converts all temperatures from the Fahrenheit scale to the Celsius scale, how is the value of the test statistic F affected?

17. a. It doesn't change.
    b. It doesn't change.
    c. It doesn't change.

8–5 Inferences about Two Means: Independent and Small Samples

In this section we present methods of inferential statistics for situations involving the means of two independent populations, but (unlike Section 8-3) at least one of the two samples is small (with $n \le 30$). We describe methods for testing hypotheses and constructing confidence intervals for situations in which the following assumptions apply.

This is the most difficult section of this chapter because we must do a preliminary F test to decide whether the population variances are the same. Figures 8-8 and 8-9 are flowcharts summarizing the complicated procedures of this section. It might be very helpful to refer to those figures early so that students can avoid confusion by referring to them as necessary.

Assumptions In testing hypotheses about the means of two populations or in constructing confidence interval estimates of the difference between those means, the methods of this section apply to cases in which

1. The two samples are *independent*.
2. The two samples are *randomly selected from normally distributed* populations.
3. At least one of the two samples is small ($n \le 30$).

When these conditions are satisfied, we use one of three different procedures corresponding to the following cases:

Case 1: The values of both population variances are known. (In reality, this case rarely occurs.)

Case 2: The two populations appear to have equal variances. (That is, based on a hypothesis test of $\sigma_1^2 = \sigma_2^2$, we fail to reject equality of the two population variances.)

Case 3: The two populations appear to have unequal variances. (That is, based on a hypothesis test of $\sigma_1^2 = \sigma_2^2$, we reject equality of the two population variances.)

This section discusses procedures that are somewhat controversial in the sense that statisticians do not universally agree with any single approach. However, the methods we present are commonly used.

Determination of the case that is appropriate may require that students conduct a preliminary *F* test. In some cases, it may be obvious that the sample standard deviations are (or are not) significantly different, so the *F* test may be skipped. If you do require a formal *F* test, be sure to inform students of that requirement.

Case 1: Both Population Variances Are Known

In reality, Case 1 doesn't occur very often. Usually, the population variances are computed from the known population data, and if we could find σ_1^2 and σ_2^2, we should be able to find the values of μ_1 and μ_2, so there would be no need to test claims or construct confidence intervals. If some strange set of circumstances allows us to know the values of σ_1^2 and σ_2^2 but not μ_1 and μ_2, hypothesis testing of claims about μ_1 and μ_2 can be done with the following test statistic.

Test Statistic: Known Population Variances

$$z = \frac{(\bar{x}_1 - \bar{x}_2) - (\mu_1 - \mu_2)}{\sqrt{\dfrac{\sigma_1^2}{n_1} + \dfrac{\sigma_2^2}{n_2}}}$$

Because this test statistic refers to the standard normal distribution, it is easy to find *P*-values: Simply use the procedure summarized in Figure 7-7.

Confidence Intervals

Confidence interval estimates of the difference $\mu_1 - \mu_2$ can be found by using

$$(\bar{x}_1 - \bar{x}_2) - E < (\mu_1 - \mu_2) < (\bar{x}_1 - \bar{x}_2) + E$$

where

$$E = z_{\alpha/2} \sqrt{\frac{\sigma_1^2}{n_1} + \frac{\sigma_2^2}{n_2}}$$

This test statistic and confidence interval reflect the property of variances that was discussed in Section 8-3: The variance of the *differences* between two independent random variables equals the variance of the first random variable *plus* the variance of the second random variable.

Because this case of known population variances is so unlikely to occur in reality, we do not illustrate it with examples. However, the calculations closely parallel those of Section 8-3, which used the same test statistic and confidence interval format.

If the three assumptions listed at the beginning of this section are satisfied and if we don't know the values of the population standard deviations, we use the following procedure.

> **Choosing between Case 2 and Case 3: Use the *F* test described in Section 8-4 to test the null hypothesis that $\sigma_1^2 = \sigma_2^2$. Use the conclusion of that test as follows.**
>
> - *Fail to reject $\sigma_1^2 = \sigma_2^2$:* **Treat the populations as if they have equal variances (Case 2).**
> - *Reject $\sigma_1^2 = \sigma_2^2$:* **Treat the populations as if they have unequal variances (Case 3).**

Cases 2 and 3 require a preliminary *F* test of the claim that $\sigma_1^2 = \sigma_2^2$, but not all statisticians agree that this preliminary *F* test should be conducted. Some argue that if we apply the *F* test with a certain level of significance and then do a *t* test at the same level, the overall result will not be at the same level of significance. Also, in addition to being sensitive to differences in population variances, the *F* statistic is very sensitive to departures from normal distributions, so it's possible to reject a null hypothesis for the wrong reason. (For one argument against the preliminary *F* test, see "Homogeneity of Variance in the Two-Sample Means Test," by Moser and Stevens, *The American Statistician*, Vol. 46, No. 1.)

Case 2: The Two Populations Appear to Have Equal Variances (Because We Fail to Reject $\sigma_1^2 = \sigma_2^2$)

If we don't have sufficient evidence to warrant rejection of equal variances (that is, we fail to reject $\sigma_1^2 = \sigma_2^2$), we calculate a **pooled estimate of σ^2** that is common to both populations; that pooled estimate is denoted by s_p^2 and is a weighted average of s_1^2 and s_2^2, as shown in the box. (Minitab and the TI-83 calculator are two of many systems using the "pooled" option of this case.)

The basic idea here is that if the populations have the same variance, we can estimate that common variance by *pooling* the two sample variances.

The format of the test statistic once again uses the same basic format of $z = (x - \mu)/\sigma$, but we adapt it here for the present circumstances.

Test Statistic (Small Independent Samples and Equal Variances)

$$t = \frac{(\bar{x}_1 - \bar{x}_2) - (\mu_1 - \mu_2)}{\sqrt{\dfrac{s_p^2}{n_1} + \dfrac{s_p^2}{n_2}}}$$

where
$$s_p^2 = \frac{(n_1 - 1)s_1^2 + (n_2 - 1)s_2^2}{(n_1 - 1) + (n_2 - 1)}$$

and the degree of freedom is given by df $= n_1 + n_2 - 2$.

Confidence Interval (Small Independent Samples and Equal Variances)

$$(\bar{x}_1 - \bar{x}_2) - E < (\mu_1 - \mu_2) < (\bar{x}_1 - \bar{x}_2) + E$$

where
$$E = t_{\alpha/2}\sqrt{\frac{s_p^2}{n_1} + \frac{s_p^2}{n_2}}$$

and s_p^2 is as given in the test statistic.

EXAMPLE   Refer to the data listed in Table 8-1 near the beginning of the chapter. Use a 0.05 significance level to test the claim that the mean amount of *nicotine* in filtered king-size cigarettes is equal to the mean amount of *nicotine* for nonfiltered king-size cigarettes. (All measurements are in milligrams, and the data are from the Federal Trade Commission.)

SOLUTION   In Figure 8-1 we used boxplots to compare the filtered and nonfiltered cigarettes. The two boxplots representing the nicotine amounts seem to indicate clearly that the nonfiltered cigarettes have more nicotine, but let's use a formal hypothesis test of the claim that the two

mean amounts are equal. We will use the sample statistics listed in the margin.

The two samples are independent (because they are separate and are not matched), the sample sizes are small, and we don't know the values of σ_1 and σ_2, so we are dealing with either Case 2 or Case 3. Using a preliminary F test (described in Section 8-4) to choose between Case 2 and Case 3, we test H_0: $\sigma_1^2 = \sigma_2^2$ with the test statistic

$$F = \frac{s_1^2}{s_2^2} = \frac{0.31^2}{0.16^2} = 3.7539$$

Nicotine (mg)	
Filtered Kings	Nonfiltered Kings
$n_1 = 21$	$n_2 = 8$
$\bar{x}_1 = 0.94$	$\bar{x}_2 = 1.65$
$s_1 = 0.31$	$s_2 = 0.16$

With $\alpha = 0.05$ in a two-tailed F test and with 20 and 7 degrees of freedom for the numerator and denominator, respectively, we use Table A-5 to find the critical F value of 4.4667. Because the computed test statistic of $F = 3.7539$ does *not* fall within the critical region, we fail to reject the null hypothesis of equal variances and proceed by using the approach outlined in Case 2.

In using the Case 2 approach, we test the claim that $\mu_1 = \mu_2$ with the following null and alternative hypotheses:

$$H_0: \mu_1 = \mu_2 \text{ (or } \mu_1 - \mu_2 = 0) \qquad H_1: \mu_1 \neq \mu_2$$

The test statistic for this case of equal variances requires the value of the pooled variance s_p^2, so we find that value first.

$$s_p^2 = \frac{(n_1 - 1)s_1^2 + (n_2 - 1)s_2^2}{(n_1 - 1) + (n_2 - 1)} = \frac{(21 - 1) \cdot 0.31^2 + (8 - 1) \cdot 0.16^2}{(21 - 1) + (8 - 1)}$$
$$= 0.078$$

We can now find the value of the test statistic:

$$t = \frac{(\bar{x}_1 - \bar{x}_2) - (\mu_1 - \mu_2)}{\sqrt{\frac{s_p^2}{n_1} + \frac{s_p^2}{n_2}}} = \frac{(0.94 - 1.65) - 0}{\sqrt{\frac{0.078}{21} + \frac{0.078}{8}}} = -6.119$$

The critical values of $t = -2.052$ and $t = 2.052$ are found from Table A-3 by referring to the column for $\alpha = 0.05$ (two tails) and to the row for df $= 27$ (the value of $21 + 8 - 2$). Figure 8-7 on the following page shows the test statistic, critical values, and critical region. We see that the test statistic does fall within the critical region, so we reject the null hypothesis that $\mu_1 = \mu_2$. There is sufficient evidence to warrant rejection of the claim that filtered king cigarettes and nonfiltered kings have the same mean amounts of nicotine. The subjective conclusion we formed by intuitively analyzing Figure 8-1 is now upheld by a formal hypothesis test.

Figure 8-7 Distribution of Differences Between Means of Nicotine Amounts in Filtered and Nonfiltered Cigarettes

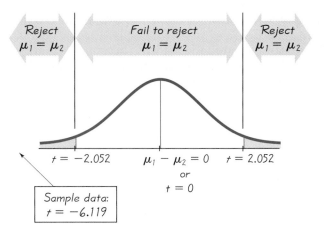

EXAMPLE Using the data in the preceding example, construct a 95% confidence interval estimate of $\mu_1 - \mu_2$.

SOLUTION It is easy to find that $\bar{x}_1 - \bar{x}_2 = 0.94 - 1.65 = -0.71$. Next, we find the value of the margin of error E.

$$E = t_{\alpha/2}\sqrt{\frac{s_p^2}{n_1} + \frac{s_p^2}{n_2}} = 2.052\sqrt{\frac{0.078}{21} + \frac{0.078}{8}} = 0.238$$

With $\bar{x}_1 - \bar{x}_2 = -0.71$ and with $E = 0.238$, we proceed to construct the confidence interval as follows.

$$(\bar{x}_1 - \bar{x}_2) - E < (\mu_1 - \mu_2) < (\bar{x}_1 - \bar{x}_2) + E$$
$$-0.71 - 0.238 < (\mu_1 - \mu_2) < -0.71 + 0.238$$
$$-0.95 < (\mu_1 - \mu_2) < -0.47$$

This final format is somewhat awkward with its negative signs, so we might interpret it by saying that we are 95% confident that nonfiltered king-size cigarettes have a mean nicotine content level that exceeds the mean for filtered kings by an amount that is between 0.47 mg and 0.95 mg.

Case 3: The Two Populations Appear to Have Unequal Variances (Because We Reject $\sigma_1^2 = \sigma_2^2$)

If we have sufficient evidence to warrant rejection of equal variances (that is, if we reject $\sigma_1^2 = \sigma_2^2$), there is no exact method for testing equality of means and constructing confidence intervals. An *approximate* method is to use the following test statistic and confidence interval.

Test Statistic (Small Independent Samples and Unequal Variances)

$$t = \frac{(\bar{x}_1 - \bar{x}_2) - (\mu_1 - \mu_2)}{\sqrt{\dfrac{s_1^2}{n_1} + \dfrac{s_2^2}{n_2}}}$$

where   df = smaller of $n_1 - 1$ and $n_2 - 1$.

Confidence Interval (Small Independent Samples and Unequal Variances)

$$(\bar{x}_1 - \bar{x}_2) - E < (\mu_1 - \mu_2) < (\bar{x}_1 - \bar{x}_2) + E$$

where $E = t_{\alpha/2}\sqrt{\dfrac{s_1^2}{n_1} + \dfrac{s_2^2}{n_2}}$

and   df = smaller of $n_1 - 1$ and $n_2 - 1$

This test statistic and confidence interval give the number of degrees of freedom as the smaller of $n_1 - 1$ and $n_2 - 1$, but this is a more conservative and simplified alternative to computing the number of degrees of freedom by using Formula 8-1.

Formula 8-1 $$df = \frac{(A + B)^2}{\dfrac{A^2}{n_1 - 1} + \dfrac{B^2}{n_2 - 1}}$$

where $A = \dfrac{s_1^2}{n_1}$ and $B = \dfrac{s_2^2}{n_2}$

More exact results are obtained by using Formula 8-1, but they continue to be only approximate. (*Note:* STATDISK uses Formula 8-1 instead of the "smaller of $n_1 - 1$ and $n_2 - 1$," so the STATDISK results will be a little better than those included in this text.)

EXAMPLE   Refer to the data listed in Table 8-1 near the beginning of this chapter. Use a 0.05 significance level to test the claim that the mean amount of tar in filtered king-size cigarettes is less than the mean amount of tar for nonfiltered king-size cigarettes. (All measurements are in milligrams, and the data are from the Federal Trade Commission.)

(*continued*)

Tar (mg)	
Filtered Kings	Nonfiltered Kings
$n_1 = 21$	$n_2 = 8$
$\bar{x}_1 = 13.3$	$\bar{x}_2 = 24.0$
$s_1 = 3.7$	$s_2 = 1.7$

SOLUTION   In Figure 8-1 we used boxplots to compare the filtered and nonfiltered cigarettes. The two boxplots representing the tar amounts seem to indicate clearly that the filtered cigarettes have less tar, but let's justify that observation with a formal hypothesis test based on the sample statistics for tar listed in the margin.

The two samples are independent (because they are separate and are not matched), the sample sizes are small, and we don't know the values of σ_1 and σ_2, so we are dealing with either Case 2 or Case 3. We use a preliminary F test to choose between Case 2 and Case 3, and we test $H_0: \sigma_1^2 = \sigma_2^2$ with the test statistic

$$F = \frac{s_1^2}{s_2^2} = \frac{3.7^2}{1.7^2} = 4.7370$$

With $\alpha = 0.05$ in a two-tailed F test and with 20 and 7 degrees of freedom for the numerator and denominator, respectively, we use Table A-5 to find the critical F value of 4.4667. Because the computed test statistic of $F = 4.7370$ does fall within the critical region, we reject the null hypothesis of equal variances and proceed by using the approach outlined in Case 3.

In using the Case 3 approach, we test the claim that $\mu_1 < \mu_2$ with the following null and alternative hypotheses:

$$H_0: \mu_1 \geq \mu_2 \,(\text{or } \mu_1 - \mu_2 \geq 0) \qquad H_1: \mu_1 < \mu_2 \,(\text{original claim})$$

The test statistic for this case of unequal variances is

$$t = \frac{(\bar{x}_1 - \bar{x}_2) - (\mu_1 - \mu_2)}{\sqrt{\dfrac{s_1^2}{n_1} + \dfrac{s_2^2}{n_2}}} = \frac{(13.3 - 24.0) - 0}{\sqrt{\dfrac{3.7^2}{21} + \dfrac{1.7^2}{8}}} = -10.630$$

The critical value of $t = -1.895$ is found from Table A-3 by referring to the column for $\alpha = 0.05$ (one tail) and to the row for df = 7 (the smaller of $21 - 1$ and $8 - 1$). Because this left-tailed test has a test statistic of $t = -10.630$ and a critical value of $t = -1.895$, the test statistic does fall within the critical region, so we reject the null hypothesis that $\mu_1 \geq \mu_2$. There is sufficient evidence to support the claim that filtered king-size cigarettes have a mean tar level that is less than the mean for nonfiltered kings.

EXAMPLE   Using the data in the preceding example, construct a 95% confidence interval estimate of $\mu_1 - \mu_2$.

SOLUTION   With $\bar{x}_1 - \bar{x}_2 = 13.3 - 24.0 = -10.7$, we proceed to find the value of the margin of error E.

$$E = t_{\alpha/2}\sqrt{\frac{s_1^2}{n_1} + \frac{s_2^2}{n_2}} = 2.365\sqrt{\frac{3.7^2}{21} + \frac{1.7^2}{8}} = 2.381$$

With $\bar{x}_1 - \bar{x}_2 = -10.7$ and with $E = 2.381$, we now construct the confidence interval as follows.

$$(\bar{x}_1 - \bar{x}_2) - E < (\mu_1 - \mu_2) < (\bar{x}_1 - \bar{x}_2) + E$$
$$-10.7 - 2.381 < (\mu_1 - \mu_2) < -10.7 + 2.381$$
$$-13.1 < (\mu_1 - \mu_2) < -8.3$$

Again, this final format is somewhat awkward with its negative signs, so we might interpret it by saying that we are 95% confident that the mean tar content for nonfiltered kings exceeds the mean tar content for filtered kings by an amount that is between 8.3 mg and 13.1 mg.

This section, Section 8-2, and Section 8-3 all deal with inferences about the means of two populations. Determining the correct procedure can be difficult because we must consider issues such as the independence of the samples, the sizes of the samples, and whether the two populations appear to have equal variances. We can avoid confusion by using the organized and systematic procedures summarized in Figures 8-8 and 8-9, found on the following two pages.

Using Computer Software and Calculators for Two Populations

The calculations in the examples of this section are complicated, but a calculator or computer can be used to simplify them. STATDISK is already programmed to automatically follow the procedures described in this section. Simply select the menu items of Analysis, Hypothesis Testing, and Mean-Two Independent Samples. When relevant, the conclusion of the prerequisite F test is included in the STATDISK display.

Minitab does not have a command to execute an F test of equality between two variances. If the two population variances appear to be equal, Minitab does allow use of a pooled estimate of the common variance. After entering the sample data in columns C1 and C2, select the options of Stat, Basic Statistics, 2-Sample t, then click on Samples in different columns and proceed to enter C1 for the first sample and C2 for the second sample. In the box identified as alternative, select the wording for the alternative hypothesis (not equal or less than or greater than), enter the

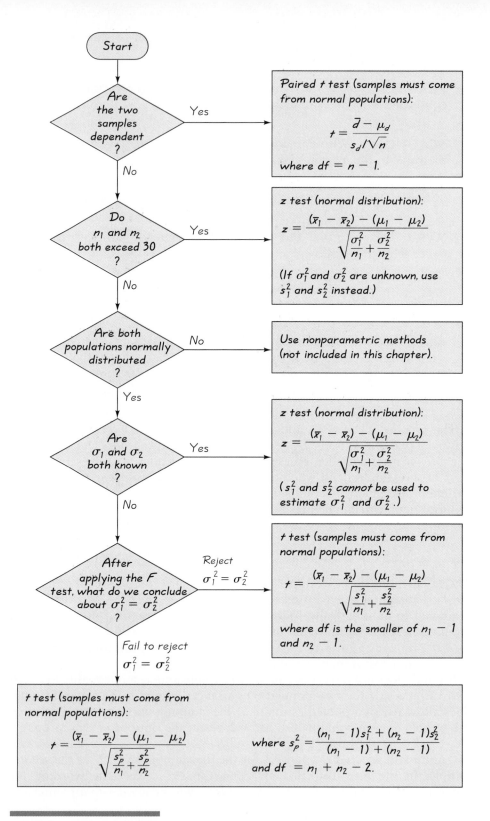

Figure 8-8 Testing Hypotheses Made about the Means of Two Populations

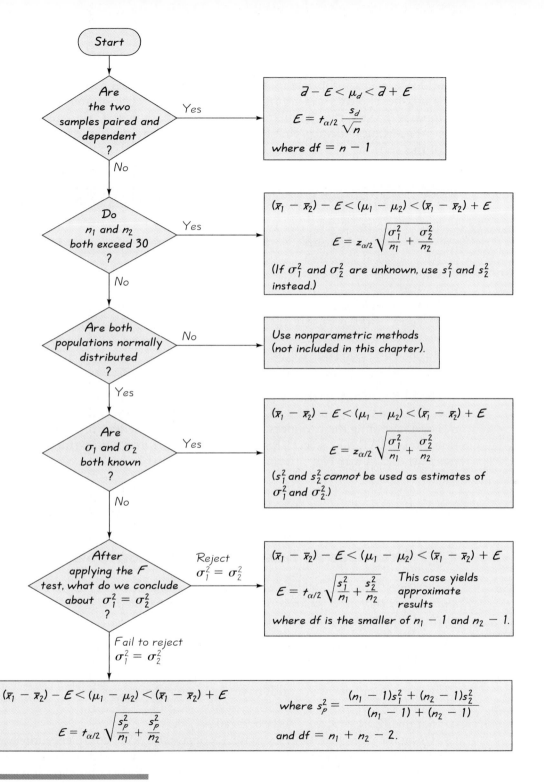

Figure 8-9 Confidence Intervals for the Difference Between Two Population Means

confidence level appropriate for the test (such as 0.95 for $\alpha = 0.05$). There will be a box next to Assume equal variances, and click on that box only if it appears that the two populations have equal variances (Case 2 of this section). The Minitab display that follows corresponds to the first example (Case 2) of this section.

MINITAB DISPLAY

```
Two sample T for C1 vs C2
           N     Mean    StDev   SE Mean
    C1    21    0.943    0.309    0.067
    C2     8    1.650    0.160    0.057


95% CI for mu C1  -  mu C2: ( -0.944, -0.470)
T-Test mu C1 = mu C2 (vs not =): T= -6.12
P=0.0000 DF= 27
Both use Pooled StDev = 0.278
```

The TI-83 calculator does a test for equality of population variances, and it does give you the option of using "pooled" variances (if you believe that $\sigma_1^2 = \sigma_2^2$) or not pooling the variances. To conduct tests of the type found in this section, press STAT, then select TESTS and choose 2-SampTTest (for a hypothesis test) or 2-SampTInt (for a confidence interval).

8-5 Exercises A: Basic Skills and Concepts

Recommended assignment: Exercises 7, 8, 11, 12, 14, 15.

In Exercises 1 and 2, test the given claim. Use a significance level of $\alpha = 0.05$, and assume that all populations are normally distributed. Use the traditional method of testing hypotheses outlined in Figure 7-4.

1. F-test results: TS: $F = 1.5625$. CV: $F = 2.8801$. Means: TS: $t = -0.990$. CV: $t = \pm 2.048$.

2. F-test results: TS: $F = 16.0000$. CV: $F = 8.6565$. Means: TS: $t = 0.889$. CV: $t = \pm 2.776$.

3. $-15 < \mu_1 - \mu_2 < 5$

4. $-21 < \mu_1 - \mu_2 < 4$ (Using STATDISK or TI-83 yields limits of -14 and 34.)

1. Claim: Populations A and B have the same mean ($\mu_1 = \mu_2$).
   Sample A: $n_1 = 10, \bar{x}_1 = 75, s_1 = 15$
   Sample B: $n_2 = 20, \bar{x}_2 = 80, s_2 = 12$
2. Claim: Populations A and B have the same mean ($\mu_1 = \mu_2$).
   Sample A: $n_1 = 15, \bar{x}_1 = 150, s_1 = 40$
   Sample B: $n_2 = 5, \bar{x}_2 = 140, s_2 = 10$
3. Using the sample data in Exercise 1, construct a 95% confidence interval estimate of the difference between the means of the two populations.
4. Using the sample data in Exercise 2, construct a 95% confidence interval estimate of the difference between the means of the two populations.

5. Equality of mean ozone concentrations for Maine and Rhode Island (see Data Set 5 in Appendix B) is tested with the Minitab results shown here. (A preliminary F test shows that there is not a significant difference between the two sample variances.) Construct a 95% confidence interval for the difference $\mu_1 - \mu_2$, where μ_1 denotes the mean ozone concentration for Maine. Do the confidence interval limits contain 0? What can you conclude about equality of the mean ozone concentrations for Maine and Rhode Island?

> ## MINITAB DISPLAY
>
> ```
> Two sample T for ME vs RI
>
>           N        Mean        StDev       SE Mean
> ME        6        2985        1352            552
> RI        6        4789        2948           1203
>
> 95% C.I. for mu ME — mu RI: (-4754, 1148)
> T-Test mu ME = mu RI (vs not =): T = -1.36
> P = 0.20 DF = 10
> Both use Pooled StDev = 2293
> ```

6. Refer to the same Minitab display for Exercise 5 and use a 0.05 significance level to test the claim that Maine and Rhode Island have the same mean ozone concentration.

7. Are severe psychiatric disorders related to biological factors that can be physically observed? One study used X-ray computed tomography (CT) to collect data on brain volumes for a group of patients with obsessive-compulsive disorders and a control group of healthy persons. Sample results (in mL) follow for volumes of the right cordate (based on data from "Neuroanatomical Abnormalities in Obsessive-Compulsive Disorder Detected with Quantitative X-Ray Computed Tomography," by Luxenberg et al., *American Journal of Psychiatry*, Vol. 145, No. 9). At the 0.01 significance level, test the claim that obsessive-compulsive patients and healthy persons have the same mean brain volume. Based on this result, does it seem that obsessive-compulsive disorders have a biological basis?

   Obsessive-compulsive patients: $n = 10$, $\bar{x} = 0.34$, $s = 0.08$
   Control group: $n = 10$, $\bar{x} = 0.45$, $s = 0.08$

8. Using the sample data from Exercise 7, construct a 99% confidence interval estimate of the difference between the mean brain volume for the patient group and the mean brain volume for the healthy control group.

5. $-4754 < \mu_1 - \mu_2 < 1148$; yes.

6. TS: $t = -1.36$. P-value: 0.20.
7. TS: $F = 1.0000$.
   Means: TS: $t = -3.075$.
   CV: $t = \pm 2.878$.
8. $-0.21 < \mu_1 - \mu_2 < -0.01$

9. $-17.16 < \mu_1 - \mu_2 < 260.40$

9. The same study cited in Exercise 7 resulted in the values summarized here for total brain volumes (in mL). Using these sample statistics, construct a 95% confidence interval for the difference between the mean brain volume of obsessive-compulsive patients and the mean brain volume of healthy persons.

Obsessive-compulsive patients: $n = 10$, $\bar{x} = 1390.03$, $s = 156.84$

Control group: $n = 10$, $\bar{x} = 1268.41$, $s = 137.97$

10. TS: $F = 1.2922$.
    CV:$F = 4.0260$.
    Means: TS: $t = 1.841$.
    CV: $t = \pm 2.101$.

10. Refer to the sample data in Exercise 9 and use a 0.05 significance level to test the claim that there is no difference between the mean for obsessive-compulsive patients and the mean for healthy persons. Based on this result, does it seem that obsessive-compulsive disorders have a biological basis?

11. TS: $F = 2.9459$.
    CV: $F = 3.0074$.
    Means: TS: $t = -1.099$.
    CV: $t = \pm 2.052$.

11. Data Set 11 in Appendix B includes a sample of 21 red M&Ms with weights having a mean of 0.9097 g and a standard deviation of 0.0275 g. The data set also includes a sample of 8 orange M&Ms with weights having a mean of 0.9251 g and a standard deviation of 0.0472 g. Use a 0.05 level of significance to test for equality of the two population means. If you were responsible for controlling production so that red and orange M&Ms have weights with the same mean, would you take corrective action?

12. TS: $F = 1.3110$.
    CV: $F = 2.9685$ Means:
    TS: $t = -0.627$.
    CV: $t = \pm 1.96$.

12. Refer to Data Set 11 in Appendix B and test the claim that yellow M&Ms and green M&Ms have the same mean weight. If you were part of a quality control team with responsibility for ensuring that yellow M&Ms and green M&Ms have the same mean weight, would you take corrective action?

13. TS: $F = 1.5155$.
    CV: $F = 2.1359$.
    Means: TS: $t = 0.261$.
    CV: $t = \pm 1.96$.

13. Refer to Data Set 10 in Appendix B and test the claim that there is no difference in length between movies rated R and movies with G, PG, or PG-13 ratings.

14. $n_1 = 20$, $\bar{x}_1 = 8344.9$, $s_1 = 15667.9$; $n_2 = 22$, $\bar{x}_2 = 16001.0$, $s_2 = 27679.4$.
    TS: $F = 3.1210$.
    CV: $F = 2.5089$
    Means: TS: $t = -1.116$.
    CV: $t = \pm 2.093$.

14. Randomly selected statistics students were given five seconds to estimate the value of a product of numbers with the results given in the accompanying table. Use a 0.05 significance level to test the claim that the two samples come from populations with the same mean. Does the order of the numbers appear to have an effect on the estimate? (See the Cooperative Group Activities at the end of Chapter 2.)

Estimates from Students Given $1 \times 2 \times 3 \times 4 \times 5 \times 6 \times 7 \times 8$

1560	169	5635	25	842	40,320	5000	500	1110	10,000	200
1252	4000	2040	175	856	42,200	49,654	560	800		

Estimates from Students Given $8 \times 7 \times 6 \times 5 \times 4 \times 3 \times 2 \times 1$

100,000	2000	42,000	1500	52,836	2050	428	372	300	225
64,582	23,410	500	1200	400	49,000	4000	1876	3600	354
750	640								

15. Use the carbon monoxide (CO) data in Table 8-1 to test the claim that filters are effective in reducing carbon monoxide in king-size cigarettes. Based on this result and the examples of this section, are cigarette filters effective, or are they just a sales gimmick with no real effect?

16. An experiment was conducted to test the effects of alcohol. The errors were recorded in a test of visual and motor skills for a treatment group of people who drank ethanol and another group given a placebo. The results are shown in the accompanying table. Use a 0.05 significance level to test the claim that the two groups come from populations with the same mean. Do these results support the common belief that drinking is hazardous for drivers, pilots, ship captains, and so on?

Treatment Group	Placebo Group	
$n_1 = 22$	$n_2 = 22$	Based on data from "Effects of Alcohol Intoxication on Risk Taking, Strategy, and Error Rate in Visuomotor Performance," by Streufert et al., *Journal of Applied Psychology*, Vol. 77, No. 4.
$\bar{x}_1 = 4.20$	$\bar{x}_2 = 1.71$	
$s_1 = 2.20$	$s_2 = 0.72$	

8-5 Exercises B: Beyond the Basics

17. An experiment was conducted to test the effects of alcohol. The breath alcohol levels were measured for a treatment group of people who drank ethanol and another group given a placebo. The results are given in the accompanying table. Use a 0.05 significance level to test the claim that the two groups come from populations with the same mean.

Treatment Group	Placebo Group	
$n_1 = 22$	$n_2 = 22$	Based on data from "Effects of Alcohol Intoxication on Risk Taking, Strategy, and Error Rate in Visuomotor Performance," by Streufert et al., *Journal of Applied Psychology*, Vol. 77, No. 4.
$\bar{x}_1 = 0.049$	$\bar{x}_2 = 0.000$	
$s_1 = 0.015$	$s_2 = 0.000$	

18. Assume that two samples have the same standard deviation and that both are independent, small, and randomly selected from normally distributed populations. Also assume that we want to test the claim that the samples come from populations with the same mean.
a. Is it necessary to conduct a preliminary F test?
b. If both samples have standard deviation s, what is the value of s_p^2 expressed in terms of s?

19. Refer to the two examples of this section that used the tar amounts in cigarettes. (See the Case 3 subsection.) How are the hypothesis test and confidence interval affected if the number of degrees of freedom is calculated with Formula 8-1, instead of using the smaller of $n_1 - 1$ and $n_2 - 1$?

15. TS: $F = 6.6835$.
CV: $F = 4.4667$.
Means: TS: $t = -3.500$.
CV: $t = -1.895$.

16. TS: $F = 9.3364$.
CV: $F = 2.4247$ (approximately).
Means: TS: $t = 5.045$.
CV: $t = \pm 2.080$.

17. No F-test results. Means: TS: $t = 15.322$. CV: $t = \pm 2.080$.

18. a. No.
b. s^2

19. df from 7 to 26; TS: same; CV: changes from -1.895 to -1.706; CI limits from $(-13.1, -8.3)$ to $(-12.8, -8.6)$.

20. 1st sample: $n = 20$, $\bar{x} =$
13.9, $s = 2.5$; TS: 2.1626;
Means: TS: $t = -10.443$.
CV: $t = -1.706$.

20. Refer to the two examples of this section that used the tar amounts in cigarettes. (See the Case 3 subsection.) How are the hypothesis test and confidence interval affected if the one extreme outlier is deleted?

8-6 Inferences about Two Proportions

This section is important because it presents methods for dealing with a very common situation: the comparison of two proportions.

Although this section appears last in the chapter, it is arguably the most important because this is where we describe methods for using two sample proportions to make inferences (hypothesis testing and confidence interval construction) about two population proportions.

When testing a hypothesis made about two population proportions—such as the proportions of cured patients in a population given some treatment and a second population given a placebo—or when constructing a confidence interval for the difference between two population proportions, we make the following assumptions and use the following notation.

Assumptions

1. We have two *independent* sets of randomly selected sample data.
2. For both samples, the conditions $np \geq 5$ and $nq \geq 5$ are satisfied. (In many cases, we will test the claim that two populations have equal proportions so that $p_1 - p_2 = 0$. Because we assume that $p_1 - p_2 = 0$, it is not necessary to specify the particular value that p_1 and p_2 have in common. In such cases, the conditions $np \geq 5$ and $nq \geq 5$ can be checked by replacing p by the estimated pooled proportion $\bar{p}$, which will be described later.)

Notation for Two Proportions

For population 1 we let

$p_1 = population$ proportion

$n_1 = $ size of the sample

$x_1 = $ number of successes in the sample

$\hat{p}_1 = \dfrac{x_1}{n_1}$ (the *sample* proportion)

$\hat{q}_1 = 1 - \hat{p}_1$

The corresponding meanings are attached to p_2, n_2, x_2, $\hat{p}_2$, and $\hat{q}_2$, which come from population 2.

Hypothesis Tests

In Section 7-5 we discussed tests of hypotheses made about a single population proportion. We will now consider tests of hypotheses made about two population proportions, but *we will be testing only claims that $p_1 = p_2$* and we will use the following pooled (or combined) estimate of the value that p_1 and p_2 have in common. (For claims that the difference between p_1 and p_2 is equal to a nonzero constant, see Exercise 17 of this section.)

Pooled Estimate of p_1 and p_2

The **pooled estimate of p_1 and p_2** is denoted by $\bar{p}$ and is given by

$$\bar{p} = \frac{x_1 + x_2}{n_1 + n_2}$$

We denote the complement of $\bar{p}$ by $\bar{q}$, so $\bar{q} = 1 - \bar{p}$.

The concept of a pooled estimate of p_1 and p_2 applies only to cases in which we assume that $p_1 = p_2$.

Test Statistic for Two Proportions

The following test statistic applies to null and alternative hypotheses that fit one of these three formats:

$$H_0: p_1 = p_2 \qquad H_0: p_1 \geq p_2 \qquad H_0: p_1 \leq p_2$$
$$H_1: p_1 \neq p_2 \qquad H_1: p_1 < p_2 \qquad H_1: p_1 > p_2$$

$$z = \frac{(\hat{p}_1 - \hat{p}_2) - (p_1 - p_2)}{\sqrt{\dfrac{\bar{p}\,\bar{q}}{n_1} + \dfrac{\bar{p}\,\bar{q}}{n_2}}}$$

where

$$p_1 - p_2 = 0$$

$$\hat{p}_1 = \frac{x_1}{n_1} \quad \text{and} \quad \hat{p}_2 = \frac{x_2}{n_2}$$

$$\bar{p} = \frac{x_1 + x_2}{n_1 + n_2}$$

$$\bar{q} = 1 - \bar{p}$$

The following example will help clarify the roles of x_1, n_1, $\hat{p}_1$, $\bar{p}$, and so on. In particular, you should recognize that under the assumption of equal proportions, the best estimate of the common proportion is obtained by pooling both samples into one big sample, so that $\bar{p}$ becomes a more obvious estimate of the common population proportion.

EXAMPLE   Johns Hopkins researchers conducted a study of pregnant IBM employees. Among 30 employees who worked with glycol ethers, 10 (or 33.3%) had miscarriages, but among 750 who were not exposed to glycol ethers, 120 (or 16.0%) had miscarriages. At the 0.01 significance level, test the claim that the miscarriage rate is greater for women exposed to glycol ethers.

Although x_1 and x_2 are technically termed numbers of "successes" in this example, this is simply a reference to the conditions being considered. Miscarriages certainly are not considered "successes" in any usual sense.

SOLUTION   For notation purposes, we stipulate that sample 1 is the group that worked with glycol ethers and sample 2 is the group not exposed, so the sample statistics can be summarized as shown here.

Exposed to Glycol Ethers	Not Exposed to Glycol Ethers
$n_1 = 30$	$n_2 = 750$
$x_1 = 10$	$x_2 = 120$
$\hat{p}_1 = \dfrac{10}{30} = 0.333$	$\hat{p}_2 = \dfrac{120}{750} = 0.160$

We will now use the traditional method of hypothesis testing, as summarized in Figure 7-4.

Step 1: The claim of a greater miscarriage rate for women exposed to glycol ethers can be represented by $p_1 > p_2$.

Step 2: If $p_1 > p_2$ is false, then $p_1 \leq p_2$.

Step 3: Because our claim of $p_1 > p_2$ does not contain equality, it becomes the alternative hypothesis, and we have

$$H_0: p_1 \leq p_2 \qquad H_1: p_1 > p_2 \text{ (original claim)}$$

Step 4: The significance level is $\alpha = 0.01$.

Step 5: We will use the normal distribution (with the test statistic previously given) as an approximation to the binomial distribution. We have two independent samples, and the conditions $np \geq 5$ and $nq \geq 5$ are satisfied for each of the two samples. To check this, we note that in conducting this test, we assume that $p_1 = p_2$, where their common value is the pooled estimate $\bar{p}$, calculated as

$$\bar{p} = \frac{x_1 + x_2}{n_1 + n_2} = \frac{10 + 120}{30 + 750} = 0.1667$$

With $\bar{p} = 0.1667$, it follows that $\bar{q} = 1 - 0.1667 = 0.8333$. We verify that $np \geq 5$ and $nq \geq 5$ for both samples as follows:

Sample 1	Sample 2
$n_1 p = (30)(0.1667) = 5 \geq 5$	$n_2 p = (750)(0.1667) = 125 \geq 5$
$n_1 q = (30)(0.8333) = 25 \geq 5$	$n_2 q = (750)(0.8333) = 625 \geq 5$

Step 6:   We can now find the value of the test statistic.

$$z = \frac{(\hat{p}_1 - \hat{p}_2) - (p_1 - p_2)}{\sqrt{\dfrac{\bar{p}\,\bar{q}}{n_1} + \dfrac{\bar{p}\,\bar{q}}{n_2}}}$$

$$= \frac{(0.333 - 0.160) - 0}{\sqrt{\dfrac{(0.1667)(0.8333)}{30} + \dfrac{(0.1667)(0.8333)}{750}}} = 2.49$$

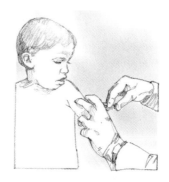

The critical value of $z = 2.33$ is found by observing that we have a right-tailed test with $\alpha = 0.01$. The value of $z = 2.33$ is found from Table A-2 as the z score corresponding to an area of $0.5 - 0.01 = 0.4900$.

Step 7:   In Figure 8-10 we see that the test statistic falls within the critical region, so we reject the null hypothesis of $p_1 \le p_2$.

Step 8:   We conclude that there is sufficient evidence to support the claim that the miscarriage rate is greater for women exposed to ethyl glycol.

These steps use the traditional approach to hypothesis testing, but it would be quite easy to use the P-value approach. In Step 6, instead of finding the critical value of z, we would find the P-value by using the procedure summarized in Figures 7-7 and 7-8. With a test statistic of $z = 2.49$ and a right-tailed test, we get

$$P\text{-value} = (\text{area to the right of } z = 2.49) = 0.0064$$

(continued)

Figure 8-10   Distribution of Differences Between Proportions for Group Exposed to Glycol Ethers and Group Not Exposed

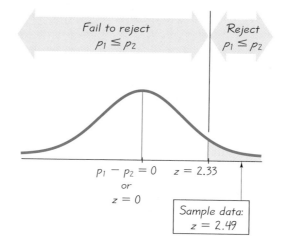

Again, we reject the null hypothesis because the *P*-value of 0.0064 is less than the significance level of $\alpha = 0.01$.

With this evidence, the Johns Hopkins researchers concluded that women employees exposed to glycol ethers "have a significantly increased risk of miscarriage." On the basis of these results, IBM warned its employees of the danger, notified the Environmental Protection Agency, and greatly reduced its use of glycol ethers.

How do we find the values of x_1 and x_2? In the preceding example we were given information from which we could easily see that $x_1 = 10$ and $x_2 = 120$. In other cases, we may have to calculate the values of x_1 and x_2, which can be found by noting that $\hat{p}_1 = x_1/n_1$ implies that

$$x_1 = n_1 \cdot \hat{p}_1$$

If we know that sample 1 consists of 500 treated patients and 30% of them were cured, the actual number of cured patients is $x_1 = 500 \cdot 0.30 = 150$. In this case, $n_1 = 500$, $\hat{p}_1 = 0.30$, and $x_1 = 150$.

Confidence Intervals

In the preceding example we found that there appears to be a larger miscarriage rate for women exposed to glycol ethers, so we now proceed to estimate the amount of the difference. Is the difference large enough to justify extensive and expensive changes in manufacturing procedures?

With the same assumptions given at the beginning of this section, a confidence interval for the difference between population proportions $p_1 - p_2$ can be constructed by evaluating the following.

$$(\hat{p}_1 - \hat{p}_2) - E < (p_1 - p_2) < (\hat{p}_1 - \hat{p}_2) + E$$

where
$$E = z_{\alpha/2} \sqrt{\frac{\hat{p}_1 \hat{q}_1}{n_1} + \frac{\hat{p}_2 \hat{q}_2}{n_2}}$$

EXAMPLE   Use the sample data given in the preceding example to construct a 99% confidence interval for the difference between the two population proportions.

SOLUTION   With a 99% degree of confidence, $z_{\alpha/2} = 2.575$ (from Table A-2). We first calculate the value of the margin of error E as shown on the following page.

$$E = z_{\alpha/2}\sqrt{\frac{\hat{p}_1\hat{q}_1}{n_1} + \frac{\hat{p}_2\hat{q}_2}{n_2}} = 2.575\sqrt{\frac{(0.333)(0.667)}{30} + \frac{(0.160)(0.840)}{750}}$$
$$= 0.2242$$

With $\hat{p}_1 = 0.333$, $\hat{p}_2 = 0.160$, and $E = 0.2242$, the confidence interval is evaluated as follows.

$$(\hat{p}_1 - \hat{p}_2) - E < (p_1 - p_2) < (\hat{p}_1 - \hat{p}_2) + E$$
$$(0.333 - 0.160) - 0.2242 < (p_1 - p_2) < (0.333 - 0.160) + 0.2242$$
$$-0.0512 < (p_1 - p_2) < 0.3972$$

The confidence interval limits contain 0, suggesting that there is not a significant difference between the two proportions. This seems to contradict the conclusion of the earlier example illustrating the hypothesis-testing procedure. This discrepancy is attributable to two factors: (1) The hypothesis test was right-tailed with a 0.01 significance level, so it corresponds to a 98% confidence interval (but the above confidence interval has a 99% degree of confidence); (2) the variance used for the hypothesis test (based on an assumption of equal proportions) is different from the variance used for the confidence interval (where the proportions are not assumed to be equal).

As always, we should be careful when interpreting confidence intervals. Because p_1 and p_2 have fixed values and are not variables, it is wrong to state that there is a 99% chance that the value of $p_1 - p_2$ falls between -0.0512 and 0.3972. It is correct to state that if we repeat the same sampling process and construct 99% confidence intervals, in the long run 99% of the intervals will actually contain the value of $p_1 - p_2$.

Why do the procedures of this section work? The test statistic given for hypothesis tests is justified by the following:

1. With $n_1 p_1 \geq 5$ and $n_1 q_1 \geq 5$, the distribution of $\hat{p}_1$ can be approximated by a normal distribution with mean p_1, standard deviation $\sqrt{p_1 q_1/n_1}$, and variance $p_1 q_1/n_1$. These conclusions are based on Sections 5-6 and 6-4 and they also apply to the second sample.

2. Because $\hat{p}_1$ and $\hat{p}_2$ are each approximated by a normal distribution, $\hat{p}_1 - \hat{p}_2$ will also be approximated by a normal distribution with mean $p_1 - p_2$ and variance

$$\sigma^2_{(\hat{p}_1 - \hat{p}_2)} = \sigma^2_{\hat{p}_1} + \sigma^2_{\hat{p}_2} = \frac{p_1 q_1}{n_1} + \frac{p_2 q_2}{n_2}$$

(In Section 8-2 we established that the variance of the differences between two independent random variables is the sum of their individual variances.)

The Lead Margin of Error

Authors Stephen Ansolabehere and Thomas Belin wrote in their article "Poll Faulting" (*Chance* magazine) that "our greatest criticism of the reporting of poll results is with the margin of error of a single proportion (usually ±3%) when media attention is clearly drawn to the *lead* of one candidate." They point out that the lead is really the *difference* between two proportions $(p_1 - p_2)$ and go on to explain how they developed the following rule of thumb: The lead is approximately $\sqrt{3}$ times larger than the margin of error for any one proportion. For a typical pre-election poll, a reported ±3% margin of error translates to about ±5% for the lead of one candidate over the other. They write that the margin of error for the lead should be reported.

3. Because the values of p_1, p_2, q_1, and q_2 are typically unknown and from the null hypothesis we assume that $p_1 = p_2$, we can pool (or combine) the sample data. The pooled estimate of the common value of p_1 and p_2 is $\bar{p} = (x_1 + x_2)/(n_1 + n_2)$. If we replace p_1 and p_2 by $\bar{p}$ and replace q_1 and q_2 by $\bar{q} = 1 - \bar{p}$, the variance from Step 2 leads to the following standard deviation.

$$\sigma_{(\hat{p}_1 - \hat{p}_2)} = \sqrt{\frac{\bar{p}\,\bar{q}}{n_1} + \frac{\bar{p}\,\bar{q}}{n_2}}$$

4. We now know that the distribution of $p_1 - p_2$ is approximately normal, with mean $p_1 - p_2$ and standard deviation as given, so that the z test statistic has the form given earlier.

The form of the confidence interval requires an expression for the variance different from the one given in Step 3. In Step 3 we are assuming that $p_1 = p_2$, but if we don't make that assumption (as in the construction of a confidence interval), we estimate the variance of $\hat{p}_1 - \hat{p}_2$ as

$$\sigma^2_{(\hat{p}_1 - \hat{p}_2)} = \sigma^2_{\hat{p}_1} + \sigma^2_{\hat{p}_2} = \frac{\hat{p}_1 \hat{q}_1}{n_1} + \frac{\hat{p}_2 \hat{q}_2}{n_2}$$

and the standard deviation becomes

$$\sqrt{\frac{\hat{p}_1 \hat{q}_1}{n_1} + \frac{\hat{p}_2 \hat{q}_2}{n_2}}$$

In the test statistic

$$z = \frac{(\hat{p}_1 - \hat{p}_2) - (p_1 - p_2)}{\sqrt{\frac{\hat{p}_1 \hat{q}_1}{n_1} + \frac{\hat{p}_2 \hat{q}_2}{n_2}}}$$

let z be positive and negative (for two tails) and solve for $p_1 - p_2$. The results are the limits of the confidence interval given earlier.

Using Computers and Calculators for Inferences with Two Proportions

We can use STATDISK for claims about two proportions. To conduct a hypothesis test, select Analysis, then Hypothesis Testing, then Proportion-Two Samples. Confidence interval limits are included with the hypothesis test results.

The use of Minitab is a bit tricky here, and we recommend that you consult the *Minitab Student Laboratory Manual and Workbook* for a procedure that will work.

The TI-83 calculator can be used for hypothesis tests and confidence intervals. Press STAT and select TESTS. Then choose the option of 2-PropZTest (for a hypothesis test) or 2-PropZInt (for a confidence interval). When test-

ing hypotheses, the TI-83 calculator will display a *P*-value instead of critical values, so the *P*-value method of testing hypotheses is used.

8-6 Exercises A: Basic Skills and Concepts

In Exercises 1 and 2, assume that you plan to use a significance level of $\alpha = 0.05$ to test the claim that $p_1 = p_2$. Use the given sample sizes and numbers of successes to find (a) the pooled estimate $\bar{p}$, (b) the z test statistic, (c) the critical z values, and (d) the P-value.

1. Sample 1   Sample 2

$n_1 = 50$ $n_2 = 100$
$x_1 = 25$ $x_2 = 55$

2. Sample 1   Sample 2

$n_1 = 500$ $n_2 = 200$
$x_1 = 300$ $x_2 = 150$

Recommended assignment: Exercises 3, 4, 6, 7, 8, 16.

1. a. 8/15
   b. −0.58
   c. ±1.96
   d. 0.5620
2. a. 0.643
   b. −3.74
   c. ±1.96
   d. 0.0002
3. TS: $z = -2.82$.
   CV: $z = -1.645$.

3. An article published in *USA Today* stated that "in a study of 200 colorectal surgery patients, 104 were kept warm with blankets and intravenous fluids; 96 were kept cool. The results show: Only 6 of those warmed developed wound infections vs. 18 who were kept cool." Use a 0.05 significance level to test the claim of the article's headline: "Warmer surgical patients recover better." If these results are verified, should surgical patients be routinely warmed?

4. A U.S. Department of Justice report (NCJ-156831) included the claim that "in spouse murder cases, wife defendants were less likely to be convicted than husband defendants." Sample data consisted of 277 convictions among 318 husband defendants, and 155 convictions among 222 wife defendants. Test the stated claim and identify one possible explanation for the result.

4. TS: $z = 4.94$.
   CV: $z = 1.645$.

5. The newly appointed head of the state mental health agency claims that a greater proportion of the crimes committed by persons younger than 21 years of age are violent crimes (when compared to the crimes committed by persons 21 years of age or older). Of 2750 randomly selected arrests of criminals younger than 21 years of age, 4.25% involve violent crimes. Of 2200 randomly selected arrests of criminals 21 years of age or older, 4.55% involve violent crimes (based on data from the Uniform Crime Reports). Construct a 95% confidence interval for the difference between the two proportions of violent crimes. Do the confidence interval limits contain 0? Does this indicate that there isn't a significant difference between the two rates of violent crimes?

5. $-0.0144 < p_1 - p_2 < 0.0086$; yes.

6. In initial tests of the Salk vaccine, 33 of 200,000 vaccinated children later developed polio. Of 200,000 children vaccinated with a placebo, 115 later developed polio. At the 0.01 level of significance, test the claim that the Salk vaccine is effective in lowering the polio rate. Does it appear that the vaccine is effective?

6. TS: $z = -6.74$.
   CV: $z = -2.33$.

7. TS: $z = 12.86$.
   CV: $z = 2.575$.

7. The *New York Times* ran an article about a study in which Professor Denise Korniewicz and other Johns Hopkins researchers subjected laboratory gloves to stress. Among 240 vinyl gloves, 63% leaked viruses. Among 240 latex gloves, 7% leaked viruses. At the 0.005 significance level, test the claim that vinyl gloves have a larger virus leak rate than latex gloves. (It might seem obvious that 63% is larger than 7%, but we must justify our conclusion by considering the sample sizes, the significance level, and the nature of the distribution.)

8. $0.469 < p_1 - p_2 < 0.651$; yes

8. Using the sample data from Exercise 7, construct the 99% confidence interval for the difference between the two proportions of gloves that leak viruses. Does that difference appear to warrant a decision to use latex gloves, even if they are more expensive?

9. TS: $z = -1.92$.
   CV: $z = \pm 1.96$.

9. A public relations expert and consultant for the airline industry is planning a strategy to influence voter perception of government regulation of airfares. In a *New York Times*/CBS News survey, it is found that 35% of 552 Democrats believe that the government should regulate airline prices, compared to 41% of the 417 Republicans surveyed. At the 0.05 significance level, test the claim that there is no difference between the proportions of Democrats and Republicans who believe in government regulation of airfares. Based on the result, should the consultant consider different approaches for Democrats and Republicans?

10. $-0.122 < p_1 - p_2 < 0.002$

10. Use the sample data from Exercise 9 to construct a 95% confidence interval for the difference between the proportions of Democrats and Republicans who believe that the government should regulate airline prices.

11. TS: $z = 3.86$.
    CV: $z = 2.33$.

11. Karl Pearson, who developed many important concepts in statistics, collected crime data in 1909. Of those convicted of arson, 50 were drinkers and 43 abstained. Of those convicted of fraud, 63 were drinkers and 144 abstained. Use a 0.01 significance level to test the claim that the proportion of drinkers among convicted arsonists is greater than the proportion of drinkers convicted of fraud. Does it seem reasonable that drinking might have an effect on the type of crime? Why?

12. $-0.271 < p_1 - p_2 < -0.089$

12. As part of a campaign for the presidency, one candidate plans to wage a voter registration drive. In considering whether to target specific age groups, this candidate finds survey results showing that among 200 randomly selected persons aged 18–24, 36.0% voted. Among 250 persons in the 25–44 age bracket, 54.0% voted (based on data from the U.S. Bureau of the Census). Construct a 95% confidence interval for the difference between the proportions of voters in the two age brackets. Are the percentages of voters in the two groups different enough that the groups should be targeted differently?

13. TS: $z = -3.81$.
    CV: $z = \pm 1.96$.

13. Using the sample data given in Exercise 12, test the claim that the proportions of voters in the two age brackets are the same. Use a 0.05 significance level.

14. TS: $z = 4.46$.
    CV: $z = \pm 2.575$.

14. Professional pollsters are becoming concerned about the growing rate of refusals among potential survey subjects. In analyzing the problem, there

is a need to know if the refusal rate is universal or if there is a difference between the rates for central-city residents and those not living in central cities. Specifically, it was found that when 294 central-city residents were surveyed, 28.9% refused to respond. A survey of 1015 residents not in a central city resulted in a 17.1% refusal rate (based on data from "I Hear You Knocking But You Can't Come In," by Fitzgerald and Fuller, *Sociological Methods and Research*, Vol. 11, No. 1). At the 0.01 significance level, test the claim that the central-city refusal rate is the same as the refusal rate in other areas.

15. A study was made of 413 children who were hospitalized as a result of motor vehicle crashes. Among 290 children who were not using seat belts, 50 were injured severely. Among 123 children using seat belts, 16 were injured severely (based on data from "Morbidity Among Pediatric Motor Vehicle Crash Victims: The Effectiveness of Seat Belts," by Osberg and Di Scala, *American Journal of Public Health*, Vol. 82, No. 3). Is there sufficient sample evidence to conclude, at the 0.05 level, that the rate of severe injuries is lower for children wearing seat belts? Based on these results, what action should be taken?

16. When games were sampled from throughout a season, it was found that the home team won 127 of 198 professional *basketball* games, and the home team won 57 of 99 professional *football* games (based on data from "Predicting Professional Sports Game Outcomes from Intermediate Game Scores," by Cooper et al., *Chance*, Vol. 5, No. 3–4). Construct a 95% confidence interval for the difference between the proportions of home wins. Do the confidence interval limits contain 0? Based on the results, is there a significant difference between the proportions of home wins? What do you conclude about the home field advantage?

15. TS: $z = 1.07$.
    CV: $z = 1.645$.

16. $-0.052 < p_1 - p_2 < 0.184$; yes.

8–6 Exercises B: Beyond the Basics

17. To test the null hypothesis that the difference between two population proportions is equal to a nonzero constant c, use the test statistic

$$z = \frac{(\hat{p}_1 - \hat{p}_2) - c}{\sqrt{\dfrac{\hat{p}_1(1 - \hat{p}_1)}{n_1} + \dfrac{\hat{p}_2(1 - \hat{p}_2)}{n_2}}}$$

As long as n_1 and n_2 are both large, the sampling distribution of the test statistic z will be approximately the standard normal distribution. Refer to the sample data included with the example presented in this section, and use a 0.05 significance level to test the claim of a medical expert that when women are exposed to glycol ethers, their percentage of miscarriages is 10 percentage points more than the percentage for women not exposed to glycol ethers.

17. TS: $z = 0.84$.
    CV: $z = \pm 1.96$.

18. a. TS: $z = 1.48$.
    CV: $z = \pm 1.96$.
    b. TS: $z = 1.63$.
    CV: $z = \pm 1.96$.
    c. TS: $z = 3.09$.
    CV: $z = \pm 1.96$.
    d. No.

19. 2135

20. Percentages cannot be
    correct.

18. Sample data are randomly drawn from three independent populations, each of size 100. The sample proportions are $\hat{p}_1 = 40/100$, $\hat{p}_2 = 30/100$, and $\hat{p}_3 = 20/100$.
    a. At the 0.05 significance level, test $H_0: p_1 = p_2$.
    b. At the 0.05 significance level, test $H_0: p_2 = p_3$.
    c. At the 0.05 significance level, test $H_0: p_1 = p_3$.
    d. In general, if hypothesis tests lead to the conclusions that $p_1 = p_2$ and $p_2 = p_3$ are reasonable, does it follow that $p_1 = p_3$ is also reasonable? Why or why not?

19. The *sample size* needed to estimate the difference between two population proportions to within a margin of error E with a confidence level of $1 - \alpha$ can be found as follows. In the expression

$$E = z_{\alpha/2} \sqrt{\frac{p_1 q_1}{n_1} + \frac{p_2 q_2}{n_2}}$$

replace n_1 and n_2 by n (assuming that both samples have the same size) and replace each of p_1, q_1, p_2, and q_2 by 0.5 (because their values are not known). Then solve for n.

  Use this approach to find the size of each sample if you want to estimate the difference between the proportions of men and women who own cars. Assume that you want 95% confidence that your error is no more than 0.03.

20. In an article from the Associated Press, it was reported that researchers "randomly selected 100 New York motorists who had been in an accident and 100 who had not. Of those in accidents, 13.7 percent owned a cellular phone, while just 10.6 percent of the accident-free drivers had a phone in the car." Analyze these results.

· ·

Vocabulary List

independent samples
dependent samples
paired samples
matched samples
F distribution

numerator degrees of freedom
denominator degrees of freedom
pooled estimate of σ^2
pooled estimate of p_1 and p_2

· ·

Review

Chapters 6 and 7 introduced two major concepts of inferential statistics: the estimation of population parameters and the methods of testing hypotheses made about population parameters. Whereas those chapters considered only cases

involving a single population, this chapter considers two samples drawn from two populations.

- Section 8-2 considered inferences made about two dependent (paired) populations.
- Section 8-3 considered inferences made about two independent populations, but the involved samples were both large ($n > 30$).
- Section 8-4 presented methods for testing claims about two population standard deviations or variances.
- Section 8-5 described procedures for making inferences about two independent populations, with at least one of the samples being small ($n \leq 30$). Sections 8-2, 8-3, and 8-5 dealt with a variety of different cases for making inferences about two population means; those cases were summarized in Figure 8-8 (for hypothesis tests) and Figure 8-9 (for confidence intervals).
- Section 8-6 discussed the procedures for making inferences about two population proportions.

Review Exercises

1. In a study of people who stop to help drivers with disabled cars, researchers hypothesized that more people would stop to help someone if they first saw another driver with a disabled car getting help. In one experiment, 2000 drivers first saw a woman being helped with a flat tire and then saw a second woman who was alone, farther down the road, with a flat tire; 2.90% of those 2000 drivers stopped to help the second woman. Among 2000 other drivers who did not see the first woman being helped, only 1.75% stopped to help (based on data from "Help on the Highway," by McCarthy, *Psychology Today*). At the 0.05 significance level, test the claim that the percentage of people who stop after first seeing a driver with a disabled car being helped is greater than the percentage of people who stop without first seeing someone else being helped.

2. Twelve Dozenol tablets are tested for solubility before and after being stored for one year. The indexes of solubility are given in the table below.

Before	472	487	506	512	489	503	511	501	495	504	494	462
After	562	512	523	528	554	513	516	510	524	510	524	508

   a. At the 0.05 significance level, test the claim that the Dozenol tablets are more soluble after the storage period.

   b. Construct a 95% confidence interval estimate of the mean difference of after − before.

3. Twelve different tablets from each of two competing cold medicines are randomly selected and tested for the amount of acetaminophen each

1. TS: $z = 2.41$.
   CV: $z = 1.645$.

2. a. TS: $t = -3.847$.
   CV: $t = -1.796$.
   b. $12.4 < \mu_d < 45.6$

3. a. TS: $F = 1.3010$.
   CV: $F =$ between 3.4296
   and 3.5257. Means: TS:
   $t = -4.337$.
   CV: $t = \pm 2.074$.
   b. $15.1 < \mu_1 - \mu_2 < 42.9$

4. TS: $F = 1.3786$.
   CV: $F = 1.8363$.

5. TS: $z = 0.78$.
   CV: $z = \pm 2.33$.

6. TS: $z = 3.67$.
   CV: $z = 1.645$.

7. TS: $t = 1.185$.
   CV: $t = \pm 2.262$.

8. TS: $z = 0.96$.
   CV: $z = 2.33$.

Winston	Barrington
$n = 18$	$n = 24$
$\bar{x} = 85.7$	$\bar{x} = 80.6$
$s = 2.8$	$s = 9.7$

contains, and the results (in milligrams) follow.

Dozenol	472	487	506	512	489	503	511	501	495	504	494	462
Niteze	562	512	523	528	554	513	516	510	524	510	524	508

a. At the 0.05 significance level, test the claim that the mean amount of acetaminophen is the same in each brand.

b. Construct a 95% confidence interval estimate of the difference between the mean for Dozenol and the mean for Niteze.

4. A study was conducted to investigate relationships between different types of standard test scores. On the Graduate Record Examination verbal test, 68 women had a mean of 538.82 and a standard deviation of 114.16, and 86 men had a mean of 525.23 and a standard deviation of 97.23. (See "Equivalencing MAT and GRE Scores Using Simple Linear Transformation and Regression Methods," by Kagan and Stock, *Journal of Experimental Education,* Vol. 49, No. 1.) At the 0.02 significance level, test the claim that the two groups come from populations with the same standard deviation.

5. Using the sample data given in Exercise 4, test the claim that the mean score for women is equal to the mean score for men. Use a 0.02 significance level.

6. A test question is considered good if it discriminates between prepared and unprepared students. The first question on a test was answered correctly by 62 of 80 prepared students and by 23 of 50 unprepared students. At the 0.05 level of significance, test the claim that this question was answered correctly by a greater proportion of prepared students.

7. In a study of techniques used to measure lung volumes, physiological data were collected for 10 subjects. The values given in the accompanying table are in liters, representing the measured forced vital capacities of the 10 subjects in a sitting position and in a supine (lying) position. At the 0.05 significance level, test the claim that the position has no effect, so the mean difference is zero.

Subject	A	B	C	D	E	F	G	H	I	J
Sitting	4.66	5.70	5.37	3.34	3.77	7.43	4.15	6.21	5.90	5.77
Supine	4.63	6.34	5.72	3.23	3.60	6.96	3.66	5.81	5.61	5.33

Based on data from "Validation of Esophageal Balloon Technique at Different Lung Volumes and Postures," by Baydur et al., *Journal of Applied Physiology,* Vol. 62, No. 1.

8. Air America is experimenting with its training program for flight attendants. With the traditional six-week program, a random sample of 60 flight attendants achieve competency test scores with a mean of 83.5 and a standard deviation of 16.3. With a new 10-day program, a random sample of 35 flight attendants achieve competency test scores with a mean of 79.8 and a standard deviation of 19.2. At the 0.01 significance level, test the claim that the 10-day program results in scores with a lower mean.

9. Researchers are testing commercial air-filtering systems made by the Winston Industrial Supply Company and the Barrington Filter Company. Random

samples are tested for each company, and the filtering efficiency is scored on a standard scale, with the results shown in the margin of the preceding page. (Higher scores correspond to better filtering.) At the 0.05 level of significance, test the claim that both systems have the same mean.

10. In an insurance study of pedestrian deaths in New York State, monthly fatalities are analyzed for two different time periods. Sample data for the first time period are summarized by these statistics: $n = 12$, $\bar{x} = 46.42$, $s = 11.07$. Sample data for the second time period are summarized by these statistics: $n = 12$, $\bar{x} = 51.00$, $s = 10.39$ (based on data from the New York State Department of Motor Vehicles).

   a. At the 0.05 significance level, test the claim that both time periods have the same mean.

   b. Construct a 95% confidence interval for the difference between the two population means.

9. TS: $F = 12.0013$.
   CV: $F = 2.5598$.
   Means: TS: $t = 2.444$.
   CV: $t = \pm 2.110$.

10. a. TS: $F = 1.1352$.
    CV: $F =$ between 3.4296
    and 3.5257. Means: TS: $t = -1.045$. CV: $t = \pm 2.074$.
    b. $-13.670 < \mu_1 - \mu_2 < 4.510$

● ●

Cumulative Review Exercises

1. The data in the accompanying table were obtained through a survey of randomly selected subjects.

   a. If one of the survey subjects is randomly selected, find the probability of getting someone ticketed for speeding.

   b. If one of the survey subjects is randomly selected, find the probability of getting a male or someone ticketed for speeding.

   c. Find the probability of getting someone ticketed for speeding, given that the selected person is a man.

   d. Find the probability of getting someone ticketed for speeding, given that the selected person is a woman.

   e. Use a 0.05 level of significance to test the claim that the percentage of women ticketed for speeding is less than the percentage for men. Can we conclude that men generally speed more than women?

1. a. 0.0707
   b. 0.369
   c. 0.104
   d. 0.054
   e. TS: $z = -2.52$.
   CV: $z = -1.645$.

	Ticketed for Speeding Within the Last Year?	
	Yes	No
Men	26	224
Women	27	473

Based on data from R.H. Bruskin Associates.

2. The Newton Scientific Instrument Company manufactures scales with its day shift and night shift. Random samples are routinely tested with errors recorded as positive amounts for readings that are too high or negative for

readings that are too low.

2. a. TS: $z = 1.95$.
   CV: $z = \pm1.96$.

a.   A random sample of 40 scales made by the day shift is tested and the errors are found to have a mean of 1.2 g and a standard deviation of 3.9 g. Use a 0.05 significance level to test the claim that the sample comes from a population with a mean error equal to 0.

b. TS: $z = -1.87$.
   CV: $z = \pm1.96$.

b.   A random sample of 33 scales made by the night shift is tested and the errors are found to have a mean of -1.4 g and a standard deviation of 4.3 g. Use a 0.05 significance level to test the claim that the sample comes from a population with a mean error equal to 0.

c. TS: $z = 2.68$;
   CV: $z = \pm1.96$.

c.   Use a 0.05 significance level to test the claim that both shifts manufacture scales with the same mean error.

d. $0.0 g < \mu < 2.4 g$

d.   Construct a 95% confidence interval for the mean error of the day shift.

e. $-2.9 g < \mu < 0.1 g$

e.   Construct a 95% confidence interval for the mean error of the night shift.

f. $0.7 g < \mu_1 - \mu_2 < 4.5 g$

f.   Construct a 95% confidence interval for the difference between the means of errors for the day shift and night shift.

g. 234

g.   If you want to estimate the mean error for the day shift, how many scales must you check if you want to be 95% confident that your mean is off by no more than 0.5 g? (*Hint:* Use the sample standard deviation in part (a) as an estimate of the population standard deviation σ.)

Computer Project

Refer to Data Set 1 in Appendix B and use STATDISK or Minitab to retrieve the weights of discarded metal and plastic, which are already stored on a disk. Use the STATDISK or Minitab programs for each of the following.

a. Test the claim that the weights of discarded metal and the weights of discarded plastic have the same mean. Record the test statistic and *P*-value, and state your conclusion. Also construct a 95% confidence interval for the difference $\mu_m - \mu_p$, where μ_m is the mean weight of discarded metal and μ_p is the mean weight of discarded plastic.

b. The weights used in part (a) are given in pounds. They can be converted to kilograms by multiplying each value by 0.4536. Convert all of those weights from pounds to kilograms and then repeat part (a). Compare the results to those originally found in part (a). In general, are the results affected if a different scale is used?

c. The first weight of discarded metal is 1.09 lb. Change that value to 109 lb by deleting the decimal point. The value of 109 lb is clearly an outlier, but it is only one wrong entry among the 62 metal weights. After making that change, repeat part (a). How are the results affected by this outlier?

Does Aspirin Help Prevent Heart Attacks?

In a recent study of 22,000 male physicians, half were given regular doses of aspirin while the other half were given placebos. The study ran for six years at a cost of $4.4 million. Among those who took the aspirin, 104 suffered heart attacks. Among those who took the placebos, 189 suffered heart attacks. (The figures are based on data from *Time* and the *New England Journal of Medicine,* Vol. 318, No. 4.) Do these results show a statistically significant decrease in heart attacks among the sample group who took aspirin? The issue is clearly important because it can affect many lives.

Use the methods of this chapter to determine whether the use of aspirin seems to help prevent heart attacks. Write a report that summarizes your findings. Include any relevant factors you can think of that might affect the validity of the study. For example, is it noteworthy that the study involved only male physicians? Is it noteworthy that aspirin sometimes causes stomach problems?

1. *Out-of-class activity:* Are Estimates Influenced by Anchoring Numbers? Refer to the related Chapter 2 Cooperative Group Activity. In Chapter 2 we noted that, according to author John Rubin, when people must estimate a value, their estimate is often "anchored" to (or influenced by) a preceding number. In that Chapter 2 activity, subjects were asked to quickly estimate the value of $8 \times 7 \times 6 \times 5 \times 4 \times 3 \times 2 \times 1$, while others are asked to quickly estimate the value of $1 \times 2 \times 3 \times 4 \times 5 \times 6 \times 7 \times 8$.

   In Chapter 2, we could compare the two sets of results by using statistics (such as the mean) and graphs (such as boxplots). The methods of Chapter 8 now allow us to compare the results with a formal hypothesis test. Specifically, collect sample data and test the claim that when we begin with larger numbers (as in $8 \times 7 \times 6$), our estimates tend to be larger.

2. *In-class activity:* In Exercise 18 from Section 8-3, we used pulse rates in Data Set 8 from Appendix B and we tested the claim that male and female statistics students have the same mean pulse rate. Divide into groups according to gender, with about 10 or 12 students in each group. Each group member should record his or her pulse rate by counting the number of heartbeats in one minute, and the group statistics $(n, \bar{x}, s)$ should be calculated. The groups should exchange their results and test the same claim given in Exercise 18 from Section 8-3. Is there a difference? Why?

3. *In-class activity:* Divide into groups of about 10 or 12 students and use the same reaction timer included with the Chapter 5 Cooperative Group Activities. Each group member should be tested for right-hand reaction time and left-hand reaction time. Using the group results, test the claim that there is no difference between the right-hand and left-hand reaction times. Compare the conclusion to the conclusion reached by other groups. Is there a difference?

9

Correlation and Regression

9-1 Overview

This chapter describes methods for dealing with relationships between two variables. The important concepts of correlation and regression are discussed.

9-2 Correlation

The relationship between two variables is investigated with a graph (called a scatter diagram) and a measure (called the linear correlation coefficient).

9-3 Regression

Linear relationships between two variables are described using the equation and graph of a straight line, called the regression line. A method for determining predicted values of a variable is also presented.

9-4 Variation and Prediction Intervals

A method is presented for analyzing the differences between the predicted values of a variable and the actual observed values. Prediction intervals, which are confidence interval estimates of predicted values, are constructed.

9-5 Multiple Regression

Methods are presented for finding a linear equation that relates three or more variables. The multiple coefficient of determination is presented as a measure of how well the sample points fit that linear equation. Because of the calculations involved, this section emphasizes the use of computer software and the interpretation of computer displays.

Chapter Problem

Can you weigh a bear with a tape measure?

Researchers have studied bears by anesthetizing them in order to obtain vital measurements, such as age, gender, length, and weight. (Do not try this at home, because checking the length or gender of a bear not fully anesthetized can be an unpleasant experience.) Because most bears are quite heavy and difficult to lift, researchers and hunters experience considerable difficulty actually weighing a bear in the wild. Can we determine the weight of a bear from other measurements that are easier to get?

See the data in Table 9-1, which represent the first eight male bears in Data Set 3 from Appen-

dix B. (For reasons of clarity, we will use this abbreviated data set, but better results could be obtained by using the more complete set of sample values.) Based on these data, does there appear to be a relationship between the length of a bear and its weight? If so, what is that relationship? If a researcher anesthetizes a bear and uses a tape measure to find that it is 71.0 in. long, how do we use that length to predict the bear's weight? These questions will be addressed in this chapter.

Table 9-1 Lengths and Weights of Male Bears

x Length (in.)	53.0	67.5	72.0	72.0	73.5	68.5	73.0	37.0
y Weight (lb)	80	344	416	348	262	360	332	34

Based on data from Minitab and Gary Alt.

9-1 Overview

In Chapter 8 we considered cases involving one variable and two populations, but in this chapter we deal with cases involving two variables in one population. We will consider cases involving two variables that correspond to a sample of *paired* data, such as the lengths and weights of the male bears represented in Table 9-1. Given such paired data, we want to determine whether there is a relationship between the two variables and, if so, identify what the relationship is. For example, using the data in Table 9-1, we want to determine whether there is a relationship between the weight of a bear and its length. If such a relationship exists, we want to identify it with an equation so that we can predict the weight of a bear by measuring its length instead of actually weighing it.

We begin in Section 9-2 by considering the concept of correlation, which is used to determine whether there is a statistically significant relationship between two variables. We investigate correlation using the scatter diagram (a graph) and the linear correlation coefficient (a measure of the strength of linear association between two variables). Section 9-3 investigates regression analysis, where we describe the relationship between two variables with an equation that relates them. Section 9-3 shows how to use that equation to predict values of a variable.

In Section 9-4 we analyze the differences between predicted values and actual observed values of a variable. Sections 9-2 through 9-4 deal with relationships between two variables, but Section 9-5 uses concepts of multiple regression to describe the relationship among three or more variables. Throughout this text we deal only with linear (or straight-line) relationships between two or more variables.

9-2 Correlation

In this section we address the issue of determining whether there appears to be a relationship between two variables. In statistics, we refer to such a relationship as a *correlation*.

DEFINITION

A **correlation** exists between two variables when one of them is related to the other in some way.

Table 9-1, for example, consists of paired data (sometimes called **bivariate data**). We will determine whether there is a correlation between the variable x (length) and the variable y (weight). This is important because the presence of a correlation could lead to a method for estimating the weight of a bear by measuring its length.

Sidebar notes:

Important changes in this chapter: Sections 9-2 and 9-3 have been rewritten to better allow for inclusion in the early part of the course, as some instructors prefer. Except for the subsection on formal hypothesis tests, Sections 9-2 and 9-3 can now follow Chapter 2.

Minimum coverage of this chapter should include linear correlation and regression, as discussed in Sections 9-2 and 9-3, which are not very difficult. Multiple regression, discussed in Section 9-5, emphasizes the use of computer software.

Consider beginning this chapter by using the chalkboard to list an unidentified table of values, such as this one:

x	78	85	92	100	85
y	89	93	99	100	84

What is the key issue if these data are test grades of subjects before and after formal instruction? ("Is the instruction effective, as indicated by higher y scores?" See Section 8-2.) What is the key issue if these data are reasoning tests for a sample of men (x) and a separate sample of women (y)? ("Does the population of men have the same mean as the population of women?" See Section 8-5.) What is the key issue if each pair represents a math reasoning score x and a starting salary y in thousands of dollars? ("Is the starting salary related to math reasoning?" See this chapter.)

As we work with sample data and develop methods of forming inferences about populations, we make the following assumptions.

Assumptions

1. The sample of paired (x, y) data is a *random* sample.
2. The pairs of (x, y) data have a **bivariate normal distribution.** (Normal distributions are discussed in Chapter 5, but this assumption basically requires that for any fixed value of x, the corresponding values of y have a distribution that is bell shaped, and for any fixed value of y, the values of x have a distribution that is bell shaped.)

The second assumption is usually difficult to check, but a partial check can be made by determining whether the values of both x and y have distributions that are basically bell shaped.

We can often form intuitive and qualitative conclusions about paired data by constructing a graph similar to the one shown in the accompanying Minitab display, which represents the data in Table 9-1. (To obtain this Minitab display, enter the x values in column C1, enter the corresponding y values in column C2, and then select the main menu item of Graph, then select Plot, and proceed to enter C2 in the box for the variable y and C1 in the box for the variable x.)

The Minitab display is an example of a **scatter diagram,** which is a plot of paired (x, y) data with a horizontal x axis and a vertical y axis. The points in the figure seem to follow a pattern, so we might conclude that there is a

Stress that although analysis of the scatter diagram is largely subjective, scatter diagrams are extremely helpful in detecting patterns that do not fit a straight line.

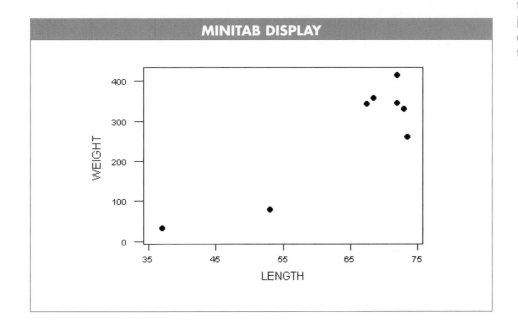

MINITAB DISPLAY

Figure 9-1   Scatter Diagrams

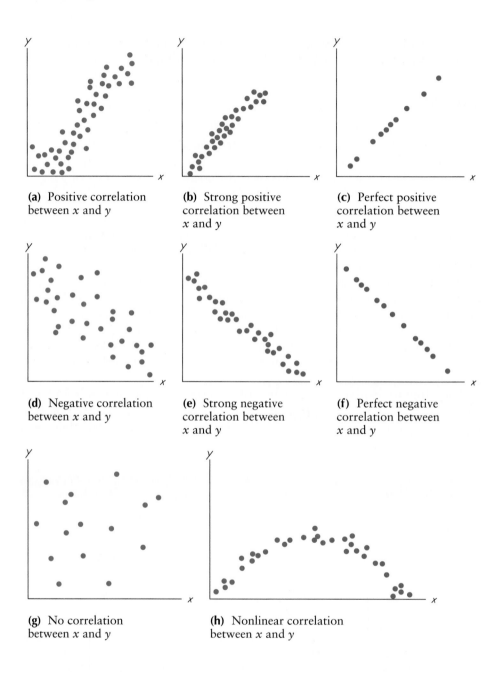

(a) Positive correlation between x and y

(b) Strong positive correlation between x and y

(c) Perfect positive correlation between x and y

(d) Negative correlation between x and y

(e) Strong negative correlation between x and y

(f) Perfect negative correlation between x and y

(g) No correlation between x and y

(h) Nonlinear correlation between x and y

relationship between the length of a bear and its weight. This conclusion is largely subjective because it is based on our perception of whether a pattern is present.

Other examples of scatter diagrams are shown in Figure 9-1. The graphs in Figure 9-1(a), (b), and (c) depict a pattern of increasing values of *y* that correspond to increasing values of *x*. As you proceed from (a) to (c), the dot pattern becomes closer to a straight line, suggesting that the relationship between *x* and *y* becomes stronger. The scatter diagrams in (d), (e), and (f) depict patterns in which the *y* values decrease as the *x* values increase. Again, as you proceed from (d) to (f), the relationship becomes stronger. In contrast to the first six graphs, the scatter diagram of (g) shows no pattern and suggests that there is no correlation (or relationship) between *x* and *y*. Finally, the scatter diagram of (h) shows a pattern, but it is not a straight-line pattern.

Because conclusions drawn from scatter diagrams tend to be subjective, we need more precise and objective methods. We will use the linear correlation coefficient, which is useful for detecting straight-line patterns, but not nonlinear patterns such as the one in Figure 9-1(h).

DEFINITION

The **linear correlation coefficient** r measures the strength of the linear relationship between the paired x and y values in a *sample*. Its value is computed by using Formula 9-1, which follows. [The linear correlation coefficient is sometimes referred to as the **Pearson product moment correlation coefficient** in honor of Karl Pearson (1857–1936), who originally developed it.]

Formula 9-1
$$r = \frac{n\Sigma xy - (\Sigma x)(\Sigma y)}{\sqrt{n(\Sigma x^2) - (\Sigma x)^2}\sqrt{n(\Sigma y^2) - (\Sigma y)^2}}$$

Because r is calculated using sample data, it is a sample statistic used to measure the strength of the linear correlation between x and y. If we had every pair of population values for x and y, the result of Formula 9-1 would be a population parameter, represented by ρ (Greek rho).

We will describe how to compute and interpret the linear correlation coefficient r given a list of paired data, but we will first identify the notation relevant to Formula 9-1. Later in this section we will present the underlying theory that led to the development of Formula 9-1.

This is the "shortcut" version of the formula for r given later in this section, along with an explanation of why it is constructed as it is. Consider doing an in-class example with six pairs of data. You might randomly select six students and get their pulse rates and heights. Stress that in computing r, they should carefully record two values for use in regression: the numerator in Formula 9-1, and the expression under the first radical in the denominator.

Notation for the Linear Correlation Coefficient
n	represents the number of pairs of data present.
Σ	denotes the addition of the items indicated.
Σx	denotes the sum of all x scores.
Σx^2	indicates that each x score should be squared and then those squares added.
$(\Sigma x)^2$	indicates that the x scores should be added and the total then squared. It is extremely important to avoid confusing Σx^2 and $(\Sigma x)^2$.
Σxy	indicates that each x score should be first multiplied by its corresponding y score. After obtaining all such products, find their sum.
r	represents the linear correlation coefficient for a *sample*.
ρ	represents the linear correlation coefficient for a *population*.

Rounding the Linear Correlation Coefficient r

Round the linear correlation coefficient r to three decimal places (so that its value can be directly compared to critical values in Table A-6). When calculating r and other statistics in this chapter, rounding in the middle of a calculation often creates substantial errors, so try using your calculator's memory to store intermediate results and round off only at the end. Many inexpensive calculators have Formula 9-1 built in so that you can automatically evaluate r after entering the sample data.

EXAMPLE   Using the data in Table 9-1, find the value of the linear correlation coefficient r. (A later example will use this value to determine whether there is a relationship between the lengths of bears and their weights.)

SOLUTION   For the sample paired data in Table 9-1, $n = 8$ because there are eight pairs of data. The other components required in Formula 9-1 are found from the calculations in Table 9-2. Note how this vertical format makes the calculations easier.

Using the calculated values and Formula 9-1, we can now evaluate r as follows:

TABLE 9-2	Finding Statistics Used to Calculate r			
Length (in.)	Weight (lb)			
x	y	$x \cdot y$	x^2	y^2
53.0	80	4,240	2809.00	6,400
67.5	344	23,220	4556.25	118,336
72.0	416	29,952	5184.00	173,056
72.0	348	25,056	5184.00	121,104
73.5	262	19,257	5402.25	68,644
68.5	360	24,660	4692.25	129,600
73.0	332	24,236	5329.00	110,224
37.0	34	1,258	1369.00	1,156
Total 516.5	2176	151,879	34,525.75	728,520
↑	↑	↑	↑	↑
Σx	Σy	Σxy	Σx^2	Σy^2

$$
\begin{aligned}
r &= \frac{n\Sigma xy - (\Sigma x)(\Sigma y)}{\sqrt{n(\Sigma x^2) - (\Sigma x)^2}\ \sqrt{n(\Sigma y^2) - (\Sigma y)^2}} \\[2mm]
&= \frac{8(151,879) - (516.5)(2176)}{\sqrt{8(34,525.75) - (516.5)^2}\ \sqrt{8(728,520) - (2176)^2}} \\[2mm]
&= \frac{91,128}{\sqrt{9433.75}\ \sqrt{1,093,184}} = 0.897
\end{aligned}
$$

If students make a point of identifying the values of 91,128 and 9433.75 as shown in the solution, they will find that the calculations required for the regression equation are much easier. For example, the slope of the regression line is 91,128/9433.75, or 9.66.

Using Computers and Calculators for Linear Correlation

STATDISK and Minitab both accept paired data as input and provide the linear correlation coefficient as output. With STATDISK, select Analysis from the main menu bar, then use the option of Correlation and Regression. With Minitab, enter the paired data in columns C1 and C2, then select Stat from the main menu bar, then choose Basic Statistics, then Correlation, and proceed to enter C1 and C2 for the columns to be used. With a TI-83 calculator, enter the paired data in lists L1 and L2, then press STAT and select TESTS. Using the option of LinRegTTest will result in several displayed values, including the value of the linear correlation coefficient r.

Interpreting the Linear Correlation Coefficient

We need to interpret a calculated value of r, such as the value of 0.897 found in the preceding example. Given the way that Formula 9-1 was developed, the value of r must always fall between -1 and $+1$ inclusive. If r is close to 0, we conclude that there is no significant linear correlation between x and y, but if r is close to -1 or $+1$, we conclude that there is a significant linear correlation between x and y. Interpretations of "close to" 0 or 1 or -1 are vague, so we use the following very specific decision criterion:

> **If the absolute value of the computed value of r exceeds the value in Table A-6, conclude that there is a significant linear correlation. Otherwise, there is not sufficient evidence to support the conclusion of a significant linear correlation.**

When there really is no linear correlation between x and y, Table A-6 lists values that are "critical" in this sense: They separate *usual* values of r from those that are *unusual*. For example, Table A-6 shows us that with $n = 8$ pairs of sample data, the critical values are 0.707 (for $\alpha = 0.05$) and 0.834 (for $\alpha = 0.01$). Critical values and the role of α are carefully described in Chapters 6 and 7. Here's how we interpret those numbers: With eight pairs of data and no linear correlation between x and y, there is a 5% chance that the absolute value of the computed linear correlation coefficient r will exceed 0.707. With $n = 8$ and no linear correlation, there is a 1% chance that $|r|$ will exceed 0.834.

EXAMPLE   Given the sample data in Table 9-1 for which $r = 0.897$, refer to Table A-6 to determine whether there is a significant linear correlation between the lengths and weights of bears. In Table A-6, use the critical value for $\alpha = 0.05$. (With $\alpha = 0.05$, we will conclude that there is a significant linear correlation only if the sample is unlikely in the sense that such a value of r occurs less than 5% of the time.)

SOLUTION   Referring to Table A-6, we locate the row for which $n = 8$ (because there are eight pairs of data). That row contains the critical values of 0.707 (for $\alpha = 0.05$) and 0.834 (for $\alpha = 0.01$). Using the critical value for $\alpha = 0.05$, we see that there is less than a 5% chance that with no linear correlation, the absolute value of the computed r will exceed 0.707. Because $r = 0.897$, its absolute value does exceed 0.707, so we conclude that there is a significant linear correlation between lengths and weights of bears.

We have already noted that the format of Formula 9-1 requires that the calculated value of r always falls between -1 and $+1$ inclusive. We list that property along with other important properties.

Power Lines Correlate with Cancer

Interesting, surprising, and useful results sometimes occur when correlations are found between variables. Several scientific studies suggest that there is a correlation between exposure to electromagnetic fields and the occurrence of cancer. Epidemiologists from Sweden's Karolinska Institute researched 500,000 Swedes who lived within 300 meters of a high-tension power line during a 25-year period. They found that the children had a higher incidence of leukemia. Their findings led Sweden's government to consider regulations that would reduce housing in close proximity to high-tension power lines. In an article on this study, *Time* magazine reported that "Although the research does not prove cause and effect, it shows an unmistakable correlation between the degree of exposure and the risk of childhood leukemia."

Properties of the Linear Correlation Coefficient *r*

1. The value of *r* is always between -1 and 1. That is,

$$-1 \leq r \leq 1$$

2. *The value of r does not change if all values of either variable are converted to a different scale.* For example, if the weights of the bears in Table 9-1 are given in kilograms instead of pounds, the value of *r* will not change.

3. *The value of r is not affected by the choice of x or y.* Interchange all *x* and *y* values and the value of *r* will not change.

4. *r measures the strength of a linear relationship.* It is not designed to measure the strength of a relationship that is not linear.

Property 2 can be very helpful in simplifying calculations, because a change in scale does not affect the value of *r*. However, a change in scale does affect the slope and *y*-intercept of the regression line, which is discussed in Section 9-3.

Common Errors Involving Correlation

We now identify three of the most common errors made in interpreting results involving correlation.

1. *We must be careful to avoid concluding that correlation implies causality.* One study showed a correlation between the salaries of statistics professors and per capita beer consumption, but those two variables are affected by the state of the economy, a third variable lurking in the background. (A **lurking variable** is formally defined as one that affects the variables being studied, but is not included in the study.)

2. *Another source of potential error arises with data based on rates or averages.* When we use rates or averages for data, we suppress the variation among the individuals or items, and this may lead to an inflated correlation coefficient. One study produced a 0.4 linear correlation coefficient for paired data relating income and education among individuals, but the linear correlation coefficient became 0.7 when regional averages were used.

3. *A third error involves the property of linearity.* The conclusion that there is no significant linear correlation does not mean that *x* and *y* are not related in any way. The data depicted in Figure 9-2 on the following page result in a value of *r* = 0, which is an indication of no *linear* correlation between the two variables. However, we can easily see from Figure 9-2 that there is a pattern reflecting a very strong *nonlinear* relationship. (Figure 9-2 is a scatter diagram that depicts the relationship between distance above ground and time elapsed for an object thrown upward.)

The first common error should be strongly emphasized in class: "Correlation does not imply causality." Here's a classic example: There is a correlation between per capita beer consumption and teachers' salaries, but teachers don't use salary increases to buy more beer.

Figure 9-2 Scatter Diagram of Distance above Ground and Time for Object Thrown Upward

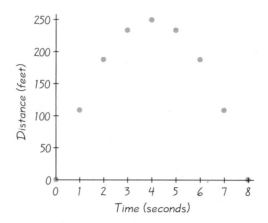

Good exercise: Ask students to identify the coordinates of the points in Figure 9-2, then use those values to calculate r. Based on the result, what do you conclude about the relationship between time and distance? What does the graph itself suggest about that relationship?

This material requires prior coverage of Chapter 7. If you are including Sections 9-2 and 9-3 early (such as following Chapter 2), skip this subsection. If you include this subsection, be sure to inform the class that you prefer Method 1 or Method 2.

Formal Hypothesis Test (Requires Coverage of Chapter 7)

We present two methods (summarized in Figure 9-3) for using a formal hypothesis test to determine whether there is a significant linear correlation between two variables. Some instructors prefer Method 1 because it reinforces concepts introduced in earlier chapters. Others prefer Method 2 because it involves easier calculations.

Figure 9-3 shows that the null and alternative hypotheses will be expressed as follows:

$$H_0: \rho = 0 \qquad \text{(No significant linear correlation)}$$
$$H_1: \rho \neq 0 \qquad \text{(Significant linear correlation)}$$

For the test statistic, we use one of the following methods.

Method 1: Test Statistic Is t   This method follows the format presented in earlier chapters. It uses the Student t distribution with a test statistic having the form $t = (r - \mu_r)/s_r$, where μ_r and s_r denote the claimed value of the mean and the sample standard deviation of r values. Because we assume that $\rho = 0$, it follows that $\mu_r = 0$. Also, it can be shown that s_r, the standard deviation of linear correlation coefficients, can be expressed as $\sqrt{(1 - r^2)/(n - 2)}$. We can therefore use the following test statistic.

Test Statistic t for Linear Correlation

$$t = \frac{r}{\sqrt{\dfrac{1 - r^2}{n - 2}}}$$

Critical values: Use Table A-3 with degrees of freedom $= n - 2$.

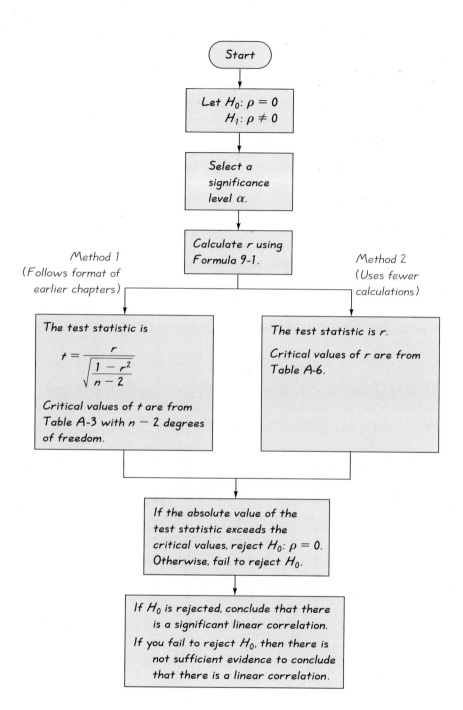

Figure 9-3 Testing for a Linear Correlation

Method 2: Test Statistic Is r   This method requires fewer calculations. Instead of calculating the test statistic just given, we use the computed value of r as the test statistic. Critical values are found in Table A-6.

Test Statistic r for Linear Correlation

Test statistic: r

Critical values: Refer to Table A-6.

Figure 9-3 shows that the decision criterion is to reject the null hypothesis of $\rho = 0$ if the absolute value of the test statistic exceeds the critical values; rejection of $\rho = 0$ means that there is sufficient evidence to support a claim of a linear correlation between the two variables. If the absolute value of the test statistic does not exceed the critical values, then we fail to reject $\rho = 0$; that is, there is not sufficient evidence to conclude that there is a linear correlation between the two variables.

● **EXAMPLE**   Using the sample data in Table 9-1, test the claim that there is a linear correlation between lengths and weights of bears. For the test statistic, use both (a) Method 1 and (b) Method 2.

SOLUTION   Refer to Figure 9-3. To claim that there is a significant linear correlation is to claim that the population linear correlation ρ is different from 0. We therefore have the following hypotheses.

$$H_0: \rho = 0 \qquad \text{(No significant linear correlation)}$$
$$H_1: \rho \neq 0 \qquad \text{(Significant linear correlation)}$$

No significance level α was specified, so use $\alpha = 0.05$.

In a preceding example we already found that $r = 0.897$. With that value, we now find the test statistic and critical value, using each of the two methods just described.

a. *Method 1:* The test statistic is

$$t = \frac{r}{\sqrt{\dfrac{1 - r^2}{n - 2}}} = \frac{0.897}{\sqrt{\dfrac{1 - 0.897^2}{8 - 2}}} = 4.971$$

The critical values of $t = -2.447$ and $t = 2.447$ are found in Table A-3, where 2.447 corresponds to 0.05 divided between two tails (with 0.025 in each tail) and the number of degrees of freedom is $n - 2 = 6$.

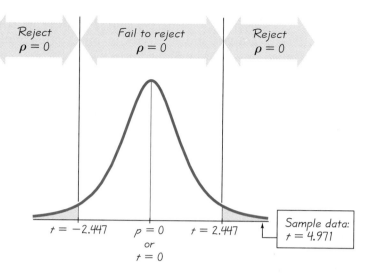

Figure 9-4   Testing H_0: $\rho = 0$ with Method 1

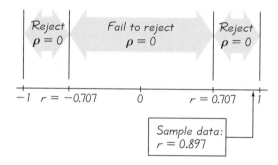

Figure 9-5   Testing H_0: $\rho = 0$ with Method 2

See Figure 9-4 for the graph that includes the test statistic and critical values.

b. *Method 2:* The test statistic is $r = 0.897$. The critical values of $r = -0.707$ and $r = 0.707$ are found in Table A-6 with $n = 8$ and $\alpha = 0.05$. See Figure 9-5 for a graph that includes this test statistic and critical values.

Using either of the two methods, we find that the absolute value of the test statistic does exceed the critical value (Method 1: $4.971 > 2.447$; Method 2: $0.897 > 0.707$); that is, the test statistic does fall within the critical region. We therefore reject H_0: $\rho = 0$. There is sufficient evidence to support the claim of a linear correlation between lengths and weights of bears. The weight of a bear does seem to correspond to its length.

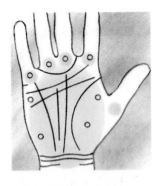

The preceding example and Figures 9-4 and 9-5 illustrate a two-tailed hypothesis test. The examples and exercises in this section will generally involve only two-tailed tests, but one-tailed tests can occur with a claim of a positive linear correlation or a claim of a negative linear correlation. In such cases, the hypotheses will be as shown here.

Claim of Negative Correlation (Left-tailed test)	Claim of Positive Correlation (Right-tailed test)
$H_0: \rho \geq 0$	$H_0: \rho \leq 0$
$H_1: \rho < 0$	$H_1: \rho > 0$

For these one-tailed tests, Method 1 can be handled as in earlier chapters. For Method 2, either calculate the critical value as described in Exercise 25 or modify Table A-6 by replacing the column headings of $\alpha = 0.05$ and $\alpha = 0.01$ by the one-sided critical values of $\alpha = 0.025$ and $\alpha = 0.005$, respectively.

Rationale: We have presented Formula 9-1 for calculating r and have illustrated its use; we will now give a justification for it. Formula 9-1 simplifies the calculations used in this equivalent formula:

$$r = \frac{\Sigma(x - \bar{x})(y - \bar{y})}{(n - 1)s_x s_y}$$

We will temporarily use this latter version of Formula 9-1 because its form relates more directly to the underlying theory. We will consider the following paired data, which are depicted in the scatter diagram shown in Figure 9-6.

x	1	1	2	4	7
y	4	5	8	15	23

Figure 9-6 includes the point $(\bar{x}, \bar{y}) = (3, 11)$, which is called the centroid of the sample points.

DEFINITION

Given a collection of paired (x, y) data, the point $(\bar{x}, \bar{y})$ is called the **centroid**.

The statistic r, sometimes called the Pearson product moment, was first developed by Karl Pearson. It is based on the product of the moments $(x - \bar{x})$ and $(y - \bar{y})$; that is, Pearson based his measure of scattering on the statistic $\Sigma(x - \bar{x})(y - \bar{y})$. In any scatter diagram, vertical and horizontal lines through the centroid $(\bar{x}, \bar{y})$ divide the diagram into four quadrants, as in Figure 9-6. If the points of the scatter diagram tend to approximate an uphill line (as in the figure), individual values of the product $(x - \bar{x})(y - \bar{y})$ tend to be

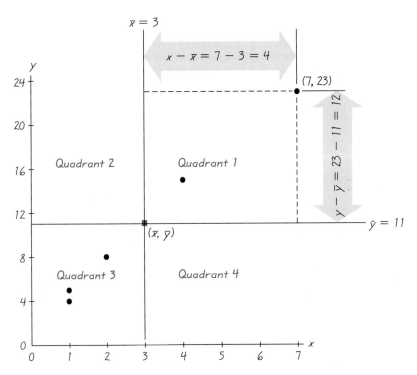

Figure 9-6 Scatter Diagram Partitioned into Quadrants

positive because most of the points are found in the first and third quadrants, where the products of $(x - \bar{x})$ and $(y - \bar{y})$ are positive. If the points of the scatter diagram approximate a downhill line, most of the points are in the second and fourth quadrants, where $(x - \bar{x})$ and $(y - \bar{y})$ are opposite in sign, so $\Sigma(x - \bar{x})(y - \bar{y})$ is negative. Points that follow no linear pattern tend to be scattered among the four quadrants, so the value of $\Sigma(x - \bar{x})(y - \bar{y})$ tends to be close to 0.

The sum $\Sigma(x - \bar{x})(y - \bar{y})$ depends on the magnitude of the numbers used. For example, if you change x from inches to feet, that sum will change. To make r independent of the particular scale used, we include the sample standard deviations as follows:

$$r = \frac{\Sigma(x - \bar{x})(y - \bar{y})}{(n - 1)s_x s_y}$$

This expression can be algebraically manipulated into the equivalent form of Formula 9-1.

Preceding chapters discussed methods of inferential statistics by addressing methods of hypothesis testing, as well as methods for constructing confidence interval estimates. A similar procedure may be used to find confidence intervals for ρ. However, because the construction of such confidence intervals involves somewhat complicated transformations, that process is presented in Exercise 28 (Beyond the Basics).

Student Ratings of Teachers

Many colleges equate high student ratings with good teaching—an equation often fostered by the fact that student evaluations are easy to administer and measure.

However, one study that compared student evaluations of teachers with the amount of material learned found a strong *negative* correlation between the two factors. Teachers rated highly by students seemed to induce less learning.

In a related study, an audience gave a high rating to a lecturer who conveyed very little information but was interesting and entertaining.

We can use the linear correlation coefficient to determine whether there is a linear relationship between two variables. Using the data in Table 9-1, we have concluded that there is a linear correlation between lengths and weights of bears. Having concluded that a relationship exists, we would like to determine what that relationship is so that we can calculate the weight of a bear when we know only its length. This next stage of analysis is addressed in the following section.

9-2 Exercises A: Basic Skills and Concepts

Recommended assignment: Exercises 1, 3, 5, 6, 8, 14, 19–22. (Exercise 5 closely parallels the text example.)

In Exercises 1 and 2, assume that a sample of n pairs of data results in the given value of r. Refer to Table A-6 (for α = 0.05) and determine whether there is a significant linear correlation between x and y. (See the decision criterion given in this section.)

1. a. Significant
   b. Significant
   c. Not significant
2. a. Not significant
   b. Not significant
   c. Significant
3. a. Significant
   b. $n = 5, \Sigma x = 25, \Sigma x^2 = 163, (\Sigma x)^2 = 625, \Sigma xy = 489, r = 0.997$
4. a. Not significant
   b. $n = 5, \Sigma x = 25, \Sigma x^2 = 163, (\Sigma x)^2 = 625, \Sigma xy = 132, r = -0.112$
5. TS: $r = 0.993$. CV: $r = \pm 0.707$.

1. a. $n = 32, r = 0.992$
   b. $n = 50, r = -0.333$
   c. $n = 17, r = 0.456$

2. a. $n = 22, r = -0.087$
   b. $n = 40, r = 0.299$
   c. $n = 25, r = -0.401$

In Exercises 3 and 4, (a) use a scatter diagram to determine whether there is a significant linear correlation between x and y, and (b) find the values of n, Σx, Σx^2, $(\Sigma x)^2$, Σxy, and the linear correlation coefficient r.

3. x	2	3	5	5	10
y	6	9	14	16	30

4. x	2	3	5	5	10
y	6	0	15	5	2

In Exercises 5–18,
a. Construct the scatter diagram.
b. Find the value of the linear correlation coefficient r.
c. Determine whether there is a significant linear correlation between the two variables. (Use only α = 0.05.)
d. Save your work because the same data will be used in the next section.

5. When bears were anesthetized, researchers measured the distances (in inches) around their chests and they weighed the bears (in pounds). The results are given below for eight male bears. Based on the results, does a bear's weight seem to be related to its chest size? Do the results change if the chest measurements are converted to feet, with each of those values divided by 12?

x Chest (in.)	26	45	54	49	41	49	44	19
y Weight (lb)	90	344	416	348	262	360	332	34

Based on data from Minitab and Gary Alt.

6. The accompanying table lists weights (in hundreds of pounds) and highway fuel usage rates (in mi/gal) for a sample of domestic new cars. Based on the result, can you expect to pay more for gas if you buy a heavier car? Do the results change if the weights are entered as 2900, 3500, . . . , 2400?

x Weight	29	35	28	44	25	34	30	33	28	24
y Fuel	31	27	29	25	31	29	28	28	28	33

Based on data from the EPA.

6. TS: $r = -0.851$.
CV: $r = \pm 0.632$.

7. The accompanying table lists weights (in pounds) of plastic discarded by a sample of households, along with the sizes of the households. Is there a significant linear correlation? This issue is important to the Census Bureau, which provided project funding, because the presence of a correlation implies that we can predict population size by analyzing discarded garbage.

Plastic (lb)	0.27	1.41	2.19	2.83	2.19	1.81	0.85	3.05
Household size	2	3	3	6	4	2	1	5

Based on data provided by Masakazu Tani and the Garbage Project at the University of Arizona.

7. TS: $r = 0.842$.
CV: $r = \pm 0.707$.

8. The paired data below consist of weights (in pounds) of discarded paper and sizes of households.

Paper	2.41	7.57	9.55	8.82	8.72	6.96	6.83	11.42
Household size	2	3	3	6	4	2	1	5

Based on data obtained from the Garbage Project at the University of Arizona.

8. TS: $r = 0.630$.
CV: $r = \pm 0.707$.

9. The paired data below consist of weights (in pounds) of discarded food and sizes of households.

Food	1.04	3.68	4.43	2.98	6.30	1.46	8.82	9.62
Household size	2	3	3	6	4	2	1	5

Based on data obtained from the Garbage Project at the University of Arizona.

9. TS: $r = 0.127$.
CV: $r = \pm 0.707$.

10. The paired data below consist of the total weights (in pounds) of discarded garbage and sizes of households.

Total weight	10.76	19.96	27.60	38.11	27.90	21.90	21.83	49.27	33.27	35.54
Household size	2	3	3	6	4	2	1	5	6	4

Based on data obtained from the Garbage Project at the University of Arizona.

10. TS: $r = 0.761$.
CV: $r = \pm 0.632$.

11. A study was conducted to investigate a relationship between age (in years) and BAC (blood alcohol concentration) measured when convicted DWI

11. TS: $r = -0.069$.
CV: $r = \pm 0.707$.

jail inmates were first arrested. Sample data are given below for randomly selected subjects. Based on the result, does the BAC level seem to be related to the age of the person tested?

Age	17.2	43.5	30.7	53.1	37.2	21.0	27.6	46.3
BAC	0.19	0.20	0.26	0.16	0.24	0.20	0.18	0.23

Based on data from the Dutchess County STOP-DWI Program.

12. The accompanying table lists the number of registered automatic weapons (in thousands), along with the murder rate (in murders per 100,000), for randomly selected states. Automatic weapons are guns that continue to fire repeatedly while the trigger is held back. Are firearm murders often committed with automatic weapons? Does a significant linear correlation imply that increased numbers of automatic weapons result in more murders?

Automatic weapons	11.6	8.3	3.6	0.6	6.9	2.5	2.4	2.6
Murder rate	13.1	10.6	10.1	4.4	11.5	6.6	3.6	5.3

Data provided by the FBI and the Bureau of Alcohol, Tobacco, and Firearms.

13. Refer to Data Set 16 in Appendix B and use the paired data for durations and intervals after eruptions of the Old Faithful geyser. Is there a significant linear correlation, suggesting that the interval after an eruption is related to the duration of the eruption?

14. Refer to Data Set 16 in Appendix B and use the paired data for intervals after eruptions and heights of eruptions of the Old Faithful geyser. Is there a significant linear correlation, suggesting that the interval after an eruption is related to the height of the eruption?

15. Refer to Data Set 4 in Appendix B and use the paired data consisting of tar and nicotine. Based on the result, does there appear to be a significant linear correlation between cigarette tar and nicotine? If so, can researchers reduce their laboratory expenses by measuring only one of these two variables?

16. Refer to Data Set 4 in Appendix B and use the paired data consisting of carbon monoxide and nicotine. Based on the result, does there appear to be a significant linear correlation between cigarette nicotine and carbon monoxide? If so, can researchers reduce their laboratory expenses by measuring only one of these two variables?

17. Refer to Data Set 7 in Appendix B and use the paired data consisting of Iowa's total annual precipitation amounts (PRECIP) and the amounts of corn produced (CORNPROD). Is there a significant linear correlation, as we might expect?

18. Refer to Data Set 7 in Appendix B and use the paired data consisting of Iowa's average annual temperatures (AVTEMP) and the amounts of corn produced (CORNPROD). Is there a significant linear correlation, as we might expect?

12. TS: $r = 0.885$.
   CV: $r = \pm 0.707$.

13. TS: $r = 0.870$.
   CV: $r = \pm 0.279$.

14. TS: $r = -0.010$.
   CV: $r = \pm 0.279$.

15. TS: $r = 0.961$.
   CV: $r = \pm 0.361$

16. TS: $r = 0.863$.
   CV: $r = \pm 0.361$.

17. TS: $r = 0.265$.
   CV: $r = \pm 0.312$.

18. TS: $r = -0.061$.
   CV: $r = \pm 0.312$.

In Exercises 19–22, describe the error in the stated conclusion. (See the list of common errors included in this section.)

19. *Given:* The paired sample data of the ages of subjects and their scores on a test of reasoning result in a linear correlation coefficient very close to 0.
    Conclusion: Younger people tend to get higher scores.

20. *Given:* There is a significant linear correlation between personal income and years of education.
    Conclusion: More education causes a person's income to rise.

21. *Given:* Subjects take a test of verbal skills and a test of manual dexterity, and those pairs of scores result in a linear correlation coefficient very close to 0.
    Conclusion: Scores on the two tests are not related in any way.

22. *Given:* There is a significant linear correlation between state average tax burdens and state average incomes.
    Conclusion: There is a significant linear correlation between individual tax burdens and individual incomes.

Exercises B: Beyond the Basics

23. How is the value of the linear correlation coefficient r affected in each of the following cases?
    a. Each x value is switched with the corresponding y value.
    b. Each x value is multiplied by the same nonzero constant.
    c. The same constant is added to each x value.

24. In addition to testing for a linear correlation between x and y, we can often use *transformations* of data to explore for other relationships. For example, we might replace each x value by x^2 and use the methods of this section to determine whether there is a linear correlation between y and x^2. Given the paired data in the accompanying table, construct the scatter diagram and then test for a linear correlation between y and each of the following. Which case results in the largest value of r?

    a. x b. x^2 c. $\log x$ d. $\sqrt{x}$ e. $1/x$

x	1.3	2.4	2.6	2.8	2.4	3.0	4.1
y	0.11	0.38	0.41	0.45	0.39	0.48	0.61

25. The critical values of r in Table A-6 are found by solving

$$t = \frac{r}{\sqrt{\dfrac{1 - r^2}{n - 2}}}$$

(continued)

19. With a linear correlation coefficient very close to 0, there does not appear to be a correlation, but the conclusion suggests that there is a correlation.

20. The presence of a significant linear correlation does not necessarily mean that one of the variables *causes* the other. Correlation does not imply causality.

21. Although there is no *linear* correlation, the variables may be related in some other *non-linear* way.

22. Averages tend to suppress variation among individuals, so a correlation among *averages* does not necessarily mean that there is a correlation among *individuals*.

23. In parts (a), (b), and (c), the value of r does not change.

24. a. 0.972
    b. 0.905
    c. 0.999 (largest)
    d. 0.992
    e. −0.984

for r to get

$$r = \frac{t}{\sqrt{t^2 + n - 2}}$$

where the t value is found from Table A-3 by assuming a two-tailed case with $n - 2$ degrees of freedom. Table A-6 lists the results for selected values of n and α. Use the formula for r given here and Table A-3 (with $n - 2$ degrees of freedom) to find the critical values of r for the given cases.

a. $H_0: \rho = 0, n = 50, \alpha = 0.05$
b. $H_1: \rho \neq 0, n = 75, \alpha = 0.10$
c. $H_0: \rho \geq 0, n = 20, \alpha = 0.05$
d. $H_0: \rho \leq 0, n = 10, \alpha = 0.05$
e. $H_1: \rho > 0, n = 12, \alpha = 0.01$

26. Use the sample data given in Exercise 5, but test the claim that there is a *positive* linear correlation between chest sizes and weights of bears. (See Exercise 25.)

27. The graph of $y = x^2$ is a parabola, not a straight line, so we might expect that the value of r would not reflect a linear correlation between x and y. Using $y = x^2$, make a table of x and y values for $x = 0, 1, 2, \ldots , 10$ and calculate the value of r. What do you conclude? How do you explain the result?

28. Given n pairs of data from which the linear correlation coefficient r can be found, use the following procedure to construct a confidence interval about the population parameter ρ.

Step a. Use Table A-2 to find $z_{\alpha/2}$ that corresponds to the desired degree of confidence.

Step b. Evaluate the interval limits w_L and w_R:

$$w_L = \frac{1}{2} \ln \left(\frac{1 + r}{1 - r} \right) - z_{\alpha/2} \cdot \frac{1}{\sqrt{n - 3}}$$

$$w_R = \frac{1}{2} \ln \left(\frac{1 + r}{1 - r} \right) - z_{\alpha/2} \cdot \frac{1}{\sqrt{n - 3}}$$

Step c. Now evaluate the confidence interval limits in the expression below.

$$\frac{e^{2w_L} - 1}{e^{2w_L} + 1} < \rho < \frac{e^{2w_R} - 1}{e^{2w_R} + 1}$$

Use this procedure to construct a 95% confidence interval for ρ, given 50 pairs of data for which $r = 0.600$.

25. a. ± 0.272
    b. ± 0.189
    c. -0.378
    d. 0.549
    e. 0.658

26. TS: $r = 0.993$.
    CV: $r = 0.621$.

27. $r = 0.963$ and $n = 11$; sig. corr. This section of the parabola can be approximated by a straight line.

28. $0.386 < \rho < 0.753$

Regression

In Section 9-2 we analyzed paired data with the goal of determining whether there is a significant linear correlation between two variables. We now want to describe the relationship by finding the graph and equation of the straight line that represents the relationship. This straight line is called the regression line, and its equation is called the regression equation. Sir Francis Galton (1822–1911) studied the phenomenon of heredity and showed that when tall or short couples have children, the heights of those children tend to *regress*, or revert to a more typical mean height. We continue to use his terminology.

Very important note about a change in notation: This edition uses the format of $\hat{y} = b_0 + b_1 x$ for the regression equation based on sample data. (The previous edition used $\hat{y} = \hat{\beta}_0 + \hat{\beta}_1 x$.) One reviewer commented on the notation from the previous edition by asking: "Did Triola think up the symbolism after a really great New Year's Eve Party?" Actually, it was after an October 18 celebration of National Statistics Day in Japan; the author really needs to get a life.

DEFINITIONS

Given a collection of paired sample data, the **regression equation**

$$\hat{y} = b_0 + b_1 x$$

describes the relationship between the two variables. The graph of the regression equation is called the **regression line** (or *line of best fit*, or *least-squares line*).

This definition expresses a relationship between x (called the **independent variable** or **predictor variable**) and $\hat{y}$ (called the **dependent variable** or **response variable**). In the preceding definition, the typical equation of a straight line ($y = mx + b$) is expressed in the format of $\hat{y} = b_0 + b_1 x$, where b_0 is the y-intercept and b_1 is the slope. The following notation box shows that b_0 and b_1 are sample statistics used to estimate the population parameters β_0 and β_1.

Notation for Regression Equation	Population Parameter	Sample Statistic
y-intercept of regression equation	β_0	b_0
Slope of regression equation	β_1	b_1
Equation of the regression line	$y = \beta_0 + \beta_1 x$	$\hat{y} = b_0 + b_1 x$

Assumptions For the regression methods given in this section, we assume that

1. We are investigating only linear relationships.
2. For each x value, y is a random variable having a normal (bell-shaped) distribution. All of these y distributions have the same variance. Also, for a

given value of x, the distribution of y values has a mean that lies on the regression line. (Results are not seriously affected if departures from normal distributions and equal variances are not too extreme.)

An important goal of this section is to use sample paired data to estimate the regression equation. Using only sample data, we can't find the exact values of the population parameters β_0 and β_1, but we can use the sample data to estimate them with b_0 and b_1, which are found by using Formulas 9-2 and 9-3.

Use the formula card to compare the forms of Formulas 9-2 and 9-3 to the form of Formula 9-1 for r. Ask the class to identify similarities and allow them to find the shortcuts.

Formula 9-2 $$b_0 = \frac{(\Sigma y)(\Sigma x^2) - (\Sigma x)(\Sigma xy)}{n(\Sigma x^2) - (\Sigma x)^2} \qquad y\text{-intercept}$$

Formula 9-3 $$b_1 = \frac{n(\Sigma xy) - (\Sigma x)(\Sigma y)}{n(\Sigma x^2) - (\Sigma x)^2} \qquad \text{(slope)}$$

These formulas might look intimidating, but they are programmed into many calculators, so the values of b_0 and b_1 can be easily found. Also, statistical software packages such as STATDISK and Minitab can be used to find b_0 and b_1. In those cases when we must use formulas instead of a calculator or computer, the required computations will be much easier if we keep the following observations in mind:

1. If the linear correlation coefficient r has been computed using Formula 9-1, the values of Σx, Σy, Σx^2, and Σxy have already been found and they can be used again in Formula 9-3. (Also, the numerator for r in Formula 9-1 is the same numerator for b_1 in Formula 9-3; the denominator for r includes the denominator for b_1. If the calculation for r is set up carefully, the calculation for b_1 requires the simple division of one known number by another.)

2. If you find the slope b_1 first, you can use Formula 9-4 to find the y-intercept b_0. [The regression line always passes through the centroid $(\bar{x}, \bar{y})$ so that $\bar{y} = b_0 + b_1\bar{x}$ must be true, and this equation can be expressed as Formula 9-4. It is usually easier to find the y-intercept b_0 by using Formula 9-4 than by using Formula 9-2.]

Formula 9-4 $$b_0 = \bar{y} - b_1\bar{x}$$

Once we have evaluated b_0 and b_1, we can identify the estimated regression equation, which has the following special property: *The regression line fits the sample points best.* (The specific criterion used to determine which line fits "best" is the least-squares property, which will be described later.) We will now briefly discuss rounding and then illustrate the procedure for finding and applying the regression equation.

Rounding the y-Intercept b_0 and the Slope b_1

It's difficult to provide a simple universal rule for rounding values of b_0 and b_1, but we usually try to round each of these values to *three significant digits* or use the values provided by a statistical software package such as STATDISK or Minitab. Because these values are very sensitive to rounding at intermediate steps of calculations, you should try to carry at least six significant digits (or use exact values) in the intermediate steps. Depending on how you round, this book's answers to examples and exercises may be slightly different from your answers.

EXAMPLE   In Section 9-2 we used the Table 9-1 data (x = lengths of bears; y = weights of bears) to find that the linear correlation coefficient of $r = 0.897$ indicates that there is a significant linear correlation. Now find the regression equation of the straight line that relates x and y.

SOLUTION   We will find the regression equation by using Formulas 9-3 and 9-4 and these values already found in Table 9-2 from Section 9-2:

$$n = 8 \qquad \Sigma x = 516.5 \qquad \Sigma y = 2176$$
$$\Sigma x^2 = 34{,}525.75 \qquad \Sigma y^2 = 728{,}520 \qquad \Sigma xy = 151{,}879$$

First find the slope b_1 by using Formula 9-3:

$$b_1 = \frac{n(\Sigma xy) - (\Sigma x)(\Sigma y)}{n(\Sigma x^2) - (\Sigma x)^2}$$

$$= \frac{8(151{,}879) - (516.5)(2176)}{8(34{,}525.75) - (516.5)^2} = \frac{91{,}128}{9433.75} = 9.65979$$

$$= 9.66 \quad \text{(rounded)}$$

Next, find the y-intercept b_0 by using Formula 9-4:

$$b_0 = \bar{y} - b_1 \bar{x}$$

$$= \frac{2176}{8} - (9.65979)\frac{516.5}{8} = -352 \quad \text{(rounded)}$$

Knowing the slope b_1 and y-intercept b_0, we can now express the estimated equation of the regression line as

$$\hat{y} = -352 + 9.66x$$

We should realize that this equation is an *estimate* of the true regression equation $y = \beta_0 + \beta_1 x$. This estimate is based on one particular set of sample data listed in Table 9-1, but another sample drawn from the same population would probably lead to a slightly different equation.

The Minitab display on the following page shows the regression line plotted on the scatter diagram. We can see that this line fits the data well.

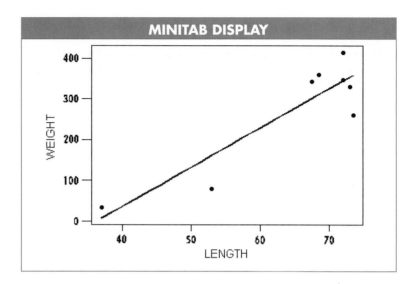

Marginal Change

We can use the regression equation to see the effect on one variable when the other variable changes by some specific amount.

DEFINITION

In working with two variables related by a regression equation, the **marginal change** in a variable is the amount that it changes when the other variable changes by exactly one unit.

The slope b_1 in the regression equation represents the marginal change resulting when x changes by one unit. For the bear data of Table 9-1, we can see that an increase in x of one unit will cause $\hat{y}$ to change by 9.66 units. That is, if a bear grows 1 inch, its predicted weight is 9.66 pounds more.

Outliers and Influential Points

The definitions of outliers and influential points are new to this edition.

A correlation/regression analysis of bivariate data should include an investigation of *outliers* and *influential points*, defined as follows.

DEFINITIONS

In a scatter diagram, an **outlier** is a point lying far away from the other data points. Paired sample data may include one or more **influential points**, which are points that strongly affect the graph of the regression line.

Here's how to determine whether a point is an influential point: Graph the regression line resulting from the data with the point included, then graph the regression line resulting from the data with the point excluded. If the graph changes by a considerable amount, the point is influential. Influential points are often found by identifying those outliers that are *horizontally* far away from the other points.

For example, refer to the preceding Minitab display. If we include an additional bear that is 35 in. long and weighs 400 lb, we have an outlier, because that point would fall in the upper-left corner of the graph and would be far away from the other points. This point would not be an influential point because the regression line would not change its position very much. If this new bear were 1 in. long and weighed 2000 lb (kind of scary, isn't it?), it would become an influential point because the graph of the regression line would change considerably, as shown by the following Minitab display.

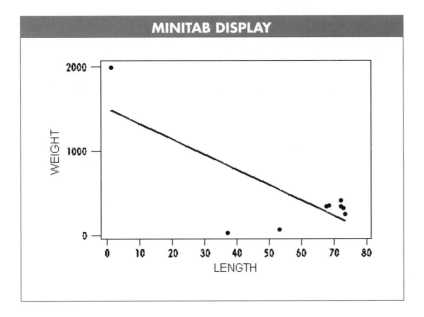

Predictions

Regression equations can be helpful when used in *predicting* the value of one variable, given some particular value for the other variable. If the regression line fits the data quite well, then it makes sense to use its equation for predictions, provided that we don't go beyond the scope of the available scores. However, *we should use the equation of the regression line only if r indicates that there is a significant linear correlation. In the absence of a significant linear correlation, we should not use the regression equation for projecting or*

Los Angeles Ozone

The South Coast Air Quality Management District monitors the ozone levels for the Los Angeles basin region. The ozone levels are affected by weather, as well as by pollutants from the infamous stream of dense traffic in the Los Angeles area. One useful indicator of an ozone problem is its level in parts per million for the worst hour of the year. Regression analysis shows a downward trend in that indicator, suggesting that despite a large increase in people and cars, ozone levels are currently at about half the level of 40 years ago. The downward trend can be seen in the "worst hour" levels for six recent consecutive years: 0.32, 0.32, 0.25, 0.21, 0.22, 022. Such statistical analyses verify the impact of clean-air legislation, and they also raise issues of the cost effectiveness of new legislation.

Figure 9-7 Predicting the Value of a Variable

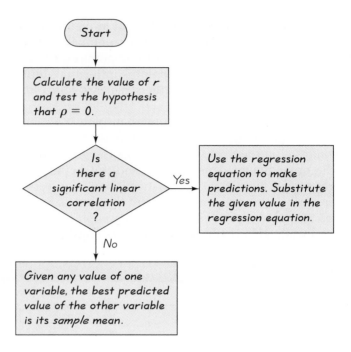

predicting; instead, our best estimate of the second variable is simply its sample mean.

In predicting a value of y based on some given value of x . . .

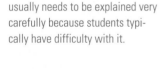
This point is very important and usually needs to be explained very carefully because students typically have difficulty with it.

1. If there is *not* a significant linear correlation, the best predicted y value is $\bar{y}$.
2. If there *is* a significant linear correlation, the best predicted y value is found by substituting the x value into the regression equation.

Figure 9-7 summarizes this process, which is easier to understand if we think of r as a measure of how well the regression line fits the sample data. If r is near -1 or $+1$, then the regression line fits the data well, but if r is near 0, then the regression line fits poorly (and should not be used for predictions).

EXAMPLE   Using the sample data in Table 9-1, we found that there is a significant linear correlation between lengths and weights of bears, and we also found that the regression equation is $\hat{y} = -352 + 9.66x$. If a bear is measured and is found to be 71.0 in. long, predict its weight.

SOLUTION   There's a strong temptation to jump in and substitute 71.0 for x in the regression equation, but we should first consider whether there is a significant linear correlation that justifies the use of that equation. In this example, we do have a significant linear correlation (with $r = 0.897$), so our predicted value is found as follows:

$$\hat{y} = -352 + 9.66x$$
$$= -352 + 9.66(71.0) = 334$$

The predicted weight of a 71.0 in. long bear is 334 lb. (If there had not been a significant linear correlation, our best predicted weight would have been $\bar{y} = 272$ lb.)

EXAMPLE   There is obviously no significant linear correlation between shoe sizes and IQs of adults. Given that an adult has a shoe size of 9, find his or her best predicted IQ score.

SOLUTION   Because there is no significant linear correlation, we do not use a regression equation. Instead, the best predicted IQ value is simply the mean IQ score, which is 100.

Carefully compare the solutions to the preceding two examples and note that we used the regression equation when there was a significant linear correlation, but in the absence of such a correlation, the best predicted value of y is simply the value of the sample mean $\bar{y}$. A common error is to use the regression equation when there is no significant linear correlation. That error violates the first of the following guidelines.

Guidelines for Using the Regression Equation

1. *If there is no significant linear correlation, don't use the regression equation to make predictions.*
2. *When using the regression equation for predictions, stay within the scope of the available sample data.* If you find a regression equation that relates women's heights and shoe sizes, it's absurd to predict the shoe size of a woman who is 10 ft tall.
3. *A regression equation based on old data is not necessarily valid now.* The regression equation relating used car prices and ages of cars is no longer usable if it's based on data from the 1970s.
4. *Don't make predictions about a population that is different from the population from which the sample data were drawn.* If we collect sample data from men and develop a regression equation relating age and TV remote

Predicting the Stock Market

Forecasting and predicting are important goals of statistics. Investment advisors seek indicators that can be used to forecast stock market performance. Some indicators are quite colorful, such as the hemline index: Rising skirt hemlines supposedly precede a rise in the Dow Jones Industrial Average. According to the Super Bowl omen, a Super Bowl victory by a team with NFL origins is followed by a year in which the New York Stock Exchange index rises; otherwise, it falls. (In 1970, the NFL and AFL merged into the current NFL.) This indicator has been correct in 27 of the past 30 years, largely due to the fact that NFL teams win much more often, and the stock market tends to follow an upward trend. Other indicators: aspirin sales, limousines on Wall Street, and elevator traffic at the New York Stock Exchange.

control usage, the results don't necessarily apply to women. If we use state *averages* to develop a regression equation relating SAT math scores and SAT verbal scores, the results don't necessarily apply to *individuals*.

Using Computers and Calculators for Correlation and Regression

Because of the messy calculations involved, the linear correlation coefficient r and the slope and y-intercept of the regression line are usually found by using a calculator or computer software. Section 9-2 described the procedures for getting the value of the linear correlation coefficient r using STATDISK and Minitab and a TI-83 calculator. Section 9-2 also included a Minitab scatter diagram for the data of Table 9-1.

To obtain the regression equation with a TI-83 calculator, enter the paired data in lists L1 and L2, then press STAT and select TESTS, then choose the option of LinRegTTest and the displayed results will include the y-intercept and slope of the regression equation. Instead of b_0 and b_1, the TI-83 display represents these values as a and b.

To obtain the regression equation with STATDISK, use the options of Analysis and Correlation and Regression. The STATDISK display will include correlation and regression results, as shown in the accompanying display.

STATDISK DISPLAY	
File Edit Analysis Data Help	
Sample Size, n	8
Degrees Freedom	6
Correlation Results:	
Correlation Coeff, r	0.89735
Critical r	±0.70673
Reject the Null Hypothesis	
Sample provides evidence that the populations are correlated	
Regression Results:	
$Y = b_0 + b_1 x$	
Y Intercept, b_0	-351.66
Slope, b_1	9.6598
Total Variation	136648
Explained Var	110035
Unexplained Var	26613
Standard Error	66.600
Coeff of Det, r^2	0.80524

To obtain the regression equation with Minitab, first enter the paired data in columns C1 and C2, then select Stat, then Regression, then Regression again. Now enter C2 for the response variable and enter C1 for the predictor variable. If you use the data in Table 9-1, the Minitab display will begin with the following.

```
The regression equation is
C2 = -352 + 9.66 C1
```

Minitab will also include other results, some of which will be discussed in the following section.

Residuals and the Least-Squares Property

We have stated that the regression equation represents the straight line that fits the data "best," and we will now describe the criterion used in determining the line that is better than all others. This criterion is based on the vertical distances between the original data points and the regression line. Such distances are called *residuals*.

> ### DEFINITION
>
> For a sample of paired (x, y) data, a **residual** is the difference $(y - \hat{y})$ between an observed sample y value and the value of y that is predicted from the regression equation.

x	1	2	4	5
y	4	24	8	32

This definition might seem as clear as tax form instructions, but you can easily understand residuals by referring to Figure 9-8 on the following page, which corresponds to the paired sample data in the margin. In Figure 9-8, the residuals are represented by the dashed lines. For a specific example, see the residual indicated as 7, which is directly above $x = 5$. If we substitute $x = 5$ into the regression equation $\hat{y} = 5 + 4x$, we get a predicted value of $\hat{y} = 25$. When $x = 5$, the *predicted* value of y is $\hat{y} = 25$, but the actual *observed* sample value is $y = 32$. The difference of $y - \hat{y} = 32 - 25 = 7$ is a residual.

The regression equation represents the line that fits the points "best" according to the *least-squares property*.

> ### DEFINITION
>
> A straight line satisfies the **least-squares property** if the sum of the squares of the residuals is the smallest sum possible.

The standard deviation is one of many different statistics that are based on a sum of squares, so it is not too surprising to see that the criterion for the line of best fit is based on a sum of squares.

Figure 9-8 Scatter Diagram with Regression Line and Residuals

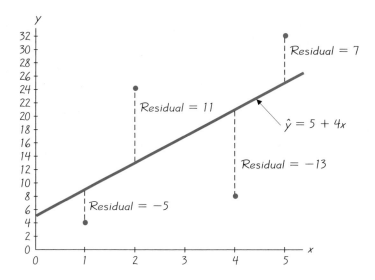

From Figure 9-8, we see that the residuals are -5, 11, -13, and 7, so the sum of their squares is

$$(-5)^2 + 11^2 + (-13)^2 + 7^2 = 364$$

Any straight line different from $\hat{y} = 5 + 4x$ will result in residuals with a sum of squares that is greater than 364.

Fortunately, we need not deal directly with the least-squares property when we want to find the equation of the regression line. Calculus has been used to build the least-squares property into Formulas 9-2 and 9-3. Because the derivations of these formulas require calculus, we don't include them in this text.

Transformations

In many cases, there is a relationship between two variables that is not linear. For example, inspection of the table shown here shows that each y value is the square of the corresponding x value, so that the two variables are related by the equation $y = x^2$ instead of a linear equation, which has the form $y = b_0 + b_1 x$.

x	2	5	4	8	10
y	4	25	16	64	100

Calculators and computers now make it easy to try nonlinear equations. The TI-83 calculator includes these options:

QuadReg: Finds the best quadratic equation ($y = ax^2 + bx + c$) that fits the sample data.

CubicReg: Finds the best third-degree polynomial ($y = ax^3 + bx^2 + cx + d$) that fits the sample data.

QuartReg: Finds the best fourth-degree polynomial ($y = ax^4 + bx^3 + cx^2 + dx + e$) that fits the sample data.

LnReg: Finds the best logarithmic equation ($y = a + b \ln x$) that fits the sample data.

ExpReg: Finds the best exponential equation ($y = ab^x$) that fits the sample data.

PwrReg: Finds the best power equation ($y = ax^b$) that fits the sample data.

Logistic: Finds the best equation of the form $y = c/(1 + ae^{-bx})$ that fits the sample data.

The best-fitting equation can be found with the help of scatter diagrams (which may reveal the general nature of the relationship) and calculator or computer displays that include values of r or other statistics that measure how well the equation fits the sample data. See Exercise 25.

9-3 Exercises A: Basic Skills and Concepts

In Exercises 1–4, use the given data to find the equation of the regression line.

1.
x	1	2	4	5
y	3	5	9	11

2.
x	5	3	2	1	0	2
y	−2	0	1	2	3	1

3.
x	2	3	5	5	10
y	6	9	14	16	30

4.
x	2	3	5	5	10
y	6	0	15	5	2

Exercises 5–18 use the same data sets as the exercises in Section 9-2. In each case, find the regression equation, letting the first variable be the independent (x) variable. Find predicted values where they are requested. **Caution: When finding predicted values, be sure to follow the prediction procedure described in this section.**

5.
x Chest (in.)	26	45	54	49	41	49	44	19
y Weight (lb)	90	344	416	348	262	360	332	34

Find the best predicted weight of a bear with a chest size of 52 in.

6.
x Weight	29	35	28	44	25	34	30	33	28	24
y Fuel	31	27	29	25	31	29	28	28	28	33

Find the best predicted fuel consumption amount for a car that weighs

Recommended assignment:
Exercises 1, 5, 6, 8, 14, 19–22.
(Exercise 5 closely parallels the text example.)

1. $\hat{y} = 1 + 2x$
2. $\hat{y} = 3 - x$
3. $\hat{y} = 0 + 3x$
4. $\hat{y} = 6.65 - 0.211x$
5. $\hat{y} = -187 + 11.3x$; 399 lb
6. $\hat{y} = 39.2 - 0.333x$;
   25.2 mi/gal

4200 lb. (Note that the table has values of x given in hundreds of pounds.)

7. Plastic (lb)

Plastic (lb)	0.27	1.41	2.19	2.83	2.19	1.81	0.85	3.05
Household size	2	3	3	6	4	2	1	5

7. $\hat{y} = 0.549 + 1.48x$; 1.3 persons

What is the best predicted size of a household that discards 0.50 lb of plastic?

8.

Paper	2.41	7.57	9.55	8.82	8.72	6.96	6.83	11.42
Household size	2	3	3	6	4	2	1	5

8. $\hat{y} = 0.152 + 0.398x$; 3.3 persons

What is the best predicted size of a household that discards 10.00 lb of paper?

9.

Food	1.04	3.68	4.43	2.98	6.30	1.46	8.82	9.62
Household size	2	3	3	6	4	2	1	5

9. $\hat{y} = 2.93 + 0.0664x$; 3.3 persons

What is the best predicted size of a household that discards 8.00 lb of food?

10.

Total weight	10.76	19.96	27.60	38.11	27.90	21.90	21.83	49.27	33.27	35.54
Household size	2	3	3	6	4	2	1	5	6	4

10. $\hat{y} = 0.183 + 0.119x$; 6.2 persons

What is the best predicted size of a household that discards a total of 50.00 lb of garbage?

11.

Age	17.2	43.5	30.7	53.1	37.2	21.0	27.6	46.3
BAC	0.19	0.20	0.26	0.16	0.24	0.20	0.18	0.23

11. $\hat{y} = 0.214 - 0.000182x$; 0.21

What is the best predicted blood alcohol level of a person 21.0 years of age who has been convicted and jailed for DWI? (The BAC level is measured when the person is first arrested.)

12.

Automatic weapons	11.6	8.3	3.6	0.6	6.9	2.5	2.4	2.6
Murder rate	13.1	10.6	10.1	4.4	11.5	6.6	3.6	5.3

12. $\hat{y} = 4.05 + 0.853x$; 12.6

What is the best predicted murder rate for a state with 10,000 registered automatic weapons? (In this table, the numbers of automatic weapons are in thousands.)

13. Refer to Data Set 16 in Appendix B and use the paired duration and interval data for eruptions of the Old Faithful geyser. What is the best predicted time before the next eruption if the previous eruption lasted for 210 seconds?

13. $\hat{y} = 41.9 + 0.179x$; 79 min

14. Refer to Data Set 16 in Appendix B and use the paired height and interval data for eruptions of the Old Faithful geyser. What is the best predicted

14. $\hat{y} = 81.9 - 0.009x$; 81 min

time before the next eruption if the previous eruption had a height of 275 ft?

15. Refer to Data Set 4 in Appendix B and use the paired data consisting of tar and nicotine. What is the best predicted amount of nicotine for a cigarette with 3 mg of tar?

16. Refer to Data Set 4 in Appendix B and use the paired data consisting of carbon monoxide and nicotine. What is the best predicted amount of nicotine for a cigarette with 18 mg of carbon monoxide?

17. Refer to Data Set 7 in Appendix B and use the paired data consisting of Iowa's total annual precipitation amounts (PRECIP) and the amounts of corn produced (CORNPROD). What is the best predicted amount of corn production in a year with 40.0 in. of precipitation?

18. Refer to Data Set 7 in Appendix B and use the paired data consisting of Iowa's average annual temperatures (AVTEMP) and the amounts of corn produced (CORNPROD). What is the best predicted amount of corn produced in a year with an average temperature of 45.00° F?

19. In each of the following cases, find the best predicted value of y given that $x = 3.00$. The given statistics are summarized from paired sample data.
   a. $r = 0.931$, $\bar{y} = 7.00$, $n = 10$, and the equation of the regression line is $\hat{y} = 4.00 + 2.00x$.
   b. $r = -0.033$, $\bar{y} = 2.50$, $n = 80$, and the equation of the regression line is $\hat{y} = 5.00 - 2.00x$.

20. In each of the following cases, find the best predicted value of y given that $x = 2.00$. The given statistics are summarized from paired sample data.
   a. $r = -0.882$, $\bar{y} = 3.57$, $n = 15$, and the equation of the regression line is $\hat{y} = 23.00 - 8.00x$.
   b. $r = 0.187$, $\bar{y} = 9.33$, $n = 60$, and the equation of the regression line is $\hat{y} = 4.00 + 8.00x$.

21. Refer to the eight pairs of data listed in Table 9-1. If we include a ninth bear that is 120 in. long and weighs 800 lb, is the new point an outlier? Is it an influential point?

22. Refer to the eight pairs of data listed in Table 9-1. If we include a ninth bear that is 120 in. long and weighs 50 lb (affectionately called "Slim"), is the new point an outlier? Is it an influential point?

9–3 Exercises B: Beyond the Basics

23. Large numbers, such as those in the accompanying table, often cause computational problems. First use the given data to find the equation of the regression line, then find the equation of the regression line after each x value has been divided by 1000. How are the results affected by the change in x? How would the results be affected if each y entry were divided by 1000? (The table is on the following page.)

Answers (right margin):

15. $\hat{y} = 0.154 + 0.0651x$; 0.3 mg

16. $\hat{y} = 0.192 + 0.0606x$; 1.3 mg

17. $\hat{y} = 420 + 18.3x$; 1015 million bushels

18. $\hat{y} = 1854 - 17.4x$; 1015 million bushels

19. a. 10.00
    b. 2.50

20. a. 7.00
    b. 9.33

21. It is an outlier but is not an influential point.

22. It is an outlier and is an influential point.

23. $\hat{y} = -182 + 0.000351x$; $\hat{y} = -182 + 0.351x$.

x	924,736	832,985	825,664	793,427	857,366
y	142	111	109	95	119

x	1	2	4	5
y	4	24	8	32

x	2.0	2.5	4.2	10.0
y	12.0	18.7	53.0	225.0

24. The residuals are -7, 10, -12, and 9, and the sum of their squares is 374.

25. $\hat{y} = -49.9 + 27.2x$; $\hat{y} = -103.2 + 134.9 \ln x$.

26. With $\beta_1 = 0$, the regression line is horizontal so that different values of x result in the same y value, and there is no correlation between x and y.

24. According to the least-squares property, the regression line minimizes the sum of the squares of the residuals. We noted that with the paired data in the margin, the regression equation is $\hat{y} = 5 + 4x$ and the sum of the squares of the residuals is 364. Show that the equation $\hat{y} = 8 + 3x$ results in a sum of squares of residuals that is greater than 364.

25. If the scatter diagram reveals a nonlinear (not a straight line) pattern that you recognize as another type of curve, you may be able to apply the methods of this section. For the data given in the margin, find the linear equation ($y = b_0 + b_1 x$) that best fits the sample data, and find the logarithmic equation ($y = a + b \ln x$) that best fits the sample data. (*Hint:* Begin by replacing each x value with $\ln x$.) Which of these two equations fits the data better? Why?

26. Explain why a test of the null hypothesis H_0: $\rho = 0$ is equivalent to a test of the null hypothesis H_0: $\beta_1 = 0$, where ρ is the linear correlation coefficient for a population of paired data, and β_1 is the slope of the regression line for that same population.

9-4 Variation and Prediction Intervals

Students with no computer software and without calculators capable of dealing with two-variable statistics will be seriously challenged by the calculations in this section.

In Section 9-2 we introduced the concept of correlation and used the linear correlation coefficient r in determining whether there is a significant linear correlation between two variables. In addition to serving as a measure of the linear correlation between two variables, the value of r can also provide us with additional information about the variation of sample points about the regression line. We begin with a sample case, which leads to an important definition.

Suppose we have a large collection of paired data, which yields the following results:

- There is a significant linear correlation.
- The equation of the regression line is $\hat{y} = 3 + 2x$.
- $\bar{y} = 9$
- One of the pairs of sample data is $x = 5$ and $y = 19$.
- The point $(5, 13)$ is one of the points on the regression line, because substitution of $x = 5$ into the regression equation yields the following.

$$\hat{y} = 3 + 2x = 3 + 2(5) = 13$$

The point $(5, 13)$ lies on the regression line, but the point $(5, 19)$ is from the original data set and does not lie on the regression line because it does not

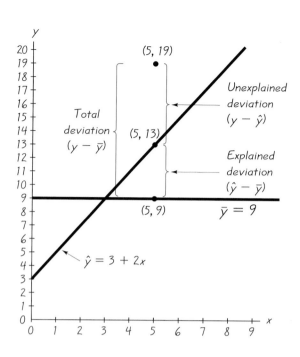

Figure 9-9   Unexplained, Explained, and Total Deviation

satisfy the regression equation. Take time to examine Figure 9-9 carefully and note the differences defined as follows.

DEFINITIONS

Assume that we have a collection of paired data containing the particular point (x, y), that $\hat{y}$ is the predicted value of y (obtained by using the regression equation), and that the mean of the sample y values is $\bar{y}$.

The **total deviation** (from the mean) of the particular point (x, y) is the vertical distance $y - \bar{y}$, which is the distance between the point (x, y) and the horizontal line passing through the sample mean $\bar{y}$.

The **explained deviation** is the vertical distance $\hat{y} - \bar{y}$, which is the distance between the predicted y value and the horizontal line passing through the sample mean $\bar{y}$.

The **unexplained deviation** is the vertical distance $y - \hat{y}$, which is the vertical distance between the point (x, y) and the regression line. (The distance $y - \hat{y}$ is also called a residual, as defined in Section 9-3.)

Predictors for Success

When a college accepts a new student, it would like to have some positive indication that the student will be successful in his or her studies. College admissions deans consider SAT scores, standard achievement tests, rank in class, difficulty of high school courses, high school grades, and extracurricular activities. In a study of characteristics that make good predictors of success in college, it was found that class rank and scores on standard achievement tests are better predictors than SAT scores. A multiple regression equation with college grade-point average predicted by class rank and achievement test score was not improved by including another variable for SAT score. This particular study suggests that SAT scores should not be included among the admissions criteria, but supporters argue that SAT scores are useful for comparing students from different geographic locations and high school backgrounds.

For the specific data under consideration, we get these results:

$$\textit{total deviation of } (5, 19) = y - \bar{y} = 19 - 9 = 10$$

$$\textit{explained deviation of } (5, 19) = \hat{y} - \bar{y} = 13 - 9 = 4$$

$$\textit{unexplained deviation of } (5, 19) = y - \hat{y} = 19 - 13 = 6$$

If we were totally ignorant of correlation and regression concepts and wanted to predict a value of y given a value of x and a collection of paired (x, y) data, our best guess would be $\bar{y}$. But we are not totally ignorant of correlation and regression concepts: We know that in this case (with a significant linear correlation) the way to predict the value of y when $x = 5$ is to use the regression equation, which yields $\hat{y} = 13$, as calculated before. We can explain the discrepancy between $\bar{y} = 9$ and $\hat{y} = 13$ by simply noting that there is a significant linear correlation best described by the regression line. Consequently, when $x = 5$, y *should be* 13 and not 9. But whereas y should be 13, *it is* 19. The discrepancy between 13 and 19 cannot be explained by the regression line and is called an unexplained deviation or a residual. The specific case illustrated in Figure 9-9 can be generalized as follows:

(total deviation)	=	(explained deviation)	+	(unexplained deviation)
or $(y - \bar{y})$	=	$(\hat{y} - \bar{y})$	+	$(y - \hat{y})$

This last expression applies to a particular point (x, y), but it can be further generalized and modified to include all of the pairs of sample data, as shown in Formula 9-5. In that formula, the **total variation** is expressed as the sum of the squares of the total deviation values, the **explained variation** is the sum of the squares of the explained deviation values, and the **unexplained variation** is the sum of the squares of the unexplained deviation values.

Formula 9-5

(total variation)	=	(explained variation)	+	(unexplained variation)
or $\Sigma(y - \bar{y})^2$	=	$\Sigma(\hat{y} - \bar{y})^2$	+	$\Sigma(y - \hat{y})^2$

The components of this last expression are used in the following important definition.

The coefficient of determination provides us with an important and meaningful interpretation, and it is quite easy to calculate its value.

DEFINITION

The **coefficient of determination** is the amount of the variation in y that is explained by the regression line. It is computed as

$$r^2 = \frac{\text{explained variation}}{\text{total variation}}$$

We can compute r^2 by using the definition just given with Formula 9-5, or we can simply square the linear correlation coefficient r, which is found by using the methods described in Section 9-2. As an example, if $r = 0.8$, then the coefficient of determination is $r^2 = 0.64$, which means that *64% of the total variation in* y *can be explained by the regression line. It follows that 36% of the total variation in* y *remains unexplained.*

EXAMPLE   Referring to the bear measurements in Table 9-1, find the percentage of the variation in y (weight) that can be explained by the regression line.

SOLUTION   Recall that Table 9-1 contains eight pairs of sample data, with each pair consisting of the lengths (in inches) and weights (in pounds) of eight male bears. In Section 9-2 we found that the linear correlation coefficient is $r = 0.897$. The coefficient of determination is $r^2 = 0.897^2 = 0.805$, indicating that the ratio of explained variation in y to total variation in y is 0.805. We can now state that 80.5% of the total variation in y can be explained by the regression line. That is, 80.5% of the total variation in weights of bears can be explained by the variation in their lengths; the other 19.5% is attributable to other factors.

Prediction Intervals

In Section 9-3 we used the Table 9-1 sample data to find the regression equation $\hat{y} = -352 + 9.66x$, where $\hat{y}$ represents the predicted bear weight and x represents the length of the bear. We then used that equation to predict the y value, given that $x = 71.0$ in. We found that the best predicted weight of a 71.0 in. bear is 334 lb. Because 334 is a single value, it is referred to as a *point estimate*. In Chapter 6 we saw that point estimates have the disadvantage of not conveying any sense of how accurate they might be. Here, we know that 334 is the best predicted value, but we don't know how accurate that value is. In Chapter 6 we developed confidence interval estimates to overcome that disadvantage, and in this section we follow that precedent. We will use a **prediction interval,** which is a confidence interval estimate of a predicted value of y.

The development of a prediction interval requires a measure of the spread of sample points about the regression line. Recall that the unexplained deviation (or residual) is the vertical distance between a sample point and the regression line, as illustrated in Figure 9-9. The standard error of estimate is a collective measure of the spread of the sample points about the regression line; it is formally defined as follows.

If using STATDISK, the value of s_e is displayed when you use the Correlation and Regression option. With Minitab, use the regression program to get a display that includes the value of s_e; it is identified as s and is on the same line as the value of R-sq and R-sq(adj).

DEFINITION

The **standard error of estimate,** denoted by s_e, is a measure of the differences (or distances) between the observed sample y values and the predicted values $\hat{y}$ that are obtained using the regression equation. It is given as

$$s_e = \sqrt{\frac{\Sigma(y - \hat{y})^2}{n - 2}}$$

where $\hat{y}$ is the predicted y value.

The development of the standard error of estimate s_e closely parallels that of the ordinary standard deviation introduced in Chapter 2. Just as the standard deviation is a measure of how scores deviate from their mean, the standard error of estimate s_e is a measure of how sample data points deviate from their regression line. The reasoning behind dividing by $n - 2$ is similar to the reasoning that led to division by $n - 1$ for the ordinary standard deviation, and we will not pursue the complex details. It is important to note that smaller values of s_e reflect points that stay close to the regression line, and larger values are from points farther away from the regression line.

Formula 9-6 can also be used to compute the standard error of estimate s_e. It is algebraically equivalent to the expression in the definition, but this form is generally easier to work with because it doesn't require that we compute each of the predicted values $\hat{y}$ by substitution in the regression equation. However, Formula 9-6 does require that we find the y-intercept b_0 and the slope b_1 of the estimated regression line.

Formula 9-6 $$s_e = \sqrt{\frac{\Sigma y^2 - b_0\Sigma y - b_1\Sigma xy}{n - 2}}$$ (standard error of estimate)

EXAMPLE   Use Formula 9-6 to find the standard error of estimate s_e for the bear measurement data listed in Table 9-1.

SOLUTION   In Section 9-2 we used the Table 9-1 data to find that:

$$n = 8 \qquad \Sigma y^2 = 728{,}520 \qquad \Sigma y = 2176 \qquad \Sigma xy = 151{,}879$$

In Section 9-3 we used the Table 9-1 data to find the y-intercept and slope of the regression line. Those values are given here with extra decimal places for greater precision.

$$b_0 = -351.660 \qquad b_1 = 9.65979$$

We can now use these values to find the standard error of estimate s_e.

$$s_e = \sqrt{\frac{\Sigma y^2 - b_0 \Sigma y - b_1 \Sigma xy}{n - 2}}$$

$$= \sqrt{\frac{728{,}520 - (-351.660)(2176) - (9.65979)(151{,}879)}{8 - 2}}$$

$$= 66.5994 = 66.6 \quad \text{(rounded)}$$

We can measure the spread of the sample points about the regression line with the standard error of estimate $s_e = 66.6$.

We can use the standard error of estimate s_e to construct interval estimates that will help us see how dependable our point estimates of y really are. Assume that for each fixed value of x, the corresponding sample values of y are normally distributed about the regression line, and those normal distributions have the same variance. The following interval estimate applies to an *individual y*. (For a confidence interval used to predict the *mean* of all y values for some given x value, see Exercise 20.)

Prediction Interval for an Individual y

Given the fixed value x_0, the *prediction interval for an individual y* is

$$\hat{y} - E < y < \hat{y} + E$$

where the margin of error E is

$$E = t_{\alpha/2} s_e \sqrt{1 + \frac{1}{n} + \frac{n(x_0 - \bar{x})^2}{n(\Sigma x^2) - (\Sigma x)^2}}$$

and x_0 represents the given value of x, $t_{\alpha/2}$ has $n - 2$ degrees of freedom, and s_e is found from Formula 9-6.

Minitab can be used to generate 95% prediction intervals, which are displayed and identified as 95% P.I. The confidence level has a default of 95%, but it can be changed as desired.

EXAMPLE   Refer to the Table 9-1 sample data listing bear lengths (x) along with their weights (y). In previous sections we have shown that

- There is a significant linear correlation (at the 0.05 significance level).
- The regression equation is $\hat{y} = -352 + 9.66x$.
- When $x = 71.0$, the predicted y value is 334 lb.

Construct a 95% prediction interval for the weight of a bear that is 71.0 in. long. This will provide a sense of how reliable the estimate of 334 lb really is.   *(continued)*

70¢

Wage Gender Gap

Many articles note that, on average, full-time female workers earn about 70¢ for each $1 earned by full-time male workers. Researchers at the Institute for Social Research at the University of Michigan analyzed the effects of various key factors and found that about one-third of the discrepancy can be explained by differences in education, seniority, work interruptions, and job choices. The other two-thirds remains unexplained by such labor factors.

SOLUTION   We have already used the Table 9-1 sample data to find the following values:

$$n = 8 \quad \bar{x} = 64.5625 \quad \Sigma x = 516.5 \quad \Sigma x^2 = 34{,}525.75$$
$$s_e = 66.5994$$

From Table A-3 we find $t_{\alpha/2} = 2.447$. (We used $8 - 2 = 6$ degrees of freedom with $\alpha = 0.05$ in two tails.) We can now calculate the margin of error E by letting $x_0 = 71.0$, because we want the prediction interval of y for $x = 71.0$.

$$\begin{aligned} E &= t_{\alpha/2}s_e \sqrt{1 + \frac{1}{n} + \frac{n(x_0 - \bar{x})^2}{n(\Sigma x^2) - (\Sigma x)^2}} \\ &= (2.447)(66.5994)\sqrt{1 + \frac{1}{8} + \frac{8(71.0 - 64.5625)^2}{8(34{,}525.75) - (516.5)^2}} \\ &= (2.447)(66.5994)(1.07710) = 176 \end{aligned}$$

With $\hat{y} = 334$ and $E = 176$, we get the prediction interval as follows.

$$\hat{y} - E < y < \hat{y} + E$$
$$334 - 176 < y < 334 + 176$$
$$158 < y < 510$$

That is, for a 71.0 in. bear, we have 95% confidence that its true weight is between 158 lb and 510 lb. That's a relatively large range. (One factor contributing to the large range is the small sample size of eight.)

In addition to knowing that the predicted weight is 334 lb, we now have a sense of how reliable that estimate really is. The 95% prediction interval found in this example shows that the 334 lb estimate can vary substantially.

Minitab can be used to generate 95% prediction intervals. Enter the x data in column C1, enter the y data in column C2, then select the options of Stat, Regression, and Regression. Enter C2 in the box labeled Response and enter C1 in the box labeled Predictors. Click on the Options box and enter 71.0 (or whatever value of x_0 is desired) in the box labeled Prediction intervals for new observations. If this procedure is used with the data in Table 9-1, Minitab will display results including (158.6, 509.8) below the heading of 95.0% P.I. This corresponds to the same prediction interval found in the preceding example, with a small discrepancy due to rounding.

Exercises A: Basic Skills and Concepts

Recommended assignment: Exercises 1-7, 11, 13, 14. (Exercises 1–4 are quick and relatively easy.)

In Exercises 1–4, use the value of the linear correlation coefficient r to find the coefficient of determination and the percentage of the total variation that can be explained by the regression line.

1. $r = 0.2$

2. $r = -0.6$

3. $r = -0.225$

4. $r = 0.837$

In Exercises 5–8, find the (a) explained variation, (b) unexplained variation, (c) total variation, (d) coefficient of determination, and (e) standard error of estimate s_e.

5. The accompanying table lists numbers x of patio tiles and costs y (in dollars) of having them manually cut to fit. (The equation of the regression line is $\hat{y} = 2 + 3x$.)

x	1	2	3	5	6
y	5	8	11	17	20

6. The paired data below consist of the chest sizes (in inches) and weights (in pounds) of a sample of male bears. (The equation of the regression line is $\hat{y} = -187.462 + 11.2713x$.)

x Chest (in.)	26	45	54	49	41	49	44	19
y Weight (lb)	90	344	416	348	262	360	332	34

7. The paired data below consist of the weights (in pounds) of discarded plastic and sizes of households. (The equation of the regression line is $\hat{y} = 0.549270 + 1.47985x$.)

Plastic (lb)	0.27	1.41	2.19	2.83	2.19	1.81	0.85	3.05
Household size	2	3	3	6	4	2	1	5

8. The paired data below consist of the total weights (in pounds) of discarded garbage and sizes of households. (The equation of the regression line is $\hat{y} = 0.182817 + 0.119423x$.)

Total weight	10.76	19.96	27.60	38.11	27.90	21.90	21.83	49.27	33.27	35.54
Household size	2	3	3	6	4	2	1	5	6	4

9. Refer to the data given in Exercise 5 and assume that the necessary conditions of normality and variance are met.
   a. For $x = 4$, find $\hat{y}$, the predicted value of y.
   b. How does the value of s_e affect the construction of the 95% prediction interval of y for $x = 4$?

10. Refer to Exercise 6 and assume that the necessary conditions of normal-

1. 0.04; 4%
2. 0.36; 36%
3. 0.051; 5.1%
4. 0.701; 70.1%

5. a. 154.8
   b. 0
   c. 154.8
   d. 1
   e. 0

6. a. 130963.47
   b. 1932.0267
   c. 132895.5
   d. 0.985462
   e. 17.944482

7. a. 13.836615
   b. 5.663385
   c. 19.5
   d. 0.70957
   e. 0.971544

8. a. 15.300055
   b. 11.099945
   c. 26.4
   d. 0.579548
   e. 1.177919

9. a. 14
   b. $s_e = 0$, $E = 0$; no interval estimate.

10. a. 399 lb
    b. 324 lb $< y <$ 473 lb

ity and variance are met.

    a. For a bear with a measured chest size of 52 in., find $\hat{y}$, the predicted weight.

    b. Find the 99% prediction interval of y for $x = 52$.

11. Refer to the data given in Exercise 7 and assume that the necessary conditions of normality and variance are met.

    a. Find the predicted size of a household that discards 2.50 lb of plastic.

    b. Find the 95% prediction interval for the size of a household that discards 2.50 lb of plastic.

12. Refer to the data given in Exercise 8 and assume that the necessary conditions of normality and variance are met.

    a. For a household discarding 20.0 lb of garbage, find the predicted household size.

    b. Find the 99% prediction interval for the size of a household that discards 20.0 lb of garbage.

In Exercises 13–16, refer to the Table 9-1 sample data. Let x represent the length of a bear (in inches), and let y represent the weight of a bear. Construct a prediction interval estimate of the weight of a bear that has the given length. Use the given degree of confidence. (See the example in this section.)

13. 50.0 in.; 95% confidence   **14.** 65.0 in.; 90% confidence

15. 49.7 in.; 90% confidence   **16.** 68.0 in.; 99% confidence

9-4 Exercises B: Beyond the Basics

17. Confidence intervals for the y-intercept β_0 and slope β_1 for a regression line $(y = \beta_0 + \beta_1 x)$ can be found by evaluating the limits in the intervals below.

$$b_0 - E < \beta_0 < b_0 + E$$

where
$$E = t_{\alpha/2} s_e \sqrt{\frac{1}{n} + \frac{\bar{x}^2}{\Sigma x^2 - \frac{(\Sigma x)^2}{n}}}$$

$$b_1 - E < \beta_1 < b_1 + E$$

where
$$E = t_{\alpha/2} \cdot \frac{s_e}{\sqrt{\Sigma x^2 - \frac{(\Sigma x)^2}{n}}}$$

In these expressions, the y-intercept b_0 and the slope b_1 are found from the sample data and $t_{\alpha/2}$ is found from Table A-3 by using $n - 2$ degrees of freedom. Using the bear data in Table 9-1, find the 95% confidence interval estimates of β_0 and β_1.

18. a. If a collection of paired data includes at least three pairs of values,

---

Answer margin notes:

11. a. 4.2 persons
   b. $1.6 < y < 6.9$

12. a. 2.6 persons
   b. $-1.7 < y < 6.8$

13. -55 lb $< y < 317$ lb
14. 139 lb $< y < 413$ lb
15. -20 lb $< y < 277$ lb
16. 42 lb $< y < 568$ lb

17. $-663 < \beta_0 < -39.9;$
   $4.91 < \beta_1 < 14.41$

what do you know about the linear correlation coefficient if $s_e = 0$?

b. If a collection of paired data is such that the total explained variation is 0, what do you know about the slope of the regression line?

19. a. Find an expression for the unexplained variation in terms of the sample size n and the standard error of estimate s_e.

b. Find an expression for the explained variation in terms of the coefficient of determination r^2 and the unexplained variation.

c. Suppose we have a collection of paired data for which $r^2 = 0.900$ and the regression equation is $\hat{y} = 3 - 2x$. Find the linear correlation coefficient.

20. From the expression given in this section for the margin of error corresponding to a prediction interval for y, we can get the expression

$$s_{\hat{y}} = s_e \sqrt{1 + \frac{1}{n} + \frac{n(x_0 - \bar{x})^2}{n(\Sigma x^2) - (\Sigma x)^2}}$$

which is the *standard error of the prediction* when predicting for a *single* y, given that $x = x_0$. When predicting for the *mean* of all values of y for which $x = x_0$, the point estimate $\hat{y}$ is the same, but $s_{\hat{y}}$ is as follows:

$$s_{\hat{y}} = s_e \sqrt{\frac{1}{n} + \frac{n(x_0 - \bar{x})^2}{n(\Sigma x^2) - (\Sigma x)^2}}$$

Use the data from Table 9-1 and extend the last example of this section to find a point estimate and a 95% confidence interval estimate of the mean weight of all bears that are 71.0 in. long.

9–5 Multiple Regression

So far, the examples and exercises in this chapter have involved the relationship between two variables, such as lengths and weights of bears. In many cases, a variable is related to two or more other variables. For example, instead of predicting a bear's weight by using only its length, perhaps we can get a better prediction by using its length, chest size, and neck size. As in the previous sections of this chapter, we will work with linear relationships only, and our goal is to develop a multiple regression equation, which expresses a dependent variable y in terms of two or more other variables.

This section emphasizes computer usage and interpretation of computer results. (The formulas required for manual calculations of multiple regression equations are monstrous.) Some instructors include this section by assigning exercises as extra-credit or out-of-class work.

DEFINITION

A **multiple regression equation** expresses a linear relationship between a dependent variable y and two or more independent variables (x_1, x_2, . . . , x_k).

18. a. $r = 1$ or $r = -1$
b. The slope is 0.

19. a. $(n-2)s_e^2$
b.
$$\frac{r^2 \cdot (\text{unexplained variation})}{1 - r^2}$$
c. $r = -0.949$

20. 334 lb; $269 < \mu_y < 399$

We will use the following notation, which follows naturally from the notation used in Section 9-3.

Notation
$\hat{y} = b_0 + b_1x_1 + b_2x_2 + \cdots + b_kx_k$ (General form of the estimated multiple regression equation)
n = sample size
k = number of *independent* variables. (The independent variables are also called **predictor variables** or x variables.)
$\hat{y}$ = predicted value of the dependent variable y (computed by using the multiple regression equation)
$x_1, x_2, \ldots, x_k$ are the independent variables
β_0 = the y-intercept, or the value of y when all of the predictor variables are 0
b_0 = estimate of β_0 based on the sample data
$\beta_1, \beta_2, \ldots, \beta_k$ are the coefficients of the independent variables $x_1, x_2, \ldots, x_k$
$b_1, b_2, \ldots, b_k$ are the sample estimates of the coefficients $\beta_1, \beta_2, \ldots, \beta_k$

Making Music with Multiple Regression

Sony manufactures millions of compact discs in Terre Haute, Indiana. At one step in the manufacturing process, a laser exposes a photographic plate so that a musical signal is transferred into a digital signal coded with 0s and 1s. This process was statistically analyzed to identify the effects of different variables, such as the length of exposure and the thickness of the photographic emulsion. Methods of multiple regression showed that among all of the variables considered, four were most significant. The photographic process was adjusted for optimal results based on the four critical variables. As a result, the percentage of defective discs dropped and the tone quality was maintained. The use of multiple regression methods led to lower production costs and better control of the manufacturing process.

Using Computer Software for Multiple Regression

In previous sections of this chapter we were able to develop procedures that could be used with a calculator, but the computations required for multiple regression are so complex that we will assume that a statistical software package is used. We will describe procedures for entering data and interpreting computer results.

The TI-83 calculator does not produce multiple regression equations, but STATDISK and Minitab both do. With STATDISK, use the options of `Analysis` and `Multiple Regression`. STATDISK will provide the multiple regression equation, the multiple coefficient of determination R^2, and the adjusted R^2, terms that will be described shortly. Minitab provides much more.

With Minitab, use the main menu item of `Statistics`, then select `Regression`, then `Regression` once again. (For details, see the *Minitab Student Laboratory Manual and Workbook*.) If you use the data from Table 9-3 and specify a response variable of `WEIGHT` and predictor variables of `HEADLEN` and `LENGTH`, the Minitab display will be as follows.

TABLE 9-3	Data from Anesthetized Male Bears								

Variable	Minitab Column	Name	Sample Data							
y	C1	WEIGHT	80	344	416	348	262	360	332	34
x_2	C2	AGE	19	55	81	115	56	51	68	8
x_3	C3	HEADLEN	11.0	16.5	15.5	17.0	15.0	13.5	16.0	9.0
x_4	C4	HEADWDTH	5.5	9.0	8.0	10.0	7.5	8.0	9.0	4.5
x_5	C5	NECK	16.0	28.0	31.0	31.5	26.5	27.0	29.0	13.0
x_6	C6	LENGTH	53.0	67.5	72.0	72.0	73.5	68.5	73.0	37.0
x_7	C7	CHEST	26	45	54	49	41	49	44	19

MINITAB DISPLAY

```
The regression equation is
WEIGHT = -374 + 18.8 HEADLEN + 5.87 LENGTH          ① Multiple regression equation

Predictor    Coef    Stdev   t-ratio        p
Constant    -374.3   134.1    -2.79     0.038
HEADLEN      18.82   23.15     0.81     0.453
LENGTH       5.875    5.065    1.16     0.299

s = 68.56   R-sq = 82.8%   R-sq(adj) = 75.9%

Analysis of Variance        $R^2 = 0.828$   ② Adjusted $R^2 = 0.759$

SOURCE        DF        SS       MS       F       p
Regression     2    113142    56571    12.03    0.012
Error          5     23506     4701
Total          7    136648
                                        ③ Overall significance of
                                          multiple regression
                                          equation
SOURCE        DF     SEQ SS
HEADLEN        1    106819
LENGTH         1      6323
```

The annotated Minitab display has three key components, identified with circled numbers:

1. Multiple regression equation:

$$WEIGHT = -374 + 18.8 \text{ HEADLEN} + 5.87 \text{ LENGTH}$$

Statistics in Court

Owners of a five-building apartment complex in New York City filed a lawsuit because of extensive damage to bricks. The damage occurred when water was absorbed by the brick face, followed by freezing and thawing cycles, causing part of the brick face to separate. With about 750,000 bricks, it was not practical to inspect each one, so methods of sampling were used instead. Statisticians used regression methods to predict the total number of damaged bricks. The independent variables included which of the five buildings was used, direction of wall exposure, height, and whether the wall faced an interior courtyard or was an exterior wall. The estimate of total damage appears to have strongly influenced the final settlements. (See "Bricks, Buildings, and the Bronx: Estimating Masonry Deterioration," by Fairley, Izenman, and Whitlock, *Chance*, Vol. 7, No. 3.)

2. Adjusted $R^2 = 0.759$

3. Overall significance of multiple regression equation: $p = 0.012$

We will now briefly address each of these three important components. Because we are focusing on interpretation of computer displays and because the underlying theory is rather complicated, we will include very little theory. For a more thorough discussion, refer to a book with more thorough coverage, such as *Elementary Business Statistics* by Triola and Franklin.

Multiple Regression Equation   The multiple regression equation of

$$\text{WEIGHT} = -374 + 18.8\ \text{HEADLEN} + 5.87\ \text{LENGTH}$$

can be expressed in our standard notation as

$$\hat{y} = -374 + 18.8x_3 + 5.87x_6$$

This equation best fits the given data (for weight, head length, and overall length) according to the same least-squares criterion described in Section 9-3. That is, when we use the Table 9-3 data for WEIGHT, HEADLEN, and LENGTH, this equation fits the data best; if we used other data, we might find another equation that fits the data better. (Later, we will discuss determination of the regression equation that fits the data best.) If the equation fits the data well, it can be used for predictions. For example, if we determine that the equation is suitable for predictions, and if we have a bear with a 14.0 in. head length and a 71.0 in. overall length, we can predict its weight by substituting those values into the regression equation to get a predicted weight of 306 lb. Also, the coefficients of $b_3 = 18.8$ and $b_6 = 5.87$ can be used to determine *marginal change*, as described in Section 9-3. For example, the coefficient of $b_3 = 18.8$ shows that when the overall length of a bear remains constant, the predicted weight increases by 18.8 lb for each 1 in. increase in the length of the head.

Adjusted R^2 R^2 denotes the **multiple coefficient of determination,** which is a measure of how well the multiple regression equation fits the sample data. A perfect fit would result in $R^2 = 1$. A very good fit results in a value near 1. A very poor fit results in a value of R^2 close to 0. The value of $R^2 = 0.828$ in the Minitab display indicates that 82.8% of the variation in bear weight can be explained by the head length x_3 and overall length x_6. *The multiple coefficient of determination R^2 is a measure of how well the regression equation fits the sample data,* but it has a serious flaw: As more variables are included, R^2 increases. (Actually, R^2 could remain the same, but it usually increases.) Although the largest R^2 is thus achieved by simply including all of the available variables, the best multiple regression equation does not necessarily use all of the available variables. Consequently, it is better to use the adjusted coefficient of determination when comparing different multiple regression equations, because it adjusts the R^2 value based on the number of variables and the sample size.

DEFINITION

The **adjusted coefficient of determination** is the multiple coefficient of determination R^2 modified to account for the number of variables and the sample size. It is calculated by using Formula 9-7.

Formula 9-7

$$\text{Adjusted } R^2 = 1 - \frac{(n-1)}{[n-(k+1)]}(1 - R^2)$$

The Minitab display shows the adjusted coefficient of determination as R-sq (adj) = 75.9%. If we use Formula 9-7 with the R^2 value of 0.828, $n = 8$, and $k = 2$, we find that the adjusted R^2 value is 0.759, confirming Minitab's displayed value of 75.9%. For the weight, head length, and length data in Table 9-3, the R^2 value of 82.8% indicates that 82.8% of the variation in weight can be explained by the head length x_3 and overall length x_6, but when we compare this multiple regression equation to others, it is better to use the adjusted R^2 of 75.9% (or 0.759).

Overall Significance: P-Value   The Minitab display shows a *P*-value of 0.012. This value is a measure of the overall significance of the multiple regression equation. In this case, the small value of 0.012 suggests that the multiple regression equation has good overall significance and is usable for predictions. That is, it makes sense to predict weights of bears based on their head lengths and overall lengths. Like the adjusted R^2, this *P*-value is a good measure of how well the equation fits the sample data. The value of 0.012 results from a test of the null hypothesis that $\beta_3 = \beta_6 = 0$. Rejection of $\beta_3 = \beta_6 = 0$ implies that at least one of β_3 and β_6 is not 0, suggesting that this regression equation is effective in determining bear weights.

A complete analysis of the Minitab results might include other important elements, such as the significance of the individual coefficients, but we will limit our discussion to the three key components previously identified.

Finding the Best Multiple Regression Equation

So far, we have considered only the multiple regression equation relating bear weights to the independent variables of head length and overall length, but there are several other variables that could be included. Perhaps some other combination of variables would result in an equation that is better in the sense that it fits the sample data better and is more likely to provide better predictions of bear weights. Table 9-4 lists a few of the combinations of variables.

Model for Alumni Contributions

One study developed a multiple regression equation that is a good predictor of alumni donations at liberal arts colleges. The college hoped to use the results of the study to improve its fund-raising strategies. The dependent variable was the amount of money donated in a year. Independent variables included income, age, whether the donor was single, whether the donor belonged to a fraternity or sorority, whether the donor was active in alumni activities, the donor's major, distance to the college, and the nation's unemployment rate (used as a measure of the economy). Other independent variables (such as whether the donor had children) were excluded for lack of statistical significance. (See "An Econometric Model of Alumni Giving: A Case Study for a Liberal Arts College," by Bruggink and Siddiqui, *The American Economist*, Vol. 39, No. 2.)

TABLE 9-4	Searching for the Best Multiple Regression Equation				
	LENGTH	CHEST	HEADLEN/ LENGTH	AGE/NECK/ LENGTH/CHEST	AGE/HEADLEN/HEADWDTH/ NECK/LENGTH/CHEST
R^2	0.805	0.983	0.828	0.999	0.999
Adjusted R^2	0.773	0.980	0.759	0.997	0.996
Overall significance	0.002	0.000	0.012	0.000	0.046

Stress that determination of the best multiple regression equation is beyond the scope of this book.

Although *determination of the best multiple regression equation is often quite difficult and beyond the scope of this book,* we can provide some help with the following guidelines.

1. *Use common sense and practical considerations to include or exclude variables.* For example, we might exclude the variable of age because inexperienced researchers might not know how to determine the age of a bear and, when questioned, bears are reluctant to reveal their ages.

2. Instead of including almost every available variable, *include relatively few independent (x) variables.* In weeding out independent variables that don't have an effect on the dependent variable, it might be helpful to find the linear correlation coefficient r for each pair of variables being considered. For example, using the data in Table 9-3, we will find that there is a 0.955 linear correlation for the paired NECK/HEADLEN data. Because there is such a high correlation between neck size and head length, there is no need to include both of those variables. In choosing between NECK and HEADLEN, we should include NECK for this reason: NECK is a better predictor of WEIGHT because the NECK/WEIGHT paired data have a linear correlation coefficient of $r = 0.971$, which is higher than $r = 0.884$ for the paired HEADLEN/WEIGHT data.

3. *Select an equation having a value of adjusted R^2 with this property: If an additional independent variable is included, the value of adjusted R^2 does not increase by a substantial amount.* For example, Table 9-4 shows that if we use only the independent variable of CHEST, the adjusted R^2 is 0.980, but when we include all six variables the adjusted R^2 increases to 0.996. Using six variables instead of only one is too high a price to pay for such a small increase in the adjusted R^2. We're better off with the single independent variable of CHEST than using all six independent variables.

4. *For a given number of independent (x) variables, select the equation with the largest value of adjusted R^2.* That is, choose those variables with the property that no other combination of the same number of independent variables will yield a larger value of adjusted R^2. For example, if we decide to use a single independent variable, we should use CHEST, because its adjusted R^2 of 0.980 is larger than the adjusted R^2 for any other single variable. Consequently, CHEST is the best *single* predictor of WEIGHT.

5. *Select an equation having overall significance, as determined by the P-value in the computer display.* For example, see the values of overall significance in Table 9-4. The use of all six independent variables results in an overall significance of 0.046, which is just barely significant at the $\alpha = 0.05$ level; we're better off with the single variable of CHEST, which has overall significance of 0.000.

Using these guidelines in an attempt to find the best equation for predicting weights of bears, we find that for the data of Table 9-3, the best regression equation uses the single independent variable of CHEST. The best regression equation appears to be

$$\text{WEIGHT} = -195 + 11.4 \text{ CHEST}$$

or
$$\hat{y} = -195 + 11.4x_7$$

For cases involving a large number of independent variables, many statistical software packages include a program for performing **stepwise regression,** whereby different combinations are tried until the best model is obtained. If we eliminate the variable AGE (as in Step 1) and then run Minitab's stepwise regression program, we will get a display suggesting that the best regression equation is the one in which CHEST is the only independent variable. (If we include all six independent variables, Minitab selects a regression equation with the independent variables of AGE, NECK, LENGTH, and CHEST, with an adjusted R^2 value of 0.997 and overall significance of 0.000.) It appears that we can estimate the weight of a bear based on its chest size, and the regression equation leads to this rule: The weight of a bear is estimated to be 11.4 times the chest size (in inches) minus 195.

When we discussed regression in Section 9-3, we listed four common errors that should be avoided when using regression equations to make predictions. These same errors should be avoided when using multiple regression equations. Be especially careful about concluding that a cause-effect relationship exists.

> Because stepwise regression can sometimes lead to very strange results, some professionals recommend trying every possible combination of variables. With many independent variables, this can be quite a project.

Exercises A: Basic Skills and Concepts

In Exercises 1–4, refer to the Minitab display at the top of the following page and answer the given questions.

1. Identify the multiple regression equation that expresses weight in terms of age, head width, and neck size.
2. Identify
   a. The *P*-value corresponding to the overall significance of the multiple regression equation
   b. The value of the multiple coefficient of determination R^2
   c. The adjusted value of R^2

> Recommended assignment: Exercises 1–8. (Exercises 1–4 are quick and relatively easy; Exercises 5–8 require STATDISK or Minitab.)
>
> 1. $\hat{y} = -285 - 1.38x_1 - 11.2x_2 + 28.6x_3$
> 2. a. 0.002
>   b. 0.969
>   c. 0.946

```
                        MINITAB DISPLAY

  The regression equation is

  WEIGHT = -285 -1.38 AGE -11.2 HEADWDTH +28.6 NECK

  Predictor       Coef       Stdev     t-ratio          p
  Constant     -285.21       78.45       -3.64      0.022
  AGE          -1.3838      0.9022       -1.53      0.200
  HEADWDTH      -11.24       20.88       -0.54      0.619
  NECK          28.594       5.870        4.87      0.008

  s = 32.49       R-sq = 96.9%   R-sq(adj) = 94.6%

  Analysis of Variance

  SOURCE        DF          SS        MS        F        p
  Regression     3      132425     44142    41.81    0.002
  Error          4        4223      1056
  Total          7      136648

  SOURCE        DF      SEQ SS
  AGE            1       90527
  HEADWDTH       1       16844
  NECK           1       25054
```

3. Yes, with a P-value of 0.002, the equation has overall significance.

4. a. 230 lb
   b. Off by 50 lb; not accurate.

5. a. $\hat{y} = -274 + 0.426x_1 + 12.1x_2$
   b. 0.928; 0.925; 0.000
   c. Yes

6. a. $\hat{y} = -265 + 5.83x_1 - 0.103x_2 + 9.36x_3$
   b. 0.937; 0.933; 0.000
   c. Yes

7. a. $\hat{y} = -235 + 0.403x_1 + 5.11x_2 - 0.555x_3 + 9.19x_4$
   b. 0.942; 0.938; 0.000
   c. Yes

8. a. $\hat{y} = -258 - 11.8x_1 + 3.95x_2 + 17.6x_3$
   b. 0.890; 0.883; 0.000
   c. Yes

3. Is the multiple regression equation usable for predicting a bear's weight based on its age, head width, and neck size? Why or why not?

4. A 32-month-old bear is found to have a head width of 5.0 in. and a neck size of 21.5 in.
   a. Find the predicted weight of the bear.
   b. The bear in question actually weighed 180 lb. How accurate is the predicted weight from part (a)?

In Exercises 5–8, refer to the bear data in Data Set 3 of Appendix B. Let the dependent variable be WEIGHT, *and let the independent variables be those given in the exercise. (The data sets are already stored on the Data/Program Disk and STATDISK.) Use software such as STATDISK or Minitab to answer these questions:*

a. *Find the multiple regression equation that expresses the dependent variable* WEIGHT *in terms of the given independent variables.*

b. *Identify the values of the multiple coefficient of determination R^2, the adjusted R^2, and (if Minitab is used) the P-value corresponding to the*

overall significance .

c. *Does the multiple regression equation seem suitable for predicting the weight of a bear based on the given independent variables?*

5. LENGTH and CHEST

6. NECK, LENGTH, and CHEST

7. AGE, NECK, LENGTH, and CHEST

8. HEADLEN, LENGTH, and NECK

In Exercises 9–14, refer to the weights of garbage in Data Set 1 of Appendix B. Let the dependent variable be HHSIZE (household size), and let the independent variables be those given in the exercise. (The data sets are already stored on the Data/Program Disk and STATDISK.) Use software such as STATDISK or Minitab to answer these questions:

a. *Find the multiple regression equation that expresses the dependent variable HHSIZE in terms of the given independent variables.*

b. *Identify the values of the multiple coefficient of determination R^2, the adjusted R^2, and (if Minitab is used) the P-value corresponding to the overall significance.*

c. *Does the multiple regression equation seem suitable for making predictions of household size based on the given independent variables?*

9. YARD

10. PLASTIC

11. PLASTIC and PAPER

12. METAL, PLASTIC, and FOOD

13. METAL and PLASTIC

14. PAPER and GLASS

15. Refer to Data Set 7 in Appendix B and find the best multiple regression equation with corn production as the dependent variable. (Don't include year as an independent variable.) Is this equation suitable for predicting Iowa's corn production? Why or why not?

16. Refer to Data Set 4 in Appendix B and find the best multiple regression equation with nicotine as the dependent variable. Is this equation suitable for predicting the amount of nicotine in a cigarette based on the amount of tar and carbon monoxide?

9-5 Exercises B: Beyond the Basics

17. In some cases, the best-fitting multiple regression equation is of the form $\hat{y} = b_0 + b_1x + b_2x^2$. The graph of such an equation is a parabola. Using the data set listed in the margin, let $x_1 = x$, let $x_2 = x^2$, and find the multiple regression equation for the parabola that best fits the given data. Based on the value of the multiple coefficient of determination, how well does this equation fit the data?

x	1	3	4	7	5
y	5	14	19	42	26

9. a. $\hat{y} = 3.56 + 0.0980x_1$
    b. 0.021; 0.005; 0.256
    c. No

10. a. $\hat{y} = 1.08 + 1.38x_1$
    b. 0.563; 0.556; 0.000
    c. Yes

11. a. $\hat{y} = 1.15 + 1.41x_1 - 0.0144x_2$
    b. 0.564; 0.549; 0.000
    c. Yes

12. a. $\hat{y} = 0.628 + 0.499x_1 + 1.13x_2 - 0.0391x_3$
    b. 0.616; 0.596; 0.000
    c. Yes

13. a. $\hat{y} = 0.554 + 0.491x_1 + 1.08x_2$
    b. 0.613; 0.600; 0.000
    c. Yes

14. a. $\hat{y} = 0.988 + 0.200x_1 + 0.222x_2$
    b. 0.343; 0.321; 0.000
    c. Yes

15. $\hat{y} = -1399 + 0.217x_1$; yes.

16. $\hat{y} = 0.154 + 0.0651x_1$; yes.

17. $\hat{y} = 2.17 + 2.44x + 0.464x^2$. Because $R^2 = 1$, the parabola fits perfectly.

· ·

Vocabulary List

correlation
bivariate data
bivariate normal distribution
scatter diagram
linear correlation coefficient
Pearson product moment
  correlation coefficient
lurking variable
centroid
regression equation
regression line
independent variable
predictor variable
dependent variable
response variable
marginal change
outlier
influential point

residual
least-squares property
total deviation
explained deviation
unexplained deviation
total variation
explained variation
unexplained variation
coefficient of determination
prediction interval
standard error of estimate
multiple regression equation
predictor variables
multiple coefficient of
  determination
adjusted coefficient of
  determination
stepwise regression

· ·

Review

In this chapter we presented basic methods for using paired data to investigate the relationship between the two corresponding variables. Throughout this chapter, we limited our discussion to linear relationships because consideration of nonlinear relationships requires more advanced mathematics.

- Section 9-2 used scatter diagrams and the linear correlation coefficient to decide whether there is a linear correlation between two variables.

- Section 9-3 presented methods for finding the equation of the regression line, which (by the least-squares criterion) best fits the paired data. When there is a significant linear correlation, the regression line can be used to predict the value of a variable, given some value of the other variable.

- Section 9-4 introduced the concept of total variation, with components of explained and unexplained variation. We defined the coefficient of determination r^2 to be the quotient obtained by dividing explained variation by total variation. We also developed methods for constructing prediction intervals, which are helpful in judging the accuracy of predicted values.

- In Section 9-5 we considered multiple regression, which allows us to investigate relationships among several variables. We discussed procedures for

obtaining a multiple regression equation, as well as the values of the multiple coefficient of determination R^2, the adjusted R^2, and a P-value for the overall significance of the equation.

• •

Review Exercises

In Exercises 1–4, use the data in the accompanying table. The data come from a study of ice cream consumption that spanned the springs and summers of three years. The ice cream consumption is in pints per capita per week, price of the ice cream is in dollars, family income of consumers is in dollars per week, and temperature is in degrees Fahrenheit.

Consumption	0.386	0.374	0.393	0.425	0.406	0.344	0.327	0.288	0.269	0.256
Price	1.35	1.41	1.39	1.40	1.36	1.31	1.38	1.34	1.33	1.39
Income	351	356	365	360	342	351	369	356	342	356
Temperature	41	56	63	68	69	65	61	47	32	24

Based on data from Kadiyala, *Econometrica*, Vol. 38.

1. **a.** Use a 0.05 significance level to test for a linear correlation between consumption and price.
   b. Find the equation of the regression line that expresses consumption (y) in terms of price (x).
   c. What is the best predicted consumption amount if the price is $1.38?
2. **a.** Use a 0.05 significance level to test for a linear correlation between consumption and income.
   b. Find the equation of the regression line that expresses consumption (y) in terms of income (x).
   c. What is the best predicted consumption amount if the income is $365?
3. **a.** Use a 0.05 significance level to test for a linear correlation between consumption and temperature.
   b. Find the equation of the regression line that expresses consumption (y) in terms of temperature (x).
   c. What is the best predicted consumption amount if the temperature is 32° F?
4. Use software such as STATDISK or Minitab to find the multiple regression equation of the form $\hat{y} = b_0 + b_1x_1 + b_2x_2 + b_3x_3$, where the dependent variable y represents consumption, x_1 represents price, x_2 represents income, and x_3 represents temperature. Also identify the value of the multiple coefficient of determination R^2, the adjusted R^2, and the P-value representing the overall significance of the multiple regression equation. Can the regression equation be used to predict ice cream consumption? Are any of the equations from Exercises 1–3 better?

1. a. TS: $r = 0.338$.
   CV: $r = \pm 0.632$.
   b. $\hat{y} = -0.488 + 0.611x$
   c. 0.3468

2. a. TS: $r = 0.116$.
   CV: $r = \pm 0.632$.
   b. $\hat{y} = 0.0657 + 0.000792x$
   c. 0.3468

3. a. TS: $r = 0.777$.
   CV: $r = \pm 0.632$.
   b. $\hat{y} = 0.193 + 0.00293x$
   c. 0.2864

4. $\hat{y} = -0.0526 + 0.747x_1 - 0.00220x_2 + 0.00303x_3$; $R^2 = 0.726$; adjusted $R^2 = 0.589$; P-value = 0.040.

In Exercises 5–8, use the sample data in the accompanying table. The table includes "minutes before midnight on the doomsday clock," a measure of the perceived threat of nuclear war as established by the Bulletin of the Atomic Scientists. The "periodicals" values are indices of print media coverage of issues related to nuclear war. People's "savings" values are expressed in terms of a standard measure of the rate of savings. The data are matched for randomly selected years.

Minutes	2.0	7.0	12.0	12.0	9.25	11.17	9.0	9.0	4.0
Periodicals	0.16	0.21	0.13	0.12	0.12	0.05	0.05	0.06	0.14
Savings	7.2	10.2	13.5	13.5	12.0	13.1	11.6	11.6	8.5

Based on data from "Saving and the Fear of Nuclear War," by Slemrod, *Journal of Conflict Resolution*, Vol. 30, No. 3.

5. a. Use a 0.05 significance level to test for a linear correlation between minutes before midnight on the doomsday clock and the index of savings.
   b. Find the equation of the regression line that expresses savings (*y*) in terms of minutes before midnight (*x*).
   c. What is the best predicted index of savings for a year in which the doomsday clock is at 5.0 min before midnight?

6. a. Use a 0.05 significance level to test for a linear correlation between the periodical index and the index of savings.
   b. Find the equation of the regression line that expresses the savings index (*y*) in terms of the periodical index (*x*).
   c. What is the best predicted index of savings for a year in which the periodical index is 0.20?

7. Use only the paired minutes/savings data. For a year in which the number of minutes before midnight on the doomsday clock is 5.0, find a 95% prediction interval estimate of the savings index.

8. Let y = savings index, x_1 = minutes before midnight, and x_2 = index of periodicals. Use software such as STATDISK or Minitab to find the multiple regression equation of the form $\hat{y} = b_0 + b_1 x_1 + b_2 x_2$. Also identify the value of the multiple coefficient of determination R^2, the adjusted R^2, and the P-value representing the overall significance of the multiple regression equation. Based on the results, should the multiple regression equation be used for making predictions? Why or why not?

5. a. TS: $r = 0.999$.
   CV: $r = \pm 0.666$.
   b. $\hat{y} = 5.92 + 0.636 x_1$
   c. 9.10

6. a. TS: $r = -0.483$.
   CV: $r = \pm 0.666$.
   b. $\hat{y} = 13.5 - 19.8 x_2$
   c. 11.2

7. $8.81 < y < 9.38$

8. $\hat{y} = 6.06 + 0.630 x_1 -$
   $0.815 x_2$; $R^2 = 0.998$; adjusted
   $R^2 = 0.998$; P-value = 0.000.

Cumulative Review Exercises

1. In 1970, the mean time between eruptions of the Old Faithful geyser was 66 minutes. Refer to the intervals (in minutes) between eruptions for the recent data listed in Data Set 16 of Appendix B.
   a. Test the claim of Yellowstone National Park geologist Rick Hutchinson that eruptions now occur at intervals that are longer than in 1970.

1. $n = 50$, $\bar{x} = 80.7$, $s = 12.0$.
   a. TS: $z = 8.66$.
   CV: $z = 1.645$.
   b. $77.3 < \mu < 84.0$

b. Construct a 95% confidence interval for the mean time between eruptions.

2. The Marc Michael Advertising Company has prepared two different television commercials for Taylor's women's jeans. One commercial is humorous and the other is serious. A test screening involves a standard scale with higher scores indicating more favorable responses. The results are listed here. Based on the results, does one commercial seem to be better? Is the issue of correlation relevant to this situation?

Consumer	A	B	C	D	E	F	G	H
Humorous commercial	26	33	19	18	29	27	23	24
Serious commercial	21	30	14	14	24	22	21	19

2. TS: $t = 10.319$.
   CV: $t = \pm 2.365$.

3. In studying the effects of heredity and environment on intelligence, it has been helpful to analyze IQs of identical twins who were separated soon after birth. Identical twins share identical genes inherited from the same fertilized egg. By studying identical twins raised apart, we can eliminate the variable of heredity and better isolate the effects of environment. The accompanying table shows the IQs of identical twins (older twins are x) raised apart. Use the sample data to determine whether there is a relationship between IQs of the twins. Write a summary statement about the effect of heredity and environment on intelligence. Note that your conclusions will be based on this relatively small sample of 12 pairs of identical twins.

3. TS: $r = 0.702$.
   CV: $r = \pm 0.576$.

x	107	96	103	90	96	113	86	99	109	105	96	89
y	111	97	116	107	99	111	85	108	102	105	100	93

Based on data from "IQs of Identical Twins Reared Apart," by Arthur Jensen, *Behavioral Genetics*.

· ·

Computer Project

Use STATDISK or Minitab to retrieve the bear data found in Data Set 3 of Appendix B. (Data Set 3 is already stored on STATDISK or the Data/Program Disk.) Using the complete data set, find the linear correlation coefficient and equation of the regression line when the dependent variable of weight is paired with (a) head length, (b) head width, (c) neck size, (d) length, and (e) chest size. Which of the preceding regression equations is best for predicting the weight of a bear given its head length, head width, neck size, length, and chest size? Also, find the multiple regression equation obtained when the dependent variable of weight is expressed in terms of the independent variables of head length, head width, neck size, and chest size. Interpret the computer results. Is the resulting multiple regression equation suitable for making predictions of a bear's weight based on its head length, head width, neck size, and chest size? Why or why not?   *(continued)*

When hiking (with a tape measure but no scale) on the Grizzly Lake trail in Yellowstone National Park, the author once encountered a bear which he wrestled to the ground and rendered temporarily unconscious. (Editor's note: The truth is that he saw a bear 250 yd away.) Given all of the preceding results, what is the best predicted weight of the bear that was measured and found to have a 12.3 in. head length, a 4.9 in. head width, a 19.0 in. neck size, a 55.3 in. length, and a 31.4 in. chest size?

•••••••• **FROM DATA TO DECISION** ••••••••

Which Computer Should You Buy?

The accompanying table lists prices (in dollars) matched with ratings of performance and features for a random sample of midrange computers. Use the methods of this chapter to analyze the sample data. Determine whether there is a correlation between price and rating. How much should a computer cost if its performance/feature rating is 105? What is the highest rating you can afford if you have only $3000? Which is the better buy: A Dell Dimension P133c computer (designated as Delc) or a Robotech P133 (designated as Robo)? Why?

Computer	Rating	Price	Computer	Rating	Price
Mem	81	2300	IBM	109	3500
NMC	80	2000	AMAX	104	4000
Blue	86	2400	Gate	103	2425
Swan	87	2600	Delx	103	2990
HP-X	85	2800	Robo	98	3300
GST	94	2500	Vekt	98	2300
Uni	90	2700	HiQ	95	2750
Sys	96	2300	Delc	100	2400
EPS	96	2300	HP-V	94	3300
Mega	88	2100	ACMA	96	3200

Based on data from *PC Magazine*.

•••••••• COOPERATIVE GROUP ACTIVITIES ••••••••

1. *Out-of-class activity:* Divide into groups of 3 or 4 people. Investigate the relationship between two variables by collecting your own paired sample data and using the methods of this chapter to determine whether there is a significant linear correlation. Also identify the regression equation and describe a procedure for predicting values of one of the variables when given values of the other variable. Suggested topics:

   - Is there a relationship between taste and cost of different brands of chocolate chip cookies (or colas)? (Taste can be measured on some scale, such as 1 to 10.)

   - Is there a relationship between salaries of professional baseball (or basketball, or football) players and their season achievements?

   - Rates versus weights: Is there a relationship between car fuel consumption rates and car weights? If so, what is it?

   - Is there a relationship between the lengths of men's (or women's) feet and their heights?

   - Is there a relationship between student grade-point averages and the amount of television watched? If so, what is it?

   - Is there a relationship between heights of fathers (or mothers) and heights of their first sons (or daughters)?

2. *In-class activity:* Divide into groups of 8 to 12 people. For each group member, *measure* his or her height and *measure* his or her arm span. For the arm span, the subject should stand with arms extended, like the wings on an airplane. It's easy to mark the height and arm span on a chalkboard, then measure the distances there. Using the paired sample data, is there a correlation between height and arm span? If so, find the regression equation with height expressed in terms of arm span. Can arm span be used as a reasonably good predictor of height?

3. *In-class activity:* Divide into groups of 8 to 12 people. For each group member, record the pulse rate by counting the number of heart beats in one minute. Also record height. Is there a relationship between pulse rate and height? If so, what is it?

4. *In-class activity:* Divide into groups of 8 to 12 people. For each group member, use a string and ruler to measure head circumference and forearm length. Is there a relationship between these two variables? If so, what is it?

interview

David Hall

Division Statistical Manager at the Boeing Commercial Airplane Group

David Hall is Division Statistical Manager, Renton Division, Boeing Commercial Airplane Group. He manages the Statistical Methods Organization, which focuses on applying statistical and other quality technology techniques to continuous quality improvement. Before joining Boeing, he worked at Battelle Pacific Northwest Laboratories, where he was manager of a statistical applications group.

How extensive is the use of statistics at Boeing, and is it increasing, decreasing, or remaining about the same?

The use of statistics is extensive and definitely increasing. I'm sure that this is true of the aircraft industry in general. People are becoming very aware of the need to improve, and to improve you must have data. Statistics is riding the wave of quality improvement. The use of control charts and, more generally, statistical process control is increasing. Designed experiments are very common. In the beginning, 99% of the useful statistical tools are very, very simple. As the processes get refined and understanding increases, more sophisticated tools are required. Regression and correlation analysis, analysis of variance, contingency tables, hypothesis testing, confidence intervals, and time series analysis— virtually all techniques are used at some time. But given the state most American manufacturing is in now, incredible gains can be made with the simplest tools, such as Pareto diagrams and run charts.

How do you sample at Boeing?

Our Quality Assurance Department uses many of the well-known sampling schemes for inspection. Currently we check the daylights out of everything. However sampling is also involved in statistical process control applications. The sampling can be quite complex and is handled on a case-by-case basis.

> **"Statistics is riding the wave of quality improvement."**

Could you cite a specific example of how the use of statistics was helpful at Boeing?

We were working with our Fabrication Division to produce more consistent hydropress-formed parts. Through the use of designed experiments, we found that the type of rubber placed over the blanks during forming could drastically reduce the part-to-part variation. That's one simple example of what goes on all the time. An example of a more sophisticated application was the use of bootstrapping to estimate the variability in wind tunnel tests.

In maintaining quality, do you aim for zero defects or is it more efficient for you to allow for defects that must be rejected or reworked?

It's never more efficient to allow defects that get inspected out. That is far too costly. We constantly preach that we should produce to target. Things have to be right, and it's been our big push to reduce variation. Also, in dealing with outside suppliers, we have a program to get suppliers to continually reduce variation about the target value. Sometimes in electronics manufacturing they allow a high defect rate, but for us that is totally unacceptable.

Do you feel job applicants are viewed more favorably if they have studied some statistics?

We naturally have a great many engineers at Boeing and almost all of them have studied some statistics. We are now expecting our managers to have more familiarity with statistics than ever before. They are expected to understand variation and how to effectively use data.

Do you have any advice for today's students?

When I get new statisticians just out of school, whether they have a B.S., M.S., or Ph.D., they still have a tremendous amount to learn before they are effective in our environment. Most of what they need involves people skills, team building, planning, and communication. Right now, American industry is crying out for people with an understanding of statistics and the ability to communicate its use.

10

Multinomial Experiments and Contingency Tables

10-1 Overview

Chapter objectives are identified. This chapter deals with sample data consisting of frequency counts arranged in one row with at least three categories (one-way frequency table), or a table with at least two rows and at least two columns (two-way frequency table).

10-2 Multinomial Experiments: Goodness-of-Fit

The goodness-of-fit procedure of hypothesis testing is presented for sample data in a one-way frequency table. This procedure is used to test claims that observed sample frequencies fit, or conform to, a particular distribution.

10-3 Contingency Tables: Independence and Homogeneity

Contingency tables (or two-way frequency tables) are defined and described. A standard method is presented for testing claims that in a contingency table, the row variable and the column variable are independent. Also presented is a method for a test of homogeneity, in which we test a claim that different populations have the same proportions of some characteristics.

Chapter Problem:

Are noncombat mortality rates the same for military personnel in a combat zone as for those military personnel not so deployed?

A study compared noncombat mortality rates for U.S. military personnel who were deployed in combat situations to those not deployed. Table 10-1 summarizes sample data for deaths in three different categories. Analysis of the table entries shows that among those who died from unintentional injuries (from cars, aircraft, explosions, and so on), 19% were deployed; among those who died from illness, 10% were deployed; among those who died from homicide or suicide, 3% were deployed. Are those differences significant?

After the military operations of Desert Shield and Desert Storm in the Persian Gulf, there were claims of dramatic increases in death rates of U.S. troops who served in that region. We can address such claims by analyzing data of the type listed in Table 10-1. Section 10-3 will consider this particular set of sample data.

Table 10-1   Noncombat Causes of Death for Deployed and Nondeployed Military Personnel

	Cause of Death		
	Unintentional Injury	Illness	Homicide or Suicide
Deployed	183	30	11
Not deployed	784	264	308

Table values are based on data from "Comparative Mortality Among U.S. Military Personnel in the Persian Gulf Region and Worldwide During Operations Desert Shield and Desert Storm," by Writer, DeFraites, and Brundage, *Journal of the American Medical Association*, Vol. 275, No. 2.

10-1 Overview

Coverage of at least Section 10-3 is strongly recommended. Analysis of survey results often involves the use of contingency tables, so the concepts of Section 10-3 are used frequently in real applications. Both Sections 10-2 and 10-3 require the χ^2 distribution, so if it has not yet been introduced, be sure to provide a thorough description of it.

Chapter 1 defined categorical (or qualitative, or attribute) data as data that can be separated into different categories (often called **cells**) that are distinguished by some nonnumeric characteristic. For example, we might separate a sample of M&Ms into the color categories of red, orange, yellow, brown, blue, and green. After finding the frequency count for each category, we might proceed to test the claim that the frequencies fit (or agree with) the color distribution claimed by the manufacturer (Mars, Inc.). In general, this chapter focuses on the analysis of categorical data consisting of frequency counts for the different categories. In Section 10-2 we will consider multinomial experiments, which consist of observed frequency counts arranged in a single row or column (called a one-way frequency table). In Section 10-3 we will consider contingency tables (or two-way frequency tables), which consist of frequency counts arranged in a table such as Table 10-1.

We will see that the two major topics of this chapter (multinomial experiments and contingency tables) use the same χ^2 (chi-square) test statistic. Recall the following important properties of the chi-square distribution:

1. Unlike the normal and Student t distributions, the chi-square distribution is not symmetric. (See Figure 10-1.)

2. The values of the chi-square distribution can be 0 or positive, but they cannot be negative. (See Figure 10-1.)

3. The chi-square distribution is different for each number of degrees of freedom. (See Figure 10-2.)

Critical values of the chi-square distribution are found in Table A-4.

In Chapters 6 and 7 the number of degrees of freedom was always $n - 1$, and we noted that we should not universally equate the number of degrees of freedom with the value of $n - 1$. In this chapter, the number of degrees of freedom is *not* $n - 1$.

10-2 Multinomial Experiments: Goodness-of-Fit

Section 4-3 defined a binomial experiment to be an experiment with a fixed number of independent trials, probabilities of outcomes that remain constant from trial to trial, and outcomes that belong to one of two categories. The requirement of *two* categories is reflected in the prefix *bi*, which begins the term *binomial*. In this section we consider multinomial experiments, in which each trial yields outcomes belonging to one of several (more than two) categories. Except for this difference, binomial and multinomial experiments are essentially the same.

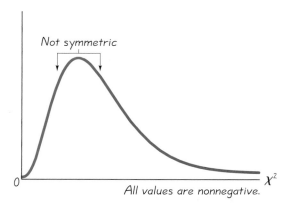

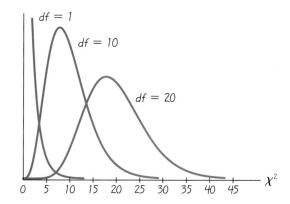

Figure 10-1   The Chi-Square Distribution

Figure 10-2   Chi-Square Distributions for 1, 10, and 20 Degrees of Freedom

---

DEFINITION

A **multinomial experiment** is an experiment that meets the following conditions.

1. The number of trials is fixed.
2. The trials are independent.
3. All outcomes of each trial must be classified into exactly one of several different categories.
4. The probabilities for the different categories remain constant for each trial.

In this section we present a method for testing a claim that in a multinomial experiment, the frequencies observed in the different categories fit a particular distribution. Because we test for how well an observed frequency distribution conforms to some theoretical distribution, this method is often called a goodness-of-fit test.

DEFINITION

A **goodness-of-fit test** is used to test the hypothesis that an observed frequency distribution fits (or conforms to) some claimed distribution.

For example, we can use a TI-83 calculator to randomly generate integers between 0 and 9 inclusive. We can test a sample of results to determine whether they fit a discrete uniform distribution in which the digits 0, 1, 2, . . ., 9 occur with the same frequency. Here are some other examples of uses of the goodness-of-fit test:

- Test the claim that New York State Lottery numbers (1, 2, . . . , 54) occur with the same frequency.
- Test Mars, Inc.'s claim that its plain M&M candies are made so that 30% are brown, 20% are yellow, 20% are red, 10% are orange, 10% are green, and 10% are blue.

Our goodness-of-fit tests will incorporate the following notation.

Notation

O represents the *observed frequency* of an outcome.

E represents the *expected frequency* of an outcome.

k represents the *number of different categories* or outcomes.

n represents the total *number of trials*.

It would be helpful to briefly review the concept of *expected value*, introduced in Section 4-2. Present a few simple and obvious examples such as this: Find the expected number of girls born in groups of 100 babies. When students respond with the correct answer of 50, ask them to describe the exact thought process that led to the answer. They will respond that they found 1/2 of 100, which can be generalized as $p \cdot n$, which leads to $E = np$. Also point out that the expected number of girls in 3 babies is 1.5, so the expected value need not be an integer.

In the typical situation requiring a goodness-of-fit test, we have *observed* frequencies (denoted by O) and must use the claimed distribution to determine the *expected* frequencies (denoted by E). In many cases, an expected frequency can be found by multiplying the probability p for a category by the number of different trials n, so

$$E = np$$

For example, if we test the claim that a die is fair by rolling it 60 times, we have $n = 60$ (because there are 60 trials) and $p = 1/6$ (because a die is fair if the six possible outcomes are equally likely with the same probability of 1/6). The expected frequency for each category or cell is therefore

$$E = np$$
$$= (60)(1/6) = 10$$

Assumptions   The following assumptions apply when we test a hypothesis that the population proportion for each of the k categories (in a multinomial experiment) is as claimed.

1. The data constitute a random sample.
2. The sample data consist of frequency counts for the k different categories.
3. For each of the k categories, the *expected* frequency is at least 5. (There is no requirement that every *observed* frequency must be at least 5.)

Emphasize that the second assumption requires that each *expected frequency* must be at least 5, but there is no requirement that *observed frequencies* must be at least 5.

We know that sample frequencies typically deviate somewhat from the values we theoretically expect, so we now present the key question: Are the differences between the actual *observed* values O and the theoretically *expected*

values E statistically significant? To answer this question we use the following test statistic, which measures the discrepancy between observed and expected frequencies.

Test Statistic for Goodness-of-Fit Tests in Multinomial Experiments

$$\chi^2 = \sum \frac{(O - E)^2}{E}$$

About the χ^2 notation: We usually use Greek letters for population parameters, but here we use χ^2 for a test statistic. Although not consistent, this is very common notation in this context. Someone once said that "consistency is the hobgoblin of small minds."

Critical Values

1. Critical values are found in Table A-4 by using $k - 1$ degrees of freedom, where k = number of categories.
2. Goodness-of-fit hypothesis tests are always right-tailed.

The form of the χ^2 test statistic is such that *close agreement* between observed and expected values will lead to a *small* value of χ^2. A large value of χ^2 will indicate strong disagreement between observed and expected values. A significantly large value of χ^2 will thus cause rejection of the null hypothesis of no difference between observed and expected frequencies. Our test is therefore right-tailed because the critical value and critical region are located at the extreme right of the distribution. Unlike previous hypothesis tests in which we had to determine whether the test was left-tailed, right-tailed, or two-tailed, these goodness-of-fit tests are all right-tailed.

Once we know how to find the values of the test statistic and critical value, we can test hypotheses by using the same procedure introduced in Chapter 7 and summarized in Figure 7-4.

Try to lead the class to a development of their own reasoning process for why the tests of this section are all right-tailed. Ask the class these questions: If there are large discrepancies between the observed frequencies and those that are expected, what do we know about the $O - E$ values? The $(O - E)^2$ values? The value of the χ^2 test statistic? Where on the χ^2 distribution do large discrepancies fall?

EXAMPLE   Many people believe that when a horse races, it has a better chance of winning if its starting line-up position is closer to the rail on the inside of the track. The starting position of 1 is closest to the inside rail, followed by position 2, and so on. Table 10-2 (on the following page) lists the numbers of wins for horses in the different starting positions. Test the claim that the probabilities of winning in the different post positions are not all the same.

SOLUTION   Table 10-2 lists results for 144 wins. If the chance of winning in each starting position is the same, the probability of winning for each position is $p = 1/8$ and the expected number of wins for each position is as follows:

$$E = np = (144)(1/8) = 18$$

(continued)

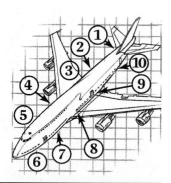

Safest Airplane Seats

Many of us believe that the rear seats are safest in an airplane crash. Safety experts do not agree that any particular part of an airplane is safer than others. Some planes crash nose first when they come down, but others crash tail first on takeoff. Matt McCormick, a survival expert for the National Transportation Safety Board, told *Travel* magazine that "There is no one safe place to sit." Goodness-of-fit tests can be used with a null hypothesis that all sections of an airplane are equally safe. Crashed airplanes could be divided into the front, middle, and rear sections. The observed frequencies of fatalities could then be compared to the frequencies that would be expected with a uniform distribution of fatalities. The χ^2 test statistic reflects the size of the discrepancies between observed and expected frequencies, and it would reveal whether some sections are safer than others.

TABLE 10-2	Starting Positions of Winning Horses							
	Starting Position							
	1	2	3	4	5	6	7	8
Number of wins	29	19	18	25	17	10	15	11

Based on data from the *New York Post*.

The observed frequencies (denoted by O) are 29, 19, 18, . . . , 11, but the expected frequencies (denoted by E) are 18, 18, 18, . . . , 18. Having identified the observed and expected frequencies, we will now follow our standard procedure for testing hypotheses.

Step 1:   The original claim is that the probabilities of winning in the different starting positions are not all the same. That is, at least one of the probabilities $p_1, p_2, \ldots, p_8$ is different from the others.

Step 2:   If the original claim is false, then all of the probabilities are the same. That is, $p_1 = p_2 = \cdots = p_8$.

Step 3:   The null hypothesis must contain the condition of equality, so we have

$$H_0: p_1 = p_2 = p_3 = p_4 = p_5 = p_6 = p_7 = p_8$$

H_1: At least one of the probabilities is different from the others.

Step 4:   No significance level was specified, so we select $\alpha = 0.05$, a very common choice.

Step 5:   Because we are testing a claim about the distribution of the wins in the different post positions, we use the goodness-of-fit test described in this section. The χ^2 distribution is used with the test statistic given earlier.

Step 6:   Using the observed frequencies O and the expected frequencies E, we compute the value of the χ^2 test statistic, as shown in Table 10-3. The test statistic is $\chi^2 = 16.333$ (rounded). The critical value is $\chi^2 = 14.067$ (found in Table A-4 with $\alpha = 0.05$ in the right tail and degrees of freedom equal to $k - 1 = 7$). The test statistic and critical value are shown in Figure 10-3.

Step 7:   Because the test statistic falls within the critical region, there is sufficient evidence to reject the null hypothesis.

Starting Position	Observed Frequency O	Expected Frequency E	$O - E$	$(O - E)^2$	$\dfrac{(O - E)^2}{E}$
1	29	18	11	121	6.7222
2	19	18	1	1	0.0556
3	18	18	0	0	0
4	25	18	7	49	2.7222
5	17	18	−1	1	0.0556
6	10	18	−8	64	3.5556
7	15	18	−3	9	0.5000
8	11	18	−7	49	2.7222

TABLE 10-3   Calculating the χ^2 Test Statistic for the Horse Race Data

144   144

(These two totals must agree.)

$$\chi^2 = \sum \frac{(O - E)^2}{E} = 16.3334$$

Step 8:   There is sufficient evidence to support the claim that the probabilities of winning in the different starting positions are not all the same. It appears that starting position should be considered when trying to select which horse will win a race.

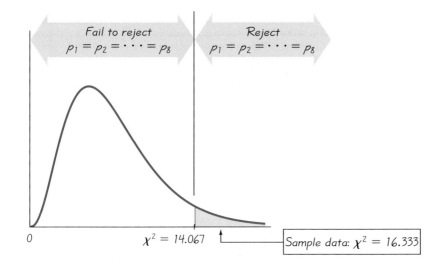

Figure 10-3   Goodness-of-Fit Test of $p_1 = p_2 = \cdots = p_8$

Fail to reject $p_1 = p_2 = \cdots = p_8$

Reject $p_1 = p_2 = \cdots = p_8$

$\chi^2 = 14.067$

Sample data: $\chi^2 = 16.333$

0

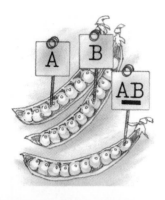

Students often have difficulty finding the expected frequencies when the proportions are not all equal, as in this example. It would be helpful to carefully explain how this is done.

The techniques in this section can be used to test that an observed frequency distribution conforms to some theoretical frequency distribution. For the winning post position horse race data, we used a goodness-of-fit test to decide whether the observed frequencies conformed to a uniform distribution, and we found that they did not. Because many statistical analyses require a normally distributed population, we can use the chi-square test in this section to determine whether given samples are drawn from normally distributed populations (see Exercise 22).

The preceding example dealt with the null hypothesis that the probabilities of winning in the different starting positions are all the same. The methods of this section can also be used when the hypothesized probabilities (or frequencies) are different, as in the next example.

EXAMPLE   Mars, Inc., claims that its M&M plain candies are distributed with the color percentages of 30% brown, 20% yellow, 20% red, 10% orange, 10% green, and 10% blue. The colors of the M&Ms listed in Data Set 11 of Appendix B are summarized in Table 10-4. Using the sample data and a 0.05 significance level, test the claim that the color distribution is as claimed by Mars, Inc.

TABLE 10-4	Frequencies of M&M Plain Candies					
	Brown	Yellow	Red	Orange	Green	Blue
Observed frequency	33	26	21	8	7	5
Expected frequency	30	20	20	10	10	10

SOLUTION   We extended Table 10-4 to include the expected frequencies, which are calculated as follows. For n, we use the total number of trials (100), which is the total number of M&Ms observed in the sample. For the probabilities, we use the decimal equivalents of the claimed percentages (30%, 20%, . . . , 10%).

Brown: $E = np = (100)(0.30) = 30$

Yellow: $E = np = (100)(0.20) = 20$

$$\vdots$$

Blue: $E = np = (100)(0.10) = 10$

In testing the given claim, Steps 1, 2, and 3 result in the following hypotheses:

H_0: $p_{br} = 0.3$ and $p_y = 0.2$ and $p_r = 0.2$ and $p_o = 0.1$ and $p_g = 0.1$ and $p_{bl} = 0.1$

H_1: At least one of the above proportions is different from the claimed value.

Steps 4, 5, and 6 lead us to use the goodness-of-fit test with a 0.05 significance level and a test statistic calculated from Table 10-5.

TABLE 10-5	Calculating the χ^2 Test Statistic for M&M Data				
Color Category	Observed Frequency O	Expected Frequency $E = np$	$O - E$	$(O - E)^2$	$\dfrac{(O - E)^2}{E}$
Brown	33	30	3	9	0.3000
Yellow	26	20	6	36	1.8000
Red	21	20	1	1	0.0500
Orange	8	10	−2	4	0.4000
Green	7	10	−3	9	0.9000
Blue	5	10	−5	25	2.5000
	100	100		$\chi^2 = \Sigma \dfrac{(O - E)^2}{E} = 5.9500$	↑

The test statistic is $\chi^2 = 5.950$. The critical value of χ^2 is 11.071, and it is found in Table A-4 (using $\alpha = 0.05$ in the right tail with $k - 1 = 5$ degrees of freedom). The test statistic and critical value are shown in Figure 10-4 (see the following page). Because the test statistic does not fall within the critical region, there is not sufficient evidence to warrant rejection of the null hypothesis. There is not sufficient evidence to warrant rejection of the claim that the colors are distributed with the percentages given by Mars, Inc.

In Figure 10-5 on the next page we graph the claimed proportions of 0.30, 0.20, . . . , 0.10, along with the observed proportions of 33/100, 26/100, 21/100, 8/100, 7/100, 5/100, so that we can visualize the discrepancy between the distribution that was claimed and the frequencies that were observed. The points along the solid line represent the claimed proportions, and the points along the broken line represent the observed proportions. The corresponding pairs of points are all fairly close, showing that all of the expected frequencies are reasonably close to the corresponding observed frequencies. In general, graphs such as Figure 10-5 are helpful in visually comparing expected frequencies and observed frequencies, as well as suggesting which categories result in the major discrepancies.

Figure 10-4 Goodness-of-Fit Test of $p_{br} = 0.3$ and $p_y = 0.2$ and $p_r = 0.2$ and $p_o = 0.1$ and $p_g = 0.1$ and $p_{bl} = 0.1$

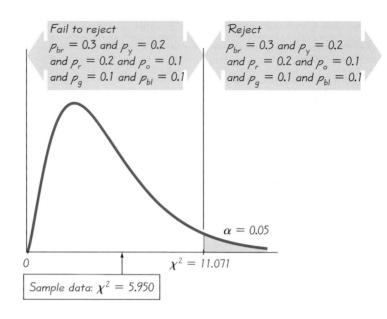

Figure 10-5 Comparison of Claimed and Observed Proportions

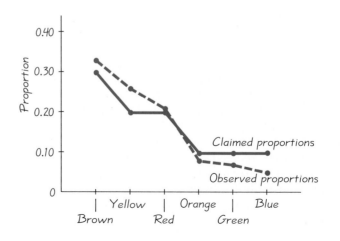

The preceding examples should be helpful in developing a sense for the role of the χ^2 test statistic. It should be clear that we want to measure the amount of disagreement between observed and expected frequencies. Simply summing the differences between observed and expected values does not result in an effective measure because that sum is always 0, as shown below.

$$\Sigma(O - E) = \Sigma O - \Sigma E = n - n = 0$$

Squaring the $O - E$ values provides a better statistic, which reflects the differences between observed and expected frequencies. (The reasons for squaring the $O - E$ values are essentially the same reasons for squaring the $x - \bar{x}$ values in the formula for standard deviation.) The value of $\Sigma(O - E)^2$ measures only the magnitude of the differences, but we need to find the magnitude of the differences relative to what was expected. This relative magnitude is found through division by the expected frequencies, as in the test statistic.

The theoretical distribution of $\Sigma(O - E)^2/E$ is a discrete distribution because the number of possible values is limited. The distribution can be approximated by a chi-square distribution, which is continuous. This approximation is generally considered acceptable, provided that all values of E are at least 5. We included this requirement with the assumptions that apply to this section. In Section 5-6 we saw that the continuous normal probability distribution can reasonably approximate the discrete binomial probability distribution, provided that np and nq are both at least 5. We now see that the continuous chi-square distribution can reasonably approximate the discrete distribution of $\Sigma(O - E)^2/E$, provided that all values of E are at least 5. (There are ways of circumventing the problem of an expected frequency that is less than 5, such as combining categories so that all expected frequencies are at least 5.)

The number of degrees of freedom reflects the fact that we can freely assign frequencies to $k - 1$ categories before the frequency for every category is determined. Although we say that we can "freely" assign frequencies to $k - 1$ categories, we cannot have negative frequencies nor can we have frequencies so large that their sum exceeds the total of the observed frequencies for all categories combined.

Delaying Death

University of California sociologist David Phillips has studied the ability of people to postpone their death until after some important event. Analyzing death rates of Jewish men who died near Passover, he found that the death rate dropped dramatically in the week before Passover, but rose the week after. He found a similar phenomenon occurring among Chinese-American women; their death rate dropped the week before their important Harvest Moon Festival, then rose the week after.

P-Values

The examples in this section used the traditional approach to hypothesis testing, but the P-value approach can also be used. P-values can be obtained by using the same methods described in Sections 7-3 and 7-6. For instance, the preceding example resulted in a test statistic of $\chi^2 = 5.950$. That example had $k = 6$ categories, so there were $k - 1 = 5$ degrees of freedom. Referring to Table A-4, we see that for the row with 5 degrees of freedom, the test statistic of 5.95 is less than the lowest right-tailed critical value of 9.236, so the P-value is greater than 0.10. If the preceding example is run on STATDISK, the display will include a P-value of 0.31111. A TI-83 calculator can be used to determine that for $\chi^2 = 5.950$ and 5 degrees of freedom, the P-value is 0.311.

The high P-value suggests that the null hypothesis should not be rejected. Remember, we reject the null hypothesis only when the P-value is equal to or less than the significance level.

Computer Software for Multinomial Experiments

STATDISK is designed to analyze multinomial experiments, but they cannot be analyzed with Minitab or the TI-83 calculator. To use STATDISK, select `Analysis` from the main menu bar, then select the option of `Multinomial Experiments`.

10-2 Exercises A: Basic Skills and Concepts

Recommended assignment: Exercises 1–3, 8, and 12 (if the Poisson distribution was covered). (Exercises 13–16 require more time because they involve the reorganization of data sets from Appendix B).

1. Test statistic: $\chi^2 = 156.500$. Critical value: $\chi^2 = 21.666$.

2. Test statistic: $\chi^2 = 11.533$. Critical value: $\chi^2 = 16.919$.

3. Test statistic: $\chi^2 = 4.600$. Critical value: $\chi^2 = 7.815$.

1. One common way to test for authenticity of data is to analyze the frequencies of digits. When people are weighed and their weights are rounded to the nearest pound, we expect the last digits 0, 1, 2, . . . , 9 to occur with about the same frequency. In contrast, if people are asked how much they weigh, the digits 0 and 5 tend to occur at higher rates. The author randomly selected 80 students, obtained their weights, and recorded only the last digits, with the results shown in the following table. At the 0.01 significance level, test the claim that the last digits occur with the same frequency. Based on the results, does it appear that the students were actually weighed or were they asked to report their weights?

Last digit	0	1	2	3	4	5	6	7	8	9
Frequency	35	0	2	1	4	24	1	4	7	2

2. A TI-83 calculator is used to generate random integers between 0 and 9 inclusive, and the following results are obtained. Use a 0.05 significance level to test the claim that the TI-83 generates digits so that they are uniformly distributed.

Digit	0	1	2	3	4	5	6	7	8	9
Frequency	32	29	37	21	41	24	33	32	25	26

3. An exercise in Section 3-4 involved four car-pooling students who missed a test and gave an excuse of a flat tire. It was noted that on the makeup test, the instructor asked the students to identify the particular tire that went flat. If they really didn't have a flat tire, would they be able to identify the same tire? The author asked 40 students to identify the tire they would select. The results are listed in the following table (except for one student who selected the spare). Use a 0.05 level of significance to test the

author's claim that the results fit a uniform distribution. What does the result suggest about the ability of students to select the same tire when they really didn't have a flat?

Tire	Left front	Right front	Left rear	Right rear
Number selected	11	15	8	6

4. It is a common belief that more fatal car crashes occur on certain days of the week, such as Friday or Saturday. A sample of motor vehicle deaths in Montana is randomly selected for a recent year. The numbers of fatalities for the different days of the week are listed below. At the 0.05 significance level, test the claim that accidents occur with equal frequency on the different days.

Day	Sun	Mon	Tues	Wed	Thurs	Fri	Sat
Number of fatalities	31	20	20	22	22	29	36

Based on data from the Insurance Institute for Highway Safety.

4. Test statistic: $\chi^2 = 9.233$.
Critical value: $\chi^2 = 12.592$.

5. Many people believe that fatal DWI crashes occur because of casual drinkers who tend to binge on Friday and Saturday nights, whereas others believe that fatal DWI crashes are caused by alcoholics who drink every day of the week. In a study of fatal car crashes, 216 cases are randomly selected from the pool in which the driver was found to have a blood alcohol content over 0.10. These cases are broken down according to the day of the week, with the results listed in the accompanying table. At the 0.05 significance level, test the claim that such fatal crashes occur on the different days of the week with equal frequency. Does the evidence support the theory that fatal DWI car crashes are due to casual drinkers or those who drink daily?

Day	Sun	Mon	Tues	Wed	Thurs	Fri	Sat
Number	40	24	25	28	29	32	38

Based on data from the Dutchess County STOP-DWI Program.

5. Test statistic: $\chi^2 = 7.417$.
Critical value: $\chi^2 = 12.592$.

6. A study was made of 147 industrial accidents that required medical attention. Among those accidents, 31 occurred on Monday, 42 on Tuesday, 18 on Wednesday, 25 on Thursday, and 31 on Friday (based on results from "Counted Data CUSUM's," by Lucas, *Technometrics,* Vol. 27, No. 2). Test the claim that accidents occur with equal proportions on the five workdays. If they are not the same, what factors might explain the differences?

6. Test statistic: $\chi^2 = 10.653$.
Critical value: $\chi^2 = 9.488$ (assuming $\alpha = 0.05$).

7. Use a 0.05 significance level and the industrial accident data from Exercise 6 to test the claim of a safety expert that accidents are distributed on workdays as follows: 30% on Monday, 15% on Tuesday, 15% on

7. Test statistic: $\chi^2 = 23.431$.
Critical value: $\chi^2 = 9.488$.

Wednesday, 20% on Thursday, and 20% on Friday. Does rejection of that claim provide any help in correcting the industrial accident problem?

8. The Gleason Supermarket's manager must decide how much of each ice cream flavor he should stock so that customer demands are satisfied but unwanted flavors don't result in waste. The ice cream supplier claims that among the four most popular flavors, customers have these preference rates: 62% prefer vanilla, 18% prefer chocolate, 12% prefer neapolitan, and 8% prefer vanilla fudge. A random sample of 200 customers produces the results below. At the $\alpha = 0.05$ significance level, test the claim that the supplier has correctly identified customer preferences.

Flavor	Vanilla	Chocolate	Neapolitan	Vanilla Fudge
Customers	120	40	18	22

Data are based on results from the International Association of Ice Cream Manufacturers.

9. The number π is an irrational number with the property that when we try to express it in decimal form, it requires an infinite number of decimal places and there is no pattern of repetition. In the decimal representation of π, the first 100 digits occur with the frequencies described in the table below. At the 0.05 significance level, test the claim that the digits are uniformly distributed.

Digit	0	1	2	3	4	5	6	7	8	9
Frequency	8	8	12	11	10	8	9	8	12	14

10. The number 22/7 is similar to π in the sense that they both require an infinite number of decimal places. However, 22/7 is a rational number because it can be expressed as the ratio of two integers, whereas π cannot. When rational numbers such as 22/7 are expressed in decimal form, there is a pattern of repetition. In the decimal representation of 22/7, the first 100 digits occur with the frequencies described in the table below. At the 0.05 significance level, test the claim that the digits are uniformly distributed. How does the result differ from that found in Exercise 9?

Digit	0	1	2	3	4	5	6	7	8	9
Frequency	0	17	17	1	17	16	0	16	16	0

11. Among drivers who have had a car crash in the last year, 88 are randomly selected and categorized by age, with the results listed in the table located at the top of the following page. If all ages have the same crash rate, we would expect (because of the age distribution of licensed drivers) the given categories to have 16%, 44%, 27%, and 13% of the subjects, respectively. At the 0.05 significance level, test the claim that the distribution of crashes conforms to the distribution of ages. Does any age group appear to have a disproportionate number of crashes?

Age	Under 25	25–44	45–64	Over 64
Drivers	36	21	12	19

Based on data from the Insurance Information Institute.

12. In analyzing hits by V-1 buzz bombs in World War II, South London was subdivided into regions, each with an area of 0.25 km^2. In Section 4-5 we presented an example and included a table of actual frequencies of hits and the frequencies expected with the Poisson distribution. Use the values listed here and test the claim that the actual frequencies fit a Poisson distribution. Use a 0.05 level of significance.

12. Test statistic: $\chi^2 = 0.976$. Critical value: $\chi^2 = 9.488$.

Number of bomb hits	0	1	2	3	4 or more
Actual number of regions	229	211	93	35	8
Expected number of regions (from Poisson distribution)	227.5	211.4	97.9	30.5	8.7

13. Refer to the Old Faithful geyser data in Data Set 16 of Appendix B. Test the claim that the time intervals are uniformly distributed among the five categories of 55–64, 65–74, 75–84, 85–94, 95–104.

13. Test statistic: $\chi^2 = 11.800$. Critical value: $\chi^2 = 9.488$ (assuming $\alpha = 0.05$).

14. Refer to Data Set 8 in Appendix B and record only the last digits of the pulse counts. Test for authenticity of the pulse counts by testing the claim that the last digits occur with equal frequency. Do the pulse counts appear to be authentic? (See Exercise 1.)

14. Test statistic: $\chi^2 = 36.851$. Critical value: $\chi^2 = 16.919$ (assuming $\alpha = 0.05$).

15. Refer to Data Set 10 in Appendix B and categorize the listed movies as poor (0.0 to 1.5 stars), fair (2.0 or 2.5 stars), good (3.0 or 3.5 stars), or excellent (4.0 stars). Movie critic James Harrington claims that movies are distributed evenly among the four categories. Test his claim at the 0.05 significance level.

15. Test statistic: $\chi^2 = 21.467$. Critical value: $\chi^2 = 7.815$.

16. Does Maryland select its lottery numbers in a way that is fair? Refer to Data Set 12 in Appendix B and use the 150 digits listed for the Pick Three lottery. Using 10 categories corresponding to the 10 possible digits, find the frequency for each category. At the 0.05 level of significance, test the claim that the lottery is fair in the sense that the 10 digits occur with the same frequency.

16. Test statistic: $\chi^2 = 4.400$. Critical value: $\chi^2 = 16.919$.

(10-2) Exercises B: Beyond the Basics

17. What do you know about the *P*-value for the hypothesis test in Exercise 16?

17. $0.10 < P$-value < 0.90 and it is close to 0.90.

18. In doing a test for goodness-of-fit as described in this section, suppose that we multiply each observed frequency by the same positive integer greater than 1. How is the critical value affected? How is the test statistic affected?

18. The critical value doesn't change, but the test statistic is multiplied by the same constant.

Critical values: The χ^2 critical value is 3.841 and it is approximately equal to the square of $z = 1.96$.

19. In this exercise we will show that a hypothesis test involving a multinomial experiment with only two categories is equivalent to a hypothesis test for a proportion (Section 7-5). Assume that a particular multinomial experiment has only two possible outcomes A and B with observed frequencies of f_1 and f_2, respectively.
   a.   Find an expression for the χ^2 test statistic, and find the critical value for a 0.05 significance level. Assume that we are testing the claim that both categories have the same frequency $(f_1 + f_2)/2$.
   b.   The test statistic

$$z = \frac{\hat{p} - p}{\sqrt{\dfrac{pq}{n}}}$$

is used to test the claim that a population proportion is equal to some value p. With the claim that $p = 0.5$, $\alpha = 0.05$, and

$$\hat{p} = \frac{f_1}{f_1 + f_2}$$

show that z^2 is equivalent to χ^2 [from part (a)]. Also show that the square of the critical z score is equal to the critical χ^2 value from part (a).

20. An observed frequency distribution is as follows:

Number of successes	0	1	2	3
Frequency	89	133	52	26

   a.   Assuming a binomial distribution with $n = 3$ and $p = 1/3$, use the binomial probability formula to find the probability corresponding to each category of the table.
   b.   Using the probabilities found in part (a), find the expected frequency for each category.
   c.   Use a 0.05 level of significance to test the claim that the observed frequencies fit a binomial distribution for which $n = 3$ and $p = 1/3$.

20. a. 0.296, 0.444, 0.222, 0.037
    b. 88.9, 133.3, 66.7, 11.1
    c. Test statistic: $\chi^2 = 23.241$.
        Critical value: $\chi^2 = 7.815$.
        Reject the claim that the observed frequencies fit a binomial distribution with $n = 3$ and $p = 1/3$.

21. Combine the last three cells.
    Test statistic: $\chi^2 = 1.012$.
    Critical value: $\chi^2 = 5.991$.
    Fail to reject H_0: The frequencies fit a Poisson distribution. There is not sufficient evidence to warrant rejection of the claim that the frequencies fit a Poisson distribution.

21. In a recent year, there were 116 homicide deaths in Richmond, Virginia (based on "A Classroom Note On the Poisson Distribution: A Model for Homicidal Deaths In Richmond, VA for 1991," *Mathematics and Computer Education*, by Winston A. Richards). If the frequencies of deaths on different days conform to a Poisson distribution, they will be as shown in the table located at the top of the following page. Use a 0.05 significance level to test the claim that the actual frequencies fit a Poisson distribution. (*Caution:* Not all of the expected frequencies are at least 5.)

	\multicolumn{5}{c}{Number of Homicides}			
	0	1	2	3	4
Actual number of days	268	79	17	1	0
Expected number of days	265.6	84.4	13.4	1.4	0.1

22. An observed frequency distribution of sample IQ scores is as follows:

IQ score	Less than 80	80–95	96–110	111–120	More than 120
Frequency	20	20	80	40	40

a. Assuming a normal distribution with $\mu = 100$ and $\sigma = 15$, use the methods given in Chapter 5 to find the probability of a randomly selected subject belonging to each class. (Use class boundaries of 79.5, 95.5, 110.5, 120.5.)

b. Using the probabilities found in part (a), find the expected frequency for each category.

c. Use a 0.01 level of significance to test the claim that the IQ scores were randomly selected from a normally distributed population with $\mu = 100$ and $\sigma = 15$.

10-3 Contingency Tables: Independence and Homogeneity

The examples and exercises in Section 10-2 involved frequencies listed in a single row (or column) according to category. This section also involves frequencies listed according to category, but here we consider cases with at least two rows and at least two columns. For example, see Table 10-1, which has two rows (deployed, nondeployed) and three columns (the three categories for causes of death). Tables similar to Table 10-1 are generally called *contingency tables*, or *two-way frequency tables*.

This section has great practical importance because contingency tables are used so often in the analysis of survey results. When selecting the topics to be included in the introductory statistics course, this section should be given a very high priority.

DEFINITIONS

A contingency table (or **two-way frequency table**) is a table in which frequencies correspond to two variables. (One variable is used to categorize rows, and a second variable is used to categorize columns.)

Contingency tables are especially important because they are frequently used to analyze survey results. Consequently, the methods presented in this section are among those used most often.

This section presents two types of hypothesis testing based on contingency tables. We first consider tests of independence, used to determine whether a contingency table's row variable is independent of its column variable. We then consider tests of homogeneity, used to determine whether different populations have the same proportions of some characteristic. Good news: Both types of hypothesis testing use the *same* basic methods. We begin with tests of independence.

Test of Independence

Home Field Advantage

In the *Chance* magazine article "Predicting Professional Sports Game Outcomes from Intermediate Game Scores," authors Harris Cooper, Kristina DeNeve, and Frederick Mosteller used statistics to analyze two common beliefs: Teams have an advantage when they play at home, and only the last quarter of professional basketball games really counts. Using a random sample of hundreds of games, they found that for the four top sports, the home team wins about 58.6% of games. Also, basketball teams ahead after 3 quarters go on to win about 4 out of 5 times, but baseball teams ahead after 7 innings go on to win about 19 out of 20 times. The statistical methods of analysis included the chi-square distribution applied to a contingency table.

> ### DEFINITION
>
> A **test of independence** tests the null hypothesis that the row variable and the column variable in a contingency table are not related. (The null hypothesis is the statement that the row and column variables are independent.)

It is very important to recognize that in this context, the word *contingency* refers to dependence, but this is only a statistical dependence and cannot be used to establish a direct cause-and-effect link between the two variables in question. For example, after analyzing the data in Table 10-1, we might conclude that there is a relationship between cause of death and whether military personnel were deployed to a combat zone, but that doesn't mean that deployment directly affects the cause of death.

Assumptions   When testing the null hypothesis of independence between the row and column variables in a contingency table, the following assumptions apply. (Note that these assumptions do not require that the parent population have a normal distribution or any other particular distribution.)

1. The sample data are randomly selected.
2. The null hypothesis H_0 is the statement that the row and column variables are *independent;* the alternative hypothesis H_1 is the statement that the row and column variables are dependent.
3. For every cell in the contingency table, the *expected* frequency E is at least 5. (There is no requirement that every *observed* frequency must be at least 5.)

Our test of independence between the row and column variables uses the following test statistic.

Test Statistic for a Test of Independence
$$\chi^2 = \sum \frac{(O - E)^2}{E}$$

It will be obvious that the test statistic given here is identical to the χ^2 test statistic given in Section 10-2. The hypothesis tests in this section are all right-tailed, as in Section 10-2. Differences between this section and Section 10-2 are found in the method used for finding expected values E and the calculation of the number of degrees of freedom.

Critical Values

1. The critical values are found in Table A-4 by using

$$\text{degrees of freedom} = (r - 1)(c - 1)$$

   where r is the number of rows and c is the number of columns.
2. Tests of independence with contingency tables involve only right-tailed critical regions.

The test statistic allows us to measure the degree of disagreement between the frequencies actually observed and those that we would theoretically expect when the two variables are independent. Small values of the χ^2 test statistic result from close agreement between observed frequencies and frequencies expected with independent row and column variables. Large values of the χ^2 test statistic are to the right of the chi-square distribution, and they reflect significant differences between observed and expected frequencies. In repeated large samplings, the distribution of the test statistic χ^2 can be approximated by the chi-square distribution, provided that all expected frequencies are at least 5. The number of degrees of freedom $(r - 1)(c - 1)$ reflects the fact that because we know the total of all frequencies in a contingency table, we can freely assign frequencies to only $r - 1$ rows and $c - 1$ columns before the frequency for every cell is determined. [However, we cannot have negative frequencies or frequencies so large that any row (or column) sum exceeds the total of the observed frequencies for that row (or column).]

In the preceding section we knew the corresponding probabilities and could easily determine the expected values, but the typical contingency table does not come with the relevant probabilities. Consequently, we need to devise a method for obtaining the corresponding expected values. We will first describe the procedure for finding the values of the expected frequencies, and then we will justify that procedure. For each cell in the frequency table, the expected frequency E can be calculated by using the following equation.

Expected Frequency for a Contingency Table
$$\text{Expected frequency} = \frac{(\text{row total})(\text{column total})}{(\text{grand total})}$$

TABLE 10-6	Observed Frequencies (and Expected Frequencies)			
	Cause of Death			
	Unintentional Injury	Illness	Homicide or Suicide	Row totals
Deployed	183	30	11	224
	(137.09)	(41.68)	(45.23)	
Not deployed	784	264	308	1356
	(829.91)	(252.32)	(273.77)	
Column totals:	967	294	319	Grand total: 1580

Here *grand total* refers to the total of all observed frequencies in the table. For example, the expected frequency for the upper left cell of Table 10-6 (a duplicate of Table 10-1 with expected frequencies inserted in parentheses) is 137.09. It is found by noting that the total of all frequencies for that row is 224, the total of the column frequencies is 967, and the total of all frequencies in the table is 1580, so we get an expected frequency of

$$E = \frac{\text{(row total)(column total)}}{\text{(grand total)}} = \frac{(224)(967)}{1580} = 137.09$$

EXAMPLE   The expected frequency for the upper left cell of Table 10-6 is 137.09. Find the expected frequency for the lower left cell, assuming independence between cause of death and whether the person was deployed.

SOLUTION   The lower left cell lies in the second row (with total 1356) and the first column (with total 967). The expected frequency is

$$E = \frac{\text{(row total)(column total)}}{\text{(grand total)}} = \frac{(1356)(967)}{1580} = 829.91$$

It would be very helpful to stop here and verify that the other expected frequencies are 41.68, 45.23, 252.32, and 273.77.

To better understand the rationale for finding expected frequencies with this procedure, let's pretend that we know only the row and column totals and

that we must fill in the cell expected frequencies by assuming independence (or no relationship) between the two variables involved—that is, pretend that we know only the row and column totals shown in Table 10-6. Let's begin with the cell in the upper left corner. Because 224 of the 1580 persons were deployed, we have $P(\text{deployed}) = 224/1580$. Similarly, 967 of the subjects died from unintentional injuries, so $P(\text{death from unintentional injury}) = 967/1580$. Because we are assuming independence between cause of death and whether the person was deployed, we can use the multiplication rule of probability to get

$$P(\text{deployed and died from unintentional injury}) = \frac{224}{1580} \cdot \frac{967}{1580}$$

This equation is an application of the multiplication rule for independent events, which is expressed in general as follows: $P(A \text{ and } B) = P(A) \cdot P(B)$. Having found an expression for the probability of being in the upper left cell, we now proceed by finding the *expected value* for that cell, which we get by multiplying the probability for that cell by the total number of subjects available, as shown in the following equation.

$$E = p \cdot n = \left[\frac{224}{1580} \cdot \frac{967}{1580} \right] \cdot 1580 = 137.09$$

The form of this product suggests a general way to obtain the expected frequency of a cell:

$$\text{expected frequency } E = \frac{(\text{row total})}{(\text{grand total})} \cdot \frac{(\text{column total})}{(\text{grand total})} \cdot (\text{grand total})$$

This expression can be simplified to

$$E = \frac{(\text{row total}) \cdot (\text{column total})}{(\text{grand total})}$$

We can now proceed to use contingency table data for testing hypotheses, as in the following example, which uses the data given in the Chapter Problem.

EXAMPLE   At the 0.05 significance level, use the data in Table 10-1 to test the claim that the cause of a noncombat death is independent of whether the military person was deployed in a combat zone.

SOLUTION   The null hypothesis and alternative hypothesis are as follows:

H_0: The cause of death is independent of whether the person was deployed in a combat zone.   *(continued)*

Instead of simply presenting the formula for the calculation of expected value E, strongly consider discussing the rationale for that expression. Refer to Table 10-1 in the Chapter Problem given at the beginning of this chapter, and ask the class these questions "in the spirit of reviewing a few of the important and basic principles of probabilities:" (1) If one of the survey subjects is randomly selected, find the probability of getting someone who was deployed; (2) Find the probability of getting someone who died from unintentional injury; (3) Find the probability of getting someone who was deployed *and* died from unintentional injury, assuming that causes of death are *independent* of whether the person was deployed.

H_1: The cause of death and whether the person was deployed are dependent.

The significance level is $\alpha = 0.05$.

Because the data are in the form of a contingency table, we use the χ^2 distribution with this test statistic:

$$\chi^2 = \sum \frac{(O - E)^2}{E}$$

$$= \frac{(183 - 137.09)^2}{137.09} + \frac{(30 - 41.68)^2}{41.68} + \frac{(11 - 45.23)^2}{45.23}$$

$$+ \frac{(784 - 829.91)^2}{829.91} + \frac{(264 - 252.32)^2}{252.32} + \frac{(308 - 273.77)^2}{273.77}$$

$$= 15.3748 + 3.2731 + 25.9052 + 2.5397 + 0.5407 + 4.2798$$

$$= 51.913$$

(The more accurate test statistic of 51.905 is obtained by carrying more decimal places in the intermediate calculations. The TI-83 calculator, STATDISK, and Minitab all agree that 51.905 is a better result.) The critical value is $\chi^2 = 5.991$, and it is found from Table A-4 by noting that $\alpha = 0.05$ in the right tail and the number of degrees of freedom is given by $(r - 1)(c - 1) = (2 - 1)(3 - 1) = 2$. The test statistic and critical value are shown in Figure 10-6. Because the test statistic falls within the critical region, we reject the null hypothesis that the cause of death is independent of whether the person was deployed in a combat zone. It appears that whether a person is deployed to a combat zone does seem to be related to the cause of death.

Figure 10-6   Test of Independence Between Noncombat Cause of Death and Whether Person was Deployed to Combat Zone

If the chi-square distribution has only 1 or 2 degrees of freedom, the shape of the distribution is as shown here.

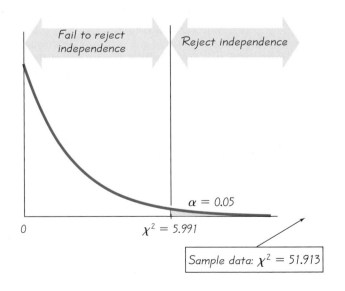

Fail to reject independence

Reject independence

$\alpha = 0.05$

0

$\chi^2 = 5.991$

Sample data: $\chi^2 = 51.913$

Using Computers and Calculators with Contingency Tables

Because of the large number of calculations required, it is often helpful to use a calculator or statistical software package for tests of the type discussed in this section. The TI-83 calculator will allow you to enter the table as a matrix, and it will calculate and display the χ^2 test statistic, P-value, and the number

---

STATDISK DISPLAY

File   Edit   Analysis   Data   Help

Degrees Freedom	2
Test Statistic, χ^2	51.905
Critical χ^2	5.9915
P-Value	0.000000

Reject the Null Hypothesis

Data provides evidence that the rows and columns are related

MINITAB DISPLAY

Expected counts are printed below observed counts

	C1	C2	C3	Total
1	183	30	11	224
	137.09	41.68	45.23	
2	784	264	308	1356
	829.91	252.32	273.77	
Total	967	294	319	1580

ChiSq =   15.372 + 3.274 + 25.901 +
            2.539 + 0.541 + 4.279 =   51.905
df = 2,   p = 0.000

of degrees of freedom. The preceding STATDISK and Minitab displays show the results obtained for the data in Table 10-1. If you are using STATDISK, select Analysis from the main menu bar, then select Contingency Tables and proceed to enter the data as requested. If using Minitab, first enter the observed frequencies in columns, then select Stat from the main menu bar. Next, select the option of Tables, then proceed to enter the names of the columns containing the observed frequencies, such as C1 C2. Minitab provides the test statistic and P-value, but not the critical value or conclusion. The STATDISK results include the test statistic, critical value, P-value, and conclusion.

P-Values

The preceding example used the traditional approach to hypothesis testing, but we can easily use the P-value approach. The TI-83 calculator, STATDISK, and Minitab all provide the same P-value of 0.000 (when rounded). Using the criterion given in Section 7-3, we reject the null hypothesis because the P-value is less than the significance level of $\alpha = 0.05$.

If we don't have a suitable calculator or statistical software package, we can estimate P-values by using the same methods introduced earlier. The preceding example resulted in a test statistic of $\chi^2 = 51.913$, and the critical value is based on 2 degrees of freedom. Refer to Table A-4 and note that for the row with 2 degrees of freedom, the test statistic of $\chi^2 = 51.913$ is larger than (or to the right of) the largest critical value of 10.597, so the P-value must be less than 0.005, which we can express as follows: P-value < 0.005. On the basis of this relatively small P-value, we again reject the null hypothesis and conclude that there is sufficient sample evidence to warrant rejection of the claim that cause of death and deployment are independent. If the P-value had been greater than the significance level of 0.05, we would have failed to reject the null hypothesis of independence.

Test of Homogeneity

Emphasize that the test of homogeneity uses the same procedures already presented in this section. Because a test of homogeneity is a test of the claim that different populations have the same proportions of some characteristics, we use data sampled from different populations, and we therefore have predetermined totals for either the rows or the columns in the contingency table.

In the preceding example, we illustrated a test of independence by using a sample of 1580 subjects drawn from a single population of military personnel. In some cases samples are drawn from different populations, and we want to determine whether those populations have the same proportions of the characteristics being considered.

DEFINITION

In a **test of homogeneity,** we test the claim that different populations have the same proportions of some characteristics.

Because a test of homogeneity uses data sampled from different populations, we have predetermined totals for either the rows or the columns in the contingency table. Consequently, a test of homogeneity involves random selections made in such a way that either the row totals are predetermined or the column totals are predetermined. In trying to distinguish between a test for homogeneity and a test for independence, we can therefore pose the following question:

> **Were predetermined sample sizes used for different populations (test of homogeneity), or was one big sample drawn so both row and column totals were determined randomly (test of independence)?**

As an example of a test of homogeneity, suppose we want to test the claim that the proportion of voters who are Republicans is the same in New York, California, Texas, and Iowa. If we choose to find the political party registrations for 200 New Yorkers, 250 Californians, 100 Texans, and 80 Iowans, then the contingency table summarizing the results will have either the row totals or the column totals (whichever represent the different states) predetermined as 200, 250, 100, 80.

In conducting a test of homogeneity, we can use the same procedures already presented in this section, as illustrated in the following example.

EXAMPLE   Does a pollster's gender have an effect on poll responses by men? In a *U.S. News & World Report* article about polls, it was stated that "On sensitive issues, people tend to give 'acceptable' rather than honest responses; their answers may depend on the gender or race of the interviewer." To support that claim, data were provided for an Eagleton Institute poll in which surveyed men were asked if they agreed with this statement: "Abortion is a private matter that should be left to the woman to decide without government intervention." We will analyze the effect of gender on male survey subjects only. Table 10-7 is based on the responses of surveyed men. Assume that the survey was designed so that male inter-

If Section 8-6 was covered, you might suggest that this hypothesis can be tested by using a z test of the equality of two proportions. See Exercise 18.

TABLE 10-7	Gender and Survey Responses	
	Gender of Interviewer	
	Man	Woman
Men who agree	560	308
Men who disagree	240	92

(continued)

viewers were instructed to obtain 800 responses from male subjects, and female interviewers were instructed to obtain 400 responses from male subjects. Using a 0.05 significance level, test the claim that the proportions of agree/disagree responses are the same for the subjects interviewed by men and the subjects interviewed by women.

SOLUTION   Because we have predetermined column totals of 800 subjects interviewed by men and 400 subjects interviewed by women, we test for homogeneity with these hypotheses:

H_0: The proportions of agree/disagree responses are the same for the subjects interviewed by men as the subjects interviewed by women.

H_1: The proportions are different.

The significance level is $\alpha = 0.05$. We use the same χ^2 test statistic described earlier, and it is calculated by using the same procedure. Instead of listing the details of that calculation, we provide the accompanying Minitab display that results from the data in Table 10-7.

MINITAB DISPLAY

```
Expected counts are printed below observed counts

              C1          C2      Total
    1        560         308      868
          578.67      289.33

    2        240          92      332
          221.33      110.67

Total        800         400     1200

ChiSq =   0.602 + 1.204 +
          1.574 + 3.149 = 6.529
df = 1,  p = 0.011
```

The Minitab display shows the expected frequencies of 578.67, 289.33, 221.33, and 110.67. The display also includes the test statistic of $\chi^2 = 6.529$ and the P-value of 0.011. Using the P-value approach to hypothesis testing, we reject the null hypothesis of equal (homogeneous) proportions. There is sufficient evidence to warrant rejection of the claim that the proportions are the same. It appears that response and the gender of the

interviewer are dependent. It seems that men are influenced by the gender of the interviewer, although this conclusion is a statement of causality that is not justified by the statistical analysis.

(10–3) Exercises A: Basic Skills and Concepts

1. Table 10-7 summarizes data for male survey subjects, but the accompanying table summarizes data for a sample of women. Using a 0.01 significance level, and assuming that the sample sizes of 800 men and 400 women are predetermined, test the claim that the proportions of agree/disagree responses are the same for the subjects interviewed by men and the subjects interviewed by women.

	Gender of Interviewer	
	Man	Woman
Women who agree	512	336
Women who disagree	288	64

Based on data from the Eagleton Institute.

2. In the judicial case *United States v. City of Chicago,* fair employment practices were challenged. A minority group (group A) and a majority group (group B) took the Fire Captain Examination. At the 0.05 significance level, use the given results to test the claim that success on the test is independent of the group.

3. Nicorette is a chewing gum designed to help people stop smoking cigarettes. Tests for adverse reactions yielded the results given in the accompanying table. At the 0.05 significance level, test the claim that the treatment (drug or placebo) is independent of the reaction (whether or not mouth or throat soreness was experienced). If you are thinking about using Nicorette as an aid to stop smoking, should you be concerned about mouth or throat soreness?

	Drug	Placebo
Mouth or throat soreness	43	35
No mouth or throat soreness	109	118

Based on data from Merrell Dow Pharmaceuticals, Inc.

4. A study was conducted of 531 persons injured in bicycle crashes, and randomly selected sample results are summarized in the table located at the top of the following page. At the 0.05 significance level, test the claim that wearing a helmet has no effect on whether facial injuries are received. Based on these results, does a helmet seem to be effective in helping to prevent facial injuries in a crash?

Recommended assignment: Exercises 1, 2, 5, 6. (Exercises 15 and 16 require more time because they involve the reorganization of data from data sets in Appendix B.)

1. Test statistic: $\chi^2 = 51.458$. Critical value: $\chi^2 = 6.635$.

	Pass	Fail
Group A	10	14
Group B	417	145

2. Test statistic: $\chi^2 = 12.321$. Critical value: $\chi^2 = 3.841$. Reject the claim that success and group are independent.

3. Test statistic: $\chi^2 = 1.174$. Critical value: $\chi^2 = 3.841$.

4. Test statistic: $\chi^2 = 10.708$. Critical value: $\chi^2 = 3.841$.

	Helmet Worn	No Helmet
Facial injuries received	30	182
All injuries nonfacial	83	236

Based on data from "A Case-Control Study of the Effectiveness of Bicycle Safety Helmets in Preventing Facial Injury," by Thompson, Thompson, Rivara, and Wolf, *American Journal of Public Health*, Vol. 80, No. 12.

5. Test statistic: $\chi^2 = 2.195$.
Critical value: $\chi^2 = 5.991$.

5. A survey was conducted to determine whether there is a gender gap in the confidence people have in police. The sample results are listed in the accompanying table. Use a 0.05 level of significance to test the claim that there is such a gender gap.

	Confidence in Police		
	Great Deal	Some	Very Little or None
Men	115	56	29
Women	175	94	31

Based on data from the U.S. Department of Justice and the Gallup Organization.

6. Test statistic: $\chi^2 = 10.814$.
Critical value: $\chi^2 = 5.991$.
(assuming $\alpha = 0.05$).

6. In a study of store checkout scanning systems, samples of purchases were used to compare the scanned prices to the posted prices. The accompanying table summarizes results for a sample of 819 items. When stores use scanners to check out items, are the error rates the same for regular-priced items as they are for advertised-special items? How might the behavior of consumers change if they believe that disproportionately more overcharges occur with advertised-special items?

	Regular-Priced Items	Advertised-Special Items
Undercharge	20	7
Overcharge	15	29
Correct price	384	364

Based on data from "UPC Scanner Pricing Systems: Are They Accurate?" by Ronald Goodstein, *Journal of Marketing*, Vol. 58.

7. Test statistic: $\chi^2 = 119.330$.
Critical value: $\chi^2 = 5.991$.

7. The accompanying table lists survey results obtained from a random sample of different crime victims. At the 0.05 level of significance, test the claim that the type of crime is independent of whether the criminal is a stranger.

	Homicide	Robbery	Assault
Criminal was a stranger	12	379	727
Criminal was acquaintance or relative	39	106	642

Based on data from the U.S. Department of Justice.

8. A study of seat-belt users and nonusers yielded the randomly selected sample data summarized in the accompanying table. Test the claim that the amount of smoking is independent of seat-belt use. A plausible theory is that people who smoke more are less concerned about their health and safety and are therefore less inclined to wear seat belts. Is this theory supported by the sample data?

| | Number of Cigarettes Smoked per Day | | | |
	0	1–14	15–34	35 and over
Wear seat belts	175	20	42	6
Don't wear seat belts	149	17	41	9

Based on data from "What Kinds of People Do Not Use Seat Belts?" by Helsing and Comstock, *American Journal of Public Health*, Vol. 67, No. 11.

8. Test statistic: $\chi^2 = 1.358$. Critical value: $\chi^2 = 7.815$. (assuming $\alpha = 0.05$).

9. Many people believe that criminals who plead guilty tend to get lighter sentences than those who are convicted in trials. The accompanying table summarizes randomly selected sample data for San Francisco defendants in burglary cases. All of the subjects had prior prison sentences. At the 0.05 significance level, test the claim that the sentence (sent to prison or not sent to prison) is independent of the plea. If you were an attorney defending a guilty defendant, would these results suggest that you should encourage a guilty plea?

	Guilty Plea	Not-Guilty Plea
Sent to prison	392	58
Not sent to prison	564	14

Based on data from "Does It Pay to Plead Guilty? Differential Sentencing and the Functioning of the Criminal Courts," by Brereton and Casper, *Law and Society Review*, Vol. 16, No. 1.

9. Test statistic: $\chi^2 = 42.557$. Critical value: $\chi^2 = 3.841$.

10. A study of randomly selected car accidents and drivers who use cellular phones provided the following sample data. At the 0.05 level of significance, test the claim that the occurrence of accidents is independent of the use of cellular phones. Based on these results, does it appear that the use of cellular phones affects driving safety?

	Had Accident in Last Year	Had No Accident in Last Year
Cellular phone user	23	282
Not cellular phone user	46	407

Based on results from AT&T and the Automobile Association of America.

10. Test statistic: $\chi^2 = 1.505$. Critical value: $\chi^2 = 3.841$.

11. Test statistic: $\chi^2 = 0.615$.
Critical value: $\chi^2 = 7.815$.

11. A study conducted to determine the rate of smoking among people from different age groups provided the randomly selected sample data summarized in the accompanying table. At the 0.05 significance level, test the claim that smoking is independent of the four age groups listed. Based on the data, does it make sense to target cigarette advertising to particular age groups?

	Age (years)			
	20–24	25–34	35–44	45–64
Smoke	18	15	17	15
Don't smoke	32	35	33	35

Based on data from the National Center for Health Statistics.

12. Test statistic: $\chi^2 = 4.737$.
Critical value: $\chi^2 = 6.251$.

12. Winning team data were collected for teams in different sports, with the results given in the accompanying table. Use a 0.10 level of significance to test the claim that home/visitor wins are independent of the sport.

	Basketball	Baseball	Hockey	Football
Home team wins	127	53	50	57
Visiting team wins	71	47	43	42

Based on data from "Predicting Professional Sports Game Outcomes from Intermediate Game Scores," by Copper, DeNeve, and Mosteller, *Chance*, Vol. 5, No. 3–4.)

13. Test statistic: $\chi^2 = 49.731$.
Critical value: $\chi^2 = 11.071$
(assuming $\alpha = 0.05$).

13. The accompanying table lists sample data that statistician Karl Pearson used in 1909. Does the type of crime appear to be related to whether the criminal drinks or abstains? Are there any crimes that appear to be associated with drinking?

	Arson	Rape	Violence	Stealing	Coining	Fraud
Drinker	50	88	155	379	18	63
Abstainer	43	62	110	300	14	144

14. Test statistic: $\chi^2 = 20.271$.
Critical value: $\chi^2 = 15.086$.

14. A study of people who refused to answer survey questions provided the randomly selected sample data in the accompanying table. At the 0.01 significance level, test the claim that the cooperation of the subject (response, refusal) is independent of the age category. Does any particular age group appear to be particularly uncooperative?

	Age					
	18–21	22–29	30–39	40–49	50–59	60 and over
Responded	73	255	245	136	138	202
Refused	11	20	33	16	27	49

Based on data from "I Hear You Knocking But You Can't Come In," by Fitzgerald and Fuller, *Sociological Methods and Research*, Vol. 11, No. 1.

15. Refer to Data Set 8 in Appendix B and test the claim that the gender of statistics students is independent of whether they smoke.
16. Refer to Data Set 8 in Appendix B and test the claim that whether statistics students exercise is independent of gender.

15. Test statistic: $\chi^2 = 0.122$.
Critical value: $\chi^2 = 3.841$ (assuming $\alpha = 0.05$).
16. Test statistic: $\chi^2 = 2.194$.
Critical value: $\chi^2 = 3.841$ (assuming $\alpha = 0.05$).

10–3 Exercises B: Beyond the Basics

17. The chi-square distribution is continuous, whereas the test statistic used in this section is discrete. Some statisticians use *Yates' correction for continuity* in cells with an expected frequency of less than 10 or in all cells of a contingency table with two rows and two columns. With Yates' correction, we replace

$$\sum \frac{(O - E)^2}{E} \quad \text{with} \quad \sum \frac{(|O - E| - 0.5)^2}{E}$$

Given the contingency table in Exercise 10, find the value of the χ^2 test statistic with and without Yates' correction. In general, what effect does Yates' correction have on the value of the test statistic?

17. With Yates' correction: $\chi^2 = 1.205$. Without Yates' correction: $\chi^2 = 1.505$. In general, the test statistic decreases with Yates' correction.

18. a. For a contingency table with two rows and two columns and frequencies of a and b in the first row and frequencies of c and d in the second row, verify that the test statistic becomes

$$\chi^2 = \frac{(a + b + c + d)(ad - bc)^2}{(a + b)(c + d)(b + d)(a + c)}$$

b. Let $\hat{p}_1 = a/(a + c)$ and let $\hat{p}_2 = b/(b + d)$. Show that the test statistic

$$z = \frac{(\hat{p}_1 - \hat{p}_2) - 0}{\sqrt{\dfrac{\overline{p}\,\overline{q}}{n_1} + \dfrac{\overline{p}\,\overline{q}}{n_2}}}$$

where

$$\overline{p} = \frac{a + b}{a + b + c + d}$$

and

$$\overline{q} = 1 - \overline{p}$$

is such that $z^2 = \chi^2$ [the same result as in part (a)]. (This result shows that the chi-square test involving a 2×2 table is equivalent to the test for the difference between two proportions, as described in Section 8-6.)

18. a. The row totals are $a + b$ and $c + d$. The column totals are $a + c$ and $b + d$. The grand total is $a + b + c + d$. Use these values to find the expected frequencies and then calculate the test statistic using $\Sigma(O - E)^2/E$ to obtain the given expression.

•••

Vocabulary List

cells
multinomial experiment
goodness-of-fit test
contingency table

two-way frequency table
test of independence
test of homogeneity

•••

Review

Section 10-2 described the goodness-of-fit test for determining whether a single row (or column) of frequencies has some claimed distribution. Section 10-3 described contingency tables and the test of independence between the row and column variables, as well as the test of homogeneity, in which different populations have the same proportions of some characteristics. The following are some key elements of the methods discussed in this chapter.

- *Section 10-2 (Test for goodness-of-fit):*

  Test statistic is $\chi^2 = \sum \dfrac{(O - E)^2}{E}$

  Test is right-tailed with $k - 1$ degrees of freedom. All expected frequencies must be at least 5.

- *Section 10-3 (Contingency table test of independence or homogeneity):*

  Test statistic is $\chi^2 = \sum \dfrac{(O - E)^2}{E}$

  Test is right-tailed with $(r - 1)(c - 1)$ degrees of freedom. All expected frequencies must be at least 5.

•••

Review Exercises

1. Test statistic: $\chi^2 = 42.004$.
Critical value: $\chi^2 = 5.991$.

1. A study of seat-belt use in taxi cabs involved 77 New York taxis, 129 Chicago taxis, and 72 Pittsburgh taxis, with the results summarized in the table at the top of the following page. A spokesperson for the taxi-cab industry argues that although it appears that very few taxis have usable seat belts, the total sample size from the three cities was only 278, so the results aren't significant. At the 0.05 level of significance, test the claim that the three cities have the same proportion of taxis with usable seat belts. That is, test for homogeneity of proportions of usable taxi seat belts in the three cities.

		New York	Chicago	Pittsburgh
Taxi has usable	Yes	3	42	2
seat belt?	No	74	87	70

Based on "The Phantom Taxi Seat Belt," by Welkon and Reisinger, *American Journal of Public Health*, Vol. 67, No. 11.

2. When *Time* magazine tracked U.S. deaths by gunfire during a one-week period, the results shown in the accompanying table were obtained. At the 0.05 significance level, test the claim that gunfire death rates are the same for the different days of the week. Is there any support for the theory that more gunfire deaths occur on weekends when more people are at home?

Weekday	Mon	Tues	Wed	Thurs	Fri	Sat	Sun
Number of deaths by gunfire	74	60	66	71	51	66	76

3. In an experiment on extrasensory perception, subjects were asked to identify the month showing on a calendar in the next room. If the results were as shown, test the claim that months were selected with equal frequencies. Assume a significance level of 0.05. If it appears that the months were not selected with equal frequencies, is the claim that the subjects have extrasensory perception supported?

Month	Jan	Feb	Mar	Apr	May	June	July	Aug	Sept	Oct	Nov	Dec
Number selected	8	12	9	15	6	12	4	7	11	11	5	20

4. The accompanying table summarizes results for 100 flights randomly selected from each of three different airline companies. Use a 0.05 level of significance to test the claim that USAir, American, and Delta have the same proportions of on-time flights.

	USAir	American	Delta
Arrived on time	80	77	76
Arrived late	20	23	24

Based on data from the Department of Transportation.

5. Clinical tests of the allergy drug Seldane yielded results summarized in the accompanying table. At the 0.05 significance level, test the claim that the occurrence of headaches is independent of the group (Seldane, placebo, control). Based on these results, should Seldane users be concerned about getting headaches?

	Seldane Users	Placebo Users	Control
Headache	49	49	24
No headache	732	616	602

Based on data from Merrell Dow Pharmaceuticals, Inc.

2. Test statistic: $\chi^2 = 6.780$.
   Critical value: $\chi^2 = 12.592$.

3. Test statistic: $\chi^2 = 22.600$.
   Critical value: $\chi^2 = 19.675$.

4. Test statistic: $\chi^2 = 0.500$.
   Critical value: $\chi^2 = 5.991$.

5. Test statistic: $\chi^2 = 7.607$.
   Critical value: $\chi^2 = 5.991$.

● ●

Cumulative Review Exercises

Exercise 1 requires Section 9-2,
Exercise 2 requires Section 8-2,
Exercise 3 requires Section 10-3,
and Exercise 4 requires Section
8-5.

TABLE 10-8

	A	B	C	D
x	66	80	82	75
y	77	89	94	84

1. (Use the linear correlation coefficient, as in Section 9-2.) Test statistic: $r = 0.978$. Critical value: $r = 0.950$ (assuming $\alpha = 0.05$).

2. (Use the test for two dependent means, as in Section 8-2.) With $\overline{d} = -10.25$ and $s_d = 1.5$, the test statistic is $t = -13.667$. The critical value is $t = -2.353$ (assuming $\alpha = 0.05$).

3. Test statistic: $\chi^2 = 0.055$. Critical value: $\chi^2 = 7.815$ (assuming $\alpha = 0.05$).

4. Test statistic is $t = -2.014$. Critical values: $t = \pm 2.447$ (assuming $\alpha = 0.05$).

1. Assume that Table 10-8 lists test scores for four people, where the x score is from a test of memory and the y score is from a test of reasoning. Test the claim that there is a relationship between the x and y scores.

2. Assume that Table 10-8 lists test scores for four people, where the x score is from a pretest taken before a training session on memory improvement and the y score is from a posttest taken after the training. Test the claim that the training session is effective in raising scores.

3. Assume that in Table 10-8, the letters A, B, C, D represent the choices on the first question of a multiple-choice quiz. Also assume that x represents men and y represents women, and the table entries are frequency counts, so that 66 men chose answer A, 77 women chose answer A, 80 men chose answer B, and so on. Test the claim that men and women choose the different answers in the same proportions.

4. Assume that in Table 10-8, the letters A, B, C, D represent different versions of the same test of reasoning. The x scores were obtained by four randomly selected men and the y scores were obtained by four randomly selected women. Test the claim that men and women have the same mean score.

● ●

Computer Project

Use your calculator to randomly generate 100 digits between 0 and 9 inclusive, then use a statistical software package (such as STATDISK or Minitab) to randomly generate another 100 digits between 0 and 9 inclusive. Record the results in the accompanying table, then test the claim that the digit selected is independent of whether you used your calculator or your statistical software package. Test the claim by using STATDISK or Minitab, and obtain a printed copy of the display. Interpret the computer results and write your conclusions.

	0	1	2	3	4	5	6	7	8	9
Calculator										
STATDISK or Minitab										

● ● ● ● ● ● ● ● **FROM DATA TO DECISION** ● ● ● ● ● ● ● ●

The table summarizes randomly selected deaths of men aged 45–64. Test the claim that smoking is independent of the cause of death. Other data show only 45% of males in the 45–64 age bracket are smokers. Using the sample data, test the claim that the proportion of deaths among smokers is $p = 0.45$. (*Hint:* See Section 7-5.) Among the 1000 males in the 45–64 age bracket who died, are smokers disproportionately represented?

	Cause of Death		
	Cancer	Heart Disease	Other
Smoker	135	310	205
Nonsmoker	55	155	140

Based on data from "Chartbook on Smoking, Tobacco, and Health," USDHEW publication CDC75-7511.

•••••••• COOPERATIVE GROUP ACTIVITIES ••••••••

1. *Out-of-class activity:* Divide into groups of four or five students. Each group member should survey at least 15 male students and 15 female students at the same college by asking two questions: (1) Which political party does the subject favor most? (2) If the subject were to make up an absence excuse of a flat tire, which tire would he or she say went flat if the instructor asked? (See Exercise 3 in Section 10-2.) Ask the subject to write the two responses on an index card, and also record the gender of the subject and whether the subject wrote with the right or left hand. Use the methods of this chapter to analyze the data collected.

2. *Out-of-class activity:* In random sampling, each member of the population has the same chance of being selected. When the author asked 60 students to "randomly" select three digits each, the results listed below were obtained. Use this sample of 180 digits to test the claim that students select digits randomly.

213	169	812	125	749	137	202	344	496	348	714	765
831	491	169	312	263	192	584	968	377	403	372	123
493	894	016	682	390	123	325	734	316	357	945	208
115	776	143	628	479	316	229	781	628	356	195	199
223	114	264	308	105	357	333	421	107	311	458	007

Were the students successful in choosing their own random numbers? Select your own sample of students and ask each one to randomly select three digits. Test your results for randomness. Also test your results to determine whether they agree with those listed below. Write a report describing your experiment and clearly state your major conclusions.

3. *In-class activity:* Divide into groups of three or four students. Each group should be given a die along with the instruction that it should be tested for "fairness." Is the die fair or is it biased? Describe the analysis and results.

4. *In-class activity:* Divide into groups of six or more students. Each group member should provide the digits of his or her social security number. (The digits should be rearranged so that privacy is maintained.) Construct a combined list of those digits and test the claim that the digits 0, 1, 2, . . . , 9 occur with the same frequency.

11

Analysis of Variance

11-1 Overview

Objectives are identified for this chapter, which introduces methods for testing the hypothesis that three or more populations have the same mean. In Chapter 8 we tested for equality between the means of two populations and used test statistics having normal or Student t distributions. The methods of this chapter, however, are very different. Because of the nature of the calculations required, this chapter will emphasize the use of computer software and the TI-83 calculator.

11-2 One-Way ANOVA

This section introduces the basic method of analysis of variance (ANOVA), which is used to test the claim that three or more populations have the same mean. One-way ANOVA involves different samples that are categorized according to a single characteristic.

11-3 Two-Way ANOVA

Section 11-2 deals with a single factor (or characteristic used to differentiate between populations), but this section considers two factors. Because the calculations are very difficult, the emphasis is on interpreting computer displays rather than performing manual calculations.

Chapter Problem:

Is Old Faithful becoming less faithful?

One of the most popular natural attractions in the United States is the Old Faithful geyser in Yellowstone National Park. This geyser was named for its predictable time intervals between eruptions, but is Old Faithful changing? Table 11-1 lists time intervals (in minutes) between eruptions for four different years. The four samples have means of 63.3 min, 74.3 min, 81.7 min, and 83.8 min, respectively, so it appears that the mean time interval between eruptions is not the same for the four years. Figure 11-1 includes the boxplots representing the four samples, and those boxplots also suggest that the four samples come from populations with means that are not all equal. But, considering that each sample consists of only 12 scores, are those differences statistically significant?

Chapter 8 included methods for testing equality between two population means. We now want to test for equality among four population means. Specifically, we want to test the claim that $\mu_{1951} = \mu_{1985} = \mu_{1995} = \mu_{1996}$. This chapter introduces methods for testing such claims, and we will consider the data included in Table 11-1.

Table 11-1 Time Intervals (in min) Between Eruptions of the Old Faithful Geyser

1951	1985	1995	1996
74	89	86	88
60	90	86	86
74	60	62	85
42	65	104	89
74	82	62	83
52	84	95	85
65	54	79	91
68	85	62	68
62	58	94	91
66	79	79	56
62	57	86	89
60	88	85	94
$n_1 = 12$	$n_2 = 12$	$n_3 = 12$	$n_4 = 12$
$\bar{x}_1 = 63.3$	$\bar{x}_2 = 74.3$	$\bar{x}_3 = 81.7$	$\bar{x}_4 = 83.8$
$s_1 = 9.4$	$s_2 = 14.2$	$s_3 = 13.7$	$s_4 = 10.9$

Based on data from geologist Rick Hutchinson and the National Park Service.

Figure 11-1

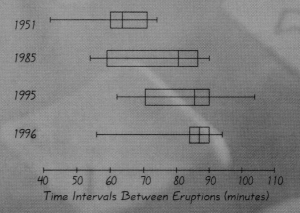

11-1 Overview

In Sections 8-2, 8-3, and 8-5 we developed procedures for testing the hypothesis that *two* population means are equal (H_0: $\mu_1 = \mu_2$). Section 11-2 introduces a procedure for testing the hypothesis that *three or more* population means are equal. (The methods used in this chapter can also be used to test for equality between two population means, but the methods used in Chapter 8 are more efficient.) A typical null hypothesis in Section 11-2 will be H_0: $\mu_1 = \mu_2 = \mu_3 = \mu_4$; the alternative hypothesis is H_1: At least one mean is different. The method we use is based on an analysis of sample variances.

DEFINITION

Analysis of variance (ANOVA) is a method of testing the equality of three or more population means by analyzing sample variances.

ANOVA is used in applications such as the following:

- When three different groups of people (such as smokers, nonsmokers exposed to environmental tobacco smoke, and nonsmokers not so exposed) are measured for cotinine (a marker of nicotine), we can test to determine whether they have the same level.

- When Kaplan, Princeton Review, and Collins Test Prep give students different SAT test preparation courses, we can test to determine whether there are differences in mean SAT scores for the three populations of students from the three different courses.

Why bother with a new procedure when we can test for equality of two means by using the methods presented in Chapter 8, where we developed tests of H_0: $\mu_1 = \mu_2$? For example, if we want to use the sample data from Table 11-1 to test the claim (at the $\alpha = 0.05$ level) that the four populations have the same mean, why not simply pair them off and do two at a time by testing H_0: $\mu_1 = \mu_2$, then H_0: $\mu_2 = \mu_3$, and so on? This approach (doing two at a time) requires six different hypothesis tests, so the degree of confidence could be as low as 0.95^6 (or 0.735). In general, as we increase the number of individual tests of significance, we increase the likelihood of finding a difference by chance alone. The risk of a type I error—finding a difference in one of the pairs when no such difference actually exists—is excessively high. The method of analysis of variance lets us avoid that particular pitfall (rejecting a true null hypothesis) by using one test for equality of several means.

F Distribution

The ANOVA methods use the F distribution, which was first introduced in Section 8-4. In Section 8-4 we noted that the F distribution has the following important properties (see Figure 11-2):

1. The F distribution is not symmetric; it is skewed to the right.

2. The values of F can be 0 or positive, but they cannot be negative.

3. There is a different F distribution for each pair of degrees of freedom for the numerator and denominator.

Critical values of F are given in Table A-5.

Analysis of variance (ANOVA) is based on a comparison of two different estimates of the variance common to the different populations. Those estimates (the *variance between samples* and the *variance within samples*) will be described in Section 11-2. The term *one-way* is used because the sample data are separated into groups according to one characteristic, or factor. For example, the times listed in Table 11-1 are separated into four different groups according to the one characteristic (or factor) of the year in which they were observed. Section 11-3 will introduce two-way analysis of variance, which allows us to compare populations separated into categories using two characteristics (or factors). For example, we might separate lengths of movies using the following two criteria: (1) their star ratings (1 star, 2 stars, 3 stars, 4 stars) and (2) their viewer discretion ratings (G, PG, PG-13, R).

Because the procedures used in this chapter require complicated calculations, we will emphasize the use of computer software, such as STATDISK and Minitab, along with the TI-83 calculator.

When introducing (or reviewing) the F distribution, consider comparing it to the normal, t, and χ^2 distributions relative to the properties of symmetry, positive/negative values, and one standard curve versus a family of curves that correspond to values of df.

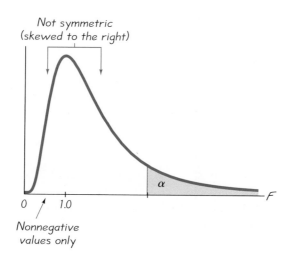

Figure 11-2   The F Distribution
There is a different F distribution for each different pair of degrees of freedom for numerator and denominator.

11-2 One-Way ANOVA

It is extremely important to consider how the topics of this chapter will be presented. These three steps do not constitute the only workable approach, but clearly inform the class of the approach you decide to use.

   Also, it may be helpful to use the TI-83 calculator, which includes a program for doing hypothesis tests with one-way ANOVA.

Because of the complexity of the calculations involved, we recommend the following approach to this section:

1. Develop a clear understanding of how to interpret a computer display that lists analysis of variance results.
2. Develop an understanding of the underlying rationale by focusing on the calculations that apply to the example in which the samples all have the same number of values.
3. Become acquainted with the nature of the SS (sum of square) and MS (mean square) values and their role in determining the F test statistic, but depend upon statistical software packages for finding those values.

In this section we consider tests of hypotheses that three or more population means are all equal, as in H_0: $\mu_1 = \mu_2 = \mu_3$.

Assumptions   The following assumptions apply when testing the hypothesis that three or more samples come from populations with the same mean:

1. The populations have normal distributions.
2. The populations have the same variance σ^2 (or standard deviation σ).
3. The samples are random and independent of each other.
4. The different samples are from populations that are categorized in only one way.

The nonparametric Kruskal-Wallis test in Section 13-5 does not require that the populations have normal distributions. It uses ranks to test the claim that independent and random samples come from identical populations. If the requirement of normally distributed populations is not met, the Kruskal-Wallis test is a good alternative.

The requirements of normality and equal variances are somewhat relaxed, as the methods in this section work reasonably well unless a population has a distribution that is very nonnormal or the population variances differ by large amounts. University of Wisconsin statistician George E. P. Box showed that as long as the sample sizes are equal (or nearly equal), the variances can differ by amounts where the largest is up to nine times the smallest and the results of ANOVA will continue to be essentially reliable. (If the samples are independent but the distributions are very nonnormal, we can use the Kruskal-Wallis test presented in Section 13-5.)

The method we use is called **one-way analysis of variance** (or **single-factor analysis of variance**) because we use a single property, or characteristic, for categorizing the populations. This characteristic is sometimes referred to as a treatment or factor.

DEFINITION

A **treatment** (or **factor**) is a property, or characteristic, that allows us to distinguish the different populations from one another.

For example, the times listed in Table 11-1 are sample data drawn from four different populations that are distinguished according to the treatment (or factor) of year in which they were observed. The term *treatment* is used because early applications of analysis of variance involved agricultural experiments in which different plots of farmland were treated with different fertilizers, seed types, insecticides, and so on.

Instead of choosing between "treatment" and "factor," consider referring to "treatment or factor" repeatedly so that it becomes known that these two terms are synonymous. Minitab and TI-83 displays include the term "factor," so it's important to recognize and understand that reference.

Using Computers and Calculators for One-Way ANOVA

Because the calculations required for analysis of variance are extremely involved, almost everyone uses computer software (except for one statistician living in a Wyoming tent). We will first describe the interpretation of computer results, then we will consider the rationale underlying the method of analysis of variance.

EXAMPLE   Given the sample data in Table 11-1 and a significance level of $\alpha = 0.05$, use STATDISK or Minitab or a TI-83 calculator to test the claim that the four samples come from populations with the same mean ($\mu_{1951} = \mu_{1985} = \mu_{1995} = \mu_{1996}$). With STATDISK, select `Analysis` from the main menu bar, then select `One-Way Analysis of Variance` and proceed to enter the sample data. With Minitab, first enter the sample data in columns C1, C2, C3, C4, then select `Stat, ANOVA, ONEWAY (UNSTACKED)`, and enter `C1   C2   C3   C4` in the box identified as `Responses` (in separate columns). With a TI-83 calculator, enter the data as lists in L1, L2, L3, and L4, then press STAT, select `TESTS`, and choose the option of `ANOVA`.

SOLUTION   The STATDISK and Minitab computer displays are as shown on page 576. A key element is the *P*-value, which is 0.00066 in STATDISK or 0.001 in Minitab. Because the *P*-value is less than or equal to the significance level of $\alpha = 0.05$, we reject the null hypothesis of equal means. There is sufficient evidence to warrant rejection of the claim that the four samples come from populations with the same mean. It appears that Old Faithful's mean time between eruptions is changing.

The lower portion of the Minitab display will also include the individual sample means, standard deviations, and graphs of the 95% confidence interval estimates of each of the four population means. These confidence intervals are computed by the same methods used in Chapter 6, except that a pooled standard deviation is used instead of the individual sample standard deviations. (The last entry in the Minitab display shows that the pooled standard deviation is 12.22.)

STATDISK DISPLAY

File   Edit   Analysis   Data   Help

Equal Length Samples

Total Num Values	48
Upper Deg Free	3
Lower Deg Free	44

SS(treatment)	3090.1
SS(error)	6571.4
SS(total)	9661.5
MS(treatment)	1030.0
MS(error)	149.35
MS(total)	205.56

Test Statistic, F	6.8967
Critical F	2.8165
P-Value	0.000660

Reject the Null Hypothesis

Data provides evidence that the sample means are unequal

MINITAB DISPLAY

Analysis of Variance

Source	DF	SS	MS	F	p
Factor	3	3090	1030	6.90	0.001
Error	44	6571	149		
Total	47	9661			

Individual 95% CIs For Mean
Based on Pooled StDev

Level	N	Mean	StDev	----+---------+---------+---------+--
C1	12	63.25	9.45	(------*------)
C2	12	74.25	14.17	(------*------)
C3	12	81.67	13.72	(------*------)
C4	12	83.75	10.91	(------*------)
				----+---------+---------+---------+--
				60 70 80 90

Pooled StDev= 12.22

If you run the Table 11-1 data through the TI-83, the ANOVA display will include the test statistic of $F = 6.896673726$, the P-value of 0.00066042992, as well as other relevant results.

Using computer software, the analysis of variance procedure is quite easy: Enter the data, run the ANOVA program, identify the P-value, and reject the null hypothesis of equal means if the P-value is less than or equal to the significance level α (otherwise, fail to reject the null hypothesis).

> If using Minitab or a TI-83 calculator, you don't get critical values, but you do get P-values, so be sure to review the P-value method for testing hypotheses. Specifically, refer to Figures 7-7 and 7-8 in Section 7-3.

Rationale

The method of analysis of variance is based on this fundamental concept: With the assumption that the populations all have the same variance σ^2, we estimate the common value of σ^2 using two different approaches. The F test statistic is the ratio of those estimates, so that a significantly *large* F test statistic (located far to the right in the F distribution graph) is evidence against equal population means. The two approaches for estimating the common value of σ^2 are as follows.

> Even if you are making heavy use of technology, a basic understanding of underlying principles is important. Carefully explain the content of this subsection.

1. The **variance between samples** (also called **variation due to treatment**) is an estimate of the common population variance σ^2 that is based on the variability among the sample *means*.

2. The **variance within samples** (also called **variation due to error**) is an estimate of the common population variance σ^2 based on the sample *variances*.

Test Statistic for One-Way ANOVA

$$F = \frac{\text{variance between samples}}{\text{variance within samples}}$$

The numerator measures variation between sample means. The estimate of variance in the denominator depends only on the sample variances and is not affected by differences among the sample means. Consequently, sample means that are close in value result in an F test statistic that is close to 1, and we conclude that there are no significant differences among the sample means. But if the value of F is excessively *large*, then we reject the claim of equal means. (The vague terms "close to 1" and "excessively large" are made objective by the use of a specific critical value, which clearly differentiates between an F test statistic that is in the critical region and one that is not.) Because excessively large values of F reflect unequal means, the test is right-tailed.

Calculations with Equal Sample Sizes

If the data sets all have the same sample size (as in Table 11-1), the required calculations aren't overwhelmingly difficult. First, find the variance between samples by evaluating $ns_{\bar{x}}^2$, where $s_{\bar{x}}^2$ is the variance of the sample means. That is, consider the sample means to be an ordinary set of scores and calculate the variance. (From the central limit theorem, $\sigma_{\bar{x}} = \sigma/\sqrt{n}$ can be solved for σ to get $\sigma = \sqrt{n} \cdot \sigma_{\bar{x}}$, so that we can estimate σ^2 with $ns_{\bar{x}}^2$. For example, the sample means in Table 11-1 are 63.3, 74.3, 81.7, and 83.8. Those four values have a standard deviation of $s_{\bar{x}} = 9.26116$, so that

$$\text{variance between samples} = ns_{\bar{x}}^2 = 12(9.26116)^2 = 1029.23$$

Next, estimate the variance within samples by calculating s_p^2, which is the pooled variance obtained by finding the mean of the sample variances. The sample standard deviations in Table 11-1 are 9.4, 14.2, 13.7, and 10.9, so that

$$\text{variance within samples} = s_p^2$$
$$= \frac{9.4^2 + 14.2^2 + 13.7^2 + 10.9^2}{4}$$
$$= 149.125$$

Finally, evaluate the F test statistic as follows.

$$F = \frac{\text{variance between samples}}{\text{variance within samples}} = \frac{1029.23}{149.125} = 6.9018$$

Carrying more decimal places would result in the more accurate test statistic of $F = 6.8967$.

The critical value of F is found by assuming a right-tailed test, because large values of F correspond to significant differences among means. With k samples each having n scores, the numbers of degrees of freedom are computed as follows:

Degrees of Freedom with k Samples of the Same Size n
numerator degrees of freedom $= k - 1$ denominator degrees of freedom $= k(n - 1)$

For the sample data in Table 11-1, $k = 4$ and $n = 12$, so the degrees of freedom are 3 for the numerator and $4(11) = 44$ for the denominator. With $\alpha = 0.05$, 3 degrees of freedom for the numerator, and 44 degrees of freedom for the denominator, the critical value is $F = 2.8387$. (Table A-5 does not include 44 degrees of freedom for the denominator, so we use the closest value, which corresponds to 40 degrees of freedom.) Note that STATDISK and Minitab display the numbers of degrees of freedom along with the calculated

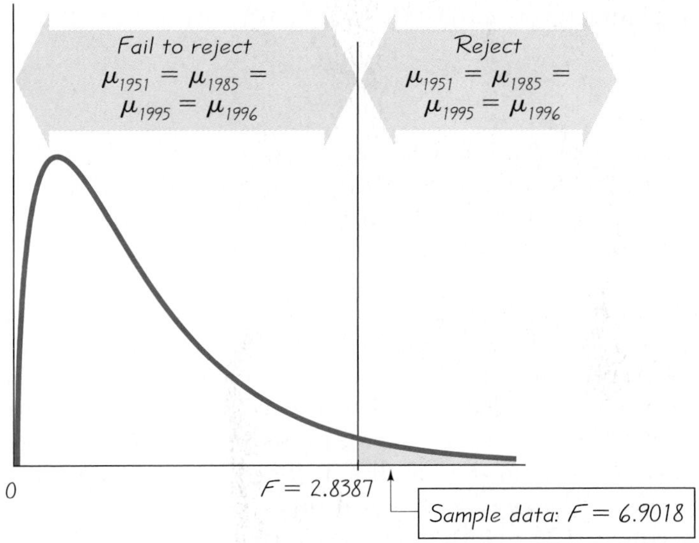

Figure 11-3   *F* Distribution for Old Faithful Data

F test statistic. Also, see Figure 11-3, which shows the test statistic and critical value. On the basis of these results, we reject the null hypothesis of equal means. There is sufficient evidence to warrant rejection of the claim that the four samples come from populations having the same mean.

Calculations with Unequal Sample Sizes

While the calculations required for cases with equal sample sizes are reasonable, they become really complicated when the sample sizes are not all the same. The same basic reasoning applies because we calculate the *F* test statistic that is the ratio of two different estimates of the common population variance σ^2, but those estimates involve weighted measures that take the sample sizes into account, as shown below.

$$F = \frac{\text{variance between samples}}{\text{variance within samples}} = \frac{\left[\dfrac{\Sigma n_i(\overline{x}_i - \overline{\overline{x}})^2}{k-1}\right]}{\left[\dfrac{\Sigma(n_i-1)s_i^2}{\Sigma(n_i-1)}\right]}$$

where $\overline{\overline{x}}$ = mean of all sample scores combined

k = number of population means being compared

n_i = number of values in the *i*th sample

N = total number of values in all samples combined

$\overline{x}_i$ = mean of values in the *i*th sample

s_i^2 = variance of values in the *i*th sample

Note that the numerator in this test statistic is really a form of the formula

$$s^2 = \frac{\Sigma(x - \bar{x})^2}{n - 1}$$

for variance that was given in Chapter 2. The factor of n_i is included so that larger samples carry more weight. The denominator of the test statistic is simply the mean of the sample variances, but it is a weighted mean with the weights based on the sample sizes.

Because calculating this test statistic can lead to large rounding errors, the various software packages typically use a different (but equivalent) expression that involves SS (for sum of squares) and MS (for mean square) notation. Although the following notation and components are complicated and involved, the basic idea is the same: The test statistic F is a ratio with a numerator reflecting variation *between* the means of the samples and a denominator reflecting variation *within* the samples. If the populations have equal means, the F ratio tends to be close to 1, but if the population means are not equal, the F ratio tends to be significantly larger than 1. Key components in our ANOVA method are identified in the following boxes.

SS(total), or total sum of squares, is a measure of the total variation (around $\bar{\bar{x}}$) in all of the sample data combined.

Formula 11-1	$SS(total) = \Sigma(x - \bar{\bar{x}})^2$

SS(total) can be broken down into the components of SS(treatment) and SS(error), described as follows:

SS(treatment) is a measure of the variation between the sample means. [In one-way ANOVA, SS(treatment) is sometimes referred to as SS(factor). Because it is a measure of variability *between* the sample means, it is also referred to as SS(between groups) or SS(between samples).]

Formula 11-2

$$SS(treatment) = n_1(\bar{x}_1 - \bar{\bar{x}})^2 + n_2(\bar{x}_2 - \bar{\bar{x}})^2 + \cdots + n_k(\bar{x}_k - \bar{\bar{x}})^2$$
$$= \Sigma n_i(\bar{x}_i - \bar{\bar{x}})^2$$

If the population means ($\mu_1, \mu_2, \ldots, \mu_k$) are equal, then the sample means $\bar{x}_1, \bar{x}_2, \ldots, \bar{x}_k$ will all tend to be close together and also close to $\bar{\bar{x}}$. The result will be a relatively small value of SS(treatment). If the population means are

not all equal, however, then at least one of $\bar{x}_1, \bar{x}_2, \ldots, \bar{x}_k$ will tend to be far apart from the others and also far apart from $\bar{\bar{x}}$. The result will be a relatively large value of SS(treatment).

SS(error) is a sum of squares representing the variability that is assumed to be common to all the populations being considered.

Formula 11-3

$$SS(error) = (n_1 - 1)s_1^2 + (n_2 - 1)s_2^2 + \cdots + (n_k - 1)s_k^2$$
$$= \Sigma(n_i - 1)s_i^2$$

Because SS(error) is a measure of the variance within groups, it is sometimes denoted as SS(within groups) or SS(within samples). Given the preceding expressions for SS(total), SS(treatment), and SS(error), the following relationship will always hold.

Formula 11-4   SS(total) = SS(treatment) + SS(error)

SS(treatment) and SS(error) are both sums of squares, and if we divide each by its corresponding number of degrees of freedom, we get mean squares, defined as follows.

MS(treatment) is a mean square for treatment, obtained as follows:

Formula 11-5 $$MS(treatment) = \frac{SS(treatment)}{k - 1}$$

MS(error) is a mean square for error, obtained as follows:

Formula 11-6 $$MS(error) = \frac{SS(error)}{N - k}$$

MS(total) is a mean square for the total variation, obtained as follows:

Formula 11-7 $$MS(total) = \frac{SS(total)}{N - 1}$$

The SS and MS values are used for determining the F test statistic that applies when the samples do not all have the same size.

Test Statistic for ANOVA with Unequal Sample Sizes

In testing the null hypothesis $H_0: \mu_1 = \mu_2 = \cdots = \mu_k$ against the alternative hypothesis that these means are not all equal, the test statistic

Formula 11-8
$$F = \frac{MS(\text{treatment})}{MS(\text{error})}$$

has an F distribution (when the null hypothesis H_0 is true) with degrees of freedom given by

$$\text{numerator degrees of freedom} = k - 1$$
$$\text{denominator degrees of freedom} = N - k$$

This test statistic is essentially the same as the one given earlier, and its interpretation is also the same as described earlier. The denominator depends only on the sample variances that measure variation within the treatments and is not affected by the differences among the sample means. In contrast, the numerator does depend on differences among the sample means. If the differences among the sample means are extreme, they will cause the numerator to be excessively large, so F will also be excessively large. Consequently, very large values of F suggest unequal means, and the ANOVA test is therefore right-tailed.

Tables are a convenient format for summarizing key results in ANOVA calculations, and Table 11-2 has a format often used in computer displays. (See the preceding Minitab display on page 576.) The entries in Table 11-2 result from the data in Table 11-1.

We use ANOVA methods to conclude that there is sufficient evidence to reject (or fail to reject) the claim of equal population means, but we cannot conclude that any particular mean is different from the others. For the data of Table 11-1, we rejected equality of means, and the boxplots in Figure 11-1 suggest that the times in 1951 have a lower mean, but we should not identify the particular means that are different. There are several other tests that can be used to make such identifications. Procedures for specifically identifying the means that are different are called **multiple comparison procedures.** Comparison of confidence intervals, the Scheffé test, the extended Tukey test, and the Bonferroni test are four common multiple comparison procedures that are usually included in more advanced texts.

As efficient and reliable as computer programs may be, they are totally worthless if you don't understand the relevant concepts. You should recognize that the methods in this section are used to test the claim that several samples

TABLE 11-2	ANOVA Table for Old Faithful Data			
Source of Variation	Sum of Squares SS	Degrees of Freedom	Mean Square MS	F Test Statistic
Treatments	3090.06250	3	1030.02083	6.8967
Error	6571.41667	44	149.350379	
Total	9661.47917	47		

come from populations with the same mean. These methods require normally distributed populations with the same variance, and the samples must be independent. We reject or fail to reject the null hypothesis of equal means by analyzing these two estimates of variance: the variance between samples and the variance within samples. MS(treatment) is an estimate of the variation between samples, and MS(error) is an estimate of the variation within samples. If MS(treatment) is significantly greater than MS(error), we reject the claim of equal means; otherwise, we fail to reject that claim.

11-2 Exercises A: Basic Skills and Concepts

1. Given below are the analysis of variance results for a Minitab display. This display results from three samples with five scores each. The three samples correspond to measured consumer reactions to television commercials from three different advertising companies. Assume that we want to use a 0.05 significance level in testing the null hypothesis that the three different companies produced commercials with the same mean reaction score.
   a. Identify the test statistic.   1. a. $F = 0.49$
   b. Find the critical value from Table A-5.   b. $F = 3.8853$
   c. Identify the P-value.   c. 0.627
   d. Based on the preceding results, should you reject or fail to reject the null hypothesis?   d. Fail to reject the null hypothesis of equal means.
   e. Write a brief statement that clearly summarizes the appropriate conclusion.

```
Source    DF      SS      MS      F       p
Factor     2     3.20    1.60    0.49    0.627
Error     12    39.51    3.29
Total     14    42.71
```

Recommended assignment: Exercises 1, 2, 3, 9, 10, 14.

If students are using Minitab, their results might not be as precise as the answers given in the back of the book. A student might use Minitab to get a test statistic of $F = 15.81$, whereas the answer in the back of the book might be $F = 15.8142$. We recommend telling students that they shouldn't be concerned about those last digits unless the test statistic is very close to the critical value—a condition which is very rare.

e. There is not sufficient evidence to warrant rejection of the claim that the three different companies produced commercials with the same mean reaction score. The means appear to be equal.

2. a. $F = 6.68$

b. $F = 3.8853$

c. 0.011

d. Reject the null hypothesis of equal means.

e. There is sufficient evidence to warrant rejection of the claim that the three different companies produced commercials with the same mean reaction score. The means appear to be different.

3. a. 0.864667

b. 0.46

c. $F = 1.8797$

d. 3.8853 (assuming $\alpha = 0.05$)

e. There is not sufficient evidence to warrant rejection of the claim that the three age-group populations have the same mean body temperature.

4. Test statistic: $F = 103.1651$. Critical F value is between 2.6049 and 2.6802. Reject the claim of equal means. Paper seems to be a particularly large part of the problem (although this cannot be concluded using the statistical methods in Section 11-2).

2. Repeat Exercise 1 assuming that the Minitab results are as shown below.

Source	DF	SS	MS	F	p
Factor	2	20.68	10.34	6.68	0.011
Error	12	18.59	1.55		
Total	14	39.27			

3. Do different age groups have different mean body temperatures? The accompanying table lists the body temperatures of five randomly selected subjects from each of three different age groups.

 a. Find the variance *between* samples by evaluating $ns_{\bar{x}}^2$. (Recall that $s_{\bar{x}}^2$ is the variance of the three sample means.)

 b. Find the variance *within* samples by evaluating the pooled variance s_p^2, which is the mean of the sample variances.

 c. Use the results from parts (a) and (b) to find the F test statistic.

 d. Find the F critical value. (Recall that with equal sample sizes, the degrees of freedom are $k - 1$ for the numerator and $k(n - 1)$ for the denominator.)

 e. Based on the preceding results, what do you conclude about the claim that the three age-group populations have the same mean body temperature?

Body Temperatures (°F) Categorized by Age

18–20	21–29	30 and older
98.0	99.6	98.6
98.4	98.2	98.6
97.7	99.0	97.0
98.5	98.2	97.5
97.1	97.9	97.3
$n_1 = 5$	$n_2 = 5$	$n_3 = 5$
$\bar{x}_1 = 97.940$	$\bar{x}_2 = 98.580$	$\bar{x}_3 = 97.800$
$s_1 = 0.568$	$s_2 = 0.701$	$s_3 = 0.752$

Based on data from Dr. Philip Mackowiak, Dr. Steven Wasserman, and Dr. Myron Levine of the University of Maryland.

4. The City Resource Recovery Company (CRRC) collects the waste discarded by households in a region. Discarded waste must be separated into categories of metal, paper, plastic, and glass. In planning for the equipment needed to collect and process the garbage, CRRC refers to the data we have summarized in Data Set 1 in Appendix B (provided by the

Garbage Project at the University of Arizona). The results (weights in pounds) are summarized in the list that follows. At the 0.05 level of significance, test the claim that the four specific populations of garbage have the same mean. Based on the results, does it appear that these four categories require the same collection and processing resources? Does any single category seem to be a particularly large part of the waste-management problem?

Metal	Paper	Plastic	Glass
$n = 62$	$n = 62$	$n = 62$	$n = 62$
$\bar{x} = 2.218$	$\bar{x} = 9.428$	$\bar{x} = 1.911$	$\bar{x} = 3.752$
$s = 1.091$	$s = 4.168$	$s = 1.065$	$s = 3.108$

For Exercises 5–8, use the data given in the table in the margin. The randomly selected data represent 30 different homes sold in Dutchess County, New York. The zones (1, 4, 7) correspond to different geographical regions of the county. The values of SP are the selling prices in thousands of dollars. The values of LA are the living areas in hundreds of square feet. The values of Acres are the lot sizes in acres, and the Taxes values are the annual tax bills in thousands of dollars. For example, the first home is in zone 1; it sold for $147,000; it has a living area of 2000 square feet; it is on a 0.50-acre lot; and the annual taxes are $1900.

5. At the 0.05 significance level, test the claim that the means of the selling prices are the same in zone 1, zone 4, and zone 7. Does any zone seem to have homes with higher selling prices?
6. At the 0.05 significance level, test the claim that the means of the living areas are the same in zone 1, zone 4, and zone 7. Does any zone seem to have larger homes?
7. At the 0.05 significance level, test the claim that the means of the lot sizes (in acres) are the same in zone 1, zone 4, and zone 7. Does any zone seem to have larger lots?
8. At the 0.05 significance level, test the claim that the means of the tax amounts are the same in zone 1, zone 4, and zone 7. Does any zone seem to have higher taxes?
9. Flammability tests were conducted on children's sleepwear. The Vertical Semirestrained Test was used, in which pieces of fabric were burned under controlled conditions. After the burning stopped, the length of the charred portion was measured and recorded. Results are given in the margin on the following page for the same fabric tested at different laboratories. Because the same fabric was used,

5. Test statistic: $F = 0.1587$.
   Critical value: $F = 3.3541$.
6. Test statistic: $F = 0.6972$.
   Critical value: $F = 3.3541$.
7. Test statistic: $F = 5.0793$.
   Critical value: $F = 3.3541$.
8. Test statistic: $F = 2.0035$.
   Critical value: $F = 3.3541$.

Zone	SP	LA	Acres	Taxes
1	147	20	0.50	1.9
1	160	18	1.00	2.4
1	128	27	1.05	1.5
1	162	17	0.42	1.6
1	135	18	0.84	1.6
1	132	13	0.33	1.5
1	181	24	0.90	1.7
1	138	15	0.83	2.2
1	145	17	2.00	1.6
1	165	16	0.78	1.4
4	160	18	0.55	2.8
4	140	20	0.46	1.8
4	173	19	0.94	3.2
4	113	12	0.29	2.1
4	85	9	0.26	1.4
4	120	18	0.33	2.1
4	285	28	1.70	4.2
4	117	10	0.50	1.7
4	133	15	0.43	1.8
4	119	12	0.25	1.6
7	215	21	3.04	2.7
7	127	16	1.09	1.9
7	98	14	0.23	1.3
7	147	23	1.00	1.7
7	184	17	6.20	2.2
7	109	17	0.46	2.0
7	169	20	3.20	2.2
7	110	14	0.77	1.6
7	68	12	1.40	2.5
7	160	18	4.00	1.8

Laboratory

1	2	3	4	5
2.9	2.7	3.3	3.3	4.1
3.1	3.4	3.3	3.2	4.1
3.1	3.6	3.5	3.4	3.7
3.7	3.2	3.5	2.7	4.2
3.1	4.0	2.8	2.7	3.1
4.2	4.1	2.8	3.3	3.5
3.7	3.8	3.2	2.9	2.8
3.9	3.8	2.8	3.2	
3.1	4.3	3.8	2.9	
3.0	3.4	3.5		
2.9	3.3			

9. Test statistic: $F = 2.9493$.
   Critical value: $F = 2.6060$.
10. Test statistic: $F = 4.0497$.
    Critical value: $F = 3.4028$.
11. Test statistic: $F = 17.8360$.
    Critical value: $F = 2.3683$
    (approximate value, assuming $\alpha = 0.05$).
12. Test statistic: $F = 6.2053$.
    Critical value: $F = 1.9105$
    (assuming $\alpha = 0.05$).
13. Test statistic: $F = 0.5083$.
    Critical value: $F = 2.2899$
    (approximately).
14. Test statistic: $F = 19.9510$.
    Critical value: $F = 4.7865$
    (approximately).

the different laboratories should have obtained the same results. Did they? Use a 0.05 significance level to test the claim that the different laboratories have the same population mean. (The data were provided by Minitab, Inc.)

10. The values in the accompanying list are measured maximum breadths of male Egyptian skulls from different epochs (based on data from *Ancient Races of the Thebaid*, by Thomson and Randall-Maciver). Changes in head shape over time suggest that interbreeding occurred with immigrant populations. Use a 0.05 significance level to test the claim that the mean is the same for the different epochs.

4000 B.C.	1850 B.C.	150 A.D.
131	129	128
138	134	138
125	136	136
129	137	139
132	137	141
135	129	142
132	136	137
134	138	145
138	134	137

11. Refer to Data Set 5 in Appendix B. Using the single factor of month, test the claim that April, May, June, July, August, and September have the same mean ozone concentrations for the states sampled.

12. Refer to the nitrate depositions from acid rain as listed in Data Set 6 in Appendix B. Using the single factor of year, test the claim that the years from 1980 through 1990 have the same mean amount.

13. Refer to Data Set 11 in Appendix B. At the 0.05 significance level, test the claim that the mean weight of M&Ms is the same for each of the six different color populations. If it is the intent of Mars, Inc., to make the candies so that the different color populations have the same mean, do these results suggest that the company has a problem that requires corrective action?

14. Refer to Data Set 9 in Appendix B. Use a 0.01 significance level to test the claim that the mean cotinine level is different for these three groups: nonsmokers not exposed to environmental tobacco smoke, nonsmokers who are exposed to tobacco smoke, and people who smoke. What do the results suggest about "second-hand" smoke?

11-2 **Exercises B: Beyond the Basics**

15. Repeat Exercise 3 after adding 2° to each temperature listed in the 18–20 age group. Compare these results to those found in Exercise 3.

16. If Exercise 3 is repeated after changing the temperature readings from the Fahrenheit scale to the Celsius scale, are the results affected? In general, how is the analysis of variance F test statistic affected by the scale used?

17. How are analysis of variance results affected in each of the following cases?
    a. The same constant is added to every sample score.
    b. Every sample score is multiplied by the same constant.
    c. The order of the samples is changed.

18. Five independent samples of 50 scores each are randomly drawn from populations that are normally distributed with equal variances. We wish to test the claim that $\mu_1 = \mu_2 = \mu_3 = \mu_4 = \mu_5$.
    a. If we used only the methods given in Chapter 8, we would test the individual claims $\mu_1 = \mu_2, \mu_1 = \mu_3, \ldots, \mu_4 = \mu_5$. How many ways could we pair off 5 means?
    b. Assume that for each test of equality between two means, there is a 0.95 probability of not making a type I error. If all possible pairs of means are tested for equality, what is the probability of making no type I errors? (Although the tests are not actually independent, assume that they are.)
    c. If we use analysis of variance to test the claim that $\mu_1 = \mu_2 = \mu_3 = \mu_4 = \mu_5$ at the 0.05 level of significance, what is the probability of not making a type I error?
    d. Compare the results of parts (b) and (c). Which approach is better in the sense of giving us a greater chance of not making a type I error?

19. In this exercise you will verify that when you have two sets of sample data, the t test for independent samples and the ANOVA method of this section are equivalent. Refer to the real-estate data used in Exercises 5–8, but exclude the zone 7 data.
    a. Use a 0.05 level and the method of Section 8-5 to test the claim that zones 1 and 4 have the same mean selling prices. (Assume that both zones have the same variance.)
    b. Use a 0.05 level and the ANOVA method of this section to test the claim made in part (a).
    c. Verify that the squares of the t test statistic and critical value from part (a) are equal to the F test statistic and critical value from part (b).

20. Complete the ANOVA table at the top of the following page if it is known that there are three samples with sizes of 5, 7, and 7, respectively.

15. a. 5.864667
    b. 0.46
    c. $F = 12.7493$
    d. 3.8853 (assuming $\alpha = 0.05$)
    e. There is sufficient evidence to warrant rejection of the claim of equal means. By adding 2° to each score in the first sample, the variance between samples increased, but the variance within samples did not change. The F test statistic increased to reflect the greater disparity among the three sample means.

16. The test statistic F is not affected by the scale used.

17. In each case, the test statistic F is not affected.

18. a. 10
    b. 0.599
    c. 0.95
    d. Analysis of variance

19. a. Test statistic: $t = 0.262$. Critical values: $t = \pm 2.101$. Fail to reject the null hypothesis that zones 1 and 4 have the same mean selling prices.
    b. Test statistic: $F = 0.0688$. Critical value: $F = 4.4139$. Fail to reject the null hypothesis that zones 1 and 4 have the same mean selling prices.
    c. Test statistics: $t^2 = 0.0686 \approx F = 0.0688$. Critical values: $t^2 = 4.4142 \approx F = 4.4139$

20. SS (treatments) = 23.45,
df (treatments) = 2,
df(error) = 16, df(total) =
18, MS(treatments) = 11.725,
MS (error) = 6.25,
F = 1.8760.

Source of Variation	Sum of Squares SS	Degrees of Freedom	Mean Square MS	Test Statistic
Treatments	?	?	?	
Error	100.00	?	?	F = ?
Total	123.45	?		

11-3 Two-Way ANOVA

If the Kruskal-Wallis test (Section 13-5) has not yet been covered, consider covering it now before getting into Section 11-3.

Section 11-2 illustrated the use of analysis of variance in deciding whether three or more populations have the same mean. That section used procedures referred to as one-way analysis of variance (or single-factor analysis of variance) because the data are categorized into groups according to a *single* factor (or treatment). Recall that a factor, or treatment, is a property that is the basis for categorizing the different groups of data. For example, in Table 11-3 the lengths of 30 movies are partitioned into three categories according to the single factor of star rating—a fair movie is rated with 2.0 or 2.5 stars, a good movie is rated with 3.0 or 3.5 stars, and an excellent movie is rated with 4.0 stars. Because the three samples in Table 11-3 are categorized according to a single factor (star rating), we can use the methods from Section 11-2 to test the claim that $\mu_1 = \mu_2 = \mu_3$. Following is the Minitab display for the data in Table 11-3.

MINITAB DISPLAY FOR ONE FACTOR OF STAR RATING

```
Analysis of Variance
Source    DF      SS      MS       F      p
Factor     2    1582     791    1.17   0.325
Error     27   18244     676
Total     29   19826

                         Individual 95% CIs for Mean
                         Based on Pooled StDev
Level      N    Mean    StDev    ----------+---------+---------+------
C1        10  108.80    18.74    (-----------*----------)
C2        10  107.30    14.81    (-----------*----------)
C3        10  123.40    38.16            (----------*-----------)
                                 ----------+---------+---------+------
Pooled StDev =  25.99                   105    .   120        135
```

From the Minitab display we can see that the test statistic is

$$F = \frac{\text{MS(treatment)}}{\text{MS(error)}} = \frac{791}{676} = 1.17$$

TABLE 11-3	Lengths (in min) of Movies Categorized by Star Ratings								
Fair (2.0–2.5 Stars)	98	100	123	92	99	110	114	96	101	155
Good (3.0–3.5 Stars)	93	94	94	105	111	115	133	106	93	129
Excellent (4.0 Stars)	103	193	168	88	121	72	120	106	104	159

The *P*-value of 0.325 indicates that at the 0.05 significance level, we fail to reject the claim of equal means. Based on our sample data, it appears that the three categories of fair, good, and excellent movies have the same mean length.

The one-way analysis of variance for Table 11-3 involves the single factor of star rating. If we were investigating lengths of movies, a more complex analysis might include additional factors such as the nature of the movie (adventure, drama, thriller, comedy, and so on), the studio that produced the movie, the budget for the movie, the actors and actresses who are featured, the director, and the musical score. **Two-way analysis of variance** involves two factors, such as the Table 11-4 factors of (1) the star rating, by which we categorize a movie as fair, good, or excellent, and (2) the Motion Picture Association of America (MPAA) rating, by which movies are separated into a category of R or another category of G, PG, or PG-13. Using the two factors of star rating and MPAA rating, we partition the data into six categories, as

TABLE 11-4	Lengths (in min) of Movies Categorized by Two Factors: Star Ratings and MPAA Ratings (G/PG/PG-13, R)		
	Fair 2.0–2.5 Stars	Good 3.0–3.5 Stars	Excellent 4.0 Stars
MPAA Rating: G/PG/PG-13	98	93	103
	100	94	193
	123	94	168
	92	105	88
	99	111	121
MPAA Rating: R	110	115	72
	114	133	120
	96	106	106
	101	93	104
	155	129	159

shown in Table 11-4. Such subcategories are often called *cells,* so Table 11-4 consists of six cells containing five scores each. Tables 11-3 and 11-4 include the same numbers, but Table 11-3 classifies the data with the single factor of star rating, whereas Table 11-4 classifies the data with the two factors of star rating and MPAA rating.

Because we've already discussed the one-way analysis of variance for the single factor of star rating, it might seem reasonable to simply proceed with another one-way ANOVA for the factor of MPAA rating. Unfortunately, that approach wastes information and totally ignores any effect from an interaction between the two factors.

DEFINITION

There is an **interaction** between two factors if the effect of one of the factors changes for different categories of the other factor.

In using ANOVA for the data of Table 11-4, we will consider the effect of an *interaction* between star ratings and MPAA ratings, as well as the effects of star ratings and the effects of MPAA ratings on movie length. The calculations are quite involved, so *we will assume that a software package is being used.* (The TI-83 calculator can be used for one-way analysis of variance, but it is not programmed for the methods of this section.) The following Minitab display shows results from the data in Table 11-4. To use Minitab, enter the 30 sample scores in column C1, enter the corresponding row numbers (1 or 2) in column C2 to identify the MPAA rating, and enter the corresponding column numbers (1 or 2 or 3) in C3 to identify the star rating. From the main menu, select `Stat`, then `ANOVA`, then `Two-Way`, and proceed to enter C1 for `Response`, enter C2 for `row factor`, and enter C3 for `column factor`, then click on OK.)

The Minitab display includes SS (sum of square) components similar to those described in Section 11-2. Because the circumstances of Section 11-2 involved only a single factor, we used SS(treatment) as a measure of the variation due to the different treatment categories and we used SS(error) as a measure of the variation due to sampling error. We now use SS(STAR) as a measure of variation among the star rating means. We use SS(MPAA) as a measure of variation among the MPAA means. We continue to use SS(error) as a measure of variation due to sampling error. Similarly, we use MS(STAR) and MS(MPAA) for the two different mean squares and continue to use MS(error) as before. Also, we use df(STAR) and df(MPAA) for the two different degrees of freedom.

Table 11-5 compares the analyses of the data in Tables 11-3 and 11-4. Because Tables 11-3 and 11-4 have the same data in the star rating categories,

MINITAB DISPLAY

```
Two-Way Analysis of Variance

Analysis of Variance for C1

Source          DF           SS           MS
MPAA             1           32           32
STAR             2         1582          791
Interaction      2         2256         1128
Error           24        15956          665
Total           29        19826
```

When using Minitab for ANOVA, it's wise to *name* variables so that you can keep track of which results go with which variables, thereby reducing confusion.

TABLE 11-5	Comparison of Components Used in One-Way ANOVA and Two-Way ANOVA

One-Way ANOVA	Two-Way ANOVA
Sample data: Table 11-3	Sample data: Table 11-4
One factor: STAR rating	Two factors: STAR rating and MPAA rating
SS(STAR) = 1582	Same
MS(STAR) = 791	Same
df(STAR) = 2	Same
SS(total) = 19,826	Same
df(total) = 29	Same
	SS(MPAA) = 32
SS(error) = 18,244	SS(interaction) = 2256
	SS(error) = 15,956
	df(MPAA) = 1
df(error) = 27	df(interaction) = 2
	df(error) = 24
	MS(MPAA) = 32
MS(error) = 676	MS(interaction) = 1128
	MS(error) = 665

the calculated values of SS(STAR), MS(STAR), and df(STAR) are also the same in both cases. But note that as we go from the one-factor case (Table 11-3) to the two-factor case (Table 11-4) by partitioning the data according to the second factor of MPAA rating, the value of SS(error) is partitioned into SS(MPAA), SS(interaction), and SS(error). (You can see that 18,244 = 32 + 2256 + 15,956.) Also, df(error) is partitioned into df(MPAA), df(interaction), and df(error), which can be easily verified by noting that 27 = 1 + 2 + 24. Similar partitioning does not apply to MS(error), where the value of MS(error) = 676 does not equal 32 + 1128 + 665.

In executing a two-way analysis of variance, we consider three effects:

1. The effect due to the *interaction* between the two factors of star rating and MPAA rating
2. The effect due to the *row* factor (MPAA rating)
3. The effect due to the *column* factor (star rating)

The following comments summarize the basic procedure for two-way analysis of variance. It is actually quite similar to the procedures presented in Section 11-2. We form conclusions about equal means by analyzing two estimates of variance, and the test statistic F is the ratio of those two estimates. A significantly large value for F indicates that there is a statistically significant difference in means. The following procedure for two-way ANOVA is summarized in Figure 11-4.

Procedure for Two-Way ANOVA

Step 1: In two-way analysis of variance, begin by testing the null hypothesis that there is no interaction between the two factors. Using Minitab for the data in Table 11-4, we get the following test statistic:

$$F = \frac{MS(\text{interaction})}{MS(\text{error})} = \frac{1128}{665} = 1.6962$$

With df(interaction) = 2 and df(error) = 24, we get a critical value of $F = 3.4028$ (assuming a 0.05 significance level). Because the test statistic does not exceed the critical value, we fail to reject the null hypothesis of no interaction between the two factors. There is not sufficient evidence to conclude that movie length is affected by an interaction between star rating and MPAA rating.

Step 2: If we do reject the null hypothesis of no interaction between factors, then we should stop now; we should not proceed with the two additional tests. (If there is an interaction between factors, we shouldn't consider the effects of either factor without considering those of the other.)

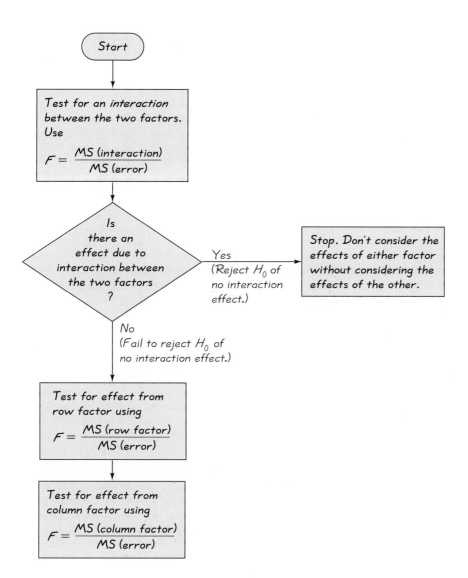

Figure 11-4 Procedure for Two-Way ANOVA

If we fail to reject the null hypothesis of no interaction between factors, then we should proceed to test the following two hypotheses:

H_0: There are no effects from the row factor (that is, the row means are equal).

H_0: There are no effects from the column factor (that is, the column means are equal).

In Step 1, we failed to reject the null hypothesis of no interaction between factors, so we proceed with the next two hypothesis tests identified in Step 2.

For the row factor,

$$F = \frac{MS(MPAA)}{MS(error)} = \frac{32}{665} = 0.0481$$

This value is not significant because the critical value, based on df(MPAA) = 1, df(error) = 24, and a 0.05 significance level, is $F = 4.2597$. We fail to reject the null hypothesis of no effects from MPAA rating. The MPAA rating does not appear to have an effect on movie length.

For the column factor,

$$F = \frac{MS(STAR)}{MS(error)} = \frac{791}{665} = 1.1895$$

This value is not significant because the critical value, based on df(STAR) = 2, df(error) = 24, and a 0.05 significance level, is $F = 3.4028$. We fail to reject the null hypothesis of no effects from star rating. The star rating of a movie does not appear to have an effect on its length.

Special Case: One Observation per Cell and No Interaction

Table 11-4 contains five observations per cell. If our sample data consist of only one observation per cell, we lose MS(interaction), SS(interaction), and df(interaction) because those values are based on sample variances computed for each individual cell. If there is only one observation per cell, there is no variation within individual cells and those sample variances cannot be calculated. Here's how we proceed when there is one observation per cell: *If it seems reasonable to assume (based on knowledge about the circumstances) that there is no interaction between the two factors, make that assumption and then proceed as before to test the following two hypotheses separately:*

H_0: There are no effects from the row factor.

H_0: There are no effects from the column factor.

There may be reason to believe that there is an interaction because there are persons who don't consider a movie to be good unless it has a rating of R resulting from some minimum level of sex and/or violence. However, the star ratings are based on professional movie critics who are much less inclined to consider the R rating when they judge the quality of movies.

As an example, suppose that we have only the first score in each cell of Table 11-4. Note that the two row means are 98.0 and 99.0. Is that difference significant, suggesting that there is an effect due to MPAA rating? The column means are 104.0, 104.0, and 87.5. Are those differences significant, suggesting that there is an effect due to star rating? If we believe the number of stars a movie receives seems to be totally unrelated to whether it is rated G, PG, PG-13, or R, we can then assume that there is no interaction between star rating and MPAA rating. (If we believe there is an interaction, the method described here does not apply.) To illustrate the method, we will make the assumption of no interaction. Following is the Minitab display for the data in Table 11-4, with only the first score from each cell.

MINITAB DISPLAY

```
Two-Way Analysis of Variance

Analysis of Variance for C1

Source        DF          SS         MS
MPAA          1            2          2
STAR          2          363        181
Error         2          793        397
Total         5         1157
```

We first use the results from the Minitab display to test the null hypothesis of no effects from the row factor of MPAA rating. We get the test statistic

$$F = \frac{MS(MPAA)}{MS(error)} = \frac{2}{397} = 0.0050$$

Assuming a 0.05 significance level, we find that the critical value is $F = 18.513$. (Refer to Table A-5 and note that the Minitab display provides degrees of freedom of 1 and 2 for numerator and denominator.) Because the test statistic of $F = 0.0050$ does not exceed the critical value of $F = 18.513$, we fail to reject the null hypothesis; it seems that the MPAA rating does not affect movie length.

We now use the Minitab display to test the null hypothesis of no effect from the column factor of star rating. The test statistic is

$$F = \frac{MS(STAR)}{MS(error)} = \frac{181}{397} = 0.4559$$

which does not exceed the critical value of $F = 19.000$. [Refer to Table A-5 with df(STAR) = 2 and df(error) = 2.] We fail to reject the null hypothesis; it seems that the star rating does not affect movie length. Again, these conclusions are based on very limited data and might not be valid when more sample data are acquired.

Randomized Block Design

When we use the one-way (or single factor) analysis of variance technique and conclude that the differences among the means are significant, we cannot necessarily conclude that the given factor is responsible for the differences. It is possible that the variation of some other unknown factor is responsible for the

interview

Lester Curtin

Chief of the Statistical Methods Staff in the Office of Research and Methodology, National Center for Health Statistics, Centers for Disease Control

Lester Curtin has a doctorate in statistics and is temporarily the Acting Director of the Division of Vital Statistics. He regularly functions as Chief of the Statistical Methods Staff at the National Center for Health Statistics, which is part of the Centers for Disease Control.

What does the National Center for Health Statistics do?

The agency is designated by Congress to collect information on the nation's health. We collect vital statistics, including births, deaths, marriages, and divorces. We also do several large-scale national health surveys, including the National Health Interview Survey, which is a personal interview survey of about 120,000 people each year. We do the National Health and Nutrition Examination Survey, which involves mobile exam centers that go around the country taking measurements on people. We get biological measurements instead of just interview responses, which are subject to measurement error. To determine calcium levels, in addition to asking people what they ate, we can also draw blood to get serum calcium levels. Our statistics are used to monitor the nation's health and to determine what changes are occurring.

Could you cite an example of how your data are used?

The National Health and Nutrition Exam Survey was being conducted in the field at about the time Congress banned lead in gasoline. There was then a movement to put lead back into gasoline. Because the surveyors were in the field over time, we could do a temporal analysis,

•••••••• COOPERATIVE GROUP ACTIVITIES ••••••••

1. *In-class activity:* Divide into groups of five or six students. This activity is a contest of reaction times to determine which group is fastest, if there is a "fastest" group. Use the same reaction timer included with the Cooperative Group Activities in Chapter 5. Test and record the reaction time using the dominant hand of each group member. (Only one try per person.) Each group should calculate the values of n, $\bar{x}$, and s and record those summary statistics on the chalkboard along with the original list of reaction times. After all groups have reported their results, identify the group with the fastest mean reaction time. But is that group actually fastest? Use ANOVA to determine whether the means are significantly different. If they are not, there is no real "winner."

2. *Out-of-class activity:* The World Almanac and Book of Facts includes a section called "Noted Personalities," with subsections comprised of architects, artists, business leaders, cartoonists, social scientists, military leaders, philosophers, political leaders, scientists, writers, composers, entertainers, and others. Design and conduct an observational study that begins with the selection of samples from select groups, to be followed by a comparison of life spans of people from the different categories. Do any particular groups appear to have life spans that are different from the other groups? Can you explain such differences?

interview

Lester Curtin

Chief of the Statistical Methods Staff in the Office of Research and Methodology, National Center for Health Statistics, Centers for Disease Control

Lester Curtin has a doctorate in statistics and is temporarily the Acting Director of the Division of Vital Statistics. He regularly functions as Chief of the Statistical Methods Staff at the National Center for Health Statistics, which is part of the Centers for Disease Control.

What does the National Center for Health Statistics do?

The agency is designated by Congress to collect information on the nation's health. We collect vital statistics, including births, deaths, marriages, and divorces. We also do several large-scale national health surveys, including the National Health Interview Survey, which is a personal interview survey of about 120,000 people each year. We do the National Health and Nutrition Examination Survey, which involves mobile exam centers that go around the country taking measurements on people. We get biological measurements instead of just interview responses, which are subject to measurement error. To determine calcium levels, in addition to asking people what they ate, we can also draw blood to get serum calcium levels. Our statistics are used to monitor the nation's health and to determine what changes are occurring.

Could you cite an example of how your data are used?

The National Health and Nutrition Exam Survey was being conducted in the field at about the time Congress banned lead in gasoline. There was then a movement to put lead back into gasoline. Because the surveyors were in the field over time, we could do a temporal analysis,

•••••••• FROM DATA TO DECISION ••••••••

Was the Experiment Designed Correctly?

An experiment was designed to compare the effects of weight loss to the effects of aerobic exercise training on coronary artery disease risk factors in men who are healthy but sedentary, obese, and middle-aged or older. (See "Effects of Weight Loss vs. Aerobic Exercise Training on Risk Factors for Coronary Disease in Healthy, Obese, Middle-aged and Older Men," by Katzel, Bleecker, et al., *Journal of the American Medical Association*, Vol. 274, No. 24.) The study involved 170 men divided into three groups: (1) 73 men were assigned to lose weight, (2) 71 men were assigned to aerobic exercise, and (3) 26 men were in a control group instructed to not lose weight or change the level of physical activity. In such a study, it is important to begin with groups that are similar in "baseline" characteristics of age, weight, body mass, body fat, and so on. In a report of the study, it was claimed that "There were no significant differences among the three groups in any of the baseline characteristics." The mean weights for the three sample groups are 94.3 kg, 93.9 kg, and 88.3 kg. Are these differences significant? Is there sufficient evidence to warrant rejection of the claim of no significant differences among the three groups? Analyze the data in the following list, form a conclusion, and write a report stating your results.

Weights (in kg) of men assigned to the weight loss group:

95	82	94	87	84	111	98	101	85	117	111	86	114	85	98	102	97
92	76	95	88	107	89	81	93	77	91	106	97	107	78	101	88	92
101	110	96	83	92	106	107	106	110	102	105	84	93	69	96	70	80
108	87	86	93	78	98	86	93	101	94	85	107	74	101	105	93	89
90	103	103	80	116												

Weights (in kg) of men assigned to the aerobic exercise group:

90	101	89	76	93	110	87	108	109	105	83	98	100	82	71	67	126
114	102	78	101	79	97	100	106	84	80	100	103	80	71	99	108	84
78	92	92	101	107	95	106	84	111	108	109	79	116	98	77	110	93
95	85	86	104	93	92	86	93	74	74	89	85	91	104	97	97	80
104	95	109														

Weights (in kg) of men assigned to the control group:

91	71	74	86	97	78	81	82	78	106	79	110	91	93	87	72	95
88	83	96	98	77	84	95	103	102								

Presidents		Popes		Kings and Queens	
Fillmore	24	Pius VIII	2	Edward VIII	36
Pierce	16	Greg XVI	15	George VI	15
Buchanan	12	Pius IX	32		
Lincoln	4	Leo XIII	25		
A. Johnson	10	Pius X	11		
Grant	17	Ben XV	8		
Hayes	16	Pius XII	17		
Garfield	0	Pius XIII	19		
Arthur	7	John XXIII	5		
Cleveland	24	Paul VI	15		
Harrison	12	John Paul I	0		
McKinley	4				
T. Roosevelt	18				
Taft	21				
Wilson	11				
Harding	2				
Coolidge	9				
Hoover	36				
F. Roosevelt	12				
Truman	28				
Kennedy	3				
Eisenhower	16				
L. Johnson	9				
Nixon	25				

Based on data from *Computer-Interactive Data Analysis*, by Lunn and McNeil, John Wiley & Sons.

2. In the United States, weights of newborn babies are normally distributed with a mean of 7.54 lb and a standard deviation of 1.09 lb (based on data from "Birth Weight and Perinatal Mortality," by Wilcox, Skjaerven, Buekens, and Kiely, *Journal of the American Medical Association*, Vol. 273, No. 9).

a. If a newborn baby is randomly selected, what is the probability that he or she weighs more than 8.00 lb?   2. a. 0.3372

b. If 16 newborn babies are randomly selected, what is the probability that their mean weight is more than 8.00 lb?   b. 0.0455

c. What is the probability that each of the next three babies will have a birth weight greater than 7.54 lb?   c. 1/8

Computer Project

For presidents, popes, and British monarchs after 1690, the accompanying table lists the numbers of years that they lived after their inauguration, election, or coronation. Use boxplots and analysis of variance to determine whether the survival times for the different groups differ. Conduct the analysis of variance by running STATDISK or Minitab or some other statistical software package. Obtain printed copies of the computer displays and write your observations and conclusions.

Presidents		Popes		Kings and Queens	
Washington	10	Alex VIII	2	James II	17
J. Adams	29	Innoc XII	9	Mary II	6
Jefferson	26	Clem XI	21	William III	13
Madison	28	Innoc XIII	3	Anne	12
Monroe	15	Ben XIII	6	George I	13
J. Q. Adams	23	Clem XII	10	George II	33
Jackson	17	Ben XIV	18	George III	59
Van Buren	25	Clem XIII	11	George IV	10
Harrison	0	Clem XIV	6	William IV	7
Tyler	20	Pius VI	25	Victoria	63
Polk	4	Pius VII	23	Edward VII	9
Taylor	1	Leo XII	6	George V	25

table continues on page 605

4. Refer to the same data used in Exercise 3 and assume that fuel consumption is not affected by an interaction between engine size and type of transmission. Use a 0.05 level of significance to test the claim that fuel consumption is not affected by engine size.

5. Refer to the same data used in Exercise 3 and assume that fuel consumption is not affected by an interaction between engine size and type of transmission. Use a 0.05 level of significance to test the claim that fuel consumption is not affected by type of transmission.

• •

Cumulative Review Exercises

1. The Rocky Mountain Brewing Company plans to launch a major media campaign. Three advertising companies prepared trial commercials in an attempt to win a $2 million contract. The commercials were tested on randomly selected consumers, whose reactions were measured; the results are summarized in the accompanying table. (Higher scores indicate more positive reactions to the commercial.)

   a. Construct a boxplot for each of the three samples. Use the same scale so that the boxplots can be compared. Do the boxplots reveal any notable differences?

   b. Find the mean and standard deviation for each of the three sets of sample data.

   c. Use the methods of Section 8-5 to test the claim that the population of Barnum scores has a mean that is equal to the population mean for Solomon & Ford scores. Use a 0.05 significance level.

   d. For each of the three samples, construct a 95% confidence interval estimate of the population mean μ. Do the results suggest any notable differences?

   e. At the 0.05 significance level, test the claim that the three populations have the same mean reaction score. If you were responsible for advertising at this brewery, which company would you select on the basis of these results? Why?

Barnum Advertising Co.	Solomon & Ford Advertising	Diaz and Florio Advertising
52 68 75 40	69 73 82 59	42 73 69 53
77 63 55 72	66 84 75 70	57 61 73 74

Answers (margin)

4. Test statistic: $F = 1.9115$. Critical value: $F = 5.1433$. The size of the engine does not appear to have an effect on fuel consumption.

5. Test statistic: $F = 3.5664$. Critical value: $F = 5.9874$. The type of transmission does not appear to have an effect on fuel consumption.

1. b. Barnum: $\bar{x} = 62.8$, $s = 12.9$; Solomon & Ford: $\bar{x} = 72.3$, $s = 8.2$; Diaz & Florio: $\bar{x} = 62.8$, $s = 11.6$

   c. F test results: Test statistic is $F = 2.4592$. Critical value: $F = 4.9949$. Fail to reject H_0: $\sigma_1^2 = \sigma_2^2$. Test of means: Test statistic is $t = -1.760$. Critical values: $t = \pm 2.145$.

   d. Barnum: $52.0 < \mu < 73.5$; Solomon & Ford: $65.4 < \mu < 79.1$; Diaz and Florio: $53.1 < \mu < 72.4$. Solomon & Ford seem slightly higher.

   e. Test statistic: $F = 1.9677$. Critical value: $F = 3.4668$.

Review Exercises

1. The Associated Insurance Institute sponsors studies of the effects of drinking and driving. In one such study, 3 groups of adult men were randomly selected for an experiment designed to measure their blood alcohol levels after consuming 5 drinks. Members of group A were tested after 1 hour, members of group B were tested after 2 hours, and members of group C were tested after 4 hours. The results are given in the accompanying table; the Minitab display for these data is also shown. At the 0.05 level of significance, test the claim that the 3 groups have the same mean level.

A	B	C
0.11	0.08	0.04
0.10	0.09	0.04
0.09	0.07	0.05
0.09	0.07	0.05
0.10	0.06	0.06
		0.04
		0.05

Analysis of Variance

Source	DF	SS	MS	F	p
Factor	2	0.0076571	0.0038286	46.90	0.000
Error	14	0.0011429	0.0000816		
Total	16	0.0088000			

2. The accompanying list shows selling prices (in thousands of dollars) for homes located on Long Beach Island in New Jersey. Different mean selling prices are expected for the different locations. Do these sample data support the claim of different mean selling prices? Use a 0.05 significance level.

Oceanside:	235	395	547	469	369	279
Oceanfront:	538	446	435	639	499	399
Bayside:	199	219	239	309	399	190
Bayfront:	695	389	489	489	599	549

3. Twelve different 4-cylinder cars were tested for fuel consumption (in mi/gal) after being driven under identical highway conditions; the results are listed in the table and accompanying Minitab display. At the 0.05 significance level, test the claim that fuel consumption is not affected by an *interaction* between engine size and transmission type.

ANALYSIS OF VARIANCE MPG

SOURCE	DF	SS	MS
TRANS	1	40.3	40.3
SIZE	2	43.2	21.6
INTERACTION	2	1.2	0.6
ERROR	6	68.0	11.3
TOTAL	11	152.7	

Highway Fuel Consumption (mi/gal) of Different 4-Cylinder Compact Cars

	Engine Size (liters)		
	1.5	2.2	2.5
Automatic Transmission	31, 32	28, 26	31, 23
Manual Transmission	33, 36	33, 30	27, 34

• •

Vocabulary List

analysis of variance (ANOVA)

one-way analysis of variance

single-factor analysis of variance

treatment

factor

variance between samples

variation due to treatment

variance within samples

variation due to error

multiple comparison procedures

two-way analysis of variance

interaction

completely randomized design

rigorously controlled design

randomized block design

block

• •

Review

In this chapter we used analysis of variance (or ANOVA) to test for equality of population means. This method requires (1) normally distributed populations, (2) populations with the same standard deviation (or variance), and (3) random samples that are independent of each other.

In Section 11-2 we considered one-way analysis of variance, characterized by sample data categorized according to a single factor. The following are key features of one-way analysis of variance:

- The F test statistic is based on the ratio of two different estimates of the common population variance σ^2, as shown below.

$$F = \frac{\text{variance between samples}}{\text{variance within samples}} = \frac{\text{MS(treatment)}}{\text{MS(error)}}$$

- Critical values of F are found in Table A-5, with the degrees of freedom (df) found as follows:

df for numerator $= k - 1$ (where $k =$ number of samples)

df for denominator $= N - k$ (where $N =$ the total number of values in all samples combined)

In Section 11-3 we considered two-way analysis of variance, characterized by data categorized according to two different factors. The method for two-way analysis of variance is summarized in Figure 11-4. We also considered two-way analysis of variance with one observation per cell and two-way analysis of variance with randomized block design. These cases use the same ANOVA methods, but in a randomized block experiment we try to control for extraneous variation by making blocks one of the factors.

Because of the nature of the calculations required throughout this chapter, we emphasized the interpretation of computer displays.

8. Test statistic: $F = 5.8007$. Critical value: $F = 5.1433$. Reject the null hypothesis of no effects from the machines. It appears that the machines do affect the numbers of support beams manufactured.

9. (The given results use the first nine values for each cell, but the pulse rate of 8 was excluded as being not feasible.) Test statistic: $F = 0.4200$. Critical value: $F = 4.1709$ (approximately). There does not appear to be an interaction between gender and smoking.

10. Test statistic: $F = 11.9141$. Critical value: $F = 4.1709$ (approximately, assuming $\alpha = 0.05$). Gender appears to have an effect on pulse rates.

11. Test statistic: $F = 0.1814$. Critical value: $F = 4.1709$ (approximately, assuming $\alpha = 0.05$). Smoking does not appear to have an effect on pulse rates. (This is not surprising, because college students haven't been smoking very long.)

12. Transposing the table does not change the results or conclusions.

13. Nothing changes.

14. The SS and MS values are multiplied by 100, but the values of the test statistics do not change.

15. Answers vary.

8. Using a 0.05 significance level, test the claim that the choice of machine has no effect on the production output. Identify the test statistic and critical value, and state the conclusion. (Refer to the instructions on page 599.)

9. Refer to Data Set 8 in Appendix B and construct a table with pulse rates categorized according to the two factors of gender and whether the individual smokes. Select nine scores for each cell and test the null hypothesis of no interaction between gender and smoking.

10. Use the same data collected for Exercise 9, assume that pulse rates are not affected by an interaction between gender and smoking, and test the null hypothesis that gender has no effect on pulse rates.

11. Use the same data collected for Exercise 9, assume that pulse rates are not affected by an interaction between gender and smoking, and test the null hypothesis that smoking has no effect on pulse rates.

11-3 Exercises B: Beyond the Basics

12. Use a statistics software package, such as Minitab or SPSS/PC, that can produce results for two-way analysis of variance. First enter the data in the table used for Exercises 1–4 and verify that the results are as given in this section. Then transpose the table by making the MPAA rating the column factor and making the star rating the row factor. Obtain the computer display for the transposed table, and compare the results to those previously given.

13. Refer to the data in the table used for Exercises 1–4 and subtract 10 from each table entry. Using a statistics software package with a two-way analysis of variance capability, determine the effects of subtracting 10 from each entry.

14. Refer to the data in the table used for Exercises 1–4 and multiply each table entry by 10. Using a statistics software package with a two-way analysis of variance capability, determine the effects of multiplying each entry by 10.

15. In analyzing Table 11-4, we concluded that movie length is not affected by an interaction between star rating and MPAA rating; it is not affected by star rating; and it is not affected by MPAA rating.
   a. Change the table entries so that there is an effect from the interaction between star rating and MPAA rating.
   b. Change the table entries so that there is no effect from the interaction between star rating and MPAA rating and there is no effect from star rating, but there is an effect from MPAA rating.
   c. Change the table entries so that there is no effect from the interaction between star rating and MPAA rating and there is no effect from MPAA rating, but there is an effect from star rating.

In Exercises 5 and 6, use only the first value from each of the eight cells in the table used for Exercises 1–4. When we use only these first values, the Minitab display is as follows.

MINITAB DISPLAY

Two-Way Analysis of Variance

Analysis of Variance for C1

Source	DF	SS	MS
MPAA	1	0	0
STAR	3	459	153
Error	3	799	266
Total	7	1258	

5. Assuming that there is no effect on movie length from the interaction between star rating and MPAA rating, test the null hypothesis that MPAA rating has no effect on movie length. Identify the test statistic and critical value, and state the conclusion. Use a 0.05 significance level.

6. Assuming that there is no effect on movie length from the interaction between star rating and MPAA rating, test the null hypothesis that star rating has no effect on movie length. Identify the test statistic and critical value, and state the conclusion. Use a 0.05 significance level.

Exercises 7 and 8 refer to the sample data in the following table and the corresponding Minitab display. The table entries are the numbers of support beams manufactured by four different operators using each of three different machines. In the Minitab results, the operators are represented by C2 and the machines are represented by C3.

MINITAB DISPLAY

Analysis of Variance for C1

Source	DF	SS	MS
C2	3	59.58	19.86
C3	2	93.17	46.58
Error	6	48.17	8.03
Total	11	200.92	

		Machine		
		1	2	3
	1	66	74	67
Operator	2	58	67	68
	3	65	71	65
	4	60	64	66

7. Using a 0.05 significance level, test the claim of the hypothesis that the four operators have the same mean production output. Identify the test statistic and critical value, and state the conclusion.

5. Test statistic: $F = 0$. Critical value: $F = 10.128$. It appears that MPAA rating does not have an effect on movie length.

6. Test statistic: $F = 0.5752$. Critical value: $F = 9.2766$. It appears that star rating does not have an effect on movie length.

7. Test statistic: $F = 2.4732$. Critical value: $F = 4.7571$. Fail to reject the null hypothesis of no effects from the operators. It appears that the operators do not affect the numbers of support beams manufactured.

(11-3) Exercises A: Basic Skills and Concepts

Recommended assignment: Exercises 1–6 and 9–11. (Exercises 1–6 are based on a computer display that is provided, but Exercises 9–11 require reorganization of data from a data set in Appendix B, as well as using software to obtain results.)

In Exercises 1–4, use the Minitab display, which corresponds to the data in the accompanying table.

MINITAB DISPLAY

```
Analysis of Variance for C1

Source           DF            SS         MS
MPAA              1            14         14
STAR              3          1049        350
Interaction       3          3790       1263
Error             8          5560        695
Total            15         10413
```

Lengths (in min) of Movies Categorized by Star Ratings and MPAA Ratings (G/PG/PG-13, R)

	Poor 0.0–1.5 Stars	Fair 2.0–2.5 Stars	Good 3.0–3.5 Stars	Excellent 4.0 Stars
MPAA Rating: G/PG/PG-13	108	98	93	103
	91	100	94	193
MPAA Rating: R	105	110	115	72
	96	114	133	120

1. a. 1263   b. 695
   c. 350   d. 14
2. Test statistic: $F = 1.8173$. Critical value: $F = 4.0662$, assuming that the significance level is $\alpha = 0.05$. There appears to be no effect on movie length that results from an interaction between MPAA rating and star rating.
3. Test statistic: $F = 0.5036$. Critical value: $F = 4.0662$, assuming that the significance level is $\alpha = 0.05$. It appears that star rating does not have an effect on movie length.
4. Test statistic: $F = 0.0201$. Critical value: $F = 5.3177$, assuming that the significance level is $\alpha = 0.05$. It appears that MPAA rating does not have an effect on movie length.

1. Identify the indicated values.
   a. MS(interaction)
   b. MS(error)
   c. MS(STAR)
   d. MS(MPAA)
2. Find the test statistic and critical value for the null hypothesis of no interaction between star rating and MPAA rating. What do you conclude?
3. Assume that the length of a movie is not affected by an interaction between its star rating and MPAA rating. Find the test statistic and critical value for the null hypothesis that star rating has no effect on movie length. What do you conclude?
4. Assume that the length of a movie is not affected by an interaction between its star rating and MPAA rating. Find the test statistic and critical value for the null hypothesis that MPAA rating has no effect on movie length. What do you conclude?

```
                        MINITAB DISPLAY

Two-way Analysis of Variance

Analysis of Variance for C1

Source          DF                  SS              MS
GAS             2                47.17           23.58
CAR             3               390.25          130.08
Error           6                11.50            1.92
Total          11               448.92
```

Considering the treatment (row) effects, we calculate the test statistic

$$F = \frac{MS(GAS)}{MS(error)} = \frac{23.58}{1.92} = 12.2813$$

From the display we get df(GAS) = 2 and df(error) = 6, and we can now refer to Table A-5. With a 0.05 significance level, the critical value is $F = 5.1433$. Because the test statistic of $F = 12.2813$ exceeds the critical value of $F = 5.1433$, we reject the null hypothesis. The grade of gas does seem to have an effect on the mileage.

For the block (column) effects, the test statistic is

$$F = \frac{MS(CAR)}{MS(error)} = \frac{130.08}{1.92} = 67.7500$$

and the critical value is $F = 4.7571$. We reject the null hypothesis of equal block means and conclude that the different cars have different mileage values. (If we interchange the roles played by blocks and treatments by transposing Table 11-6, we obtain the same results.)

In this section we have briefly discussed an important branch of statistics. We have emphasized the interpretation of computer displays while omitting the manual calculations and formulas, which are quite formidable. More advanced texts typically discuss this topic in much greater detail, but our intent here is to give some general insight into the nature of two-way analysis of variance.

```
                    MINITAB DISPLAY

  Two-Way Analysis of Variance

  Analysis of Variance for C1

  Source            DF            SS            MS
  MPAA              1             2             2
  STAR              2             363           181
  Error             2             793           397
  Total             5             1157
```

We first use the results from the Minitab display to test the null hypothesis of no effects from the row factor of MPAA rating. We get the test statistic

$$F = \frac{MS(MPAA)}{MS(error)} = \frac{2}{397} = 0.0050$$

Assuming a 0.05 significance level, we find that the critical value is $F = 18.513$. (Refer to Table A-5 and note that the Minitab display provides degrees of freedom of 1 and 2 for numerator and denominator.) Because the test statistic of $F = 0.0050$ does not exceed the critical value of $F = 18.513$, we fail to reject the null hypothesis; it seems that the MPAA rating does not affect movie length.

We now use the Minitab display to test the null hypothesis of no effect from the column factor of star rating. The test statistic is

$$F = \frac{MS(STAR)}{MS(error)} = \frac{181}{397} = 0.4559$$

which does not exceed the critical value of $F = 19.000$. [Refer to Table A-5 with df(STAR) = 2 and df(error) = 2.] We fail to reject the null hypothesis; it seems that the star rating does not affect movie length. Again, these conclusions are based on very limited data and might not be valid when more sample data are acquired.

Randomized Block Design

When we use the one-way (or single factor) analysis of variance technique and conclude that the differences among the means are significant, we cannot necessarily conclude that the given factor is responsible for the differences. It is possible that the variation of some other unknown factor is responsible for the

differences. One way to reduce the effect of the extraneous factors is to design the experiment so that it has a **completely randomized design,** in which each element is given the same chance of belonging to the different categories, or treatments. Another way to reduce the effect of extraneous factors is to use a **rigorously controlled design,** in which elements are carefully chosen so that all other factors have no variability. That is, select elements that are the same in every characteristic except for the single factor being considered.

Yet another way to control for extraneous variation is to use a **randomized block design,** in which a measurement is obtained for each treatment on each of several individuals that are matched according to similar characteristics. A **block** is a group of similar individuals. See the following example, where we use four different cars to test the mileage produced by three different grades (regular, extra, premium) of a gasoline. Using a randomized block design as a way to control differences among cars, we burn each grade of gas in each of the four cars, randomly selecting the order in which this is done. We have three treatments corresponding to the three grades of gas; we have four blocks corresponding to the four cars used.

EXAMPLE   When we burn each of three different grades of gas in each of four different cars, we obtain the results shown in Table 11-6. Use two-way analysis of variance to test the claim that the grade of gasoline does not affect mileage, and test the claim that the different cars do not affect mileage.

SOLUTION   The method we use is the same as the method used for the special case of one observation per cell, but we need not use special knowledge of the circumstances to assume that there is no interaction. We can assume that there is no interaction because of the way we designed the experiment. The Minitab display for the data in Table 11-6 is as follows.

TABLE 11-6	Mileage (mi/gal) for Different Grades of Gasoline			
		Block		
		Car 1	Car 2	Car 3	Car 4
	Regular Gas	19	33	23	27
Treatment	Extra Gas	19	34	26	29
	Premium Gas	22	39	26	34

and we saw that lead levels in people were dropping. This conclusion was fought by the gasoline industry's experts. The government had its experts, and the two groups went round and round. We used data from the National Health and Nutrition Exam Survey to present congressional testimony in favor of keeping the lead out of gasoline. The validity of our data was upheld and was very influential in saving the congressional law to keep the lead out of gasoline.

It's a lot of fun dealing with health statistics. We're measuring serum cotinine as a way to see the effects of passive smoking. We used data on monetary expenditures for mathematical modeling of health costs used for the health care reform package that President Clinton proposed. We're looking for what might make a difference in terms of dietary habits on future health.

In your field, is the use of statistics increasing, decreasing, or remaining about the same?

The National Center for Health Statistics once had a mandate that our group should simply collect and disseminate data. Over the last few years we've become much more involved with the analytic aspects of data. We are progressing more with the use of sophisticated statistical methods.

What statistical methods do you use?

We do just about everything. We do a lot of survey design work. We use a lot of quality control in coding and editing the data. We study the cognitive aspect of survey design and how people think when they answer a question. We might ask a question two or three different ways to see if that has an impact on how people are responding. We use confidence intervals, hypothesis testing, regression, time series analysis, ARIMA modeling, logistic regression on risk factors, and analysis of variance. The National Center for Health Statistics employs about 200 statisticians and another 50 people with general quantitative statistical backgrounds. We have employees with bachelor's degrees, master's degrees, and Ph.Ds.

How do you convince people of the validity of your data?

We have a basic policy of not presenting an estimate unless we present some measure of a sampling error. We also have a policy that when we publish a report from the National Center for Health Statistics and we say that one statistic is different from another, we have to back that up with a statistical test.

12

Statistical Process Control

12-1 Overview

The pattern of data over time is discussed as an important characteristic in a data set. The importance of quality control and monitoring processes is discussed.

12-2 Control Charts for Variation and Mean

A definition is presented for "process" data. Analysis of process data is described with run charts and control charts. Illustrations are provided of the construction and interpretation of (1) run charts (used

to monitor the behavior of individual process elements), (2) control charts for R (used to monitor process variation), and (3) control charts for $\bar{x}$ (used to monitor the process mean).

12-3 Control Charts for Attributes

This section describes the construction of p charts, which are used to monitor the proportion of items having some attribute, such as being defective.

Chapter Problem:

An axial load of an aluminum can is the maximum weight supported by its sides. It is important to have an axial load high enough so that the can isn't crushed when the top lid is pressed into place. The sample data in Table 12-1 come from a population of cans with a 0.0109 in. thickness. In this chapter we will analyze the manufacturing process by focusing on the pattern of axial loads over time.

Table 12-1   *Axial Loads (in pounds) of Aluminum Cans*

Day	Axial Load (pounds)							$\bar{x}$	Median	Range	s
1	270	273	258	204	254	228	282	252.7	258	78	27.6
2	278	201	264	265	223	274	230	247.9	264	77	29.7
3	250	275	281	271	263	277	275	270.3	275	31	10.6
4	278	260	262	273	274	286	236	267.0	273	50	16.3
5	290	286	278	283	262	277	295	281.6	283	33	10.7
6	274	272	265	275	263	251	289	269.9	272	38	11.8
7	242	284	241	276	200	278	283	257.7	276	84	31.4
8	269	282	267	282	272	277	261	272.9	272	21	7.9
9	257	278	295	270	268	286	262	273.7	270	38	13.5
10	272	268	283	256	206	277	252	259.1	268	77	25.9
11	265	263	281	268	280	289	283	275.6	280	26	10.1
12	263	273	209	259	287	269	277	262.4	269	78	25.3
13	234	282	276	272	257	267	204	256.0	267	78	27.8
14	270	285	273	269	284	276	286	277.6	276	17	7.3
15	273	289	263	270	279	206	270	264.3	270	83	27.0
16	270	268	218	251	252	284	278	260.1	268	66	22.3
17	277	208	271	208	280	269	270	254.7	270	72	32.2
18	294	292	289	290	215	284	283	278.1	289	79	28.1
19	279	275	223	220	281	268	272	259.7	272	61	26.5
20	268	279	217	259	291	291	281	269.4	279	74	25.9
21	230	276	225	282	276	289	288	266.6	276	64	27.2
22	268	242	283	277	285	293	248	270.9	277	51	19.3
23	278	285	292	282	287	277	266	281.0	282	26	8.4
24	268	273	270	256	297	280	256	271.4	270	41	14.3
25	262	268	262	293	290	274	292	277.3	274	31	14.1

12-1 Overview

So far, we have considered data sets by analyzing central tendency, variation, and distribution shape. But those characteristics sometimes change over time, so that the population does not have fixed properties. This chapter focuses on the analysis of such patterns over time. The concepts of this chapter are used extensively by companies striving to succeed with goods or services that are monitored with the goal of maintaining high quality.

In Chapter 2 we noted that data sets generally have three important characteristics: (1) the nature or shape of the *distribution*, such as bell shaped; (2) a *representative value*, such as the mean; and (3) a measure of the scattering or *variation*, such as the standard deviation. In this chapter we consider a fourth characteristic of some data that is often critically important: its *pattern over time*. By monitoring this characteristic, we are better able to control the production of goods and services, thereby ensuring better quality.

There is currently a strong trend toward trying to improve the quality of American goods and services, and the methods presented in this chapter are being used by growing numbers of businesses. Evidence of the increasing importance of quality is found in its greater role in advertising and the growing number of books and articles that focus on the issue of quality. In many cases, job applicants enjoy a distinct advantage when they can tell employers that they have studied statistics and methods of quality control. This chapter will present some of the basic tools commonly used to monitor quality.

Minitab and other software packages include programs for automatically generating charts of the type discussed in this chapter, and we will include several examples of Minitab displays. The use of such software will save much time and effort in achieving the objective of improving quality by monitoring a process.

12-2 Control Charts for Variation and Mean

This section considers (1) run charts, (2) control charts to monitor process variation, and (3) control charts to monitor a process mean. Section 12-3 considers control charts for monitoring some qualitative attribute, such as whether the process items are defective.

In this section we focus on the pattern of data over time. Such data are often referred to as *process data*.

DEFINITION

Process data are data arranged according to some time sequence. They are measurements of a characteristic of goods or services that results from some combination of equipment, people, materials, methods, and conditions.

For example, Table 12-1 includes process data consisting of the measured axial loads of aluminum cans over 25 consecutive days of production. Each day, seven cans were randomly selected and tested. Because the data in Table 12-1 are arranged according to the time at which they were selected, they are process data. It is very important to recognize this point:

Important characteristics of process data can change over time.

In making aluminum cans, a manufacturer might use a machine that begins with desired properties, but if the machine wears with use, those desired properties might deteriorate to unacceptable levels. Companies have gone bankrupt because they allowed manufacturing processes to continue without constant monitoring. There are various methods that can be used to monitor a process to ensure that the important desired characteristics don't change—the *run chart* is one such method.

DEFINITION

A **run chart** is a sequential plot of *individual* data values over time. One axis (usually the vertical axis) is used for the data values, and the other axis (usually the horizontal axis) is used for the time sequence.

EXAMPLE   Treating the 175 axial loads in Table 12-1 as a string of consecutive measurements, construct a run chart by using a vertical axis for the axial loads and a horizontal axis to identify the order of the sample data.

SOLUTION   Figure 12-1 is the Minitab-generated run chart for the data of Table 12-1. The vertical scale is designed to be suitable for axial loads ranging from 200 lb to 297 lb, which are the minimum and maximum values in Table 12-1. The horizontal scale is designed to include the 175 values arranged in sequence. The first point represents the first value of 270 lb, the second point represents the second value of 273 lb, and so on.   *(continued)*

Because Minitab has excellent programs for generating run charts and control charts, this chapter uses Minitab displays. If Minitab is not available, consider using STATDISK or a TI-83 calculator to generate graphs as *scatter diagrams.* To generate the run chart for the data of Table 12-1, use the paired data (270, 1), (273, 2), . . . , (292, 175) where the first coordinates are the data and the second coordinates are consecutive positive integers. Manual construction of graphs is more time consuming, but students do become much more personally involved.

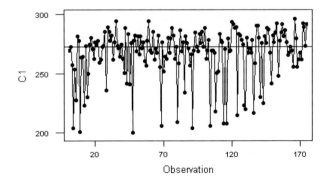

Figure 12-1   Run Chart of Axial Loads in Table 12-1

In Figure 12-1, the horizontal scale identifies the sample number, so the number 20 indicates the 20th sample item. The vertical scale represents the measured axial load, so 250 indicates a load of 250 lb. Now examine Figure 12-1 and observe that there aren't any dramatic patterns that jump out begging for attention. Figure 12-1 does not appear to reveal any problems that need correction.

Using Computer Software to Generate Run Charts

Minitab can be used to construct run charts, such as the one shown in Figure 12-1. Begin by entering all of the sample data in column C1. Select the options of Stat, then Quality Tools, then Run Chart. In the indicated boxes, enter C1 for the single column variable, enter 1 for the subgroup size, then click on OK.

Whether we construct the run chart manually or by using computer software, it is important to interpret the chart carefully in order to learn how the process is behaving.

DEFINITION

A process is **statistically stable** (or **within statistical control**) if it has only natural variation, with no patterns, cycles, or any unusual points.

Only when a process is statistically stable can its data be treated as if they came from a population with a constant mean, standard deviation, distribution, and other characteristics. Figure 12-2 illustrates these common patterns, indicating that the process of filling 16-oz soup cans is not statistically stable:

- **Figure 12-2(a):** There is an obvious *upward trend* that corresponds to values that are increasing over time. If the filling process were to follow this type of pattern, the cans would be filled with more and more soup until they began to overflow, eventually leaving the employees swimming in soup.

- **Figure 12-2(b):** There is an obvious *downward trend* that corresponds to steadily decreasing values. The cans would be filled with less and less soup until they were extremely underfilled. Such a process would require a complete reworking of the cans in order to get them full enough for distribution to consumers.

- **Figure 12-2(c):** There is an *upward shift*. A run chart such as this one might result from an adjustment to the filling process, making all subsequent values higher.

Consider discussing a class example of controlling the process of class absences. How is the course affected if attendance varies by large amounts? How is the course affected if absences gradually increase over the semester? If absences lack statistical stability, what can instructors do to fix the problem so that they are within statistical control?

Consider asking students to identify possible *causes* for each of the patterns in Figure 12-2. For example, the upward trend in Figure 12-2(a) might result from a nozzle with a hole that is growing larger through wear. As the soup makes the nozzle opening larger, more soup is poured in the same amount of time, resulting in an upward trend. This reinforces the importance of using statistics to help in making practical changes so that the process continues to have a high level of quality.

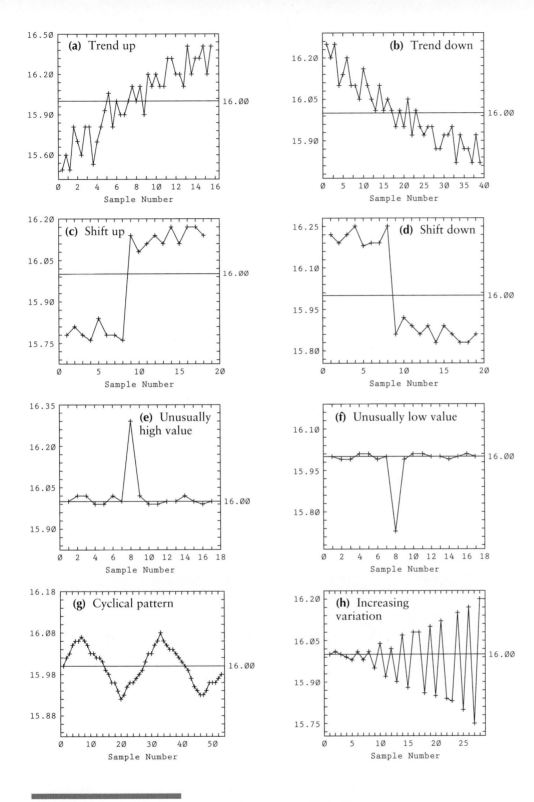

Figure 12-2 Processes with Patterns That Aren't Statistically Stable

- **Figure 12-2(d)**: There is a *downward shift*—the first few values are relatively stable, and then something happened so that the last several values are relatively stable, but at a much lower level.

- **Figure 12-2(e)**: The process is stable except for one *exceptionally high value*. The cause of that unusual value should be investigated. Perhaps the cans became temporarily stuck and one particular can was filled twice instead of once.

- **Figure 12-2(f)**: There is an *exceptionally low value*.

- **Figure 12-2(g)**: There is a *cyclical pattern* (or repeating cycle). This pattern is clearly nonrandom and therefore reveals a statistically unstable process. Perhaps periodic overadjustments are being made to the machinery, with the effect that some desired value is continually being chased but never quite captured.

- **Figure 12-2(h)**: The *variation is increasing over time*. This is a common problem in quality control. The net effect is that products vary more and more until almost all of them are worthless. For example, some soup cans will be overflowing with wasted soup and some will be underfilled and unsuitable for distribution to consumers.

A common goal of many different methods of quality control is this: To *reduce variation* in a product or service. For example, Ford became concerned with variability when it found that its transmissions required significantly more warranty repairs than the same type of transmissions made by Mazda in Japan. A study showed that the Mazda transmissions had substantially less variation in the gearboxes; that is, crucial gearbox measurements varied much less in the Mazda transmissions. Although the Ford transmissions were built within the allowable limits, the Mazda transmissions were more reliable because of their lower variability. Variation in a process can be described in terms of two types of causes.

DEFINITIONS

Random variation is due to chance; it is the type of variation inherent in any process that is not capable of producing every good or service exactly the same way every time.

Assignable variation results from causes that can be identified.

As an example, let's again consider the axial loads of aluminum cans. Figure 12-1 shows that the manufacturing process appears to be stable. The axial loads are all well above the 165-lb maximum load applied when the top lid is pressed into place. We are dealing with a high-quality product. The axial loads do vary, but the variations are small and are the result of random variation. If

the machinery begins to wear so that the axial loads gradually decrease over time, then we have assignable variation attributable to the machinery. Later in the chapter we will look at objective ways to distinguish between assignable variation and random variation.

The run chart is one tool for monitoring the stability of a process. We will now consider control charts, which are also extremely useful for that same purpose.

DEFINITIONS

A **control chart** of a process characteristic (such as mean or variation) consists of values plotted sequentially over time, and it includes a **center line** as well as a **lower control limit** (LCL) and an **upper control limit** (UCL). The center line represents a central value of the characteristic measurements, whereas the control limits are boundaries used to separate and identify any points considered to be unusual.

We will assume that the population standard deviation σ is not known as we consider only two of several different types of *control charts:* **R charts** (or **range charts**) used to monitor variation and **$\overline{x}$ charts** used to monitor means. When using control charts to monitor a process, it is common to consider R charts and $\overline{x}$ charts together, because a statistically unstable process may be the result of increasing variation or changing means or both.

It's important to monitor both the variation and the center of a process because an out-of-control situation could be the result of statistically unstable variation, a statistically unstable mean, or both. In many companies, R charts and $\overline{x}$ charts are generated together in pairs.

Control Chart for Monitoring Variation: The R Chart

In developing a control chart for monitoring variation, it makes sense to use some measure of variation, such as standard deviation or range. Although the use of standard deviations has some theoretical advantage, we will use ranges because they are currently used more often. Before the age of calculators and computers, control charts for ranges had the distinct advantage of requiring much simpler calculations than control charts for standard deviations. Consequently, control charts based on ranges were used much more often, and this preference has continued to current applications, even though the calculations are no longer a serious obstacle.

To construct a control chart for monitoring *variation,* we plot sample *ranges,* instead of individual values. Using the data of Table 12-1, for example, we could plot the 25 sample ranges corresponding to the 25 different days in which 7 cans were sampled. The center line would be located at $\overline{R}$, which denotes the mean of all sample ranges. Following is a summary of the notation we will be using.

Notation

Given: Process data consist of a sequence of samples all of the same size, and the distribution of the process data is essentially normal.

n = size of each sample, or *subgroup*

$\bar{\bar{x}}$ = mean of the sample means (that is, the sum of the sample means, divided by the number of samples), which is equivalent to the mean of all sample scores combined

$\bar{R}$ = mean of the sample ranges (that is, the sum of the sample ranges, divided by the number of samples)

The basic format of the R chart will include the sample ranges plotted in sequence; there will be a center line and upper and lower control limits. The center line and control limits can be described as follows.

Monitoring Process Variation: Control Chart for R

Points plotted: Sample ranges

Center line: $\bar{R}$

Upper control limit (UCL): $D_4\bar{R}$

Lower control limit (LCL): $D_3\bar{R}$

where the values of D_4 and D_3 are found in Table 12-2.

Upper and lower control limits correspond closely to the concept of confidence interval limits presented in Chapter 6.

The values of D_4 and D_3 were computed by quality control experts, and they are intended to simplify calculations. The upper and lower control limits of $D_4\bar{R}$ and $D_3\bar{R}$ are values that are roughly equivalent to 99.7% confidence interval limits. It is therefore highly unlikely that values would fall beyond those limits. If a value does fall beyond the control limits, it's very likely that something is wrong with the process.

● **EXAMPLE**   Refer to the axial loads in Table 12-1. Using samples of size $n = 7$ collected each working day, construct a control chart for R.

SOLUTION
We begin by finding the value of $\bar{R}$, the mean of the sample ranges.

$$\bar{R} = \frac{78 + 77 + \cdots + 31}{25} = 54.96$$

TABLE 12-2	Control Chart Constants					
	$\bar{x}$		s		$\bar{R}$	
Observations in Subgroup, n	A_2	A_3	B_3	B_4	D_3	D_4
2	1.880	2.659	0.000	3.267	0.000	3.267
3	1.023	1.954	0.000	2.568	0.000	2.574
4	0.729	1.628	0.000	2.266	0.000	2.282
5	0.577	1.427	0.000	2.089	0.000	2.114
6	0.483	1.287	0.030	1.970	0.000	2.004
7	0.419	1.182	0.118	1.882	0.076	1.924
8	0.373	1.099	0.185	1.815	0.136	1.864
9	0.337	1.032	0.239	1.761	0.184	1.816
10	0.308	0.975	0.284	1.716	0.223	1.777
11	0.285	0.927	0.321	1.679	0.256	1.744
12	0.266	0.886	0.354	1.646	0.283	1.717
13	0.249	0.850	0.382	1.618	0.307	1.693
14	0.235	0.817	0.406	1.594	0.328	1.672
15	0.223	0.789	0.428	1.572	0.347	1.653
16	0.212	0.763	0.448	1.552	0.363	1.637
17	0.203	0.739	0.466	1.534	0.378	1.622
18	0.194	0.718	0.482	1.518	0.391	1.608
19	0.187	0.698	0.497	1.503	0.403	1.597
20	0.180	0.680	0.510	1.490	0.415	1.585
21	0.173	0.663	0.523	1.477	0.425	1.575
22	0.167	0.647	0.534	1.466	0.434	1.566
23	0.162	0.633	0.545	1.455	0.443	1.557
24	0.157	0.619	0.555	1.445	0.451	1.548
25	0.153	0.606	0.565	1.435	0.459	1.541

Source: Adapted from *ASTM Manual on the Presentation of Data and Control Chart Analysis,* © 1976 ASTM, pp. 134–136. Reprinted with permission of American Society for Testing and Materials.

Don't Tamper!

Nashua Corp. had trouble with its paper-coating machine and considered spending a million dollars to replace it. The machine was working well with a stable process, but samples were taken every so often and, based on the results, adjustments were made. These overadjustments, called *tampering,* caused shifts away from the distribution that had been good. The effect was an increase in defects. When statistician and quality expert W. Edwards Deming studied the process, he recommended that no adjustments be made unless warranted by a signal that the process had shifted or had become unstable. The company was better off with no adjustments than with the tampering that took place.

The center line for our R chart is therefore located at $\bar{R} = 54.96$. To find the upper and lower control limits, we must first find the values of D_3 and D_4. Referring to Table 12-2 for $n = 7$, we get $D_3 = 0.076$ and $D_4 = 1.924$, so the control limits are as follows: *(continued)*

$$\text{Upper control limit: } D_4\overline{R} = (1.924)(54.96) = 105.74$$
$$\text{Lower control limit: } D_3\overline{R} = (0.076)(54.96) = 4.18$$

Using a center line value of $\overline{R} = 54.96$ and control limits of 105.74 and 4.18, we now proceed to plot the sample ranges. The result is shown in the Minitab display that follows. (There is a very small discrepancy between the control limits calculated here and those shown in the Minitab display.)

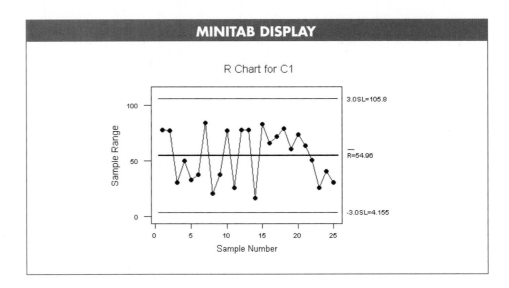

MINITAB DISPLAY

R Chart for C1

Using Computer Software to Generate *R* Charts

To use Minitab to construct a control chart for *R* with the data in Table 12-1, first enter the 175 individual axial loads in column C1. Next, select the options of Stat, Control Charts, and R. Enter C1 in the "single column" box, enter 7 in the box for the subgroup size, and click on estimate. Select Rbar. (Selection of the *R* bar estimate causes the variation of the population distribution to be estimated with the sample ranges instead of the sample standard deviations, which is the default.) Click OK twice.

Interpreting Control Charts

The fact that our analyses are based on *actual* behavior and not the *desired* behavior is extremely important and should be clarified and stressed.

Here is an extremely important point:

Upper and lower control limits of a control chart are based on the *actual* behavior of the process, not the *desired* behavior.

For example, control charts might suggest that 16-oz soup cans are being filled with a statistically stable process, but if they are being filled to a level of

only 8 oz, then there is a major problem because the statistically stable actual behavior is very different from the desired behavior. Also, we should clearly understand the specific criteria for determining whether a process is in statistical control (that is, whether it is statistically stable). So far, we have noted that a process is not statistically stable if its pattern resembles any of those shown in Figure 12-2. This criterion is included with some others in the following list.

Criteria for Using Control Charts to Determine That a Process Is Not Statistically Stable (Out of Statistical Control)

1. There is a pattern, trend, or cycle that is obviously not random (such as those depicted in Figure 12-2).

2. There is a point lying beyond the upper or lower control limits.

3. **Run of 8 Rule:** There are eight consecutive points all above or all below the center line. (With a statistically stable process, there is a 0.5 probability that a point will be above or below the center line, so it is very unlikely that eight consecutive points will all be above the center line or all below it.)

We will use only the three out-of-control criteria listed, but some businesses use additional criteria such as these:

- There are six consecutive points all increasing or all decreasing.

- There are 14 consecutive points all alternating between up and down (such as up, down, up, down, and so on).

- Two out of three consecutive points are beyond control limits that are 1 standard deviation away from the center line.

- Four out of five consecutive points are beyond control limits that are 2 standard deviations away from the center line.

● **EXAMPLE**   Examine the *R* chart shown in the Minitab display and determine whether the process variation is within statistical control.

SOLUTION   We can interpret control charts for *R* by applying the three out-of-control criteria just listed. Applying the three criteria to the Minitab display of the *R* chart, we conclude that variation in this process is within statistical control because of the following.

1. There is no pattern, trend, or cycle that is obviously not random.

2. No point lies beyond the upper or lower control limits.

3. There are not 8 consecutive points all above or all below the center line.

(continued)

INSULATION 2.912 mm

Quality Control at Perstorp

Perstorp Components, Inc., uses a computer that automatically generates control charts to monitor the thicknesses of the floor insulation the company makes for Ford Rangers and Jeep Grand Cherokees. The $20,000 cost of the computer was offset by a first-year savings of $40,000 in labor, which had been used to manually generate control charts to ensure that insulation thicknesses were between the specifications of 2.912 mm and 2.988 mm. Through the use of control charts and other quality-control methods, Perstorp reduced its waste by more than two-thirds.

Bribery Detected with Control Charts

Control charts were used to help convict a person who bribed Florida jai alai players to lose. (See "Using Control Charts to Corroborate Bribery in Jai Alai," by Charnes and Gitlow, *The American Statistician*, Vol. 49, No. 4.) An auditor for one jai alai facility noticed that abnormally large sums of money were wagered for certain types of bets, and some contestants didn't win as much as expected when those bets were made. R charts and $\bar{x}$ charts were used in court as evidence of highly unusual patterns of betting. Examination of the control charts clearly shows points well beyond the upper control limit, indicating that the process of betting was out of statistical control. The statistician was able to identify a date at which assignable variation appeared to stop, and prosecutors knew that it was the date of the suspect's arrest.

We therefore conclude that the variation (not necessarily the mean) of the process is within statistical control. No action is required to correct the *variation* among the axial loads of the cans.

Control Chart for Monitoring Means: The $\bar{x}$ Chart

In constructing a control chart for monitoring means, we can use various approaches to determine the locations of the control limits. One approach applies the central limit theorem by locating the control limits at

$$\bar{\bar{x}} + \frac{3\bar{s}}{\sqrt{n}} \quad \text{and} \quad \bar{\bar{x}} - \frac{3\bar{s}}{\sqrt{n}}$$

(The use of "3" reflects the very low probability of having a sample mean differ from a population mean by more than 3 standard deviations; the symbol $\bar{s}$ denotes the mean of the sample standard deviations.) We will use a similar approach, common in business and industry, in which the center line and control limits are based on ranges instead of standard deviations.

Monitoring Process Mean: Control Chart for $\bar{x}$
Points plotted: Sample means
Center line: $\bar{\bar{x}}$
Upper control limit (UCL): $\bar{\bar{x}} + A_2\bar{R}$
Lower control limit (LCL): $\bar{\bar{x}} - A_2\bar{R}$
where the values of A_2 are found in Table 12-2.

● **EXAMPLE**   Refer to the axial loads in Table 12-1. Using samples of size $n = 7$ collected each working day, construct a control chart for $\bar{x}$. Based on the control chart for $\bar{x}$ only, determine whether the process mean is within statistical control.

SOLUTION   Before plotting the 25 points corresponding to the 25 values of $\bar{x}$, we must first find the value for the center line and the values for the control limits. We get

$$\bar{\bar{x}} = \frac{252.7 + 247.9 + \cdots + 277.3}{25} = 267.12$$

$$\bar{R} = \frac{78 + 77 + \cdots + 31}{25} = 54.96$$

Referring to Table 12-2, we find that for $n = 7$, $A_2 = 0.419$. Knowing the values of $\overline{\overline{x}}$, A_2, and $\overline{R}$, we are now able to find the control limits.

Upper control limit: $\overline{\overline{x}} + A_2\overline{R} = 267.12 + (0.419)(54.96) = 290.15$
Lower control limit: $\overline{\overline{x}} - A_2\overline{R} = 267.12 - (0.419)(54.96) = 244.09$

The resulting control chart for $\overline{x}$ will be as shown in the accompanying Minitab display. Examination of the control chart shows that the process mean is within statistical control because none of the three out-of-control criteria are satisfied. Again, no corrective action is required.

MINITAB DISPLAY

X-bar Chart for C1

3.0SL=290.2

$\overline{\overline{X}}$=267.1

-3.0SL=244.1

Sample Mean — Sample Number

Using Computer Software for $\overline{x}$ Charts

To use Minitab with the data of Table 12-1 for generating a control chart for $\overline{x}$, first enter the 175 individual axial loads in column C1. Next, select the options of `Stat`, `Control Charts`, and `Xbar`. Enter C1 in the "single column" box, enter 7 in the `subgroup size` box, and click on `estimate`, then select `Rbar`. Click OK twice.

Recommended assignment: Exercises 1–6.

12–2 Exercises A: Basic Skills and Concepts

In Exercises 1–3, use the following information:

The Old Faithful Geyser was monitored for 25 recent and consecutive years. In each year, six intervals (in minutes) between eruptions were recorded with the results given in the accompanying table.

The Flynn Effect: Upward Trend in IQ Scores

A run chart or control chart of IQ scores would reveal that they exhibit an upward trend, because IQ scores have been steadily increasing since they began to be used about 70 years ago. The trend is worldwide, and it is the same for different types of IQ tests, even those that rely heavily on abstract and nonverbal reasoning with minimal cultural influence. This upward trend has been named the *Flynn effect*, because political scientist James R. Flynn discovered the trend in his studies of U.S. military recruits. The amount of the increase is quite substantial: Based on a current mean IQ score of 100, it is estimated that the mean IQ in 1920 would be about 77. The typical student of today is therefore brilliant when compared to his or her grandparents. So far, there is no generally accepted explanation for the Flynn effect.

1. There is an upward trend indicating that the intervals between eruptions are increasing, with the result that tourists must wait longer. There is increasing variation, with the result that predicted times of eruptions are becoming less reliable.

2.

There is a point lying beyond the upper control limit, there are 8 points below the center line, and there is an upward trend, so the process variation is out of statistical control. The prediction of eruptions is becoming less reliable.

3. There are points lying beyond the upper control limit and there is an upward trend, so the process mean is out of statistical control. An out-of-control process mean indicates that we cannot make accurate predictions.

Intervals Between Eruptions

Year	Interval (min)						Mean	Range
1	65	72	60	69	65	67	66.3	12
2	74	65	60	69	68	59	65.8	15
3	68	66	69	64	70	73	68.3	9
4	73	65	71	77	63	77	71.0	14
5	79	67	64	61	81	77	71.5	20
6	74	76	65	69	76	64	70.7	12
7	70	73	74	77	65	73	72.0	12
8	71	68	70	79	75	82	74.2	14
9	62	63	61	48	59	77	61.7	29
10	60	74	77	57	52	78	66.3	26
11	67	73	47	81	92	57	69.5	45
12	79	84	79	72	61	80	75.8	23
13	83	78	83	74	61	68	74.5	22
14	57	68	72	75	56	79	67.8	23
15	59	76	78	86	64	72	72.5	27
16	63	63	71	77	81	65	70.0	18
17	67	84	72	75	70	70	73.0	17
18	93	83	85	79	90	74	84.0	19
19	81	74	80	65	70	84	75.7	19
20	83	67	71	67	97	88	78.8	30
21	62	61	57	86	70	77	68.8	29
22	67	75	67	89	93	81	78.7	26
23	86	65	70	74	83	74	75.3	21
24	74	67	99	75	41	83	73.2	58
25	97	93	73	81	85	90	86.5	24

1. Construct a run chart for the 150 values. Does there appear to be a pattern suggesting that the process is not within statistical control? What are the implications for tourists wishing to see Old Faithful erupt?
2. Construct an R chart and determine whether the process variation is within statistical control. If it is not, identify which of the three out-of-control criteria lead to rejection of statistically stable variation. How would tourists be affected by out-of-control variation?
3. Construct an $\bar{x}$ chart and determine whether the process mean is within statistical control. If it is not, identify which of the three out-of-control criteria lead to rejection of a statistically stable mean. How would tourists be affected by an out-of-control mean?

In Exercises 4–6, use the following information:

The U.S. Mint has a goal of making quarters with a weight of 5.670 g, but any weight between 5.443 g and 5.897 g is considered acceptable. A new minting machine is placed into service and the weights are recorded for a quarter randomly selected every 12 min for 20 consecutive hours. The results are listed in the accompanying table.

Weights (in grams) of Minted Quarters

Hour	Weight (grams)					$\bar{x}$	s	Range
1	5.639	5.636	5.679	5.637	5.691	5.6564	0.0265	0.055
2	5.655	5.641	5.626	5.668	5.679	5.6538	0.0211	0.053
3	5.682	5.704	5.725	5.661	5.721	5.6986	0.0270	0.064
4	5.675	5.648	5.622	5.669	5.585	5.6398	0.0370	0.090
5	5.690	5.636	5.715	5.694	5.709	5.6888	0.0313	0.079
6	5.641	5.571	5.600	5.665	5.676	5.6306	0.0443	0.105
7	5.503	5.601	5.706	5.624	5.620	5.6108	0.0725	0.203
8	5.669	5.589	5.606	5.685	5.556	5.6210	0.0545	0.129
9	5.668	5.749	5.762	5.778	5.672	5.7258	0.0520	0.110
10	5.693	5.690	5.666	5.563	5.668	5.6560	0.0534	0.130
11	5.449	5.464	5.732	5.619	5.673	5.5874	0.1261	0.283
12	5.763	5.704	5.656	5.778	5.703	5.7208	0.0496	0.122
13	5.679	5.810	5.608	5.635	5.577	5.6618	0.0909	0.233
14	5.389	5.916	5.985	5.580	5.935	5.7610	0.2625	0.596
15	5.747	6.188	5.615	5.622	5.510	5.7364	0.2661	0.678
16	5.768	5.153	5.528	5.700	6.131	5.6560	0.3569	0.978
17	5.688	5.481	6.058	5.940	5.059	5.6452	0.3968	0.999
18	6.065	6.282	6.097	5.948	5.624	6.0032	0.2435	0.658
19	5.463	5.876	5.905	5.801	5.847	5.7784	0.1804	0.442
20	5.682	5.475	6.144	6.260	6.760	6.0642	0.5055	1.285

4. Construct a run chart for the 100 values. Does there appear to be a pattern suggesting that the process is not within statistical control? What are the practical implications of the run chart?

5. Construct an R chart and determine whether the process variation is within statistical control. If it is not, identify which of the three out-of-control criteria lead to rejection of statistically stable variation.

6. Construct an $\bar{x}$ chart and determine whether the process mean is within statistical control. If it is not, identify which of the three out-of-control criteria lead to rejection of a statistically stable mean. Does this process need corrective action?

4.

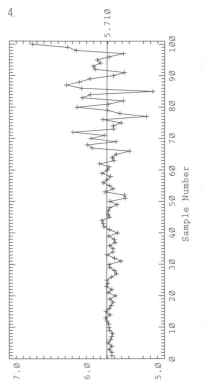

There is a pattern of increasing variation, so the process is out of statistical control.

6.

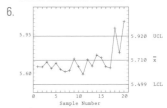

There is a pattern of increasing variation, there are points lying beyond the upper control limit, and there are eight consecutive points all below the center line, so the process mean is out of statistical control. This process does need corrective action.

7.

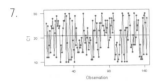

The process appears to be within statistical control.

8.

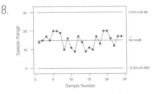

The process appears to be within statistical control.

In Exercises 7–9, use the following information:

Pisa Pizza provides home delivery and advertises that the order is free if it is not delivered in 30 minutes or less. To monitor the quality of delivery, the shop records delivery times for six orders randomly selected each day for 24 consecutive days. The delivery times (in minutes) are listed in the accompanying table.

Pizza Delivery Times

Sample	Time (min)						Mean	Range
1	26	12	25	12	17	23	19.17	14.0
2	14	25	24	28	24	13	21.33	15.0
3	22	27	12	29	25	27	23.67	17.0
4	24	25	10	17	14	21	18.50	15.0
5	24	27	27	30	10	29	24.50	20.0
6	18	10	11	30	28	11	18.00	20.0
7	12	27	19	15	10	29	18.67	19.0
8	17	24	16	14	14	19	17.33	10.0
9	16	17	20	14	14	30	18.50	16.0
10	23	18	21	13	20	24	19.83	11.0
11	15	18	20	19	11	11	15.67	9.0
12	27	12	23	23	29	18	22.00	17.0
13	26	15	12	23	26	16	19.67	14.0
14	26	30	25	25	23	21	25.00	9.0
15	27	22	24	27	16	19	22.50	11.0
16	28	20	25	28	22	30	25.50	10.0
17	27	11	13	17	18	10	16.00	17.0
18	11	13	17	23	23	10	16.17	13.0
19	30	29	13	10	23	24	21.50	20.0
20	10	24	19	30	29	25	22.83	20.0
21	16	17	14	21	17	30	19.17	16.0
22	17	20	29	23	25	27	23.50	12.0
23	12	17	11	12	11	28	15.17	17.0
24	13	29	30	26	13	28	23.17	17.0

7. Construct a run chart for the 144 values. Does there appear to be a pattern suggesting that the process is not within statistical control?

8. Construct an *R* chart and determine whether the process variation is within statistical control. If it is not, identify which of the three out-of-control criteria lead to rejection of statistically stable variation.

9. Construct an $\bar{x}$ chart and determine whether the process mean is within statistical control. Does this process appear to be working well, or does it require correction? If the $\bar{x}$ chart were to have a downward trend, as in Figure 12-2(b), should the process be modified to make the means stable?

In Exercises 10–12, use the following information:

The Medassist Pharmaceutical Supply Company provides drug stores with bulk packages of phenobarbital, all labeled as containing 5 lb. Each day, a sample of five packages is randomly selected and weighed. The results for 20 consecutive days are given in the table.

Weight (in pounds) of Phenobarbital

Sample	Weight (lb)					Mean	Range
1	4.98	5.02	4.98	4.88	4.99	4.970	0.14
2	5.04	5.03	5.03	5.02	5.03	5.030	0.02
3	4.96	4.89	5.05	5.02	4.98	4.980	0.16
4	5.05	4.95	4.96	4.96	4.95	4.974	0.10
5	5.04	4.96	4.96	5.01	4.98	4.990	0.08
6	5.00	4.93	4.96	4.95	5.02	4.972	0.09
7	4.99	5.01	5.08	5.01	5.03	5.024	0.09
8	4.96	4.93	5.00	5.02	5.04	4.990	0.11
9	4.91	4.93	4.97	4.92	4.99	4.944	0.08
10	4.96	4.97	5.03	4.98	5.08	5.004	0.12
11	4.97	5.07	5.05	5.01	4.94	5.008	0.13
12	4.83	4.44	4.94	5.22	5.73	5.032	1.29
13	4.92	5.19	4.70	5.02	4.91	4.948	0.49
14	5.38	4.71	5.10	4.85	5.73	5.154	1.02
15	5.18	4.71	5.95	5.03	5.55	5.284	1.24
16	4.68	4.84	6.15	4.85	5.30	5.164	1.47
17	4.56	5.59	5.03	4.81	4.81	4.960	1.03
18	5.53	5.20	5.26	4.91	5.40	5.260	0.62
19	5.96	5.50	5.25	5.23	4.54	5.296	1.42
20	5.39	5.28	5.90	5.31	5.10	5.396	0.80

10. Construct a run chart for the 100 values. Is there a pattern indicating that the process is not statistically stable?
11. Construct an R chart and determine whether the process variation is within statistical control. If it is not, identify which of the three out-of-control criteria lead to rejection of statistically stable variation.

9. The process mean is within statistical control and the process appears to be working well. A downward trend would have indicated that the process is out of statistical control, but it is *improving* because the delivery times are decreasing. Such a downward trend should be investigated to ensure that the reduced times continue.

10.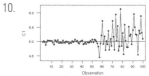

The process is out of control because there is increasing variation and an upward trend.

11. The process is out of control because there is a shift up, there are 8 consecutive points below the center line, there are 8 consecutive points below the center line, and there are points beyond the upper control limit.

12.

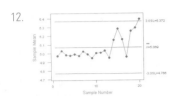

The process is out of control
because there is an upward
trend, there are 8 consecutive
points below the center line,
and there is a point lying beyond
the upper control limit.

13. Using Table 12-1 values, $\bar{s}$ =
20.05 and LCL = 2.366.

14.

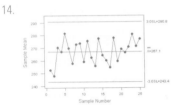

The result and conclusions are
very similar to those for the $\bar{x}$
chart based on sample ranges.

The examples and exercises in this
section assume that all samples
have the same size n. There are
ways to deal with unequal sample
sizes, but we do not consider them
in this text. (See the *Minitab Reference Manual*.)

12. Construct an $\bar{x}$ chart and determine whether the process mean is within statistical control. If it is not, identify a consequence of not correcting the process to make it statistically stable.

12-2 Exercises B: Beyond the Basics

13. In this section we described control charts for R and $\bar{x}$ based on ranges. Control charts for monitoring variation and center (mean) can also be based on standard deviations. An *s chart* for monitoring variation is made by plotting sample standard deviations with a center line at $\bar{s}$ (the mean of the sample standard deviations) and control limits at $B_4\bar{s}$ and $B_3\bar{s}$, where B_4 and B_3 are found in Table 12-2. Construct an *s* chart for the data of Table 12-1. Compare the result to the R chart given in this section.

14. An $\bar{x}$ chart based on standard deviations (instead of ranges) is made by plotting sample means with a center line at $\bar{\bar{x}}$ and control limits at $\bar{\bar{x}} + A_3\bar{s}$ and $\bar{\bar{x}} - A_3\bar{s}$, where A_3 is found in Table 12-2 and $\bar{s}$ is the mean of the sample standard deviations. Use the data in Table 12-1 to construct an $\bar{x}$ chart based on standard deviations. Compare the result to the $\bar{x}$ chart based on sample ranges (as shown in this section).

12-3 Control Charts for Attributes

In this section we construct a control chart for an attribute, such as the proportion p of defective items. Instead of monitoring the *quantitative* characteristics of variation and mean (as in Section 12-2), we now consider the *qualitative* attribute of whether an item has some particular characteristic, such as being defective, being acceptable, weighing under 16 oz, or being nonconforming. A good or a service is nonconforming if it doesn't meet specifications or requirements. (Nonconforming goods are sometimes discarded, repaired, or called "seconds" and sold at reduced prices.) As in Section 12-2, we select samples of size n at regular time intervals and plot points in a sequential graph with a center line and control limits. (There are ways to deal with samples of different sizes, but we don't consider them here.) The **control chart for p** (or **p chart**) is a control chart used to monitor the proportion p; the relevant notation and control chart values are as shown on page 629 (where the attribute of "defective" can be replaced by any other relevant attribute).

We use $\bar{p}$ for the center line because it is the best estimate of the proportion of defects from the process. The expressions for the control limits correspond to 3 standard deviations away from the center line. In Section 6-4 we noted that the standard deviation of sample proportions is $\sqrt{pq/n}$, and this is expressed in the control limits shown in the box.

Notation

$\bar{p}$ = pooled estimate of the proportion of defective items in the process

$\phantom{\bar{p}}$ = $\dfrac{\text{total number of defects found among all items sampled}}{\text{total number of items sampled}}$

$\bar{q}$ = pooled estimate of the proportion of process items that are *not* defective

$\phantom{\bar{q}}$ = $1 - \bar{p}$

n = size of each sample (not the number of samples)

Control Chart for p

Center line: $\bar{p}$

Upper control limit: $\bar{p} + 3\sqrt{\dfrac{\bar{p}\,\bar{q}}{n}}$

Lower control limit: $\bar{p} - 3\sqrt{\dfrac{\bar{p}\,\bar{q}}{n}}$

(If the calculation for the lower control limit results in a negative value, use 0 instead. If the calculation for the upper control limit exceeds 1, use 1 instead.)

These control limits correspond very closely to 99.7% confidence interval limits.

● **EXAMPLE**   Physicians report that infectious diseases should be carefully monitored over time because they are much more likely to have sudden changes in trends than are other diseases, such as cancer. In each of 13 consecutive and recent years, 100,000 subjects were randomly selected and the number who died from respiratory tract infections is recorded, with the results given here (based on data from "Trends in Infectious Diseases Mortality in the United States," by Pinner et al., *Journal of the American Medical Association*, Vol. 275, No. 3). Construct a control chart for p and determine whether the process is within statistical control. If not, identify which of the three out-of-control criteria apply.

Number of deaths: 25 24 22 25 27 30 31 30 33 32 33 32 31

(continued)

Motorola and the Six-Sigma (σ) Quality Standard

From the 1950s to the 1970s, U.S. manufacturers sought to make 99.8% of their components good, with only 0.2% defective. This rate corresponds to a "three-sigma" level of tolerance. This would seem to be a high level of quality, but Motorola recognized that when many such components are *combined*, the chance of getting a defective product is quite high. The company sought to reduce the variation in all processes by one-half. This standard became known as the "six-sigma" program. A six-sigma level corresponds to about two defects per *billion* items. Motorola found that it could achieve this level with many components and processes. The extra cost was more than offset by reduced costs of repair, scrap, and warranty work. In 1988 Motorola won the first Malcolm Baldridge National Quality Award.

SOLUTION

The center line for our control chart is located by the value of $\bar{p}$:

$$\bar{p} = \frac{\text{total number of deaths from all samples combined}}{\text{total number of subjects sampled}}$$

$$= \frac{25 + 24 + 22 + \cdots + 31}{13 \cdot 100{,}000} = \frac{375}{1{,}300{,}000} = 0.000288$$

Because $\bar{p} = 0.000288$, it follows that $\bar{q} = 1 - \bar{p} = 0.999712$. Using $\bar{p} = 0.000288$, $\bar{q} = 0.999712$, and $n = 100{,}000$, we find the control limits as follows:

Upper control limit:

$$\bar{p} + 3\sqrt{\frac{\bar{p}\,\bar{q}}{n}} = 0.000288 + 3\sqrt{\frac{(0.000288)(0.999712)}{100{,}000}} = 0.000449$$

Lower control limit:

$$\bar{p} - 3\sqrt{\frac{\bar{p}\,\bar{q}}{n}} = 0.000288 - 3\sqrt{\frac{(0.000288)(0.999712)}{100{,}000}} = 0.000127$$

Having found the values for the center line and control limits, we can proceed to plot the yearly proportion of deaths from respiratory tract infections. The Minitab control chart for p is shown in the accompanying display.

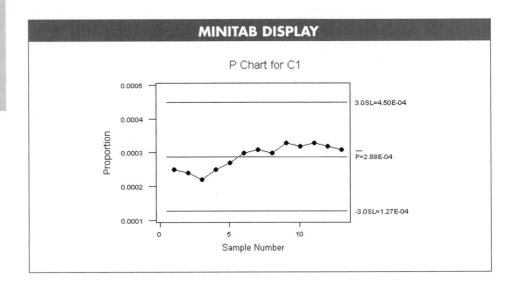

MINITAB DISPLAY

P Chart for C1

We can interpret the control chart for p by considering the same three out-of-control criteria listed in Section 12-2. Using those criteria, we conclude that this process is out of statistical control for these reasons: There appears to be an upward trend, and there are eight consecutive points all lying above the center line (Run of 8 Rule). Based on these data, public health policies affecting respiratory tract infections should be modified to cause a decrease in the death rate.

Using Computer Software for p Charts

To obtain a p chart using Minitab, enter the numbers of defects (or items with any particular attribute) in column C1. Select the options of Stat, then Control Charts, then P. Enter C1 in the box identified as variable, and enter the size of the samples in the box identified as subgroup size, then click on OK. Minitab also allows you to generate an np chart, described as follows.

If Minitab is not available, students can use the *scatter diagram* capability of STATDISK or a TI-83 calculator, or they can manually construct p charts.

A variation of the control chart for p is the **np chart** in which the *actual numbers* of defects are plotted instead of the *proportions* of defects. The np chart will have a center line value of $n\overline{p}$, and the control limits will have values of $n\overline{p} + 3\sqrt{n\overline{p}\,\overline{q}}$ and $n\overline{p} - 3\sqrt{n\overline{p}\,\overline{q}}$. The p chart and the np chart differ only in the scale of values used for the vertical axis.

⟨12-3⟩ Exercises A: Basic Skills and Concepts

In Exercises 1–4, examine the given control chart for p and determine whether the process is within statistical control. If it is not, identify which of the three out-of-control criteria apply.

Recommended assignment: Exercises 1–6.

1.

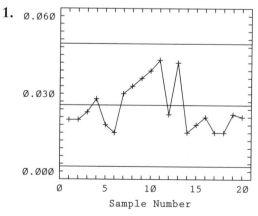

2.

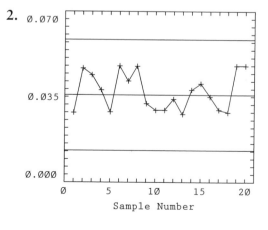

1. Process appears to be within statistical control.
2. Process appears to be within statistical control.

3. Process appears to be out of statistical control because there is a pattern of an upward trend and there is a point that lies beyond the upper control limit.

4. Process appears to be out of statistical control because there are points lying beyond the control limits.

3.

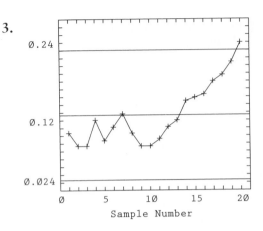

4.

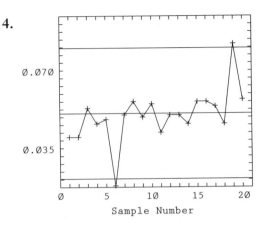

5.

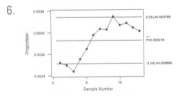

The process is out of statistical control because there is a downward trend and there are eight consecutive points all lying below the center line. This downward trend is good and its causes should be identified so that it can continue.

6.

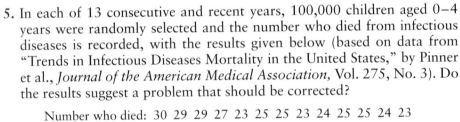

The process is out of control because there is a shift up and there are points beyond the control limits.

In Exercises 5–8, use the given process data to construct a control chart for p. *In each case, use the three out-of-control criteria listed in Section 12-2 and determine whether the process is within statistical control. If it is not, identify which of the three out-of-control criteria apply.*

5. In each of 13 consecutive and recent years, 100,000 children aged 0–4 years were randomly selected and the number who died from infectious diseases is recorded, with the results given below (based on data from "Trends in Infectious Diseases Mortality in the United States," by Pinner et al., *Journal of the American Medical Association*, Vol. 275, No. 3). Do the results suggest a problem that should be corrected?

Number who died: 30 29 29 27 23 25 25 23 24 25 25 24 23

6. In each of 13 consecutive and recent years, 100,000 adults 65 years of age or older were randomly selected and the number who died from infectious diseases is recorded, with the results given below (based on data from "Trends in Infectious Diseases Mortality in the United States," by Pinner et al., *Journal of the American Medical Association*, Vol. 275, No. 3).

Number who died: 270 264 250 278 302 334 348
347 377 357 362 351 343

7. In each of 20 consecutive and recent years, 1000 adults were randomly selected and surveyed. Each value below is the number who were victims of violent crime (based on data from the U.S. Department of Justice, Bureau of Justice Statistics). Do the data suggest a problem that should be corrected?

29 33 24 29 27 33 36 22 25 24 31 31 27 23 30 35 26 31 32 24

8. Columbia House runs a membership club from Terre Haute, Indiana, and uses control charts for *p* to monitor proportions of CD and tape orders that are incorrectly filled. Suppose that 300 orders are randomly selected and monitored each day; the results for 21 consecutive days are listed below. If further investigation revealed that on the 6th and 7th days tem-

porary employees were hired to fill in for vacationing employees, what would you recommend?

$$3\ 2\ 4\ 7\ 3\ 15\ 18\ 2\ 6\ 4\ 3\ 5\ 4\ 6\ 5\ 2\ 4\ 3\ 6\ 1\ 5$$

12-3 Exercises B: Beyond the Basics

9. Construct the np chart for the example given in this section. Compare the result with the control chart for p given in this section.
10. a. Identify the locations of the center line and control limits for a p chart representing a process that has been having a 5% rate of nonconforming items, based on samples of size 100.
    b. Repeat part (a) after changing the sample size to 300.
    c. Compare the two sets of results. What are an advantage and a disadvantage of using the larger sample size? Which chart would be better in detecting a shift from 5% to 10%?

• •

Vocabulary List

process data	lower control limit
run chart	upper control limit
statistically stable	R chart
within statistical control	range chart
random variation	$\bar{x}$ chart
assignable variation	control chart for p
control chart	p chart
center line	np chart

• •

Review

Whereas earlier chapters focused on the important data characteristics of distribution shape, representative value, and variation, this chapter focused on pattern over time. Process data were defined to be data arranged according to some time sequence, and such data can be analyzed with run charts and control charts. Control charts have a center line, an upper control limit, and a lower control limit. A process is statistically stable (or within statistical control) if it has only natural variation with no patterns, cycles, or unusual points. Decisions about statistical stability are based on how a process is actually behaving, not how we might like it to behave because of such factors as manufacturer specifications.

• A run chart is a sequential plot of *individual* data values over time.

8.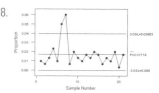

The process is out of statistical control because there are points lying beyond the upper control limit. If temporary employees were hired on the 6th and 7th days, that could easily explain the jump in defective filled orders. The company should consider staggering vacations so that the negative effect of temporary employees can be reduced.

9.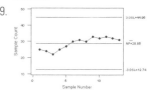

Except for the scale used, the charts are identical.

10. a. LCL = 0; center line is at 0.0500; UCL = 0.1154.
    b. LCL = 0.01225; center line is at 0.0500; UCL = 0.08775.
    c. As the sample size n increases, the control lines get closer. The advantage is that smaller shifts in the process are detected. A disadvantage is that more items must be sampled. The chart in part (b) would be better for detecting a change from 5% to 10%.

1.

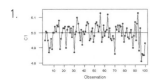

The process appears to be within statistical control.

2.

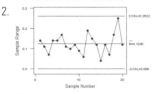

The process appears to be within statistical control.

3.

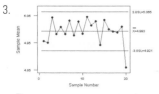

The process mean appears to be out of statistical control because there is a point lying beyond the lower control limit. The process should be corrected because too little epoxy cement hardener is being put into some containers.

- An R chart is a control chart that uses ranges in an attempt to monitor the variation in a process.
- An $\bar{x}$ chart is a control chart used to determine whether the process mean is within statistical control.
- A p chart is a control chart used to monitor the proportion of some process attribute, such as whether items are defective.

Review Exercises

In Exercises 1–3, use the following information:

The York Chemical Company supplies epoxy cement hardener in containers that are designed to hold 5 oz of the liquid. Too much or too little of the hardener results in a mixture with little strength. Each hour, a sample of five containers of hardener is carefully measured; the table shows the results for 20 consecutive hours of operation.

Amounts of Epoxy Cement Hardener

Sample	Amount (oz)					Mean	Range
1	4.95	5.01	5.00	4.95	4.87	4.956	0.14
2	4.90	4.96	4.89	5.00	5.00	4.950	0.11
3	5.04	5.01	5.05	5.04	5.08	5.044	0.07
4	4.89	4.96	5.00	5.03	5.03	4.982	0.14
5	5.04	4.90	5.02	5.04	5.04	5.008	0.14
6	5.00	5.03	4.97	5.04	4.87	4.982	0.17
7	4.99	5.01	5.10	5.06	5.00	5.032	0.11
8	4.95	5.02	4.92	5.01	4.99	4.978	0.10
9	5.01	5.05	5.07	5.06	4.95	5.028	0.12
10	4.96	5.01	5.03	4.94	4.98	4.984	0.09
11	5.06	5.07	5.05	5.01	5.04	5.046	0.06
12	5.13	4.94	4.97	5.05	4.98	5.014	0.19
13	5.01	5.02	5.12	4.97	5.03	5.030	0.15
14	4.96	4.95	4.98	4.96	4.86	4.942	0.12
15	5.04	5.02	5.05	5.01	5.05	5.034	0.04
16	5.04	4.98	5.06	4.94	4.98	5.000	0.12
17	5.00	5.03	4.97	4.96	5.00	4.992	0.07
18	5.02	4.97	4.90	4.98	5.07	4.988	0.17
19	4.95	5.06	5.13	4.88	5.03	5.010	0.25
20	4.81	4.86	4.85	4.85	4.93	4.860	0.12

1. Construct a run chart for the 100 values. Does there appear to be a pattern suggesting that the process is not within statistical control?
2. Construct an R chart and determine whether the process variation is within statistical control. If it is not, identify which of the three out-of-control criteria lead to rejection of statistically stable variation.
3. Construct an $\bar{x}$ chart and determine whether the process mean is within statistical control. Does this process appear to be working well, or does it require correction?
4. The Paris Cosmetics Company manufactures lipstick tubes that dispense (you guessed it) lipstick. Each day, 200 tubes are randomly selected and tested for defects. The results are given below for 20 consecutive days. Construct a control chart for p (proportion of defects), then determine whether the process is within statistical control. If not, identify which criteria lead to rejection of statistical stability.

    Number of defects: 4 2 3 6 1 4 3 0 5 5 2 1 9 3 4 1 0 2 6 3

4.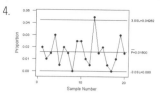

The process is out of statistical control because there is a point lying beyond the upper control limit.

Cumulative Review Exercises

1. The Telektronic Company produces 20-amp fuses used to protect car radios from too much electrical power. Each day 400 fuses are randomly selected and tested; the results (numbers of defects per 400 fuses tested) for 20 consecutive days are as follows:

    10 8 7 6 6 9 12 5 4 7 9 6 11 4 6 5 10 5 9 11

    a. Use a control chart for p to verify that the process is within statistical control, so the data can be treated as coming from a population with fixed variation and mean.
    b. Using all of the data combined, construct a 95% confidence interval for the proportion of defects.
    c. Using a 0.05 significance level, test the claim that the rate of defects is 1% or less.
2. When interpreting control charts, one of the three out-of-control criteria is based on eight consecutive points that are all above or all below the center line. For a statistically stable process, there is a 0.5 probability that a point will be above the center line and there is a 0.5 probability that a point will be below the center line. In each of the following, assume that sample values are independent and the process is statistically stable.
    a. Find the probability that when eight consecutive points are randomly selected, they are all above the center line.
    b. Find the probability that when eight consecutive points are randomly selected, they are all below the center line.
    c. Find the probability that when eight consecutive points are randomly selected, they are all above or all below the center line.

1. a.

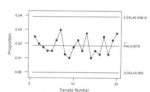

The process is in statistical control.

b. $0.0158 < p < 0.0217$

c. Test statistic: $z = 5.77$. Critical value: $z = 1.645$. Reject H_0: $p \le 0.01$. There is sufficient evidence to warrant rejection of the claim that the rate of defects is 1% or less.

2. a. 1/256
   b. 1/256
   c. 1/128

Computer Project

a. Simulate 20 days of manufacturing heart pacemakers with a 1% rate of defective units. If you are using Minitab, sample 200 pacemakers each day as follows. From the main menu bar, select `Calc`, then `Random  Data`, then `Integer`. Enter 200 in the box for the number of rows of data, enter C1 as the column to be used for storing the data, enter 1 for the minimum value, and enter 100 for the maximum value. Consider the occurrence of "1" to be a defective pacemaker, whereas any other integer from 2 to 100 represents an acceptable pacemaker. Repeat this procedure until results for 20 days have been simulated.

   If you are using STATDISK, select `Data` from the main menu bar, then select `Uniform  Generator` and proceed to generate 200 scores with a minimum of 1 and a maximum of 100. Display the ranked data by using the `Format` menu in `Sample Editor`. Consider an outcome beginning with 1 to be defective and all other outcomes to be acceptable. Repeat this procedure until results for 20 days have been simulated.

b. Construct a p chart for the proportion of defective pacemakers, and determine whether the process is within statistical control. Since we know the process is actually stable with $p = 0.01$, the conclusion that it is not stable would be a type I error; that is, we would have a false positive signal, causing us to believe that the process needed to be adjusted when in fact it should be left alone.

c. The result from part (a) is a simulation of 20 days. Now simulate another 10 days of manufacturing pacemakers, but modify these last 10 days so that the nonconforming rate is 3% instead of 1%. (Modify the preceding Minitab and STATDISK commands for the change from 1% to 3%.)

d. Combine the data generated from parts (a) and (c) to represent a total of 30 days of sample results. Construct a p chart for this combined data set. Is the process out of control? If we concluded that the process was not out of control, we would be making a type II error; that is, we would believe that the process was okay when in fact it should be repaired or adjusted to correct the shift to the 3% nonconforming rate.

•••••••• FROM DATA TO DECISION ••••••••

Is the Process of Manufacturing Cans Proceeding as It Should?

The Chapter Problem listed process data from a New York company that manufactures 0.0109-in. thick aluminum cans for a major beverage supplier, and those data were analyzed in Section 12-2. Refer to Data Set 15 in Appendix B and conduct an analysis of the process data for the cans that are 0.0111 in. thick. The values in the data set are the measured axial loads of cans, and the top lids are pressed into place with pressures that vary between 158 pounds and 165 pounds.

Should you take any corrective action? Write a report summarizing your conclusions. Address not only the issue of statistical stability, but also the ability of the cans to withstand the pressures applied when the top lids are pressed into place. Also compare the behavior of the 0.0111-in. cans to the behavior of the 0.0109-in. cans and recommend which thickness should be used.

•••••••• COOPERATIVE GROUP ACTIVITIES ••••••••

1. *Out-of-class activity:* Collect your own set of process data and analyze them using the methods of this section. It would be ideal to collect data from a real manufacturing process, but that is usually difficult to implement. Instead, consider using a simulation or referring to published data, such as those found in an almanac. Here are some suggestions:

   • Shoot five crumpled sheets of paper into a wastebasket and record the number of shots made, then repeat this procedure 20 times and use a p chart to test for statistical stability in the proportion of shots made.

   • Select five candies from a bag of M&Ms and record the number that are brown. Repeat this procedure 20 times and pretend that the 20 samples were selected on 20 consecutive days of production. Use a p chart to test for statistical stability of the proportion of brown M&Ms.

   • The *Information Please Almanac* lists the divorce rate in terms of divorces per 1000 population for several years. Assume that in each year 1000 people were randomly selected and surveyed to determine whether they became divorced. Use a p chart to test for statistical stability of the divorce rate. (Other possible rates: marriage, birth, death, accident fatality.)

   Obtain a printed copy of computer results, and write a report summarizing your conclusions.

2. *In-class activity:* If the instructor can distribute the numbers of absences for each class meeting, groups of three or four students can analyze them and make recommendations based on the conclusions.

interview

Bill James

Bill James is a recognized baseball expert who specializes in the analysis of baseball statistics. He has written many books, including *The Baseball Book, Bill James Presents . . . Stats Major League Handbook, Bill James Presents . . . Stats Minor League Handbook, STATS 1997 Batter vs. Pitcher Match-Ups, The Politics of Glory: How Baseball's Hall of Fame Really Works, The Bill James Player Ratings Book, Bill James Historical Baseball Abstract,* and *Bill James Guide to Baseball Managers: From 1870 to Today.*

How did you become involved with the statistics of baseball?

I wanted to know things like which catchers were best in preventing stolen bases. I sat down with a stack of box scores and counted the stolen bases in different games with different catchers. Sometimes, available statistics don't meet your true objectives, and it becomes necessary to find the statistics that are appropriate for answering your real questions.

What is *sabermetrics*?

I coined the word, which stands for Society for American Baseball Research, and "metrics," which refers to measurements. With sabermetrics I try to answer objective questions, such as "Which catcher is most effective in preventing stolen bases," or "which batter on the Yankees created the most runs?"

Why isn't the batting average a perfectly good statistic for measuring the effectiveness of a batter?

The real question involves what the hitter is attempting to do, and that is to create runs, because runs win games. If you want to understand the value of a hitter, you start with the number of runs created, then you can factor in things like the ballpark and the quality of pitching. There are some players who are extremely effective in creating runs and winning games, even though they have low batting averages.

> "Sometimes, available statistics don't meet your true objectives, and it becomes necessary to find the statistics that are appropriate for answering your real questions."

Do you use formal methods of statistics?

I am not a statistician by training, but I sometimes use some formal methods. For example, I occasionally use the Pearson product moment in correlation. (Author's note: The Pearson product moment is the *linear correlation* described in Chapter 9.) I use distributions all the time, and I've never found anything in baseball that had a normal distribution. Nothing in baseball is normally distributed because only the very best players are in the major leagues. A typical batter hits .260, and there are thousands of players who could hit .160 but very few who could hit .360. Any calculations in baseball that begin with the assumption of a normal distribution will go in a wrong direction.

I just wrote a magazine article in which I began by asking which is the largest number: (1) the number of seconds that have passed since the formation of the earth; (2) the number of inches in a light year; (3) the number of pennies in the national debt; (4) the number of possible ways a manager can choose a nine-man lineup from a 25-man roster. The answer is the fourth number. (Author's note: If you calculate the fourth number, see Section 3-6 and consider that the nine-man lineup has both a batting order and an order for field positions.)

Can your methods be applied to fields other than sports?

I suppose so. I get many letters from people who claim that they have applied my methods to other fields, and that's nice, but in reality most of my methods are applied *from* other fields. Most of what I've done is to apply to baseball the methods that I was taught in economics, psychology, and other fields. When I was a student of economics, I would always take what I was taught and immediately try to apply it to baseball. So, yes, my methods apply outside of baseball, but they originated outside of baseball in the first place.

One individual who works for the IRS wrote to tell me that he created a computer program to help spot potential problems with returns. He said that he got the idea from one of my books that discussed profiling techniques. I use profiling techniques to find the player that best fits specific requirements. The IRS employee used profiling to identify returns that should be examined more closely.

What would you recommend to today's college students? Would you recommend that they take a statistics course?

Certainly, yes. Also, I would say that all useful knowledge comes from crossbreeding of ideas. If you really want to have original insight into baseball or any other field, you need to study something different. You then have to figure out how to apply that knowledge to the questions that you are interested in.

13

Nonparametric Statistics

Chapter Problem:

Will our radio message to outer space be mistaken for "random noise"?

Dr. Frank Drake of Cornell University developed a radio message to send to intelligent life beyond our solar system. The message, transmitted as a series of pulses and gaps, is a sequence of 1271 0s and 1s, as listed in Figure 13-1. If we factor 1271 into the prime numbers of 41 and 31 and then make a 41 × 31 grid and put a dot at those positions corresponding to a 1 in the message, we get the pattern shown in Figure 13-1. This pattern contains information including the location of Earth in the solar system; the symbols for hydrogen, carbon, and oxygen; and drawings of a man, woman, child, fish, and water. Suppose we send this message and it is intercepted by extraterrestrial intelligent life. Will they think that the pattern is random? In Section 13-7 we will consider a test for determining whether such a sequence is random.

Figure 13-1 Radio Message
Message shows water, fish, man, child, and woman; symbols for hydrogen, carbon, and oxygen; and the location of Earth.

```
10000000000000000000000000000000000000000001
00001110000000000010000010000001000001000000
00010001000000000000000000000000000000000000
00010001000010000000000000000000100000000000
00010001000001000001000000001000100010001000
00001110000000000000001000000000001000000000
00000000000000000000000000000000000000000000
00000000000000000100001000000100000010000000
11000100000000000000000000000000000000000000
00000000001100001100001100001100001100001000
00000000010010010010010010010010010010010010
10100100100011000110000110000110000110001100
00000001000000000001111101000000000000000000
00000010000000000010000010000000000000010110
11100100000000000001111101000000000000000000
00000000000000100000000000000010001000100111
00000000000010100000000000000010100100000001
10010101110010100000000000000010100100000001
00000000001001001001001001001001001000000000
00000000000111110000000000000011111000000001
11010100000010101000000000001010100000000001
00000000000100010100000000001010000100000000
00000000010001001000100010010011011001111101
10110100000100010001010101010001000001000001
00000000001000100010010010010010001000000001
00000000001110000111110000011110000011000001
11110100000101010000010100000100001000000001
00000000001000001000011100001000001000001000
10000000001000001000100010001000001000001000
00001100001000001000100010001000001000001000
10000000001100000110110001101100000011000111
```

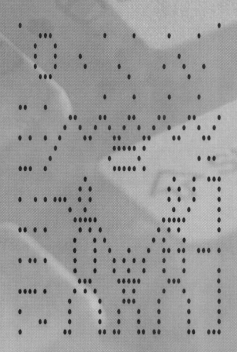

13-1 Overview

This chapter can be covered as one unit devoted to nonparametric methods, or its individual sections can be covered along with the related parametric methods as indicated below.

- 13-2/13-3 ↔ 8-2
- 13-4   ↔ 8-3/8-5
- 13-5   ↔ 11-2
- 13-6   ↔ 9-2

Section 13-7 has been rewritten so that most of it can now be covered at any time throughout the course. Consider covering Section 13-7 very early, such as following Chapter 1.

Most of the methods of inferential statistics covered thus far in this text can be called *parametric methods* because they are based on sampling from a population with specific parameters, such as the mean μ, standard deviation σ, or proportion p. Those parametric methods usually must conform to some fairly strict conditions, such as a requirement that the sample data come from a normally distributed population. This chapter introduces nonparametric methods, which do not have such strict requirements.

DEFINITIONS

Parametric tests require assumptions about the nature or shape of the populations involved; **nonparametric tests** do not require such assumptions. Consequently, nonparametric tests of hypotheses are often called **distribution-free tests**.

Although the term *nonparametric* strongly suggests that the test is not based on a parameter, there are some nonparametric tests that do depend on a parameter such as the median, but don't require a particular distribution. Although *distribution-free* is a more accurate description, the term *nonparametric* is more commonly used. The following are major advantages and disadvantages of nonparametric methods.

Advantages of Nonparametric Methods

1. Nonparametric methods can be applied to a wide variety of situations because they do not have the more rigid requirements associated with their parametric counterparts. In particular, nonparametric methods do not require normally distributed populations.
2. Unlike parametric methods, nonparametric methods can often be applied to nonnumerical data, such as the genders of survey respondents.
3. Nonparametric methods usually involve simpler computations than the corresponding parametric methods and are therefore easier to understand.

Disadvantages of Nonparametric Methods

1. Nonparametric methods tend to waste information because exact numerical data are often reduced to a qualitative form. For example, in the nonparametric sign test (described in Section 13-2), weight losses by dieters are recorded simply as negative signs; the actual magnitudes of the weight losses are ignored.

| TABLE 13-1 | Comparison of Parametric and Nonparametric Tests |

Application	Parametric Test	Nonparametric Test	Efficiency of Nonparametric Test with Normal Population
Two dependent samples (paired data)	t test or z test	Sign test Wilcoxon signed-ranks test	0.63 0.95
Two independent samples	t test or z test	Wilcoxon rank-sum test	0.95
Several independent samples	Analysis of variance (F test)	Kruskal-Wallis test	0.95
Correlation	Linear correlation	Rank correlation test	0.91
Randomness	No parametric test	Runs test	No basis for comparison

2. Nonparametric tests are not as efficient as parametric tests, so with a non-parametric test we generally need stronger evidence (such as a larger sample or greater differences) before we reject a null hypothesis.

When the requirements of population distributions are satisfied, nonparametric tests are generally less efficient than their parametric counterparts, but the reduced efficiency can be compensated for by an increased sample size. For example, Section 13-6 will deal with a concept called *rank correlation,* which has an efficiency rating of 0.91 when compared to the linear correlation presented in Chapter 9. This means that with all other things being equal, nonparametric rank correlation requires 100 sample observations to achieve the same results as 91 sample observations analyzed through parametric linear correlation, assuming the stricter requirements for using the parametric method are met. Table 13-1 lists the nonparametric methods covered in this chapter, along with the corresponding parametric approach and **efficiency** rating. Table 13-1 shows that several nonparametric tests have efficiency ratings above 0.90, so the lower efficiency might not be a critical factor in choosing between parametric or nonparametric methods. However, because parametric tests do have higher efficiency ratings than their nonparametric counterparts, it's generally better to use the parametric tests when their required assumptions are satisfied.

Ranks

Some earlier procedures (such as finding percentiles) required that data be ranked (or arranged in order). In this chapter we sometimes use the ranks themselves as the data. We will now briefly describe ranks so that they can be used in Sections 13-3 through 13-6.

Parametric tests typically require populations with normal distributions of data, but that requirement cannot be met when original data consist of ranks. Such data sets are better analyzed with nonparametric methods, such as those presented in this chapter.

> ### DEFINITION
>
> Data are *ranked* when they are arranged according to some criterion, such as smallest to largest or best to worst. A **rank** is a number assigned to an individual sample item according to its order in the ranked list. The first item is assigned a rank of 1, the second item is assigned a rank of 2, and so on.

EXAMPLE   The numbers 5, 3, 40, 10, and 12 can be arranged from lowest to highest as 3, 5, 10, 12, and 40, and these numbers have ranks of 1, 2, 3, 4, and 5, respectively:

| 5 | 3 | 40 | 10 | 12 | Original scores |
| 3 | 5 | 10 | 12 | 40 | Scores arranged in order |

$\uparrow$ $\uparrow$ $\uparrow$ $\uparrow$ $\uparrow$

| 1 | 2 | 3 | 4 | 5 | Ranks |

This is one method for dealing with ties in ranks. For other approaches, see Exercise 14 in Section 13-2.

Handling ties in ranks: If a tie in ranks occurs, the usual procedure is to find the mean of the ranks involved and then assign this mean rank to each of the tied items, as in the following example.

EXAMPLE   The numbers 3, 5, 5, 10, and 12 are given ranks of 1, 2.5, 2.5, 4, and 5, respectively. In this case, ranks 2 and 3 were tied, so we found the mean of 2 and 3 (which is 2.5) and assigned it to the scores that created the tie:

| 3 | 5 | 5 | 10 | 12 | Original scores |

$\uparrow$ $\quad$ $\uparrow$ $\quad$ $\uparrow$ $\quad$ $\uparrow$ $\quad$ $\uparrow$

| 1 | 2.5 | 2.5 | 4 | 5 | Ranks |

2 and 3 are tied

The sign test, summarized in the flowchart of Figure 13-2, can be covered along with Section 8-2. Sections 13-2 and 8-2 both deal with tests of claims about *two dependent samples.*

13-2 Sign Test

The **sign test** is one of the easiest nonparametric tests to use because it is based on plus and minus signs, which are easily found. It is applicable to several different situations, including (1) claims involving two dependent samples; (2) claims involving nominal data; and (3) claims made about the median of a single population. We will address each of these three applications.

Claims Involving Two Dependent Samples

When using the sign test with two dependent populations, we begin by subtracting each value of the second variable from the corresponding value of the first variable, but we record only the *sign* of the difference. In our sign test procedure, we exclude any ties (represented by 0s). (For other ways to handle ties, see Exercise 14.) The next step consists of recording the positive and negative signs. The key concept underlying the sign test is this:

> **If the two sets of data have equal medians, the number of positive signs should be approximately equal to the number of negative signs.**

For consistency and ease, we will stipulate the following notation.

Notation for the Sign Test

x = the test statistic representing the number of times the *less frequent* sign occurs

n = the total number of positive and negative signs combined

Table A-7 lists critical values for the sign test. The entries in Table A-7 were found by using the binomial probability distribution (Section 4-3). [The binomial probability distribution applies because the results fall into two categories (positive sign, negative sign) and we have a fixed number of independent cases (or pairs of values)].

EXAMPLE   Table 13-2 consists of sample data obtained when 14 subjects are tested for reaction times with their left and right hands. (Only right-handed subjects were used.) Use a 0.05 significance level to test the claim of no difference between the right- and left-hand reaction times.

TABLE 13-2	Reaction Times (in thousandths of a second) of 14 Subjects												
Right	191	97	116	165	116	129	171	155	112	102	188	158	121	133
Left	224	171	191	207	196	165	177	165	140	188	155	219	177	174
Sign of difference (right − left)	−	−	−	−	−	−	−	−	−	−	+	−	−	−

Product Testing

The United States Testing Company in Hoboken, New Jersey, is the world's largest independent product-testing laboratory. It's often hired to verify advertising claims. A vice president has said that the most difficult part of his job is "telling a client when his product stinks. But if we didn't do that, we'd have no credibility." He says that there have been a few clients who wanted positive results fabricated, but most want honest results. The United States Testing Company evaluates laundry detergents, cosmetics, insulation materials, zippers, pantyhose, football helmets, toothpastes, fertilizers, and a wide variety of other products.

SOLUTION   Here's the basic idea: If people generally have the same reaction times with their right and left hands, the numbers of positive and negative signs should be approximately equal—but in Table 13-2 we have 13 negative signs and 1 positive sign. Are the numbers of positive and negative signs approximately equal, or are they significantly different? We follow the same basic steps for testing hypotheses as outlined in Figure 7-4.

Steps 1, 2, 3:   The null hypothesis is the claim of no difference between the right- and left-hand reaction times, and the alternative hypothesis is the claim that there is a difference.

H_0:   There is no difference.
(The median of the differences is equal to 0.)
H_1:   There is a difference.
(The median of the differences is not equal to 0.)

Step 4:   The significance level is $\alpha = 0.05$.

Step 5:   We are using the nonparametric sign test.

Step 6:   The test statistic x is the number of times the less frequent sign occurs. Table 13-2 includes differences with 13 negative signs and 1 positive sign. We let x equal the smaller of 13 and 1, so $x = 1$. Also, $n = 14$ (the total number of positive and negative signs combined). Our test is two-tailed with $\alpha = 0.05$, and reference to Table A-7 shows that the critical value is 2.

Step 7:   With a test statistic of $x = 1$ and a critical value of 2, we reject the null hypothesis of no difference. The critical value of 2 indicates that the number of the less frequent sign must be less than or equal to 2 in order to be significant. (See Note 2 included with Table A-7: "The null hypothesis is rejected if the number of the less frequent sign (x) is less than or equal to the value in the table." Because $x = 1$ is less than or equal to 2, we reject the null hypothesis.)

Step 8:   There is sufficient evidence to warrant rejection of the claim that the median of the differences is equal to 0; that is, there is sufficient evidence to warrant rejection of the claim that there is no difference between the right- and left-hand reaction times. This is the same conclusion reached in Section 8-2 where we used a parametric t test, but sign test results do not always agree with parametric test results.

Using Computer Software for the Sign Test

The TI-83 calculator is not programmed for the sign test, but computer software packages such as STATDISK and Minitab can be used. With STATDISK,

select `Analysis` from the main menu bar, then select `Sign Test`. The displayed result will include the critical value for both a one-tailed test and a two-tailed test, as well as conclusions for one- and two-tailed tests. A Minitab display for the preceding example follows. Minitab's P-value of 0.0018 is less than the significance level of $\alpha = 0.05$, so we reject the null hypothesis of no difference. See the *Minitab Student Laboratory Manual and Workbook* for details on using Minitab for a sign test.

MINITAB DISPLAY

```
Sign test of median = 0.00000 versus N.E. 0.00000

      N    BELOW    EQUAL    ABOVE    P-VALUE    MEDIAN
C3   14      13        0        1     0.0018     -41.50
```

Caution: When applying the sign test in a one-tailed test, be very careful to avoid making the wrong conclusion when one sign occurs significantly more often than the other, but the sample data are *consistent with* the null hypothesis. If the sense of the data is consistent with (instead of conflicting with) the null hypothesis, then fail to reject the null hypothesis and don't proceed with the sign test. Figure 13-2 on page 648 summarizes the procedure for the sign test and includes this check: Do the sample data "support" (in the sense of being consistent with) H_0? If the answer is yes, fail to reject the null hypothesis. *It is always important to think about the data and to avoid relying on blind calculations or computer results.*

In the preceding example we found the critical value in Table A-7, but that table includes values of n up to 25 only. When $n > 25$, we use a normal approximation to obtain critical values. For $n > 25$, the test statistic x is converted to a z score as shown in the following box. (After the next example, we will justify the format of the test statistic for $n > 25$.)

Test Statistic for the Sign Test

For $n \leq 25$: x (the number of times the less frequent sign occurs)

For $n > 25$: $z = \dfrac{(x + 0.5) - \left(\dfrac{n}{2}\right)}{\dfrac{\sqrt{n}}{2}}$

Critical values:
For $n \leq 25$, critical x values are in Table A-7.
For $n > 25$, critical z values are in Table A-2.

Important point to clarify: We should *never* apply procedures blindly without *thinking* about what we are doing. This text explains that in applying the sign test, it is important to check that the sense of the sample data conflicts with the null hypothesis. Whenever sample data support the null hypothesis, we can terminate the test by failing to reject the null hypothesis. For example, you need not formally test a claim that a coin favors heads if 100 tosses result in 52 tails. There's no way we could ever support the claim of heads being *favored* if we get heads in fewer than 50% of the trials.

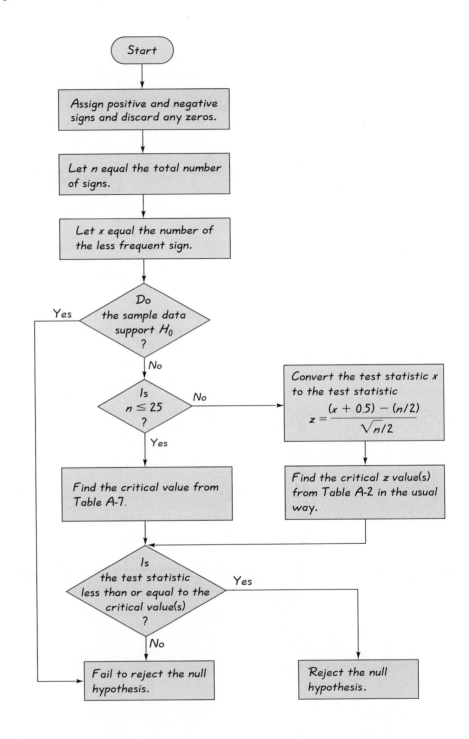

Figure 13-2 Sign Test Procedure

Claims Involving Nominal Data

The following example illustrates the use of nonparametric methods with nominal data. Note that we obtain positive and negative signs by representing women with + signs and men with − signs. (Those signs are chosen arbitrarily, honest.) Also note the procedure for handling cases in which $n > 25$.

It might be helpful to review the concept of nominal data that was introduced in Section 1-2. Nominal data consist of names or categories only, such as gender (man, woman) or political party (Democrat, Republican, Independent, and so on).

EXAMPLE   The Malloy and Turner Advertising Company claims that its hiring practices are fair: "We do not discriminate on the basis of gender, and the fact that 40 of the last 50 new employees are men is just a fluke." The company acknowledges that applicants are about half men and half women, and all have met the basic job-qualification standards. Test the null hypothesis that men and women are hired equally by this company. Use a significance level of 0.05.

SOLUTION   The null and alternative hypotheses are

H_0: $p_1 = p_2$ (the proportions of men and women are equal)
H_1: $p_1 \neq p_2$

Denoting hired women by + and hired men by −, we have 10 positive signs and 40 negative signs. Refer now to the flowchart in Figure 13-2. The test statistic x is the smaller of 10 and 40, so $x = 10$. This test involves two tails because a disproportionately low number of either gender will cause us to reject the claim of equality. The sample data do not support the null hypothesis because 10 and 40 are not precisely equal. (That is, the sample data are not consistent with the null hypothesis.) Continuing with the procedure in Figure 13-2, we note that the value of $n = 50$ is above 25, so the test statistic x is converted (using a correction for continuity) to the test statistic z as follows:

$$z = \frac{(x + 0.5) - \left(\dfrac{n}{2}\right)}{\dfrac{\sqrt{n}}{2}}$$

$$= \frac{(10 + 0.5) - \left(\dfrac{50}{2}\right)}{\dfrac{\sqrt{50}}{2}} = -4.10$$

(continued)

Figure 13-3 Testing the Claim
That Hiring Practices Are Fair

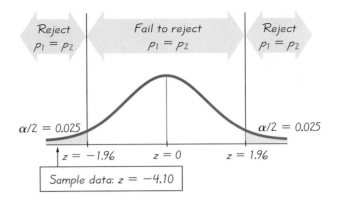

With $\alpha = 0.05$ in a two-tailed test, the critical values are $z = \pm 1.96$. The test statistic $z = -4.10$ is less than these critical values (see Figure 13-3), so we reject the null hypothesis of equality. There is sufficient sample evidence to warrant rejection of the claim that the hiring practices are fair. This company is in trouble.

Rationale for the test statistic used when $n > 25$: When finding critical values for the sign test, we use Table A-7 only for n up to 25, so we need another procedure for finding critical values when $n > 25$. When $n > 25$, the test statistic z is based on a normal approximation to the binomial probability distribution with $p = q = 1/2$. Recall that in Section 5-6 we saw that the normal approximation to the binomial distribution is acceptable when both $np \geq 5$ and $nq \geq 5$. Recall also that in Section 4-4 we saw that $\mu = np$ and $\sigma = \sqrt{npq}$ for binomial experiments. Because this sign test assumes that $p = q = 1/2$, we meet the $np \geq 5$ and $nq \geq 5$ prerequisites whenever $n \geq 10$. Also, with the assumption that $p = q = 1/2$, we get $\mu = np = n/2$ and $\sigma = \sqrt{npq} = \sqrt{n/4} = \sqrt{n}/2$, so

$$z = \frac{x - \mu}{\sigma} \quad \text{becomes} \quad z = \frac{x - \left(\dfrac{n}{2}\right)}{\dfrac{\sqrt{n}}{2}}$$

Finally, we replace x by $x + 0.5$ as a correction for continuity. That is, the values of x are discrete, but because we are using a continuous probability distribution, a discrete value such as 10 is actually represented by the interval from 9.5 to 10.5. Because x represents the less frequent sign, we act conservatively by concerning ourselves only with $x + 0.5$; we thus get the test statistic z, as given in the equation and in Figure 13-2.

Claims About the Median of a Single Population

The previous examples involved applications of the sign test to comparisons of two sets of data, but we can also use the sign test for a claim made about the median of one population. In the next example, negative and positive signs are based on the claimed value of the median.

EXAMPLE   In Chapter 7 we used sample data to test the claim that the mean body temperature of healthy adults is not 98.6° F. With the same data, use the sign test to test the claim that the median is less than 98.6° F. The data set used in Chapter 7 has 106 subjects—68 subjects with temperatures below 98.6° F, 23 subjects with temperatures above 98.6° F, and 15 temperatures equal to 98.6° F.

SOLUTION   The claim that the median is less than 98.6° F is the alternative hypothesis, while the null hypothesis is the claim that the median is at least 98.6° F.

H_0:   Median is at least 98.6° F. (Median $\geq$ 98.6° F)
H_1:   Median is less than 98.6° F. (Median $<$ 98.6° F)

Following the procedure outlined in Figure 13-2, we discard the 15 zeros, we use the negative sign $-$ to denote each temperature that is below 98.6° F, and we use the positive sign $+$ to denote each temperature that is above 98.6° F. We therefore have 68 negative signs and 23 positive signs, so $n = 91$ and $x = 23$ (the number of the less frequent sign). The sample data thus conflict with the null hypothesis, because fewer than half of the 91 temperatures are at least 98.6° F. We must now proceed to determine whether this conflict is significant. (If the sample data did not conflict with the null hypothesis, we could immediately terminate the test by concluding that we fail to reject the null hypothesis.) The value of n exceeds 25, so we convert the test statistic x to the test statistic z.

$$z = \frac{(x + 0.5) - \left(\dfrac{n}{2}\right)}{\dfrac{\sqrt{n}}{2}}$$

$$= \frac{(23 + 0.5) - \left(\dfrac{91}{2}\right)}{\dfrac{\sqrt{91}}{2}} = -4.61$$

(continued)

Figure 13-4 Testing the Claim That the Median Is Less Than 98.6° F

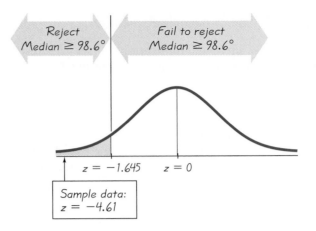

In this one-tailed test with $\alpha = 0.05$, we use Table A-2 to get the critical z value of -1.645. From Figure 13-4 we can see that the test statistic of $z = -4.61$ does fall within the critical region. We therefore reject the null hypothesis. On the basis of the available sample evidence, we support the claim that the median body temperature of healthy adults is less than 98.6° F.

In this sign test of the claim that the median is below 98.6° F, we get a test statistic of $z = -4.61$, but a parametric test of the claim that $\mu < 98.6$° F results in a test statistic of $z = -6.64$. Because -4.61 isn't as extreme as -6.64, we see that the sign test isn't as sensitive as the parametric test. Both tests lead to rejection of the null hypothesis, but the sign test doesn't consider the sample data to be as extreme, partly because the sign test uses only information about the direction of the differences between pairs of data, ignoring the magnitudes of those differences. The next section introduces the Wilcoxon signed-ranks test, which largely overcomes that disadvantage.

13–2 Exercises A: Basic Skills and Concepts

Recommended assignment:
Exercises 1, 2, 4, 5, 6.

In Exercises 1–12, use the sign test.

1. Does it pay to take preparatory courses for standardized tests such as the SAT? Using a 0.05 significance level, test the claim that the Allan Preparation Course has no effect on SAT scores. Use the sample data in the table at the top of the following page.

Subject	A	B	C	D	E	F	G	H	I	J
SAT score before course	700	840	830	860	840	690	830	1180	930	1070
SAT score after course	720	840	820	900	870	700	800	1200	950	1080

Based on data from the College Board and "An Analysis of the Impact of Commercial Test Preparation Courses on SAT Scores," by Sesnowitz, Bernhardt, and Knain, *American Educational Research Journal*, Vol. 19, No. 3.

2. Captopril is a drug designed to lower systolic blood pressure. When subjects were tested with this drug, their systolic blood pressure readings (in mm of mercury) were measured before and after the drug was taken, with the results given below. Use a 0.05 significance level to test the claim that the drug has no effect on systolic blood pressure readings.

Before	200	174	198	170	179	182	193	209	185	155	169	210
After	191	170	177	167	159	151	176	183	159	145	146	177

Based on data from "Essential Hypertension: Effect of an Oral Inhibitor of Angiotensin-Converting Enzyme," by MacGregor et al., *British Medical Journal*, Vol. 2.

3. A television commercial advertises that 7 out of 10 dentists surveyed prefer Covariant toothpaste over the leading competitor. Assume that 10 randomly selected dentists are surveyed; 7 do prefer Covariant, and 3 favor the other brand. Is this a reasonable basis for making the claim that most (more than half) dentists favor Covariant toothpaste? Use a significance level of 0.05.

4. A college aptitude test was given to 100 randomly selected high school seniors. After a period of intensive training, a similar test was given to the same students; 59 students received higher grades, 36 received lower grades, and 5 received the same grades. At the 0.05 level of significance, test the claim that the training is effective.

5. The Diaz Manufacturing Company claims that hiring is done without any gender bias. If 18 of the last 25 new employees are women, and job applicants are about half men and half women who are all qualified, is there sufficient evidence to charge gender bias? Use a 0.01 level of significance, because we don't want to make such a serious charge unless there is very strong evidence.

6. Refer to Data Set 16 in Appendix B. Test the claim that the intervals between eruptions of the Old Faithful geyser have a median greater than 77 min, which was the median about 20 years ago.

7. Using the weights of only the *brown* M&Ms listed in Data Set 11 of Appendix B, test the claim that the median is greater than 0.9085 g. (For the 1498 M&Ms to produce a total package weight of 1361 g, the mean weight must be at least 0.9085 g.) Use a 0.05 significance level. Based on the result, does it appear that the package is labeled correctly?

8. Refer to Data Set 2 in Appendix B. Use the paired data consisting of body temperatures of women at 8:00 A.M. and at 12:00 A.M. on Day 2. Using a

1. The test statistic $x = 2$ is not less than or equal to the critical value of 1.

2. The test statistic $x = 0$ is less than or equal to the critical value of 2.

3. The test statistic $x = 3$ is not less than or equal to the critical value of 1.

4. Convert $x = 36$ to the test statistic $z = -2.26$. The critical value is $z = -1.645$.

5. The test statistic $x = 7$ is not less than or equal to the critical value of 5.

6. (Instead of a right-tailed test to determine whether $x = 37$ is large enough to be significant, use a left-tailed test to determine whether $x = 13$ is small enough to be significant.) Convert $x = 13$ to the test statistic $z = -3.25$. Critical value: $z = -1.645$ (assuming $\alpha = 0.05$).

7. (Instead of a right-tailed test to determine whether $x = 18$ is large enough to be significant, use a left-tailed test to determine whether $x = 15$ is small enough to be significant.) Convert $x = 15$ to the test statistic $z = -0.35$. Critical value: $z = -1.645$.

8. The test statistic $x = 2$ is not less than or equal to the critical value of 1.

9. There are 38 weights below 5.670 g, 2 equal to 5.670 g, and 10 above 5.670 g, so use $x = 10$ and $n = 48$. Test statistic: $z = -3.90$. Critical values: $z = \pm 1.96$ (assuming $\alpha = 0.05$).

10. Convert $x = 6$ to the test statistic $z = -2.83$. The critical values are $z = \pm 2.575$.

11. The test statistic $x = 2$ is not less than or equal to the critical value of 1.

12. With 11 scores below 1.0 mg and 11 scores above 1.0 mg, the sample data tend to support the null hypothesis that the median is at least 1.0 mg, so fail to reject that null hypothesis. There is not sufficient evidence to support the claim that the median is less than 1.0 mg.

13. With $k = 3$ we get $6 < M < 17$.

0.05 level of significance, test the claim that for those temperatures, the median difference is 0. Based on the result, do morning and night body temperatures appear to be about the same?

9. Refer to Data Set 13 in Appendix B. If the weights of the sampled quarters are consistent with current manufacturing specifications of the U.S. Mint, the median should be 5.670 g. Use the data to test the claim that the sampled quarters have weights with that median.

10. The Life Trust Insurance Company funded a university study of drinking and driving. After 30 randomly selected drivers were tested for reaction times, they were given two drinks and tested again, with the result that 22 had slower reaction times, 6 had faster reaction times, and 2 received the same scores as before the drinks. At the 0.01 significance level, test the claim that the drinks had no effect on the reaction times. Based on these very limited results, does it appear that the insurance company is justified in charging higher rates for those who drink and drive?

11. Mental measurements of young children are often made by giving them blocks and telling them to build a tower as tall as possible. One experiment of block building was repeated a month later, with the times (in seconds) listed below. Use a 0.01 level of significance to test the claim that there is no difference between the two times.

First trial	30	19	19	23	29	178	42	20	12	39	14	81	17	31	52
Second trial	30	6	14	8	14	52	14	22	17	8	11	30	14	17	15

Based on data from "Tower Building," by Johnson and Courtney, *Child Development*, Vol. 3.

12. Refer to the nicotine contents of cigarettes listed in Data Set 4 of Appendix B. Use the sample data to test the claim that the population median is less than 1.0 mg.

13-2 Exercises B: Beyond the Basics

13. Given n sample scores sorted in ascending order $(x_1, x_2, \ldots, x_n)$, if we wish to find the approximate $1 - \alpha$ confidence interval for the population median M, we get

$$x_{k+1} < M < x_{n-k}$$

Here k is the critical value (Table A-7) for the number of signs in a two-tailed hypothesis test conducted at the significance level of α. Find the approximate 95% confidence interval for the sample scores listed below.

3   8   6   2   1   7   9   11   17   23   25   10   14   8   30

14. In the sign test procedure described in this section, we excluded ties (represented by 0 instead of a sign of $+$ or $-$). A second approach is to treat half of the 0s as positive signs and half as negative signs. (If the number of 0s is odd, exclude one so that they can be divided equally.) With a third approach, in two-tailed tests make half of the 0s positive and half nega-

tive; in one-tailed tests make all 0s either positive or negative, whichever supports the null hypothesis. Assume that in using the sign test on a claim that the median score is at least 100, we get 60 scores below 100, 40 scores above 100, and 21 scores equal to 100. Identify the test statistic and conclusion for the three different ways of handling differences of 0. Assume a 0.05 significance level in all three cases.

15. Table A-7 lists critical values for limited choices of α. Use Table A-1 to add a new column in Table A-7 (down to $n = 15$) that represents a significance level of 0.03 in one tail or 0.06 in two tails. For any particular n we use $p = 0.5$, as the sign test requires the assumption that

$$P(\text{positive sign}) = P(\text{negative sign}) = 0.5$$

The probability of x or fewer like signs is the sum of the probabilities for values up to and including x.

14. First approach: $z = -1.90$; reject H_0.
Second approach: $z = -1.73$; reject H_0.
Third approach: $z = 0$; fail to reject H_0.

15. *, *, *, *, *, 0, 0, 0, 1, 1, 1, 2, 2, 3, 3

Wilcoxon Signed-Ranks Test for Two Dependent Samples

In Section 13-2 we used the sign test to analyze the differences between sample paired data. The sign test used only the signs of the differences and did not use their actual magnitudes (how large the numbers are). This section introduces the **Wilcoxon signed-ranks test,** which is also used with sample paired data. By using ranks, this test takes the magnitudes of the differences into account. Because the Wilcoxon signed-ranks test incorporates and uses more information than the sign test, it tends to yield conclusions that better reflect the true nature of the data. The Wilcoxon signed-ranks test requires the following assumption.

Like the previous section, this section addresses two dependent samples. This section could be covered along with Section 8-2, which includes a t test for claims about two dependent samples.

Assumption   In using the Wilcoxon signed-ranks test for two dependent (paired) samples, we assume that the population of differences (found from the pairs of data) has a distribution that is approximately symmetric.

In Section 2-4 we defined a distribution to be *symmetric* if the left half of its histogram is roughly a mirror image of its right half. Unlike the Student t test for paired data (described in Section 8-2), the Wilcoxon signed-ranks test does *not* require that the data have a normal distribution.

In general, the null hypothesis is the claim that both samples come from populations that have the same distribution. The null and alternative hypotheses can therefore be generalized as follows:

H_0:   The two samples come from populations with the same distribution.
H_1:   The two samples come from populations with different distributions.

The test statistic is based on ranks of the differences between the pairs of data. The rationale for the test statistic will be discussed later in this section; for now we present the steps for finding its value.

When working with more than a few scores, ranking data can be a very annoying task. To ease the ranking process, consider using the TI-83 calculator (Sort) or stem-and-leaf plots.

Procedure for Finding the Value of the Test Statistic

Step 1:   For each pair of data, find the difference d by subtracting the second score from the first score. Keep the signs, but discard any pairs for which $d = 0$.

Step 2:   *Ignore the signs of the differences,* then rank the differences from lowest to highest. When differences have the same numerical value, assign to them the mean of the ranks involved in the tie. (See Section 13-1 for the method of ranking data.)

Step 3:   Attach to each rank the sign of the difference from which it came. That is, insert those signs that were ignored in Step 2.

Step 4:   Find the sum of the absolute values of the negative ranks. Also find the sum of the positive ranks.

Step 5:   Let T be the *smaller* of the two sums found in Step 4. Either sum could be used, but for a simplified procedure we arbitrarily select the smaller of the two sums. (See the notation for T in the box below.)

Step 6:   Let n be the number of pairs of data for which the difference d is not 0.

Step 7:   Determine the test statistic and critical values based on the sample size, as shown in the Test Statistic box.

Step 8:   When forming the conclusion, reject the null hypothesis if the sample data lead to a test statistic that is in the critical region—that is, the test statistic is less than or equal to the critical value(s). Otherwise, fail to reject the null hypothesis.

Notation

$T = $ the smaller of the following two sums:

1. The sum of the absolute values of the negative ranks
2. The sum of the positive ranks

Test Statistic for the Wilcoxon Signed-Ranks Test for Two Dependent Samples

For $n \leq 30$: T

For $n > 30$: $z = \dfrac{T - \dfrac{n(n + 1)}{4}}{\sqrt{\dfrac{n(n + 1)(2n + 1)}{24}}}$

Critical values:

If $n \leq 30$, the critical T value is found in Table A-8.
If $n > 30$, the critical z values are found in Table A-2.

TABLE 13-3	Reaction Times (in thousandths of a second) of 14 Subjects													
Right	191	97	116	165	116	129	171	155	112	102	188	158	121	133
Left	224	171	191	207	196	165	177	165	140	188	155	219	177	174
Differences d	−33	−74	−75	−42	−80	−36	−6	−10	−28	−86	+33	−61	−56	−41
Ranks of differences	4.5	11	12	8	13	6	1	2	3	14	4.5	10	9	7
Signed ranks	−4.5	−11	−12	−8	−13	−6	−1	−2	−3	−14	+4.5	−10	−9	−7

EXAMPLE   Table 13-3 consists of sample data obtained when 14 subjects are tested for reaction times with their left and right hands. (Only right-handed subjects were used.) Use the Wilcoxon signed-ranks test to test the claim of no difference between reaction times with right and left hands. Use a significance level of $\alpha = 0.05$.

SOLUTION   The null and alternative hypotheses are as follows:

H_0:   There is no difference between right- and left-hand reaction times.
H_1:   There is a difference between right- and left-hand reaction times.

The significance level is $\alpha = 0.05$. We are using the Wilcoxon signed-ranks test procedure, so the test statistic is calculated by using the eight-step procedure presented earlier in this section.

Step 1:   In Table 13-3, the row of differences is obtained by computing
d = Right − Left
for each pair of data.

Step 2:   Ignoring their signs, we rank the absolute differences from lowest to highest. Note that the differences of −33 and +33 are tied for ranks of 4 and 5, so we assign a rank of 4.5 to each of them.

(continued)

Step 3: The bottom row of Table 13-3 is created by attaching to each rank the sign of the corresponding difference. If there is no difference between right- and left-hand reaction times, we expect the number of positive ranks to be approximately equal to the number of negative ranks.

Step 4: We now find the sum of the absolute values of the negative ranks, and we also find the sum of the positive ranks.

Sum of absolute values of negative ranks: 100.5, which is

$$4.5 + 11 + 12 + 8 + 13 + 6 + 1 + 2 + 3 + 14 + 10 + 9 + 7$$

Sum of positive ranks: 4.5

Step 5: Letting T be the smaller of the two sums found in Step 4, we find that $T = 4.5$.

Step 6: Letting n be the number of pairs of data for which the difference d is not 0, we have $n = 14$.

Step 7: Because $n = 14$, we have $n \leq 30$, so we use a test statistic of $T = 4.5$ (and we do not calculate a z test statistic). Also, because $n \leq 30$, we use Table A-8 to find the critical value of 21.

Step 8: The test statistic $T = 4.5$ is less than or equal to the critical value of 21, so we reject the null hypothesis. It appears that there is a difference between right- and left-hand reaction times. Examining the sample data, we see that there are 13 negative signs and only 1 positive sign. Because negative signs result from lower values with the right hand, it appears that right-handed people have faster reaction times with their right hands.

If we use the sign test with the preceding example, we will arrive at the same conclusion. Although the sign test and the Wilcoxon signed-ranks test agree in this particular case, there are other cases in which they do not agree. [For example, see Cumulative Review Exercise 2, parts (a) and (b).]

Using Computer Software for the Wilcoxon Signed-Ranks Test

The TI-83 calculator is not programmed for the Wilcoxon signed-ranks test, but computer software such as STATDISK and Minitab can be used. With STATDISK, select `Analysis` from the main menu bar, then select `Wilcoxon Tests` and proceed to use the Wilcoxon signed-ranks test for *dependent* samples. The STATDISK display will include the test statistic, critical value, and conclusion. With Minitab, enter the paired data in columns C1 and C2, then enter the command LET C3 = C1 − C2, then select the options of `Stat`,

Nonparametrics, and 1-Sample Wilcoxon. Enter C3 for the variable and click on the circle for Test Median. The Minitab display will be as shown. The displayed *P*-value of 0.003 is less than the significance level of $\alpha = 0.05$, indicating that we should reject the null hypothesis of no difference.

MINITAB DISPLAY

TEST OF MEDIAN = 0.000000 VERSUS MEDIAN N.E. 0.000000

	N	N FOR TEST	WILCOXON STATISTIC	P-VALUE	ESTIMATED MEDIAN
C3	14	14	4.5	0.003	-44.50

In this example the unsigned ranks of 1 through 14 have a total of 105, so if there are no significant differences, each of the two signed-rank totals should be around $105 \div 2$, or 52.5. The table of critical values shows that at the 0.05 level of significance with 14 pairs of data, a 21-84 split represents a significant departure from the null hypothesis, and any split that is farther apart (such as 20-85 or 19-86 or 4.5-100.5) will also represent a significant departure from the null hypothesis. Conversely, splits like 22-83 do not represent significant departures away from a 52.5-52.5 split, and they would not be a basis for rejecting the null hypothesis. The Wilcoxon signed-ranks test is based on the lower rank total, so instead of analyzing both numbers constituting the split, we consider only the lower number.

The sum $1 + 2 + 3 + \cdots + n$ of all the ranks is equal to $n(n + 1)/2$; if this is a rank sum to be divided equally between two categories (positive and negative), each of the two totals should be near $n(n + 1)/4$, which is half of $n(n + 1)/2$. Recognition of this principle helps us understand the test statistic used when $n > 30$. The denominator in that expression represents a standard deviation of T and is based on the principle that

$$1^2 + 2^2 + 3^3 + \cdots + n^2 = \frac{n(n + 1)(2n + 1)}{6}$$

The Wilcoxon signed-ranks test can only be used for dependent (matched) data. The next section will describe a rank-sum test that can be applied to two sets of independent data that are not paired.

13–3 Exercises A: Basic Skills and Concepts

In Exercises 1–4, refer to the sample data for the given exercises in Section 13-2. Instead of the sign test, use the Wilcoxon signed-ranks test to test the

Recommended assignment:
Exercises 1, 2, 5, 6.

1. Test statistic: $T = 9.5$. Critical value: $T = 2$.
2. Test statistic: $T = 0$. Critical value: $T = 14$.
3. Test statistic: $T = 6.5$. Critical value: $T = 8$.
4. Test statistic: $T = 5.5$. Critical value: $T = 13$.
5. Test statistic: $T = 1$. Critical value: $T = 0$.
6. Test statistic: $T = 0$. Critical value: $T = 8$.
7. Test statistic: $T = 1$. Critical value: $T = 14$ (assuming $\alpha = 0.05$).
8. Test statistic: $T = 1$. Critical value: $T = 0$.

claim that both samples come from populations having the same distribution. Use the significance level α that is given below.

1. Exercise 1; $\alpha = 0.01$ **2.** Exercise 2; $\alpha = 0.05$
3. Exercise 8; $\alpha = 0.05$ **4.** Exercise 11; $\alpha = 0.01$

In Exercises 5–8, use the Wilcoxon signed-ranks test.

5. A study was conducted to investigate the effectiveness of hypnotism in reducing pain. Results for randomly selected subjects are given below. The measurements are in centimeters on a pain scale. At the 0.01 significance level, test the claim that hypnotism has no effect.

Subject	A	B	C	D	E	F	G	H
Before hypnosis	6.6	6.5	9.0	10.3	11.3	8.1	6.3	11.6
After hypnosis	6.8	2.4	7.4	8.5	8.1	6.1	3.4	2.0

Based on "An Analysis of Factors That Contribute to the Efficacy of Hypnotic Analgesia," by Price and Barber, *Journal of Abnormal Psychology*, Vol. 96, No. 1.

6. In a study of techniques used to measure lung volumes, physiological data were collected for 10 subjects. The values given in the table are in liters and represent the measured functional residual capacities of the 10 subjects both in a sitting position and in a supine (lying) position. At the 0.05 significance level, test the claim that there is no significant difference between the measurements taken in the two positions.

Sitting	2.96	4.65	3.27	2.50	2.59	5.97	1.74	3.51	4.37	4.02
Supine	1.97	3.05	2.29	1.68	1.58	4.43	1.53	2.81	2.70	2.70

Based on "Validation of Esophageal Balloon Technique at Different Lung Volumes and Postures," by Baydur, Cha, and Sassoon, *Journal of Applied Physiology*, Vol. 62, No. 1.

7. The Acton Paper Company gives new employees a test for mathematics anxiety. The test is given twice: once before and once after an intensive program in which required mathematical skills are identified and the company's hiring and firing policies are discussed. Sample scores from the math anxiety test are given below. Test the claim that the program has no effect on the anxiety level.

Before program	104	97	60	88	39	82	91	87	41	43	82	58	67
After program	83	64	58	72	40	70	87	79	25	43	64	40	52

8. The Malloy and Turner Advertising Company prepared two different television commercials for Taylor women's jeans. One commercial is humorous, and the other is serious. A test screening of both commercials involved eight randomly selected consumers who were tested for their reactions; the results are listed in the table. At the 0.01 significance level,

test the claim that there is no difference between the reactions to the commercials. Based on the result, which commercial seems better?

Consumer	A	B	C	D	E	F	G	H
Humorous commercial	26.2	20.3	25.4	19.6	21.5	28.3	23.7	24.0
Serious commercial	24.1	21.3	23.7	18.0	20.1	25.8	22.4	21.4

13-3 Exercises B: Beyond the Basics

9. Assume that the Wilcoxon signed-ranks test is being used for a two-tailed hypothesis test with a significance level of 0.05.
   a. With $n = 10$ pairs of data, find the lowest and highest possible values of T.   9. a. 0, 27.5
   b. With $n = 50$ pairs of data, find the lowest and highest possible values of T.   b. 0, 637.5
   c. If there are $n = 100$ pairs of data with no differences of 0 and no tied ranks, find the critical value of T.   c. 1954

10. The Wilcoxon signed-ranks test can be used to test the claim that a sample comes from a population with a specified median. The procedure used is the same as the one described in this section, except that the differences (Step 1) are obtained by subtracting the value of the hypothesized median from each score. Use the sample data consisting of the 106 body temperatures listed in Table 7-1. (See the Chapter Problem at the beginning of Chapter 7.) At the 0.05 level of significance, test the claim that healthy adults have a median body temperature that is equal to 98.6° F.

10. Convert $T = 710$ to test statistic $z = -5.47$. Critical values: $z = \pm 1.96$.

13-4 Wilcoxon Rank-Sum Test for Two Independent Samples

Section 13-3 presented the Wilcoxon signed-ranks test as a method for analyzing two dependent (or paired) samples. This section introduces the **Wilcoxon rank-sum test,** which is a nonparametric test that can be applied to data from two samples that are *independent* and not paired.

The Wilcoxon rank-sum test is equivalent to the **Mann-Whitney U test,** which is included in some other textbooks (see Exercise 9) and software packages (such as Minitab). The basis for the procedure used in the Wilcoxon rank-sum test is the principle that if two samples are drawn from identical populations and the individual scores are all ranked as one combined collection of values, then the high and low ranks should fall evenly between the two samples. If the low ranks are found predominantly in one sample and the high ranks are found predominantly in the other sample, we suspect that the two populations are not identical.

Because this topic involves two independent samples, it could be covered along with Sections 8-3 and 8-5. The method described in this section is equivalent to the Mann-Whitney *U* test described in some other textbooks.

Assumptions

Once again, the requirement of a normal distribution or any other particular distribution is *not* included among the requirements, so the title of a "distribution-free" method is appropriate.

1. We have two independent samples.
2. We are testing the null hypothesis that the two independent samples come from the same distribution; the alternative hypothesis is the claim that the two distributions are different in some way.
3. Each of the two samples has more than 10 scores. (For samples with 10 or fewer values, special tables are available in reference books, such as *CRC Standard Probability and Statistics Tables and Formulae*, published by CRC Press.)

Note that unlike the corresponding hypothesis tests in Sections 8-3 and 8-5, the Wilcoxon rank-sum test does *not* require normally distributed populations. Also, the Wilcoxon rank-sum test can be used with data at the ordinal level of measurement, such as data consisting of ranks. In contrast, the parametric methods of Sections 8-3 and 8-5 cannot be used with data at the ordinal level of measurement. In Table 13-1 we noted that the Wilcoxon rank-sum test has a 0.95 efficiency rating when compared with the parametric t test or z test. Because this test has such a high efficiency rating and involves easier calculations, it is often preferred over the parametric tests presented in Sections 8-3 and 8-5, even when the condition of normality is satisfied.

The procedure for using the Wilcoxon rank-sum test begins with these steps:

1. Rank all the sample data combined.
2. Find the sum of the ranks for one of the two samples.

In the following notation, either sample can be used as sample 1.

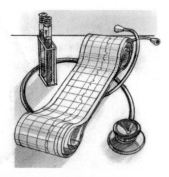

Gender Gap in Drug Testing

A study of the relationship between heart attacks and doses of aspirin involved 22,000 male physicians. This study, like many others, excluded women. The General Accounting Office recently criticized the National Institutes of Health for not including both sexes in many studies because results of medical tests on males do not necessarily apply to females. For example, women's hearts are different from men's in many important ways. When forming conclusions based on sample results, we should be wary of an inference that extends to a population larger than the one from which the sample was drawn.

Notation for the Wilcoxon Rank-Sum Test
n_1 = size of sample 1
n_2 = size of sample 2
R_1 = sum of ranks for sample 1
R_2 = sum of ranks for sample 2
R = same as R_1 (sum of ranks for sample 1)
μ_R = mean of the sample R values that is expected when the two populations are identical
σ_R = standard deviation of the sample R values that is expected when the two populations are identical

If testing the null hypothesis of identical populations and if both sample sizes are greater than 10, then the sampling distribution of R is approximately normal with mean μ_R and standard deviation σ_R, and the test statistic is as follows. (Because the test statistic is based on the normal distribution, critical values can be found in Table A-2.)

Test Statistic for the Wilcoxon Rank-Sum Test for Two Independent Samples

$$z = \frac{R - \mu_R}{\sigma_R}$$

where $\mu_R = \dfrac{n_1(n_1 + n_2 + 1)}{2}$

where $\sigma_R = \sqrt{\dfrac{n_1 n_2(n_1 + n_2 + 1)}{12}}$

$n_1 =$ size of the sample from which the rank sum R is found

$n_2 =$ size of the other sample

$R =$ sum of ranks of the sample with size n_1

The expression for μ_R is a variation of the following result of mathematical induction: The sum of the first n positive integers is given by $1 + 2 + 3 + \cdots + n = n(n + 1)/2$. The expression for σ_R is a variation of a result that states that the integers $1, 2, 3, \ldots, n$ have standard deviation $\sqrt{(n^2 - 1)/12}$.

EXAMPLE Samples of M&M plain candies are randomly selected, and the red and yellow M&Ms are weighed, with the results listed in the margin on the following page (from Data Set 11 in Appendix B). Use a 0.05 level of significance to test the claim that weights of red M&Ms and yellow M&Ms have the same distribution.

Class Attendance Does Help

In a study of 424 undergraduates at the University of Michigan, it was found that students with the worst attendance records tended to get the lowest grades. (Is anybody surprised?) Those who were absent less than 10% of the time tended to receive grades of B or above. The study also showed that students who sit in the front of the class tend to get significantly better grades.

SOLUTION The null and alternative hypotheses are as follows:

H_0: Red and yellow M&M plain candies have weights with identical populations.

H_1: The two populations are not identical.

Rank all 47 weights combined, beginning with a rank of 1 (assigned to the lowest weight of 0.868 g). Ties in ranks are handled as described in Section 13-1: Find the mean of the ranks involved and assign this mean rank to each of the tied values. The ranks corresponding to the individual sample values are shown in parentheses in the table. R denotes the sum of the ranks for the sample we choose as sample 1. If we choose the red M&Ms, we get

(continued)

Weights (in grams) of M&Ms			
Red		Yellow	
0.870	(2)	0.906	(19)
0.933	(35)	0.978	(45)
0.952	(42)	0.926	(34)
0.908	(21)	0.868	(1)
0.911	(24.5)	0.876	(5)
0.908	(21)	0.968	(44)
0.913	(27)	0.921	(30)
0.983	(46)	0.893	(15)
0.920	(29)	0.939	(38)
0.936	(37)	0.886	(10.5)
0.891	(13)	0.924	(32)
0.924	(32)	0.910	(23)
0.874	(4)	0.877	(6)
0.908	(21)	0.879	(7.5)
0.924	(32)	0.941	(40)
0.897	(16)	0.879	(7.5)
0.912	(26)	0.940	(39)
0.888	(12)	0.960	(43)
0.872	(3)	0.989	(47)
0.898	(17)	0.900	(18)
0.882	(9)	0.917	(28)
		0.911	(24.5)
		0.892	(14)
		0.886	(10.5)
		0.949	(41)
		0.934	(36)
$n_1 = 21$		$n_2 = 26$	
$R_1 = 469.5$		$R_2 = 658.5$	

$$R = 2 + 35 + 42 + \cdots + 9 = 469.5$$

Because there are 21 red M&Ms, we have $n_1 = 21$. Also, $n_2 = 26$, because there are 26 yellow M&Ms. We can now determine the values of μ_R, σ_R, and the test statistic z.

$$\mu_R = \frac{n_1(n_1 + n_2 + 1)}{2} = \frac{21(21 + 26 + 1)}{2} = 504$$

$$\sigma_R = \sqrt{\frac{n_1 n_2(n_1 + n_2 + 1)}{12}} = \sqrt{\frac{(21)(26)(21 + 26 + 1)}{12}} = 46.73$$

$$z = \frac{R - \mu_R}{\sigma_R} = \frac{469.5 - 504}{46.73} = -0.74$$

The test is two-tailed because a large positive value of z would indicate that the higher ranks are found disproportionately in the first sample, and a large negative value of z would indicate that the first sample had a disproportionate share of lower ranks. In either case, we would have strong evidence against the claim that the two samples come from identical populations.

The significance of the test statistic z can be treated in the same manner as in previous chapters. We are now testing (with $\alpha = 0.05$) the hypothesis that the two populations are the same, so we have a two-tailed test with critical z values of 1.96 and -1.96. The test statistic of $z = -0.74$ does not fall within the critical region, so we fail to reject the null hypothesis that red and yellow M&Ms have the same weights.

We can verify that if we interchange the two sets of weights and consider the sample of yellow M&Ms to be first, $R = 658.5$, $\mu_R = 624$, $\sigma_R = 46.73$, and $z = 0.74$, so the conclusion is the same.

Using Computer Software for the Wilcoxon Rank-Sum Test

Computer software such as STATDISK and Minitab can be used for the Wilcoxon rank-sum test. If using STATDISK, select Analysis from the main menu bar, then select Wilcoxon Tests and proceed to use the Wilcoxon rank-sum test for *independent* samples. The STATDISK display will include the rank sums, sample size, test statistic, critical value, and conclusion.

If using Minitab, first enter the two sets of sample data in columns C1 and C2. Then select the options of Stat, Nonparametrics, and Mann-Whitney, and proceed to enter C1 for the first sample and C2 for the second sample. The confidence level of 95.0 corresponds to a significance level of $\alpha = 0.05$, and the "alternate: not equal" box refers to the alternative hypothesis, where "not equal" corresponds to a two-tailed hypothesis test. The

Minitab display will be as shown. Note the inclusion of the test statistic of 469.5, the *P*-value of 0.4668, and the comment "Cannot reject (the null hypothesis) at alpha = 0.05."

```
                        MINITAB DISPLAY

RED             N = 21          Median = 0.90800
YELLOW          N = 26          Median = 0.91400
Point estimate for ETA1-ETA2 is      -0.00600
95.2 Percent C.I. for ETA1-ETA2 is (-0.02600,0.01200)
W = 469.5
Test of ETA1 = ETA2 vs. ETA1 ~= ETA2 is significant at 0.4669
The test is significant at 0.4668 (adjusted for ties)

Cannot reject at alpha = 0.05
```

13-4 Exercises A: Basic Skills and Concepts

In Exercises 1–8, use the Wilcoxon rank-sum test.

1. The example in this section tested the null hypothesis that red and yellow M&M plain candies have weights with identical populations. Refer to Data Set 11 in Appendix B and test the claim that red and brown M&M plain candies have weights with identical populations. Use a 0.05 level of significance.

2. Listed below are time intervals (in minutes) between eruptions of the Old Faithful geyser in Yellowstone National Park. Using a 0.05 significance level, test the claim that the times are the same for both years.

1951	74	60	74	42	74	52	65	68	62	66	62	60
1996	88	86	85	89	83	85	91	68	91	56	89	94

Based on data from geologist Rick Hutchinson and the National Park Service.

3. Flammability tests were conducted on children's sleepwear. The Vertical Semirestrained Test was used, in which pieces of fabric were burned under controlled conditions. After the burning stopped, the length of the charred portion was measured and recorded. The table at the top of the following page shows results for the same fabric tested at two different laboratories. Because the same fabric was used, the different laboratories should have obtained the same results. Did they? Use a 0.05 significance level to test the claim that the two laboratories have identical populations.

Recommended assignment: Exercises 1, 2, 3, 4, 8.

1. $\mu_R = 577.5$, $\sigma_R = 56.358$, $R = 569$, $z = -0.15$. Test statistic: $z = -0.15$. Critical values: $z = \pm 1.96$.

2. $\mu_R = 150$, $\sigma_R = 17.3205$, $R = 91.5$, $z = -3.38$. Test statistic: $z = -3.38$. Critical values: $z = \pm 1.96$.

3. $\mu_R = 126.5$, $\sigma_R = 15.2288$, $R = 103$, $z = -1.54$. Test statistic: $z = -1.54$. Critical values: $z = \pm 1.96$.

4. $\mu_R = 150$, $\sigma_R = 17.321$,
$R = 96.5$, $z = -3.09$. Test
statistic: $z = -3.09$. Critical
values: $z = \pm 2.575$.

5. $\mu_R = 162$, $\sigma_R = 19.442$,
$R = 89.5$, $z = -3.73$. Test
statistic: $z = -3.73$. Critical
values: $z = \pm 1.96$.

6. $\mu_R = 188.5$, $\sigma_R = 21.71$,
$R = 182$, $z = -0.30$. Test
statistic: $z = -0.30$. Critical
values: $z = \pm 1.96$.

Lab 1	2.9	3.1	3.1	3.7	3.1	4.2	3.7	3.9	3.1	3.0	2.9
Lab 2	2.7	3.4	3.6	3.2	4.0	4.1	3.8	3.8	4.3	3.4	3.3

Data provided by Minitab, Inc.

4. Are severe psychiatric disorders related to biological factors that can be physically observed? One study used X-ray computed tomography (CT) to collect data on brain volumes for a group of patients with obsessive-compulsive disorders and a control group of healthy persons. The accompanying list shows sample results (in milliliters) for volumes of the right cordate (based on data from "Neuroanatomical Abnormalities in Obsessive-Compulsive Disorder Detected with Quantitative X-Ray Computed Tomography," by Luxenberg et al., *American Journal of Psychiatry*, Vol. 145, No. 9).

Obsessive-compulsive patients				Control group			
0.308	0.210	0.304	0.344	0.519	0.476	0.413	0.429
0.407	0.455	0.287	0.288	0.501	0.402	0.349	0.594
0.463	0.334	0.340	0.305	0.334	0.483	0.460	0.445

At the 0.01 significance level, test the claim that obsessive-compulsive patients and healthy persons have the same brain volumes. Based on this result, can we conclude that obsessive-compulsive disorders have a biological basis?

5. The sample data in the following list show BAC (blood alcohol concentration) levels at arrest of randomly selected jail inmates who were convicted of DWI or DUI offenses. The data are categorized by the type of drink consumed (based on data from the U.S. Department of Justice).

Beer				Liquor			
0.129	0.146	0.148	0.152	0.220	0.225	0.185	0.182
0.154	0.155	0.187	0.212	0.253	0.241	0.227	0.205
0.203	0.190	0.164	0.165	0.247	0.224	0.226	0.234
				0.190	0.257		

At the 0.05 significance level, test the claim that beer drinkers and liquor drinkers have the same BAC levels. Based on these results, do both groups seem equally dangerous, or is one group more dangerous than the other?

6. Sample data were collected in a study of calcium supplements and their effects on blood pressure. A placebo group and a calcium group began the study with measures of blood pressures. At the 0.05 significance level, test the claim that the two sample groups come from populations with the same blood pressure levels. (The data are based on "Blood Pressure and Metabolic Effects of Calcium Supplementation in Normotensive White

and Black Men," by Lyle et al., *Journal of the American Medical Association*, Vol. 257, No. 13.)

Placebo group

124.6	104.8	96.5	116.3
106.1	128.8	107.2	123.1
118.1	108.5	120.4	122.5
113.6			

Calcium group

129.1	123.4	102.7	118.1
114.7	120.9	104.4	116.3
109.6	127.7	108.0	124.3
106.6	121.4	113.2	

7. The arrangement of test items was studied for its effect on anxiety. Sample results are as follows:

Easy to difficult

24.64	39.29	16.32	32.83
28.02	33.31	20.60	21.13
26.69	28.90	26.43	24.23
7.10	32.86	21.06	28.89
28.71	31.73	30.02	21.96
25.49	38.81	27.85	30.29
30.72			

Difficult to easy

33.62	34.02	26.63	30.26
35.91	26.68	29.49	35.32
27.24	32.34	29.34	33.53
27.62	42.91	30.20	32.54

At the 0.05 level of significance, test the claim that the two samples come from populations with the same scores. (The data are based on "Item Arrangement, Cognitive Entry Characteristics, Sex and Test Anxiety as Predictors of Achievement Examination Performance," by Klimko, *Journal of Experimental Education*, Vol. 52, No. 4.)

8. Refer to Data Set 9 in Appendix B. Use a 0.01 significance level to test the claim that the population of cotinine levels is the same for these two groups: Nonsmokers who are not exposed to environmental tobacco smoke (labeled as NOETS), and nonsmokers who are exposed to tobacco smoke (labeled as ETS). What do the results suggest about "second-hand" smoke?

13-4 Exercises B: Beyond the Basics

9. The Mann-Whitney U test is equivalent to the Wilcoxon rank-sum test for independent samples in the sense that they both apply to the same situations and always lead to the same conclusions. In the Mann-Whitney U test we calculate

---

7. $\mu_R = 525$, $\sigma_R = 37.417$, $R = 437$, $z = -2.35$. Test statistic: $z = -2.35$. Critical values: $z = \pm 1.96$.

8. $\mu_R = 2525$, $\sigma_R = 145.05746$, $R = 1825$, $z = -4.83$. Test statistic: $z = -4.83$. Critical values: $z = \pm 2.575$.

9. The denominators are the same, so you need to consider only the numerators. First set $R - \mu_R = -U + n_1 n_2/2$, and then replace U by $n_1 n_2 + n_1(n_1 + 1)/2 - R$ and replace μ_R by $n_1(n_1 + n_2 + 1)/2$. Remove parentheses and simplify to see that both sides are equal.

10. a

ABAB	4
ABBA	5
BBAA	7
BAAB	5
BABA	6

b.

R	p
3	1/6
4	1/6
5	2/6
6	1/6
7	1/6

c. No.

Rank	Rank sum for treatment A
1 2 3 4	
A A B B	3

$$z = \frac{U - \dfrac{n_1 n_2}{2}}{\sqrt{\dfrac{n_1 n_2 (n_1 + n_2 + 1)}{12}}}$$

where

$$U = n_1 n_2 + \frac{n_1(n_1 + 1)}{2} - R$$

Show that if the expression for U is substituted into the preceding expression for z, we get the same test statistic (with opposite sign) used in the Wilcoxon rank-sum test for two independent samples.

10. Assume that we have two treatments (A and B) that produce measurable results, and we have only two observations for treatment A and two observations for treatment B. We cannot use the test statistic given in this section because both sample sizes do not exceed 10.

   a. Complete the accompanying table by listing the five rows corresponding to the other five cases, and enter the corresponding rank sums for treatment A.

   b. List the possible values of R, along with their corresponding probabilities. (Assume that the rows of the table from part (a) are equally likely.)

   c. Is it possible, at the 0.10 significance level, to reject the null hypothesis that there is no difference between treatments A and B? Explain.

13-5 Kruskal-Wallis Test

Because this section deals with comparisons of three or more sets of sample data, it could be covered along with Section 11-2 (ANOVA). If the class lacks time or energy to get into analysis of variance, this section might be used as a quicker and easier alternative.

In Section 11-2 we used one-way analysis of variance (ANOVA) to test hypotheses that differences in means among several samples were significant. That method uses an F test that requires that all of the involved populations must have normal distributions with variances that are approximately equal. This section introduces the nonparametric **Kruskal-Wallis test** (also called the H **test**), which is used to test hypotheses that different samples come from the same or identical populations, but this test does not require normal distributions. Also, the Kruskal-Wallis test can be used with data at the ordinal level of measurement, such as data consisting of ranks. The assumptions for the Kruskal-Wallis test are as follows.

Assumptions

1. We have at least three samples, all of which are random.

2. We want to test the null hypothesis that the samples come from the same or identical populations.

3. Each sample has at least five observations. (If samples have fewer than five observations, refer to special tables of critical values, such as *CRC Standard Probability and Statistics Tables and Formulae*, published by CRC Press.)

4. The Kruskal-Wallis test requires equal variances, so this test shouldn't be used if the different samples have variances that are very far apart.

In applying the Kruskal-Wallis test, we compute the **test statistic H, which has a distribution that can be approximated by the chi-square distribution as long as each sample has at least five observations.** When we use the chi-square distribution in this context, the number of degrees of freedom is $k - 1$, where k is the number of samples. (For a quick review of the key features of the chi-square distribution, see Section 6-5.) The relevant notation and test statistic are as follows.

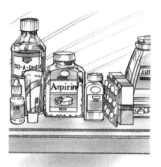

Notation for the Kruskal-Wallis Test

N = total number of observations in all samples combined

k = number of samples

R_1 = sum of ranks for sample 1

n_1 = number of observations in sample 1

For sample 2, the sum of ranks is R_2 and the number of observations is n_2, and similar notation is used for the other samples.

Test Statistic for the Kruskal-Wallis Test

$$H = \frac{12}{N(N + 1)}\left(\frac{R_1^2}{n_1} + \frac{R_2^2}{n_2} + \cdots + \frac{R_k^2}{n_k}\right) - 3(N + 1)$$

where   degrees of freedom = $k - 1$

Drug Approval

The Pharmaceutical Manufacturing Association has reported that the development and approval of a new drug costs around $87 million and takes about eight years. Extensive laboratory testing is followed by FDA approval for human testing, which is done in three phases. Phase I human testing involves about 80 people, and phase II involves about 250 people. In phase III, between 1000 and 3000 volunteers are used. Overseeing such a complex, extensive, and time-consuming process would be enough to give anyone a headache, but the process does protect us from dangerous or worthless drugs.

The Kruskal-Wallis test includes these key steps:

1. Considering all observations combined, assign a rank to each one. (Rank from lowest to highest, and in cases of ties, assign to each observation the mean of the ranks involved.)

2. For each sample, find the sum of the ranks and the sample size.

3. Using the results from Step 2, calculate the test statistic H, which has a distribution approximated by the chi-square distribution with $k - 1$ degrees

of freedom. The critical value is found from Table A-4, the table of χ^2 values.

The test statistic H is basically a measure of the variance of the rank sums R_1, $R_2, \ldots, R_k$. If the ranks are distributed evenly among the sample groups, then H should be a relatively small number. If the samples are very different, then the ranks will be excessively low in some groups and high in others, with the net effect that H will be large. Consequently, only large values of H lead to rejection of the null hypothesis that the samples come from identical populations. **The Kruskal-Wallis test is therefore a right-tailed test.**

● **EXAMPLE**   The Chapter Problem for Chapter 11 used the data in Table 13-4, and we tested the null hypothesis that the mean time interval between eruptions of the Old Faithful geyser is the same for the four different years. Now use the Kruskal-Wallis test to test the null hypothesis that the different years have time intervals with identical populations.

TABLE 13-4	Time Intervals (in minutes) Between Eruptions of the Old Faithful Geyser						
1951		**1985**		**1995**		**1996**	
74	(21)	89	(40)	86	(34.5)	88	(37.5)
60	(8)	90	(42)	86	(34.5)	86	(34.5)
74	(21)	60	(8)	62	(12)	85	(30.5)
42	(1)	65	(15.5)	104	(48)	89	(40)
74	(21)	82	(26)	62	(12)	83	(27)
52	(2)	84	(28)	95	(47)	85	(30.5)
65	(15.5)	54	(3)	79	(24)	91	(43.5)
68	(18.5)	85	(30.5)	62	(12)	68	(18.5)
62	(12)	58	(6)	94	(45.5)	91	(43.5)
66	(17)	79	(24)	79	(24)	56	(4)
62	(12)	57	(5)	86	(34.5)	89	(40)
60	(8)	88	(37.5)	85	(30.5)	94	(45.5)
$n_1 = 12$		$n_2 = 12$		$n_3 = 12$		$n_4 = 12$	
$R_1 = 157$		$R_2 = 265.5$		$R_3 = 358.5$		$R_4 = 395$	

Data from geologist Rick Hutchinson and the National Park Service.

SOLUTION   We will follow the same hypothesis testing procedure summarized in Figure 7-4.

Steps 1, 2, and 3:   The null and alternative hypotheses are as follows.

H_0:   The four years have time intervals with identical populations.

H_1:   The four populations are not identical.

Step 4:   No significance level was specified. In the absence of any over-riding circumstances, we use $\alpha = 0.05$.

Step 5:   In Chapter 11 we used an F test for analysis of variance, but here we illustrate the method of the Kruskal-Wallis test.

Step 6:   In determining the value of the test statistic H, we must first rank all of the data. We begin with the lowest value of 42 min, which is assigned a rank of 1. Ranks are shown in parentheses with the original data in Table 13-4. Next we find the sample size, n, and sum of ranks, R, for each sample; these values are shown at the bottom of Table 13-4. Because the total number of observations is 48, we have $N = 48$. We can now evaluate the test statistic as follows:

$$H = \frac{12}{N(N + 1)} \left(\frac{R_1^2}{n_1} + \frac{R_2^2}{n_2} + \cdots + \frac{R_k^2}{n_k} \right) - 3(N + 1)$$

$$= \frac{12}{48(48 + 1)} \left(\frac{157^2}{12} + \frac{265.5^2}{12} + \frac{358.5^2}{12} + \frac{395^2}{12} \right) - 3(48 + 1)$$

$$= 14.431$$

Because each sample has at least five observations, the distribution of H is approximately a chi-square distribution with $k - 1$ degrees of freedom. The number of samples is $k = 4$, so we have $4 - 1 = 3$ degrees of freedom. Refer to Table A-4 to find the critical value of 7.815, which corresponds to 3 degrees of freedom and a 0.05 significance level.

Step 7:   The test statistic $H = 14.431$ is in the critical region bounded by 7.815, so we reject the null hypothesis of identical populations.

Step 8:   The time intervals between eruptions do not appear to have the same distribution in the four years. (When we considered the same data in Section 11-2, we rejected the claim that the four samples come from populations having the same mean.)

The test statistic H, as presented earlier, is the rank version of the test statistic F used in the analysis of variance discussed in Chapter 11. When we deal with ranks R instead of raw scores x, many components are predetermined.

For example, the sum of all ranks can be expressed as $N(N + 1)/2$, where N is the total number of scores in all samples combined. The expression

$$H = \frac{12}{N(N + 1)}\Sigma n_i(\overline{R}_i - \overline{\overline{R}})^2$$

where

$$\overline{R}_i = \frac{R_i}{n_i} \qquad \overline{\overline{R}} = \frac{\Sigma R_i}{\Sigma n_i}$$

combines weighted variances of ranks to produce the test statistic H given here. This expression for H is algebraically equivalent to the expression for H given earlier as the test statistic. The earlier form of H (not the one given here) is easier to work with. In comparing the procedures of the parametric F test for analysis of variance and the nonparametric Kruskal-Wallis test, we see that in the absence of computer software, the Kruskal-Wallis test is much simpler to apply. We need not compute the sample variances and sample means. We do not require normal population distributions. Life becomes so much easier. However, the Kruskal-Wallis test is not as efficient as the F test, and it may require more dramatic differences for the null hypothesis to be rejected.

Using Computer Software for the Kruskal-Wallis Test

STATDISK and Minitab both execute the Kruskal-Wallis test. If using STAT-DISK, select Analysis from the main menu bar, then select Kruskal-Wallis Test. STATDISK will display the sum of the ranks for each sample, the H test statistic, the critical value, and the conclusion. If using Minitab, refer to the *Minitab Student Laboratory Manual and Workbook* for the procedure required to use the options of Stat, Nonparametrics, and Kruskal-Wallis. Following is the Minitab display for the preceding example. Note that the Minitab display includes the P-value of 0.002, which suggests that we should reject the null hypothesis of identical populations.

MINITAB DISPLAY				
LEVEL	NOBS	MEDIAN	AVE. RANK	Z VALUE
1	12	63.50	13.1	−3.26
2	12	80.50	22.1	−0.68
3	12	85.50	29.9	1.54
4	12	87.00	32.9	2.40
OVERALL	48		24.5	

$H = 14.43$ d.f. $= 3$ $p = 0.002$
$H = 14.48$ d.f. $= 3$ $p = 0.002$ (adjusted for ties)

13-5 Exercises A: Basic Skills and Concepts

In Exercises 1–8, use the Kruskal-Wallis test.

Recommended assignment: Exercises 2, 3, 6.

1. The accompanying table lists body temperatures of five randomly selected subjects from each of three different age groups. Using a significance level of $\alpha = 0.05$, test the claim that the three age-group populations of body temperatures are identical.

   1. Test statistic: $H = 2.180$. Critical value: $\chi^2 = 5.991$.

   Body Temperatures (°F) Categorized by Age

18–20	21–29	30 and older
98.0	99.6	98.6
98.4	98.2	98.6
97.7	99.0	97.0
98.5	98.2	97.5
97.1	97.9	97.3

   Data from Dr. Philip Mackowiak, Dr. Steven Wasserman, and Dr. Myron Levine of the University of Maryland.

2. The accompanying values are measured maximum breadths of male Egyptian skulls from different epochs. Changes in head shape over time suggest that interbreeding occurred with immigrant populations. Use a 0.05 significance level to test the claim that the three samples come from identical populations.

   2. Test statistic: $H = 6.631$. Critical value: $\chi^2 = 5.991$.

4000 B.C.	1850 B.C.	150 A.D.
131	129	128
138	134	138
125	136	136
129	137	139
132	137	141
135	129	142
132	136	137
134	138	145
138	134	137

   Based on data from *Ancient Races of the Thebaid,* by Thomson and Randall-Maciver.

3. Test statistic: $H = 13.075$. Critical value: $\chi^2 = 7.815$.
4. Test statistic: $H = 9.129$. Critical value: $\chi^2 = 9.488$ (assuming $\alpha = 0.05$).
5. Test statistic: $H = 2.075$. Critical value: $\chi^2 = 11.071$.
6. Test statistic: $H = 84.31$. Critical value: $\chi^2 = 9.210$.

Laboratory

1	2	3	4	5
2.9	2.7	3.3	3.3	4.1
3.1	3.4	3.3	3.2	4.1
3.1	3.6	3.5	3.4	3.7
3.7	3.2	3.5	2.7	4.2
3.1	4.0	2.8	2.7	3.1
4.2	4.1	2.8	3.3	3.5
3.7	3.8	3.2	2.9	2.8
3.9	3.8	2.8	3.2	
3.1	4.3	3.8	2.9	
3.0	3.4	3.5		
2.9	3.3			

Data provided by Minitab, Inc.

7. Test statistic: $H = 3.489$. Critical value: $\chi^2 = 5.991$.
8. Test statistic: $H = 122.500$. Critical value: $\chi^2 = 7.815$. P-value: 0.000.

3. The accompanying table shows selling prices (in thousands of dollars) for homes located on Long Beach Island in New Jersey. Do these sample data support the claim of different populations of selling prices in the four regions?

Oceanside	235	395	547	469	369	279
Oceanfront	538	446	435	639	499	399
Bayside	199	219	239	309	399	190
Bayfront	695	389	489	489	599	549

4. Flammability tests were conducted on children's sleepwear. The Vertical Semirestrained Test was used, in which pieces of fabric were burned under controlled conditions. After the burning stopped, the length of the charred portion was measured and recorded. Results are given in the margin for the same fabric tested at different laboratories. Because the same fabric was used, the different laboratories should have obtained the same results. Did they?

5. Refer to Data Set 11 in Appendix B. At the 0.05 significance level, test the claim that the weights of M&Ms are the same for each of the six different color populations. If it is the intent of Mars, Inc., to make the candies so that the different color populations are the same, do your results suggest that the company has a problem that requires corrective action?

6. Refer to Data Set 9 in Appendix B. Use a 0.01 significance level to test the claim that the populations of cotinine levels are identical for these three groups: nonsmokers who are not exposed to environmental tobacco smoke, nonsmokers who are exposed to tobacco smoke, and people who smoke. What do the results suggest about "second-hand" smoke?

7. Refer to Data Set 10 in Appendix B and use a 0.05 significance level to test the claim that the populations of movie lengths (in minutes) are the same for these three categories of star ratings: 0.0–2.5 stars, 3.0–3.5 stars, and 4.0 stars. Based on the results, does it appear that bad movies are longer than good movies, or does it just seem that way?

8. The City Resource Recovery Company (CRRC) collects the waste discarded by households in a region. Discarded waste must be separated into categories of metal, paper, plastic, and glass. In planning for the equipment needed to collect and process the garbage, CRRC refers to the data in Data Set 1 in Appendix B (provided by the Garbage Project at the University of Arizona) and uses Minitab to obtain the results shown in the display at the top of the following page. At the 0.05 level of significance, test the claim that the four specific populations of garbage (metal, paper, plastic, glass) are the same. Based on the results, does it appear that these four categories require the same collection and processing resources? Does any single category seem to be a particularly large part of the waste management problem?

```
LEVEL     NOBS     MEDIAN     AVE. RANK     Z VALUE
  1        62      1.975        90.4         -4.32
  2        62      9.300       206.2         10.36
  3        62      1.645        76.3         -6.11
  4        62      2.980       125.1          0.07
OVERALL   248                  124.5

H = 122.50   d.f. = 3     p = 0.000
H = 122.50   d.f. = 3     p = 0.000 (adjusted for ties)
```

13-5 Exercises B: Beyond the Basics

9. a. In general, how is the value of the test statistic H affected if a constant is added to (or subtracted from) each score?

  b. In general, how is the value of the test statistic H affected if each score is multiplied (or divided) by a positive constant?

10. For three samples, each of size 5, what are the largest and smallest possible values of H?

11. In using the Kruskal-Wallis test, there is a correction factor that should be applied whenever there are many ties: Divide H by

$$1 - \frac{\Sigma T}{N^3 - N}$$

Here $T = t^3 - t$, where t is the number of observations that are tied for a group of tied scores. That is, find t for each group of tied scores, then compute the value of T, then add the T values to get ΣT. For the example presented in this section, use this procedure to find the corrected value of H. Does the corrected value of H differ substantially from the value of 14.431 that was found in this section?

12. Show that for the case of two samples, the Kruskal-Wallis test is equivalent to the Wilcoxon rank-sum test. This can be done by showing that for the case of two samples, the test statistic H equals the square of the test statistic z used in the Wilcoxon rank-sum test. Also note that with 1 degree of freedom, the critical values of χ^2 correspond to the square of the critical z score.

9. a. The test statistic H does not change.

b. The test statistic H does not change.

10. 12.5, 0

11. $\Sigma T = 366$. Dividing H by the correction factor results in $14.431335 \div 0.99668910$, which is rounded to 14.479. Even though 35 of the 48 scores are involved in ties, the correction factor does not change H by a substantial amount.

13-6 Rank Correlation

Section 9-2 introduced the concept of correlation and included the *linear correlation coefficient r* as a measure of the strength of the association between two variables. In this section we study rank correlation, the nonparametric

Because this section deals with the correlation between two sets of paired data, it could be covered along with Section 9-2.

Direct Link Between Smoking and Cancer

When we find a statistical correlation between two variables, we must be extremely careful to avoid the mistake of concluding that there is a cause-effect link. The tobacco industry has consistently emphasized that correlation does not imply causality. However, Dr. David Sidransky of Johns Hopkins University now says that "we have such strong molecular proof that we can take an individual cancer and potentially, based on the patterns of genetic change, determine whether cigarette smoking was the cause of that cancer." Based on his findings, he also said that "The smoker had a much higher incidence of the mutation, but the second thing that nailed it was the very distinct pattern of mutations . . . so we had the smoking gun." Although statistical methods cannot prove that smoking *causes* cancer, such proof can be established with physical evidence of the type described by Dr. Sidransky.

version of that parametric measure. In Chapter 9 we used paired sample data to compute values for the linear correlation coefficient r; in this section we consider the **rank correlation coefficient,** which uses ranks as the basis for measuring the strength of the association between two variables. Rank correlation has some distinct advantages over the parametric methods discussed in Chapter 9:

1. The nonparametric method of rank correlation can be used in a wider variety of circumstances than the parametric method of linear correlation. With rank correlation, we can analyze paired data that can be ranked but not measured. Methods using linear correlation require normal distributions and therefore do not apply to data consisting of ranks. For example, if two judges rank 30 different gymnasts, we can use those ranks to test for a correlation between the two judges, but we cannot test for a correlation using linear correlation.

2. Rank correlation can be used to detect some relationships that are not linear; an example will be given later in this section.

3. The computations for rank correlation are much simpler than those for linear correlation, as can be readily seen by comparing the formulas used to compute these statistics. If calculators or computers are not available, the rank correlation coefficient is easier to compute.

A disadvantage of rank correlation is its efficiency rating of 0.91, as described in Section 13-1. This efficiency rating indicates that with all other circumstances being equal, the nonparametric approach of rank correlation requires 100 pairs of sample data to achieve the same results as only 91 pairs of sample observations analyzed through the parametric approach, assuming the stricter requirements of the parametric approach are met.

We will use the following notation, which closely parallels the notation used in Chapter 9 for linear correlation. (Recall from Chapter 9 that r denotes the linear correlation coefficient for sample paired data, ρ denotes the linear correlation coefficient for all paired data in a population, and n denotes the number of pairs of data.)

Notation

r_s = rank correlation coefficient for sample paired data (r_s is a sample statistic)

ρ_s = rank correlation coefficient for all the population data (ρ_s is a population parameter)

n = number of pairs of data

d = difference between ranks for the two observations within a pair

---

We use the notation r_s for the rank correlation coefficient so that we don't confuse it with the linear correlation coefficient r. The subscript s has nothing to do with standard deviation; it is used in honor of Charles Spearman (1863–1945), who originated the rank correlation approach. In fact, r_s is often called **Spearman's rank correlation coefficient.**

Given a collection of sample paired data, if we want to test for a relationship between the two variables, we can use the null hypothesis H_0: $\rho_s = 0$, which is the claim of no correlation. The test statistic is as follows:

Test Statistic for the Rank Correlation Coefficient

$$r_s = 1 - \frac{6\Sigma d^2}{n(n^2 - 1)}$$

where each value of d is a difference between the ranks for a pair of sample data

Critical values:

If $n \le 30$, refer to Table A-9.
If $n > 30$, use Formula 13-1 (see below).

Ties in ranks of the original sample values can be handled as in the preceding sections of this chapter: Find the mean of the ranks involved in the tie, and then assign the mean rank to each of the tied items. This test statistic yields the exact value of r_s only if there are no ties. With a relatively small number of ties, the test statistic is a good approximation of r_s. (When ties occur, we can get an exact value of r_s by ranking the data and using Formula 9-1 for the linear correlation coefficient; after finding the value of r_s, we can proceed with the methods of this section.)

See Figure 13-5 for a flowchart summary of the hypothesis testing method that uses rank correlation. Figure 13-5 shows that when the number of ranks n exceeds 30, we get the critical values by evaluating

Formula 13-1 $r_s = \dfrac{\pm z}{\sqrt{n - 1}}$ (critical values when $n > 30$)

where   the value of z corresponds to the significance level.

Point out that Tables A-6 and A-9 are somewhat similar and they should not be confused. Comment that Table A-6 is used with the method in Chapter 9, whereas Table A-9 is used for the method in this section.

EXAMPLE   *Business Week* magazine ranked business schools two different ways. Corporate rankings were based on surveys of corporate recruiters and graduate rankings were based on surveys of MBA graduates. Table 13-5 is based on the results for 10 schools. Is there a correlation between the corporate rankings and the graduate rankings? The

(continued)

Figure 13-5 Rank Correlation
for Testing H_0: $\rho_s = 0$.

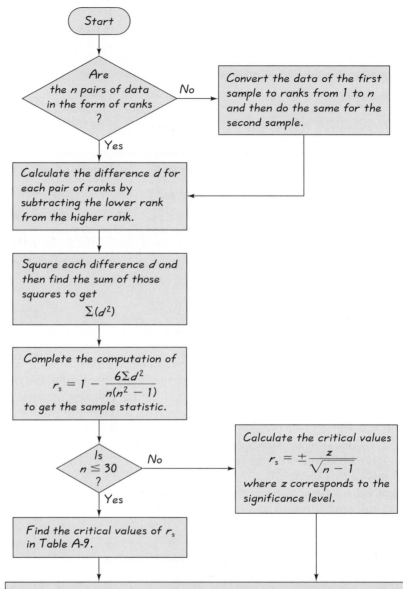

TABLE 13-5	Rankings of Business Schools									
School	PA	NW	Chi	Sfd	Hvd	MI	IN	Clb	UCLA	MIT
Corporate ranking	1	2	4	5	3	6	8	7	10	9
Graduate ranking	3	5	4	1	10	7	6	8	2	9
d	2	3	0	4	7	1	2	1	8	0
d^2	4	9	0	16	49	1	4	1	64	0 →Total = 148

linear correlation coefficient r (Section 9-2) should not be used because it requires normal distributions, but the data consist of ranks that are not normally distributed. Instead, use the rank correlation coefficient to test the claim that there is a relationship between corporate and graduate rankings (that is, $\rho_s \neq 0$). Use a significance level of $\alpha = 0.05$.

SOLUTION   We follow the same basic steps for testing hypotheses as outlined in Figure 7-4.

Step 1:   The claim of a correlation is expressed symbolically as $\rho_s \neq 0$.

Step 2:   The negation of the claim in Step 1 is $\rho_s = 0$.

Step 3:   Because the null hypothesis must contain the condition of equality, we have

$$H_0: \rho_s = 0$$
$$H_1: \rho_s \neq 0 \quad \text{(Original claim)}$$

Step 4:   The significance level is $\alpha = 0.05$.

Step 5:   As we noted, we cannot use the linear correlation approach of Section 9-2 because ranks do not satisfy the requirement of a normal distribution. We will use the rank correlation approach instead.

Step 6:   We now find the value of the test statistic. Table 13-5 shows the calculation of the differences d and their squares d^2, which results in a value of $\Sigma d^2 = 148$. With $n = 10$ (for 10 pairs of data) and $\Sigma d^2 = 148$, we can find the value of the test statistic r_s as follows:

(continued)

$$r_s = 1 - \frac{6\Sigma d^2}{n(n^2 - 1)} = 1 - \frac{6(148)}{10(10^2 - 1)}$$

$$= 1 - \frac{888}{990} = 0.103$$

Now refer to Table A-9 to determine that the critical values are ± 0.648 (based on $\alpha = 0.05$ and $n = 10$). Because the test statistic $r_s = 0.103$ does not exceed the critical value of 0.648, we fail to reject the null hypothesis. There is not sufficient evidence to support the claim of a correlation between corporate and graduate rankings of business schools. It appears that corporate recruiters and business school graduates have different perceptions of the qualities of the schools.

Manatees Saved

Manatees are large mammals that like to float just below the water's surface, where they are in danger from powerboat propellers. A Florida study of the number of powerboat registrations and the numbers of accidental manatee deaths confirmed that there was a significant positive correlation. As a result, Florida created coastal sanctuaries where powerboats are prohibited so that manatees could thrive. (See Program 1 from the series *Against All Odds: Inside Statistics* for a discussion of this case.) This is one of many examples of the beneficial use of statistics.

EXAMPLE   Assume that the preceding example is expanded by including a total of 40 business schools and that the test statistic r_s is found to be 0.291. If the significance level is $\alpha = 0.05$, what do you conclude about the correlation between corporate and graduate rankings?

SOLUTION   Because there are 40 pairs of data, we have $n = 40$. Because n exceeds 30, we find the critical values from Formula 13-1 instead of Table A-9. With $\alpha = 0.05$ in two tails, we let $z = 1.96$ to get

$$r_s = \frac{\pm 1.96}{\sqrt{40 - 1}} = \pm 0.314$$

The test statistic of $r_s = 0.291$ does not exceed the critical value of 0.314, so we fail to reject the null hypothesis. There is not sufficient evidence to support the claim of a correlation between the corporate rankings and graduate rankings.

The next example is intended to illustrate the principle that rank correlation can sometimes be used to detect relationships that are not linear.

EXAMPLE   Ten students study for a test; the following table lists the number of hours studied (x) and the corresponding number of correct answers (y).

Hours studied (x)	5	9	17	1	2	21	3	29	7	100
Correct answers (y)	6	16	18	1	3	21	7	20	15	22

At the 0.05 level of significance, use Spearman's rank correlation approach to determine whether there is a relationship between hours studied and the number of correct answers.

SOLUTION

We will test the null hypothesis of no rank correlation ($\rho_s = 0$).

$$H_0: \rho_s = 0$$
$$H_1: \rho_s \neq 0$$

Refer to Figure 13-5, which we follow in this solution. The given data are not ranks, so we convert them into ranks as shown in the table. (Section 13-1 describes the procedure for converting scores into ranks.)

Ranks for hours studied	4	6	7	1	2	8	3	9	5	10
Ranks for correct answers	3	6	7	1	2	9	4	8	5	10
d	1	0	0	0	0	1	1	1	0	0
d^2	1	0	0	0	0	1	1	1	0	0 → Total = 4

After expressing all data as ranks, we calculate the differences, d, and then square them. The sum of the d^2 values is 4. We now calculate

$$r_s = 1 - \frac{6\Sigma d^2}{n(n^2 - 1)} = 1 - \frac{6(4)}{10(10^2 - 1)}$$

$$= 1 - \frac{24}{990} = 0.976$$

Proceeding with Figure 13-5, we have $n = 10$, so we answer yes when asked if $n \leq 30$. We use Table A-9 to get the critical values of ± 0.648. Finally, the sample statistic of 0.976 exceeds 0.648, so we conclude that there is significant correlation. More hours of study appear to be associated with higher grades. (You didn't really think we would suggest otherwise, did you?)

In the preceding example, if we compute the linear correlation coefficient r (using Formula 9-1) for the original data, we get $r = 0.629$, which leads to the conclusion that there is not enough evidence to support the claim of a significant linear correlation at the 0.05 level of significance. If we examine the scatter diagram in Figure 13-6 on the following page, we can see that there does seem to be a relationship, but it's not linear. This last example illustrates these two advantages of the nonparametric approach over the parametric approach: (1) With rank correlation, we can sometimes detect relationships

Figure 13-6 Scatter Diagram for Hours Studied and Correct Answers

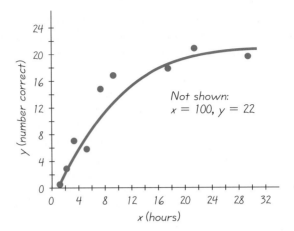

Not shown:
x = 100, y = 22

that are not linear, and (2) Spearman's rank correlation coefficient r_s is less sensitive to an outlier, such as the 100 hours in the preceding data.

Using Calculators and Computer Software for Rank Correlation

If you are using a calculator with 2-variable statistics, you can find the value of r_s as follows: (1) Replace each sample value by its corresponding rank, then (2) calculate the value of the linear correlation coefficient r with the same procedures used in Section 9-2.

STATDISK and Minitab can be used to obtain the rank correlation coefficient. If you are using STATDISK, select Analysis from the main menu bar, then select Rank Correlation. If you are using Minitab, enter the paired data in columns C1 and C2. If the data are not already ranks, use Minitab's Manip and Rank options to convert the data to ranks, then select Stat, followed by Basic Statistics, followed by Correlation. Minitab will display a one-line statement of the value for r_s, such as the following:

```
Correlation of C1 and C2 = 0.976
```

Recommended assignment: Exercises 1–4, 10, 15.

1. a. $r_s = 1$ and there appears to be a correlation between x and y.
 b. $r_s = -1$ and there appears to be a correlation between x and y.
 c. $r_s = 0$ and there does not appear to be a correlation between x and y.

13-6 Exercises A: Basic Skills and Concepts

1. For each of the following samples of paired ranks, sketch a scatter diagram, find the value of r_s, and state whether there appears to be a correlation between x and y.

a.

x	1	3	5	4	2
y	1	3	5	4	2

b.

x	1	2	3	4	5
y	5	4	3	2	1

c.

x	1	2	3	4	5
y	2	5	3	1	4

2. Find the critical value(s) for r_s by using either Table A-9 or Formula 13-1, as appropriate. Assume two-tailed cases, where α represents the level of significance and n represents the number of pairs of data.

 a. $n = 20, \alpha = 0.05$ **b.** $n = 50, \alpha = 0.05$
 c. $n = 40, \alpha = 0.02$ **d.** $n = 15, \alpha = 0.01$
 e. $n = 82, \alpha = 0.04$

In Exercises 3–18, use the rank correlation coefficient to test the claim of no correlation between the two variables. Use a significance level of $\alpha = 0.05$.

3. The accompanying table lists salary rankings and stress rankings for randomly selected jobs. Does it appear that salary increases as stress increases?

Job	Salary Rank	Stress Rank
Stockbroker	2	2
Zoologist	6	7
Electrical engineer	3	6
School principal	5	4
Hotel manager	7	5
Bank officer	10	8
Occ. safety inspector	9	9
Home economist	8	10
Psychologist	4	3
Airline pilot	1	1

Based on data from *The Jobs Rated Almanac.*

4. Exercise 3 includes paired salary and stress level ranks for 10 randomly selected jobs. The physical demands of the jobs were also ranked; the salary and physical demand ranks are given below. Does there appear to be a relationship between the salary of a job and its physical demands?

Salary	2	6	3	5	7	10	9	8	4	1
Physical demand	5	2	3	8	10	9	1	7	6	4

Based on data from *The Jobs Rated Almanac.*

5. Ten jobs were randomly selected and ranked according to stress level and physical demand, with the results given below. Does there appear to be a relationship between the stress levels of jobs and their physical demands?

Stress level	2	7	6	4	5	8	9	10	3	1
Physical demand	5	2	3	8	10	9	1	7	6	4

Based on data from *The Jobs Rated Almanac.*

6. The following table ranks eight states according to teachers' salaries and students' SAT scores.

2. a. ± 0.450
   b. ± 0.280
   c. ± 0.373
   d. ± 0.689
   e. ± 0.228

3. $r_s = 0.855$. Critical values: $r_s = \pm 0.648$. Significant correlation.

4. $r_s = 0.261$. Critical values: $r_s = \pm 0.648$. No significant correlation.

5. $r_s = -0.067$. Critical values: $r_s = \pm 0.648$. No significant correlation.

6. $r_s = 0.571$. Critical values: $r_s = \pm 0.738$. No significant correlation.

	NY	CA	FL	NJ	TX	NC	MD	OR
Teacher's salary	1	2	7	4	6	8	3	5
SAT score	4	3	5	6	7	8	2	1

Based on data from the U.S. Department of Education.

7. $r_s = 0.715$. Critical values: $r_s = \pm 0.591$. Significant correlation.

7. In studying the effects of heredity and environment on intelligence, scientists have learned much by analyzing the IQ scores of identical twins who were separated soon after birth. Identical twins have identical genes, which they inherited from the same fertilized egg. By studying identical twins raised apart, we can eliminate the variable of heredity and better isolate the effects of environment. Following are the IQ scores of identical twins (first-born twins are x) raised apart.

x	107	96	103	90	96	113	86	99	109	105	96	89
y	111	97	116	107	99	111	85	108	102	105	100	93

Based on data from "IQ's of Identical Twins Reared Apart," by Arthur Jensen, *Behavioral Genetics*.

8. $r_s = 0.190$. Critical values: $r_s = \pm 0.738$. No significant correlation.

8. *Consumer Reports* tested VHS tapes used in VCRs. Following are performance scores and prices (in dollars) of randomly selected tapes.

Performance	91	92	82	85	87	80	94	97
Price	4.56	6.48	5.99	7.92	5.36	3.32	7.32	5.27

9. $r_s = 0.363$. Critical values: $r_s = \pm 0.738$.

9. Researchers have studied bears by anesthetizing them in order to obtain vital measurements. The data in the table represent the first eight male bears in Data Set 3 from Appendix B.

10. $r_s = 0.994$. Critical values: $r_s = \pm 0.738$. Significant correlation.

x Length (in.)	53.0	67.5	72.0	72.0	73.5	68.5	73.0	37.0
y Weight (lb)	80	344	416	348	262	360	332	34

Based on data from Gary Alt.

11. $r_s = -0.758$. Critical values: $r_s = \pm 0.648$. Significant correlation.

10. When bears were anesthetized, researchers measured the distances (in inches) around their chests and weighed the bears (in pounds). The results are given for eight male bears. Based on the results, does a bear's weight seem to be related to its chest size?

x Chest (in.)	26	45	54	49	41	49	44	19
y Weight (lb)	80	344	416	348	262	360	332	34

Based on data from Gary Alt.

11. The accompanying table lists weights (in hundreds of pounds) and highway fuel consumption amounts (in mi/gal) for a sample of domestic new

cars. Based on the result, can you expect to pay more for gas if you buy a heavier car? How do the results change if the weights are entered as 2900, 3500, . . . , 2400?

x Weight	29	35	28	44	25	34	30	33	28	24
y Fuel	31	27	29	25	31	29	28	28	28	33

Based on data from the EPA.

12. The accompanying table lists weights (in pounds) of plastic discarded by a sample of households, along with the sizes of the households. Is there a correlation? This issue is important to the Census Bureau, which provided project funding, because the presence of a correlation implies that we can predict population size by analyzing discarded garbage.

Plastic (lb)	0.27	1.41	2.19	2.83	2.19	1.81	0.85	3.05
Household size	2	3	3	6	4	2	1	5

Based on data provided by Masakazu Tani and the Garbage Project at the University of Arizona.

13. Refer to Data Set 16 in Appendix B and use the paired data for durations and intervals after eruptions of the Old Faithful geyser. Is there a correlation, suggesting that the interval after an eruption is related to the duration of the eruption?

14. Refer to Data Set 16 in Appendix B and use the paired data for intervals after eruptions and heights of eruptions of the Old Faithful geyser. Is there a correlation, suggesting that the interval after an eruption is related to the height of the eruption?

15. Refer to Data Set 4 in Appendix B and use the paired data consisting of tar and nicotine. Based on the result, does there appear to be a correlation between cigarette tar and nicotine? If so, can researchers reduce their laboratory expenses by measuring only one of these two variables?

16. Refer to Data Set 4 in Appendix B and use the paired data consisting of carbon monoxide and nicotine. Based on the result, does there appear to be a correlation between cigarette nicotine and carbon monoxide? If so, can researchers reduce their laboratory expenses by measuring only one of these two variables?

17. Refer to Data Set 7 in Appendix B and use the paired data consisting of Iowa's total annual precipitation amounts (PRECIP) and the amounts of corn produced (CORNPROD). Is there a correlation, as we might expect?

18. Refer to Data Set 7 in Appendix B and use the paired data consisting of Iowa's average annual temperatures (AVTEMP) and the amounts of corn produced (CORNPROD). Is there a correlation, as we might expect?

12. $r_s = 0.869$. Critical values: $r_s = \pm 0.738$. Significant correlation.

13. $r_s = 0.786$. Critical values: $r_s = \pm 0.280$. Significant correlation.

14. $r_s = -0.036$. Critical values: $r_s = \pm 0.280$. No significant correlation.

15. $r_s = 0.918$. Critical values: $r_s = \pm 0.370$. Significant correlation.

16. $r_s = 0.739$. Critical values: $r_s = \pm 0.370$. Significant correlation.

17. $r_s = 0.265$. Critical values: $r_s = \pm 0.310$. No significant correlation.

18. $r_s = -0.076$. Critical values: $r_s = \pm 0.310$. No significant correlation.

13-6 Exercises B: Beyond the Basics

19. a. ±0.707

  b. ±0.514

  c. ±0.361

  d. ±0.463

  e. ±0.834

19. One alternative to using Table A-9 involves an approximation of critical values for r_s given as

$$r_s = \pm \sqrt{\frac{t^2}{t^2 + n - 2}}$$

Here t is the t score from Table A-3 corresponding to the significance level and $n - 2$ degrees of freedom. Apply this approximation to find critical values of r_s for the following cases.

  a. $n = 8$, $\alpha = 0.05$ **b.** $n = 15$, $\alpha = 0.05$

  c. $n = 30$, $\alpha = 0.05$ **d.** $n = 30$, $\alpha = 0.01$

  e. $n = 8$, $\alpha = 0.01$

20. a. r_s does not change.

  b. r_s changes sign.

  c. r_s does not change.

  d. Both values will be the same.

  e. r_s does not change.

20. a. How is r_s affected if the scale for one of the variables is changed from feet to inches?

  b. How is r_s affected if one variable is ranked from low to high while the other variable is ranked from high to low?

  c. How is r_s affected if the two variables are interchanged?

  d. One researcher ranks both variables from low to high, while another researcher ranks both variables from high to low. How will their values of r_s compare?

  e. How is r_s affected if each value of x is replaced by log x?

13-7 Runs Test for Randomness

Important: This section (up to, but not including the subsection "Large Sample Cases") has been rewritten so that it can be covered *anytime*, including at the beginning of the course.

Many of the statistical procedures used in this book require the random selection of data. In this section we describe the runs test for randomness, which is based on the number of runs in a sequence of data. The runs test applies only if we can categorize each data value into one of *two* separate categories.

DEFINITIONS

The **runs test** is a systematic and standard procedure for testing the randomness of data.

A **run** is a sequence of data that exhibit the same characteristic; the sequence is preceded and followed by different data or no data at all.

As an example, consider the cola choice of consumers in a market research project. We let D denote a consumer who prefers *diet* cola and R denote a consumer who prefers *regular* cola. The following sequence contains exactly four runs.

$$\underbrace{D\ D\ D\ D}_{\text{1st run}}\quad \underbrace{R\ R}_{\text{2nd run}}\quad \underbrace{D\ D\ D}_{\text{3rd run}}\quad \underbrace{R}_{\text{4th run}}$$

Given these sample data, we can use the runs test to test for the randomness with which diet and regular occur. Let's use common sense to see how runs relate to randomness. Examine the following sequence, and then stop to consider how randomly diet and regular occur. Also count the number of runs.

$$D\quad D\quad D\quad D\quad R\quad R\quad R\quad R\quad R$$

It is reasonable to conclude that diet and regular occur in a sequence that is *not* random. Note that in this sequence of 9 data, there are only two runs. This pattern suggests that *if the number of runs is very low, randomness is lacking.* Now consider the following sequence of 20 data. Try again to form your own conclusion about randomness before you continue reading.

$$D\quad R\quad D\quad R\quad D\quad R\quad D\quad R\quad D\quad R\quad D\quad R\quad D\quad R\quad D\quad R\quad D\quad R\quad D\quad R$$

It should be apparent that the sequence of diet and regular is again *not* random because there is a distinct, predictable pattern. In this sample, the number of runs is 20. This example suggests that *if the number of runs is very high, randomness is lacking.*

Note that **this test for randomness is based on the *order* in which the data occur. This runs test is *not* based on the *frequency* of the data.** For example, a particular sequence containing 3 men and 20 women might lead to the conclusion that the sequence is random. The issue of whether 3 men and 20 women constitute a *biased* sample is not addressed by the runs test.

The two sequences given in this section are obvious in their lack of randomness, but most sequences are not so obvious, so we need more sophisticated techniques for analysis. We begin by introducing some notation.

The issue of bias could be considered with other methods, such as the sign test or a parametric test (as in Section 7-5 or 8-6).

Notation

n_1 = Number of elements in the sequence that have one particular characteristic. (The characteristic chosen for n_1 is arbitrary.)

n_2 = Number of elements in the sequence that have the other characteristic

G = Number of runs

For example, the sequence

$$D\quad D\quad D\quad D\quad R\quad R\quad R\quad R\quad R$$

results in the following values for n_1, n_2, and G.

$n_1 = 4$ because there are 4 diet cola consumers

$n_2 = 5$ because there are 5 regular cola consumers

$G = 2$ because there are 2 runs

On the basis of the preceding discussion, we now proceed to reject randomness of the data if the number of runs G is too small or too large. But how do we determine exactly which values of G are too small or too large? Our criterion is the following:

5% Cutoff Criterion:
Reject randomness of the data if the number of runs G is so small or so large that in repeated samplings, a value at least as extreme as G will occur less than 5% of the time.

Although this criterion wins no prizes for simplicity, it's quite easy to apply if we use Table A-10. Table A-10 identifies those values of G that are so small or so large that they belong to the category of exceptional sequences that happen less than 5% of the time. Using Table A-10, the 5% criterion can be simply restated as follows:

Simplified 5% Cutoff Criterion:
Reject randomness if the number of runs G is less than or equal to the smaller Table A-10 entry or greater than or equal to the larger entry in that table.

Recall that the sequence

D   D   D   D   R   R   R   R   R

results in these values: $n_1 = 4$, $n_2 = 5$, and $G = 2$ runs. With $n_1 = 4$ and $n_2 = 5$, Table A-10 indicates that we should reject randomness if the number of runs is 2 or less or 9 or greater. With $G = 2$ runs, we therefore reject randomness according to the simplified 5% cutoff criterion based on Table A-10.

How was Table A-10 constructed? Let's consider the cutoff values of 2 and 9 that correspond to $n_1 = 4$ and $n_2 = 5$. If we had an abundance of time and patience, we could list all possible sequences of 4 diet colas and 5 regular colas. Examination of that list would reveal these facts:

- There are 126 different possible sequences of 4 diet and 5 regular colas.
- Among the 126 different possible cases, 2 cases have 2 runs, 7 cases have 3 runs, and so on, as summarized in Table 13-6.

Based on the 5% cutoff criterion, and based on the 126 cases summarized in Table 13-6, we should reject randomness if the number of runs G is 2 or 9 because, with randomly selected data, we will get 2 runs or 9 runs only 2.38% of the time. (With only 2 cases having 2 runs and with only 1 case having 9 runs, the number of runs G is excessively low or high in 3 cases, which is 2.38% of the total number of cases. We arrived at 2.38% by converting

TABLE 13-6

Number of runs	2	3	4	5	6	7	8	9
Frequency in 126 cases	2	7	24	30	36	18	8	1

$3 \div 126$ to 0.0238, which is equivalent to 2.38%.) It is easy to get 3, 4, 5, 6, 7, or 8 runs, because these values occur more than 95% of the time; but it is unusual to get 2 or 9 runs because these values occur only 2.38% of the time. By analyzing the 126 cases summarized in Table 13-6, we can see how to find the cutoff values of 2 and 9 that are included in Table A-10. Although this more detailed analysis for finding cutoff values is perfectly valid, it's too awkward to apply very often, so we use Table A-10 instead.

Table A-10 applies when the following three conditions are all met:

1. We are using 5% as the cutoff for sequences that have too few or too many runs,
2. $n_1 \leq 20$, and
3. $n_2 \leq 20$.

EXAMPLE   The president of the Mutual Fidelity investment firm has observed that men and women have been hired in the following sequence: M, M, M, W, M, M, M, M, W, W, W, M. Use the (simplified) 5% cutoff criterion to test the personnel officer's claim that the sequence of men and women is random. (Note that we are not testing for a *bias* in favor of one gender over the other. There are 8 men and 4 women, but we are investigating only the *randomness* in the way they appear in the given sequence.)

SOLUTION
We must first find the values of n_1, n_2, and the number of runs G.

$$n_1 = \text{number of men} = 8$$
$$n_2 = \text{number of women} = 4$$
$$G = \text{number of runs} = 5$$

With $G = 5$, we refer to Table A-10 to find the cutoff values of 3 and 10. Because $G = 5$ is not less than or equal to 3, nor is it greater than or equal to 10, *we do not reject randomness*. (Perhaps the hiring is not being done randomly with respect to gender, but we do not have sufficient evidence to make that charge.)

Random Walk and Stocks

In the book *A Random Walk Down Wall Street*, Burton Malkiel states, "When the term (random walk) is applied to the stock market, it means that short-run changes in stock prices cannot be predicted. Investment advisory services, earnings predictions, and complicated chart patterns are useless." He also writes that "taken to its logical extreme, it means that a blindfolded monkey throwing darts at a newspaper's financial pages could select a portfolio that would do just as well as one carefully selected by the experts." This suggests that investors could save money by no longer securing the expensive advice of financial consultants. They can simply select stocks in a way that is random.

Large Sample Cases

So far we have discussed only cases in which $n_1 \leq 20$ and $n_2 \leq 20$ and $\alpha = 0.05$. If we wish to use the runs test for randomness but $n_1 > 20$ or $n_2 > 20$ or $\alpha \neq 0.05$, we use the property that the number of runs G has a distribution that is approximately normal with mean μ_G and standard deviation σ_G described as follows:

Formula 13-2
$$\mu_G = \frac{2n_1 n_2}{n_1 + n_2} + 1$$

Formula 13-3
$$\sigma_G = \sqrt{\frac{(2n_1 n_2)(2n_1 n_2 - n_1 - n_2)}{(n_1 + n_2)^2(n_1 + n_2 - 1)}}$$

After finding the values of μ_G and σ_G, the test statistic can be computed as $z = (G - \mu_G)/\sigma_G$. The normal approximation (with test statistic z) is quite good. If the entire table of critical values (Table A-10) had been computed using this normal approximation, no critical value would be off by more than one unit.

Test Statistic for the Runs Test for Randomness

If $\alpha = 0.05$ and $n_1 \leq 20$ and $n_2 \leq 20$, the test statistic is G.
If $\alpha \neq 0.05$ or $n_1 > 20$ or $n_2 > 20$, the test statistic is

$$z = \frac{G - \mu_G}{\sigma_G} \quad \text{(where } \mu_G \text{ and } \sigma_G \text{ are from Formulas 13-2 and 13-3)}$$

Critical values:

- If the test statistic is G, critical values are found in Table A-10.
- If the test statistic is z, critical values are found in Table A-2 by using the same procedures introduced in Chapter 6.

Figure 13-7 summarizes the procedures for the runs test for randomness and includes cases in which the test statistic is G as well as cases in which the test statistic is z.

EXAMPLE   Refer to the sequence of 1s and 0s given in Figure 13-1. Recall that the sequence was designed to be a radio message that could be translated into a picture packed with information about us earthlings. If an intelligent extraterrestrial life form intercepted the message and used the runs test for randomness, would the sequence of 1s and 0s appear to

Figure 13-7   Runs Test for Randomness

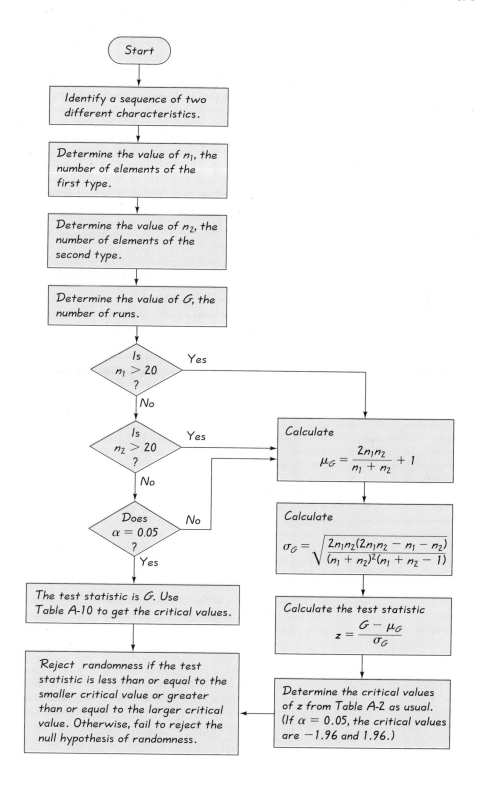

be random and the message mistaken for "random noise"? Use a 0.05 significance level in testing the sequence for randomness.

SOLUTION   The null and alternative hypotheses are as follows:

H_0:   The sequence is random.

H_1:   The sequence is not random.

The significance level is $\alpha = 0.05$; we are using the runs test for randomness. The test statistic is obtained by first finding the number of 1s, the number of 0s, and the number of runs. After painstakingly analyzing the sequence, we find that

$$n_1 = \text{number of 1s} = 249$$
$$n_2 = \text{number of 0s} = 1022$$
$$G = \text{number of runs} = 353$$

As we follow Figure 13-7, we answer yes to "Is $n_1 > 20$?" We therefore need to evaluate μ_G and σ_G before we can determine the test statistic. Using Formulas 13-2 and 13-3 we get

$$\mu_G = \frac{2n_1 n_2}{n_1 + n_2} + 1 = \frac{2(249)(1022)}{249 + 1022} + 1 = 401.437$$

$$\sigma_G = \sqrt{\frac{(2n_1 n_2)(2n_1 n_2 - n_1 - n_2)}{(n_1 + n_2)^2(n_1 + n_2 - 1)}}$$

$$= \sqrt{\frac{(2)(249)(1022)[2(249)(1022) - 249 - 1022]}{(249 + 1022)^2(249 + 1022 - 1)}} = 11.223$$

We can now find the test statistic.

$$z = \frac{G - \mu_G}{\sigma_G} = \frac{353 - 401.437}{11.223} = -4.32$$

Because the significance level is $\alpha = 0.05$ and we have a two-tailed test, the critical values are $z = -1.96$ and $z = 1.96$. The test statistic of $z = -4.32$ does fall within the critical region, so we reject the null hypothesis of randomness. The given sequence does not appear to be random, which is good news for those who developed the radio message because it implies that the message would not be mistaken for random noise.

In each of the preceding examples, the data clearly fit into two categories, but we can also test for randomness in the way numerical data fluctuate above or below a mean or median. To test for randomness above and below the median, for example, use the sample data to find the value of the median, then replace each individual score with the letter A if it is *above* the median, and

replace it with B if it is *below* the median. Delete any values that are equal to the median. It is helpful to write the As and Bs directly above the numbers they represent because this makes checking easier and also reduces the chance of having the wrong number of letters. After finding the sequence of A and B letters, we can proceed to apply the runs test as described earlier. Economists use the runs test for randomness above and below the median in an attempt to identify trends or cycles. An upward economic trend would contain a predominance of Bs at the beginning and As at the end, so the number of runs would be small. A downward trend would have As dominating at the beginning and Bs at the end, with a low number of runs. A cyclical pattern would yield a sequence that systematically changes, so the number of runs would tend to be large. (See Exercise 10.)

Using Computer Software for the Runs Test

STATDISK is programmed for the runs test but, because of the nature of the data, you must first determine the values of n_1, n_2, and the number of runs G. To use STATDISK, select `Analysis` from the main menu bar, then select `Runs Test`. The STATDISK display will include the upper and lower critical values, the mean and standard deviation for the number of runs expected with a random sequence, and the conclusion. Minitab will do a runs test with a sequence of numerical data only. See the *Minitab Student Laboratory Manual and Workbook* for the procedure required to use the options of `Stat`, `Nonparametrics`, and `Runs Test`. The TI-83 calculator is not programmed for the runs test.

13-7 Exercises A: Basic Skills and Concepts

In Exercises 1–4, use the given sequence to determine the values of n_1, n_2, the number of runs G, and the 5% cutoff values from Table A-10.

1. A A A B B B B B A A A A A A

2. T F T F T F T F T F T T T T T T T T T T

3. O O O O O O O O O O O E E O O E E

4. M F F F F M F F M M M F F F F F

In Exercises 5–12, use the runs test of this section to determine whether the given sequence is random. Use a significance level of $\alpha = 0.05$. (All data are listed in order by row.)

5. In conducting research for this book, the author recorded the outcomes of a roulette wheel in the Stardust Casino. (Yes, it was hard work, but somebody had to do it.) Test for randomness of odd (O) and even (E) numbers

Recommended assignment: Exercises 1, 7, 8, 9, 10, 16. (Exercises 1–12 use Table A-10, but Exercises 13–16 require the calculation of μ_G and σ_G.)

1. $n_1 = 10$, $n_2 = 5$, $G = 3$, 5% cutoff values: 3, 12

2. $n_1 = 15$, $n_2 = 5$, $G = 11$, 5% cutoff values: 4, 12

3. $n_1 = 12$, $n_2 = 4$, $G = 4$, 5% cutoff values: 3, 10

4. $n_1 = 5$, $n_2 = 11$, $G = 6$, 5% cutoff values: 4, 12

5. $n_1 = 12$, $n_2 = 8$, $G = 10$, 5% cutoff values: 6, 16.

6. $n_1 = 18$, $n_2 = 4$, $G = 4$, 5% cutoff values: 4, 10.

7. $n_1 = 10$, $n_2 = 14$, $G = 4$, 5% cutoff values: 7, 18.

8. $n_1 = 15$, $n_2 = 11$, $G = 15$, 5% cutoff values: 8, 19.

9. $n_1 = 17$, $n_2 = 14$, $G = 14$, 5% cutoff values: 10, 23.

10. Median = 1061.5, $n_1 = 15$, $n_2 = 15$, $G = 2$, 5% cutoff values: 10, 22.

for the results given in the following sequence. What would a lack of randomness mean to the author? To the casino?

O O E E E E O O E O E O O O O O O E O E

6. The Niko Music Company uses a machine to produce compact discs that must meet certain specifications. When a sample of discs is selected in sequence from the assembly line, they are examined and judged to be defective (D) or acceptable (A), with the following results.

A A A A A A A A D D A A A A A A A A D D

7. A statistics quiz consists of true-false questions with the answers listed below. If the answers don't appear to be random, is there a noticeable pattern that students could use?

T T T T F F F F F F F T T T T T F F F F F F

8. Test the claim that the sequence of World Series wins by American League and National League teams is random. Given below are recent results with American and National league teams represented by A and N, respectively. What does the result suggest about the abilities of the two leagues?

A N A A A N N A A N N N N A A A N A N A N A A A N A

9. For a recent sequence of presidential elections, the political party of the winner is indicated by D for Democrat and R for Republican. Does it appear that we elect Democrat and Republican candidates in a sequence that is random?

R R D R D R R R R D D R R R D D
D D D R R D D R R R D R R R D D

10. Trends in business and economics applications are often analyzed with the runs test. The accompanying list shows (in order by row) the annual high points of the Dow Jones Industrial Average for a recent sequence of years. First find the median of the values, then replace each value by A if it is above the median and B if it is below the median. Then apply the runs test to the resulting sequence of As and Bs. What does the result suggest about the stock market as an investment consideration?

943	985	969	842	951	1036	1052	892	882	1015
1000	908	898	1000	1024	1071	1287	1287	1553	1956
2722	2184	2791	3000	3169	3413	3794	3978	5667	6624

11. Men were once drafted into the U.S. Army by using a process that was supposed to randomly select birthdays. Suppose the first few selections are

as listed below. Test the sequence for randomness before and after the middle of the year.

Nov. 27	July 7	Aug. 3	Oct. 19	Dec. 19	Sept. 21	Apr. 1
Mar. 5	June 10	May 21	June 27	Jan. 5		

12. Listed below are numbers selected in consecutive drawings from the Maryland Pick Three lottery. Test for randomness of odd (O) and even (E) numbers. What would lack of randomness mean for those who play the lottery?

0 0 0 7 1 3 3 6 4 6 8 6 2 4 7 7 6 9
6 2 5 6 7 7 6 1 1 3 3 3 8 2 2

13. The Restaurant Market Research Company hires Diana Washington to survey adults about their dining habits. A check of her first 40 results shows that their genders are represented in the following sequence. Use the runs test to determine whether the subjects are randomly selected with respect to gender.

F F F F F M M M M M F F F F M M M M F F
F F F M M M M M M M M M F F F F F F

14. A *New York Times* article about the calculation of decimal places of π noted that "mathematicians are pretty sure that the digits of π are indistinguishable from any random sequence." Given below are the first 100 decimal places of π. Test for randomness of odd (O) and even (E) digits.

1 4 1 5 9 2 6 5 3 5 8 9 7 9 3 2 3 8 4 6
2 6 4 3 3 8 3 2 7 9 5 0 2 8 8 4 1 9 7 1
6 9 3 9 9 3 7 5 1 0 5 8 2 0 9 7 4 9 4 4
5 9 2 3 0 7 8 1 6 4 0 6 2 8 6 2 0 8 9 9
8 6 2 8 0 3 4 8 2 5 3 4 2 1 1 7 0 6 7 9

15. Use the 100 decimal digits for π given in Exercise 14. Test for randomness above (A) and below (B) the value of 4.5.

16. Test the claim that the sequence of World Series wins by American League and National League teams is random. Given below are recent results, with American and National League teams represented by A and N, respectively.

A N A N N N A A A A N A A A A N A N A N
N A A N N A A A A N A N N A A A A A
N A N A N A N A A A A A A N N A N
A N N A A N N N A N A N A N A A A N
N A A N N N N A A A N A N A N A N A A A
N A

13-7 Exercises B: Beyond the Basics

17. Minimum is 2, maximum is 4. Critical values of 1 and 6 can never be realized so that the null hypothesis of randomness can never be rejected.

18. b. The 84 sequences yield two runs of 2, seven runs of 3, twenty runs of 4, twenty-five runs of 5, twenty runs of 6, and ten runs of 7.

   c. With $P(2 \text{ runs}) = 2/84$, $P(3 \text{ runs}) = 7/84$, $P(4 \text{ runs}) = 20/84$, $P(5 \text{ runs}) = 25/84$, $P(6 \text{ runs}) = 20/84$, and $P(7 \text{ runs}) = 10/84$, the G values of 3, 4, 5, 6, 7 can easily occur by chance, whereas $G = 2$ is unlikely because $P(2 \text{ runs})$ is less than 0.025. The lower critical G value is therefore 2, and there is no upper critical value that can be equalled or exceeded.

   d. Critical value of $G = 2$ agrees with Table A-10. The table lists 8 as the upper critical value, but it is impossible to get 8 runs using the given elements.

17. Using the elements A, A, B, B, what is the minimum number of possible runs that can be arranged? What is the maximum number of runs? Now refer to Table A-10 to find the 5% cutoff G values for $n_1 = n_2 = 2$. What do you conclude about this case?

18. a. Using all of the elements A, A, A, B, B, B, B, B, B, list the 84 different possible sequences.
   b. Find the number of runs for each of the 84 sequences.
   c. Use the results from parts (a) and (b) to find your own 5% cutoff values for G.
   d. Compare your results to those given in Table A-10.

· ·

Vocabulary List

parametric tests
nonparametric tests
distribution-free tests
efficiency
rank
sign test
Wilcoxon signed-ranks test
Wilcoxon rank-sum test

Mann-Whitney U test
Kruskal-Wallis test
H test
rank correlation coefficient
Spearman's rank correlation
   coefficient
runs test
run

· ·

Review

In this chapter we examined six different nonparametric methods for analyzing sample data. When compared to their corresponding parametric methods, nonparametrics have the following major advantages:

• Nonparametric methods do not require normally distributed populations.
• Nonparametric methods can often be applied to nonnumerical data.
• Nonparametric methods usually involve simpler computations.

A major disadvantage is that nonparametric tests are not as efficient as parametric tests, so we generally need stronger evidence before we reject a null hypothesis.

   Table 13-7 lists the nonparametric tests presented in this chapter, along with their functions. The table also lists the corresponding parametric tests.

TABLE 13-7	Summary of Nonparametric Tests	
Nonparametric Test	Function	Parametric Test
Sign test (Section 13-2)	Test for claimed value of average with one sample	t test or z test (Sections 7-2, 7-3, 7-4)
	Test for difference between two dependent samples	t test or z test (Section 8-2)
Wilcoxon signed-ranks test (Section 13-3)	Test for difference between two dependent samples	t test or z test (Section 8-2)
Wilcoxon rank-sum test (Section 13-4)	Test for difference between two independent samples	t test or z test (Sections 8-3, 8-5)
Kruskal-Wallis test (Section 13-5)	Test for more than two independent samples coming from identical populations	Analysis of variance (Section 11-2)
Rank correlation (Section 13-6)	Test for relationship between two variables	Linear correlation (Section 9-2)
Runs test (Section 13-7)	Test for randomness of sample data	(No parametric test)

· ·

Review Exercises

In Exercises 1–12, use a 0.05 significance level with the indicated test. If no particular test is specified, use the appropriate nonparametric test from this chapter.

1. The accompanying table shows samples of data from presidents, popes, and British monarchs. The values are the numbers of years the subjects lived after their inauguration, election, or coronation. Test the claim that the three populations are not all the same.

Presidents	26	15	25	1	16	17	0	12	18	21	2	36
Popes	3	18	25	6	15	11	19					
Monarchs	17	13	12	33	10	7	25					

Based on data from *Computer-Interactive Data Analysis,* by Lunn and McNeil, John Wiley & Sons.

2. The accompanying table shows the heights of presidents matched with the heights of the candidates they beat. All heights are in inches, and only the

1. Test statistic: $H = 0.144$. Critical value: $\chi^2 = 5.991$.

2. Test statistic: $T = 10$. Critical value: $T = 2$.

second-place candidates are included. Use the Wilcoxon signed-ranks test to test the claim that both samples come from populations with the same distribution.

Winner	76	66	70	70	74	71.5	73	74
Runner-up	64	71	72	72	68	71	69.5	74

3. $n_1 = 6$, $n_2 = 24$, $G = 12$, $\mu_G = 10.6$, $\sigma_G = 1.6873$. Test statistic: $z = 0.83$. Critical values: $z = \pm 1.96$.

3. The Dallas Manufacturing Company uses a machine to produce surgical knives that must meet certain specifications. A sample of knives is randomly selected and each knife is judged to be defective (D) or acceptable (A), with the results given below. Use the runs test for randomness to determine if the sequence of defective and acceptable knives is random.

$$D\ A\ A\ A\ A\ D\ A\ A\ A\ A\ D\ A\ A\ A\ A$$
$$D\ A\ A\ A\ A\ D\ A\ A\ A\ A\ D\ A\ A\ A\ A$$

4. $r_s = 0.810$. Critical values: $r_s = \pm 0.738$. Significant correlation.

4. The paired data in the accompanying table consist of weights (in pounds) of discarded paper and sizes of households. Test the claim that there is a correlation between discarded paper and household size.

Paper	2.41	7.57	9.55	8.82	8.72	6.96	6.83	11.42
Household size	2	3	3	6	4	2	1	5

Based on data from the Garbage Project at the University of Arizona.

5. The test statistic $x = 1$ is less than or equal to the critical value of 1.

5. Air America experimented with two different reservation systems and recorded the times (in seconds) required to process randomly selected passenger requests. The results are listed in the accompanying table. Use the sign test to test the claim that there is no difference between the two systems. If you had to select the reservation system to be used by Air America, how would you decide which system to adopt?

Passenger	A	B	C	D	E	F	G	H	I	J	K
MicroAir Software	21	23	25	27	27	29	31	32	30	41	47
Flight Services Software	18	20	21	26	27	24	22	33	27	34	34

6. $\mu_R = 201.5$, $\sigma_R = 23.894$, $R = 163$, $z = -1.61$. Test statistic: $z = -1.61$. Critical values: $z = \pm 1.96$.

6. The Medassist Pharmaceutical Company conducted a study to determine whether a drug affects eye movements. A standardized scale was developed, and the drug was administered to one group, while a control group was given a placebo that produces no effects. The eye movement ratings of subjects are listed below. Test the claim that the drug has no effect on eye movements. Use the Wilcoxon rank-sum test.

Drugged group	Control group
652 512 711 621 508 603 787	674 676 821 830 565 821 837 652 549
747 516 624 627 777 729	668 772 563 703 789 800 711 598

7. To test the effectiveness of the use of technology in a statistics course, four different sections were taught with different combinations of calculators and computer software, with final course averages listed below for randomly selected students from each section. Test the claim that the samples come from populations that are not all the same.

Calculators only	Computers only	Calculators and computers	No Calculators and no computers
74 85 91 62	82 87 60 71	78 89 82 64	66 78 83 55
73 87 66 80	77 63 84 70	77 91 73 85	72 57 81 65

8. A teacher develops a true/false test with the answers given below. Test the claim that the sequence of answers is random.

> F T T T F T T T F T T T F T T
> F F F F F T T T T T F F F F F

9. A consumer investigator obtained prices from mail order companies and computer stores. The accompanying lists show the prices (in dollars) quoted for cartons of floppy disks from various manufacturers. Use the Wilcoxon rank-sum test to test the claim that there is no difference between mail order and store prices.

Mail order				Computer store			
23.00	26.00	27.99	31.50	30.99	33.98	37.75	38.99
32.75	27.00	27.98	24.50	35.79	33.99	34.79	32.99
24.75	28.15	29.99	29.99	29.99	33.00	32.00	

10. A study was conducted to investigate some effects of physical training. Sample data (weights in kilograms) are listed in the margin. (See "Effect of Endurance Training on Possible Determinants of VO_2 During Heavy Exercise," by Casaburi et al., *Journal of Applied Physiology,* Vol. 62, No. 1.) Use the sign test to test the claim that the training has no effect on weight.

11. Do Exercise 10 using the Wilcoxon signed-ranks test.

12. The following table ranks eight states according to SAT scores and cost per student. Is there a correlation between SAT score and cost per student?

State	NY	CA	FL	NJ	TX	NC	MD	OR
SAT score	4	3	5	6	7	8	2	1
Cost per student	1	5	6	2	7	8	3	4

Based on data from the U.S. Department of Education.

Pre-training	Post-training
99	94
57	57
62	62
69	69
74	66
77	76
59	58
92	88
70	70
85	84

7. Test statistic: $H = 4.096$. Critical value: $\chi^2 = 7.815$.

8. $n_1 = 14$, $n_2 = 16$, $G = 11$, 5% cutoff values: 10, 22.

9. $\mu_R = 144$, $\sigma_R = 16.25$, $R = 84$, $z = -3.69$. Test statistic: $z = -3.69$. Critical values: $z = \pm 1.96$.

10. The test statistic $x = 0$ is less than or equal to the critical value of 0.

11. Test statistic: $T = 0$. Critical value: $T = 1$.

12. $r_s = 0.524$. Critical values: $r_s = \pm 0.738$. No significant correlation.

Cumulative Review Exercises

1. The Diaz Opinion Research Organization assigned a pollster to collect data from 30 randomly selected adults. As the data were submitted to the company, the genders of the interviewed subjects were noted. The sequence obtained is shown in the accompanying list.
   a. At the 0.05 significance level, test the claim that the sequence is random.
   b. At the 0.05 significance level, test the claim that the proportion of women is different from 0.5.
   c. Use the sample data to construct a 95% confidence interval for the proportion of women.
   d. What do the preceding results suggest? Is the sample biased against either gender? Was the sample obtained in a random sequence? If you are the manager, do you have any problems with these results?

   M M F M M F M M F M F M M F M
   F M F M M F M M F M M F M M M

2. The Allan Preparation Course is designed to help students achieve better scores on the SAT exam. The accompanying table lists results for randomly selected students. Use the indicated test with a 0.05 level of significance to test the hypothesis that the course has no effect on SAT scores.
   a. Sign test
   b. Wilcoxon signed-ranks test
   c. t test for a claim about two dependent samples (Section 8-2)
   d. How do the preceding results support the statement that nonparametric tests lack the sensitivity of parametric tests (so stronger evidence is required before a null hypothesis is rejected)?

Subject	A	B	C	D	E	F	G	H	I	J
SAT score before course	700	840	830	860	840	690	830	1180	930	1070
SAT score after course	800	840	820	980	980	800	800	1270	1080	1220

Computer Project

In Section 13-4 we saw that the Wilcoxon rank-sum test can be used to test the null hypothesis that two independent samples come from the same distribution, and the alternative hypothesis is the claim that the two distributions differ in some way. Use STATDISK, Minitab, or any other statistical software package to run the Wilcoxon rank-sum test and determine whether it detects the difference in the two

distributions. The two samples are from populations with the same mean of 100 and the same standard deviation of 15, but the distributions are different (normal, uniform).

Sample randomly selected from a *normally* distributed population with mean 100 and standard deviation 15:

82	92	84	107	104	130	84	125	110	96
98	83	87	99	108	97	84	122	106	114
78	117	105	100	94	114	135	99	109	79
106	83	72	105	108	102	100	106	109	119
95	124	95	86	113	83	93	95	99	122
77	106	87	94	92	95	112	95	101	65
69	108	125	105	132	107	101	94	122	80
101	111	108	107	96	134	105	51	85	141
94	102	74	84	101	99	109	83	116	119
118	114	129	98	101	117	92	96	83	114

Sample randomly selected from a *uniformly* distributed population with a minimum of 74 and a maximum of 126 (such a population has a mean of 100 and a standard deviation of 15):

111	104	96	109	93	78	88	85	88	91
92	108	124	114	82	110	100	104	84	113
111	116	95	121	82	120	110	115	93	124
118	94	102	106	104	126	114	123	115	82
84	100	118	76	82	110	81	123	108	96
107	91	89	122	92	100	122	108	81	107
89	123	81	111	118	97	94	86	109	74
87	113	97	78	115	95	117	77	125	109
77	82	111	80	117	108	107	102	94	93
114	94	112	94	116	111	88	101	81	110

········ FROM DATA TO DECISION ········

Was the Draft Lottery Random?

In 1970, a lottery was used to determine who would be drafted into the U.S. Army. The 366 dates in the year were placed in individual capsules. First, the 31 January capsules were placed in a box; then the 29 February capsules were added and the two months were mixed. Then the 31 March capsules were added and the three months were mixed. This process continued until all months were included. The first capsule selected was September 14, so men born on that date were drafted first. The accompanying list shows the 366 dates in the order of selection.

a. Use the runs test to test the sequence for randomness above and below the median of 183.5.

b. Use the Kruskal-Wallis test to test the claim that the 12 months had priority numbers drawn from the same population.

c. Calculate the 12 monthly means. Then plot those 12 means on a graph. (The horizontal scale lists the 12 months, and the vertical scale ranges from 100 to 260.) Note any pattern suggesting that the original priority numbers were not randomly selected.

d. Based on the results from parts (a), (b), and (c), decide whether this particular draft lottery was fair. Write a statement either supporting your position that it was fair or explaining why you believe that it was not fair. If you decided that this lottery was unfair, describe a process for selecting lottery numbers that would have been fair.

Jan:	305	159	251	215	101	224	306	199	194	325	329	221	318	238	017	121
	235	140	058	280	186	337	118	059	052	092	355	077	349	164	211	
Feb:	086	144	297	210	214	347	091	181	338	216	150	068	152	004	089	212
	189	292	025	302	363	290	057	236	179	365	205	299	285			
Mar:	108	029	267	275	293	139	122	213	317	323	136	300	259	354	169	166
	033	332	200	239	334	265	256	258	343	170	268	223	362	217	030	
Apr:	032	271	083	081	269	253	147	312	219	218	014	346	124	231	273	148
	260	090	336	345	062	316	252	002	351	340	074	262	191	208		
May:	330	298	040	276	364	155	035	321	197	065	037	133	295	178	130	055
	112	278	075	183	250	326	319	031	361	357	296	308	226	103	313	
Jun:	249	228	301	020	028	110	085	366	335	206	134	272	069	356	180	274
	073	341	104	360	060	247	109	358	137	022	064	222	353	209		
Jul:	093	350	115	279	188	327	050	013	277	284	248	015	042	331	322	120
	098	190	227	187	027	153	172	023	067	303	289	088	270	287	193	
Aug:	111	045	261	145	054	114	168	048	106	021	324	142	307	198	102	044
	154	141	311	344	291	339	116	036	286	245	352	167	061	333	011	
Sep:	225	161	049	232	082	006	008	184	263	071	158	242	175	001	113	207
	255	246	177	063	204	160	119	195	149	018	233	257	151	315		
Oct:	359	125	244	202	024	087	234	283	342	220	237	072	138	294	171	254
	288	005	241	192	243	117	201	196	176	007	264	094	229	038	079	
Nov:	019	034	348	266	310	076	051	097	080	282	046	066	126	127	131	107
	143	146	203	185	156	009	182	230	132	309	047	281	099	174		
Dec:	129	328	157	165	056	010	012	105	043	041	039	314	163	026	320	096
	304	128	240	135	070	053	162	095	084	173	078	123	016	003	100	

COOPERATIVE GROUP ACTIVITIES

1. *In-class activity:* Use the existing class seating arrangement and apply the runs test to determine whether the students are arranged randomly according to gender. After recording the seating arrangement, analysis can be done in subgroups of 3 or 4 students.

2. *In-class activity:* Divide into groups of 8 to 12 people. For each group member, *measure* his or her height and *measure* his or her arm span. For the arm span, the subject should stand with arms extended, like the wings on an airplane. It's easy to mark the height and arm span on a chalkboard, then measure the distances there. Divide the following tasks among subgroups of 3 or 4 people.

    a. Use rank correlation with the paired sample data to determine whether there is a correlation between height and arm span.

    b. Use the sign test to test for a difference between the two variables.

    c. Use the Wilcoxon signed-ranks test to test for a difference between the two variables.

3. *In-class activity:* Do activity 2 using pulse rate instead of arm span. Measure pulse rates by counting the number of heartbeats in one minute.

4. *In-class activity:* Divide into groups of about 10 or 12 students and use the same reaction timer included with the Chapter 5 Cooperative Group Activities. Each group member should be tested for right-hand reaction time and left-hand reaction time. Analyze the results using methods from this chapter. State the methods used and the conclusions reached.

5. *Out-of-class activity:* See the preceding "From Data to Decision" project, which involves analysis of the 1970 lottery used for drafting men into the U.S. Army. Because the 1970 results raised concerns about the randomness of selecting draft priority numbers, design a new procedure for generating the 366 priority numbers. Use your procedure to generate the 366 numbers and test your results by using the techniques suggested in parts (a), (b), and (c) of the "From Data to Decision" project. How do your results compare to those obtained in 1970? Does your random selection process appear to be better than the one used in 1970? Write a report that clearly describes the process you designed. Also include your analyses and conclusions.

14

Project and Epilogue

14-1 A Statistics Group Project

A great way to conclude an introductory statistics course is to do a project in which methods of statistics are actually applied to real data collected by students. A list of suggested topics follows. Some of these topics can be addressed by actually conducting experiments, whereas others might require researching results already obtained. For example, testing the effectiveness of air bags by actually crashing cars is strongly discouraged, but destructive taste tests of chocolate chip cookies can be an easy and somewhat enjoyable experiment. In addition to a 5- to 10-minute-long oral report, students should submit a type-written report, which includes the following components:

1. List of data collected.
2. Description of the method of analysis.
3. Relevant graphs and/or statistics, including STATDISK or Minitab displays.
4. Statement of conclusions.
5. Reasons why the results might not be correct, along with a description of ways in which the study could be improved, given more time and money.

Suggested Topics

1. Are the ages of student cars different than those of faculty? If so, how?
2. Is the proportion of foreign cars the same for students and faculty?

3. Car ages in the parking lot of a discount store compared to car ages in the parking lot of an upscale department store

4. Are husbands older than their wives?

5. Are husband/wife age differences the same for young married couples as for older married couples?

6. Analysis of the ages of books in the college library

7. How do the ages of books in the college library compare to those in the library of a nearby college?

8. Comparison of the ages of science books and English books in the college library

9. How many hours do students study each week?

10. Is there a relationship between hours studied and grades earned?

11. Is there a relationship between hours worked and grades earned?

12. A study of *reported* heights compared to *measured* heights

13. A study of the accuracy of wrist watches

14. Is there a relationship between taste and cost of different brands of chocolate chip cookies?

15. Is there a relationship between taste and cost of different brands of peanut butter?

16. Is there a relationship between taste and cost of different brands of cola?

17. Is there a relationship between salaries of professional baseball (or basketball, or football) players and their season achievements?

18. Rates versus weights: Is there a relationship between car fuel-consumption rates and car weights? If so, what is it?

19. Is there a relationship between the lengths of men's (or women's) feet and their heights?

20. Are there differences in taste between ordinary tap water and different brands of bottled water?

21. What is the probability that a dropped tack will land with the pointed end up?

22. Were auto fatality rates affected by laws requiring the use of seat belts?

23. Were auto fatality rates affected by the presence of air bags?

24. Is there a difference in taste between Coke and Pepsi?

25. Is there a relationship between student grade-point averages and the amount of television watched? If so, what is it?

26. Is there a relationship between the selling price of a home and its living area (in square feet), lot size (in acres), number of rooms, number of baths, and the annual tax bill?

Note to instructors:

Here are some helpful hints that should facilitate the inclusion of a cooperative project in the statistics course.

- Point out that the project involves cooperation within a group, computer usage, preparation of a report, and an oral presentation. All of these components require skills that employers generally consider to be extremely valuable.

- Select groups and describe the project approximately midway through the course. (At the first class meeting, announce that a project will be given later in the course.)

- Aim for groups of four, although some groups of three or five may be necessary. For a group of five, include the student with the worst attendance record. For a group of three, try to include students with regular attendance records.

- Select groups with one student from each quartile (based on the midterm grade).

- Avoid having close friends in the same group. Try to achieve a good mixture of ages, genders, and personality types.

- Have each group member write a sentence describing the participation of the other group members. If some group members contribute nothing to the project, that should be stated and the members in question should be graded accordingly.

- Deadline: An absolute deadline should be the fourth class from the end. This will allow for more than a full class of group presentations and at least a class or two devoted to review for the final exam.

27. A comparison of the numbers of keys carried by males and females

28. A comparison of the numbers of credit cards carried by males and females

29. Are murderers now younger than they were in the past?

30. Do people who exercise vigorously tend to have lower pulse rates than those who do not?

31. Do people who exercise vigorously tend to have reaction times that are different than those who do not?

32. Do people who smoke tend to have higher pulse rates than those who do not?

33. For people who don't exercise, how is pulse rate affected by climbing a flight of stairs?

34. Do statistics students tend to have pulse rates that are different than people not studying statistics?

35. A comparison of GPAs of statistics students versus students not taking statistics

36. Do left-handed people tend to be involved in more car crashes?

37. Do men have more car crashes than women?

38. Do young drivers have more car crashes than older drivers?

39. Are drivers who get tickets more likely to be involved in crashes?

40. Do smokers tend to be involved in more car crashes?

41. Do people with higher pulse rates tend to be involved in more/fewer car crashes?

42. A comparison of reaction times measured with right and left hands

43. Are the proportions of male and female smokers equal?

44. Do statistics students tend to smoke more (or less) than the general population?

45. Are people more likely to smoke if their parents smoked?

46. Evidence to support/refute the belief that smoking tends to stunt growth

47. Does a sports team have an advantage by playing at home instead of away?

48. Analysis of service times (in seconds) for a car drive-up window at a bank

49. A comparison of service times for car drive-up windows at two different banks

50. Analysis of times that McDonalds' patrons are seated at a table

51. Analysis of times that McDonalds' patrons wait in line

52. Analysis of times that cars require for refueling

53. Is the state lottery a wise investment?

54. Comparison of casino games: craps versus roulette

55. Starting with $1, is it easier to win a million dollars by playing casino craps or by playing a state lottery?

56. Bold versus cautious strategies of gambling: When gambling with $100, does it make any difference if you bet $1 at a time or if you bet the whole $100 at once?

57. Designing and analyzing results from a test for extrasensory perception

58. Analyzing paired data consisting of heights of fathers (or mothers) and heights of their first sons (or daughters)

14-2 Which Procedure Applies?

When trying to actually use methods of statistics with some collection of data, one of the most difficult tasks is determining the specific procedure that is most appropriate. This text includes a wide variety of procedures that apply to many different circumstances. We generally begin by clearly identifying the questions that need to be answered and by evaluating the quality of the sampling procedure. We must also answer questions such as these:

- What is the level of measurement (nominal, ordinal, interval, ratio) of the data?
- Does the study involve one, two, or more populations?
- Is there a claim to be tested or a parameter to be estimated?
- What is the relevant parameter (mean, standard deviation, proportion)?
- Is the sample large ($n > 30$) or small?
- Is there reason to believe that the population is normally distributed?

In Figure 14-1 on the following page we list the major methods included in this book, along with a scheme for determining which of those methods should be used. To use Figure 14-1, start at the extreme left side of the figure and begin by identifying the level of measurement of the data. Proceed to follow the path suggested by the level of measurement, the number of populations, and the claim or parameter being considered.

14-3 A Perspective

No one expects a single introductory statistics course to transform anyone into an expert statistician. After studying several of the chapters in this book, it is natural for students to feel that they have not yet mastered the material to the extent necessary for using statistics in real applications. Many important topics (such as factor analysis and discriminate analysis) are not included in this text because they are too advanced for this introductory level. Some easier

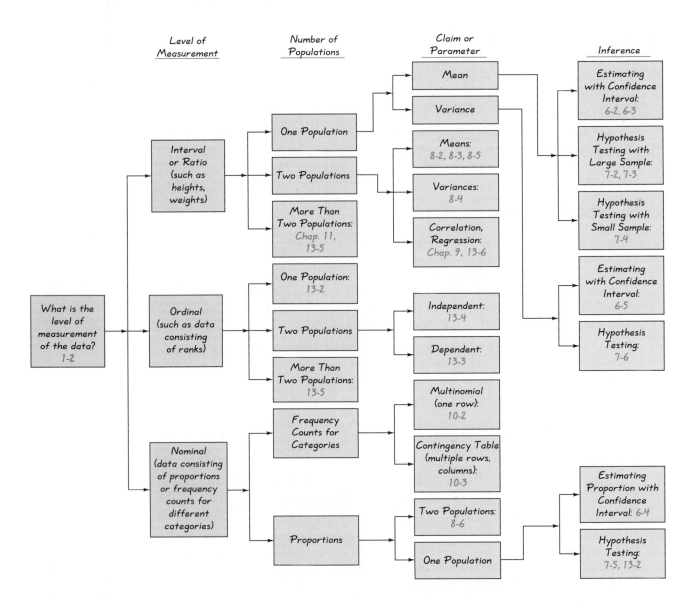

Figure 14-1   Inferential Statistics: Applicable Text Sections

Note: This figure applies to a fixed population. If the data are from a process that may change over time, construct a control chart (see Chapter 12) to determine whether the process is statistically stable. This figure applies to process data only if the process is statistically stable.

topics (such as time series) have been excluded for other reasons. It is important to know that professional help is available from expert statisticians, and this introductory statistics course will help you open a dialogue with one of these experts.

Although this course is not designed to make you an expert statistician, it is designed to make you a better-educated person with improved job marketability. You should know and understand the basic concepts of probability and chance. You should know that in attempting to gain insight into a set of data, it is important to investigate measures of central tendency (such as mean and median), measures of dispersion (such as range and standard deviation), and the nature of the distribution (via a frequency table or graph). You should know and understand the importance of estimating population parameters (such as a mean, standard deviation, and proportion), as well as testing claims made about population parameters. You should realize that the nature and configuration of the data can have a dramatic effect on the particular statistical procedures that are used.

Throughout this text we have emphasized the importance of good sampling. You should recognize that a bad sample may be beyond repair by even the most expert statisticians, using the most sophisticated techniques. There are many mail, magazine, and telephone call-in surveys that allow respondents to be "self-selected." The results of such surveys are generally worthless when judged according to the criteria of sound statistical methodology. Keep this in mind when you are exposed to self-selected surveys, so that you don't let them affect your beliefs and decisions. You should also recognize, however, that many surveys and polls obtain very good results, even though the sample sizes might seem to be relatively small. Although many people refuse to believe it, a nationwide survey of only 1700 voters can provide good results if the sampling is carefully planned and executed.

At one time a person was considered educated if he or she could read, but our society has become much more demanding. A modern education typically provides students with specific skills, such as the ability to read, write, understand the significance of the Renaissance, operate a computer, and do algebra. The larger picture involves several disciplines that use different approaches to a common goal—seeking the truth. The study of statistics helps us see the truth that is sometimes distorted by others or concealed by data that are disorganized or perhaps not yet collected. Understanding statistics is essential for a growing number of future employees, employers, and citizens.

Appendices

Appendix A: Tables

TABLE A-1	Binomial Probabilities

								p								
n	x	.01	.05	.10	.20	.30	.40	.50	.60	.70	.80	.90	.95	.99		x
2	0	980	902	810	640	490	360	250	160	090	040	010	002	0+		0
	1	020	095	180	320	420	480	500	480	420	320	180	095	020		1
	2	0+	002	010	040	090	160	250	360	490	640	810	902	980		2
3	0	970	857	729	512	343	216	125	064	027	008	001	0+	0+		0
	1	029	135	243	384	441	432	375	288	189	096	027	007	0+		1
	2	0+	007	027	096	189	288	375	432	441	384	243	135	029		2
	3	0+	0+	001	008	027	064	125	216	343	512	729	857	970		3
4	0	961	815	656	410	240	130	062	026	008	002	0+	0+	0+		0
	1	039	171	292	410	412	346	250	154	076	026	004	0+	0+		1
	2	001	014	049	154	265	346	375	346	265	154	049	014	001		2
	3	0+	0+	004	026	076	154	250	346	412	410	292	171	039		3
	4	0+	0+	0+	002	008	026	062	130	240	410	656	815	961		4
5	0	951	774	590	328	168	078	031	010	002	0+	0+	0+	0+		0
	1	048	204	328	410	360	259	156	077	028	006	0+	0+	0+		1
	2	001	021	073	205	309	346	312	230	132	051	008	001	0+		2
	3	0+	001	008	051	132	230	312	346	309	205	073	021	001		3
	4	0+	0+	0+	006	028	077	156	259	360	410	328	204	048		4
	5	0+	0+	0+	0+	002	010	031	078	168	328	590	774	951		5
6	0	941	735	531	262	118	047	016	004	001	0+	0+	0+	0+		0
	1	057	232	354	393	303	187	094	037	010	002	0+	0+	0+		1
	2	001	031	098	246	324	311	234	138	060	015	001	0+	0+		2
	3	0+	002	015	082	185	276	312	276	185	082	015	002	0+		3
	4	0+	0+	001	015	060	138	234	311	324	246	098	031	001		4
	5	0+	0+	0+	002	010	037	094	187	303	393	354	232	057		5
	6	0+	0+	0+	0+	001	004	016	047	118	262	531	735	941		6
7	0	932	698	478	210	082	028	008	002	0+	0+	0+	0+	0+		0
	1	066	257	372	367	247	131	055	017	004	0+	0+	0+	0+		1
	2	002	041	124	275	318	261	164	077	025	004	0+	0+	0+		2
	3	0+	004	023	115	227	290	273	194	097	029	003	0+	0+		3
	4	0+	0+	003	029	097	194	273	290	227	115	023	004	0+		4
	5	0+	0+	0+	004	025	077	164	261	318	275	124	041	002		5
	6	0+	0+	0+	0+	004	017	055	131	247	367	372	257	066		6
	7	0+	0+	0+	0+	0+	002	008	028	082	210	478	698	932		7
8	0	923	663	430	168	058	017	004	001	0+	0+	0+	0+	0+		0
	1	075	279	383	336	198	090	031	008	001	0+	0+	0+	0+		1
	2	003	051	149	294	296	209	109	041	010	001	0+	0+	0+		2
	3	0+	005	033	147	254	279	219	124	047	009	0+	0+	0+		3
	4	0+	0+	005	046	136	232	273	232	136	046	005	0+	0+		4
	5	0+	0+	0+	009	047	124	219	279	254	147	033	005	0+		5
	6	0+	0+	0+	001	010	041	109	209	296	294	149	051	003		6
	7	0+	0+	0+	0+	001	008	031	090	198	336	383	279	075		7
	8	0+	0+	0+	0+	0+	001	004	017	058	168	430	663	923		8

NOTE: 0+ represents a positive probability less than 0.0005.

(continued)

n	x	.01	.05	.10	.20	.30	.40	.50	.60	.70	.80	.90	.95	.99	x
9	0	914	630	387	134	040	010	002	0+	0+	0+	0+	0+	0+	0
	1	083	299	387	302	156	060	018	004	0+	0+	0+	0+	0+	1
	2	003	063	172	302	267	161	070	021	004	0+	0+	0+	0+	2
	3	0+	008	045	176	267	251	164	074	021	003	0+	0+	0+	3
	4	0+	001	007	066	172	251	246	167	074	017	001	0+	0+	4
	5	0+	0+	001	017	074	167	246	251	172	066	007	001	0+	5
	6	0+	0+	0+	003	021	074	164	251	267	176	045	008	0+	6
	7	0+	0+	0+	0+	004	021	070	161	267	302	172	063	003	7
	8	0+	0+	0+	0+	0+	004	018	060	156	302	387	299	083	8
	9	0+	0+	0+	0+	0+	0+	002	010	040	134	387	630	914	9
10	0	904	599	349	107	028	006	001	0+	0+	0+	0+	0+	0+	0
	1	091	315	387	268	121	040	010	002	0+	0+	0+	0+	0+	1
	2	004	075	194	302	233	121	044	011	001	0+	0+	0+	0+	2
	3	0+	010	057	201	267	215	117	042	009	001	0+	0+	0+	3
	4	0+	001	011	088	200	251	205	111	037	006	0+	0+	0+	4
	5	0+	0+	001	026	103	201	246	201	103	026	001	0+	0+	5
	6	0+	0+	0+	006	037	111	205	251	200	088	011	001	0+	6
	7	0+	0+	0+	001	009	042	117	215	267	201	057	010	0+	7
	8	0+	0+	0+	0+	001	011	044	121	233	302	194	075	004	8
	9	0+	0+	0+	0+	0+	002	010	040	121	268	387	315	091	9
	10	0+	0+	0+	0+	0+	0+	001	006	028	107	349	599	904	10
11	0	895	569	314	086	020	004	0+	0+	0+	0+	0+	0+	0+	0
	1	099	329	384	236	093	027	005	001	0+	0+	0+	0+	0+	1
	2	005	087	213	295	200	089	027	005	001	0+	0+	0+	0+	2
	3	0+	014	071	221	257	177	081	023	004	0+	0+	0+	0+	3
	4	0+	001	016	111	220	236	161	070	017	002	0+	0+	0+	4
	5	0+	0+	002	039	132	221	226	147	057	010	0+	0+	0+	5
	6	0+	0+	0+	010	057	147	226	221	132	039	002	0+	0+	6
	7	0+	0+	0+	002	017	070	161	236	220	111	016	001	0+	7
	8	0+	0+	0+	0+	004	023	081	177	257	221	071	014	0+	8
	9	0+	0+	0+	0+	001	005	027	089	200	295	213	087	005	9
	10	0+	0+	0+	0+	0+	001	005	027	093	236	384	329	099	10
	11	0+	0+	0+	0+	0+	0+	0+	004	020	086	314	569	895	11
12	0	886	540	282	069	014	002	0+	0+	0+	0+	0+	0+	0+	0
	1	107	341	377	206	071	017	003	0+	0+	0+	0+	0+	0+	1
	2	006	099	230	283	168	064	016	002	0+	0+	0+	0+	0+	2
	3	0+	017	085	236	240	142	054	012	001	0+	0+	0+	0+	3
	4	0+	002	021	133	231	213	121	042	008	001	0+	0+	0+	4
	5	0+	0+	004	053	158	227	193	101	029	003	0+	0+	0+	5
	6	0+	0+	0+	016	079	177	226	177	079	016	0+	0+	0+	6
	7	0+	0+	0+	003	029	101	193	227	158	053	004	0+	0+	7
	8	0+	0+	0+	001	008	042	121	213	231	133	021	002	0+	8
	9	0+	0+	0+	0+	001	012	054	142	240	236	085	017	0+	9
	10	0+	0+	0+	0+	0+	002	016	064	168	283	230	099	006	10
	11	0+	0+	0+	0+	0+	0+	003	017	071	206	377	341	107	11
	12	0+	0+	0+	0+	0+	0+	0+	002	014	069	282	540	886	12

NOTE: 0+ represents a positive probability less than 0.0005.

(continued)

n	x	.01	.05	.10	.20	.30	.40	.50	.60	.70	.80	.90	.95	.99	x
13	0	878	513	254	055	010	001	0+	0+	0+	0+	0+	0+	0+	0
	1	115	351	367	179	054	011	002	0+	0+	0+	0+	0+	0+	1
	2	007	111	245	268	139	045	010	001	0+	0+	0+	0+	0+	2
	3	0+	021	100	246	218	111	035	006	001	0+	0+	0+	0+	3
	4	0+	003	028	154	234	184	087	024	003	0+	0+	0+	0+	4
	5	0+	0+	006	069	180	221	157	066	014	001	0+	0+	0+	5
	6	0+	0+	001	023	103	197	209	131	044	006	0+	0+	0+	6
	7	0+	0+	0+	006	044	131	209	197	103	023	001	0+	0+	7
	8	0+	0+	0+	001	014	066	157	221	180	069	006	0+	0+	8
	9	0+	0+	0+	0+	003	024	087	184	234	154	028	003	0+	9
	10	0+	0+	0+	0+	001	006	035	111	218	246	100	021	0+	10
	11	0+	0+	0+	0+	0+	001	010	045	139	268	245	111	007	11
	12	0+	0+	0+	0+	0+	0+	002	011	054	179	367	351	115	12
	13	0+	0+	0+	0+	0+	0+	0+	001	010	055	254	513	878	13
14	0	869	488	229	044	007	001	0+	0+	0+	0+	0+	0+	0+	0
	1	123	359	356	154	041	007	001	0+	0+	0+	0+	0+	0+	1
	2	008	123	257	250	113	032	006	001	0+	0+	0+	0+	0+	2
	3	0+	026	114	250	194	085	022	003	0+	0+	0+	0+	0+	3
	4	0+	004	035	172	229	155	061	014	001	0+	0+	0+	0+	4
	5	0+	0+	008	086	196	207	122	041	007	0+	0+	0+	0+	5
	6	0+	0+	001	032	126	207	183	092	023	002	0+	0+	0+	6
	7	0+	0+	0+	009	062	157	209	157	062	009	0+	0+	0+	7
	8	0+	0+	0+	002	023	092	183	207	126	032	001	0+	0+	8
	9	0+	0+	0+	0+	007	041	122	207	196	086	008	0+	0+	9
	10	0+	0+	0+	0+	001	014	061	155	229	172	035	004	0+	10
	11	0+	0+	0+	0+	0+	003	022	085	194	250	114	026	0+	11
	12	0+	0+	0+	0+	0+	001	006	032	113	250	257	123	008	12
	13	0+	0+	0+	0+	0+	0+	001	007	041	154	356	359	123	13
	14	0+	0+	0+	0+	0+	0+	0+	001	007	044	229	488	869	14
15	0	860	463	206	035	005	0+	0+	0+	0+	0+	0+	0+	0+	0
	1	130	366	343	132	031	005	0+	0+	0+	0+	0+	0+	0+	1
	2	009	135	267	231	092	022	003	0+	0+	0+	0+	0+	0+	2
	3	0+	031	129	250	170	063	014	002	0+	0+	0+	0+	0+	3
	4	0+	005	043	188	219	127	042	007	001	0+	0+	0+	0+	4
	5	0+	001	010	103	206	186	092	024	003	0+	0+	0+	0+	5
	6	0+	0+	002	043	147	207	153	061	012	001	0+	0+	0+	6
	7	0+	0+	0+	014	081	177	196	118	035	003	0+	0+	0+	7
	8	0+	0+	0+	003	035	118	196	177	081	014	0+	0+	0+	8
	9	0+	0+	0+	001	012	061	153	207	147	043	002	0+	0+	9
	10	0+	0+	0+	0+	003	024	092	186	206	103	010	001	0+	10
	11	0+	0+	0+	0+	001	007	042	127	219	188	043	005	0+	11
	12	0+	0+	0+	0+	0+	002	014	063	170	250	129	031	0+	12
	13	0+	0+	0+	0+	0+	0+	003	022	092	231	267	135	009	13
	14	0+	0+	0+	0+	0+	0+	0+	005	031	132	343	366	130	14
	15	0+	0+	0+	0+	0+	0+	0+	0+	005	035	206	463	860	15

NOTE: 0+ represents a positive probability less than 0.0005.

From Frederick C. Mosteller, Robert E. K. Rourke, and George B. Thomas, Jr., *Probability with Statistical Applications*, 2nd ed., © 1970 Addison-Wesley Publishing Co., Reading, MA. Reprinted with permission.

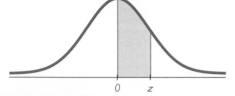

TABLE A-2 Standard Normal (z) Distribution

z	.00	.01	.02	.03	.04	.05	.06	.07	.08	.09
0.0	.0000	.0040	.0080	.0120	.0160	.0199	.0239	.0279	.0319	.0359
0.1	.0398	.0438	.0478	.0517	.0557	.0596	.0636	.0675	.0714	.0753
0.2	.0793	.0832	.0871	.0910	.0948	.0987	.1026	.1064	.1103	.1141
0.3	.1179	.1217	.1255	.1293	.1331	.1368	.1406	.1443	.1480	.1517
0.4	.1554	.1591	.1628	.1664	.1700	.1736	.1772	.1808	.1844	.1879
0.5	.1915	.1950	.1985	.2019	.2054	.2088	.2123	.2157	.2190	.2224
0.6	.2257	.2291	.2324	.2357	.2389	.2422	.2454	.2486	.2517	.2549
0.7	.2580	.2611	.2642	.2673	.2704	.2734	.2764	.2794	.2823	.2852
0.8	.2881	.2910	.2939	.2967	.2995	.3023	.3051	.3078	.3106	.3133
0.9	.3159	.3186	.3212	.3238	.3264	.3289	.3315	.3340	.3365	.3389
1.0	.3413	.3438	.3461	.3485	.3508	.3531	.3554	.3577	.3599	.3621
1.1	.3643	.3665	.3686	.3708	.3729	.3749	.3770	.3790	.3810	.3830
1.2	.3849	.3869	.3888	.3907	.3925	.3944	.3962	.3980	.3997	.4015
1.3	.4032	.4049	.4066	.4082	.4099	.4115	.4131	.4147	.4162	.4177
1.4	.4192	.4207	.4222	.4236	.4251	.4265	.4279	.4292	.4306	.4319
1.5	.4332	.4345	.4357	.4370	.4382	.4394	.4406	.4418	.4429	.4441
1.6	.4452	.4463	.4474	.4484	.4495 *	.4505	.4515	.4525	.4535	.4545
1.7	.4554	.4564	.4573	.4582	.4591	.4599	.4608	.4616	.4625	.4633
1.8	.4641	.4649	.4656	.4664	.4671	.4678	.4686	.4693	.4699	.4706
1.9	.4713	.4719	.4726	.4732	.4738	.4744	.4750	.4756	.4761	.4767
2.0	.4772	.4778	.4783	.4788	.4793	.4798	.4803	.4808	.4812	.4817
2.1	.4821	.4826	.4830	.4834	.4838	.4842	.4846	.4850	.4854	.4857
2.2	.4861	.4864	.4868	.4871	.4875	.4878	.4881	.4884	.4887	.4890
2.3	.4893	.4896	.4898	.4901	.4904	.4906	.4909	.4911	.4913	.4916
2.4	.4918	.4920	.4922	.4925	.4927	.4929	.4931	.4932	.4934	.4936
2.5	.4938	.4940	.4941	.4943	.4945	.4946	.4948	.4949 *	.4951	.4952
2.6	.4953	.4955	.4956	.4957	.4959	.4960	.4961	.4962	.4963	.4964
2.7	.4965	.4966	.4967	.4968	.4969	.4970	.4971	.4972	.4973	.4974
2.8	.4974	.4975	.4976	.4977	.4977	.4978	.4979	.4979	.4980	.4981
2.9	.4981	.4982	.4982	.4983	.4984	.4984	.4985	.4985	.4986	.4986
3.0	.4987	.4987	.4987	.4988	.4988	.4989	.4989	.4989	.4990	.4990
3.10 and higher	.4999									

NOTE: For values of z above 3.09, use 0.4999 for the area.

*Use these common values that result from interpolation:

z score	Area
1.645	0.4500
2.575	0.4950

From Frederick C. Mosteller and Robert E. K. Rourke, *Sturdy Statistics,* 1973, Addison-Wesley Publishing Co., Reading, MA. Reprinted with permission of Frederick Mosteller.

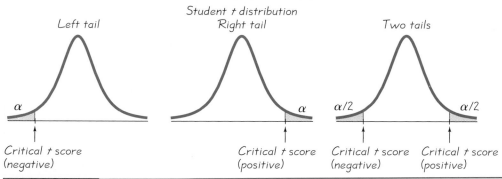

Student *t* distribution

Left tail | Right tail | Two tails

α — Critical *t* score (negative)

α — Critical *t* score (positive)

$\alpha/2$ — Critical *t* score (negative) | $\alpha/2$ — Critical *t* score (positive)

TABLE A-3   *t* Distribution

	α					
Degrees of freedom	.005 (one tail) .01 (two tails)	.01 (one tail) .02 (two tails)	.025 (one tail) .05 (two tails)	.05 (one tail) .10 (two tails)	.10 (one tail) .20 (two tails)	.25 (one tail) .50 (two tails)
1	63.657	31.821	12.706	6.314	3.078	1.000
2	9.925	6.965	4.303	2.920	1.886	.816
3	5.841	4.541	3.182	2.353	1.638	.765
4	4.604	3.747	2.776	2.132	1.533	.741
5	4.032	3.365	2.571	2.015	1.476	.727
6	3.707	3.143	2.447	1.943	1.440	.718
7	3.500	2.998	2.365	1.895	1.415	.711
8	3.355	2.896	2.306	1.860	1.397	.706
9	3.250	2.821	2.262	1.833	1.383	.703
10	3.169	2.764	2.228	1.812	1.372	.700
11	3.106	2.718	2.201	1.796	1.363	.697
12	3.054	2.681	2.179	1.782	1.356	.696
13	3.012	2.650	2.160	1.771	1.350	.694
14	2.977	2.625	2.145	1.761	1.345	.692
15	2.947	2.602	2.132	1.753	1.341	.691
16	2.921	2.584	2.120	1.746	1.337	.690
17	2.898	2.567	2.110	1.740	1.333	.689
18	2.878	2.552	2.101	1.734	1.330	.688
19	2.861	2.540	2.093	1.729	1.328	.688
20	2.845	2.528	2.086	1.725	1.325	.687
21	2.831	2.518	2.080	1.721	1.323	.686
22	2.819	2.508	2.074	1.717	1.321	.686
23	2.807	2.500	2.069	1.714	1.320	.685
24	2.797	2.492	2.064	1.711	1.318	.685
25	2.787	2.485	2.060	1.708	1.316	.684
26	2.779	2.479	2.056	1.706	1.315	.684
27	2.771	2.473	2.052	1.703	1.314	.684
28	2.763	2.467	2.048	1.701	1.313	.683
29	2.756	2.462	2.045	1.699	1.311	.683
Large (*z*)	2.575	2.327	1.960	1.645	1.282	.675

TABLE A-4	Chi-Square (χ^2) Distribution

Area to the Right of the Critical Value

Degrees of freedom	0.995	0.99	0.975	0.95	0.90	0.10	0.05	0.025	0.01	0.005
1	—	—	0.001	0.004	0.016	2.706	3.841	5.024	6.635	7.879
2	0.010	0.020	0.051	0.103	0.211	4.605	5.991	7.378	9.210	10.597
3	0.072	0.115	0.216	0.352	0.584	6.251	7.815	9.348	11.345	12.838
4	0.207	0.297	0.484	0.711	1.064	7.779	9.488	11.143	13.277	14.860
5	0.412	0.554	0.831	1.145	1.610	9.236	11.071	12.833	15.086	16.750
6	0.676	0.872	1.237	1.635	2.204	10.645	12.592	14.449	16.812	18.548
7	0.989	1.239	1.690	2.167	2.833	12.017	14.067	16.013	18.475	20.278
8	1.344	1.646	2.180	2.733	3.490	13.362	15.507	17.535	20.090	21.955
9	1.735	2.088	2.700	3.325	4.168	14.684	16.919	19.023	21.666	23.589
10	2.156	2.558	3.247	3.940	4.865	15.987	18.307	20.483	23.209	25.188
11	2.603	3.053	3.816	4.575	5.578	17.275	19.675	21.920	24.725	26.757
12	3.074	3.571	4.404	5.226	6.304	18.549	21.026	23.337	26.217	28.299
13	3.565	4.107	5.009	5.892	7.042	19.812	22.362	24.736	27.688	29.819
14	4.075	4.660	5.629	6.571	7.790	21.064	23.685	26.119	29.141	31.319
15	4.601	5.229	6.262	7.261	8.547	22.307	24.996	27.488	30.578	32.801
16	5.142	5.812	6.908	7.962	9.312	23.542	26.296	28.845	32.000	34.267
17	5.697	6.408	7.564	8.672	10.085	24.769	27.587	30.191	33.409	35.718
18	6.265	7.015	8.231	9.390	10.865	25.989	28.869	31.526	34.805	37.156
19	6.844	7.633	8.907	10.117	11.651	27.204	30.144	32.852	36.191	38.582
20	7.434	8.260	9.591	10.851	12.443	28.412	31.410	34.170	37.566	39.997
21	8.034	8.897	10.283	11.591	13.240	29.615	32.671	35.479	38.932	41.401
22	8.643	9.542	10.982	12.338	14.042	30.813	33.924	36.781	40.289	42.796
23	9.260	10.196	11.689	13.091	14.848	32.007	35.172	38.076	41.638	44.181
24	9.886	10.856	12.401	13.848	15.659	33.196	36.415	39.364	42.980	45.559
25	10.520	11.524	13.120	14.611	16.473	34.382	37.652	40.646	44.314	46.928
26	11.160	12.198	13.844	15.379	17.292	35.563	38.885	41.923	45.642	48.290
27	11.808	12.879	14.573	16.151	18.114	36.741	40.113	43.194	46.963	49.645
28	12.461	13.565	15.308	16.928	18.939	37.916	41.337	44.461	48.278	50.993
29	13.121	14.257	16.047	17.708	19.768	39.087	42.557	45.722	49.588	52.336
30	13.787	14.954	16.791	18.493	20.599	40.256	43.773	46.979	50.892	53.672
40	20.707	22.164	24.433	26.509	29.051	51.805	55.758	59.342	63.691	66.766
50	27.991	29.707	32.357	34.764	37.689	63.167	67.505	71.420	76.154	79.490
60	35.534	37.485	40.482	43.188	46.459	74.397	79.082	83.298	88.379	91.952
70	43.275	45.442	48.758	51.739	55.329	85.527	90.531	95.023	100.425	104.215
80	51.172	53.540	57.153	60.391	64.278	96.578	101.879	106.629	112.329	116.321
90	59.196	61.754	65.647	69.126	73.291	107.565	113.145	118.136	124.116	128.299
100	67.328	70.065	74.222	77.929	82.358	118.498	124.342	129.561	135.807	140.169

From Donald B. Owen, *Handbook of Statistical Tables*, © 1962 Addison-Wesley Publishing Co., Reading, MA. Reprinted with permission of the publisher.

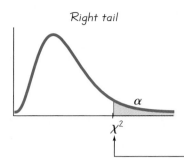

Right tail

To find this value, use the column with the area α given at the top of the table.

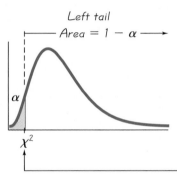

Left tail

Area $= 1 - \alpha$

To find this value, determine the area of the region to the right of this boundary (the unshaded area) and use the column with this value at the top. If the left tail has area α, use the column with the value of $1 - \alpha$ at the top of the table.

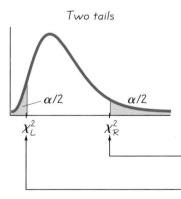

Two tails

To find this value, use the column with area $\alpha/2$ at the top of the table.

To find this value, use the column with area $1 - \alpha/2$ at the top of the table.

0.01

F

TABLE A-5   F Distribution (α = 0.01 in the right tail)

Numerator degrees of freedom (df$_1$)

df$_2$	1	2	3	4	5	6	7	8	9
1	4052.2	4999.5	5403.4	5624.6	5763.6	5859.0	5928.4	5981.1	6022.5
2	98.503	99.000	99.166	99.249	99.299	99.333	99.356	99.374	99.388
3	34.116	30.817	29.457	28.710	28.237	27.911	27.672	27.489	27.345
4	21.198	18.000	16.694	15.977	15.522	15.207	14.976	14.799	14.659
5	16.258	13.274	12.060	11.392	10.967	10.672	10.456	10.289	10.158
6	13.745	10.925	9.7795	9.1483	8.7459	8.4661	8.2600	8.1017	7.9761
7	12.246	9.5466	8.4513	7.8466	7.4604	7.1914	6.9928	6.8400	6.7188
8	11.259	8.6491	7.5910	7.0061	6.6318	6.3707	6.1776	6.0289	5.9106
9	10.561	8.0215	6.9919	6.4221	6.0569	5.8018	5.6129	5.4671	5.3511
10	10.044	7.5594	6.5523	5.9943	5.6363	5.3858	5.2001	5.0567	4.9424
11	9.6460	7.2057	6.2167	5.6683	5.3160	5.0692	4.8861	4.7445	4.6315
12	9.3302	6.9266	5.9525	5.4120	5.0643	4.8206	4.6395	4.4994	4.3875
13	9.0738	6.7010	5.7394	5.2053	4.8616	4.6204	4.4410	4.3021	4.1911
14	8.8616	6.5149	5.5639	5.0354	4.6950	4.4558	4.2779	4.1399	4.0297
15	8.6831	6.3589	5.4170	4.8932	4.5556	4.3183	4.1415	4.0045	3.8948
16	8.5310	6.2262	5.2922	4.7726	4.4374	4.2016	4.0259	3.8896	3.7804
17	8.3997	6.1121	5.1850	4.6690	4.3359	4.1015	3.9267	3.7910	3.6822
18	8.2854	6.0129	5.0919	4.5790	4.2479	4.0146	3.8406	3.7054	3.5971
19	8.1849	5.9259	5.0103	4.5003	4.1708	3.9386	3.7653	3.6305	3.5225
20	8.0960	5.8489	4.9382	4.4307	4.1027	3.8714	3.6987	3.5644	3.4567
21	8.0166	5.7804	4.8740	4.3688	4.0421	3.8117	3.6396	3.5056	3.3981
22	7.9454	5.7190	4.8166	4.3134	3.9880	3.7583	3.5867	3.4530	3.3458
23	7.8811	5.6637	4.7649	4.2636	3.9392	3.7102	3.5390	3.4057	3.2986
24	7.8229	5.6136	4.7181	4.2184	3.8951	3.6667	3.4959	3.3629	3.2560
25	7.7698	5.5680	4.6755	4.1774	3.8550	3.6272	3.4568	3.3239	3.2172
26	7.7213	5.5263	4.6366	4.1400	3.8183	3.5911	3.4210	3.2884	3.1818
27	7.6767	5.4881	4.6009	4.1056	3.7848	3.5580	3.3882	3.2558	3.1494
28	7.6356	5.4529	4.5681	4.0740	3.7539	3.5276	3.3581	3.2259	3.1195
29	7.5977	5.4204	4.5378	4.0449	3.7254	3.4995	3.3303	3.1982	3.0920
30	7.5625	5.3903	4.5097	4.0179	3.6990	3.4735	3.3045	3.1726	3.0665
40	7.3141	5.1785	4.3126	3.8283	3.5138	3.2910	3.1238	2.9930	2.8876
60	7.0771	4.9774	4.1259	3.6490	3.3389	3.1187	2.9530	2.8233	2.7185
120	6.8509	4.7865	3.9491	3.4795	3.1735	2.9559	2.7918	2.6629	2.5586
∞	6.6349	4.6052	3.7816	3.3192	3.0173	2.8020	2.6393	2.5113	2.4073

Denominator degrees of freedom (df$_2$)

From Maxine Merrington and Catherine M. Thompson, "Tables of Percentage Points of the Inverted Beta (F) Distribution," Biometrika 33 (1943): 80–84. Reproduced with permission of the Biometrika Trustees.

(continued)

TABLE A-5   F Distribution ($\alpha = 0.01$ in the right tail) (continued)

Numerator degrees of freedom (df_1)

Denominator degrees of freedom (df_2)	10	12	15	20	24	30	40	60	120	∞
1	6055.8	6106.3	6157.3	6208.7	6234.6	6260.6	6286.8	6313.0	6339.4	6365.9
2	99.399	99.416	99.433	99.449	99.458	99.466	99.474	99.482	99.491	99.499
3	27.229	27.052	26.872	26.690	26.598	26.505	26.411	26.316	26.221	26.125
4	14.546	14.374	14.198	14.020	13.929	13.838	13.745	13.652	13.558	13.463
5	10.051	9.8883	9.7222	9.5526	9.4665	9.3793	9.2912	9.2020	9.1118	9.0204
6	7.8741	7.7183	7.5590	7.3958	7.3127	7.2285	7.1432	7.0567	6.9690	6.8800
7	6.6201	6.4691	6.3143	6.1554	6.0743	5.9920	5.9084	5.8236	5.7373	5.6495
8	5.8143	5.6667	5.5151	5.3591	5.2793	5.1981	5.1156	5.0316	4.9461	4.8588
9	5.2565	5.1114	4.9621	4.8080	4.7290	4.6486	4.5666	4.4831	4.3978	4.3105
10	4.8491	4.7059	4.5581	4.4054	4.3269	4.2469	4.1653	4.0819	3.9965	3.9090
11	4.5393	4.3974	4.2509	4.0990	4.0209	3.9411	3.8596	3.7761	3.6904	3.6024
12	4.2961	4.1553	4.0096	3.8584	3.7805	3.7008	3.6192	3.5355	3.4494	3.3608
13	4.1003	3.9603	3.8154	3.6646	3.5868	3.5070	3.4253	3.3413	3.2548	3.1654
14	3.9394	3.8001	3.6557	3.5052	3.4274	3.3476	3.2656	3.1813	3.0942	3.0040
15	3.8049	3.6662	3.5222	3.3719	3.2940	3.2141	3.1319	3.0471	2.9595	2.8684
16	3.6909	3.5527	3.4089	3.2587	3.1808	3.1007	3.0182	2.9330	2.8447	2.7528
17	3.5931	3.4552	3.3117	3.1615	3.0835	3.0032	2.9205	2.8348	2.7459	2.6530
18	3.5082	3.3706	3.2273	3.0771	2.9990	2.9185	2.8354	2.7493	2.6597	2.5660
19	3.4338	3.2965	3.1533	3.0031	2.9249	2.8442	2.7608	2.6742	2.5839	2.4893
20	3.3682	3.2311	3.0880	2.9377	2.8594	2.7785	2.6947	2.6077	2.5168	2.4212
21	3.3098	3.1730	3.0300	2.8796	2.8010	2.7200	2.6359	2.5484	2.4568	2.3603
22	3.2576	3.1209	2.9779	2.8274	2.7488	2.6675	2.5831	2.4951	2.4029	2.3055
23	3.2106	3.0740	2.9311	2.7805	2.7017	2.6202	2.5355	2.4471	2.3542	2.2558
24	3.1681	3.0316	2.8887	2.7380	2.6591	2.5773	2.4923	2.4035	2.3100	2.2107
25	3.1294	2.9931	2.8502	2.6993	2.6203	2.5383	2.4530	2.3637	2.2696	2.1694
26	3.0941	2.9578	2.8150	2.6640	2.5848	2.5026	2.4170	2.3273	2.2325	2.1315
27	3.0618	2.9256	2.7827	2.6316	2.5522	2.4699	2.3840	2.2938	2.1985	2.0965
28	3.0320	2.8959	2.7530	2.6017	2.5223	2.4397	2.3535	2.2629	2.1670	2.0642
29	3.0045	2.8685	2.7256	2.5742	2.4946	2.4118	2.3253	2.2344	2.1379	2.0342
30	2.9791	2.8431	2.7002	2.5487	2.4689	2.3860	2.2992	2.2079	2.1108	2.0062
40	2.8005	2.6648	2.5216	2.3689	2.2880	2.2034	2.1142	2.0194	1.9172	1.8047
60	2.6318	2.4961	2.3523	2.1978	2.1154	2.0285	1.9360	1.8363	1.7263	1.6006
120	2.4721	2.3363	2.1915	2.0346	1.9500	1.8600	1.7628	1.6557	1.5330	1.3805
∞	2.3209	2.1847	2.0385	1.8783	1.7908	1.6964	1.5923	1.4730	1.3246	1.0000

From Maxine Merrington and Catherine M. Thompson, "Tables of Percentage Points of the Inverted Beta (F) Distribution," *Biometrika 33* (1943): 80–84. Reproduced with permission of the Biometrika Trustees.

(continued)

0.025

TABLE A-5 F Distribution ($\alpha = 0.025$ in the right tail)

Denominator degrees of freedom (df_2)

	Numerator degrees of freedom (df_1)								
	1	2	3	4	5	6	7	8	9
1	647.79	799.50	864.16	899.58	921.85	937.11	948.22	956.66	963.28
2	38.506	39.000	39.165	39.248	39.298	39.331	39.335	39.373	39.387
3	17.443	16.044	15.439	15.101	14.885	14.735	14.624	14.540	14.473
4	12.218	10.649	9.9792	9.6045	9.3645	9.1973	9.0741	8.9796	8.9047
5	10.007	8.4336	7.7636	7.3879	7.1464	6.9777	6.8531	6.7572	6.6811
6	8.8131	7.2599	6.5988	6.2272	5.9876	5.8198	5.6955	5.5996	5.5234
7	8.0727	6.5415	5.8898	5.5226	5.2852	5.1186	4.9949	4.8993	4.8232
8	7.5709	6.0595	5.4160	5.0526	4.8173	4.6517	4.5286	4.4333	4.3572
9	7.2093	5.7147	5.0781	4.7181	4.4844	4.3197	4.1970	4.1020	4.0260
10	6.9367	5.4564	4.8256	4.4683	4.2361	4.0721	3.9498	3.8549	3.7790
11	6.7241	5.2559	4.6300	4.2751	4.0440	3.8807	3.7586	3.6638	3.5879
12	6.5538	5.0959	4.4742	4.1212	3.8911	3.7283	3.6065	3.5118	3.4358
13	6.4143	4.9653	4.3472	3.9959	3.7667	3.6043	3.4827	3.3880	3.3120
14	6.2979	4.8567	4.2417	3.8919	3.6634	3.5014	3.3799	3.2853	3.2093
15	6.1995	4.7650	4.1528	3.8043	3.5764	3.4147	3.2934	3.1987	3.1227
16	6.1151	4.6867	4.0768	3.7294	3.5021	3.3406	3.2194	3.1248	3.0488
17	6.0420	4.6189	4.0112	3.6648	3.4379	3.2767	3.1556	3.0610	2.9849
18	5.9781	4.5597	3.9539	3.6083	3.3820	3.2209	3.0999	3.0053	2.9291
19	5.9216	4.5075	3.9034	3.5587	3.3327	3.1718	3.0509	2.9563	2.8801
20	5.8715	4.4613	3.8587	3.5147	3.2891	3.1283	3.0074	2.9128	2.8365
21	5.8266	4.4199	3.8188	3.4754	3.2501	3.0895	2.9686	2.8740	2.7977
22	5.7863	4.3828	3.7829	3.4401	3.2151	3.0546	2.9338	2.8392	2.7628
23	5.7498	4.3492	3.7505	3.4083	3.1835	3.0232	2.9023	2.8077	2.7313
24	5.7166	4.3187	3.7211	3.3794	3.1548	2.9946	2.8738	2.7791	2.7027
25	5.6864	4.2909	3.6943	3.3530	3.1287	2.9685	2.8478	2.7531	2.6766
26	5.6586	4.2655	3.6697	3.3289	3.1048	2.9447	2.8240	2.7293	2.6528
27	5.6331	4.2421	3.6472	3.3067	3.0828	2.9228	2.8021	2.7074	2.6309
28	5.6096	4.2205	3.6264	3.2863	3.0626	2.9027	2.7820	2.6872	2.6106
29	5.5878	4.2006	3.6072	3.2674	3.0438	2.8840	2.7633	2.6686	2.5919
30	5.5675	4.1821	3.5894	3.2499	3.0265	2.8667	2.7460	2.6513	2.5746
40	5.4239	4.0510	3.4633	3.1261	2.9037	2.7444	2.6238	2.5289	2.4519
60	5.2856	3.9253	3.3425	3.0077	2.7863	2.6274	2.5068	2.4117	2.3344
120	5.1523	3.8046	3.2269	2.8943	2.6740	2.5154	2.3948	2.2994	2.2217
∞	5.0239	3.6889	3.1161	2.7858	2.5665	2.4082	2.2875	2.1918	2.1136

(continued)

TABLE A-5 F Distribution (α = 0.025 in the right tail) (continued)

Numerator degrees of freedom (df_1)

df_2	10	12	15	20	24	30	40	60	120	∞
1	968.63	976.71	984.87	993.10	997.25	1001.4	1005.6	1009.8	1014.0	1018.3
2	39.398	39.415	39.431	39.448	39.456	39.465	39.473	39.481	39.490	39.498
3	14.419	14.337	14.253	14.167	14.124	14.081	14.037	13.992	13.947	13.902
4	8.8439	8.7512	8.6565	8.5599	8.5109	8.4613	8.4111	8.3604	8.3092	8.2573
5	6.6192	6.5245	6.4277	6.3286	6.2780	6.2269	6.1750	6.1225	6.0693	6.0153
6	5.4613	5.3662	5.2687	5.1684	5.1172	5.0652	5.0125	4.9589	4.9044	4.8491
7	4.7611	4.6658	4.5678	4.4667	4.4150	4.3624	4.3089	4.2544	4.1989	4.1423
8	4.2951	4.1997	4.1012	3.9995	3.9472	3.8940	3.8398	3.7844	3.7279	3.6702
9	3.9639	3.8682	3.7694	3.6669	3.6142	3.5604	3.5055	3.4493	3.3918	3.3329
10	3.7168	3.6209	3.5217	3.4185	3.3654	3.3110	3.2554	3.1984	3.1399	3.0798
11	3.5257	3.4296	3.3299	3.2261	3.1725	3.1176	3.0613	3.0035	2.9441	2.8828
12	3.3736	3.2773	3.1772	3.0728	3.0187	2.9633	2.9063	2.8478	2.7874	2.7249
13	3.2497	3.1532	3.0527	2.9477	2.8932	2.8372	2.7797	2.7204	2.6590	2.5955
14	3.1469	3.0502	2.9493	2.8437	2.7888	2.7324	2.6742	2.6142	2.5519	2.4872
15	3.0602	2.9633	2.8621	2.7559	2.7006	2.6437	2.5850	2.5242	2.4611	2.3953
16	2.9862	2.8890	2.7875	2.6808	2.6252	2.5678	2.5085	2.4471	2.3831	2.3163
17	2.9222	2.8249	2.7230	2.6158	2.5598	2.5020	2.4422	2.3801	2.3153	2.2474
18	2.8664	2.7689	2.6667	2.5590	2.5027	2.4445	2.3842	2.3214	2.2558	2.1869
19	2.8172	2.7196	2.6171	2.5089	2.4523	2.3937	2.3329	2.2696	2.2032	2.1333
20	2.7737	2.6758	2.5731	2.4645	2.4076	2.3486	2.2873	2.2234	2.1562	2.0853
21	2.7348	2.6368	2.5338	2.4247	2.3675	2.3082	2.2465	2.1819	2.1141	2.0422
22	2.6998	2.6017	2.4984	2.3890	2.3315	2.2718	2.2097	2.1446	2.0760	2.0032
23	2.6682	2.5699	2.4665	2.3567	2.2989	2.2389	2.1763	2.1107	2.0415	1.9677
24	2.6396	2.5411	2.4374	2.3273	2.2693	2.2090	2.1460	2.0799	2.0099	1.9353
25	2.6135	2.5149	2.4110	2.3005	2.2422	2.1816	2.1183	2.0516	1.9811	1.9055
26	2.5896	2.4908	2.3867	2.2759	2.2174	2.1565	2.0928	2.0257	1.9545	1.8781
27	2.5676	2.4688	2.3644	2.2533	2.1946	2.1334	2.0693	2.0018	1.9299	1.8527
28	2.5473	2.4484	2.3438	2.2324	2.1735	2.1121	2.0477	1.9797	1.9072	1.8291
29	2.5286	2.4295	2.3248	2.2131	2.1540	2.0923	2.0276	1.9591	1.8861	1.8072
30	2.5112	2.4120	2.3072	2.1952	2.1359	2.0739	2.0089	1.9400	1.8664	1.7867
40	2.3882	2.2882	2.1819	2.0677	2.0069	1.9429	1.8752	1.8028	1.7242	1.6371
60	2.2702	2.1692	2.0613	1.9445	1.8817	1.8152	1.7440	1.6668	1.5810	1.4821
120	2.1570	2.0548	1.9450	1.8249	1.7597	1.6899	1.6141	1.5299	1.4327	1.3104
∞	2.0483	1.9447	1.8326	1.7085	1.6402	1.5660	1.4835	1.3883	1.2684	1.0000

Denominator degrees of freedom (df_2)

From Maxine Merrington and Catherine M. Thompson, "Tables of Percentage Points of the Inverted Beta (F) Distribution," *Biometrika 33* (1943): 80–84. Reproduced with permission of the Biometrika Trustees.

(continued)

0.05

TABLE A-5 F Distribution (α = 0.05 in the right tail)

Numerator degrees of freedom (df_1)

	1	2	3	4	5	6	7	8	9
1	161.45	199.50	215.71	224.58	230.16	233.99	236.77	238.88	240.54
2	18.513	19.000	19.164	19.247	19.296	19.330	19.353	19.371	19.385
3	10.128	9.5521	9.2766	9.1172	9.0135	8.9406	8.8867	8.8452	8.8123
4	7.7086	6.9443	6.5914	6.3882	6.2561	6.1631	6.0942	6.0410	6.9988
5	6.6079	5.7861	5.4095	5.1922	5.0503	4.9503	4.8759	4.8183	4.7725
6	5.9874	5.1433	4.7571	4.5337	4.3874	4.2839	4.2067	4.1468	4.0990
7	5.5914	4.7374	4.3468	4.1203	3.9715	3.8660	3.7870	3.7257	3.6767
8	5.3177	4.4590	4.0662	3.8379	3.6875	3.5806	3.5005	3.4381	3.3881
9	5.1174	4.2565	3.8625	3.6331	3.4817	3.3738	3.2927	3.2296	3.1789
10	4.9646	4.1028	3.7083	3.4780	3.3258	3.2172	3.1355	3.0717	3.0204
11	4.8443	3.9823	3.5874	3.3567	3.2039	3.0946	3.0123	2.9480	2.8962
12	4.7472	3.8853	3.4903	3.2592	3.1059	2.9961	2.9134	2.8486	2.7964
13	4.6672	3.8056	3.4105	3.1791	3.0254	2.9153	2.8321	2.7669	2.7144
14	4.6001	3.7389	3.3439	3.1122	2.9582	2.8477	2.7642	2.6987	2.6458
15	4.5431	3.6823	3.2874	3.0556	2.9013	2.7905	2.7066	2.6408	2.5876
16	4.4940	3.6337	3.2389	3.0069	2.8524	2.7413	2.6572	2.5911	2.5377
17	4.4513	3.5915	3.1968	2.9647	2.8100	2.6987	2.6143	2.5480	2.4943
18	4.4139	3.5546	3.1599	2.9277	2.7729	2.6613	2.5767	2.5102	2.4563
19	4.3807	3.5219	3.1274	2.8951	2.7401	2.6283	2.5435	2.4768	2.4227
20	4.3512	3.4928	3.0984	2.8661	2.7109	2.5990	2.5140	2.4471	2.3928
21	4.3248	3.4668	3.0725	2.8401	2.6848	2.5727	2.4876	2.4205	2.3660
22	4.3009	3.4434	3.0491	2.8167	2.6613	2.5491	2.4638	2.3965	2.3419
23	4.2793	3.4221	3.0280	2.7955	2.6400	2.5277	2.4422	2.3748	2.3201
24	4.2597	3.4028	3.0088	2.7763	2.6207	2.5082	2.4226	2.3551	2.3002
25	4.2417	3.3852	2.9912	2.7587	2.6030	2.4904	2.4047	2.3371	2.2821
26	4.2252	3.3690	2.9752	2.7426	2.5868	2.4741	2.3883	2.3205	2.2655
27	4.2100	3.3541	2.9604	2.7278	2.5719	2.4591	2.3732	2.3053	2.2501
28	4.1960	3.3404	2.9467	2.7141	2.5581	2.4453	2.3593	2.2913	2.2360
29	4.1830	3.3277	2.9340	2.7014	2.5454	2.4324	2.3463	2.2783	2.2229
30	4.1709	3.3158	2.9223	2.6896	2.5336	2.4205	2.3343	2.2662	2.2107
40	4.0847	3.2317	2.8387	2.6060	2.4495	2.3359	2.2490	2.1802	2.1240
60	4.0012	3.1504	2.7581	2.5252	2.3683	2.2541	2.1665	2.0970	2.0401
120	3.9201	3.0718	2.6802	2.4472	2.2899	2.1750	2.0868	2.0164	1.9588
∞	3.8415	2.9957	2.6049	2.3719	2.2141	2.0986	2.0096	1.9384	1.8799

Denominator degrees of freedom (df_2)

(continued)

TABLE A-5 F Distribution ($\alpha = 0.05$ in the right tail) (continued)

Numerator degrees of freedom (df_1)

df_2	10	12	15	20	24	30	40	60	120	∞
1	241.88	243.91	245.95	248.01	249.05	250.10	251.14	252.20	253.25	254.31
2	19.396	19.413	19.429	19.446	19.454	19.462	19.471	19.479	19.487	19.496
3	8.7855	8.7446	8.7029	8.6602	8.6385	8.6166	8.5944	8.5720	8.5494	8.5264
4	5.9644	5.9117	5.8578	5.8025	5.7744	5.7459	5.7170	5.6877	5.6581	5.6281
5	4.7351	4.6777	4.6188	4.5581	4.5272	4.4957	4.4638	4.4314	4.3985	4.3650
6	4.0600	3.9999	3.9381	3.8742	3.8415	3.8082	3.7743	3.7398	3.7047	3.6689
7	3.6365	3.5747	3.5107	3.4445	3.4105	3.3758	3.3404	3.3043	3.2674	3.2298
8	3.3472	3.2839	3.2184	3.1503	3.1152	3.0794	3.0428	3.0053	2.9669	2.9276
9	3.1373	3.0729	3.0061	2.9365	2.9005	2.8637	2.8259	2.7872	2.7475	2.7067
10	2.9782	2.9130	2.8450	2.7740	2.7372	2.6996	2.6609	2.6211	2.5801	2.5379
11	2.8536	2.7876	2.7186	2.6464	2.6090	2.5705	2.5309	2.4901	2.4480	2.4045
12	2.7534	2.6866	2.6169	2.5436	2.5055	2.4663	2.4259	2.3842	2.3410	2.2962
13	2.6710	2.6037	2.5331	2.4589	2.4202	2.3803	2.3392	2.2966	2.2524	2.2064
14	2.6022	2.5342	2.4630	2.3879	2.3487	2.3082	2.2664	2.2229	2.1778	2.1307
15	2.5437	2.4753	2.4034	2.3275	2.2878	2.2468	2.2043	2.1601	2.1141	2.0658
16	2.4935	2.4247	2.3522	2.2756	2.2354	2.1938	2.1507	2.1058	2.0589	2.0096
17	2.4499	2.3807	2.3077	2.2304	2.1898	2.1477	2.1040	2.0584	2.0107	1.9604
18	2.4117	2.3421	2.2686	2.1906	2.1497	2.1071	2.0629	2.0166	1.9681	1.9168
19	2.3779	2.3080	2.2341	2.1555	2.1141	2.0712	2.0264	1.9795	1.9302	1.8780
20	2.3479	2.2776	2.2033	2.1242	2.0825	2.0391	1.9938	1.9464	1.8963	1.8432
21	2.3210	2.2504	2.1757	2.0960	2.0540	2.0102	1.9645	1.9165	1.8657	1.8117
22	2.2967	2.2258	2.1508	2.0707	2.0283	1.9842	1.9380	1.8894	1.8380	1.7831
23	2.2747	2.2036	2.1282	2.0476	2.0050	1.9605	1.9139	1.8648	1.8128	1.7570
24	2.2547	2.1834	2.1077	2.0267	1.9838	1.9390	1.8920	1.8424	1.7896	1.7330
25	2.2365	2.1649	2.0889	2.0075	1.9643	1.9192	1.8718	1.8217	1.7684	1.7110
26	2.2197	2.1479	2.0716	1.9898	1.9464	1.9010	1.8533	1.8027	1.7488	1.6906
27	2.2043	2.1323	2.0558	1.9736	1.9299	1.8842	1.8361	1.7851	1.7306	1.6717
28	2.1900	2.1179	2.0411	1.9586	1.9147	1.8687	1.8203	1.7689	1.7138	1.6541
29	2.1768	2.1045	2.0275	1.9446	1.9005	1.8543	1.8055	1.7537	1.6981	1.6376
30	2.1646	2.0921	2.0148	1.9317	1.8874	1.8409	1.7918	1.7396	1.6835	1.6223
40	2.0772	2.0035	1.9245	1.8389	1.7929	1.7444	1.6928	1.6373	1.5766	1.5089
60	1.9926	1.9174	1.8364	1.7480	1.7001	1.6491	1.5943	1.5343	1.4673	1.3893
120	1.9105	1.8337	1.7505	1.6587	1.6084	1.5543	1.4952	1.4290	1.3519	1.2539
∞	1.8307	1.7522	1.6664	1.5705	1.5173	1.4591	1.3940	1.3180	1.2214	1.0000

Denominator degrees of freedom (df_2)

TABLE A-6	Critical Values of the Pearson Correlation Coefficient r

n	$\alpha = .05$	$\alpha = .01$
4	.950	.999
5	.878	.959
6	.811	.917
7	.754	.875
8	.707	.834
9	.666	.798
10	.632	.765
11	.602	.735
12	.576	.708
13	.553	.684
14	.532	.661
15	.514	.641
16	.497	.623
17	.482	.606
18	.468	.590
19	.456	.575
20	.444	.561
25	.396	.505
30	.361	.463
35	.335	.430
40	.312	.402
45	.294	.378
50	.279	.361
60	.254	.330
70	.236	.305
80	.220	.286
90	.207	.269
100	.196	.256

NOTE: To test H_0: $\rho = 0$ against H_1: $\rho \neq 0$, reject H_0 if the absolute value of r is greater than the critical value in the table.

| TABLE A-7 | Critical Values for the Sign Test |

	α			
	.005 (one tail) .01 (two tails)	.01 (one tail) .02 (two tails)	.025 (one tail) .05 (two tails)	.05 (one tail) .10 (two tails)
n				
1	*	*	*	*
2	*	*	*	*
3	*	*	*	*
4	*	*	*	*
5	*	*	*	0
6	*	*	0	0
7	*	0	0	0
8	0	0	0	1
9	0	0	1	1
10	0	0	1	1
11	0	1	1	2
12	1	1	2	2
13	1	1	2	3
14	1	2	2	3
15	2	2	3	3
16	2	2	3	4
17	2	3	4	4
18	3	3	4	5
19	3	4	4	5
20	3	4	5	5
21	4	4	5	6
22	4	5	5	6
23	4	5	6	7
24	5	5	6	7
25	5	6	7	7

NOTES:
1. * indicates that it is not possible to get a value in the critical region.
2. Reject the null hypothesis if the number of the less frequent sign *(x)* is less than or equal to the value in the table.
3. For values of *n* greater than 25, a normal approximation is used with

$$z = \frac{(x + 0.5) - \left(\frac{n}{2}\right)}{\frac{\sqrt{n}}{2}}$$

TABLE A-8 Critical Values of T for the Wilcoxon Signed-Ranks Test

	α			
n	.005 (one tail) .01 (two tails)	.01 (one tail) .02 (two tails)	.025 (one tail) .05 (two tails)	.05 (one tail) .10 (two tails)
5	*	*	*	1
6	*	*	1	2
7	*	0	2	4
8	0	2	4	6
9	2	3	6	8
10	3	5	8	11
11	5	7	11	14
12	7	10	14	17
13	10	13	17	21
14	13	16	21	26
15	16	20	25	30
16	19	24	30	36
17	23	28	35	41
18	28	33	40	47
19	32	38	46	54
20	37	43	52	60
21	43	49	59	68
22	49	56	66	75
23	55	62	73	83
24	61	69	81	92
25	68	77	90	101
26	76	85	98	110
27	84	93	107	120
28	92	102	117	130
29	100	111	127	141
30	109	120	137	152

NOTES:

1. * indicates that it is not possible to get a value in the critical region.
2. Reject the null hypothesis if the test statistic T is less than or equal to the critical value found in this table. Fail to reject the null hypothesis if the test statistic T is greater than the critical value found in this table.

From *Some Rapid Approximate Statistical Procedures,* Copyright © 1949, 1964 Lederle Laboratories Division of American Cyanamid Company. Reprinted with the permission of the American Cyanamid Company.

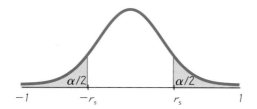

TABLE A-9	Critical Values of Spearman's Rank Correlation Coefficient r_s			
n	$\alpha = 0.10$	$\alpha = 0.05$	$\alpha = 0.02$	$\alpha = 0.01$
5	.900	—	—	—
6	.829	.886	.943	—
7	.714	.786	.893	—
8	.643	.738	.833	.881
9	.600	.683	.783	.833
10	.564	.648	.745	.794
11	.523	.623	.736	.818
12	.497	.591	.703	.780
13	.475	.566	.673	.745
14	.457	.545	.646	.716
15	.441	.525	.623	.689
16	.425	.507	.601	.666
17	.412	.490	.582	.645
18	.399	.476	.564	.625
19	.388	.462	.549	.608
20	.377	.450	.534	.591
21	.368	.438	.521	.576
22	.359	.428	.508	.562
23	.351	.418	.496	.549
24	.343	.409	.485	.537
25	.336	.400	.475	.526
26	.329	.392	.465	.515
27	.323	.385	.456	.505
28	.317	.377	.448	.496
29	.311	.370	.440	.487
30	.305	.364	.432	.478

NOTE: For $n > 30$ use $r_s = \pm z/\sqrt{n-1}$, where z corresponds to the level of significance. For example, if $\alpha = 0.05$, then $z = 1.96$.

To test H_0: $\rho_s = 0$
against H_1: $\rho_s \neq 0$

From "Distribution of sums of squares of rank differences to small numbers of individuals," *The Annals of Mathematical Statistics,* Vol. 9, No. 2. Reprinted with permission of the Institute of Mathematical Statistics.

TABLE A-10 — Critical Values for Number of Runs G

Value of n_2

Value of n_1	2	3	4	5	6	7	8	9	10	11	12	13	14	15	16	17	18	19	20
2	1	1	1	1	1	1	1	1	1	1	2	2	2	2	2	2	2	2	2
	6	6	6	6	6	6	6	6	6	6	6	6	6	6	6	6	6	6	6
3	1	1	1	1	2	2	2	2	2	2	2	2	2	3	3	3	3	3	3
	6	8	8	8	8	8	8	8	8	8	8	8	8	8	8	8	8	8	8
4	1	1	1	2	2	2	3	3	3	3	3	3	3	3	4	4	4	4	4
	6	8	9	9	9	10	10	10	10	10	10	10	10	10	10	10	10	10	10
5	1	1	2	2	3	3	3	3	3	4	4	4	4	4	4	4	5	5	5
	6	8	9	10	10	11	11	12	12	12	12	12	12	12	12	12	12	12	12
6	1	2	2	3	3	3	3	4	4	4	4	5	5	5	5	5	5	6	6
	6	8	9	10	11	12	12	13	13	13	13	14	14	14	14	14	14	14	14
7	1	2	2	3	3	3	4	4	5	5	5	5	5	6	6	6	6	6	6
	6	8	10	11	12	13	13	14	14	14	14	15	15	15	16	16	16	16	16
8	1	2	3	3	3	4	4	5	5	5	6	6	6	6	6	7	7	7	7
	6	8	10	11	12	13	14	14	15	15	16	16	16	16	17	17	17	17	17
9	1	2	3	3	4	4	5	5	5	6	6	6	7	7	7	7	8	8	8
	6	8	10	12	13	14	14	15	16	16	16	17	17	18	18	18	18	18	18
10	1	2	3	3	4	5	5	5	6	6	7	7	7	7	8	8	8	8	9
	6	8	10	12	13	14	15	16	16	17	17	18	18	18	19	19	19	20	20
11	1	2	3	4	4	5	5	6	6	7	7	7	8	8	8	9	9	9	9
	6	8	10	12	13	14	15	16	17	17	18	19	19	19	20	20	20	21	21
12	2	2	3	4	4	5	6	6	7	7	7	8	8	8	9	9	9	10	10
	6	8	10	12	13	14	16	16	17	18	19	19	20	20	21	21	21	22	22
13	2	2	3	4	5	5	6	6	7	7	8	8	9	9	9	10	10	10	10
	6	8	10	12	14	15	16	17	18	19	19	20	20	21	21	22	22	23	23
14	2	2	3	4	5	5	6	7	7	8	8	9	9	9	10	10	10	11	11
	6	8	10	12	14	15	16	17	18	19	20	20	21	22	22	23	23	23	24
15	2	3	3	4	5	6	6	7	7	8	8	9	9	10	10	11	11	11	12
	6	8	10	12	14	15	16	18	18	19	20	21	22	22	23	23	24	24	25
16	2	3	4	4	5	6	6	7	8	8	9	9	10	10	11	11	11	12	12
	6	8	10	12	14	16	17	18	19	20	21	21	22	23	23	24	25	25	25
17	2	3	4	4	5	6	7	7	8	9	9	10	10	11	11	11	12	12	13
	6	8	10	12	14	16	17	18	19	20	21	22	23	23	24	25	25	26	26
18	2	3	4	5	5	6	7	8	8	9	9	10	10	11	11	12	12	13	13
	6	8	10	12	14	16	17	18	19	20	21	22	23	24	25	25	26	26	27
19	2	3	4	5	6	6	7	8	8	9	10	10	11	11	12	12	13	13	13
	6	8	10	12	14	16	17	18	20	21	22	23	23	24	25	26	26	27	27
20	2	3	4	5	6	6	7	8	9	9	10	10	11	12	12	13	13	13	14
	6	8	10	12	14	16	17	18	20	21	22	23	24	25	25	26	27	27	28

NOTE:

1. The entries in this table are the critical G values, assuming a two-tailed test with a significance level of $\alpha = 0.05$.
2. The null hypothesis of randomness is rejected if the total number of runs G is less than or equal to the smaller entry or greater than or equal to the larger entry.

From "Tables for testing randomness of groupings in a sequence of alternatives," *The Annals of Mathematical Statistics*, Vol. 14, No. 1. Reprinted with permission of the Institute of Mathematical Statistics.

Data Set 1: Weights (in pounds) of Household Garbage for One Week

HHSIZE = household size
METAL = weight of discarded metals
PAPER = weight of discarded paper goods
PLAS = weight of discarded plastic goods
GLASS = weight of discarded glass products
FOOD = weight of discarded food items
YARD = weight of discarded yard waste
TEXT = weight of discarded textile goods
OTHER = weight of discarded goods not included in the above
categories
TOTAL = total weight of discarded materials

STATDISK: Variable names are HHSIZE, METAL, PAPER, PLAS, GLASS, FOOD, YARD, TEXT, OTHER, and TOTAL.

Minitab: Enter RETRIEVE 'GARBAGE' to get columns C1 through C10, which correspond to the 10 columns of data.

HOUSEHOLD	HHSIZE	METAL	PAPER	PLAS	GLASS	FOOD	YARD	TEXT	OTHER	TOTAL
1	2	1.09	2.41	0.27	0.86	1.04	0.38	0.05	4.66	10.76
2	3	1.04	7.57	1.41	3.46	3.68	0.00	0.46	2.34	19.96
3	3	2.57	9.55	2.19	4.52	4.43	0.24	0.50	3.60	27.60
4	6	3.02	8.82	2.83	4.92	2.98	0.63	2.26	12.65	38.11
5	4	1.50	8.72	2.19	6.31	6.30	0.15	0.55	2.18	27.90
6	2	2.10	6.96	1.81	2.49	1.46	4.58	0.36	2.14	21.90
7	1	1.93	6.83	0.85	0.51	8.82	0.07	0.60	2.22	21.83
8	5	3.57	11.42	3.05	5.81	9.62	4.76	0.21	10.83	49.27
9	6	2.32	16.08	3.42	1.96	4.41	0.13	0.81	4.14	33.27
10	4	1.89	6.38	2.10	17.67	2.73	3.86	0.66	0.25	35.54
11	4	3.26	13.05	2.93	3.21	9.31	0.70	0.37	11.61	44.44
12	7	3.99	11.36	2.44	4.94	3.59	13.45	4.25	1.15	45.17
13	3	2.04	15.09	2.17	3.10	5.36	0.74	0.42	4.15	33.07
14	5	0.99	2.80	1.41	1.39	1.47	0.82	0.44	1.03	10.35
15	6	2.96	6.44	2.00	5.21	7.06	6.14	0.20	14.43	44.44
16	2	1.50	5.86	0.93	2.03	2.52	1.37	0.27	9.65	24.13
17	4	2.43	11.08	2.97	1.74	1.75	14.70	0.39	2.54	37.60
18	4	2.97	12.43	2.04	3.99	5.64	0.22	2.47	9.20	38.96
19	3	1.42	6.05	0.65	6.26	1.93	0.00	0.86	0.00	17.17
20	3	3.60	13.61	2.13	3.52	6.46	0.00	0.96	1.32	31.60
21	2	4.48	6.98	0.63	2.01	6.72	2.00	0.11	0.18	23.11
22	2	1.36	14.33	1.53	2.21	5.76	0.58	0.17	1.62	27.56
23	4	2.11	13.31	4.69	0.25	9.72	0.02	0.46	0.40	30.96
24	1	0.41	3.27	0.15	0.09	0.16	0.00	0.00	0.00	4.08
25	4	2.02	6.67	1.45	6.85	5.52	0.00	0.68	0.03	23.22

(continued)

Data Set 1 (continued)

HOUSEHOLD	HHSIZE	METAL	PAPER	PLAS	GLASS	FOOD	YARD	TEXT	OTHER	TOTAL
26	6	3.27	17.65	2.68	2.33	11.92	0.83	0.28	4.03	42.99
27	11	4.95	12.73	3.53	5.45	4.68	0.00	0.67	19.89	51.90
28	3	1.00	9.83	1.49	2.04	4.76	0.42	0.54	0.12	20.20
29	4	1.55	16.39	2.31	4.98	7.85	2.04	0.20	1.48	36.80
30	3	1.41	6.33	0.92	3.54	2.90	3.85	0.03	0.04	19.02
31	2	1.05	9.19	0.89	1.06	2.87	0.33	0.01	0.03	15.43
32	2	1.31	9.41	0.80	2.70	5.09	0.64	0.05	0.71	20.71
33	2	2.50	9.45	0.72	1.14	3.17	0.00	0.02	0.01	17.01
34	4	2.35	12.32	2.66	12.24	2.40	7.87	4.73	0.78	45.35
35	6	3.69	20.12	4.37	5.67	13.20	0.00	1.15	1.17	49.37
36	2	3.61	7.72	0.92	2.43	2.07	0.68	0.63	0.00	18.06
37	2	1.49	6.16	1.40	4.02	4.00	0.30	0.04	0.00	17.41
38	2	1.36	7.98	1.45	6.45	4.27	0.02	0.12	2.02	23.67
39	2	1.73	9.64	1.68	1.89	1.87	0.01	1.73	0.58	19.13
40	2	0.94	8.08	1.53	1.78	8.13	0.36	0.12	0.05	20.99
41	3	1.33	10.99	1.44	2.93	3.51	0.00	0.39	0.59	21.18
42	3	2.62	13.11	1.44	1.82	4.21	4.73	0.64	0.49	29.06
43	2	1.25	3.26	1.36	2.89	3.34	2.69	0.00	0.16	14.95
44	2	0.26	1.65	0.38	0.99	0.77	0.34	0.04	0.00	4.43
45	3	4.41	10.00	1.74	1.93	1.14	0.92	0.08	4.60	24.82
46	6	3.22	8.96	2.35	3.61	1.45	0.00	0.09	1.12	20.80
47	4	1.86	9.46	2.30	2.53	6.54	0.00	0.65	2.45	25.79
48	4	1.76	5.88	1.14	3.76	0.92	1.12	0.00	0.04	14.62
49	3	2.83	8.26	2.88	1.32	5.14	5.60	0.35	2.03	28.41
50	3	2.74	12.45	2.13	2.64	4.59	1.07	0.41	1.14	27.17
51	10	4.63	10.58	5.28	12.33	2.94	0.12	2.94	15.65	54.47
52	3	1.70	5.87	1.48	1.79	1.42	0.00	0.27	0.59	13.12
53	6	3.29	8.78	3.36	3.99	10.44	0.90	1.71	13.30	45.77
54	5	1.22	11.03	2.83	4.44	3.00	4.30	1.95	6.02	34.79
55	4	3.20	12.29	2.87	9.25	5.91	1.32	1.87	0.55	37.26
56	7	3.09	20.58	2.96	4.02	16.81	0.47	1.52	2.13	51.58
57	5	2.58	12.56	1.61	1.38	5.01	0.00	0.21	1.46	24.81
58	4	1.67	9.92	1.58	1.59	9.96	0.13	0.20	1.13	26.18
59	2	0.85	3.45	1.15	0.85	3.89	0.00	0.02	1.04	11.25
60	4	1.52	9.09	1.28	8.87	4.83	0.00	0.95	1.61	28.15
61	2	1.37	3.69	0.58	3.64	1.78	0.08	0.00	0.00	11.14
62	2	1.32	2.61	0.74	3.03	3.37	0.17	0.00	0.46	11.70

Data provided by Masakazu Tani, the Garbage Project, University of Arizona.

Data Set 2: Body Temperatures (in degrees Fahrenheit) of Healthy Adults

This is the only Appendix B data set not available on disk.

SUBJECT	AGE	SEX	SMOKE	Temperature Day 1		Temperature Day 2	
				8 AM	12 AM	8 AM	12 AM
1	22	M	Y	98.0	98.0	98.0	98.6
2	23	M	Y	97.0	97.6	97.4	—
3	22	M	Y	98.6	98.8	97.8	98.6
4	19	M	N	97.4	98.0	97.0	98.0
5	18	M	N	98.2	98.8	97.0	98.0
6	20	M	Y	98.2	98.8	96.6	99.0
7	27	M	Y	98.2	97.6	97.0	98.4
8	19	M	Y	96.6	98.6	96.8	98.4
9	19	M	Y	97.4	98.6	96.6	98.4
10	24	M	N	97.4	98.8	96.6	98.4
11	35	M	Y	98.2	98.0	96.2	98.6
12	25	M	Y	97.4	98.2	97.6	98.6
13	25	M	N	97.8	98.0	98.6	98.8
14	35	M	Y	98.4	98.0	97.0	98.6
15	21	M	N	97.6	97.0	97.4	97.0
16	33	M	N	96.2	97.2	98.0	97.0
17	19	M	Y	98.0	98.2	97.6	98.8
18	24	M	Y	—	—	97.2	97.6
19	18	F	N	—	—	97.0	97.7
20	22	F	Y	—	—	98.0	98.8
21	20	M	Y	—	—	97.0	98.0
22	30	F	Y	—	—	96.4	98.0
23	29	M	N	—	—	96.1	98.3
24	18	M	Y	—	—	98.0	98.5
25	31	M	Y	—	98.1	96.8	97.3
26	28	F	Y	—	98.2	98.2	98.7
27	27	M	Y	—	98.5	97.8	97.4
28	21	M	Y	—	98.5	98.2	98.9
29	30	M	Y	—	99.0	97.8	98.6
30	27	M	N	—	98.0	99.0	99.5
31	32	M	Y	—	97.0	97.4	97.5
32	33	M	Y	—	97.3	97.4	97.3
33	23	M	Y	—	97.3	97.5	97.6
34	29	M	Y	—	98.1	97.8	98.2
35	25	M	Y	—	—	97.9	99.6
36	31	M	N	—	97.8	97.8	98.7

(continued)

Data Set 2 (continued)

SUBJECT	AGE	SEX	SMOKE	Temperature Day 1 8 AM	Temperature Day 1 12 AM	Temperature Day 2 8 AM	Temperature Day 2 12 AM
37	25	M	Y	—	99.0	98.3	99.4
38	28	M	N	—	97.6	98.0	98.2
39	30	M	Y	—	97.4	—	98.0
40	33	M	Y	—	98.0	—	98.6
41	28	M	Y	98.0	97.4	—	98.6
42	22	M	Y	98.8	98.0	—	97.2
43	21	F	Y	99.0	—	—	98.4
44	30	M	N	—	98.6	—	98.6
45	22	M	Y	—	98.6	—	98.2
46	22	F	N	98.0	98.4	—	98.0
47	20	M	Y	—	97.0	—	97.8
48	19	M	Y	—	—	—	98.0
49	33	M	N	—	98.4	—	98.4
50	31	M	Y	99.0	99.0	—	98.6
51	26	M	N	—	98.0	—	98.6
52	18	M	N	—	—	—	97.8
53	23	M	N	—	99.4	—	99.0
54	28	M	Y	—	—	—	96.5
55	19	M	Y	—	97.8	—	97.6
56	21	M	N	—	—	—	98.0
57	27	M	Y	—	98.2	—	96.9
58	29	M	Y	—	99.2	—	97.6
59	38	M	N	—	99.0	—	97.1
60	29	F	Y	—	97.7	—	97.9
61	22	M	Y	—	98.2	—	98.4
62	22	M	Y	—	98.2	—	97.3
63	26	M	Y	—	98.8	—	98.0
64	32	M	N	—	98.1	—	97.5
65	25	M	Y	—	98.5	—	97.6
66	21	F	N	—	97.2	—	98.2
67	25	M	Y	—	98.5	—	98.5
68	24	M	Y	—	99.2	97.0	98.8
69	25	M	Y	—	98.3	97.6	98.7
70	35	M	Y	—	98.7	97.5	97.8
71	23	F	Y	—	98.8	98.8	98.0
72	31	M	Y	—	98.6	98.4	97.1
73	28	M	Y	—	98.0	98.2	97.4
74	29	M	Y	—	99.1	97.7	99.4
75	26	M	Y	—	97.2	97.3	98.4
76	32	M	N	—	97.6	97.5	98.6
77	32	M	Y	—	97.9	97.1	98.4

(continued)

Data Set 2 (continued)

SUBJECT	AGE	SEX	SMOKE	Temperature Day 1 8 AM	Temperature Day 1 12 AM	Temperature Day 2 8 AM	Temperature Day 2 12 AM
78	21	F	Y	—	98.8	98.6	98.5
79	20	M	Y	—	98.6	98.6	98.6
80	24	F	Y	—	98.6	97.8	98.3
81	21	F	Y	—	99.3	98.7	98.7
82	28	M	Y	—	97.8	97.9	98.8
83	27	F	N	98.8	98.7	97.8	99.1
84	28	M	N	99.4	99.3	97.8	98.6
85	29	M	Y	98.8	97.8	97.6	97.9
86	19	M	N	97.7	98.4	96.8	98.8
87	24	M	Y	99.0	97.7	96.0	98.0
88	29	M	N	98.1	98.3	98.0	98.7
89	25	M	Y	98.7	97.7	97.0	98.5
90	27	M	N	97.5	97.1	97.4	98.9
91	25	M	Y	98.9	98.4	97.6	98.4
92	21	M	Y	98.4	98.6	97.6	98.6
93	19	M	Y	97.2	97.4	96.2	97.1
94	27	M	Y	—	—	96.2	97.9
95	32	M	N	98.8	96.7	98.1	98.8
96	24	M	Y	97.3	96.9	97.1	98.7
97	32	M	Y	98.7	98.4	98.2	97.6
98	19	F	Y	98.9	98.2	96.4	98.2
99	18	F	Y	99.2	98.6	96.9	99.2
100	27	M	N	—	97.0	—	97.8
101	34	M	Y	—	97.4	—	98.0
102	25	M	N	—	98.4	—	98.4
103	18	M	N	—	97.4	—	97.8
104	32	M	Y	—	96.8	—	98.4
105	31	M	Y	—	98.2	—	97.4
106	26	M	N	—	97.4	—	98.0
107	23	M	N	—	98.0	—	97.0

Data provided by Dr. Steven Wasserman, Dr. Philip Mackowiak, and Dr. Myron Levine of the University of Maryland.

Data Set 3: Bears (wild bears anesthetized)

Age: Months

Month: Month of measurement (1 = Jan., 2 = Feb., etc.)

Sex: 1 = male, 2 = female

HEADLEN: Length (inches) of head

HEADWTH: Width (inches) of head

NECK: Distance (inches) around neck

LENGTH: Length (inches) of body

CHEST: Distance (inches) around the chest

WEIGHT: Measured weight (pounds)

STATDISK: Variable names are BEARAGE, MONTH, BEARSEX, HEADLEN, HEADWTH, NECK, BEARLEN, CHEST, BEARWT

Minitab: Enter RETRIEVE 'BEARS' to get columns C1 through C9, which correspond to the nine columns of data.

AGE	MONTH	SEX	HEADLEN	HEADWTH	NECK	LENGTH	CHEST	WEIGHT
19	7	1	11.0	5.5	16.0	53.0	26.0	80
55	7	1	16.5	9.0	28.0	67.5	45.0	344
81	9	1	15.5	8.0	31.0	72.0	54.0	416
115	7	1	17.0	10.0	31.5	72.0	49.0	348
104	8	2	15.5	6.5	22.0	62.0	35.0	166
100	4	2	13.0	7.0	21.0	70.0	41.0	220
56	7	1	15.0	7.5	26.5	73.5	41.0	262
51	4	1	13.5	8.0	27.0	68.5	49.0	360
57	9	2	13.5	7.0	20.0	64.0	38.0	204
53	5	2	12.5	6.0	18.0	58.0	31.0	144
68	8	1	16.0	9.0	29.0	73.0	44.0	332
8	8	1	9.0	4.5	13.0	37.0	19.0	34
44	8	2	12.5	4.5	10.5	63.0	32.0	140
32	8	1	14.0	5.0	21.5	67.0	37.0	180
20	8	2	11.5	5.0	17.5	52.0	29.0	105
32	8	1	13.0	8.0	21.5	59.0	33.0	166
45	9	1	13.5	7.0	24.0	64.0	39.0	204
9	9	2	9.0	4.5	12.0	36.0	19.0	26
21	9	1	13.0	6.0	19.0	59.0	30.0	120
177	9	1	16.0	9.5	30.0	72.0	48.0	436
57	9	2	12.5	5.0	19.0	57.5	32.0	125
81	9	2	13.0	5.0	20.0	61.0	33.0	132
21	9	1	13.0	5.0	17.0	54.0	28.0	90
9	9	1	10.0	4.0	13.0	40.0	23.0	40
45	9	1	16.0	6.0	24.0	63.0	42.0	220
9	9	1	10.0	4.0	13.5	43.0	23.0	46
33	9	1	13.5	6.0	22.0	66.5	34.0	154
57	9	2	13.0	5.5	17.5	60.5	31.0	116
45	9	2	13.0	6.5	21.0	60.0	34.5	182
21	9	1	14.5	5.5	20.0	61.0	34.0	150
10	10	1	9.5	4.5	16.0	40.0	26.0	65
82	10	2	13.5	6.5	28.0	64.0	48.0	356
70	10	2	14.5	6.5	26.0	65.0	48.0	316
10	10	1	11.0	5.0	17.0	49.0	29.0	94
10	10	1	11.5	5.0	17.0	47.0	29.5	86
34	10	1	13.0	7.0	21.0	59.0	35.0	150
34	10	1	16.5	6.5	27.0	72.0	44.5	270
34	10	1	14.0	5.5	24.0	65.0	39.0	202
58	10	2	13.5	6.5	21.5	63.0	40.0	202
58	10	1	15.5	7.0	28.0	70.5	50.0	365
11	11	1	11.5	6.0	16.5	48.0	31.0	79
23	11	1	12.0	6.5	19.0	50.0	38.0	148
70	10	1	15.5	7.0	28.0	76.5	55.0	446
11	11	2	9.0	5.0	15.0	46.0	27.0	62
83	11	2	14.5	7.0	23.0	61.5	44.0	236
35	11	1	13.5	8.5	23.0	63.5	44.0	212
16	4	1	10.0	4.0	15.5	48.0	26.0	60
16	4	1	10.0	5.0	15.0	41.0	26.0	64
17	5	1	11.5	5.0	17.0	53.0	30.5	114
17	5	2	11.5	5.0	15.0	52.5	28.0	76
17	5	2	11.0	4.5	13.0	46.0	23.0	48
8	8	2	10.0	4.5	10.0	43.5	24.0	29
83	11	1	15.5	8.0	30.5	75.0	54.0	514
18	6	1	12.5	8.5	18.0	57.3	32.8	140

Data from Gary Alt and Minitab, Inc.

Data Set 4: Cigarette Tar, Nicotine, and Carbon Monoxide

All measurements are in milligrams per cigarette and all cigarettes are 100 mm long, filtered, and not menthol or light types.

STATDISK: Variable names are TAR, NICOTINE, CO.

Minitab: Enter RETRIEVE 'CIGARET' to get columns C1 through C3, which correspond to the three columns of data.

Brand	TAR	NICOTINE	CO
American Filter	16	1.2	15
Benson & Hedges	16	1.2	15
Camel	16	1.0	17
Capri	9	0.8	6
Carlton	1	0.1	1
Cartier Vendome	8	0.8	8
Chelsea	10	0.8	10
GPC Approved	16	1.0	17
Hi-Lite	14	1.0	13
Kent	13	1.0	13
Lucky Strike	13	1.1	13
Malibu	15	1.2	15
Marlboro	16	1.2	15
Merit	9	0.7	11
Newport Stripe	11	0.9	15
Now	2	0.2	3
Old Gold	18	1.4	18
Pall Mall	15	1.2	15
Players	13	1.1	12
Raleigh	15	1.0	16
Richland	17	1.3	16
Rite	9	0.8	10
Silva Thins	12	1.0	10
Tareyton	14	1.0	17
Triumph	5	0.5	7
True	6	0.6	7
Vantage	8	0.7	11
Viceroy	18	1.4	15
Winston	16	1.1	18

Based on data from the Federal Trade Commission.

Data Set 5: Ozone Concentrations (in parts per billion) for Northeastern States for a Recent Summer

STATDISK: All of the values are stored in the file named OZONE.

Minitab: Enter RETRIEVE 'OZONE' to get columns C1 through C12, which correspond to the 12 columns listed below.

	CT	DE	ME	MD	MA	NH	NJ	NY	NC	OH	PA	RI
April	1384	1721	1623	1915	1479	1439	1379	2024	4947	2329	1916	1583
May	4832	8802	3088	8084	4525	3596	8416	6073	7109	6843	6351	5017
June	7667	12632	4041	11748	6053	3933	12658	6557	9276	9031	10208	8546
July	9057	17527	2523	16032	6391	4253	16322	7571	12787	7860	10208	7604
August	3953	8035	5010	8390	4314	4782	6796	5107	10254	6835	6377	4508
September	1193	3696	1628	1628	1098	1161	2620	1421	4452	2783	1737	1473

Based on data from the U.S. Department of Agriculture.

Data Set 6: Acid Rain
Nitrate and Sulfate Depositions (kg/hectare) for the
Northeastern United States, State Averages for
July–September, 1980–1990

STATDISK: Variable names are the state abbreviations affixed to NIT (for nitrate) or SUL (for sulfate), as in CTNIT (for the Connecticut nitrate levels), DENIT (for the Delaware nitrate levels), . . . , RISUL (for the Rhode Island sulfate levels).

Minitab: Enter RETRIEVE 'NITRATE' to get columns C1 through C12, which correspond to the nitrate amounts in the 12 different states. Enter RETRIEVE 'SULFATE' to get columns C1 through C12, which correspond to the sulfate amounts in the 12 different states.

Nitrate

	CT	DE	ME	MD	MA	NH	NJ	NY	NC	OH	PA	RI
1980	6.29	6.32	3.97	7.98	6.40	5.84	5.67	6.57	3.41	8.04	8.38	6.45
1981	5.18	4.86	5.47	5.25	5.21	5.20	5.09	5.26	3.29	4.40	5.23	4.93
1982	4.78	5.08	4.16	5.27	4.66	4.14	5.33	4.43	3.98	4.20	4.92	5.21
1983	4.44	4.96	3.89	5.55	5.24	4.98	4.82	5.23	3.50	4.28	5.73	5.07
1984	6.47	7.86	5.59	6.95	6.96	5.25	8.36	6.77	5.46	5.10	7.08	7.47
1985	5.19	5.07	4.21	6.33	5.53	4.73	4.72	5.23	4.56	4.64	6.11	6.12
1986	7.18	6.79	5.75	7.28	8.23	7.05	6.97	8.34	3.49	3.61	7.02	8.99
1987	7.26	5.25	4.29	4.88	6.80	5.32	7.24	6.25	3.22	4.01	6.48	7.39
1988	5.21	4.74	3.82	4.56	5.78	5.22	4.87	5.11	3.43	4.40	5.46	5.78
1989	6.17	6.50	4.79	6.34	6.00	4.88	6.38	5.49	4.85	5.38	5.81	6.53
1990	5.86	5.07	5.78	5.70	5.41	4.17	6.31	5.25	3.77	5.64	6.53	6.29

Sulfate

	CT	DE	ME	MD	MA	NH	NJ	NY	NC	OH	PA	RI
1980	11.94	13.09	7.96	17.29	12.12	11.80	11.53	13.47	6.77	14.24	17.59	11.54
1981	11.28	10.88	12.84	13.87	11.21	11.37	11.72	12.17	7.65	9.26	12.84	9.80
1982	10.38	12.19	7.38	13.64	9.95	8.88	13.49	9.52	8.18	7.85	11.55	11.10
1983	8.00	10.75	7.26	12.37	8.77	9.72	10.54	11.37	7.04	7.05	12.72	8.10
1984	12.12	17.21	10.12	15.73	11.68	9.59	18.36	14.02	10.40	8.34	14.33	11.06
1985	10.27	10.26	8.89	13.21	9.71	8.92	10.14	10.57	10.60	9.25	12.23	11.19
1986	14.80	15.49	11.60	17.94	15.59	14.15	15.56	16.71	6.79	7.18	15.16	15.34
1987	13.52	11.61	9.02	11.22	13.05	10.45	15.32	11.88	6.48	8.11	14.08	13.03
1988	10.55	10.53	7.78	10.57	11.77	10.76	10.12	10.54	8.12	8.70	10.85	11.40
1989	9.81	12.50	8.70	13.29	9.37	9.00	12.36	10.39	9.43	9.91	11.87	8.86
1990	11.27	9.94	10.50	11.28	10.54	8.75	12.37	11.22	7.82	10.07	12.93	11.53

Based on data from the U.S. Department of Agriculture.

Data Set 7: Iowa Weather/Corn Data for 1950–1990

PRECIP values are the amounts of total annual precipitation (in inches).

AVTEMP values are the average annual temperatures (in °F).

CORNPROD values are the amounts of corn produced (in millions of bushels).

ACREHARV are the amounts of harvested acres (in thousands of acres).

STATDISK: Variable names are PRECIP, AVTEMP, CORNPROD, ACREHARV.

Minitab: Enter RETRIEVE 'IOWA' to get columns C1 through C5, which correspond to the five columns listed.

YEAR	PRECIP	AVTEMP	CORNPROD	ACREHARV
1950	29.1	46.04	456	9396
1951	42.3	45.60	421	9680
1952	29.8	49.08	653	10449
1953	26.0	50.37	573	10811
1954	34.3	50.54	546	10014
1955	22.7	49.25	499	10293
1956	24.2	49.39	504	9413
1957	31.6	48.77	611	9860
1958	25.6	48.10	646	9782
1959	37.9	48.03	789	12139
1960	33.9	47.21	773	12166
1961	37.4	47.82	753	9976
1962	31.3	47.53	745	9677
1963	27.0	48.80	852	10645
1964	31.5	49.61	755	9738
1965	39.9	47.80	815	9933
1966	25.3	47.51	902	10132
1967	30.4	47.78	986	11145
1968	32.7	48.77	909	9775
1969	35.0	47.18	945	9549
1970	33.8	48.03	866	10077
1971	29.4	48.51	1178	11550
1972	37.1	46.53	1230	10600
1973	42.9	49.73	1207	11280
1974	32.2	48.53	968	12100
1975	30.5	48.31	1118	12420
1976	23.7	48.26	1174	12900
1977	37.0	49.21	1092	12700
1978	34.1	46.20	1478	12850
1979	34.4	45.93	1664	13100
1980	29.3	48.86	1463	13300
1981	32.4	49.80	1731	13850
1982	41.8	46.93	1578	13150
1983	36.5	48.47	744	8550
1984	37.5	48.28	1445	12900
1985	31.6	46.67	1707	13550
1986	40.3	49.31	1627	12050
1987	33.3	51.87	1320	10150
1988	21.6	49.34	899	10700
1989	24.7	47.06	1446	12250
1990	39.4	49.88	1562	12400

Sex: 1 = male, 2 = female

Age: Years

Height: Inches

Coins: Value (in cents) of coins in possession of respondent

Keys: Number of keys in possession of respondent

Credit: Number of credit cards in possession of respondent

Pulse: Count of heartbeats in one minute

Exercise: 1 = yes, 2 = no (for vigorous exercise consisting of at least 20 minutes at least twice a week)

Smoke: 1 = yes, 2 = no

Color: 1 = yes, 2 = no ("Are you color blind?")

Hand: 1 = left-handed, 2 = right-handed, 3 = ambidextrous

 STATDISK: Variable names are STATSEX, STATAGE, STATHT, COINS, KEYS, CREDIT, PULSE, EXERCISE, SMOKE, COLOR, HAND.

Minitab: Enter RETRIEVE 'STATSURV' to get columns C1 through C11, which correspond to the 11 columns of data.

SEX	AGE	HEIGHT	COINS	KEYS	CREDIT	PULSE	EXERCISE	SMOKE	COLOR	HAND
2	19	64	0	3	0	97	2	2	2	2
2	28	67.5	100	5	0	88	1	2	2	2
1	19	68	0	0	1	69	1	2	1	2
1	20	70.5	23	4	2	67	1	2	2	3
2	18	65	35	5	5	83	1	2	2	2
2	17	63	185	6	0	77	1	2	2	2
1	18	75	0	3	0	66	2	2	2	2
2	48	64	0	3	0	60	2	2	2	3
2	19	68.75	43	3	0	78	2	2	2	3
2	17	57	35	3	0	73	1	1	2	1
2	35	63	250	10	2	8	1	2	2	2
2	18	64	178	5	10	67	1	1	2	2
1	19	72	10	2	1	55	1	2	2	2
2	28	67	90	5	0	72	1	1	2	2
2	24	62.5	0	8	14	82	1	1	2	1
2	30	63	200	1	10	70	1	2	2	2
1	21	69	0	5	0	47	1	2	2	1
1	19	68	40	4	2	63	1	1	2	2
1	19	68	73	2	0	52	2	1	2	2
1	24	68	20	2	1	55	1	2	2	2
2	22	5	500	4	1	67	1	2	2	1
1	21	69	0	2	1	75	2	2	2	1
1	19	69	0	3	3	76	1	2	2	2

(continued)

SEX	AGE	HEIGHT	COINS	KEYS	CREDIT	PULSE	EXERCISE	SMOKE	COLOR	HAND
2	19	60	35	10	0	60	2	2	2	2
1	20	69	130	3	0	84	1	2	2	2
1	30	73	62	10	1	40	1	2	2	2
1	33	74	5	7	8	64	1	2	2	2
1	19	67	0	3	0	72	2	2	2	2
1	18	70	0	5	1	72	1	2	2	2
1	20	70	0	3	5	75	1	2	1	2
2	18	76	0	3	0	80	2	2	2	2
1	20	68	32	2	4	63	1	2	2	2
1	50	72	74	8	4	72	2	2	2	2
2	20	65	14	4	4	90	1	1	2	2
1	18	68	25	2	0	70	2	2	2	2
2	18	64	0	2	0	100	2	2	2	2
2	18	64	25	1	0	69	1	2	2	2
1	22	68	0	5	5	64	1	2	2	2
2	21	64	27	2	2	80	2	2	2	2
1	41	72	76	2	0	60	2	1	2	2
1	18	68	160	3	0	66	2	1	2	2
1	21	68	34	26	0	78	1	2	2	2
2	17	60	75	3	0	60	2	1	2	2
2	40	64	20	10	5	68	1	2	2	2
1	19	74	0	1	3	72	1	2	2	2
1	19	69	0	5	0	60	1	2	2	2
2	28	68	453	5	7	88	1	1	2	2
2	28	64	0	3	0	58	1	2	2	2
2	19	63	79	6	0	88	1	2	2	1
2	41	63	100	6	3	80	2	2	2	2
1	18	71	25	2	1	61	1	2	2	2
1	18	73	181	6	0	67	2	2	2	2
1	21	71	72	3	4	60	1	1	2	2
1	18	73	0	5	0	80	1	2	2	2
1	22	69	75	12	5	60	1	1	2	2
2	22	69	0	3	5	80	1	2	2	2
1	21	72	0	4	0	68	1	1	2	2
2	26	60	25	15	0	78	1	2	2	2
1	21	72	97	2	0	54	1	2	2	2
1	20	54	0	5	5	81	2	2	2	2
2	19	65	0	8	0	67	2	2	2	2
1	22	66	30	8	4	70	1	2	2	2
1	20	76	0	6	1	63	1	2	2	2
1	19	71	0	7	0	90	1	2	1	2
1	36	73	18	11	4	70	2	2	2	2
1	20	71	0	5	1	69	2	2	1	2
1	19	71	50	7	0	69	2	1	2	2
2	18	67.5	0	4	0	75	2	1	2	2
2	19	64	0	3	1	80	1	1	2	2
1	52	71.7	51	4	5	92	2	2	2	2
2	41	68	800	7	4	72	2	2	2	2
1	20	69	0	3	3	63	1	1	2	2
1	20	72	85	3	1	60	1	2	2	2

(continued)

SEX	AGE	HEIGHT	COINS	KEYS	CREDIT	PULSE	EXERCISE	SMOKE	COLOR	HAND
2	30	63	111	5	0	78	2	2	2	2
1	21	73	77	5	0	77	1	2	2	2
1	21	70	35	5	4	71	1	1	2	2
2	34	65.5	300	4	2	15	2	2	2	2
1	19	69	36	5	0	83	2	1	2	2
1	20	69	0	2	0	80	1	2	2	2
1	20	69	0	1	1	71	1	2	2	2
2	20	66	45	4	0	86	2	2	2	2
2	36	64.5	116	3	8	65	1	2	2	2
1	19	68	52	4	2	70	1	2	2	2
2	20	67	358	7	7	76	1	2	2	2
1	19	71	0	4	0	78	1	1	2	2
1	19	72	15	10	1	63	1	2	2	3
1	19	71	1	0	0	52	1	1	2	2
1	25	71	25	4	1	78	2	2	2	1
2	29	64	26	6	3	92	2	2	2	2
1	19	81	0	4	3	48	1	2	2	2
2	17	67.5	0	3	0	68	1	2	2	2
2	19	68	0	8	1	85	2	2	2	2
1	24	73	0	14	0	64	1	2	2	1
2	19	63	0	3	4	65	2	2	2	2
2	18	64	25	3	0		1	2	2	2
2	23	69	0	3	26		2	1	2	1
1	19	60	50	4	0		1	2	2	2
1	19	72	83	6	2		1	1	2	2
2	21	67	50	3	8		1	2	2	2
1	20	74	0	6	0		2	1	2	1

Data Set 9: Passive and Active Smoke

All values are measured levels of serum cotinine (in ng/ml), a metabolite of nicotine. (When nicotine is absorbed by the body, cotinine is produced.)

NOETS: Subjects are nonsmokers who have no environmental tobacco smoke (ETS) exposure at home or work.

ETS: Subjects are nonsmokers who are exposed to environmental tobacco smoke at home or work.

SMOKERS: Subjects report tobacco use.

STATDISK: Variable names are NOETS, ETS, and SMOKERS.

Minitab: Enter RETRIEVE 'COTININE' to get columns C1, C2, and C3 which correspond to the three columns of data.

NOETS	ETS	SMOKERS
0.03	0.03	0.08
0.05	0.07	0.14
0.05	0.08	0.27
0.06	0.08	0.44
0.06	0.09	0.51
0.06	0.09	1.78
0.07	0.10	2.55
0.08	0.11	3.03
0.08	0.12	3.44
0.08	0.12	4.98
0.08	0.14	6.87
0.08	0.17	11.12
0.08	0.20	12.58
0.08	0.23	13.73

NOETS	ETS	SMOKERS
0.08	0.27	14.42
0.08	0.28	18.22
0.09	0.30	19.28
0.09	0.33	20.16
0.10	0.37	23.67
0.10	0.38	25.00
0.10	0.44	25.39
0.10	0.49	29.41
0.12	0.51	30.71
0.13	0.51	32.54
0.13	0.68	32.56
0.15	0.82	34.21
0.15	0.97	36.73
0.16	1.12	37.73
0.16	1.23	39.48
0.18	1.37	48.58
0.19	1.40	51.21
0.20	1.67	56.74
0.20	1.98	58.69
0.20	2.33	72.37
0.22	2.42	104.54
0.24	2.66	114.49
0.25	2.87	145.43
0.28	3.13	187.34
0.30	3.54	226.82
0.32	3.76	267.83
0.32	4.58	328.46
0.37	5.31	388.74
0.41	6.20	405.28
0.46	7.14	415.38
0.55	7.25	417.82
0.69	10.23	539.62
0.79	10.83	592.79
1.26	17.11	688.36
1.58	37.44	692.51
8.56	61.33	983.41

Data are based on measurements from the National Health and Nutrition Examination Survey (National Institutes of Health).

Data Set 10: Movie Ratings and Lengths

STATDISK: Variable names are MOVSTAR and MOVLEN for the star ratings and movie lengths.

Minitab: Enter RETRIEVE 'MOVIES' to get columns C1 and C2, which correspond to the star ratings and movie lengths.

	Movie	Star Rating	Length (min)	Rating
1	After the Rehearsal	4.0	72	R
2	All of Me	3.5	93	PG
3	Amityville II	2.0	110	R
4	At Close Range	3.5	115	R
5	Best Little Whorehouse in Texas	2.0	114	R
6	Beverly Hills Cop II	1.0	105	R
7	Birdy	4.0	120	R
8	The Blues Brothers	3.0	133	R
9	Breakin' 2—Electric Boogaloo	3.0	94	PG
10	Bugsy Malone	3.5	94	G
11	Cannonball Run II	0.5	108	PG
12	Choose Me	3.5	106	R
13	The Class of 1984	3.5	93	R
14	Conan the Barbarian	3.0	129	R
15	Crocodile Dundee	2.0	98	PG-13
16	Dead of Winter	2.5	100	PG-13
17	Desert Hearts	2.5	96	R
18	The Dogs of War	3.0	102	R
19	Drive, He Said	3.0	90	R
20	The Elephant Man	2.0	123	PG
21	Extreme Prejudice	3.0	104	R
22	Father of the Bride	3.0	105	PG
23	Flashdance	1.5	96	R
24	Frances	3.5	139	R
25	The Gauntlet	3.0	111	R
26	Godzilla	1.0	91	PG
27	Gremlins	3.0	111	PG
28	Hardcore	4.0	106	R
29	Heat	2.0	101	R
30	Honkeytonk Man	3.0	123	PG
31	Infra-Man	2.5	92	PG
32	Jo Jo Dancer	3.0	97	R
33	The King of Marvin Gardens	3.0	104	R
34	Last House on the Left	3.5	82	R

(continued)

Data Set 10 (continued)

	Movie	Star Rating	Length (min)	Rating
35	The Little Drummer Girl	2.0	155	R
36	Love Letters	3.5	98	R
37	The Waterdance	3.5	107	R
38	Melvin and Howard	3.5	95	R
39	Mona Lisa	4.0	104	R
40	The Muppets Take Manhattan	3.0	94	G
41	Nashville	4.0	159	R
42	1984	3.5	117	R
43	Oh, God!	3.5	104	PG
44	Ordinary People	4.0	125	R
45	Peggy Sue Got Married	4.0	103	PG-13
46	Plenty	3.5	119	R
47	Pretty in Pink	3.0	96	PG-13
48	Q	2.5	92	R
49	Ran	4.0	160	R
50	The Right Stuff	4.0	193	PG
51	The Rose	3.0	134	R
52	Scenes from a Marriage	4.0	168	PG
53	Silent Movie	4.0	88	PG
54	A Soldier's Story	2.5	99	PG
55	Spaceballs	2.5	100	PG
56	Star Wars	4.0	121	PG
57	Stepping Out	2.0	108	PG
58	Superman	4.0	144	PG
59	Sweetie	3.5	100	R
60	Testament	4.0	90	PG

Data Set 11: Weights (in grams) of a Sample of M&M Plain Candies

STATDISK: Variable names are RED, ORANGE, YELLOW, BROWN, BLUE, GREEN.

Minitab: Enter RETRIEVE 'M&M' to get columns C1 through C6, which correspond to the six columns of data.

RED	ORANGE	YELLOW	BROWN	BLUE	GREEN
0.870	0.903	0.906	0.932	0.838	0.911
0.933	0.920	0.978	0.860	0.875	1.002
0.952	0.861	0.926	0.919	0.870	0.902
0.908	1.009	0.868	0.914	0.956	0.930
0.911	0.971	0.876	0.914	0.968	0.949
0.908	0.898	0.968	0.904		0.890
0.913	0.942	0.921	0.930		0.902
0.983	0.897	0.893	0.871		
0.920		0.939	1.033		
0.936		0.886	0.955		
0.891		0.924	0.876		
0.924		0.910	0.856		
0.874		0.877	0.866		
0.908		0.879	0.858		
0.924		0.941	0.988		
0.897		0.879	0.936		
0.912		0.940	0.930		
0.888		0.960	0.923		
0.872		0.989	0.867		
0.898		0.900	0.965		
0.882		0.917	0.902		
		0.911	0.928		
		0.892	0.900		
		0.886	0.889		
		0.949	0.875		
		0.934	0.909		
			0.976		
			0.921		
			0.898		
			0.897		
			0.902		
			0.920		
			0.909		

Data Set 12: Maryland Lottery Results

STATDISK: Variable name for the 150 Pick Three numbers is PICK3.

Minitab: Enter RETRIEVE 'LOTTO' to get the 6 Lotto columns of C1, C2, . . . , C6. Enter RETRIEVE 'PICK' to get the 3 Pick Three columns of C1, C2, and C3.

	Maryland Lotto: 50 Consecutive Drawings							Maryland Pick Three: 50 Consecutive Drawings		
1	2	17	21	38	44	49	1	0	0	0
2	24	31	34	40	41	49	2	7	1	3
3	1	8	13	19	27	33	3	3	6	4
4	2	21	23	27	35	45	4	6	8	6
5	1	8	13	38	40	48	5	2	4	7
6	13	21	26	36	40	48	6	7	6	9
7	1	5	10	29	30	46	7	6	2	5
8	7	8	14	29	34	43	8	6	7	7
9	5	11	12	13	26	32	9	6	1	1
10	8	18	20	24	37	47	10	3	3	3
11	7	17	22	31	41	42	11	8	2	2
12	2	7	15	21	25	38	12	1	9	7
13	1	14	19	20	25	33	13	7	6	9
14	5	16	27	44	47	48	14	8	7	4
15	11	13	16	32	38	39	15	7	1	5
16	2	19	27	30	44	45	16	1	0	3
17	26	28	30	31	37	47	17	6	4	5
18	11	20	24	30	42	47	18	8	0	0
19	9	15	27	29	38	43	19	6	5	0
20	1	11	27	35	36	48	20	9	5	1
21	7	20	32	37	38	44	21	5	4	9
22	9	19	25	27	44	49	22	2	6	2
23	12	13	15	16	23	31	23	1	2	1
24	12	26	33	39	41	43	24	5	6	0
25	1	10	21	24	39	44	25	0	3	6
26	22	25	35	38	40	49	26	3	8	2
27	1	5	19	21	31	36	27	9	2	6
28	1	24	29	33	40	45	28	9	5	3
29	3	8	19	36	45	46	29	0	3	9
30	1	7	21	23	28	33	30	7	4	4
31	5	6	8	23	29	34	31	7	8	2
32	9	11	37	39	43	44	32	5	0	4
33	4	13	33	38	43	48	33	8	5	5
34	2	4	25	42	43	45	34	5	5	1

(continued)

Data Set 12 (continued)

	Maryland Lotto: 50 Consecutive Drawings							Maryland Pick Three: 50 Consecutive Drawings		
35	11	15	19	27	38	41	35	4	3	8
36	8	16	17	22	33	42	36	9	4	4
37	8	9	16	17	29	36	37	1	1	8
38	9	22	24	37	40	47	38	1	7	1
39	1	5	7	16	31	49	39	0	6	6
40	1	9	17	24	32	33	40	3	5	6
41	8	9	12	19	24	39	41	3	4	5
42	1	5	19	21	23	47	42	9	9	9
43	1	7	9	14	28	41	43	4	4	1
44	15	34	35	39	48	49	44	8	2	9
45	12	14	24	28	32	36	45	7	3	0
46	3	4	9	29	36	47	46	7	5	7
47	7	13	25	37	38	49	47	7	5	7
48	17	22	27	32	40	49	48	5	3	3
49	12	15	23	28	34	44	49	8	7	3
50	16	19	22	32	39	41	50	8	1	6

Data Set 13: Weights (in grams) of Quarters

STATDISK: Variable name is QUARTERS.

Minitab: Enter RETRIEVE 'QUARTERS' to get column C1, which contains the 50 weights.

5.60	5.63	5.58	5.56	5.66	5.58	5.57	5.59	5.67	5.61	5.84
5.73	5.53	5.58	5.52	5.65	5.57	5.71	5.59	5.53	5.63	5.68
5.62	5.60	5.53	5.58	5.60	5.58	5.59	5.66	5.73	5.59	5.63
5.66	5.67	5.60	5.74	5.57	5.62	5.73	5.60	5.60	5.57	5.71
5.62	5.72	5.57	5.70	5.60	5.49					

Data Set 14: Weights (in milligrams) of Bufferin Tablets

STATDISK: Variable name is ASPIRIN.

Minitab: Enter RETRIEVE 'ASPIRIN' to get column C1, which contains the 30 weights.

672.2	679.2	669.8	672.6	672.2	662.2	662.7	661.3	654.2
667.4	667.0	670.7	665.5	672.9	664.8	655.1	669.1	663.6
655.2	657.5	655.7	662.5	665.6	684.7	659.5	660.5	658.1
671.3	662.1	667.0						

Data Set 15: Aluminum Cans
Axial loads (in pounds) of aluminum cans with 0.0109 in.
thickness and 0.0111 in. thickness

STATDISK: Variable names are CANS109 and CANS111.

Minitab: Enter RETRIEVE 'CANS109' to get columns C1 through C7, which correspond to the seven columns of values listed for the cans with thickness 0.0109 in. Enter RETRIEVE 'CANS111' to get columns C1 through C7, which correspond to the seven columns of values listed for the cans with thickness 0.0111 in.

Sample	Aluminum cans 0.0109 in. Load (pounds)							Sample	Aluminum cans 0.0111 in. Load (pounds)						
1	270	273	258	204	254	228	282	1	287	216	260	291	210	272	260
2	278	201	264	265	223	274	230	2	294	253	292	280	262	295	230
3	250	275	281	271	263	277	275	3	283	255	295	271	268	225	246
4	278	260	262	273	274	286	236	4	297	302	282	310	305	306	262
5	290	286	278	283	262	277	295	5	222	276	270	280	288	296	281
6	274	272	265	275	263	251	289	6	300	290	284	304	291	277	317
7	242	284	241	276	200	278	283	7	292	215	287	280	311	283	293
8	269	282	267	282	272	277	261	8	285	276	301	285	277	270	275
9	257	278	295	270	268	286	262	9	290	288	287	282	275	279	300
10	272	268	283	256	206	277	252	10	293	290	313	299	300	265	285
11	265	263	281	268	280	289	283	11	294	262	297	272	284	291	306
12	263	273	209	259	287	269	277	12	263	304	288	256	290	284	307
13	234	282	276	272	257	267	204	13	273	283	250	244	231	266	504
14	270	285	273	269	284	276	286	14	284	227	269	282	292	286	281
15	273	289	263	270	279	206	270	15	296	287	285	281	298	289	283
16	270	268	218	251	252	284	278	16	247	279	276	288	284	301	309
17	277	208	271	208	280	269	270	17	284	284	286	303	308	288	303
18	294	292	289	290	215	284	283	18	306	285	289	292	295	283	315
19	279	275	223	220	281	268	272	19	290	247	268	283	305	279	287
20	268	279	217	259	291	291	281	20	285	298	279	274	205	302	296
21	230	276	225	282	276	289	288	21	282	300	284	281	279	255	210
22	268	242	283	277	285	293	248	22	279	286	293	285	288	289	281
23	278	285	292	282	287	277	266	23	297	314	295	257	298	211	275
24	268	273	270	256	297	280	256	24	247	279	303	286	287	287	275
25	262	268	262	293	290	274	292	25	243	274	299	291	281	303	269

Data Set 16: Old Faithful Geyser

Durations (in seconds), time intervals (in minutes) to the next eruption, and heights (in feet) of eruptions of the Old Faithful geyser in Yellowstone National Park.

 STATDISK: Variable names are DURATION, INTERVAL, GEYSERHT.

Minitab: Enter RETRIEVE 'OLDFAITH' to get columns C1, C2, and C3, which correspond to the three columns of data listed here.

DURATION	INTERVAL	HEIGHT
240	86	140
237	86	154
122	62	140
267	104	140
113	62	160
258	95	140
232	79	150
105	62	150
276	94	160
248	79	155
243	86	125
241	85	136
214	86	140
114	58	155
272	89	130
227	79	125
237	83	125
238	82	139

DURATION	INTERVAL	HEIGHT
203	84	125
270	82	140
218	78	140
226	91	135
250	89	141
245	79	140
120	57	139
267	100	110
103	62	140
270	87	135
241	70	140
239	88	135
233	82	140
238	83	139
102	56	100
271	81	105
127	74	130
275	102	135
140	61	131
264	83	135
134	73	153
268	97	155
124	67	140
270	90	150
249	84	153
237	82	120
235	81	138
228	78	135
265	89	145
120	69	130
275	98	136
241	79	150

Data courtesy of the National Park Service and research geologist Rick Hutchinson.

Appendix C: The Internet

There is now a Web site that could be used by students and/or instructors using this book. It includes information about the book, its supplements, and other items such as a collection of data sets, and links to other relevant sites. This site is listed along with other Internet sites that may be helpful. Because of the changing nature of the Internet, these sites may no longer exist or they may exist in different formats.

http://hepg.awl.com   Internet site with a link to the resources for this book

http://www.minitab.com   Select the option of "Statistics Resources" to gain access to these "on-line libraries and data sources":

- Dr. B's Wide World of Web Data
- DASL, the Data Story and Library
- Statlib
- JASA data (where JASA is the *Journal of the American Statistical Association*)
- Virtual Statistics Library

http//ticalc.org   This is the web site for Texas Instruments calculators, including information about the TI-83.

http://www2.ncsu.edu/ncsu/pams/stat/info/jse   Use the above address or simply use a search engine for *Journal of Statistics Education,* which is a "refereed *electronic* journal of postsecondary teaching of statistics." This journal exists on the Internet, but is not printed.

- Current issue
- Data Archive
- Index
- Search JSE
- JSE Information Service

http://www.geom.umn.edu/locate/chance   The *Chance* magazine database home page

http://it.stlawu.edu/~rlock/datasurf.html#textbook   This is the web site of Robin Lock at St. Lawrence University, who has developed an excellent search tool for "data surfing."

http://www.stat.ucla.edu   Select "Case Studies and Data Sets."

User groups: **sci.stat.edu**
               sci.stat.math

Appendix D: Glossary

Absolute deviation   The measure of dispersion equal to the sum of the deviations of each score from the mean, divided by the number of scores

Addition rule   Rule for determining the probability that, on a single trial, either event A occurs, or event B occurs, or they both occur

Adjusted coefficient of determination   Multiple coefficient of determination R^2 modified to account for the number of variables and sample size

Alpha (α)   Symbol used to represent the probability of a type I error. *See also* significance level.

Alternative hypothesis   Statement that is equivalent to the negation of the null hypothesis; denoted by H_1

Analysis of variance   Method of analyzing population variance in order to make inferences about the population

ANOVA   *See* analysis of variance.

Arithmetic mean   Sum of a set of scores divided by the number of scores; usually referred to as the mean

Assignable variation   Type of variation in a process that results from causes that can be identified

Attribute data   Data that can be separated into different categories distinguished by some nonnumeric characteristic

Average   Any one of several measures designed to reveal the central tendency of a collection of data

Beta (β)   Symbol used to represent the probability of a type II error

Bimodal   Having two modes

Binomial experiment   Experiment with a fixed number of independent trials, where each outcome falls into exactly one of two categories

Binomial probability formula   Expression used to calculate probabilities in a binomial experiment (see Formula 4-5 in Section 4-3)

Bivariate data   Data arranged in pairs

Bivariate normal distribution   Distribution of paired data in which, for any fixed value of one variable, the values of the other variable are normally distributed

Blinding   Procedure used in experiments whereby the subject doesn't know whether he or she is receiving a treatment or a placebo

Block   In analysis of variance, a group of similar individuals

Box-and-whisker diagram   *See* boxplot.

Boxplot   Graphical representation of the spread of a set of data

Categorical data   Data that can be separated into different categories that are distinguished by some nonnumeric characteristic

Cell   Category used to separate qualitative (or attribute) data

Census   Collection of data from every element in a population

Center line   Line used in a control chart to represent a central value of the characteristic measurements

Central limit theorem   Theorem stating that sample means tend to be normally distributed with mean μ and standard deviation $\sigma/\sqrt{n}$

Centroid   The point $(\bar{x}, \bar{y})$ determined from a collection of bivariate data

Chebyshev's theorem   Theorem that uses the standard deviation to provide information about the distribution of data

Chi-square distribution   A continuous probability distribution (first introduced in Section 6-5)

Class boundaries   Values obtained from a frequency table by increasing the upper class limits and decreasing the lower class limits by the same amount so that there are no gaps between consecutive classes

Classical approach to probability   Approach in which the probability of an event is determined by dividing the number of ways the event can occur by the total number of possible outcomes

Classical method of testing hypotheses   Method of testing hypotheses based on a comparison of the test statistic and critical values

Class marks   Midpoints of the classes in a frequency table

Class width   The difference between two consecutive lower class limits in a frequency table

Cluster sampling   Dividing the population area into sections (or clusters), then randomly selecting a few of those sections, and then choosing *all* the members from those selected sections

Coefficient of determination   Amount of the variation in y that is explained by the regression line

Combinations rule   Rule for determining the number of different combinations of selected items

Complement of an event   All outcomes in which the original event does not occur

Completely randomized design   Procedure in an experiment whereby each element is given the same chance of belonging to the different categories or treatments

Compound event Combination of simple events

Conditional probability The probability of an event, given that some other event has already occurred

Confidence coefficient *See* degree of confidence.

Confidence interval Range of values used to estimate some population parameter with a specific level of confidence; also called an interval estimate

Confidence interval limits Two numbers that are used as the high and low boundaries of a confidence interval

Confounding A situation that occurs when the effects from two or more variables cannot be distinguished from each other

Contingency table Table of observed frequencies where the rows correspond to one variable of classification and the columns correspond to another variable of classification; also called a two-way table

Continuity correction Adjustment made when a discrete random variable is being approximated by a continuous random variable (Section 5-6)

Continuous data Data resulting from infinitely many possible values that can be associated with points on a continuous scale in such a way that there are no gaps or interruptions

Continuous random variable A random variable with infinite values that can be associated with points on a continuous line interval

Control chart Any one of several types of charts (Chapter 12) depicting some characteristic of a process in order to determine whether there is statistical stability

Control group A group of subjects in an experiment who are not given a particular treatment

Control limit Boundary used in a control chart for identifying unusual points

Convenience sampling Sampling in which data are selected because they are readily available

Correlation Statistical association between two variables

Correlation coefficient Measurement of the strength of the relationship between two variables

Critical region The set of all values of the test statistic that would cause rejection of the null hypothesis

Critical value Value separating the critical region from the values of the test statistic that would not lead to rejection of the null hypothesis

Cumulative frequency Sum of the frequencies for a class and all preceding classes

Cumulative frequency table Frequency table in which each class and frequency represents cumulative data up to and including that class

Data Numbers or information describing some characteristic

Decile The nine values that divide ranked data into ten groups with approximately 10% of the scores in each group

Degree of confidence Probability that a population parameter is contained within a particular confidence interval; also called level of confidence

Degrees of freedom Number of values that are free to vary after certain restrictions have been imposed on all values

Denominator degrees of freedom Degrees of freedom corresponding to the denominator of the F test statistic

Density curve Graph of a continuous probability distribution

Dependent events Events for which the occurrence of any one event affects the probabilities of the occurrences of the other events

Dependent sample Sample whose values are related to the values in another sample

Dependent variable y variable in a regression or multiple regression equation

Descriptive statistics Methods used to summarize the key characteristics of known population data

Deviation Amount of difference between a score and the mean; expressed as $x - \bar{x}$

Discrete data Data resulting from either a finite number of possible values or a countable number of possible values

Discrete random variable Random variable with either a finite number of values or a countable number of values

Distribution-free tests Tests not requiring a particular distribution, such as the normal distribution. *See also* nonparametric tests.

Dotplot Graph in which each data value is plotted as a point (or dot) along a scale of values

Double-blind Procedure used in an experiment whereby the subject doesn't know whether he or she is receiving a treatment or placebo, and the person administering the treatment also does not know

Efficiency Measure of the sensitivity of a nonparametric test in comparison to a corresponding parametric test

Empirical rule Rule that uses standard deviation to provide information about data with a bell-shaped distribution (Section 2-5)

Estimate Specific value or range of values used to approximate some population parameter

Estimator Sample statistic (such as the sample mean $\bar{x}$) used to approximate a population parameter

Event Result or outcome of an experiment

Expected frequency Theoretical frequency for a cell of a contingency table or multinomial table

Expected value For a discrete random variable, the average value of the outcomes

Experiment Process for collecting observations

Experimental units Subjects in an experiment

Explained deviation For one pair of values in a collection of bivariate data, the difference between the predicted y value and the mean of the y values

Explained variation Sum of the squares of the explained deviations for all pairs of bivariate data in a sample

Exploratory data analysis (EDA) Branch of statistics emphasizing the investigation of data

Factor In analysis of variance, a property or characteristic that allows us to distinguish the different populations from one another

Factorial rule Rule stating that n different items can be arranged $n!$ different ways

F distribution Continuous probability distribution first introduced in Section 8-4

Finite population correction factor Factor for correcting the standard error of the mean when a sample size exceeds 5% of the size of a finite population

Five-number summary Minimum score, maximum score, median, and the first and third quartiles of a set of data

Fractiles Numbers that partition data into parts that are approximately equal in size

Frequency polygon Graphical representation of the distribution of data using connected straight-line segments

Frequency table List of categories of scores along with their corresponding frequencies

Fundamental counting rule Rule stating that, for a sequence of two events in which the first event can occur m ways and the second can occur n ways, the events together can occur a total of $m \cdot n$ ways

Goodness-of-fit test Test for how well some observed frequency distribution fits some theoretical distribution

Histogram Graph of vertical bars representing the frequency distribution of a set of data

H test *See* Kruskal-Wallis test.

Hypothesis Statement or claim about some property of a population

Hypothesis test Method for testing claims made about populations; also called test of significance

Independent events Events for which the occurrence of any one of the events does not affect the probabilities of the occurrences of the other events

Independent sample Sample whose values are not related to the values in another sample

Independent variable The x variable in a regression equation, or one of the x variables in a multiple regression equation

Inferential statistics Methods involving the use of sample data to make generalizations or inferences about a population

Influential point Point that strongly affects the graph of a regression line

Interaction In two-way analysis of variance, the effect when one of the factors changes for different categories of the other factor

Interquartile range The difference between the first and third quartiles

Interval Level of measurement of data; characterizes data that can be arranged in order and for which differences between data values are meaningful

Interval estimate *See* confidence interval.

Kruskal-Wallis test Nonparametric hypothesis test used to compare three or more independent samples; also called an H test

Least-squares property Property stating that, for a regression line, the sum of the squares of the vertical deviations of the sample points from the regression line is the smallest sum possible

Left-tailed test Hypothesis test in which the critical region is located in the extreme left area of the probability distribution

Level of confidence *See* degree of confidence.

Linear correlation coefficient Measure of the strength of the relationship between two variables

Lower class limits Smallest numbers that can actually belong to the different classes in a frequency table

Lower control limit Boundary used in a control chart to separate points that are unusually low

Lurking variable Variable that affects the variables being studied, but is not itself included in the study

Mann-Whitney U test Hypothesis test equivalent to the Wilcoxon rank-sum test for two independent samples

Marginal change For variables related by a regression equation, the amount of change in the dependent variable when one of the independent variables changes by one unit and the other independent variables remain constant

Margin of error Maximum likely (with probability $1 - \alpha$) difference between the observed sample statistic and the true value of the population parameter

Matched samples *See* paired samples.

Maximum error of estimate *See* margin of error.

Mean The sum of a set of scores divided by the number of scores

Mean deviation Measure of variation equal to the sum of the deviations of each score from the mean, divided by the number of scores

Measure of central tendency Value intended to indicate the center of the values in a collection of data

Measure of dispersion Any of several measures designed to reflect the amount of variation for a set of values

Median Middle value of a set of scores arranged in order of magnitude

Midquartile One-half of the sum of the first and third quartiles

Midrange One-half the sum of the highest and lowest scores

Mode Score that occurs most frequently

MS(error) Mean square for error; used in analysis of variance

MS(total) Mean square for total variation; used in analysis of variance

MS(treatment) Mean square for treatments; used in analysis of variance

Multimodal Having more than two modes

Multinomial experiment Experiment with a fixed number of independent trials, where each outcome falls into exactly one of several categories

Multiple coefficient of determination Measure of how well a multiple regression equation fits the sample data

Multiple comparison procedures Procedures for identifying which particular means are different, after concluding that three or more means are not all equal

Multiple regression Study of linear relationships among three or more variables

Multiple regression equation Equation that expresses a linear relationship between a dependent variable y and two or more independent variables $(x_1, x_2, \ldots, x_k)$

Multiplication rule Rule for determining the probability that event A will occur on one trial and event B will occur on a second trial

Mutually exclusive events Events that cannot occur simultaneously

Negatively skewed Skewed to the left

Nominal Level of measurement of data; characterizes data that consist of names, labels, or categories only

Nonparametric tests Statistical procedures for testing hypotheses or estimating parameters, where there are no required assumptions about the nature or shape of population distributions; also called distribution-free tests

Nonsampling errors Errors from external factors not related to sampling

Normal distribution Bell-shaped probability distribution described algebraically by Formula 5-1 in Section 5-1

np chart Control chart in which numbers of defects are plotted so that a process can be monitored

Null hypothesis Claim made about some population characteristic, usually involving the case of no difference; denoted by H_0

Numerator degrees of freedom Degrees of freedom corresponding to the numerator of the F test statistic

Numerical data Data consisting of numbers representing counts or measurements

Observational study Study in which we observe and measure specific characteristics, but don't attempt to manipulate or modify the subjects being studied

Observed frequency Actual frequency count recorded in one cell of a contingency table or multinomial table

Odds against Ratio of the probability of an event not occurring to the event occurring, usually expressed in the form of $a{:}b$ where a and b are integers having no common factors

Odds in favor Ratio of the probability of an event occurring to the event not occurring, usually expressed as the ratio of two integers with no common factors

Ogive Graphical representation of a cumulative frequency table

One-way analysis of variance Analysis of variance involving data classified into groups according to a single criterion only

Ordinal   Level of measurement of data; characterizes data that may be arranged in order, but differences between data values either cannot be determined or are meaningless

Outliers   Values that are very unusual in the sense that they are very far away from most of the data

Paired samples   Two samples that are dependent in the sense that the data values are matched by pairs

Parameter   Measured characteristic of a population

Parametric tests   Statistical procedures, based on population parameters, for testing hypotheses or estimating parameters

Pareto chart   Bar graph for qualitative data, with the bars arranged in order according to frequencies

p chart   Control chart used to monitor the proportion p for some attribute in a process

Pearson's product moment correlation coefficient   *See* linear correlation coefficient.

Percentile   The 99 values that divide ranked data into 100 groups with approximately 1% of the scores in each group

Permutations rule   Rule for determining the number of different arrangements of selected items

Pie chart   Graphical representation of data in the form of a circle containing wedges

Placebo effect   Effect that occurs when an untreated subject incorrectly believes that he or she is receiving a real treatment and reports an improvement in symptoms

Point estimate   Single value that serves as an estimate of a population parameter

Poisson distribution   Discrete probability distribution that applies to occurrences of some event over a specified interval of time, distance, area, volume, or some similar unit

Pooled estimate of p_1 and p_2   Probability obtained by combining the data from two sample proportions and dividing the total number of successes by the total number of observations

Pooled estimate of σ^2   Estimate of the variance σ^2 that is common to two populations, found by computing a weighted average of the two sample variances

Population   Complete and entire collection of elements to be studied

Positively skewed   Skewed to the right

Power of a test   Probability $(1 - \beta)$ of rejecting a false null hypothesis

Predicted values   Values of a dependent variable found by using values of independent variables in a regression equation

Prediction interval   Confidence interval estimate of a predicted value of y

Predictor variables   Independent variables in a regression equation

Probability   Measure of the likelihood that a given event will occur; expressed as a number between 0 and 1

Probability distribution   Collection of values of a random variable along with their corresponding probabilities

Probability histogram   Histogram with outcomes listed along the horizontal axis and probabilities listed along the vertical axis

Probability value   *See* P-value.

Process data   Data, arranged according to some time sequence, that measure a characteristic of goods or services resulting from some combination of equipment, people, materials, methods, and conditions

P-value   Probability that a test statistic in a hypothesis test is at least as extreme as the one actually obtained

Qualitative data   Data that can be separated into different categories distinguished by some nonnumeric characteristic

Quantitative data   Data consisting of numbers representing counts or measurements

Quartiles   The three values that divide ranked data into four groups with approximately 25% of the scores in each group

Randomized block design   Design in which a measurement is obtained for each treatment on each of several individuals matched according to similar characteristics

Random sample   Sample selected in a way that allows every member of the population to have the same chance of being chosen

Random selection   Selection of sample elements in such a way that all elements available for selection have the same chance of being selected

Random variable   Variable (typically represented by x) that has a single numerical value (determined by chance) for each outcome of an experiment

Random variation   Type of variation in a process that is due to chance; the type of variation inherent in any process not capable of producing every good or service exactly the same way every time

Range   The measure of variation that is the difference between the highest and lowest scores

Range chart   Control chart based on sample ranges; used to monitor variation in a process

Range rule of thumb   Rule stating that the range of a set of data is approximately 4 standard deviations (4*s*) wide

Rank   Numerical position of an item in a sample set arranged in order

Rank correlation coefficient   Measure of the strength of the relationship between two variables; based on the ranks of the values

Ratio   Level of measurement of data; characterizes data that can be arranged in order, for which differences between data values are meaningful, and there is an inherent zero starting point

R chart   *See* range chart.

Regression equation   Algebraic equation describing the relationship among variables

Regression line   Straight line that best fits a collection of points representing paired sample data

Relative frequency   Frequency for a class, divided by the total of all frequencies

Relative frequency approximation of probability   Estimated value of probability based on actual observations

Relative frequency histogram   Variation of the basic histogram in which frequencies are replaced by relative frequencies

Relative frequency table   Variation of the basic frequency table in which the frequency for each class is divided by the total of all frequencies

Replication   Requirement that sample sizes be large enough so that effects of chance sample variation are reduced

Residual   Difference between an observed sample *y* value and the value of *y* that is predicted from a regression equation

Right-tailed test   Hypothesis test in which the critical region is located in the extreme right area of the probability distribution

Rigorously controlled design   Design of experiment in which all factors are forced to be constant so effects of extraneous factors are eliminated

Run   Sequence of data exhibiting the same characteristic; used in runs test for randomness

Run chart   Sequential plot of individual data values over time, where one axis (usually the vertical axis) is used for the data values and the other axis (usually the horizontal axis) is used for the time sequence

Runs test   Nonparametric method used to test for randomness

Sample   Subset of a population

Sample size   Number of items in a sample

Sample space   Set of all possible outcomes or events in an experiment that cannot be further broken down

Sampling distribution of sample means   Distribution of the sample means that is obtained when we repeatedly draw samples of the same size from the same population

Sampling error   Difference between a sample result and the true population result; results from chance sample fluctuations

Scatter diagram   Graphical display of paired (*x*, *y*) data

***s* chart**   Control chart, based on sample standard deviations, that is used to monitor variation in a process

Self-selected survey   Survey in which the respondents themselves decide whether to be included

Semi-interquartile range   One-half of the difference between the first and third quartiles

Significance level   Probability of making a type I error when conducting a hypothesis test

Sign test   Nonparametric hypothesis test used to compare samples from two populations

Simple event   Experimental outcome that cannot be further broken down

Simple random sample   Sample of a particular size selected so that every possible sample of the same size has the same chance of being chosen

Simulation   Process that behaves in a way that is similar to some experiment so that similar results are produced

Single factor analysis of variance   *See* one-way analysis of variance.

Skewed   Not symmetric and extending more to one side than the other

Slope   Measure of steepness of a straight line

Spearman's rank correlation coefficient   *See* rank correlation coefficient.

SS(error)   Sum of squares representing the variability that is assumed to be common to all the populations being considered; used in analysis of variance

SS(total)   Measure of the total variation (around $\bar{\bar{x}}$) in all of the sample data combined; used in analysis of variance

SS(treatment)   Measure of the variation between the sample means; used in analysis of variance

Standard deviation   Measure of variation equal to the square root of the variance

Standard error of estimate   Measure of spread of sample points about the regression line

Standard error of the mean Standard deviation of all possible sample means $\bar{x}$

Standard normal distribution Normal distribution with a mean of 0 and a standard deviation equal to 1

Standard score Number of standard deviations that a given value is above or below the mean; also called z score

Statistic Measured characteristic of a sample

Statistically stable process Process with only natural variation and no patterns, cycles, or unusual points

Statistical process control (SPC) Use of statistical techniques such as control charts to analyze a process or its outputs so as to take appropriate actions to achieve and maintain a state of statistical control and to improve the process capability

Statistics Collection of methods for planning experiments, obtaining data, organizing, summarizing, presenting, analyzing, interpreting, and drawing conclusions based on data

Stem-and-leaf plot Method of sorting and arranging data to reveal the distribution

Stepwise regression Process of using different combinations of variables until the best model is obtained; used in multiple regression

Stratified sampling Sampling in which samples are drawn from each stratum (class)

Student t distribution *See* t distribution.

Subjective probability Guess or estimate of a probability based on knowledge of relevant circumstances

Symmetric Property of data for which the distribution can be divided into two halves that are approximately mirror images by drawing a vertical line through the middle

Systematic sampling Sampling in which every kth element is selected

t distribution Bell-shaped distribution usually associated with small sample experiments; also called the Student t distribution

10–90 percentile range Difference between the 10th and 90th percentiles

Test of homogeneity Test of the claim that different populations have the same proportion of some characteristic

Test of independence Test of the null hypothesis that for a contingency table, the row variable and column variable are not related

Test of significance *See* hypothesis test.

Test statistic Sample statistic based on the sample data; used in making the decision about rejection of the null hypothesis

Total deviation Sum of the explained deviation and unexplained deviation for a given pair of values in a collection of bivariate data

Total variation Sum of the squares of the total deviation for all pairs of bivariate data in a sample

Traditional method of testing hypotheses Method of testing hypotheses based on a comparison of the test statistic and critical values

Treatment Property or characteristic that allows us to distinguish the different populations from one another; used in analysis of variance

Treatment group Group of subjects given some treatment in an experiment

Tree diagram Graphical depiction of the different possible outcomes in a compound event

Two-tailed test Hypothesis test in which the critical region is divided between the left and right extreme areas of the probability distribution

Two-way analysis of variance Analysis of variance involving data classified according to two different factors

Two-way table *See* contingency table.

Type I error Mistake of rejecting the null hypothesis when it is true

Type II error Mistake of failing to reject the null hypothesis when it is false

Unbiased estimator Sample statistic that tends to target the population parameter that it is used to estimate

Unexplained deviation For one pair of values in a collection of bivariate data, the difference between the y coordinate and the predicted value

Unexplained variation Sum of the squares of the unexplained deviations for all pairs of bivariate data in a sample

Uniform distribution Probability distribution in which every value of the random variable is equally likely

Upper class limits Largest numbers that can belong to the different classes in a frequency table

Upper control limit Boundary used in a control chart to separate points that are unusually high

Variance Measure of dispersion found by using Formula 2-5 in Section 2-5

Variance between samples In analysis of variance, the variation among the different samples

Variation due to error *See* variation within samples.

Variation due to treatment *See* variance between samples.

Variation within samples In analysis of variance, the variation that is due to chance

Weighted mean Mean of a collection of scores that have been assigned different degrees of importance

Wilcoxon rank-sum test Nonparametric hypothesis test used to compare two independent samples

Wilcoxon signed-ranks test Nonparametric hypothesis test used to compare two dependent samples

Within statistical control *See* statistically stable process.

$\bar{x}$ **chart** Control chart used to monitor the mean of a process

y-**intercept** Point at which a straight line crosses the *y*-axis

z **score** Number of standard deviations that a given score is above or below the mean

Appendix E: Bibliography

*An asterisk denotes a book recommended for reading. Other books are recommended as reference texts.

Andrews D., and A. Herzberg. 1985. *Data: A Collection of Problems from Many Fields for the Student and Research Worker.* New York: Springer.

Beyer, W. 1991. *CRC Standard Probability and Statistics Tables and Formulae.* Boca Raton, Fla.: CRC Press.

*Campbell, S. 1974. *Flaws and Fallacies in Statistical Thinking.* Englewood Cliffs, N.J.: Prentice-Hall.

*Crossen, C. 1994. *Tainted Truth: The Manipulation of Fact in America.* New York: Simon & Schuster.

Devore, J., and R. Peck. 1986. *Statistics: The Exploration and Analysis of Data.* St. Paul, Minn.: West Publishing.

*Fairley, W., and F. Mosteller. 1977. *Statistics and Public Policy.* Reading, Mass.: Addison-Wesley.

Fisher, R. 1966. *The Design of Experiments.* 8th ed. New York: Hafner.

*Freedman, D., R. Pisani, R. Purves, and A. Adhikari. 1991. *Statistics.* 2nd ed. New York: Norton.

*Gonick, L., and W. Smith. 1993. *The Cartoon Guide to Statistics.* New York: HarperCollins.

Hoaglin, D., F. Mosteller, and J. Tukey, eds. 1983. *Understanding Robust and Exploratory Data Analysis.* New York: Wiley.

*Hollander, M., and F. Proschan. 1984. *The Statistical Exorcist: Dispelling Statistics Anxiety.* New York: Marcel Dekker.

*Holmes, C. 1990. *The Honest Truth About Lying with Statistics.* Springfield, Ill.: Charles C Thomas.

*Hooke, R. 1983. *How to Tell the Liars from the Statisticians.* New York: Marcel Dekker.

*Huff, D. 1993. *How to Lie with Statistics.* New York: Norton.

*Jaffe, A., and H. Spirer. 1987. *Misused Statistics.* New York: Marcel Dekker.

*Kimble, G. 1978. *How to Use (and Misuse) Statistics.* Englewood Cliffs, N.J.: Prentice-Hall.

Kotz, S., and D. Stroup. 1983. *Educated Guessing— How to Cope in an Uncertain World.* New York: Marcel Dekker.

*Loyer, M. 1998. *Student Solutions Manual to Accompany Elementary Statistics.* 7th ed. Reading, Mass.: Addison-Wesley.

*Moore, D. 1997. *Statistics: Concepts and Controversies.* 4th ed. San Francisco: Freeman.

*Morgan, L. 1998. *The TI-83 Companion to Accompany Elementary Statistics 7th ed.* Reading, Mass.: Addison-Wesley.

Mosteller, F., R. Rourke, and G. Thomas, Jr. 1970. *Probability with Statistical Applications.* 2nd ed. Reading, Mass.: Addison-Wesley.

Ott, L., and W. Mendenhall. 1994. *Understanding Statistics.* 6th ed. Boston: Duxbury Press.

Owen, D. 1962. *Handbook of Statistical Tables.* Reading, Mass.: Addison-Wesley.

*Paulos, J. 1988. *Innumeracy: Mathematical Illiteracy and Its Consequences.* New York: Hill and Wang.

*Reichard, R. 1974. *The Figure Finaglers.* New York: McGraw-Hill.

*Reichmann, W. 1962. *Use and Abuse of Statistics.* New York: Oxford University Press.

*Rossman, A. 1996. *Workshop Statistics: Discovery with Data.* New York: Springer.

Ryan, T., B. Joiner, and B. Ryan. 1995. *MINITAB Handbook.* 3rd ed. Boston: Duxbury.

Schaeffer, R., M. Gnanadesikan, A. Watkins, and J. Witmer. 1996. *Activity-Based Statistics: Student Guide.* New York: Springer.

Schmid, C. 1983. *Statistical Graphics.* New York: Wiley.

Simon, J. 1992. *Resampling: The New Statistics.* Belmont, Calif.: Duxbury Press.

Smith, G. 1995. *Statistical Process Control and Quality Improvement.* 2nd ed. Columbus: Merrill.

*Stigler, S. 1986. *The History of Statistics.* Cambridge, Mass.: Harvard University Press.

*Tanur, J., ed. 1989. *Statistics: A Guide to the Unknown.* 3rd ed. Belmont, Calif.: Wadsworth.

Triola, M. 1998. *Minitab Student Laboratory Manual and Workbook.* 7th ed. Reading, Mass.: Addison-Wesley.

Triola, M. 1998. *STATDISK 7.0 Student Laboratory Manual and Workbook.* 7th ed. Reading, Mass.: Addison-Wesley.

Triola, M., and L. Franklin. 1994. *Business Statistics.* Reading, Mass.: Addison-Wesley.

*Tufte, E. 1983. *The Visual Display of Quantitative Information.* Cheshire, Conn.: Graphics Press.

Tukey, J. 1977. *Exploratory Data Analysis.* Reading, Mass.: Addison-Wesley.

Utts, J. 1996. *Seeing Through Statistics.* Belmont, Calif.: Wadsworth.

Appendix F: Answers to Odd-Numbered Exercises (and ALL Review Exercises and Cumulative Review Exercises)

Section 1-2

1. Continuous   3. Discrete   5. Discrete   7. Continuous
9. Ordinal   11. Nominal   13. Interval   15. Nominal
17. Ordinal
19. Interval. Differences between the years can be determined and are meaningful, but there is no inherent starting point since time did not begin in the year zero.

Section 1-3

1. People with unlisted numbers and people without telephones are excluded.
3. A study sponsored by the citrus industry is much more likely to reach conclusions favorable to that industry.
5. Because the respondents are self-selected, the survey results are not likely to be valid at all.
7. 62% of 8% of 1875 is only 93.
9. Mothers who eat lobsters tend to be wealthier and can afford better health care.
11. A maker of shoe polish has an obvious interest in the importance of the product and there are many ways in which this could affect the survey results.
13. J. Douglas Carroll wrote in a letter to the editor of the *New York Times* that the mean of 69.5 for all males is measured from birth, whereas males don't become conductors until they have already survived for about 30 years. When this is taken into account, the mean of 73.4 years is not significant.
15. The wording of the question is biased and tends to encourage negative responses. The sample size of 20 is too small. Survey respondents are self-selected instead of being selected by the newspaper. If 20 readers respond, the percentages should be multiples of 5; 87% and 13% are not possible results.

Section 1-4

1. Observational study   3. Experiment   5. Convenience
7. Stratified   9. Random   11. Stratified   13. Cluster
15. Random

17. a. An advantage of open questions is that they provide the subject and the interviewer with a much wider variety of responses; a disadvantage is that open questions can be very difficult to analyze.
    b. An advantage of closed questions is that they reduce the chance of misinterpreting the topic; a disadvantage is that closed questions prevent the inclusion of valid responses the pollster might not have considered.
    c. Closed questions are easier to analyze with formal statistical procedures.

Section 1-5

1. 2.636   3. 3.6055513   5. 1067.1111   7. 5005
9. STATDISK: 0.838 0.875 0.870
   Minitab: 3.22 1 1

Chapter 1 Review Exercises

1. a. Continuous   b. Ratio   c. Stratified
   d. Observational study
   e. The products that use the batteries may be damaged.
2. a. Ratio   b. Ordinal   c. Ordinal   d. Interval
   e. Nominal
3. Because it is a mail survey, the respondents are self-selected and will likely include those with strong opinions about the issue. Self-selected respondents do not necessarily represent the views of everyone who invests.
4. a. Discrete   b. Continuous   c. Continuous
5. a. Systematic   b. Random   c. Cluster   d. Stratified
   e. Convenience
6. Respondents often tend to round off to a nice even number like 50.
7. The sample could be biased by excluding those who work, those who don't eat at school, those who commute, and so on.
8. The figure is very precise, but it is probably not very accurate. The use of such a precise number may incorrectly suggest that it is also accurate.

Chapter 1 Cumulative Review Exercises

1. The second version of the question is substantially less confusing because it doesn't include a double negative. One possibility for a better question:
   "Which of the following two statements do you agree with more?
   - The Nazi extermination of Jews never happened.
   - The Nazi extermination of Jews definitely happened."
2. Answer varies

Section 2-2

1. Class width: 6. Class marks: 2.5, 8.5, 14.5, 20.5, 26.5. Class boundaries: −0.5, 5.5, 11.5, 17.5, 23.5, 29.5.
3. Class width: 2.0. Class marks: 0.95, 2.95, 4.95, 6.95, 8.95. Class boundaries: −0.05, 1.95, 3.95, 5.95, 7.95, 9.95.

5.

Absences	Relative Frequency
0–5	0.195
6–11	0.205
12–17	0.190
18–23	0.200
24–29	0.210

7.

Weight (kg)	Relative Frequency
0.0–1.9	0.133
2.0–3.9	0.213
4.0–5.9	0.327
6.0–7.9	0.207
8.0–9.9	0.120

9.

Absences	Cumulative Frequency
Less than 6	39
Less than 12	80
Less than 18	118
Less than 24	158
Less than 30	200

11.

Weight (kg)	Cumulative Frequency
Less than 2.0	20
Less than 4.0	52
Less than 6.0	101
Less than 8.0	132
Less than 10.0	150

13. In Exercise 1, the numbers of absences are (approximately) evenly spread over the five classes, but the absences in Exercise 2 have frequencies that start relatively low, increase to a maximum in the middle class, then decrease to the last class.
15. 0.26–0.75

17.

Weight (lb)	Frequency
0–49	6
50–99	10
100–149	10
150–199	7
200–249	8
250–299	2
300–349	4
350–399	3
400–449	3
450–499	0
500–549	1

19.

Time (min)	Frequency
56–63	8
64–71	3
72–79	9
80–87	17
88–95	8
96–103	4
104–111	1

21. Relative frequencies for men: 0.019, 0.071, 0.118, 0.171, 0.087, 0.273, 0.142, 0.118. Relative frequencies for women: 0.010, 0.072, 0.173, 0.265, 0.042, 0.279, 0.060, 0.100. The distributions are very similar except that there are disproportionately more women in the 3.0–3.9 class and disproportionately fewer women in the 10.0–14.9 class.
23. The third guideline is violated because the class width varies. The fifth guideline is also violated because the number of classes is not between 5 and 20.

Section 2-3

1. Only one eruption out of 200 took as long as 109 min from the time of the preceding eruption, so allocating 100 min would please almost all bus trip participants.

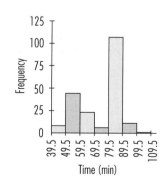

3. Although the posted limit is 30 mi/h, it appears that the police only issue tickets to those traveling at least 42 mi/h.

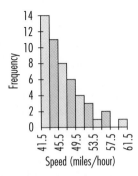

5. 570   571   577   581   583   583   584   589
   594   595   596   596   597   598   602   603

7.

9.
3	67
4	00133
4	667889
5	022334
5	778999
6	0011123333444
6	556778
7	00222233
7	56

11.
5.4	9
5.5	2333
5.5	677777888889999
5.6	00000001222333
5.6	5666778
5.7	01123334
5.7	
5.8	4

13. Networking appears to dominate as the most effective approach in getting a job.

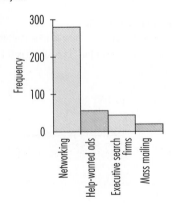

15.

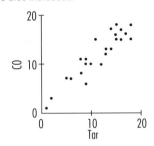

17. As the amount of tar in cigarettes increases, the amount of carbon monoxide also increases.

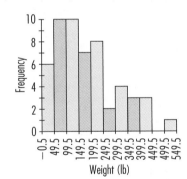

19. Skewed

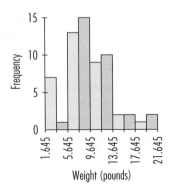

21. Bell-shaped

23. 10,000/422,000: 2.4%

25. 13,000 (from 37,000 to 24,000)

27.

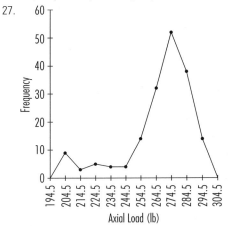

Axial Load (lb)

29. The heights of the bars will be approximately halved. Also, depending on the nature of the data, the general shape may be changed.

31.

Actors	Stem	Actresses
	2	1466678
998776532221	3	001133444555778
8876543322100	4	111249
66531	5	0
210	6	011
6	7	4
	8	0

Section 2-4

1. $\bar{x}$ = 19.3 oz; median = 19.5 oz; mode = 20 oz; midrange = 19.0 oz; no

3. $\bar{x}$ = 6.020; median = 5.780; mode: none; midrange = 6.445

5. Jefferson Valley: $\bar{x}$ = 7.15; median = 7.20; mode = 7.7; midrange = 7.10
   Providence: same results as Jefferson Valley

7. 4000 B.C.: $\bar{x}$ = 128.7; median = 128.5; mode = 131; midrange = 128.5
   150 A.D.: $\bar{x}$ = 133.3; median = 133.5; mode = 126; midrange = 133.5

9. $\bar{x}$ = 98.13; median = 98.20; mode = 97.4, 98.0, 98.2, 98.8; midrange = 97.80

11. $\bar{x}$ = 182.9; median = 150.0; mode = 140, 150, 166, 202, 204, 220; midrange = 270.0

13. 74.3 min   15. 46.7 mi/h   17. 82.0

19. a. $\bar{x}$ = 193,000; median = 206,000; mode = 236,000; midrange = 172,000
    b. Each result is increased by k.
    c. Each result is multiplied by k.
    d. 5.269 ≠ 5.286, so they're not equal.

21. 1.092

23. a. 7.0   b. 7.1   c. 7.3
    In this case, the open-ended class doesn't have too much effect on the mean. The mean is likely to be around 7.1, give or take about 0.1.

25. a. 182.9 lb   b. 171.0 lb   c. 159.2 lb
    The results differ by substantial amounts, suggesting that the mean of the original set of weights is strongly affected by extreme scores.

Section 2-5

1. range = 4.0 oz; s^2 = 1.2 oz^2; s = 1.1 oz

3. range = 3.570; s^2 = 1.030; s = 1.015

5. Jefferson Valley: range = 1.20; s^2 = 0.23; s = 0.48
   Providence: range = 5.80; s^2 = 3.32; s = 1.82

7. 4000 B.C.: range = 19.0; s^2 = 21.5; s = 4.6
   150 A.D.: range = 15.0; s^2 = 25.2; s = 5.0

9. 0.76   11. 121.8   13. 14.7   15. 4.3

17. The population of batteries with σ = 1 month are much more consistent, so they have a smaller chance of failing much earlier than expected.

19. Use $s \approx$ (tallest − shortest) ÷ 4

21. a. 68%   b. 95%   c. 50, 110

23. a. range = 128,000; s = 53,122
    b. The results will be the same.
    c. The range and standard deviation will be multiplied by k.
    d. The standard deviation of the log x values is 0.1409, but log s = 4.7253.
    e. $\bar{x}$ = 36.78; s = 0.34

25. Section 1: range = 19.0; s = 5.7
    Section 2: range = 17.0; s = 6.7
    The ranges suggest that Section 2 has less variation, but the standard deviations suggest that Section 1 has less variation.

27. I = −0.80; there is not significant skewness.

Section 2-6

1. a. −2.14   b. 5.71   c. 0.26   3. 2.43   5. 2.56; unusual
7. −2.43; no   9. 2.00; 2.50; 250 is better
11. 1.50; 1.54; 1.47; 398 is highest
13. 17   15. 60   17. 279   19. 276.5   21. 282   23. 230
25. 44   27. 80   29. 344   31. 86   33. 360   35. 150
37. a. 20   b. 272   c. 59   d. Yes; yes   e. No
39. 222.9, 311.3

Section 2-7

1.

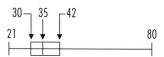

3. Actors:

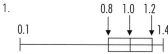

Actresses:

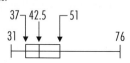

Oscar-winning actresses tend to be younger than actors.

5. Male:

Female:

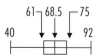

The two groups do not appear to be different.

7. Red:

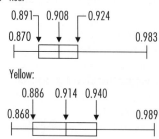

Yellow:

Red and yellow M&Ms appear to have weights that are about the same.

9.

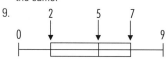

The lottery numbers appear to have a uniform distribution, as they should.

11. Select the batteries represented by the boxplot farthest to the left. They have the best combination of highest mean and lowest variation.

Chapter 2 Review Exercises

1.

Time	Frequency
235–334	4
335–434	9
435–534	11
535–634	9
635–734	9
735–834	6
835–934	8
935–1034	2
1035–1134	2

2.

Time	Relative Frequency
235–334	0.067
335–434	0.150
435–534	0.183
535–634	0.150
635–734	0.150
735–834	0.100
835–934	0.133
935–1034	0.033
1035–1134	0.033

3.

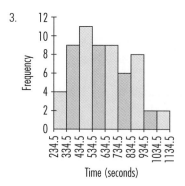

4. a. 447.5   b. 575.5   c. 7
5. $s \approx (1128 - 235)/4 = 223$ 6. 621.2; 210.7

7.
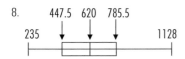

2	35 40 92
3	25 35 37 45 63 78 96 96
4	04 20 43 47 48 57 74 83 94 95
5	03 06 14 40 52 64 87
6	09 15 25 26 27 66 70 70 76 88 93
7	00 04 23 48 56 78 93 94
8	20 52 53 60 61 62 71
9	15 29 91
10	23 70
11	28

8.
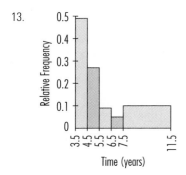

447.5 620 785.5

235   1128

9. a. 124.4   b. 121.0   c. 135   d. 148.0   e. 158.0   f. 48.6
   g. 2364.5
10. a. 54.9   b. 55.0   c. 51   d. 55.5   e. 27.0   f. 6.3
   g. 39.6   h. 51   i. 51   j. 57
11. a. No, the z score is 1.5, so 260 is within 2 standard deviations
   of the mean.
   b. −0.38
   c. About 95% of the scores should fall between 120 and 280.
   d. 220   e. 40
12. 5.15; 1.67; 8 years is not unusual because it is within 2 standard
   deviations of the mean.

13.
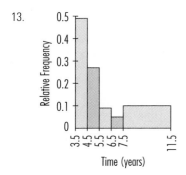

14. The score of 450 is better because its z score of −0.63 is greater
   than the other z score of −0.75.
15. The mean appears to be greater in 150 A.D., so there does appear
   to be an upward shift in the maximum skull breadth.

16.
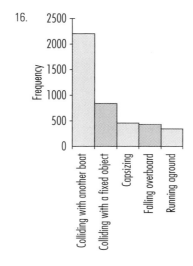

Chapter 2 Cumulative Review Exercises

1. a. $\bar{x} = 2.40$; median = 2.20; mode: 0.0, 2.0, 2.1, 2.4, 4.4;
   midrange = 2.25
   b. $s = 1.29$; $s^2 = 1.67$; range = 4.50
   c. Continuous
   d. Ratio
2. a. Mode, because the other measures of central tendencies
   require calculations that cannot (or should not) be done with
   data at the nominal level of measurement.
   b. Convenience
   c. Cluster
3. No, the 50 values should be weighted according to the corresponding populations. A weighted mean should be computed
   using the populations as weights.

Section 3-2

1. −0.2, 3/2, $\sqrt{2}$ 3. 1/2   5. 0.210   7. 0.860; yes
9. 0.320   11. 0.450
13. a. 1/365   b. 6/73 (reduced from 30/365)   15. 0.130
17. 0.340; yes   19. 0.0896   21. 0.200
23. a. bb, bg, gb, gg   b. 1/4   c. 1/2
25. b. 1/8   c. 1/8   d. 1/2
27. 4:1
29. a. 37:1   b. 35:1   c. The casino pays off at 35:1 so that it will
   make a profit, instead of paying off at 37:1, which would not
   yield a profit.
31. 3/13   33. 0.600   35. a. 4/1461   b. 400/146,097

Section 3-3

1. a. No   b. No   c. Yes   3. a. 3/5   b. 0.487
5. a. 4/13 (reduced from 16/52)   b. 2/13 (reduced from 8/52)
7. 0.698   9. 0.290   11. 0.580   13. 0.471   15. 0.740
17. 0.550   19. 0.220   21. 0.140   23. 0.180
25. a. 17/60   b. $P(A \text{ or } B) = 0.9$ c. $P(A \text{ or } B) < 0.9$
27. $P(A \text{ or } B) = P(A) + P(B) - 2P(A \text{ and } B)$

Section 3-4

1. a. Dependent   b. Independent   c. Dependent
3. 0.010   5. 0.00000369   7. 0.999992   9. 1/64
11. a. 0.0542   b. 0.799
13. 0.0138; he's probably lying, unless his rate is above 70%.
15. 0.000000250   17. 0.787   19. 0.00000410; fraud
21. 0.0728
23. a. 0.210   b. 0.0530   c. 0.593   d. 0.0637 (not 0.0638)
25. 0.100   27. 0.145
29. a. 0.679   b. 0.0541
    c. Given a positive test result, the likelihood of actually being
       HIV infected changes dramatically, depending on whether the
       subject is in the at-risk population.
31. a. 0.431   b. 0.569   33. 0.0192

Section 3-5

1. 0.240; the simulated result is too high.
3. 0.200; the result is a little low, but it's not dramatically different.
5. Approximately 16/94 = 0.170
7. a. 1.64   b. 0.98

Section 3-6

1. 720   3. 970,200   5. 720   7. 15   9. 79,833,600
11. 3,838,380   13. 1   15. $n!$ 17. a. 10,000   b. 13.9 hours
19. 1/1,000,000,000   21. 43,758   23. 5040
25. a. 3,838,380   b. 1/3,838,380   c. 3,838,379:1
27. 14,348,907   29. a. 230,300   b. 1/230,300
31. 259,459,200   33. a. 100,000   b. 1/100,000
35. 1/76,904,685; yes, if all other factors are equal.
37. 1/479,001,600; yes   39. 144
41. 2,095,681,645,538 (about 2 trillion)
43. a. Calculator: 3.0414093×10^{64}; approximation: 3.0363452×10^{64}
    b. 615

Chapter 3 Review Exercises

1. 0.650   2. 0.805   3. 0.0359   4. 0.0550   5. 0.655
6. 0.274   7. 0.711
8. 0.208; no, smokers have a greater probability of having cancer.
9. 0.877   10. 0.0483; should not speed.
11. a. 1/56   b. 336   12. 0.979   13. a. 9/19   b. 10:9   c. $5
14. 1/120   15. 0.000000531; no.   16. 1/12,870

Chapter 3 Cumulative Review Exercises

1. a. 43.2   b. 24.1   c. 0.350   d. 0.580   e. 0.00163
2. a. 1/4   b. 3/4   c. 1/16
3. a. 11.55 cm   b. 11.30 cm   c. 0.86 cm   d. 0.74 cm^2
   e. Ratio   f. 1/5   g. 0.0571   h. 0.0711   i. 2/3

Section 4-2

1. Continuous   3. Discrete
5. Probability distribution with $\mu = 0.7$, $\sigma^2 = 0.8$, $\sigma = 0.9$
7. Not a probability distribution because $\Sigma P(x) = 0.475 \neq 1$.
9. Probability distribution with $\mu = 0.8$, $\sigma^2 = 0.5$, $\sigma = 0.7$
11. Probability distribution with $\mu = 4.2$, $\sigma^2 = 4.5$, $\sigma = 2.1$
13. $-26¢$; $5.26¢$ 15. $-\$106$
17. $\mu = 1.5$, $\sigma^2 = 0.8$, $\sigma = 0.9$; minimum $= -0.3$ and maxi-
    mum $= 3.3$, but reality dictates that the minimum and maxi-
    mum are 0 and 3.
19. $\mu = 0.4$, $\sigma^2 = 0.3$, $\sigma = 0.5$
21. a. Yes   b. No, $\Sigma P(x) > 1$ c. Yes   d. Yes
23. a. $\mu = 4.5$, $\sigma = 2.872281323$ b. $\mu = 0$, $\sigma = 1$

Section 4-3

1. Not binomial; more than two outcomes   3. Binomial
5. Not binomial; more than two outcomes
7. Not binomial; trials are not independent
9. 0.243   11. 0.075   13. 0.141   15. 0.238   17. 0.3087
19. 0.5812   21. a. 0.205   b. 0.828   c. 0.120   23. 0.250
25. 0.007; study
27. The probability of 0.006 is too low to attribute to bad luck; there
    seems to be a problem in Newport.
29. 0.392 suggests that the defect rate could easily be 10%; there isn't
    sufficient evidence to conclude that the newly instituted measures
    were effective in lowering the defect rate.
31. 0.001; the results are not likely to occur by chance.
33. 0.0524   35. 0.000535

Section 4-4

1. $\mu = 32.0$, $\sigma^2 = 16.0$, $\sigma = 4.0$
3. $\mu = 267.0$, $\sigma^2 = 200.3$, $\sigma = 14.2$
5. $\mu = 12.5$, $\sigma^2 = 6.3$, $\sigma = 2.5$ 7. $\mu = 13.2$, $\sigma = 3.6$
9. a. $\mu = 1140.0$, $\sigma = 7.5$ b. Yes
11. a. $\mu = 338.0$, $\sigma = 17.1$ b. No
13. a. $\mu = 1200.0$, $\sigma = 29.0$
    b. 1272 is unusually high, probably due to quality teams or particularly weak programs on other channels.
15. a. $\mu = 745.0$, $\sigma = 25.2$ b. Yes, unusually high
17. a. $\mu = 44.6$, $\sigma = 6.1$ b. 32.4, 56.9

Section 4-5

1. 0.180   3. 0.153   5. a. 0.105   b. 0.237   c. 0.113
7. a. 0.819   b. 0.164   c. 0.0164   9. 0.594
11. a. 0.497   b. 0.348   c. 0.122   d. 0.0284   e. 0.00497
    The expected frequencies of 139, 97, 34, 8, and 1.4 compare reasonably well to the actual frequencies of 144, 91, 32, 11, and 2. The Poisson distribution does provide good results.
13. Table: 0.130; Poisson formula: 0.129

Chapter 4 Review Exercises

1. a. A random variable is a variable that has a single numerical value (determined by chance) for each outcome of an experiment.
   b. A probability distribution gives the probability for each value of the random variable.
   c. Yes, because each probability value is between 0 and 1 and the sum of the probabilities is 1.   d. $\mu = 3.4$ e. $\sigma = 0.7$
2. a. 7.5   b. 7.5   c. 2.5
   d. Yes, using the range rule of thumb, the number is usually between 2.5 and 12.5, so 15 is unusually high.   e. 0.123
3. a. 0.103   b. 0.150   c. $\mu = 3.0$, $\sigma = 1.4$
4. a. 0.394   b. 0.00347   c. 0.000142   d. 0.0375
5. a. 0.85   b. 0.84   c. Yes
6. a. $\mu = 0.230$ b. 0.0210

Chapter 4 Cumulative Review Exercises

1. a.

x	r.f.
4	0.146
5	0.247
6	0.225
7	0.382

   b. Yes, each cumulative frequency is a value between 0 and 1, and the sum of the frequencies is 1.
   c. 0.854   d. 0.143   e. 5.8   f. 1.1
   g. 5.8 games; about 175,000 hot dogs

2. a. $\bar{x} = 9.7$, $s = 2.9$ b. 0.4 isn't close to 0.0278.   c. 0.431
   d. Claim that the sample of 20 results is too small to obtain meaningful results.

Section 5-2

1. 0.6   3. 0.4   5. 0.4987   7. 0.4901   9. 0.0049
11. 0.0183   13. 0.0863   15. 0.1203   17. 0.5319
19. 0.9890   21. 0.9545   23. 0.8412   25. 0.0099
27. 0.9759   29. 1.28°   31. −0.67°   33. 1.75°
35. −2.05°
37. a. 0.92   b. 0.41   c. 0.72   d. −0.68   e. −0.23
39. The heights at $x = 0$, 1 are 0.4000 and 0.2434. Using a trapezoid, we approximate the area as 0.3217. The value from Table A-2 is 0.3413.

Section 5-3

1. 0.2123   3. 0.9861   5. 0.1844   7. 0.1379   9. 0.2238
11. 0.0038; either a very rare event has occurred or the husband is not the father.
13. 0.7823   15. a. 0.0179   b. 1343
17. 0.2562   19. 96.32%
21. a. Normal distribution   b. $\bar{x} = 0.9147$, $s = 0.0369$
    c. 0.0104
23. a. Normal distribution   b. $\bar{x} = 32.473$, $s = 5.601$
    c. 0.9099

Section 5-4

1. 66.2 in.   3. 61.0 in.   5. 9.1 years   7. 7.6 lb
9. 242 days   11. 135   13. a. 0.7054   b. 238.3 mg/100 mL
15. a. 98.74%   b. 57.8 in., 69.4 in.   17. 11.9   19. 131.6; 782

Section 5-5

1. a. 0.1554   b. 0.4918   3. a. 0.4364   b. 0.1292
5. 0.1251   7. 0.8716   9. a. 0.3446   b. 0.0838
   c. Because the original population has a normal distribution, samples of any size will yield sample means that are normally distributed.
11. 0.0668   13. 0.9345   15. 0.0069; level is acceptable
17. a. 0.2981   b. 0.0038
    c. Yes, because it is highly unlikely (with a probability of only 0.0038) that the mean will be that low because of chance.
19. a. 0.5675   b. 0.9999   c. Yes   21. 0.0057
23. ≈ 0.1949 (using $s \approx 22.4$)

Section 5-6

1. The area to the right of 35.5
3. The area to the left of 41.5
5. The area to the left of 72.5
7. The area between 124.5 and 150.5
9. a. 0.183   b. 0.1817
11. a. 0.996   b. Normal approximation is not suitable.
13. 0.1841   15. 0.1020   17. 0.0454
19. 0.0668; not *very* unusual   21. 0.0001; no   23. 0.0526; no
25. 0.0075; something seems wrong with the sample.
27. 0.6368   29. 0.2946
31. a. $0.4129 - 0.3264 = 0.0865$
    b. $0.3192 - 0.2643 = 0.0549$
    c. $0.0256 - 0.0228 = 0.0028$
    As n gets larger, the difference becomes smaller.

Chapter 5 Review Exercises

1. 64.4 in., 73.6 in.   2. 3.67%   3. 0.9876   4. 0.0081
5. 0.6940   6. a. 0.3944   b. 0.3085   c. 0.8599
    d. 0.6247   e. 0.2426   7. 78.8   8. 671   9. 0.0091
10. a. 0.9049   b. 0.8133   c. 27,563 mi

Chapter 5 Cumulative Review Exercises

1. a. 0.001   b. 0.271
    c. The requirement that $np \geq 5$ is not satisfied, indicating that the normal approximation would result in errors that are too large.
    d. 5.0   e. 2.1
    f. No, 8 is within two standard deviations of the mean and is within the range of values that could easily occur by chance.
2. a. 200.46 ms   b. 200.45 ms   c. 200.5 ms
    d. 0.29 ms   e. 0.14   f. 7.5%
    g. 3.14% (using $\bar{x} = 200.46$ and $s = 0.29$)   h. Yes

Section 6-2

1. 2.575   3. 2.33   5. a. 0.7 in.   b. 62.7 in. $< \mu <$ 64.1 in.
7. a. 1.9   b. 75.7 $< \mu <$ 79.5   9. 133.3 mm $< \mu <$ 135.7 mm
11. 5.08 yr $< \mu <$ 5.22 yr   13. 543
15. 5.600 g $< \mu <$ 5.644 g; no, the quarters become lighter as they wear down.
17. 432   19. 190.0 $< \mu <$ 193.4; no   21. 236   23. 782
25. $n = 62, \bar{x} = 9.428, s = 4.168$; 8.391 lb $< \mu <$ 10.466 lb
27. $n = 33, \bar{x} = 0.9128, s = 0.0395$; 0.8979 $< \mu <$ 0.9277; no, the sample is small.
29. 147   31. 0.5 mg

Section 6-3

1. 3.250   3. 2.528   5. a. 1.7 in.   b. 61.7 in. $< \mu <$ 65.1 in.
7. a. 6.2   b. 71.4 $< \mu <$ 83.8   9. $1066 $< \mu <$ $2506
11. 1.79 h $< \mu <$ 3.01 h   13. 3.84 $< \mu <$ 4.04
15. $29,835 $< \mu <$ $48,707
17. $\bar{x} = 7.96, s = 1.60$; 7.11 $< \mu <$ 8.81
19. $n = 100, \bar{x} = 0.9147$ g, $s = 0.0369$ g; 0.9075 g $< \mu <$ 0.9219 g; there is reasonable agreement with the result from Exercise 18.
21. The confidence interval limits are closer than they should be.

Section 6-4

1. 0.0300   3. 0.0174   5. 0.720 $< p <$ 0.780
7. 0.385 $< p <$ 0.417   9. 2401   11. 664
13. a. 64.0%   b. 61.3% $< p <$ 66.7%   15. a. 474   b. 601
17. 84.1% $< p <$ 88.7%   19. 9.53% $< p <$ 18.5%; yes
21. 2944   23. 82.8% $< p <$ 85.2%; yes
25. a. 0.00916 $< p <$ 0.0233   b. 4229
27. 45.9% $< p <$ 70.8%; yes   29. 1853   31. 89%
33. $p > 0.818$; 81.8%

Section 6-5

1. 13.120, 40.646   3. 43.188, 79.082   5. 1.7 in. $< \sigma <$ 4.4 in.
7. 11.0 $< \sigma <$ 20.4   9. 20   11. 171
13. 38.2 mL $< \sigma <$ 95.7 mL; no, the fluctuation appears to be too high.
15. 3.0 mm $< \sigma <$ 5.0 mm
17. $s = 0.47667832$; 0.33 min $< \sigma <$ 0.87 min
19. a. $s \approx (42.9 - 21.6)/4 = 5.325$
    b. Use $s = 5.6006261$; 4.33 in. $< \sigma <$ 7.78 in.   c. Yes
21. a. 98%   b. 27.0

Chapter 6 Review Exercises

1. a. 12.27 in. $< \mu <$ 13.63 in.   b. 1.79 in. $< \sigma <$ 2.86 in.
    c. 398
2. 2944   3. 5.47 yr $< \mu <$ 8.55 yr   4. 2.92 yr $< \sigma <$ 5.20 yr
5. 0.296 $< p <$ 0.344; 2.4 percentage points   6. 221
7. $n = 16, \bar{x} = 72.6625, s = 2.5809236$; 71.29 cm $< \mu <$ 74.04 cm
8. 36.5 $< \mu <$ 44.9   9. 404   10. 21.5% $< p <$ 26.5%

Chapter 6 Cumulative Review Exercises

1. a. 70.13 in.   b. 70.10 in.
    c. 69.2 in., 69.9 in., 70.0 in., 70.2 in., 70.8 in.   d. 69.80 in.

e. 6.00 in.   f. 1.82 in.^{2} g. 1.35 in.   h. 69.3 in.
i. 70.1 in.   j. 70.8 in.   k. ratio
l.

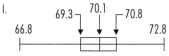

m. Answer varies, depending on the choices for the number of classes and the starting point. The histogram is approximately bell shaped.
n. 69.36 in. $< \mu <$ 70.90 in.   o. 0.97 in. $< \sigma <$ 2.13 in.
p. 303
q. The sample mean of $\bar{x} = 70.13$ in. appears to be reasonably close to the population mean of 69.7 in., but the sample standard deviation of $s = 1.35$ in. appears to be considerably smaller than the population standard deviation of $\sigma = 2.8$ in.

2. a. 0.0089   b. $0.260 < p < 0.390$
    c. Because the confidence interval limits do not contain 0.25, it is unlikely that the expert is correct.

Section 7-2

1. a. $\mu \neq 650$ mg   b. $H_0: \mu = 650$ mg   c. $H_1: \mu \neq 650$ mg
    d. Two-tailed
    e. The error of rejecting the claim that the mean equals 650 mg when it really does equal 650 mg.
    f. The error of failing to reject the claim that the mean is equal to 650 mg when it is really different from 650 mg.
    g. There is sufficient evidence to support the claim that the mean is different from 650 mg.
    h. There is not sufficient evidence to support the claim that the mean is different from 650 mg.

3. a. $p > 0.5$ b. $H_0: p \leq 0.5$ c. $H_1: p > 0.5$ d. Right-tailed
    e. The error of rejecting the claim that she is favored by at most 1/2 of the voters when she really is favored by at most 1/2 of them.
    f. The error of failing to reject the claim that she is favored by at most 1/2 of the voters when she is really favored by more than 1/2 of them.
    g. There is sufficient evidence to support the claim that she is favored by more than 1/2 of the voters.
    h. There is not sufficient evidence to support the claim that she is favored by more than 1/2 of the voters.

5. a. $p = 0.15$ b. $H_0: p = 0.15$ c. $H_1: p \neq 0.15$ d. Two-tailed
    e. The error of rejecting the claim that it gets 15% of the viewers when NBC really does get 15% of them.
    f. The error of failing to reject the claim that it gets 15% of the viewers when NBC does not get 15% of them.

g. There is sufficient evidence to warrant rejection of the claim that NBC gets 15% of the viewers.
h. There is not sufficient evidence to warrant rejection of the claim that NBC gets 15% of the viewers.

7. a. $\mu > 30$ mi/gal   b. $H_0: \mu \leq 30$ mi/gal
    c. $H_1: \mu > 30$ mi/gal   d. Right-tailed
    e. The error of rejecting the claim that the mean is at most 30 mi/gal when it really is at most 30 mi/gal.
    f. The error of failing to reject the claim that the mean is at most 30 mi/gal when it is actually greater than 30 mi/gal.
    g. There is sufficient evidence to support the claim that the mean is greater than 30 mi/gal.
    h. There is not sufficient evidence to support the claim that the mean is greater than 30 mi/gal.

9. 1.645   11. -2.33 13. ± 1.645 15. -1.96

17. There are no finite values of z for which $\alpha = 0$. The null hypothesis will never be rejected, so the test loses its usefulness.

Section 7-3

1. Test statistic: $z = 2.00$. Critical values: $z = \pm 1.96$. Reject $H_0: \mu = 75$. There is sufficient evidence to warrant rejection of the claim that the mean equals 75. P-value: 0.0456.

3. Test statistic: $z = -0.50$. Critical value: $z = -2.05$. Fail to reject $H_0: \mu \geq 2.50$. There is not sufficient evidence to support the claim that the mean is less than 2.50. P-value: 0.3085.

5. Test statistic: $z = -0.60$. Critical values: $z = \pm 1.96$. Fail to reject $H_0: \mu = 92.84$ in. There is not sufficient evidence to support the claim that the mean is different from 92.84 in. These balls do not appear to be "juiced." P-value: 0.5486.

7. Test statistic: $z = 14.61$. Critical value: $z = 2.33$. Reject $H_0: \mu \leq 0.21$. There is sufficient evidence to support the claim that the mean is greater than 0.21. Compulsive buyers do seem to get substantially higher scores than the general population. P-value: 0.0001.

9. Test statistic: $z = -38.82$. Critical value: $z = -2.33$. Reject $H_0: \mu \geq 1.39$ days. There is sufficient evidence to support the claim that the mean is less than 1.39 days. The seat belts appear to be very helpful in reducing the time spent in hospitals. P-value: 0.0001.

11. Test statistic: $z = 1.37$. Critical value: $z = 1.645$. Fail to reject $H_0: \mu \leq 0$. There is not sufficient evidence to support the claim that the course is effective (with a mean increase greater than 0). People should not take this course until it can be shown to be effective. P-value: 0.0853.

13. Test statistic: $z = -1.31$. Critical value: $z = -1.645$. Fail to reject H_0: $\mu \geq 7.5$ yr. There is not sufficient evidence to support the manager's claim that the mean is less than 7.5 yr. P-value: 0.0951.

15. Test statistic: $z = -4.22$. Critical values: $z = \pm 2.575$. There is sufficient evidence to warrant rejection of the claim that the mean is equal to 600 mg. Don't buy this cold medicine because it is not as described on the label. P-value: 0.0002.

17. Test statistic: $z = 2.44$. Critical value: $z = 2.575$. Fail to reject H_0: $\mu \leq \$40,000$. There is not sufficient evidence to support the claim that the mean is greater than $40,000. P-value: 0.0073.

19. Test statistic: $z = -8.63$. Critical value: $z = 1.645$. Fail to reject H_0: $\mu \leq 420$ hr. There is not sufficient evidence to support the claim of improved reliability. (In fact, it appears that quality has deteriorated by a significant amount.) P-value: 0.9999.

21. Test statistic: $z = -4.78$. Critical value: $z = -2.33$. Reject H_0: $\mu \geq 35$ lb. There is sufficient evidence to support the claim that the mean is less than 35 lb. P-value: 0.0001.

23. Test statistic: $z = -8.93$. P-value: 0.0000. Reject H_0: $\mu = 4.5$. There is sufficient evidence to warrant rejection of the claim that the mean is 4.5. The last digits of the river lengths suggest that those values were not measured and/or reported accurately.

25. 11.0   27. a. $\beta = 0.6178$ b. $\beta = 0.0868$

Section 7-4

1. a. ± 2.447 b. -1.796 c. 2.896
3. Test statistic: $t = -1.991$. Critical values: $t = \pm 2.201$. Fail to reject H_0: $\mu = 64.8$. There is not sufficient evidence to support the claim that the mean is not equal to 64.8.
5. Test statistic: $t = 2.23$. P-value: 0.025. Fail to reject H_0: $\mu \leq$ 12.00 kg/hectare (because the P-value is greater than 0.01). There is not sufficient evidence to support the claim that the mean is greater than 12.00 kg/hectare.
7. Test statistic: $t = 2.746$. Critical value: $t = 2.821$. Fail to reject H_0: $\mu \leq 40$ mg. There is not sufficient evidence to support the editor's belief that the mean is greater than 40 mg. She should not charge that the mean is greater than 40 mg.
9. Test statistic: $t = -3.214$. Critical values: $t = \pm 2.064$. Reject H_0: $\mu = 98.6$. There is sufficient evidence to warrant rejection of the claim that the mean is equal to 98.6.
11. Test statistic: $t = 3.344$. Critical value: $t = 2.718$. Reject H_0: $\mu \leq$ $1800. There is sufficient evidence to support the claim that the mean exceeds $1800.

13. Test statistic: $z = 1.00$. Critical value: $z = 1.96$. Fail to reject H_0: $\mu \leq \$2000$. There is not sufficient evidence to support the claim that the mean is greater than $2000. The new monitoring system will not be implemented.

15. Test statistic: $t = -3.408$. Critical value: $t = -2.602$. Reject H_0: $\mu \geq 10.00$. There is sufficient evidence to support the claim that the mean rating is less than 10.00.

17. Test statistic: $z = 0.80$. Critical value: $z = 1.28$. Fail to reject H_0: $\mu \leq 5$ yr. There is not sufficient evidence to support the claim that the mean is greater than 5 years.

19. Test statistic: $t = -0.277$. Critical value: $t = -2.132$. Fail to reject H_0: $\mu \geq 0.9085$. There is not sufficient evidence to warrant rejection of the claim that the mean is at least 0.9085 g. We cannot conclude that the package contents disagree with the claimed weight printed on the label.

21. $n = 7, \bar{x} = 10601.2857$, $s = 1026.7175$. Test statistic: $t = -1.027$. Critical values: $t = \pm 2.447$. Fail to reject H_0: $\mu = 11,000$ kWh. There is not sufficient evidence to warrant rejection of the claim that the mean is equal to 11,000 kWh.

23. $n = 15, \bar{x} = 69.5$, $s = 19.3$. Test statistic: $t = 1.264$ (1.271 is more accurate). Critical value: $t = 1.761$ (assuming that $\alpha = 0.05$). Fail to reject H_0: $\mu \leq 63.2$ s. There is not sufficient evidence to support the claim that the mean is greater than 63.2 s. Her new show does not appear to be significantly better.

25. a. $0.01 < $ P-value $ < 0.025$ b. $0.10 < $ P-value $ < 0.25$
    c. P-value $ < 0.01$ d. P-value $ = 0.0026$

27. With $z = 1.645$, the table and the approximation both result in $t = 1.833$.

Section 7-5

1. Test statistic: $z = 2.19$. Critical values: $z = \pm 1.96$. Reject H_0: $p = 0.01$. There is sufficient evidence to warrant rejection of the claim that with scanners, 1% of sales are overcharges. Based on these results, scanners appear to increase overcharges instead of helping to avoid them. P-value: 0.0286.

3. Test statistic: $z = 1.25$. Critical value: $z = 1.645$. Fail to reject H_0: $p \leq 0.04$. There is not sufficient evidence to warrant rejection of the claim that production is within control. No corrective actions appear to be necessary. P-value: 0.1056.

5. Test statistic: $z = 1.33$. Critical value: $z = 1.645$. Fail to reject H_0: $p \leq 0.5$. There is not sufficient evidence to support the claim that most students don't know what the Holocaust is. P-value: 0.0918.

7. Test statistic: $z = 9.56$. Critical value: $z = 2.575$. Reject H_0: $p \leq$ 0.043. There is sufficient evidence to support the claim that the

clients are audited at a rate that is above the 4.3% rate for the general population. This tax company is not a good choice for those who earn above $100,000 because the high audit rate suggests problems that are attracting IRS scrutiny. *P*-value: 0.0001.

9. Test statistic: $z = -1.79$. Critical value: $z = -1.645$. Reject H_0: $p \geq 0.5$. There is sufficient evidence to support the claim that fewer than half of all adults are annoyed by the violence shown on television. *P*-value: 0.0367.

11. Test statistic: $z = 5.96$. Critical values: $z = \pm 2.33$. Reject H_0: $p = 0.10$. There is sufficient evidence to warrant rejection of the claim that 10% of drivers use telephones. *P*-value: 0.0002.

13. Test statistic: $z = 0.99$. Critical values: $z = \pm 1.645$. Fail to reject H_0: $p = 0.08$. There is not sufficient evidence to warrant rejection of the claim that 8% of Seldane users experience drowsiness. *P*-value: 0.3222.

15. Test statistic: $z = -4.37$. Critical value: $z = -2.33$. Reject H_0: $p \geq 0.10$. There is sufficient evidence to support the claim that fewer than 10% of medical students prefer pediatrics. *P*-value: 0.0001.

17. Test statistic: $z = 0.83$. Critical value: $z = 1.28$. Fail to reject H_0: $p \leq 0.5$. There is not sufficient evidence to support the claim that the majority are smoking a year after therapy. Although many of the smokers did not stop with the nicotine patch therapy, there are many who did stop smoking, so the therapy is effective for them. *P*-value: 0.2033.

19. Test statistic: $z = -14.00$. Critical values: $z = \pm 2.575$. Reject H_0: $p = 0.5$. There is sufficient evidence to reject the claim that half of all adults eat their fruitcakes. Fruitcake producers should consider changes. *P*-value: 0.0002.

21. From the table of binomial probabilities with $n = 15$ and $p = 0.1$, we get $P(0) = 0.206$. If p is really 0.1, then there is a good chance (0.206) that none of the 15 residents will believe that the mayor is doing a good job. Because that result could easily occur by chance, there is not sufficient evidence to reject the reporter's claim. We would reject the claim only if the probability were found to be less than 0.05.

23. a. $\hat{p} = 21/100$. Test statistic: $z = 0.25$. Critical values: $z = \pm 1.96$. Fail to reject H_0: $p = 0.20$. There is not sufficient evidence to warrant rejection of the claim that 20% of the candies are red.

b. The *P*-value of 0.8026 exceeds the significance level of 0.05, so fail to reject H_0: $p = 0.20$. There is not sufficient evidence to warrant rejection of the claim that 20% of the candies are red.

c. $0.130 < p < 0.290$; Because the confidence interval limits contain the claimed value of 0.20, fail to reject H_0: $p = 0.20$.

There is not sufficient evidence to warrant rejection of the claim that 20% of the candies are red.

Section 7-6

1. a. 1.735, 23.589   b. 45.642   c. 10.851

3. Test statistic: $\chi^2 = 114.586$. Critical values: $\chi^2 = 57.153$, 106.629. Reject H_0: $\sigma = 43.7$ ft. There is sufficient evidence to support the claim that the standard deviation is different from 43.7 ft. Because the standard deviation seems to be larger than it was, the new production method results in less consistency and is therefore worse.

5. Test statistic: $\chi^2 = 9.016$. Critical value: $\chi^2 = 13.848$. Reject H_0: $\sigma \geq 6.2$ min. There is sufficient evidence to support the claim that there is lower variation with a single line. Although the waiting times are more consistent, the wait is not necessarily shorter.

7. More consistent implies a lower standard deviation. Test statistic: $\chi^2 = 24.576$. Critical value: $\chi^2 = 43.188$. Reject H_0: $\sigma \geq 0.75$ lb. There is sufficient evidence to support the claim that the weights are more consistent than in the past.

9. Test statistic: $\chi^2 = 44.800$. Critical value: $\chi^2 = 51.739$. Reject H_0: $\sigma \geq 0.15$ oz. There is sufficient evidence to support the claim that the new machine fills bottles with lower variation. Based on the sample variation, the new machine should be purchased.

11. Test statistic: $\chi^2 = 69.135$. Critical value: $\chi^2 = 67.505$ (approximately). Reject H_0: $\sigma \leq 19.7$. There is sufficient evidence to support the claim that women have more variation.

13. $s = 2.340$. Test statistic: $\chi^2 = 15.628$. Critical values: $\chi^2 = 12.401, 39.364$ (assuming $\alpha = 0.05$). Fail to reject H_0: $\sigma = 2.9$ in. There is not sufficient evidence to support the claim that the standard deviation is different from 2.9 in.

15. a. $0.01 < P\text{-value} < 0.02$ b. $P\text{-value} < 0.005$
c. $0.005 < P\text{-value} < 0.01$

17. a. Estimated values: 74.216, 129.565; Table A-4 values: 74.222, 129.561
b. $\chi^2 = 117.093, 184.690$

Chapter 7 Review Exercises

1. a. $t = -1.833$ b. $z = \pm 2.575$ c. $\chi^2 = 8.907, 32.852$
2. a. $z = \pm 1.645$ b. $\chi^2 = 19.023$ c. $z = -2.33$
3. a. $\mu = \$90,000$ b. Two-tailed
c. Rejecting the claim that the mean equals $90,000 when it actually does equal that amount.
d. Failing to reject the claim that the mean equals $90,000 when it is actually different from that amount.
e. 0.05

4. a. $\sigma \geq 15$ s
   b. Left-tailed
   c. Rejecting the claim that the standard deviation is at least 15 s when it actually is at least 15 s.
   d. Failing to reject the claim that the standard deviation is at least 15 s when it is actually less than 15 s.
   e. 0.01

5. Test statistic: $z = 3.14$. Critical value: $z = 2.33$. Reject H_0: $p \leq 0.5$. There is sufficient evidence to support the claim that the majority of gun owners favor stricter gun control laws.

6. Test statistic: $z = -2.73$. Critical value: $z = -2.33$. Reject H_0: $\mu \geq 5.00$. There is sufficient evidence to support the claim that the mean radiation dosage is below 5.00 milliroentgens.

7. Test statistic: $\chi^2 = 20.429$. The critical value is between 14.954 and 22.164, but it should be close to 18.559. Fail to reject H_0: $\sigma \geq 2.50$ milliroentgens. There is not sufficient evidence to support the claim that the standard deviation is less than 2.50 milliroentgens. If the standard deviation is too high, there may be some machines that emit dangerously high levels of radiation.

8. Test statistic: $z = -2.36$. Critical value: $z = -2.33$. Reject H_0: $p \geq 0.10$. There is sufficient evidence to support the claim that the true percentage is less than 10%. The sample of 8% is significantly below 10%, so the phrase "almost 1 out of 10" is not justified.

9. Test statistic: $t = -2.321$. Critical value: $t = -2.500$. Fail to reject H_0: $\mu \geq 12$ oz. There is not sufficient evidence to support the claim that the company is cheating consumers.

10. $\bar{x} = 23.29$, $s = 4.53$. Test statistic: $t = 3.240$. Critical values: $t = \pm 2.861$. Reject H_0: $\mu = 20$ mg. There is sufficient evidence to warrant rejection of the claim that the mean equals 20 mg. These pills are unacceptable because their contents do not correspond to the label.

Chapter 7 Cumulative Review Exercises

1. a. 0.2358
   b. 0.0020

2. a. $\bar{x} = 124.23$, $s = 22.52$
   b. Test statistic: $t = 1.675$. Critical values: $t = \pm 2.132$. Fail to reject H_0: $\mu = 114.8$. There is not sufficient evidence to warrant rejection of the claim that the mean is equal to 114.8.
   c. $112.23 < \mu < 136.23$; yes
   d. Test statistic: $\chi^2 = 44.339$. Critical values: $\chi^2 = 6.262$, 27.488. Reject H_0: $\sigma = 13.1$. There is sufficient evidence to warrant rejection of the claim that the standard deviation is equal to 13.1.

e. Although the mean does not appear to change by a significant amount, the variation among the values appears to be significantly larger.

3. a. 6.3
   b. 2.2
   c. 0.0024 (or 0.0019 if unrounded statistics are used)
   d. Based on the low probability value in part (c), reject H_0: $p = 0.25$. There is sufficient evidence to reject the claim that the subject made random guesses.
   e. 423

4. a.

9.	0005
10.	00000
11.	00055555
12.	05555
13.	000000005555555
14.	00555
15.	055555
16.	00055
17.	0

   b. The distribution appears to be approximately normal.
   c. A histogram could also be used to test for a normal distribution.
   d. The last digits are all 0s and 5s, suggesting that the measurements were rounded to the nearest half inch. It is unlikely that the bears were asked their head-length measurements and they reported rounded results.

Section 8-2

1. a. $\bar{d} = 2.8$
   b. $s_d = 3.6$
   c. $t = 1.757$
   d. $t = \pm 2.776$

3. $-1.6 < \mu_d < 7.2$

5. Test statistic: $t = -1.718$. Critical values: $t = \pm 2.262$. Fail to reject H_0: $\mu_d = 0$. There is not sufficient evidence to support the claim that the course has an effect.

7. $9.5 < \mu_d < 27.6$

9. Test statistic: $t = 3.036$. Critical value: $t = 1.895$. Reject H_0: $\mu_d \leq 0$. There is sufficient evidence to support the claim that the sensory measurements are lower after hypnotism.

11. Test statistic: $t = 2.631$. Critical values: $t = \pm 2.977$. Fail to reject H_0: $\mu_d = 0$. There is not sufficient evidence to warrant rejection of the claim that there is no difference between the two times.

13. Test statistic: $t = 2.301$. Critical values: $t = \pm 2.262$. Reject H_0: $\mu_d = 0$. There is sufficient evidence to warrant rejection of the claim that the mean pretraining weight equals the mean post-training weight.

15. $-1.40 < \mu_d < -0.17$

17. Test statistic: $t = -0.41$. P-value $= 0.69$. Fail to reject H_0: $\mu_d = 0$. There is not sufficient evidence to support the claim that astemizole has an effect. Based on the available evidence, do not take astemizole because it doesn't appear to help.

19. $-48.8 < \mu_d < 33.8$; yes

21. The hypothesis test results are not affected. If each of the original values is multiplied by a constant (such as 0.001), the confidence interval limits will also be multiplied by the same constant.

Section 8-3

1. Test statistic: $z = 0.79$. Critical values: $z = \pm 1.96$. Fail to reject H_0: $\mu_1 = \mu_2$ There is not sufficient evidence to warrant rejection of the claim that the two samples come from populations with the same mean.

3. $-3 < \mu_1 - \mu_2 < 7$; yes. The control group and experimental group do not appear to be significantly different.

5. Test statistic: $z = 2.88$. Critical value: $z = 2.33$. Reject H_0: $\mu_1 \leq \mu_2$. There is sufficient evidence to support the claim that the mean amount recalled is less for those exposed to stress (but we cannot conclude that it was *caused* by stress).

7. Test statistic: $z = 2.06$. Critical values: $z = \pm 2.575$. Fail to reject H_0: $\mu_1 = \mu_2$. There is not sufficient evidence to warrant rejection of the claim that the two populations have the same mean. Advertise, because there does not appear to be a difference.

9. Test statistic: $z = 4.91$. Critical value: $z = 1.645$. Reject H_0: $\mu_1 \leq \mu_2$. There is sufficient evidence to support the claim that student cars are older than faculty cars.

11. Test statistic: $z = 8.56$. Critical value: $z = 2.33$. Reject H_0: $\mu_1 \leq \mu_2$. There is sufficient evidence to support the claim that older men come from a population with a mean weight that is less than the mean for the population of younger men.

13. Test statistic: $z = 2.64$. Critical values: $z = \pm 1.645$. Reject H_0: $\mu_1 = \mu_2$. There is sufficient evidence to warrant rejection of the claim that American and TWA have the same mean salary for flight attendants. Because the means are significantly different, salary might well be an important factor in selecting an airline for which to work.

15. Test statistic: $z = -0.23$. Critical values: $z = \pm 1.96$ (assuming $\alpha = 0.05$). Fail to reject H_0: $\mu_1 = \mu_2$. There is not sufficient evi-

dence to warrant rejection of equality of the two population means. The discrepancy does not appear to be significant.

17. Group aged 18–24: $n = 37$, $\bar{x} = 98.224324$, $s = 0.659716$. Group aged 25 and over: $n = 56$, $\bar{x} = 98.057143$, $s = 0.635283$.

a. Test statistic: $z = 1.21$. Critical values: $z = \pm 1.96$. Fail to reject H_0: $\mu_1 = \mu_2$. There is not sufficient evidence to warrant rejection of the claim that both age groups have the same mean body temperature.

b. P-value $= 0.2262$. Because the P-value is not less than or equal to the significance level of $\alpha = 0.05$, conclusions are the same as in part (a).

c. $-0.10 < \mu_1 - \mu_2 < 0.44$; because the interval contains 0, the conclusions are the same as in part (a).

d. All three parts suggest that there is not sufficient evidence to warrant rejection of equal means.

19. The boxplot for the 0.0111 in. cans uses a 5-number summary of 205, 275, 285, 294, 317. Before the deletion of the outlier, the 5-number summary was 205, 275, 285, 295, 504, so the only major change is shortening of the right whisker. In the hypothesis test, the test statistic changes from -5.48 to -5.66, so the conclusions will be the same. After deleting the outlier of 504, the confidence interval limits change from $(-21.6, -7.8)$ to $(-19.5, -7.3)$, so those changes are not substantial.

21. a. 50/3   b. 2/3   c. 52/3

e. The range of the *x-y* values equals the range of the *x* values plus the range of the *y* values.

Section 8-4

1. Test statistic: $F = 2.2222$. Critical value: $F = 2.2090$. Reject H_0: $\sigma_1^2 = \sigma_2^2$. There is sufficient evidence to support the claim that the two population variances are different.

3. Test statistic: $F = 1.2949$. Critical value: $F = 1.6928$ (approximately). Fail to reject H_0: $\sigma_1 = \sigma_2$. There is not sufficient evidence to support the claim that the samples come from populations with different standard deviations.

5. Test statistic: $F = 1.6804$. Critical value of F is less than 1.5330. Reject H_0: $\sigma_1 \geq \sigma_2$. There is sufficient evidence to support the claim that the older men come from a population with a standard deviation less than that for the younger men.

7. Test statistic: $F = 1.0722$. Critical value: $F = 1.7444$. Fail to reject H_0: $\sigma_1 = \sigma_2$. There is not sufficient evidence to warrant rejection of the claim that the standard deviations for American and TWA are the same.

9. Test statistic: $F = 1.3478$. Critical value: $F = 2.6171$. Fail to reject H_0: $\sigma_1^2 = \sigma_2^2$. There is not sufficient evidence to warrant

rejection of the claim that both groups come from populations with the same variance.

11. Test statistic: $F = 1.2478$. Critical value: $F = 2.5342$. Fail to reject H_0: $\sigma_1 = \sigma_2$. There is not sufficient evidence to warrant rejection of the claim that both groups come from populations with the same standard deviation.

13. a. Reject H_0: $\sigma_1^2 = \sigma_2^2$.
    b. Fail to reject H_0: $\sigma_1^2 = \sigma_2^2$.
    c. Reject H_0: $\sigma_1^2 = \sigma_2^2$.

15. a. $F_L = 0.2484$, $F_R = 4.0260$
    b. $F_L = 0.2315$, $F_R = 5.5234$
    c. $F_L = 0.1810$, $F_R = 4.3197$
    d. $F_L = 0.3071$, $F_R = 4.7290$
    e. $F_L = 0.2115$, $F_R = 3.2560$

17. a. It doesn't change.
    b. It doesn't change.
    c. It doesn't change.

Section 8-5

1. F-test results: Test statistic is $F = 1.5625$. Critical value: $F = 2.8801$. Fail to reject $\sigma_1^2 = \sigma_2^2$. Test of means: Test statistic is $t = -0.990$. Critical values: $t = \pm2.048$. Fail to reject H_0: $\mu_1 = \mu_2$. There is not sufficient evidence to warrant rejection of the claim that the two populations have equal means.

3. $-15 < \mu_1 - \mu_2 < 5$

5. $-4754 < \mu_1 - \mu_2 < 1148$; yes. There is not sufficient evidence to warrant rejection of equality of the two means.

7. F-test results: Test statistic is $F = 1.0000$, so fail to reject H_0: $\sigma_1^2 = \sigma_2^2$. Test of means: Test statistic is $t = -3.075$. Critical values: $t = \pm2.878$. Reject H_0: $\mu_1 = \mu_2$. There is sufficient evidence to warrant rejection of equality of the two population means. It appears that obsessive-compulsive disorders have a biological basis.

9. $-17.16 < \mu_1 - \mu_2 < 260.40$

11. F-test results: Test statistic is $F = 2.9459$. Critical value: $F = 3.0074$. Fail to reject H_0: $\sigma_1^2 = \sigma_2^2$. Test of means: Test statistic is $t = -1.099$. Critical values: $t = \pm2.052$. Fail to reject H_0: $\mu_1 = \mu_2$. There is not sufficient evidence to warrant rejection of the claim that red and orange M&Ms have the same mean weight. No corrective action is necessary.

13. F-test results: Test statistic is $F = 1.5155$. Critical value: $F = 2.1359$. Fail to reject H_0: $\sigma_1^2 = \sigma_2^2$. Test of means: Test statistic is $t = 0.261$. Critical values: $t = \pm1.96$. Fail to reject H_0: $\mu_1 = \mu_2$.

There is not sufficient evidence to warrant rejection of the claim that there is no difference in length between movies rated R and movies with G, PG, or PG-13 ratings.

15. Filtered: $n = 21$, $\bar{x} = 12.857$, $s = 3.071$. Nonfiltered: $n = 8$, $\bar{x} = 15.625$, $s = 1.188$. F-test results: Test statistic is $F = 6.6835$. Critical value: $F = 4.4667$ (assuming $\alpha = 0.05$). Reject H_0: $\sigma_1^2 = \sigma_2^2$. Test of means: Test statistic is $t = -3.500$. Critical value: $t = -1.895$ (assuming $\alpha = 0.05$). Reject H_0: $\mu_1 \geq \mu_2$. There is sufficient evidence to support the claim that filters are effective in reducing the mean carbon monoxide level in cigarettes.

17. F-test results: The test statistic doesn't exist, because it would be calculated with division by 0. However, the standard deviations do not appear to be equal, so reject H_0: $\sigma_1^2 = \sigma_2^2$. Test of means: Test statistic is $t = 15.322$. Critical values: $t = \pm2.080$. Reject H_0: $\mu_1 = \mu_2$. There is sufficient evidence to warrant rejection of the claim that the two groups come from populations with the same mean. (There is a serious question about the requirement that the samples come from normally distributed populations, because the 22 scores in the second sample appear to be the same.)

19. The number of degrees of freedom changes from 7 to 26. The test statistic doesn't change, the critical value changes from $t = -1.895$ to $t = -1.706$, but the conclusions remain the same. The confidence interval limits change from $(-13.1, -8.3)$ to $(-12.8, -8.6)$.

Section 8-6

1. a. 8/15
   b. -0.58
   c. ±1.96
   d. 0.5620

3. Test statistic: $z = -2.82$. Critical value: $z = -1.645$. Reject H_0: $p_1 \geq p_2$. There is sufficient evidence to support the claim that warmer surgical patients recover better. It appears that surgical patients should be routinely warmed.

5. $-0.0144 < p_1 - p_2 < 0.0086$; yes. There does not appear to be a significant difference between the two rates of violent crimes.

7. Test statistic: $z = 12.86$. Critical value: $z = 2.575$. Reject H_0: $p_1 \leq p_2$. There is sufficient evidence to support the claim that vinyl gloves have a larger virus leak rate than latex gloves.

9. Test statistic: $z = -1.92$. Critical values: $z = \pm1.96$. Fail to reject H_0: $p_1 = p_2$. There is not sufficient evidence to warrant rejection of the claim that there is no difference between the pro-

portions of Democrats and Republicans who believe in government regulation of airfares. Don't use different approaches for Democrats and Republicans.

11. Test statistic: $z = 3.86$. Critical value: $z = 2.33$. Reject H_0: $p_1 \leq p_2$. There is sufficient evidence to support the claim that arson is committed by a proportion of drinkers that is greater than the proportion of drinkers convicted of fraud. It does seem reasonable that drinking would have an effect on the type of crime.

13. Test statistic: $z = -3.81$. Critical values: $z = \pm1.96$. Reject H_0: $p_1 = p_2$. There is sufficient evidence to warrant rejection of the claim that the proportions of voters in the two age brackets are the same.

15. Test statistic: $z = 1.07$. Critical value: $z = 1.645$. Fail to reject H_0: $p_1 \leq p_2$. There is not sufficient evidence to support the claim that the rate of severe injuries is lower for children wearing seat belts. Based on these results, no action should be taken.

17. Test statistic: $z = 0.84$. Critical values: $z = \pm1.96$. Fail to reject H_0: $p_1 - p_2 = 0.1$. There is not sufficient evidence to warrant rejection of the claim that when women are exposed to glycol ethers, their percentage of miscarriages is 10 percentage points more than the percentage for women not exposed to glycol ethers.

19. 2135

Chapter 8 Review Exercises

1. Test statistic: $z = 2.41$. Critical value: $z = 1.645$. Reject H_0: $p_1 \leq p_2$. There is sufficient evidence to support the claim that the percentage of people who stop after first seeing someone else being helped is greater than the percentage of people who stop without first seeing someone else being helped.

2. a. Test statistic: $t = -3.847$. Critical value: $t = -1.796$. Reject H_0: $\mu_d \geq 0$. There is sufficient evidence to support the claim that the Dozenol tablets are more soluble after the storage period.
   b. $12.4 < \mu_d < 45.6$

3. a. F-test results: Test statistic is $F = 1.3010$. Critical value of F is between 3.4296 and 3.5257. Fail to reject H_0: $\sigma_1^2 = \sigma_2^2$. Test of means: Test statistic is $t = -4.337$. Critical values: $t = \pm2.074$. Reject H_0: $\mu_1 = \mu_2$. There is sufficient evidence to warrant rejection of the claim that the mean amount of acetaminophen is the same in each brand.
   b. $15.1 < \mu_1 - \mu_2 < 42.9$

4. Test statistic: $F = 1.3786$. Critical value: $F = 1.8363$. Fail to reject H_0: $\sigma_1 = \sigma_2$. There is not sufficient evidence to warrant rejection of the claim that the two groups come from populations with the same standard deviation.

5. Test statistic: $z = 0.78$. Critical values: $z = \pm2.33$. Fail to reject H_0: $\mu_1 = \mu_2$. There is not sufficient evidence to warrant rejection of the claim that the mean score for women is equal to the mean score for men.

6. Test statistic: $z = 3.67$. Critical value: $z = 1.645$. Reject H_0: $p_1 \leq p_2$. There is sufficient evidence to support the claim that the question was answered correctly by a greater proportion of prepared students.

7. Test statistic: $t = 1.185$. Critical values: $t = \pm2.262$. Fail to reject H_0: $\mu_d = 0$. There is not sufficient evidence to warrant rejection of the claim that the position has no effect.

8. Test statistic: $z = 0.96$. Critical value: $z = 2.33$. Fail to reject H_0: $\mu_1 \leq \mu_2$. There is not sufficient evidence to support the claim that the 10-day program results in scores with a lower mean.

9. F-test results: Test statistic is $F = 12.0013$. Critical value: $F = 2.5598$. Reject H_0: $\sigma_1^2 = \sigma_2^2$. Test of means: Test statistic is $t = 2.444$. Critical values: $t = \pm2.110$. Reject H_0: $\mu_1 = \mu_2$. There is sufficient evidence to warrant rejection of the claim that both systems have the same mean.

10. a. F-test results: Test statistic is $F = 1.1352$. Critical F value is between 3.4296 and 3.5257. Fail to reject H_0: $\sigma_1^2 = \sigma_2^2$. Test of means: Test statistic is $t = -1.045$. Critical values: $t = \pm2.074$. Fail to reject H_0: $\mu_1 = \mu_2$. There is not sufficient evidence to warrant rejection of the claim that both time periods have the same mean.
    b. $-13.670 < \mu_1 - \mu_2 < 4.510$

Chapter 8 Cumulative Review Exercises

1. a. 0.0707
   b. 0.369
   c. 0.104
   d. 0.054
   e. Test statistic: $z = -2.52$. Critical value: $z = -1.645$. Reject H_0: $p_1 \geq p_2$. There is sufficient evidence to support the claim that women are ticketed at a lower rate. We cannot conclude that men generally speed more, only that they are ticketed more. Perhaps men drive more or perhaps they are more likely to get a ticket when speeding.

2. a. Test statistic: $z = 1.95$. Critical values: $z = \pm1.96$. Fail to reject H_0: $\mu = 0$. There is not sufficient evidence to warrant rejection of the claim that the sample comes from a population with a mean equal to 0.

b. Test statistic: $z = -1.87$. Critical values: $z = \pm 1.96$. Fail to reject H_0: $\mu = 0$. There is not sufficient evidence to warrant rejection of the claim that the sample comes from a population with a mean equal to 0.

c. Test statistic is $z = 2.68$. Critical values: $z = \pm 1.96$. Reject H_0: $\mu_1 = \mu_2$. There is sufficient evidence to warrant rejection of the claim that both shifts manufacture scales with the same mean error.

d. $0.0\,g < \mu < 2.4\,g$

e. $-2.9\,g < \mu < 0.1\,g$

f. $0.7\,g < \mu_1 - \mu_2 < 4.5\,g$

g. 234

Section 9-2

1. a. Significant linear correlation
   b. Significant linear correlation
   c. No significant linear correlation

3. a. Significant linear correlation
   b. $n = 5$, $\Sigma x = 25$, $\Sigma x^2 = 163$, $(\Sigma x)^2 = 625$, $\Sigma xy = 489$, $r = 0.997$

5. Test statistic: $r = 0.993$. Critical values: $r = \pm 0.707$. Reject the claim of no significant linear correlation. There does appear to be a linear correlation between chest size and weight. The results do not change if the chest measurements are converted to feet.

7. Test statistic: $r = 0.842$. Critical values: $r = \pm 0.707$. Reject the claim of no significant linear correlation. There does appear to be a linear correlation between weight of discarded plastic and household size.

9. Test statistic: $r = 0.127$. Critical values: $r = \pm 0.707$. Fail to reject the claim of no significant linear correlation. There does not appear to be a linear correlation between weight of discarded food and household size.

11. Test statistic: $r = -0.069$. Critical values: $r = \pm 0.707$. Fail to reject the claim of no significant linear correlation. There does not appear to be a linear correlation between BAC level and the age of the person.

13. Test statistic: $r = 0.870$. Critical values: $r = \pm 0.279$. Reject claim of no significant linear correlation. There does appear to be a linear correlation between the interval after an eruption and the duration of the eruption.

15. Test statistic: $r = 0.961$. Critical values: $r = \pm 0.361$ (approximately). Reject the claim of no significant linear correlation. There does appear to be a linear correlation between the amounts of tar and nicotine in cigarettes. Laboratory expenses could be reduced by measuring only one of the variables.

17. Test statistic: $r = 0.265$. Critical values: $r = \pm 0.312$ (approximately). Fail to reject the claim of no significant linear correlation. There does not appear to be a linear correlation between the annual precipitation amount and the amount of corn production.

19. With a linear correlation coefficient very close to 0, there does not appear to be a correlation, but the conclusion suggests that there is a correlation.

21. Although there is no *linear* correlation, the variables may be related in some other *nonlinear* way.

23. In parts (a), (b), and (c), the value of r does not change.

25. a. ± 0.272
   b. ± 0.189
   c. -0.378
   d. 0.549
   e. 0.658

27. With $r = 0.963$ and $n = 11$, it is reasonable to conclude that there is a significant linear correlation. This section of the parabola can be approximated by a straight line.

Section 9-3

1. $\hat{y} = 1 + 2x$ 3. $\hat{y} = 0 + 3x$ 5. $\hat{y} = -187 + 11.3x$; 399 lb

7. $\hat{y} = 0.549 + 1.48x$; 1.3 persons

9. $\hat{y} = 2.93 + 0.0664x$; 3.3 persons

11. $\hat{y} = 0.214 - 0.000182x$; 0.21   13. $\hat{y} = 41.9 + 0.179x$; 79 min

15. $\hat{y} = 0.154 + 0.0651x$; 0.3 mg

17. $\hat{y} = 420 + 18.3x$; 1015 million bushels

19. a. 10.00   b. 2.50

21. It is an outlier but is not an influential point. The regression line changes to $\hat{y} = -346 + 9.57x$, which is not a dramatic change from $\hat{y} = -352 + 9.66x$.

23. $\hat{y} = -182 + 0.000351x$; $\hat{y} = -182 + 0.351x$. The slope is multiplied by 1000 and the y-intercept doesn't change. If each y entry is divided by 1000, the slope and y-intercept are both divided by 1000.

25. The linear equation $\hat{y} = -49.9 + 27.2x$ is better because it has $r = 0.997$, which is higher than $r = 0.963$ for $\hat{y} = -103.2 + 134.9 \ln x$.

Section 9-4

1. 0.04; 4%   3. 0.051; 5.1%

5. a. 154.8   b. 0   c. 154.8   d. 1   e. 0

7. a. 13.836615   b. 5.663385   c. 19.5   d. 0.70957
   e. 0.971544

9. a. 14

b. Because $s_e = 0$, $E = 0$ and there is no "interval" estimate.

11. a. 4.2 persons   b. $1.6 < y < 6.9$

13. $-55 \text{ lb} < y < 317 \text{ lb}$ 15. $-20 \text{ lb} < y < 277 \text{ lb}$

17. $-663 < \beta_0 < -39.9$; $4.91 < \beta_1 < 14.41$

19. a. $(n-2)s_e^2$

b. $\dfrac{r^2 \cdot \text{(unexplained variation)}}{1 - r^2}$ c. $r = -0.949$

Section 9-5

1. $\hat{y} = -285 - 1.38x_1 - 11.2x_2 + 28.6x_3$

3. Yes, with a P-value of 0.002, the equation has overall significance.

5. a. $\hat{y} = -274 + 0.426x_1 + 12.1x_2$ b. 0.928; 0.925; 0.000
   c. Yes

7. a. $\hat{y} = -235 + 0.403x_1 + 5.11x_2 - 0.555x_3 + 9.19x_4$
   b. 0.942; 0.938; 0.000   c. Yes

9. a. $\hat{y} = 3.56 + 0.0980x_1$ b. 0.021; 0.005; 0.256   c. No

11. a. $\hat{y} = 1.15 + 1.41x_1 - 0.0144x_2$ b. 0.564; 0.549; 0.000
    c. Yes

13. a. $\hat{y} = 0.554 + 0.491x_1 + 1.08x_2$ b. 0.613; 0.600; 0.000
    c. Yes

15. $\hat{y} = -1399 + 0.217x_1$ (where x_1 represents ACREHARV); yes, because adjusted $R^2 = 0.630$ and P-value $= 0.000$.

17. $\hat{y} = 2.17 + 2.44x + 0.464x^2$. Because $R^2 = 1$, the parabola fits perfectly.

Chapter 9 Review Exercises

1. a. Test statistic: $r = 0.338$. Critical values: $r = \pm 0.632$. Fail to reject the claim of no significant linear correlation. There does not appear to be a linear correlation between consumption and price.
   b. $\hat{y} = -0.488 + 0.611x$ c. 0.3468

2. a. Test statistic: $r = 0.116$. Critical values: $r = \pm 0.632$. Fail to reject the claim of no significant linear correlation. There does not appear to be a linear correlation between consumption and income.
   b. $\hat{y} = 0.0657 + 0.000792x$ c. 0.3468

3. a. Test statistic: $r = 0.777$. Critical values: $r = \pm 0.632$. Reject the claim of no significant linear correlation. There does appear to be a linear correlation between consumption and temperature.
   b. $\hat{y} = 0.193 + 0.00293x$ c. 0.2864

4. $\hat{y} = -0.0526 + 0.747x_1 - 0.00220x_2 + 0.00303x_3$; $R^2 = 0.726$; adjusted $R^2 = 0.589$; P-value $= 0.040$. This equation is a good predictor of ice cream consumption, but the equation from Exercise 3 is better.

5. a. Test statistic: $r = 0.999$. Critical values: $r = \pm 0.666$. Reject the claim of no significant linear correlation. There does appear to be a linear correlation between minutes before midnight on the doomsday clock and the index of savings.
   b. $\hat{y} = 5.92 + 0.636x_1$ c. 9.10

6. a. Test statistic: $r = -0.483$. Critical values: $r = \pm 0.666$. Fail to reject the claim of no significant linear correlation. There does not appear to be a linear correlation between the periodical index and the index of savings.
   b. $\hat{y} = 13.5 - 19.8x_2$
   c. 11.2

7. $8.81 < y < 9.38$

8. $\hat{y} = 6.06 + 0.630x_1 - 0.815x_2$; $R^2 = 0.998$; adjusted $R^2 = 0.998$; P-value $= 0.000$. The multiple regression equation is usable for predicting the savings index, because the P-value indicates overall significance of 0.000 and the adjusted value of R^2 is very close to 1.

Chapter 9 Cumulative Review Exercises

1. $n = 50$, $\bar{x} = 80.7$, $s = 12.0$.
   a. Test statistic: $z = 8.66$. Critical value: $z = 1.645$ (assuming $\alpha = 0.05$). Reject H_0: $\mu \le 66$. There is sufficient evidence to support the claim that the mean interval is now longer than 66 min.
   b. $77.3 < \mu < 84.0$

2. (Use the methods of Section 8-2 for two matched samples.) Test statistic: $t = 10.319$. Critical values: $t = \pm 2.365$ (assuming $\alpha = 0.05$). Reject H_0: $\mu_d = 0$. There does appear to be a significant difference between the scores on the humorous commercial and those on the serious commercial. The humorous commercial appears to have higher scores and is therefore better. The issue of correlation is not relevant here; the issue is whether either commercial gets significantly higher scores than the other, not whether the scores are related.

3. Test for a significant linear correlation between the two variables: Test statistic: $r = 0.702$. Critical values: $r = \pm 0.576$ (assuming $\alpha = 0.05$). Reject the claim of no significant linear correlation. There does appear to be a significant linear correlation between the IQs of identical twins who were raised apart. Because $r^2 = 0.493$, we can conclude that 49.3% of the total variation in the IQs of the younger twins can be explained by the variation in IQs of the older twins. This suggests that 49.3% of the variation in IQs can be explained by heredity, and the other 50.7% is attributable to other factors.

Section 10-2

1. Test statistic: $\chi^2 = 156.500$. Critical value: $\chi^2 = 21.666$. Reject H_0: The last digits occur with the same frequency. There is sufficient evidence to warrant rejection of the claim that the last digits occur with equal frequency. It appears that the students reported their weights.

3. Test statistic: $\chi^2 = 4.600$. Critical value: $\chi^2 = 7.815$. Fail to reject H_0: The results fit a uniform distribution. There is not sufficient evidence to warrant rejection of the claim that the results fit a uniform distribution. The results suggest that students generally cannot select the same tire.

5. Test statistic: $\chi^2 = 7.417$. Critical value: $\chi^2 = 12.592$. Fail to reject H_0: Fatal crashes occur on the different days with equal frequency. There is not sufficient evidence to warrant rejection of the claim that fatal crashes occur on the different days with equal frequency.

7. Test statistic: $\chi^2 = 23.431$. Critical value: $\chi^2 = 9.488$. Reject H_0: Accidents are distributed with the given percentages. There is sufficient evidence to warrant rejection of the claim that accidents are distributed with the given percentages. Rejection of the claim does not provide help in correcting the problem.

9. Test statistic: $\chi^2 = 4.200$. Critical value: $\chi^2 = 16.919$. Fail to reject H_0: The digits are uniformly distributed. There is not sufficient evidence to warrant rejection of the claim that the digits are uniformly distributed.

11. Test statistic: $\chi^2 = 53.051$. Critical value: $\chi^2 = 7.815$. Reject H_0: The distribution of crashes conforms to the distribution of ages. There is sufficient evidence to warrant rejection of the claim that the distribution of crashes conforms to the distribution of ages. The under-25 age group appears to have a disproportionate number of crashes.

13. Test statistic: $\chi^2 = 11.800$. Critical value: $\chi^2 = 9.488$ (assuming $\alpha = 0.05$). Reject H_0: The time intervals are uniformly distributed among the five categories. There is sufficient evidence to warrant rejection of the claim that the time intervals are uniformly distributed among the five categories.

15. Test statistic: $\chi^2 = 21.467$. Critical value: $\chi^2 = 7.815$. Reject H_0: The movies are distributed evenly among the four categories. There is sufficient evidence to warrant rejection of the claim that the movies are distributed evenly among the four categories.

17. $0.10 < P\text{-value} < 0.90$ and it is close to 0.90.

19. a.
$$\chi^2 = \frac{\left(f_1 - \frac{f_1 + f_2}{2}\right)^2}{\frac{f_1 + f_2}{2}} + \frac{\left(f_2 - \frac{f_1 + f_2}{2}\right)^2}{\frac{f_1 + f_2}{2}}$$
$$= \frac{(f_1 - f_2)^2}{f_1 + f_2}$$

b. $z^2 = \dfrac{\left(\dfrac{f_1}{f_1 + f_2} - 0.5\right)^2}{\dfrac{1/4}{f_1 + f_2}} = \dfrac{(f_1 - f_2)^2}{f_1 + f_2}$

Critical values: The χ^2 critical value is 3.841 and it is approximately equal to the square of $z = 1.96$.

21. Combine the last three cells. Test statistic: $\chi^2 = 1.012$. Critical value: $\chi^2 = 5.991$. Fail to reject H_0: The frequencies fit a Poisson distribution. There is not sufficient evidence to warrant rejection of the claim that the frequencies fit a Poisson distribution.

Section 10-3

1. Test statistic: $\chi^2 = 51.458$. Critical value: $\chi^2 = 6.635$. Reject the claim (H_0) that the proportions of agree/disagree responses are the same for the subjects interviewed by men and the subjects interviewed by women.

3. Test statistic: $\chi^2 = 1.174$. Critical value: $\chi^2 = 3.841$. Fail to reject the claim that the treatment is independent of the reaction. There is not enough evidence to be concerned.

5. Test statistic: $\chi^2 = 2.195$. Critical value: $\chi^2 = 5.991$. Fail to reject H_0 that confidence is independent of gender. There is not sufficient evidence to support the claim that there is a gender gap in the confidence people have in police.

7. Test statistic: $\chi^2 = 119.330$. Critical value: $\chi^2 = 5.991$. Reject the claim that the type of crime is independent of whether the criminal is a stranger.

9. Test statistic: $\chi^2 = 42.557$. Critical value: $\chi^2 = 3.841$. Reject the claim that the sentence is independent of the plea. The data suggest that you should encourage a guilty plea.

11. Test statistic: $\chi^2 = 0.615$. Critical value: $\chi^2 = 7.815$. Fail to reject the claim that smoking is independent of age group. Based on the given data, it does not make sense to target cigarette advertising to particular age groups.

13. Test statistic: $\chi^2 = 49.731$. Critical value: $\chi^2 = 11.071$ (assuming $\alpha = 0.05$). Reject the claim that the type of crime is independent of whether the criminal drinks. Fraud appears to be unique in the

sense that it is associated with drinkers at a much lower rate than the other crimes.

15. Test statistic: $\chi^2 = 0.122$. Critical value: $\chi^2 = 3.841$ (assuming $\alpha = 0.05$). Fail to reject the claim that gender of statistics students is independent of whether they smoke.

17. With Yates' correction: $\chi^2 = 1.205$. Without Yates' correction: $\chi^2 = 1.505$. In general, the test statistic decreases with Yates' correction.

Chapter 10 Review Exercises

1. Test statistic: $\chi^2 = 42.004$. Critical value: $\chi^2 = 5.991$. Reject the claim that the three cities have the same proportion of taxis with usable seat belts.

2. Test statistic: $\chi^2 = 6.780$. Critical value: $\chi^2 = 12.592$. Fail to reject the claim that gunfire death rates are the same for the different days of the week. There is not enough evidence to support the theory of more gunfire deaths on weekends.

3. Test statistic: $\chi^2 = 22.600$. Critical value: $\chi^2 = 19.675$. Reject the claim that months were selected with equal frequencies. This result does not support extrasensory perception; we need data showing the frequency of correct selections.

4. Test statistic: $\chi^2 = 0.500$. Critical value: $\chi^2 = 5.991$. Fail to reject the claim that the three airline companies have the same proportions of on-time flights.

5. Test statistic: $\chi^2 = 7.607$. Critical value: $\chi^2 = 5.991$. Reject the claim that the occurrence of headaches is independent of the group. The Seldane users had a lower headache rate than the placebo group, so headaches don't seem to be a problem.

Chapter 10 Cumulative Review Exercises

1. (Use the linear correlation coefficient, as in Section 9-2.) Test statistic: $r = 0.978$. Critical value: $r = 0.950$ (assuming $\alpha = 0.05$). Reject the claim of no significant linear correlation. There does appear to be a relationship between the memory test score and the reasoning test score.

2. (Use the test for two dependent means, as in Section 8-2.) With $\bar{d} = -10.25$ and $s_d = 1.5$, the test statistic is $t = -13.667$. The critical value is $t = -2.353$ (assuming $\alpha = 0.05$). Reject H_0: $\mu_d \geq 0$. There is sufficient evidence to support the claim that the scores are higher after the training session. It appears that the training session is effective in raising scores.

3. (Use a test of homogeneity in a contingency table, as in Section 10-3.) Test statistic: $\chi^2 = 0.055$. Critical value: $\chi^2 = 7.815$ (assuming $\alpha = 0.05$). Fail to reject the claim that men and

women choose the different answers in the same proportions. The letter selections appear to be the same for men and women.

4. (Use a test for equality between two independent means, as in Section 8-5.) Preliminary F test: Test statistic is $F = 1.0344$. Critical value: $F = 15.439$ (assuming $\alpha = 0.05$). Fail to reject $\sigma_1^2 = \sigma_2^2$. Test of means: Test statistic is $t = -2.014$. Critical values: $t = \pm 2.447$ (assuming $\alpha = 0.05$). Fail to reject H_0: $\mu_1 = \mu_2$. There is not sufficient evidence to warrant rejection of the claim that men and women have the same mean score.

Section 11-2

1. a. $F = 0.49$ b. $F = 3.8853$ c. 0.627
   d. Fail to reject the null hypothesis of equal means.
   e. There is not sufficient evidence to warrant rejection of the claim that the three different companies produced commercials with the same mean reaction score. The means appear to be equal.

3. a. 0.864667   b. 0.46   c. $F = 1.8797$
   d. 3.8853 (assuming $\alpha = 0.05$)
   e. There is not sufficient evidence to warrant rejection of the claim that the three age-group populations have the same mean body temperature.

5. Test statistic: $F = 0.1587$. Critical value: $F = 3.3541$. Fail to reject the claim of equal means. The zones appear to have the same mean selling price.

7. Test statistic: $F = 5.0793$. Critical value: $F = 3.3541$. Reject the claim of equal means. Zone 7 appears to have the largest lots, although such a conclusion is not justified by the ANOVA procedure. Also, the Zone 7 variance is so much larger than the others that the assumption of equal variances is very questionable here.

9. Test statistic: $F = 2.9493$. Critical value: $F = 2.6060$. Reject the claim of equal means. The different laboratories do not appear to have the same mean.

11. Test statistic: $F = 17.8360$. Critical value: $F = 2.3683$ (approximate value, assuming $\alpha = 0.05$). Reject the claim of equal means. The means appear to be different in the different months.

13. Test statistic: $F = 0.5083$. Critical value: $F = 2.2899$ (approximately). Fail to reject the claim of equal means. The mean weights appear to be the same for the different colors. No corrective action is required.

15. a. 5.864667   b. 0.46   c. $F = 12.7493$
   d. 3.8853 (assuming $\alpha = 0.05$)
   e. There is sufficient evidence to warrant rejection of the claim of equal means. By adding $2°$ to each score in the first sample,

the variance between samples increased, but the variance within samples did not change. The F test statistic increased to reflect the greater disparity among the three sample means.

17. In each case, the test statistic F is not affected.

19. a. Test statistic: $t = 0.262$. Critical values: $t = \pm 2.101$. Fail to reject the null hypothesis that zones 1 and 4 have the same mean selling prices.

   b. Test statistic: $F = 0.0688$. Critical value: $F = 4.4139$. Fail to reject the null hypothesis that zones 1 and 4 have the same mean selling prices.

   c. Test statistics: $t^2 = 0.0686 \approx F = 0.0688$.
   Critical values: $t^2 = 4.4142 \approx F = 4.4139$

Section 11-3

1. a. 1263   b. 695   c. 350   d. 14

3. Test statistic: $F = 0.5036$. Critical value: $F = 4.0662$, assuming that the significance level is $\alpha = 0.05$. It appears that star rating does not have an effect on movie length.

5. Test statistic: $F = 0$. Critical value: $F = 10.128$. It appears that MPAA rating does not have an effect on movie length.

7. Test statistic: $F = 2.4732$. Critical value: $F = 4.7571$. Fail to reject the null hypothesis of no effects from the operators. It appears that the operators do not affect the numbers of support beams manufactured.

9. (The given results use the first nine values for each cell, but the pulse rate of 8 was excluded as being not feasible.) Test statistic: $F = 0.4200$. Critical value: $F = 4.1709$ (approximately). There does not appear to be an interaction between gender and smoking.

11. Test statistic: $F = 0.1814$. Critical value: $F = 4.1709$ (approximately, assuming $\alpha = 0.05$). Smoking does not appear to have an effect on pulse rates. (This is not surprising, because college students haven't been smoking very long.)

13. Nothing changes.

15. Answers vary.

Chapter 11 Review Exercises

1. Test statistic: $F = 46.90$. Critical value: $F = 3.7389$. Reject the null hypothesis of equal means.

2. Test statistic: $F = 9.4827$. Critical value: $F = 3.0984$. Reject the null hypothesis of equal means. There is sufficient evidence to support the claim of different mean selling prices.

3. Test statistic: $F = 0.0531$. Critical value: $F = 5.1433$. Fuel consumption does not appear to be affected by an interaction between transmission type and engine size.

4. Test statistic: $F = 1.9115$. Critical value: $F = 5.1433$. The size of the engine does not appear to have an effect on fuel consumption.

5. Test statistic: $F = 3.5664$. Critical value: $F = 5.9874$. The type of transmission does not appear to have an effect on fuel consumption.

Chapter 11 Cumulative Review Exercises

1. a.

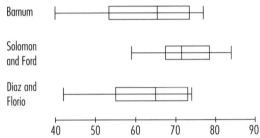

Barnum Advertising and Diaz and Florio Advertising appear to be very similar, but Solomon & Ford Advertising seems to be somewhat higher.

   b. Barnum: $\bar{x} = 62.8$, $s = 12.9$; Solomon & Ford: $\bar{x} = 72.3$, $s = 8.2$; Diaz & Florio: $\bar{x} = 62.8$, $s = 11.6$

   c. F test results: Test statistic is $F = 2.4592$. Critical value: $F = 4.9949$. Fail to reject H_0: $\sigma_1^2 = \sigma_2^2$. Test of means: Test statistic is $t = -1.760$. Critical values: $t = \pm 2.145$. Fail to reject H_0: $\mu_1 = \mu_2$. There is not sufficient evidence to warrant rejection of the claim that the population of Barnum scores has a mean equal to the population mean for Solomon & Ford scores.

   d. Barnum: $52.0 < \mu < 73.5$; Solomon & Ford: $65.4 < \mu < 79.1$; Diaz and Florio: $53.1 < \mu < 72.4$. Solomon & Ford seem slightly higher.

   e. Test statistic: $F = 1.9677$. Critical value: $F = 3.4668$. Fail to reject the null hypothesis of equal means. There is not sufficient evidence to warrant rejection of the claim that the three populations have the same mean reaction score. No single company stands out as being significantly better than the others.

2. a. 0.3372   b. 0.0455   c. 1/8

Section 12-2

1.

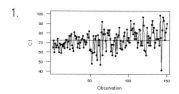

There is an upward trend indicating that the intervals between eruptions are increasing, with the result that tourists must wait longer. There is increasing variation, with the result that predicted times of eruptions are becoming less reliable.

3.

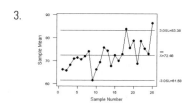

There are points lying beyond the upper control limit and there is an upward trend, so the process mean is out of statistical control. An out-of-control process mean indicates that we cannot make accurate predictions.

5.

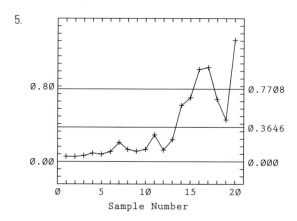

There is a pattern of increasing ranges, there are points beyond the upper control limit, and there are eight consecutive points lying below the center line, so the process variation is out of statistical control.

7.

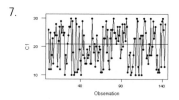

The process appears to be within statistical control.

9.

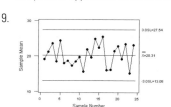

The process mean is within statistical control and the process appears to be working well. A downward trend would have indicated that the process is out of statistical control, but it is *improving* because the delivery times are decreasing. Such a downward trend should be investigated to ensure that the reduced times continue.

11.

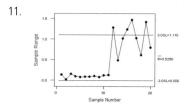

The process is out of control because there is a shift up, there are 8 consecutive points below the center line, and there are points beyond the upper control limit.

13.

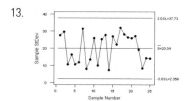

Using Table 12-1 values, $\bar{s} = 20.05$ and LCL = 2.366. The result and conclusions are very similar to those for the R chart.

Section 12-3

1. Process appears to be within statistical control.
3. Process appears to be out of statistical control because there is a

pattern of an upward trend and there is a point that lies beyond the upper control limit.

5.

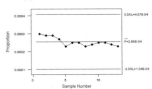

The process is out of statistical control because there is a downward trend and there are eight consecutive points all lying below the center line. This downward trend is good and its causes should be identified so that it can continue.

7.

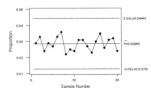

The process is within statistical control, but this describes the actual behavior of the process, not the desired behavior. People generally feel that there is too much violent crime, and the stable levels are too high.

9.

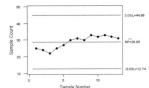

Except for the scale used, the charts are identical.

Review Exercises

1.

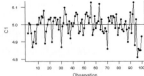

The process appears to be within statistical control.

2.

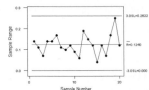

The process appears to be within statistical control.

3.

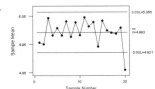

The process mean appears to be out of statistical control because there is a point lying beyond the lower control limit. The process should be corrected because too little epoxy cement hardener is being put into some containers.

4.

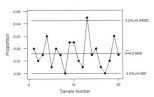

The process is out of statistical control because there is a point lying beyond the upper control limit.

Cumulative Review Exercises

1. a.

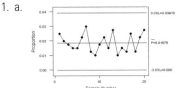

The process is in statistical control.

b. $0.0158 < p < 0.0217$

c. Test statistic: $z = 5.77$. Critical value: $z = 1.645$. Reject $H_0: p \leq 0.01$. There is sufficient evidence to warrant rejection of the claim that the rate of defects is 1% or less.

2. a. 1/256

b. 1/256

c. 1/128

Section 13-2

1. The test statistic $x = 2$ is not less than or equal to the critical value of 1. Fail to reject the claim that the course has no effect on SAT scores. The course does not appear to be effective.

3. The test statistic $x = 3$ is not less than or equal to the critical value of 1. Fail to reject the null hypothesis that Covariant is favored by at most half of all dentists. The sample data do not provide a reasonable basis for making the claim that most dentists favor Covariant toothpaste.

5. The test statistic $x = 7$ is not less than or equal to the critical value of 5. Fail to reject the null hypothesis that the proportions of men and women are equal. There is not sufficient evidence to charge gender bias.

7. (Instead of a right-tailed test to determine whether $x = 18$ is large enough to be significant, use a left-tailed test to determine whether $x = 15$ is small enough to be significant.) Convert $x = 15$ to the test statistic $z = -0.35$. Critical value: $z = -1.645$. Fail to reject H_0. There is not sufficient evidence to support the claim that the median is greater than 0.9085 g. We cannot conclude that the package is labeled incorrectly, because the median could be equal to 0.9085 g.

9. There are 38 weights below 5.670 g, 2 equal to 5.670 g, and 10 above 5.670 g, so use $x = 10$ and $n = 48$. Test statistic: $z = -3.90$. Critical values: $z = \pm 1.96$ (assuming $\alpha = 0.05$). Reject the null hypothesis that the median is 5.670 g. There is sufficient evidence to warrant rejection of the claim that quarters have a median of 5.670 g.

11. The test statistic $x = 2$ is not less than or equal to the critical value of 1. Fail to reject the null hypothesis of no difference. There is not sufficient evidence to warrant rejection of the claim of no difference between the two times.

13. With $k = 3$ we get $6 < M < 17$.

15. *, *, *, *, *, 0, 0, 0, 1, 1, 1, 2, 2, 3, 3

Section 13-3

1. Test statistic: $T = 9.5$. Critical value: $T = 2$. Fail to reject the null hypothesis that both samples come from the same population distribution.

3. Test statistic: $T = 6.5$. Critical value: $T = 8$. Reject the null hypothesis that both samples come from the same population distribution.

5. Test statistic: $T = 1$. Critical value: $T = 0$. Fail to reject the null hypothesis that both samples come from the same population distribution. Based on the available results, hypnotism does not appear to be effective.

7. Test statistic: $T = 1$. Critical value: $T = 14$ (assuming $\alpha = 0.05$). Reject the null hypothesis that both samples come from the same population distribution. There is sufficient evidence to warrant rejection of the claim that the program has no effect.

9. a. 0, 27.5
   b. 0, 637.5
   c. 1954

Section 13-4

1. $\mu_R = 577.5$, $\sigma_R = 56.358$, $R = 569$, $z = -0.15$. Test statistic: $z = -0.15$. Critical values: $z = \pm 1.96$. Fail to reject the null hypothesis that red and brown M&Ms have weights with identical populations. They appear to be the same.

3. $\mu_R = 126.5$, $\sigma_R = 15.2288$, $R = 103$, $z = -1.54$. Test statistic: $z = -1.54$. Critical values: $z = \pm 1.96$. Fail to reject the null hypothesis that the two laboratories have identical populations.

5. $\mu_R = 162$, $\sigma_R = 19.442$, $R = 89.5$, $z = -3.73$. Test statistic: $z = -3.73$. Critical values: $z = \pm 1.96$. Reject the null hypothesis that beer and liquor drinkers have the same BAC levels. It appears that liquor drinkers are more dangerous.

7. $\mu_R = 525$, $\sigma_R = 37.417$, $R = 437$, $z = -2.35$. Test statistic: $z = -2.35$. Critical values: $z = \pm 1.96$. Reject the null hypothesis that the two samples come from populations with the same scores.

9. The denominators are the same, so you need to consider only the numerators. First set $R - \mu_R = -U + n_1 n_2/2$, and then replace U by $n_1 n_2 + n_1(n_1 + 1)/2 - R$ and replace μ_R by $n_1(n_1 + n_2 + 1)/2$. Remove parentheses and simplify to see that both sides are equal.

Section 13-5

1. Test statistic: $H = 2.180$. Critical value: $\chi^2 = 5.991$. Fail to reject the null hypothesis that the three age-group populations of body temperatures are identical.

3. Test statistic: $H = 13.075$. Critical value: $\chi^2 = 7.815$. Reject the null hypothesis that the four samples come from identical populations. The sample data support the claim of different populations of selling prices in the four regions.

5. Test statistic: $H = 2.075$. Critical value: $\chi^2 = 11.071$. Fail to reject the null hypothesis that the samples come from identical populations. No corrective action is necessary.

7. Test statistic: $H = 3.489$. Critical value: $\chi^2 = 5.991$. Fail to reject the null hypothesis of identical populations. The three categories of movies appear to have the same lengths.

9. a. The test statistic H does not change.
   b. The test statistic H does not change.

11. $\Sigma T = 366$. Dividing H by the correction factor results in $14.431335 \div 0.99668910$, which is rounded to 14.479. Even though 35 of the 48 scores are involved in ties, the correction factor does not change H by a substantial amount.

Section 13-6

1. a. $r_s = 1$ and there appears to be a correlation between x and y.
   b. $r_s = -1$ and there appears to be a correlation between x and y.
   c. $r_s = 0$ and there does not appear to be a correlation between x and y.

3. $r_s = 0.855$. Critical values: $r_s = \pm0.648$. Significant correlation. There appears to be a correlation between salary and stress.

5. $r_s = -0.067$. Critical values: $r_s = \pm0.648$. No significant correlation. There does not appear to be a correlation between stress level and physical demand.

7. $r_s = 0.715$. Critical values: $r_s = \pm0.591$. Significant correlation. There appears to be a correlation between the IQ scores of identical twins separated at birth.

9. $r_s = 0.363$. Critical values: $r_s = \pm0.738$. No significant correlation. There does not appear to be a correlation between length and weight.

11. $r_s = -0.758$. Critical values: $r_s = \pm0.648$. Significant correlation. There appears to be a correlation between weights of cars and fuel consumption amounts. The results don't change if each weight is multiplied by 100.

13. $r_s = 0.786$. Critical values: $r_s = \pm0.280$. Significant correlation. There appears to be a correlation between durations and intervals between eruptions.

15. $r_s = 0.918$. Critical values: $r_s = \pm0.370$. Significant correlation. There appears to be a correlation between amounts of cigarette tar and nicotine. Yes, but the results will not be exact.

17. $r_s = 0.265$. Critical values: $r_s = \pm0.310$. No significant correlation. There does not appear to be a correlation between Iowa's annual precipitation amounts and the amounts of corn produced.

19. a. ±0.707
    b. ±0.514
    c. ±0.361
    d. ±0.463
    e. ±0.834

Section 13-7

1. $n_1 = 10$, $n_2 = 5$, $G = 3$, 5% cutoff values: 3, 12

3. $n_1 = 12$, $n_2 = 4$, $G = 4$, 5% cutoff values: 3, 10

5. $n_1 = 12$, $n_2 = 8$, $G = 10$, 5% cutoff values: 6, 16. Fail to reject the null hypothesis of randomness. A lack of randomness could result in large gains for the author and large losses to the casino. Fat chance.

7. $n_1 = 10$, $n_2 = 14$, $G = 4$, 5% cutoff values: 7, 18. Reject the null hypothesis of randomness. There is a noticeable pattern.

9. $n_1 = 17$, $n_2 = 14$, $G = 14$, 5% cutoff values: 10, 23. Fail to reject the null hypothesis of randomness. It appears that we elect Democrats and Republicans in a sequence that is random.

11. The median corresponds to July 2 so the sequence becomes A A A A A B B B B B B with $n_1 = 6$, $n_2 = 6$, $G = 2$. 5% cutoff values are 3, 11. Reject the null hypothesis of randomness.

13. $n_1 = 21$, $n_2 = 19$, $G = 7$, $\mu_G = 20.95$, $\sigma_G = 3.11346$. Test statistic: $z = -4.48$. Critical values: $z = \pm1.96$. Reject the null hypothesis of randomness.

15. $n_1 = 49$, $n_2 = 51$, $G = 54$, $\mu_G = 50.98$, $\sigma_G = 4.9727$. Test statistic: $z = 0.61$. Critical values: $z = \pm1.96$. Fail to reject the null hypothesis of randomness.

17. Minimum is 2, maximum is 4. Critical values of 1 and 6 can never be realized so that the null hypothesis of randomness can never be rejected.

Chapter 13 Review Exercises

1. Test statistic: $H = 0.144$. Critical value: $\chi^2 = 5.991$. Fail to reject the null hypothesis that the three populations are the same. There is not sufficient evidence to support the claim that the three populations are not all the same.

2. Test statistic: $T = 10$. Critical value: $T = 2$. Fail to reject the null hypothesis that both samples come from the same population distribution.

3. $n_1 = 6$, $n_2 = 24$, $G = 12$, $\mu_G = 10.6$, $\sigma_G = 1.6873$. Test statistic: $z = 0.83$. Critical values: $z = \pm1.96$. Fail to reject randomness.

4. $r_s = 0.810$. Critical values: $r_s = \pm0.738$. Significant correlation. There is sufficient evidence to support the claim that there is a correlation between weight of discarded paper and household size.

5. The test statistic $x = 1$ is less than or equal to the critical value of 1. Reject the claim that there is no difference between the two systems. Flight Services Software appears to be better because the processing times are generally less.

6. $\mu_R = 201.5$, $\sigma_R = 23.894$, $R = 163$, $z = -1.61$. Test statistic: $z = -1.61$. Critical values: $z = \pm1.96$. Fail to reject the null hypothesis that the drug has no effect on eye movements.

7. Test statistic: $H = 4.096$. Critical value: $\chi^2 = 7.815$. Fail to reject the null hypothesis that the four populations are identical. There is not sufficient evidence to support the claim that the samples come from populations that are not all the same.

8. $n_1 = 14$, $n_2 = 16$, $G = 11$, 5% cutoff values: 10, 22. Fail to reject randomness.

9. $\mu_R = 144$, $\sigma_R = 16.25$, $R = 84$, $z = -3.69$. Test statistic: $z = -3.69$. Critical values: $z = \pm1.96$. Reject the null hypothe-

sis of no difference between mail order and store prices.

10. The test statistic $x = 0$ is less than or equal to the critical value of 0. Reject the claim that the training has no effect on weight. There does appear to be an effect.

11. Test statistic: $T = 0$. Critical value: $T = 1$. Reject the null hypothesis that training has no effect on weight. There does appear to be an effect.

12. $r_s = 0.524$. Critical values: $r_s = \pm 0.738$. No significant correlation.

Cumulative Review Exercises

1. a. $G = 21$, $n_1 = 20$, $n_2 = 10$. Critical values: 9, 20. Reject randomness.
   b. Test statistic: $z = -1.83$. Critical values: $z = \pm 1.96$. Fail to reject null hypothesis that $p = 0.5$. There is not sufficient evidence to support the claim that the proportion of women is different from 0.5.
   c. $0.165 < p < 0.502$
   d. There isn't sufficient evidence to support a claim of bias against either gender, but the sequence does not appear to be random. It's possible that the pollster is using selection methods that are unacceptable.

2. a. The test statistic $x = 2$ is not less than or equal to the critical value of 1. Fail to reject the claim that the course has no effect on SAT scores. The course does not appear to have an effect on SAT scores.
   b. Test statistic: $T = 3$. Critical value: $T = 6$. Reject the null hypothesis that the course has no effect on SAT scores. It does appear to have an effect.
   c. Test statistic: $t = -3.753$. Critical values: $t = \pm 2.262$. Reject the null hypothesis that the course has no effect. It does appear to have an effect.
   d. The sign test failed to detect a difference, whereas the t test resulted in the conclusion of a significant difference. Unlike the sign test, the Wilcoxon signed-ranks test did use the magnitudes of the differences, so it is more sensitive than the sign test and it did result in the same conclusion as the t test. (The book does not describe a method for finding P-values with the Wilcoxon signed-ranks test, but the P-value is 0.024, whereas the P-value for the t test is 0.0045. Comparison of these P-values shows that the Wilcoxon signed-ranks test is not as sensitive as the parametric t test.)

Index

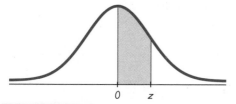

TABLE A-2	Standard Normal (z) Distribution								

z	.00	.01	.02	.03	.04	.05	.06	.07	.08	.09
0.0	.0000	.0040	.0080	.0120	.0160	.0199	.0239	.0279	.0319	.0359
0.1	.0398	.0438	.0478	.0517	.0557	.0596	.0636	.0675	.0714	.0753
0.2	.0793	.0832	.0871	.0910	.0948	.0987	.1026	.1064	.1103	.1141
0.3	.1179	.1217	.1255	.1293	.1331	.1368	.1406	.1443	.1480	.1517
0.4	.1554	.1591	.1628	.1664	.1700	.1736	.1772	.1808	.1844	.1879
0.5	.1915	.1950	.1985	.2019	.2054	.2088	.2123	.2157	.2190	.2224
0.6	.2257	.2291	.2324	.2357	.2389	.2422	.2454	.2486	.2517	.2549
0.7	.2580	.2611	.2642	.2673	.2704	.2734	.2764	.2794	.2823	.2852
0.8	.2881	.2910	.2939	.2967	.2995	.3023	.3051	.3078	.3106	.3133
0.9	.3159	.3186	.3212	.3238	.3264	.3289	.3315	.3340	.3365	.3389
1.0	.3413	.3438	.3461	.3485	.3508	.3531	.3554	.3577	.3599	.3621
1.1	.3643	.3665	.3686	.3708	.3729	.3749	.3770	.3790	.3810	.3830
1.2	.3849	.3869	.3888	.3907	.3925	.3944	.3962	.3980	.3997	.4015
1.3	.4032	.4049	.4066	.4082	.4099	.4115	.4131	.4147	.4162	.4177
1.4	.4192	.4207	.4222	.4236	.4251	.4265	.4279	.4292	.4306	.4319
1.5	.4332	.4345	.4357	.4370	.4382	.4394	.4406	.4418	.4429	.4441
1.6	.4452	.4463	.4474	.4484	.4495 ✱	.4505	.4515	.4525	.4535	.4545
1.7	.4554	.4564	.4573	.4582	.4591	.4599	.4608	.4616	.4625	.4633
1.8	.4641	.4649	.4656	.4664	.4671	.4678	.4686	.4693	.4699	.4706
1.9	.4713	.4719	.4726	.4732	.4738	.4744	.4750	.4756	.4761	.4767
2.0	.4772	.4778	.4783	.4788	.4793	.4798	.4803	.4808	.4812	.4817
2.1	.4821	.4826	.4830	.4834	.4838	.4842	.4846	.4850	.4854	.4857
2.2	.4861	.4864	.4868	.4871	.4875	.4878	.4881	.4884	.4887	.4890
2.3	.4893	.4896	.4898	.4901	.4904	.4906	.4909	.4911	.4913	.4916
2.4	.4918	.4920	.4922	.4925	.4927	.4929	.4931	.4932	.4934	.4936
2.5	.4938	.4940	.4941	.4943	.4945	.4946	.4948	.4949 ✱	.4951	.4952
2.6	.4953	.4955	.4956	.4957	.4959	.4960	.4961	.4962	.4963	.4964
2.7	.4965	.4966	.4967	.4968	.4969	.4970	.4971	.4972	.4973	.4974
2.8	.4974	.4975	.4976	.4977	.4977	.4978	.4979	.4979	.4980	.4981
2.9	.4981	.4982	.4982	.4983	.4984	.4984	.4985	.4985	.4986	.4986
3.0	.4987	.4987	.4987	.4988	.4988	.4989	.4989	.4989	.4990	.4990
3.10 and higher	.4999									

NOTE: For values of z above 3.09, use 0.4999 for the area.
*Use these common values that result from interpolation:

z score	Area
1.645	0.4500
2.575	0.4950

From Frederick C. Mosteller and Robert E. K. Rourke, *Sturdy Statistics,* 1973, Addison-Wesley Publishing Co., Reading, MA. Reprinted with permission of Frederick Mosteller.